1st International Conference on Tropical Studies and Its Application (ICTROPS 2017)

IOP Conference Series: Earth and Environmental Science
Volume 144

Samarinda, Indonesia
9 November 2017

ISBN: 978-1-5108-6278-4
ISSN: 1755-1307

Printed from e-media with permission by:

Curran Associates, Inc.
57 Morehouse Lane
Red Hook, NY 12571

proceedings.com

Some format issues inherent in the e-media version may also appear in this print version.

Printed by Curran Associates, Inc. (2018)

For permission requests, please contact the Institute of Physics
at the address below.

Institute of Physics
Dirac House, Temple Back
Bristol BS1 6BE UK

Phone: 44 1 17 929 7481
Fax: 44 1 17 920 0979

techtracking@iop.org

Additional copies of this publication are available from:

Curran Associates, Inc.
57 Morehouse Lane
Red Hook, NY 12571 USA
Phone: 845-758-0400
Fax: 845-758-2633
Email: curran@proceedings.com
Web: www.proceedings.com

Table of contents

Volume 144

1st International Conference on Tropical Studies and Its Application (ICTROPS)

9 November 2017, Samarinda, East Kalimantan, Indonesia

Accepted papers received: 9 April 2018
Published online: 8 May 2018

Preface

Papers

Natural and Basic Science

Ecosystem and Environmental Engineering

Foreword from the Chairman of Organizing Committee

The International Conference on Tropical Studies and Its Application (ICTROPS) annual conference in November 9, 2017, is organized by Universitas Mulawarman in collaboration with Islamic Development Bank (IDB) and Ministry of Research, Technology, and Higher Education of The Republic of Indonesia. ICTROPS 2017 offers generous opportunities for renowned researchers and Industries across the globe to share ideas and knowledge especially in the area of Tropical Studies. The conference also aims to exchange information and networking among researchers and industries, in particular the member of the 4 in 1 consortium (Universitas Mulawarman, Universitas Sultan Ageng Tirtayasa, Universitas Jember and Universitas Negeri Malang).

ICTROPS 2017 with a theme "Tropical Studies and Application for Better Life" covers all key areas in Germ Plasm (Diversity), Ecosystem and Environmental Engineering, Ecological Integrity and Environmental Issue, Biodiversity Mapping, Natural and Basic Science in Tropical Studies, Biomaterial, Biotechnology and Renewable Energy, and Forestry.

For ICTROPS 2017, we received a large number of abstracts and there are 126 accepted abstracts for 103 oral presentations and 22 poster presentations. After peer review process, 68 papers were selected for publication in IOP conference series: Earth and Environmental Science.

On behalf of the organizing committee, I would like to thank all the reviewers for the effort, time and expertise in reviewing the papers to ensure the high quality and standard of peer-reviewed papers. Lastly, we also like to greatly thank all of the proceeding team for their hard work and constant dedication in preparing the proceeding through a long time process.

Dr. Rahmat Gunawan, M.Si
Chair of Organizing Committe

IOP Conf. Series: Earth and Environmental Science **144** (2018) 011002 doi:10.1088/1755-1315/144/1/011002

STEERING COMMITTEE

Prof. Dr. H. Masjaya, M.Si
Dr. Sc. Mustaid Yusuf, M.Si
Dr. Bohari Yusuf, M.Si
Dr. Anton Rahmadi, M.Sc

PERSON IN CHARGE

Dr. Zeni Haryanto, M.Pd

ORGANIZING COMMITTEE CHAIRMAN

Dr. Rahmat Gunawan, M.Si

VICE CHAIRMAN

Dr. Esti Handayani Hardi, M.Si

SECRETARY

Dr. R. R. Dirgarini J. N. Subagyono, M.Sc

MEMBER

Widi Sunaryo, S.P., M.Si., Ph.D
Erwin, S.Hut., M.P., M.Pd
Veliyana Londong Allo, S.Si., M.Si
Muhammad Amin, S.Kom., M.Si
Rexy Anderson Tindangen, S.Pd
Yunisa Asrianie, S.T., M.Sc
Yuliadini Puspita, S.Tp
Achdiat Farid, S.H.
Nur Aini Haryati, S.Si
Agung Yusuf, S.Sos
Ayu Aditriya
Elma Yossiana
Rodatul Jannah
Dwi Lestarima Jaya Simanullang
Putri Maghfirah
Muhammad Nur Hadiyansyah
Dr. Rudianto Amirta, M.P.
Irawan Wijaya Kusuma, S.Hut., M.P., Ph.D
Uni Wahyuni Sagena, M.Si, Ph.D
Syahril, S.P.
Masykur Kurniawan, S.E.
Asri Widiastuti, S.Hut
Desy Nursayekti, S.Tp
Kamalia Olfah, S.P.
Khairunnisa, S.Pd
Caesar Nataniel, S.Kom
Sari Mayaningrum, B.Sc
Sabran, S.E.
Jahlil Ahad, S.E.
Supriadi
Aditya Ferry Noor
Robin Sulistio

IOP Conf. Series: Earth and Environmental Science **144** (2018) 011002 doi:10.1088/1755-1315/144/1/011002

BOARD OF REVIEWER

Opening Ceremony of ICTROPS 2017

All participants of ICTROPS 2017

First Keynote Speaker: Prof. Alan L. Chaffee (Monash University, Australia)

Second Keynote Speaker: Dr. Thi My Lien Do (Saigon University, Vietnam)

IOP Conf. Series: Earth and Environmental Science **144** (2018) 011003 doi:10.1088/1755-1315/144/1/011003

Conference gift from the Vice Rector IV of Universitas Mulawarman to Dr. Thi My Lien Do

Awards for the best oral and poster presentation

Plenary session

All invited speakers and the representatives of the member of the 4 in 1 consortium (Universitas Mulawarman, Universitas Sultan Ageng Tirtayasa, Universitas Jember and Universitas Negeri Malang)

Parallel Session: Room 1

Parallel Session: Room 2

Parallel Session: Room 3

Parallel Session: Room 4

IOP Conf. Series: Earth and Environmental Science **144** (2018) 011003 doi:10.1088/1755-1315/144/1/011003

Judging process for the best poster (1)

Judging process for the best poster (2)

IOP Conf. Series: Earth and Environmental Science **144** (2018) 011003 doi:10.1088/1755-1315/144/1/011003

The organizing committee of ICTROPS 2017

Traditional dance for the closing ceremony

IOP Conf. Series: Earth and Environmental Science **144** (2018) 011004 doi:10.1088/1755-1315/144/1/011004

SUPPORTING AGENCIES

IOP Conf. Series: Earth and Environmental Science **144** (2018) 011005 doi:10.1088/1755-1315/144/1/011005

Peer review statement

All papers published in this volume of *IOP Conference Series: Earth and Environmental Science* have been peer reviewed through processes administered by the proceedings Editors. Reviews were conducted by expert referees to the professional and scientific standards expected of a proceedings journal published by IOP Publishing.

Geophysical survey for groundwater potential investigation in peat land area, Riau, Indonesia

N Islami*, M Irianti, Azhar, M Nor and Fakhrudin

Physics Education – FKIP, University of Riau
Jl. HR Soebrantas, Km. 12.5, Pekanbaru, 28293, Indonesia

*Corresponding Author : nurislami@lecturer.unri.ac.id

Abstract. Tropical forests, especially peat lands, are particularly vulnerable to forest fires. Fires are the most common disasters in peat lands in the dry season, especially in Riau Province, Indonesia. In the process of extinguishing the peat fire, several substantial problems arise to stop peat fires during this period. This study aims to determine the possibility of using ground water as a source of water to anticipate the early mitigation of peat land fires disaster. The geoelectrical resistivity surveys were used to predict the subsurface geological data including peat thickness and depth of aquifers. The geometry of peat land was determined using geostatistics based on geoelectrical resistivity interpretation data. Peat land thickness varies up to 4 m in the north, and is thinner to the south. A shallower and deeper aquifer is available at a depth of 13 m to 18 m and 70 m to 90 m respectively. In general, the potential of groundwater in the shallow aquifer is predicted to be sufficient for peat land watering anytime.

1. Introduction

The depth and thickness of sedimentary rocks that are not consolidated vary for certain places even within the same area. This depends on the conditions under which the rocks are formed and also depends on the weathering process and the geological precipitation processes occurring in the area. In groundwater exploration, porosity and sediment thickness variables determine the amount of groundwater reserves beneath the surface. The thickness of the unconsolidated sediment depends on how far the bedrock is found in the area. If the bedrock is available at relatively shallow depths, this causes a thinner sediment thickness that can store groundwater.

Almost the area of Central Sumatera basin is located in Riau Province. One of the rivers that crosses this basin is the Siak River. The upper surface of the Central Sumatra Basin consists primarily of peat that is vulnerable to peat fires [1]. Meanwhile, alteration of peat land functions for agriculture occurs almost in all peat lands in Indonesia [2,3]. In this regard, the main problem occurring in the Sumatran basin, especially along the Siak plains and coastal areas is forest fires as a result of illegal agricultural and agricultural activities [4]. Wildfires are often out of control due to lack of water to stop them [5]. This is because peat soils have a specific structure [6,7]. The need for water to stop fires in the forest increases dramatically in case of fire on peat lands. This problem can be better addressed if water sources for forest fires are extracted from groundwater in the area.

Geophysical survey as the one of in-destructive, fastest and low cost survey method is widely used to explore the subsurface condition for the several problems. The use of resistivity survey has been reported that the method is possible to differentiate contrast between the area that has been applied with fertilization for long duration and non-fertilization [8,9,10]. This method was also used to investigate geomorphology of gravity-flow channels in a rift basin using seismic survey [11] and Interpretation of the deep-water geomorphology of the glaciated Irish margin from high-resolution marine geophysical data [12]. Geophysical data also gave the better result when combine it with the Lansat images in the interpretation [13]. In this study the use of geoelectrical resistivity will be discussed for the subsurface mapping and groundwater investigation in the Siak River delta, Indonesia. Findings of this study are very important in development of the knowledge especially for peat land, hydrogeology and geomorphology, and to develop the subsurface geological model. A part on that, particularly, for the local government is also important as the guidance to manage the groundwater resources in term of the peat fire rescue.

The location of the research area is given in Figure 1. The study area has an altitude between 8 to 12 meters above mean sea level. The yellow lines in the figure representing the direction of the resistivity data, but the length of the line does not indicate the length of the path, whereas the blue circle is the location of the wellbore. The research area is bounded at the eastern part with the Siak River, where this area is part of the Biosfere Giam Siak Kecil Reserves area, Bengkalis Regency, Riau.

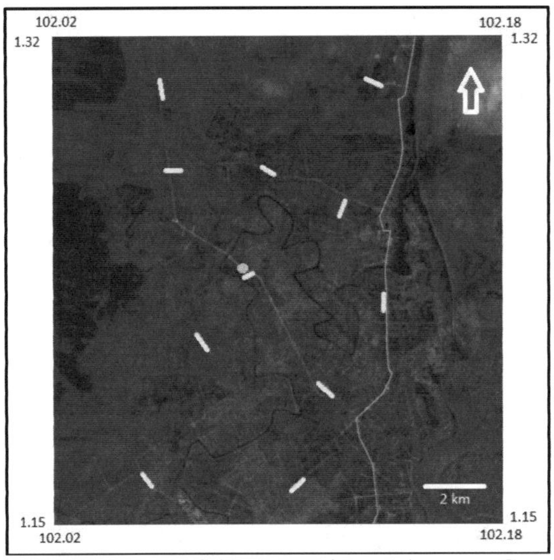

Figure 1. Satellite image of the research area (source: Google Earth)

2. Methodology

In this research, the geoelectrical resistivity survey and soil property analysis were used to study the groundwater and peat land characteristic. The geolectrical resistivity was calibrated with a geological data obtained from drilled well to attain a more accurate interpretation and with the peat soil sample. Uninterrupted peat lands were sampled using homemade equipment with a volume of 500 ml from a depth of 50 cm in some locations. Peat soil samples were then dried and the dried peat samples were watered to obtain total porosity[14]. The porosity of peat land is then used to predict water requirements for the entire study area.

The geoelectrical resistivity method with Schlumberger configuration was used to survey the study area of about 130 km^2. The data collection used an in-house resistivity meter with 4 electrodes for each measurement. The electrodes spacing was placed from a distance of 50 cm until more than hundred meter for covering the high resolution at the near surface. Inversion of the raw data from the field in order to obtain resistivity data was done using Progress1D inversion program. Dominant resistivity value at each depth was used as an input data in generating a three dimensional resistivity slice maps. Kriging technique was employed to produce contour of resistivity data.

Based on the Ohm's law, the potential difference is equal to the multiplication of current and material constraints. When combined with material properties, material durability will depend on the type of resistance material, then the resistance of the material type can be determined if the measurement geometry is known. Finally the potential difference can be determined using Equation 1. Fig. 2 is a moving electrical ray in the earth media of one electrode, and the negative electrode is assumed to lie at a great distance, which will produce potential at point h.

$$dV = iR_{shell} = i\rho\frac{dr}{A} = \frac{dr}{2\pi r^2} \quad (1)$$

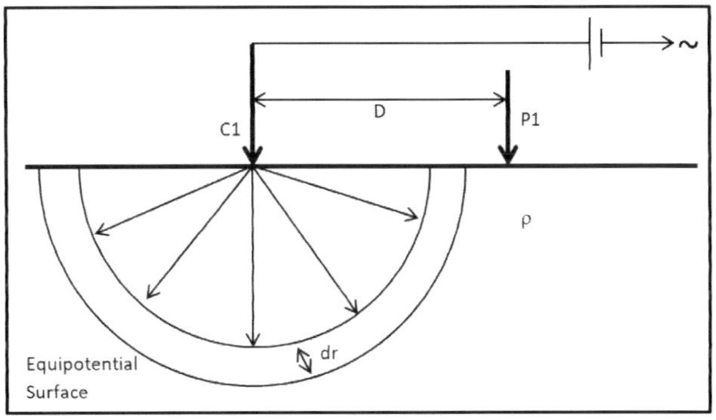

Figure 2.An illustration of electric current propagation from an electrode [15]

For the Schlumberger configuration (Fig. 3), the distance between the electrodes is as in the figure, so that the sub-surface resistivity is (Equation 2)

$$\rho_a \approx \frac{\pi}{2}\frac{L^2}{l}\frac{\Delta V}{i} \quad (2)$$

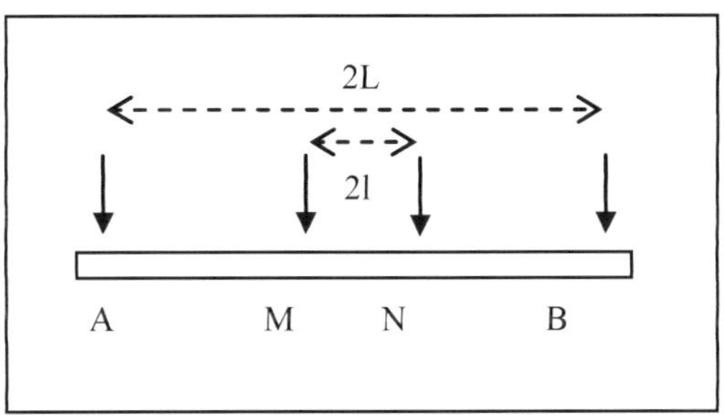

Figure 3. Schlumberger configuration (modified from Telford 1991)

A new well was drilled to obtain geological data for calibration purposes with geoelectrical resistivity data. Soil samples obtained directly from drilling were taken and analyzed in term of their grain size and color. These data were used to control the resistivity interpretation for whole data.

Semivariogram was used for the data modeling prior to Krigging simulation. The semivariogram is one of the significant functions to show the spatial relations in observations measured at the sample sites. Generally it is given as a graph that shows the variance in the measurement of the distance between all pairs of sample locations with distance d. A graph is helpful for constructing a mathematical model that describes the measured variable of a sample with another sample location.

In the modeling process, the relationship between the sample locations to show the variability of measurement by the separation distance is called semi-variogrammodeling. It is applied in calculations that involve an estimation of an unknown new location value [16,17]. With a simple semi-variogram can be described as a plot between the semi-variant of the distance, which is measured at every distance with difference in lag. The semi-variogram is actually can be expressed mathematically as in the following Equation 3:

$$\gamma_h = \frac{\sum_i^{n-h}(X_i - X_{i+h})^2}{2n} \tag{3}$$

Further stage, semi-variogram data were a requirement for the Krigging Gaussian simulations process [16]. The krigging process was done after the semi-variogram data was obtained. Simply krigging is a prediction the unknown data at a certain location by considering the direction of whole data trends, the amount of data around the unknown data and distance to the surrounding data.

Mathematically krigging can be illustrated by the equation below. Gama is a semi-variogram, h is a distance and W is a weighting. Finally the data predictions at location p can be obtained [18].

$$\hat{Y}_p = \sum W_i Y_i \tag{4}$$

Where Yp is the value at the location p which will be in prediction by weighting of the value at the surrounding location based on the given semi-variogram data.

3. Results and Discussion

Table 1.is the peat land porosity data obtained from 6 locations. It is seen clearly, the porosity of peat in this research area is quite big with an average of about 41.5%. Although the smallest porosity is obtained from peat lands that are closer to the Siak river (SK2 and SK5), however, the value of porosity at these locations differed only slightly with the porosity of peat land that is relatively far from the Siak River.

Table 1. Porosity of peat soil sample from a depth of 50cm

No	Sample ID	Long (°)	Lat (°)	Porosity (%)
1	SK1	102.050094	1.181669	42
2	SK2	102.116613	1.202139	40
3	SK3	102.072606	1.239806	42
4	SK4	102.063108	1.239455	42
5	SK5	102.129552	1.25016	40
6	SK6	102.073659	1.265075	43
		Average		41.5

Figure 4 shows the result of subsurface resistivity obtained from the inversion process for line 1 and line 8. The apparent resistivity data from some locations is then converted to the required data format, so it can be processed to obtain the actual subsurface resistivity value. In Figure 4, the dot of the circle is the data obtained from the field and the line is the calculated resistivity, whereas the blocked graph is the inverse interpretation value. The table to the right is the depth and true resistivity value obtained from the interpretation, where the interpretation is based on the trend curve, yielding the raw data point curve. The table at the bottom shows the configuration and raw data obtained from the field, and error estimation in the inversion process. The column on the lower right side shows the vertical interpretation of the subsurface based on the table above it. In the figure, it can be seen that the resistivity value ranges from 15ohm.mto more than 180 ohm.m. The interpretation of the results of this data is that the subsurface layer consists of 6 main layers, interpreted as peat layer (layer 1), very fine sediments contaminated by marine sediment (layer 2), coarse and coarse sand (layer 3), very fine sediments with relatively white (layer 4) and a deep aquifer zone with coarse sand (layer 5) and the latter is finer clay sediment mixed with sand (layer 6).The well data provides the lithology condition that was taken from direct drilling at survey location 8. This location was chosen because it is in the middle position of the research area, so it is expected to be able as a represent for the overall subsurface geological condition. This well data is used to calibrate the inversion resistivity data so that it can be interpreted in the form of lithology, rather than in the form of resistivity.

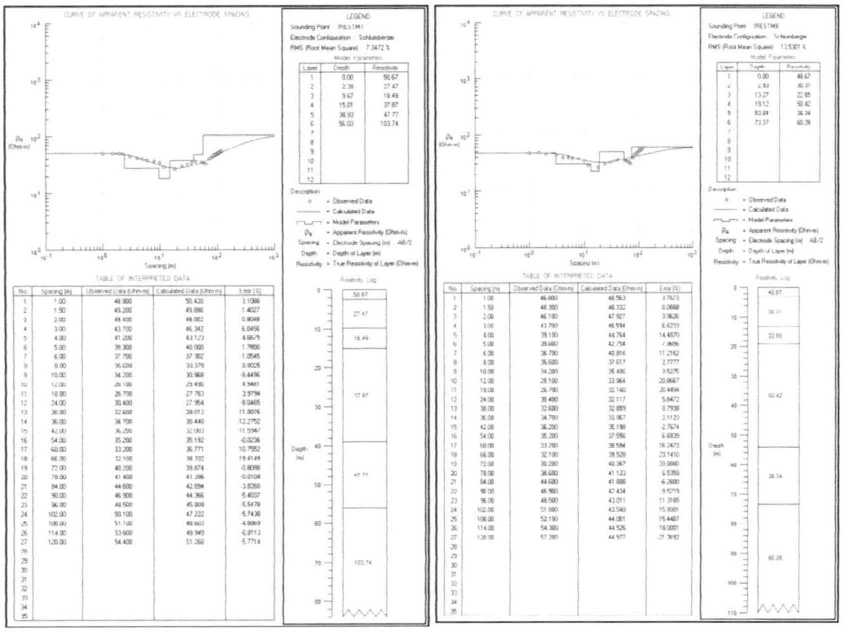

Figure 4. Inversion of resistivity data at survey line 1 and line 8, and subsurface interpretation

The average resistivity value of the first layer is 48 ohm.m with thickness of 2 meters to 4 meters, this layer is the top layer interpreted as peatland. This can be known through the supporting data obtained from direct resistivity measurement, which the average peat resistivity value is about 45-52 ohm.m. Figure 5 shows contour of the peat thickness obtained from the interpretation of the geoelectical resistivity survey data. This peat layer is not evenly distributed in this research area. In the northern part, the peat layer is found with a thickness of approximately 4 meters. But in the south, it is only with a depth of 2 meters. While in the middle of the research area, peat layer is almost not found. The peat layer resistivity in general has a decreasing trend towards the north. This is due to residual effects of sea water remaining in the peat.

The second layer is the resistivity value zone of about 25 ohm.m with a thickness of 7 meters to 9 meters, interpreted as clay layer with marine deposit. The third layer is the zone with the resistivity of 18 ohm.mto 25 ohm.m with a thickness of 8 meters on average. This layer is interpreted as sand layer which is estimated to contain brackish groundwater (aquifer). The fourth layer is the zone with average resistivity value of 40 ohm.m with average thickness of around 25 meters, this layer is interpreted as Clay layer. The fifth zone is the layer with average resistivity of around 40 ohm.m with the average thickness of 18 meters, this layer is interpreted as deep aquifer. And the last layer is the clay which is the deepest zone with a resistivity value of 70 meter.

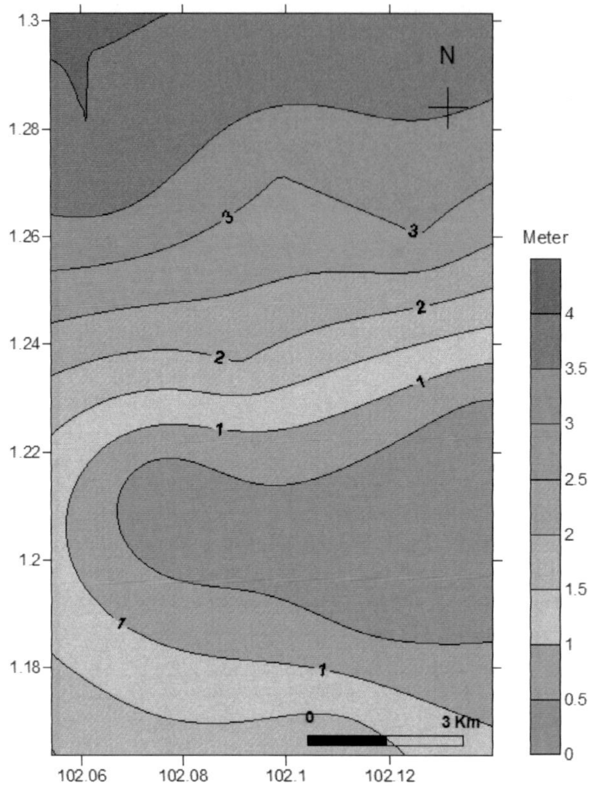

Figure 5. Thickness of the peat layer based on the geoelectrical resistivity interpretation data

Figure 6 shows a subsurface resistivity depth slice for several depth resulting from the kriggring simulation of all available resistivity data. In the krigging process, each location at the same depth in the sample data was used for the interpolating process. Semi-variogram modelling was done before krigging simulation. The result of semi-variogram analysis provides that the data trend direction is from the north of an angle about 10^0 to the west. At the center of the study area (Figure 5), at the 0 m deep, a relatively large resistivity value of around 200 ohm.m can be observed. These values are collected from direct resistivity measurement at the zone of relatively dry claysoil. It is a representative of more dense of clay soil with low water content in the pore of the soil. Around this site, peat soil was not found.

In the southeastern part, it can be seen that the resistivity value is relatively high at the depth ranging from 0 m. This value indicates the presence of relatively dry clay. This interpretation is supported by the presence of dry clays in the data collection zone in this area. In addition, the interpretation is also proved by calibration at wells indicating slightly dry clay having a resistivity value of more than 200 ohm.m

The clay soil on the surface spreads to the south with the depth of 5 m deep. Overall it is seen that the upper layer sediments are thickening towards the north with a thickness of about 5 m in the south and 30 m in the north. It can be seen also that the shallow groundwater potential is greater in the northwest. This is indicated from the relatively low resistivity of around 50 ohm.m in the area.

The possibility of deeper aquifers can be found at a depth of 60 m in the south part and has a slope towards the northwest at a depth of 100 m in this zone.Furthermore, the volume of peat soil can be calculated by providing the upper and lower boundary as well as other lateral boundary, so that the peat volume is found as much as 146.2 million cubic meters. If it is assumed that the peat porosity is 41.5% on average, then the water requirement for the whole area is 60.7 million cubic meters. Overall, the thickness of shallow aquifer is obtained 8 meter. Based on the soil obtained from the well data, the aquifer consists of the coarse-grain sand and the aquifer porosity is approximately of 25% then the volume of water present in the shallow aquifer is calculated as much as 195 million cubic meters. It is highly visible that a shallow aquifer is able to inundate the entire peat in the study area.

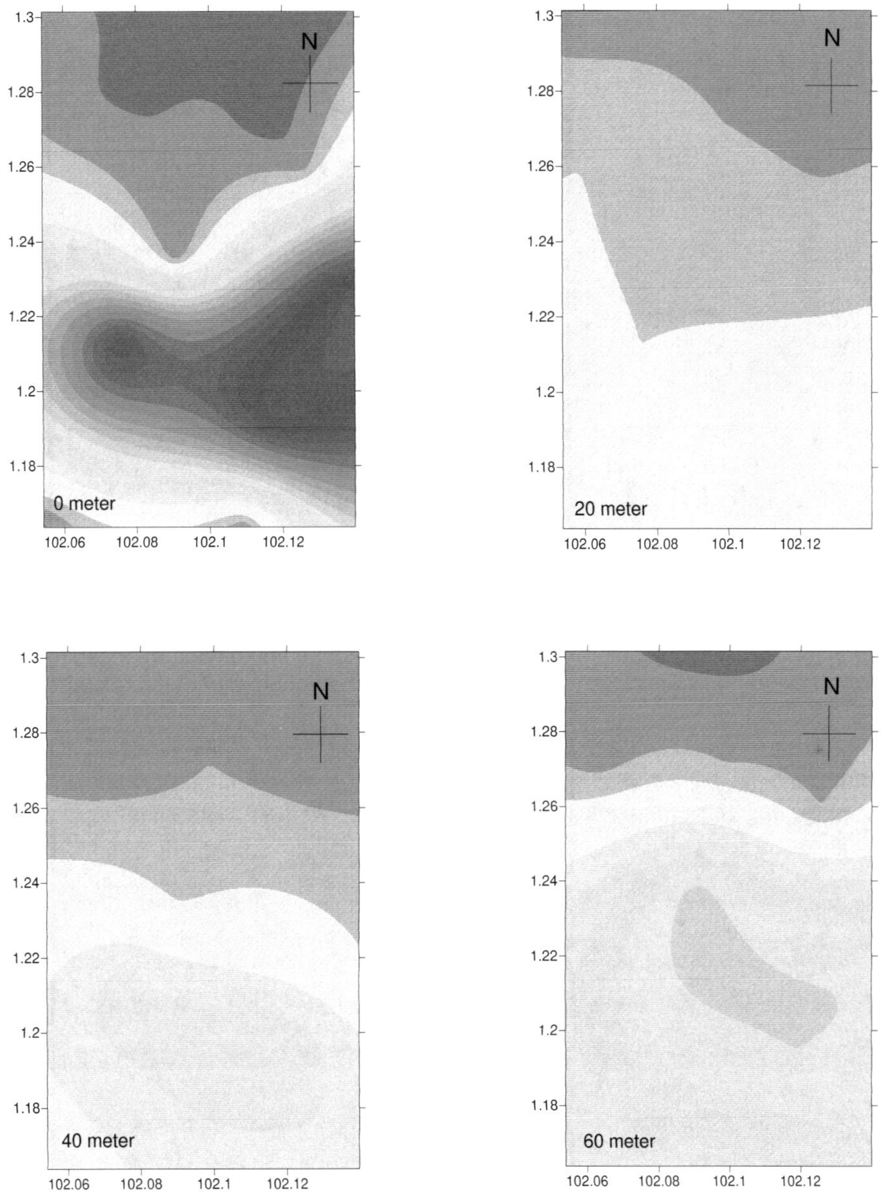

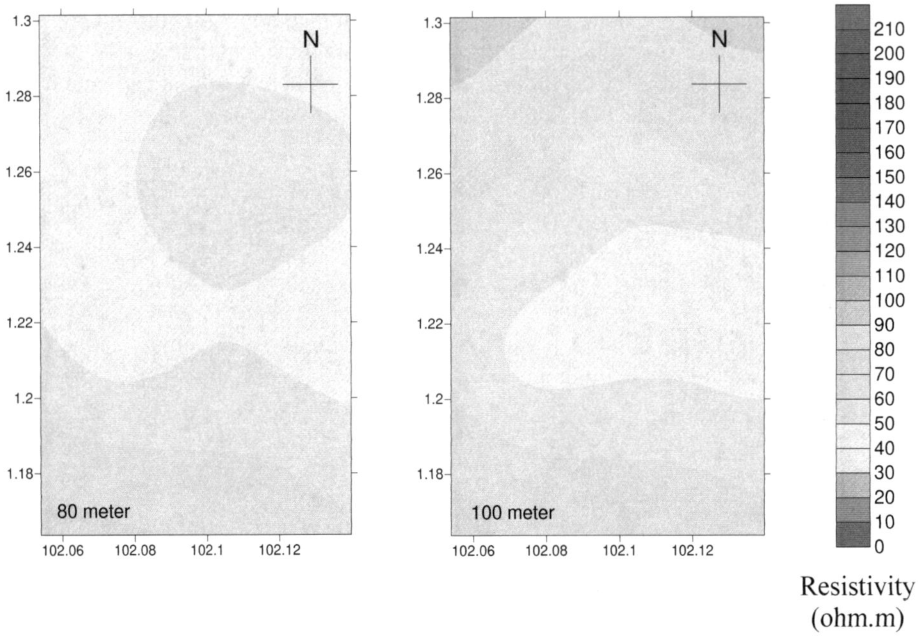

Figure 6.The geoelectrical resistivity in the depth slice that is generated using Krigging process

4. Conclusion

The geoelectrical resistivity method is well examined for detecting subsurface conditions in term of the groundwater resources. It is due to the geoelectrical resistivity is very sensitive to the granular type and the fluid in the pore. Unconsolidated sediments are not dense and have pores and it is connected to each other. This causes the flow of electric current easily if filled by a conductive fluid. This property will cause this sediment to have a small enough resistivity. Peat soil is found in the middle to the north zone of the study area. While in the southern and south eastern parts, peat lands are hardly seen. The aquifers in this study area are found in the shallow depth as well as at the deeper depth. In general, sedimentary rocks are thickening towards the northwest as well as with shallow aquifers and deep aquifers. The groundwater volume is predicted more than enough to watering the peat land in order to rescue peat fire.

Acknowledgments

Thank you to the Institute for Research and Community Service (LPPM) University of Riau, which has funded this research. Acknowledgments are also shared with field colleagues who assist in obtaining data during the survey.

References

[1] Supardi A D, Subekty, Sandra G, Neuzil 1993General geology and peat resources of the Siak Kanan and Bengkalis Island peat deposits, Sumatra, Indonesia, *Geological Society of America Special Papers***286**45-62

[2] RossieW N and Sudarmadji 2014 Emisi CO2 Tanah Akibat Alih Fungsi Rawa Gambut di Kalimantan Barat *Jurnal Manusia dan Lingkungan* **21(3)** 264-276

[3] Wahyunto, Ritung and Suprato 2015 Sebaran Gambut dan kandungan Karbon di Sumatra dan Kalimantan. Program dan Wildlife Habitat Canada, Bogor

[4] Dian A, CarolienK and Asmadi S 2016 Indonesia palm oil production without deforestation and peat conversion by 2050 *Science of The Total Environment* **557–558** 562-570

[5] Majalah Riaupos 2015 Solusi Kabut Asap MasaDepan *Riaupos* Edisi 139

[6] Fereidoun R, Jonathan S P, *et al* 2016 Structure of peat soils and implications for water storage,

flow and solute transport: A review update for geochemists *Chemical Geology* **429** 75-84
[7] YunitaL and HaryonoS 2015 Dampak pembangunan hutan industri di Lahan Gambut terhadap tingkat kematangan permukaan tanah *Jurnal Manusia dan Lingkungan* **22(2)** 179-186
[8] Islami N 2010 Geolectrical resistivity and hydrogeochemical contrast between the area that has been applied with fertilization for long duration and non-fertilization *ITB J. Eng Sci.***42(2)** 151-165
[9] Islami N, Taib S, Yusoff I and Ghani A A 2011 Time lapse chemical fertilizer monitoring in agriculture sandy soil, *International Journal of Environmental Science & Technology* **8(4)** 765-780
[10] Islami N, Taib S, Yusoff I and Ghani A A 2012 Integrated geoelectrical resistivity, hydrochemical and soil property analysis methods to study shallow groundwater in the agriculture area, Machang, Malaysia *Environmental Earth Sciences* **65(3)** 699-712
[11] Huaqing L, Zhong H,*et al*2016Sedimentary characteristics and seismic geomorphology of gravity-flow channels in a rift basin: Oligocene Shahejie Formation, Qinan Slope, Huanghua Depression of Bohai Bay Basin, China, *Marine and Petroleum Geology*, In Press, Corrected Proof, Available online 16 February 2016
[12] Sacchetti F, Benetti S, *et al* 2012Deep-water geomorphology of the glaciated Irish margin from high-resolution marine geophysical data *Marine Geology* **291–294** 113-131
[13] El Sayed A, El Gammal, *et al* 2013Applications of geomorphology, tectonics, geology and geophysical interpretation of, East KomOmbo depression, Egypt, using Landsat images, *The Egyptian Journal of Remote Sensing and Space Science* Volume **16(2)** 171-187
[14] Jing Guo, Haibo Jiang, HongfengBian, Chunguang He, YingzhiGao, 2016, Effects of hydrologic mediation and plantation of Carexschmidtii Meinsh on *peatland* restoration in China's Changbai Mountain region, Ecological Engineering, Volume 96, Pages 187-193,
[15] Telford 1991 Applied Geophysics*Cambridge University Press*, Pages 770
[16] Falk H, Vladyslav P, Steffen S, Sabine A 2014 Generating random fields with a truncated power-law variogram: A comparison of several numerical methods *Environmental Modelling & Software,* **55** 32-48
[17] Bachmaier, M and Backes, M 2008 Variogram or semivariogram? Understanding the variances in a variogram. Article doi:10.1007/s11119-008-9056-2, *Precision Agriculture, Springer-Verlag, Berlin, Heidelberg, New York.*
[18] Kigobe M M, Kizza 2006 Dealing with spatial variability under limited hydrogeological data. Case study: hydrological parameter estimation in Mpigi-Wakiso, *Proceedings from the International Conference on Advances in Engineering and Technology* 211-220

Insect-based protein: future promising protein source for fish cultured

R A Nugroho* and F M Nur

Animal Physiology, Development, and Molecular Laboratory, Department of Biology, Faculty of Mathematics and Natural Sciences Mulawarman University. Samarinda, East Kalimantan, Indonesia.

*Corresponding author: rudyagung.nugroho@fmipa.unmul.ac.id

Abstract. As one of the vital component feed used in fisheries, fishmeal (FM) is generally added to the fish diet to enhance fish growth, digestive performance and absorption of nutrients. This addition contributes significantly to the variable production cost in the aquaculture industry. Expanded production of carnivorous species requiring high protein, high-energy feeds will further tax global fish meal. Thus, research based on the low-cost budget for feed operating cost should be strategized to assist aquaculturists in enhancing fish productivity. Moreover, suitable alternative feed ingredients will have to be utilized to provide the essential nutrients and energy needed to fuel the growth of aquaculture production. To this effect, the use of insect-based protein sources to replace FM that often scarce, expensive, limited availability, and leads to high fish production costs is alternative ways and has been gaining momentum. Currently, Insects have been proposed as one of the potential future protein sources of protein because of the production of insects is highly sustainable. Farming insects is characterized by higher food conversion efficiencies, lower environmental impact, and higher potential to be grown on waste streams.

1. Introduction

As fish demand continues to grow, the strategies to increase fish productivity need to be developed. The research based on eco-friendly and sustainable commercial aims should also be strategized to assist aquaculturists in increasing fish productivity. To this effect, the use of fish meal in the fish diet in production systems to improve productivity has been gaining momentum. Fish meal plays an important role ingredient for fish feed [1] . Fish meal is a prime protein source and known as a high-quality, very digestible feed ingredient [2, 3]. Fish meal has large unit energy per unit weight which contains protein, lipids (oils), minerals, and vitamins. However, there is a limit amount of carbohydrate in fishmeal.

Fish meal is a generic term for a nutrient-rich feed ingredient used primarily in diets for domestic animals, sometimes used as a high-quality organic fertilizer. Fish meal can be made from almost any type of seafood but is generally manufactured from wild-caught, small marine fish that contain a high percentage of bones and oil, and usually deemed not suitable for direct human consumption. These fishes are considered 'industrial' since most of them are caught for the sole purpose of fishmeal and fish oil production. A small percentage of fishmeal is rendered from the by-catch of other fisheries, and by-products or trimmings created during processing (e.g., fish filleting and cannery operations) of various seafood products destined for direct human consumption.

The fish meal is one of the few major animal industries existing today that still relies greatly on a "hunting-and-gathering" technique. Most fish rendered into a meal is captured at sea. Millions of tons of fishmeal are produced worldwide. Contrary to recent that beneficial to increase weight gain, specific growth rate, feed conversion ratio, protein utilization efficiency, and carcass enhancement, more than tons of fish are used to produce fish meal.

According to United States Department of Agriculture [4], Thailand has the highest production of fish meal in Asian region, reaching up to 430.000 metric tons while Japan around 190.000 metric tons and Indonesia about 16.000 metric tons (figure 1). In 2016, it is estimated that the production annual growth rate of fish meal is up to 3.70% in Vietnam, followed by Japan 2.70%.

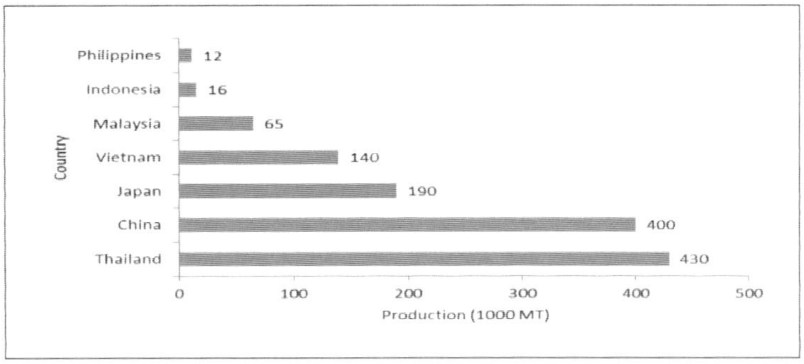

Figure 1. Fish meal production by country [4]

On the other hand, there is a reducing of fish production in related to the fish meal. The use of fish from nature to produce fish meal is a competition with human needs. Further, the decrease in the availability, sustainability, and the increase in the prices of the fish meal have stimulated the search for sustainable alternatives for aquaculture feeds. It is also stated that feeding fish to fish is useless and wasteful that because more than 6 kg of wild fish is taken up to produce 1 kg of farmed fish [5]. To this regard, the use of protein-based plant is an alternative way to satisfy the increasing demands of the aquaculture feed industry, to partially or totally replace of fish meal in fish diet.

Some attempts have been made to explore the possibilities on the use of plant-based protein to replace total or partial replacement of fish meal. Previous research stated that many protein-based plants have a good nutritional value including protein content that beneficial to fish growth and physiology (table 1).

Table 1. The use of plant-based protein and its effects on the growth and physiology of fish

No	Plant-based Protein	Effects	References
1	seaweed *Gracilaria arcuate*	Better weight gain, specific growth rate, and feed utilization on African catfish, *Clarias gariepinus*	[6]
2	pea protein concentrate	Amino acid profiles of the diets, body composition, and growth performance of juveniles of tench (*Tinca tinca* L.)	[7]
3	Soybean meal	Increase marine fish growth	[8]
4	Lupin	Better apparent digestibility and coefficient of protein (ADC-P) on the Juvenile barramundi (*Lates calcarifer* Bloch 1970)	[9]
5	Soybean meal, peas, corn gluten, and wheat	Skeletal muscle growth, flesh texture of *Senegalese sole*	[10]
6	Canola meal	Growth, feed utilisation, plasma biochemistry, histology of digestive organs and hepatic gene expression of barramundi (Asian sea bass; *Lates calcarifer*)	[11]
7	Cotton Seed Meal without lysine supplementation	Growth performance, body composition and digestive enzyme activities.	[12]

No	Plant-based Protein	Effects	References
8	Rice polish	Growth performance of Nile tilapia (*Oreochromis niloticus*)	[13]
9	Wheat gluten	Growth in the range of 40-130 g was superior to the fishmeal control with the diet containing wheat gluten gilthead seabream, *Sparus aurata* L	[14]

Due to the reason mentioned above, fishmeal is still quite commonly used in animal feed diets. Nevertheless, overfishing, the low availability and the increase in the prices of fish meal which is reflected by increasing market prices over the last decades have stimulated the search for environmentally friendly and eco-sustainable alternatives for animal feeds. Alternative protein sources of comparable value are therefore urgently needed. Thus, the potential of insect-based protein in animal feed diets has attracted much attention

2. Why insect?

It is predicted by Food and Agricultural Organization (FAO) the world need to enhance food production by 70% by 2050 that would be given to whole population in the world which cover 9 billion people. Almost 80% of the world's agricultural land is occupied for grazing and feeding farm animals. Meanwhile, global meat consumption, especially in relation to fish production is expected to rise by 50% above 2006 levels to meet expected demand by 2050 [21].

In order to balance that situation above, a comparable value of alternative protein sources is urgently needed to produce protein source production. Currently, the potential of insect-based protein has attracted much attention not only farmer but also researcher. It seems there are few reasons that can consider in related to the use of insect.

The insects such as crickets, caterpillars and silkworms could be the good source of food in the future. Insects, an edible food that contains protein, vitamins and important amino acids are efficient to be reared. Insects only need six times less feed than cattle and can be cultivated by using organic waste. As consequences, insects produce fewer greenhouse gasses emission (100-1000x) [22]. The others reasons that could be used as consideration are insects has high feed conversion rate (FCR), low use of water and energy, and nutritious as source of essential protein as well as amino acids for animal feed (table 2). However, to ensure the less cost production of insect-based protein, the insect must be able to reproduce in short time [23].

Table 2. Examples of nutrient contents of some insects

Nutrients	House Cricket (*Acheta domesticus*)	Meal worm (*Tenebrio molitor*)	Silk worm (*Bombyx mori*)	Mormon cricket (*Anabrus simplex*)	Black fly soldier (*Hermetia illucens*)
Crude protein*	55-67	47-60	52-71	60.3^	42++
Fats*)	10-22	31-43	6-37	12.9^	35++
Methionine+	1.4	1.5	46**	1.4	2.1
Cysteine+	0.8	0.8	14**	0.1	0.1
Lysine+	5.4	5.4	75**	5.9	6.6

Source: *Crude protein and ether extracts of fats (% of dry matter) in house cricket, silkworm and mealworm. +Amino acid content (g/16 g N) in insects produced as animal feed [24] **Amino acid composition of pupae of the silkworm (mg/g crude protein) [25]. ++protein and fats content [26]. ^Proximate analysis (percent, dry weight) of adult Mormon crickets [27].

3. Insect-based protein in aquaculture

Traditionally, the application of insect in aquaculture fields has been seen in a recent decade. In Uganda, a research regarding on the use of ingredient such as vegetables, grass, cereals, cereal brands, oil seed cakes, industrial and kitchen wastes fishmeal and insects by fish farmer was reported by Rutaisire [28] (Figure 2). Unfortunately, the availability of most of these ingredients is seasonal. In addition, termites for feeding fish that used by fish farmer up to 5% has been collected directly or purchased from collectors at a cost of US$0.27/kg – only at certain month, March to April and from August to

September. The availability of insect is also depending on the number and size of termite hills on the farm, moonlight intensity and termite species. Meanwhile in Southeast Asia region, fluorescent light commonly used as an insect attractant that place above fish ponds. The insect that attracted by light falls into the pond and can be eaten by fish. Wingless grasshoppers and crickets, another example of insect which cannot float, are also made as fish bait [29].

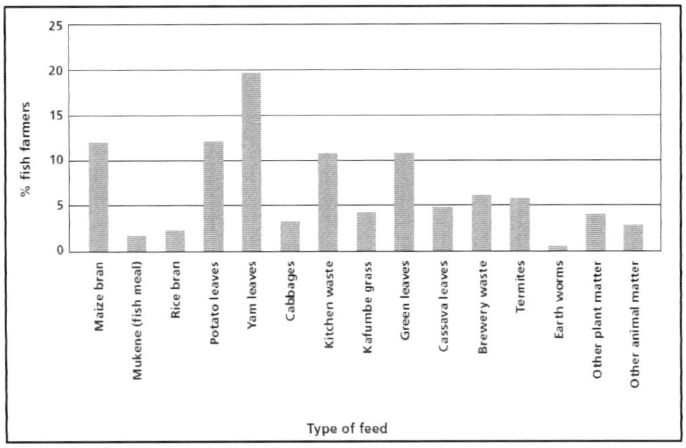

Figure 2. Several of feed types that is used by fish farmer

In current year, fish meal is very popular in aquaculture. Almost 63% of the fish meal is used in aquaculture field [30] . However, due to the low availability and the increase in the prices of fish meal have stimulated the search for environmentally friendly and eco-sustainable alternatives for aquaculture feeds [1]. To this concern, the research to find out new ingredients to replace fish meal has gained a momentum and growth importance specifically in aquaculture field. Recently, the aquaculture field has shown a significant improve and in future the amount of global productions is a key point to provide a pivotal point in supply of seafood to satisfy the growing consumers demand.

Table 3. Insect-based meal and its effect on the growth and physiological of fish

No	Insect-based Protein	Effects	References
1	*Bombyx mori*	Growth, Survival, body composition, hematological parameters of rainbow trout *Oncorhynchus mykiss*	[31,32]
2	*Tenebrio molitor*	Growth performance, whole body composition and in vivo apparent digestibility on European sea bass (*Dicentrarchus labrax* L.) juveniles	[33]
3	*Musca domestica*	Growth and nutrient utilization indices on African catfish (*Clarias gariepinus*)	[34]
4	Grasshoppers	Growth of *Clarias gariepinus* fingerlings	[35]
5	Cricket (*Gryllus bimaculatus*)	Growth performance, antioxidant enzyme activities, and haematological response of African catfish (*Clarias gariepinus*)	[36]
6	Black Soldier Fly *Hermetia illucens* (Linnaeus, 1758) Prepupae	Growth of *Clarias gariepinus* (Burchell, 1822)	[37]

Previous research stated that, feeding trials on several fish using ratio 25:50 % of fish meal was replaced with acridid (a genus of grasshoppers) meal produced results similar to the control diet, containing 100 percent fish meal. In addition, growth parameters measured for the selected fish found significantly higher fish fed acridid meal than for fish fed commercial diets. This finding indicates that acridids can be successfully implemented as fish meal replacement [38]. Another successful implementation research of insect as fish meal replacement can be seen in Table 3 above.

Though insect based protein has many advantages in animal feed specifically aquaculture fields, there are other reasons that need to be considered as the disadvantage of insect meal. Some opponent stated that the production costs for insect-based meal are still relatively high; the effects of insect nutrients on the health of animal still need to be further investigated. The legislation will also need to be paid to cove the safe use of substrates such as vegetable and domestic waste and manure that usually is economically made as insect's substrate [39].

4. Conclusion

The fish meal will stay as a pivotal point in future diets of fish. However, the concerns about the sustainability of these ingredients and the increasing price on the global market are boosting the fish farmer to use a more efficient of fish meal and find alternative feed sources such as insect-based meal. Next research needs to be conducted in order to determine the further effects on the use of insect-based meal on the physiology, immunology, and molecular basis of the animal.

Acknowledgement
The author grateful to Department of Biology, Faculty of Mathematics and natural Sciences, Mulawarman University for any kinds of support to present this paper.

References

[1] Caruso G 2015 Use of Plant Products as Candidate Fish Meal Substitutes: An Emerging Issue in Aquaculture Productions *Fisheries and Aquaculture Journal* **6** 1-3
[2] Miles R and Chapman F 2006 The Benefits of Fish Meal in Aquaculture Diets *IFAS Extension* **FA122** 1-6
[3] Tantikitti C, Chookird D and Phongdara A 2016 Effects of Fishmeal Quality on Growth Performance, Protein Digestibility and Trypsin Gene Expression in Pacific White Shrimp (*Litopenaeus vannamei*) *Songklanakarin Journal of Science and Technology* **38** 73-82
[4] USDA 2017 Fish Meal Production by Country in 1000 MT. Index Mundi. https://www.indexmundi.com/Agriculture/?commodity=fish-meal&graph=production. accessed July 2017.
[5] Schipp G 2008 Is the Use of Fishmeal and Fish Oil in Aquaculture Diets Sustainable? *Technote* **124** 2-15
[6] Al-Asgah N A, Younis E-S M, Abdel-Warith A-W A and Shamlol F S 2016 Evaluation of red seaweed Gracilaria arcuata as dietary ingredient in African catfish, Clarias gariepinus *Saudi Journal of Biological Sciences* **23** 205-10
[7] González-Rodríguez Á, Celada J D, Carral J M, Sáez-Royuela M and Fuertes J B 2016 Evaluation of Pea Protein Concentrate as Partial Replacement of Fish Meal in Practical Diets for Juvenile Tench (*Tinca tinca* L.) *Aquaculture Research* **47** 2825-34
[8] Mo W Y, Lau R S S, Kwok A C K and Wong M H 2016 Use of Soybean Meal and Papain to Partially Replace Animal Protein for Culturing Three Marine Fish Species: Fish Growth and Water Quality *Environmental Pollution* **In Press: Corrected Proof** 1-6
[9] Ilham, Fotedar R and Munilkumar S 2016 Effects of Organic Selenium Supplementation on Growth, Glutathione Peroxidase Activity and Histopathology in Juvenile Barramundi (*Lates calcarifer* Bloch 1970) Fed High Lupin Meal-Based Diets *Aquaculture* **457** 15-23
[10] Valente L M P, Cabral E M, Sousa V, Cunha L M and Fernandes J M O 2016 Plant Protein Blends in Diets for *Senegalese sole* Affect Skeletal Muscle Growth, Flesh Texture and The Expression of Related Genes *Aquaculture* **453** 77-85
[11] Ngo D T, Wade N M, Pirozzi I and Glencross B D 2016 Effects of canola meal on growth, feed utilisation, plasma biochemistry, histology of digestive organs and hepatic gene expression of barramundi (Asian seabass; Lates calcarifer) *Aquaculture* **464** 95-105

[12] Yu D H, Gong S Y, Lin Y C and Yuan Y C 2014 Partial Replacement of Fish Meal by Several Plant Proteins With or Without Iron and Lysine Supplement in Diets for Juvenile Chinese Sucker, *Myxocyprinus asiaticus Aquaculture Nutrition* **20** 205-12
[13] Khan M S K, Siddique M A M and Zamal H 2013 Replacement of Fish Meal by Plant Protein Sources in Nile Tilapia (*Oreochromis niloticus*) Diet: Growth Performance and Utilization *Iranian Journal of Fisheries Sciences* **12** 864-72
[14] Kissil G W and Lupatsch I 2004 Successful Replacement of Fishmeal by Plant Proteins in Diets for The Gilthead Seabream, *Sparus aurata* L *The Israeli Journal of Aquaculture* **56** 188-99
[15] Glencross B, Evans D, Hawkins W and Jones B 2004 Evaluation of Dietary Inclusion of Yellow Lupin (*Lupinus luteus*) Kernel Meal on The Growth, Feed Utilisation and Tissue Histology of Rainbow Trout (*Oncorhynchus mykiss*) *Aquaculture* **235** 411-22
[16] Francis G, Makkar H P S and Becker K 2001 Antinutritional Factors Present in Plant-Derived Alternate Fish Feed Ingredients and Their Effects in Fish *Aquaculture* **199** 197-227
[17] Rolland M 2014 Effects of Dietary Methionine on Feed Utilization, Plasma Amino Acid Profiles and Gene Expression in Rainbow Trout (*Oncorhynchus Mykiss*). In: *National Institute of Aquatic Resources,* (Denmark: Technical University of Denmark) p 173
[18] Chung E 2016 Shift to Plant-Based Fish Feed Could Hurt Health Environment. (Canada: CBCNews Technology and Science)
[19] Glencross B, Evans D, Rutherford N, Hawkins W, McCafferty P, Dods K, Jones B, Harris D, Morton L and Sweetingham M 2006 The Influence of The Dietary Inclusion of The Alkaloid Gramine, on Rainbow Trout (*Oncorhynchus mykiss*) Growth, Feed Utilisation and Gastrointestinal Histology *Aquaculture* **253** 512-22
[20] Hemre G, Amlund H, Aursand M, Bakke A M, Olsen R E, Ringø E and Svihus B 2009 *Opinion of the Panel on Animal Feed of The Norwegian Scientific Committee for Food Safety: Criteria for Safe Use of Plant Ingredients in Diets for Aquacultured Fish* vol 7/604 (Norwegia: Norwegian Scientific Committee for Food Safety)
[21] Laaninen T 2016 Insects – Soon to be A Regulated Food? ed M R Service (Europe: European Parliamentary Research Service) pp 1-2
[22] Frangoul A 2016 Insects: Food of The Future? (US: CNBC LLC)
[23] Józefiak D, Józefiak A, Kierończyk B, Rawski M, Świątkiewicz S, Długosz J and Engberg R M 2016 Insects: A Natural Nutrient Source for Poultry–A Review *Annals of Animal Science* **16** 297-313
[24] Makkar H P, Tran G, Heuzé V and Ankers P 2014 State of The Art on Use of Insects As Animal Feed *Animal Feed Science and Technology* **197** 1-33
[25] Zhou J and Han D 2006 Proximate, Amino Acid and Mineral Composition of Pupae of The Silkworm *Antheraea pernyi* in China *Journal of Food Composition and Analysis* **19** 850-3
[26] Stamer A, Wessels S, Neidigk R and Hoerstgen-Schwark G 2014 Black Soldier Fly (*Hermetia illucens*) Larvae-Meal as An Example for A New Feed Ingredients Class in Aquaculture Diets. In: *The 4th ISOFAR Scientific Conference,* ed G Rahman and U Aksoy (Istanbul, Turkey pp 1043-6
[27] DeFoliart G, Finke M and Sunde M 1982 Potential Value of The Mormon Cricket (Orthoptera: Tettigoniidae) Harvested as A High-Protein Feed for Poultry *Journal of Economic Entomology* **75** 848-52
[28] Rutaisire J 2007 Analysis of Feeds and Fertilizers for Sustainable Aquaculture Development in Uganda *FAO Fisheries Technical Paper* **497** 471
[29] Van Huis A, Van Itterbeeck J, Klunder H, Mertens E, Halloran A, Muir G and Vantomme P 2013 *Edible Insects: Future Prospects for Food and Feed Security* vol 171 (Rome: FAO)
[30] Boyd C E 2015 *Feed and Feeding Practices in Aquaculture,* (UK: Woodhead Publishing) p 21
[31] Shakoori M, Gholipour H and Naseri S 2015 Effect of Replacing Dietary Fish Meal With Silkworm (*Bombyx mori*) Pupae on Hematological Parameters of Rainbow Trout *Oncorhynchus mykiss Comparative Clinical Pathology* **24** 139-43

[32] Shakoori M, Gholipour H, Naseri S and Khara H 2016 Growth, Survival, and Body Composition of Rainbow Trout, *Oncorhynchus mykiss,* when Dietary Fish Meal is Replaced with Silkworm (*Bombyx mori*) pupae *Archives of Polish Fisheries* **24** 53-7

[33] Gasco L, Henry M, Piccolo G, Marono S, Gai F, Renna M, Lussiana C, Antonopoulou E, Mola P and Chatzifotis S 2016 *Tenebrio molitor* Meal in Diets for European Sea Bass (*Dicentrarchus labrax* L.) Juveniles: Growth Performance, Whole Body Composition and In Vivo Apparent Digestibility *Animal Feed Science and Technology* **220** 34-45

[34] Aniebo A O, Erondu E S and Owen O J 2009 Replacement of Fish Meal with Maggot Meal in African Catfish (*Clarias gariepinus*) Diets *Revista Cientifica UDO Agricola* **9** 666-71

[35] Grace O I 2015 Effects of Grasshopper Meal in The Diet of *Clarias Gariepinus* Fingerlings *Journal of Aquaculture Research and Development* **6** 1-3

[36] Taufek N M, Aspani F, Muin H, Raji A A, Razak S A and Alias Z 2016 The Effect of Dietary Cricket Meal (*Gryllus bimaculatus*) on Growth Performance, Antioxidant Enzyme Activities, and Haematological Response of African Catfish (*Clarias gariepinus*) *Fish Physiology and Biochemistry* **42** 1143-55

[37] Adeniyi O V and Folorunsho C Y 2015 Performance of *Clarias gariepinus* (Burchell, 1822) Fed Dietary Levels of Black Soldier Fly, *Hermetia illucens* (Linnaeus, 1758) Prepupae Meal as A Protein Supplement *International Journal Fish Aquaculture* **5** 89-93

[38] Haldar P 2012 Evaluation of Nutritional Value of Short-Horn Grasshoppers (Acridids) and Their Farm-Based Mass Production as A Possible Alternative Protein Source for Human and Livestock. In: *Expert Consultation Meeting on Assessing The Potential of Insects as Food and Feed in Assuring Food Security,* (Rome: FAO) pp 23-6

[39] Fitches E 2014 Why Are Insects Not Allowed in Animal Feed? In: *International Magazine on Animal Nutrition, Processing and Feed Management,* (UK: Reed Business Media) pp 1-9

Penile erection responses of *Nigella sativa* seed extract on isolated rat corpus cavernosum

M Aminyoto*[1] and S Ismail[2,3]

[1]Department of Community Medicine, Faculty of Medicine, Mulawarman University, Jl. Kerayan Kampus Gn. Kelua, Samarinda, Indonesia
[2]Research Center on Drugs and Public Health, University of Mulawarman, Jl. Kerayan Kampus Gn. Kelua, Samarinda, Indonesia
[3]Pharmacology Laboratory, Faculty of Medicine, Mulawarman University, Jl. Kerayan Kampus Gn. Kelua, Samarinda, Indonesia

*Corresponding author: meiliati.aminyoto@gmail.com

Abstract. *Nigella sativa* L. (NS) from Ranunculaceae family is known as black cumin in Indonesia. The seed has been used as an aphrodisiac in ethnobotanical studies and reported to have pharmacological activities such as antihypertensive through the relaxant effect of vascular smooth muscles but the direct effect to the blood vessels of the corpus cavernosum is still unknown. The purpose of this study was to examine the response of NS seed extract on penile erection in vitro. NS seeds were macerated in ethanol solvent for three days in room temperature and repeated for two times. Penile erection responses was assessed using isolated rat corpus cavernosum in Krebs-Henseleit solution, temperature 37°C, pH 7.4, aerated with carbogen gas. After acclimation, corpus cavernosum was contracted with a phenylephrine solution. Ethanolic extract of NS seeds or control solution were given after reaching the plateu phase of the highest contraction. This study showed that the contraction response of the corpus cavernosum decreased after addition of NS extract and this action was increased with the addition of the extract concentration. This study concluded that NS seed ethanol extract affects the penile erection response directly through the relaxation of blood vessels in the corpus cavernosum.

1. Introduction

Nigella sativa L. (NS) from Ranunculaceae family is well known in Southern Europe, North Africa, Middle East, Middle Asia and Far East as a natural remedy for several disorders. The seed also called as black seed, black cumin, habbatussauda, habbatul barakah. In Indonesia, it also calls as jintan hitam. Many studies have been conducted on the effect of *Nigella sativa* seed extract on various body systems and reported on several pharmacologic activities including anti-hypertensive, anti-inflammatory, antimicrobial, antidiabetic, antihistaminic, insulinotropic, hepatoprotective, antiulcer, and bronchodilator. Most of the pharmacological properties of the *Nigella sativa* seed extract mainly due to quinine constituents, which the major one is thymoquinone. The seed also contain essential oil, protein, alkaloids and saponin [1,2]

Nigella sativa seed extract also reported to have benefits in male infertility. The increased levels of testosterone and follicle stimulating hormone (FSH) in testiscular tissue after *Nigella sativa* seed extract administered have been shown to increase sperm concentration, enhancing reproductive performance, sperm motility and sperm quality [3].

Aphrodisiac is any food or drug that arouses sexual desire, increases pleasure and sexual performance. These substances are derives from plants, animals or minerals [4]. Medicinal plants that acts as aphrodisiac has a potential to use in the treatment of erectile dysfunction. Erectile dysfunction can cause relationship difficulties among the couples, characterized by inability to develop or maintain an adequate erection of the penis [5].

World Health Organization (WHO) have been reported that traditional medicine or non-convensional medicine is found in almost every country in the world with an increasing number of users and facilities. Even in some countries where the conventional heath-care system is quite well established, more than 70% of the population stil commonly use traditional medicine [6]. The demand of herbal medicine is increasing because they are considered safer and well tolerated compared to the allopathic drugs [7]. Therefore, it is necessary to scientifically study on the herbal or traditional medicine in order to develop an evidence based medicine.

There is currently not much evidence regarding the effects of *Nigella sativa* seed as an aphrodisiac. Thus, the primary endpoint of this research was to examine the response of *Nigella sativa* seed extract on penile erection in vitro.

Figure 1. *Nigella sativa* seed

2. Material and Methods

2.1. Study design

This study was conducted as a post-test only control group trial. All the experimental procedures were conducted in Pharmacology Laboratory, Faculty of Medicine, Mulawarman University. Ethical approval was granted by the Health Research Ethics Committee, Faculty of Medicine, Mulawarman University with the reference number 211/KEPK-FK/XI/2016.

2.2. Sample preparation

Nigella sativa seeds were purchased from a local herbal shop in Samarinda, East Kalimantan Province. Identification of *Nigella sativa* seeds was done by plant taxonomists from the Faculty of Forestry, Mulawarman University (reference 78/UN17.4.3.08/LL2016).

2.3. Sample maceration and extraction

The seeds were ground into powder then added to the maserator. One part of the seed powder was diluted with 10 parts of absolute ethanol solvent. It was soaked for 6 hours and stirred occationally with an orbital shaker 20 rpm for 10 minutes then stayed for three days in room temperature. The maceration solution was filtered using Whatman filter paper. All the process was repeated two times. The macerates were collected and evaporated with rotavapor vacuum, temperature 50°C, which yields a concentrated extract. The extract was dried by putting into a dessicator containing blue silica gel in an oven with temperature 50°C for one week [8].

2.4. Animal maintenance and preparation

This study used isolated rat corpus cavernosum to prove the efficacy of *Nigella sativa* seed extract as an aphrodisiac. The rats were euthanized by overdose ketamine injection then their penis excised. Dorsal penile vein and spongiose tissues were removed carefully. Strips of corpus cavernosum were mounted in organ bath containing physiologic solution (Krebs-Henseleit) at 37°C temperature, pH 7.4, continously aerated with carbogen gas and connected to an isometric transducer (resting tension 250 mg).

After 60 minutes stabilization period, corpus cavernosum was contracted with 10 μM phenylephrine solution to test the smooth muscles integrity. The smooth muscles considered in a good condition if there was a contraction response. Then it rinsed with the physiologic solution every 15 minutes until the muscle tone back to the basal condition. Further, corpus cavernosum were induced by the addition of 10 μM phenylephrine. After reaching the plateau tonic phase, corpus cavernosum was exposed to the cumulative concentrations of *Nigella sativa* seed extract (0.03, 0.1, 0.3, 1, and 3 mg/ml). Same amount of solvent solution were used as a control. The resulting relaxation of the smooth muscle was recorded and evaluated. The same protocol was repeated 5 times. Negative results indicate a relaxation activity and otherwise a positive value indicates contraction activity.

2.5. Statistical analysis

Data obtained were analyzed using the Statistical Package for the Social Sciences (SPSS) software. Activity of the smooth muscles are expressed as mean ± SE and showed in dose response curve. All the data examined with the t-test after assessed as normal distributed and normal variance. Significance was set at $p < 0.05$.

3. Results and Discussion

Extract of *Nigella sativa* seed have vasodilation activity to the corpus cavernosum characterized by negative contraction persentage. The expected effect of *Nigella sativa* seed extract addition to the isolated rat corpus cavernosum was a vasodilation of blood vessels to prove that *Nigella sativa* seed extract effective as an aphrodisiac to maintain penile erection. The percentage of corpus cavernosum contraction response after given *Nigella sativa* seed extract is shown in table 1. Positive values indicated the opposite effect than expected resulting the flaccid state of the penis. Vasodilation activity was initially recorded at the concentration level 0.1 mg/ml with the persentage contraction value -1.57 ± 0.81 and this effect was strengthened with the addition of cumulative dose. Significant differences between the effects of *Nigella sativa* seed extracts compared to control solution were also found starting from the organ exposure to concentration level 0.1 mg/ml ($p<0.05$). Emax value was obtained from the experimental data -23.84 ± 1.24 ($p<0.05$).

Table 1. Contraction response percentage of corpus cavernosum after exposed to NS seed extract

Concentration of extract (mg/ml)	Contraction percentage of control solution (Mean ± SE)	Contraction percentage of NS seed extract (Mean ± SE)
0.03	2.18 ± 0.32	1.02 ± 0.78
0.1	1.45 ± 0.37	-1,57 ± 0.81*
0.3	0.31 ± 0.44	-5.27 ± 0.52*
1	-0.75 ± 0.43	-9.92 ± 0.69*
3	-1.82 ± 0.53	-16.10 ± 0.90*

n=5 rats . Statistical analysis with *t*-test. *significantly different compared to control

Figure 2 shows the corpus cavernosum contraction activity dose response curve of *Nigella sativa* seed extract compared to control solution at the same concentration. Increased concentrations of control fluid or solvent may result in a mildly negative contraction percentage, -1.82 ± 0.53 at the concentration

3 mg/ml, meaning that the solvent causing mild vasodilation of the blood vessels within the rat corpus cavernosum.

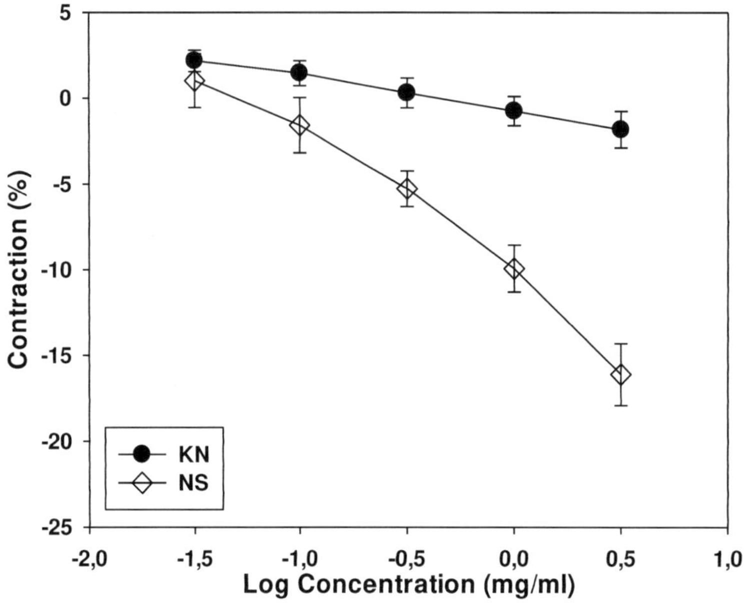

Figure 2. Dose-response curve of *Nigella sativa* seed extract and control solution effects to the corpus cavernosum contraction

Our results confirmed negative contraction response of the penile after administration of *Nigella sativa* extract. Mechanism of action of vasodilation in the corpus cavernosum needs further investigation whether influenced by vascular endothelium via eNOS (endothelial nitric oxide synthase) or by inhibition of phosphodiesterase enzyme. Activation of eNOS by an agonist in the vascuar endothelium will release nitric oxide (NO) which can diffuse easily into the smooth mucle of the blood vessels that present in the corpus cavernosum of the penis. NO will activate GC (guanylate cyclase) then GC are responsible for the conversion of guanosine triphosphate (GTP) into cyclic guanosine monophosphate (cGMP). cGMP stimulates calcium ions release out from the blood vessel smooth muscle resulting the penile relaxation. NO can also be release if there is sexual stimulation or the effect of some drugs [9].

Decreased calsium ions in the blood vessel smooth muscle within the corpus cavernosum can also because of the inhibition of phosphodiesterase enzyme (PDE). This enzyme plays role to convert cAMP (cyclic adenosine monophosphate) or cGMP into AMP or GMP. Inhibition to this enzyme will cause cGMP levels remain high then the blood vessels in corpus cavernosum remains in relaxation state allowing more blood filled into the corpus cavernosum, the penis will remain erect or increasingly tense [9-11].

Several studies on *Nigella sativa* seeds related to its benefit on male infertility, none as we know have studied the direct effect on penile erection. *Nigella sativa* have been reported increasing reproductive parameters (seminiferous tubules thickness and diameters, account of spermatogonia, primary and secondary spermatocytes, spermatids, free spermatozoa, account of sertoli and Leydig cells, diameter of Leydig cells and the height of epithelial cells entirely covered epididymal caudal), stimulates testosterone and follicle stimulating hormones as well as protein concentration. The results need to be confirmed with further experimental study as this was set in vitro [12].

4. Conclusion

The present results confirmed that Nigella sativa seed extract affects the penile erection response directly through the relaxation of the blood vessels in the corpus cavernosum of male rats. This result may support the use of Nigella sativa seed as sexual enhancing remedies. Further investigations are required to explore the mechanism of action of Nigella sativa seed as an aphrodisiac.

Acknowledgement

The authors gratefully acknowledge the support of Mulawarman University for the fully funding source of this research.

References

[1] Kamal B, Khalid K, Yassine C, Miloud E K and Youssef R 2014 *J. Chem. Pharm. Res.* **6(8)** 9-11
[2] Sabira S, Hafiz M A, Naheed A, Asif I, Haleema N and Riaz U R 2015 *J. Pharm. Pharmacol.* **4(4)** 103-6
[3] Ng C P, Noor H H and Durriyah S H A 2014 *Evid. Based Complement. Alternat. Med.* (2014)007
[4] Neelesh M, Sanjay J, Vipin B G and Savita V 2011 *Acta Pol. Pharm.* **68(1)** 3-8
[5] Asif M, Jas K, Irwin N and Manit A 2014 *BMJ* **348**
[6] WHO 2013 *Traditional medicine strategy: 2014-2023* (Geneva: WHO Press) pp 15-27
[7] Ramandeep S, Sarabjeet S, Jeyabalan G, Ashraf A 2012 *J. Pharmacogn. Phytochem.* **1(4)** 43-56
[8] MoH R I 2008 *Indonesian Herbal Pharmacopoeia* (Jakarta: Ministry of Health Republic of Indonesia)
[9] Cartledge J J, Minhas S, Eardley I and Morrison J F B 2000 *Int. J. Impot. Res.* **12** 213-221
[10] Robert C D and Tom F L 2005 *Urol Clin North Am.* **32(4)** 379-95
[11] Ashish B 2009 *J. Res. Educ. Indian Med.* (2008)15-24
[12] Sayed M B K, Reihaneh J K, Maryam H and Elham Y 2016 *BLM* **8** 4

Scientific approach as an understanding and applications of hydrological concepts of tropical rainforest

Z Haryanto* and I Setyasih

Faculty of Teacher Training and Education, Mulawarman University, Samarinda, Indonesia

*Corresponding author : zeni1968@gmail.com

Abstract. East Kalimantan has a variety of biomes, one of which is tropical rain forests. Tropical rain forests have enormous hydrological potential, so it is necessary to provide understanding to prospective teachers. Hydrology material cannot be separated from the concept of science, for it is needed the right way of learning so students easily understand the material. This research uses descriptive method with research subject is geography education students taking hydrology course at Faculty of Teacher Training and Education, Mulawarman University. The results showed that the students were able to observe, ask question, collect data, give reason, and communicate the hydrological conditions of tropical rain forest biomes, especially related to surface ground water and groundwater conditions. Tropical rainforests are very influenced by the hydrological conditions of the region and the availability of water is affected by the forest area as a catchment area. Therefore, the tropical rainforest must be maintained in condition and its duration, so that there is no water crisis and hydrological related disasters.

1. Introduction

East Kalimantan is a province that has the largest tropical rain forest in Indonesia. The forest consists of conservation forests (1.70466 million ha), protected forests (2,848,243 ha), and production forests (9,123,225 ha). Tropical rain forests of East Kalimantan consist of national parks, Forest Park (Tahura), and forest of natural attractions. Forests are natural resources that can provide double benefit, directly or indirectly benefit. Benefits of the forest are as sources of various types of goods, such as wood, sap, bark, leaves, roots, fruits, flowers and others. One of the functions of tropical rainforest is hydrology. Function of hydrology is very important for human life in surrounding woods and for germ plasm [1].

Hydrological means the forest is a warehouse of water and a place to absorb rain water and dew that will eventually flow into the springs and rivers. Tropical rain forests play an important role in maintaining the circulation of the hydrological cycle. Tropical rainforests can absorb water in large volumes because the trees in the tropical rainforest have long roots and have large trunks of trees that can store large amounts of water. Tropical rainforests are a system of continuously hydrological cycles of water below, on the surface, and on the earth. Analysis on the land-cover changes to the infiltration rates revealed that the older the forest stands, the greater the forest absorbs water from the soil. This proves that the forest stands very well in absorbing water from the soil [2].

In hydrological discussion of tropical rainforest it is required understanding and examples of concrete applications of many found scientific concepts in hydrological material. Understanding and application of the concepts should be taught to the students of geography education who take hydrological courses. Therefore, we need an appropriate learning approach so that students easily understand the hydrological material of tropical rain forest. The approach of learning, in accordance with the material, must also be in accordance with the characteristics of students and involve the student environment as a source of learning. Approach of learning that is considered in accordance with the material, learning objectives, characteristics of students, and environment is the approach of science. The science approach has many advantages over other learning approaches. The advantages are: 1) the students construct the concept through the stages of observing, questioning, collecting data, reasoning, and communicating, 2) providing material understanding through scientific approach, and 3) information not only from lecturers but also from students themselves actively looking for learning resources [3]. The scientific approach helps students arguing that prediction is the ultimate goal that sciences seek [4].

To a degree, we know that we understand (at some level) an event when we can predict the occurrence of that event. Prediction may also permit a substantial amount of control. When events can be predicted accurately, preparation in anticipation of the event can occur, and help us to perfect predictions of our behaviors and thoughts. It is important to note that these assumptions of science are not thought of as true or false, provable or unprovable. As scientists, we make certain assumptions to see where they take us in terms of achieving our goals.

The purpose of this research is to know the effectiveness of scientific approach to improve understanding and application of concept of science on hydrological material of tropical rain forest. In addition to that, this study aims to determine the response of students learning using a scientific approach.

2. Method

This research method uses descriptive methods,which aims to know the understanding and application of concept of hydrological matter of tropical rain forest. The subjects of this study were geography education students Faculty of Teacher Training and Education, Mulawarman University, who took hydrological courses with a total of 31 people. The variables to be studied in this research are: (1) scientific approach; (2) understanding and application of science concept, and (3) moderator variable that is learning style and logic ability.

3. Results and Discussion

One of the study materials in geography education is hydrology. The study material has been adjusted to the basic scientific pattern (PIP) established by the university, namely tropical rainforests and the environment. Hydrological studies include hydrometeorology, potamology, limnology (relatively calm surface water such as geohydrological lakes), and cryology (solid water like ice and snow) and water quality [5]. To be in accordance with PIP Mulawarman University, the hydrological study materials also study the hydrology of tropical rain forest and its impact on human life.

In the hydrological material it is found many concepts of science, especially the mechanical properties of fluid, which study the behaviour of water flow in micro or macro. Fluid Mechanics lays the foundations of hydrological theory focused on the engineering of fluid properties. The topic of hydrology extends in many aspects of science and engineering disciplines, including concepts such as closed flow (pipeline), dam design, pumps, turbines, hydropower, fluid dynamics counts, flow measurements, and open channel flow behaviour such as streams and gullies [6]. The many concepts of science on hydrological material require appropriate learning methods so that students can understand the concept of science and its application. To improve the understanding and application of the concept of science on hydrological material, learning can be done with a scientific approach. The scientific approach matches not only the characteristics of the hydrological material but also the characteristics of the student and student environment. The student centered learning is built on the principles of learning which consists of active and constructive process as well as social activity, requires metal reflection, uses prior knowledge, and takes of reflection, idea, action, and result [7]. Scientific approach is student centered learning, so students do a lot of activities and this requires high level thinking. The following data are related to the characteristics of students of research subjects.

Table 1. Data subject research

Indegeus	**Gender**		**Learning Style**			**Logic Abilities**		
	M	**F**	**Auditori**	**Kinestetic**	**Visual**	**Concrete**	**Transition**	**Formal**
Banjar		3	2	1		3		
Bugis		2			2	2		
Buton	1		1			1		
Dayak	5	2	1	2	4	3	2	2
Flores		1			1	1		
Java	3	6	3		6	4	4	1
Kutai	4	1	1	1	3	4	1	
Paser		1			1		1	
Sunda	1				1		1	
Toraja		1	1			1		
Amount	14	17	9	4	18	19	9	3
Total	31		31			31		

Based on the data of the study subjects, it is shown that subject of women was 55% and that of male was 45%. Most students came from Javanese tribe (29%) and at least 3% students came from Paser, Sunda, and Toraja tribes, respectively. Majority learning style is visual (58%) and the next used learning styles are auditory (29%) and kinestetik (13%). Majority student logic ability is at concrete level (61%) and the next student logic abilities are at transition (29%) and formal (10%). Here are the result of interaction data analysis between learning style and logic ability as a moderator variable.

Table 2. Results of data analysis moderat variable

		Value	**Asymp. Std. Error[a]**	**Approx. T[b]**	**Approx. Sig.**
Learning Style * Logic Ability	Kendall's tau-c	.197	.120	1.637	.102
N of Valid Cases		31			

Based on data analysis on moderate variables, it turns out the learning style does not affect the ability of logic. The result analysis of student activeness when learning with scientific approach is presented in the following table.

Table 3. Recapitulation of student activity level at learning

Subject	**Level Activity**						**Amount Subject**
	Less Active		**Active**		**Very Active**		
	F	%	F	%	F	%	
Scientific approach	-	-	5	16	26	84	31
Amount	-	-	5		26		31

When studying hydrology of tropical rainforests using a scientific approach, majority students were 84% very active and 16% active. The results indicate that the students are very active with a scientific approach.

Here is the result of the analysis of the influence of the scientific approach as the independent variable and the moderate style (learning style and logical ability) to the understanding and application of the concept of science as the dependent variable.

Table 4. Influence analysis of independent and moderate variables against tied variables

Group	Mean Difference	Std. Error	Sig.[a]	95% Confidence Interval for Difference[a]	
				Lower Bound	Upper Bound
SA*Auditori	6.927*	2.039	.001	2.900	10.953
SA*Kinestetik	7.609*	2.422	.002	2.826	12.392
SA*Visual	6.055*	1.724	.001	2.650	9.461
SA*Concrete	6.737*	2.165	.002	2.462	11.013
SA*Transision	-6.927*	2.039	.001	-10.953	-2.900
SA*Formal	-6.055*	1.724	.001	-9.461	-2.650

The results of the analysis show that the scientific approach, learning style, and logic ability influence the understanding and application of the concept of science on the hydrological material of tropical rain forest. Significant results are influenced by learning activities in a scientific approach that prioritizes student activities in learning. In a scientific approach, students observe, ask questions, collect data, give reason, and communicate results.

Observation activities were carried out by observing the hydrological conditions in Benanga Dam, Karangmumus River, Mahakam River, Wain River Balikpapan, and artesian wells of residents in the area L2 Kutai Kartanegara. Students understood the hydrological conditions and scientific concepts contained in the hydrological material, i.e.: flow velocity, water pressure, fluid, and capillarity. During the observation, the students also questioned the managers and residents around the hydrological conditions of Wain River Balikpapan and the artesian wells of residents in L2 for 5 - 10 years ago. Thus, students grasped on how the hydrological conditions change over time. In addition to observation and inquire, students also collected river flow data, as well as the distribution and depth of artesian wells owned by residents in L2. Based on observations and data collected by students, the students determined the factors impacting on the hydrological conditions in the tropical rain forest area. After that the students communicate or explain their data and findings on the condition of tropical rain forests and their impact on hydrology and human life, as well as the influence of human activities on the condition of tropical rainforests and hydrology.

Scientific approach is very effective to improve the understanding of science concept to the students, because students construct their own knowledge through scientific activities.The constructive development of new knowledge is based on empiricism and research. Such knowledge leads to the exact reasoning necessary to solve the problems posed by modern life [8]. Modern society requires efficient methods of learning, with emphasis on scientific method [9].

In addition to increase liveliness, the scientific approach also significantly affects the understanding and application of the concept of science [10]. If students understand the concept of science in hydrological material, then students will easily understand the application of the concept of hydrology in everyday life such as fluid, pressure, speed, and energy. This is because students make observations, such as the speed of river flow discharge in the student environment during field lectures. In addition to field observation, the lecturers' learning uses the media of images and video to show and explain the concepts of hydrological material.

The scientific approach is not only done with explanations from lecturers or the use of drawing media, but students also do practicum. This kind of learning activity involves not only one sense but also almost all the students' senses. Students not only hear the explanation of lecturers or see pictures displayed by lecturers, but also student do hydrological practicum. With the involvement of all the five senses of students, the students will better understand the concept of science and its application on the hydrological material of tropical rain forest. By learning such scientific measures, according to Edgar Dale's cone theory of experience [11], the student will remember and understand more if he does with according to everyday life.

In accordance with the purpose of hydrology learning in geography education program, that is knowledge, skill, and application of hydrological values in daily life to the students, the scientific approach gives meaningful learning to the students. The use of learning media, outdoor study, and the early knowledge of students on a scientific approach, help students understand the concept of science and its application with its cognitive structure. This is in line with Piaget's cognitive learning theory, Ausubel's meaningful learning theory, and Vygotsky's social interaction theory [12].

Learning using a scientific approach will help students to remember and understand the material learned, because students play an active role in learning. The findings of this research are supported by study of Adisyah stating that geography learning using scientific approach has significant effect on learning outcomes and geography skills of high school students [13].

4. Conclusions

Based on data analysis and discussion, it is concluded that scientific approach has an effect on understanding and application of tropical rainforest material science concepts to students who have different learning styles and logic abilities. Understanding and application of the concept of science on hydrological material will be easily understood by the students if the learning is adapted to the material characteristics, student characteristics, environment, and learning resources. Students can understand the hydrological condition of the tropical rainforest and its impact on human life. In addition, students understand if the destruction of tropical rainforests will result in reduced and even loss of hydrological functions of tropical rainforests.

References

[1] Jayapercunda, S. 2002. *Hutan dan Kehutanan Indonesia: Dari Masa Ke Masa*. Bogor: IPB Press.

[2] Mulyana, N. 2000. *Pengaruh hutan pinus terhadap karakteristik hidrologi di sub daerah aliran sungai Ciwulan Hulu KPH Tasikmalaya Perum Perhutani Unit III Jawa Barat (Kajian menggunakan Model POWERSIM-PINUS Ver. 3.1).* Tesis Program Pascasarjana Institut Pertanian Bogor, Bogor.

[3] Wieman, C., and Gilbert, S.2014. The teaching practices inventory: a new tool for characterizing college and university teaching in mathematics and science. *CBELife Sci. Edu.* **13:**552–569.

[4] Lawson E. A. 2009. *Basic Inferences of Scientific Reasoning, Argumentation, and Discovery*, Wiley Periodicals, Inc: Science Education, **94**, 336-364, available at http://www.interscience.wiley.com, 2009.

[5] Utaya, S. 2005. *Pengantar Hidrologi.* Malang, Fakultas MIPA Universitas Negeri Malang

[6] Gupta, B. 1979. *Water Resources Engineering and Hydrology*. New Delhi: Standard Publishers Distributors.

[7] Harsono. 2008. Student Centered Learning di Perguruan Tinggi. *Jurnal Pendidikan Kedokteran dan Profesi Kesehatan Indonesia* 3(**1**)

[8] Hodson D. 1986. Laboratory Work as Scientific Method. *Journal of Curriculum Studies* 28. 115-135

[9] Mirko G. N., Dusanka Z. O, Segedinac, M. 2012. *Effective Teaching of Physics and Scientific Method*. TEM Journal – Volume 1/Number **2**/2012. 85 www.temjournal.com

[10] Lawson, A, E. 2001. Using the Learning Cycle to Teach Biology Concepts and Reasoning Patterns. *Journal of Biological Education* ***35*** *(4)* pp 165-9

[11] Silberman, M. 1996. *Active Learning, 101 StretegiPembelajaranAktif*. Yogyakarta: Yappendis
[12] Slavin, R.E. 2000. *Educational Psychology: Theory and Practice*. 6th. Boston: Allyn and Bacon
[13] Adisyah, Bakti. Utaya,Bidijanto. 2016. *The Effect of Scientific Approach to Students Geographical Skill and Learning Outcomes of Geography in Senior High School*. *The* International Institute of Science, Technology and Education (IISTE).

Rainfall prediction using fuzzy inference system for preliminary micro-hydro power plant planning

B Suprapty[1], R Malani*[1], J Minardi[2]

[1]Department of Information Technology, State Polytechnic of Samarinda, Samarinda, East Kalimantan, Indonesia
[2]Department of Information System, Faculty of Science and Technology, Nahdlatul Ulama of Islamic University, Jepara, Central Java, Indonesia

*Corresponding Author : anaogie@gmail.com

Abstract. East Kalimantan is a very rich area with water sources, in the form of river streams that branch to the remote areas. The conditions of natural potency like this become alternative solution for area that has not been reached by the availability of electric energy from State Electricity Company. The river water in selected location (catchment area) which is channelled to the canal, pipeline or penstock can be used to drive the waterwheel or turbine. The amount of power obtained depends on the volume/water discharge and headwater (the effective height between the reservoir and the turbine). The water discharge is strongly influenced by the amount of rainfall. Rainfall is the amount of water falling on the flat surface for a certain period measured, in units of mm^3, above the horizontal surface in the absence of evaporation, run-off and infiltration. In this study, the prediction of rainfall is done in the area of East Kalimantan which has 13 watersheds which, in principle, have the potential for the construction of Micro Hydro Power Plant. Rainfall time series data is modelled by using AR (Auto Regressive) Model based on FIS (Fuzzy Inference System). The FIS structure of the training results is then used to predict the next two years rainfall.

1. Introduction

Hydropower is a system that utilizes water flow energy to be converted into electrical energy or mechanical energy. There are various ways to harness the energy of the water flow, one of which is the run of river system that does not require a large reservoir. This system is known as Microhydro. The power plant system using the microhydro is called as the Micro Hydro Power Plant (MHPP). MHPP is an alternative renewable energy system that meets energy conservation policies, especially for areas that have great potential for water energy sources.

East Kalimantan is a very rich area with water sources, in the form of river streams that branch to the remote areas. The conditions of natural potency like this become alternative solution for the area that has not been reached by the availability of electric energy from State Electricity Company. The river water in the selected location (catchment area) which is channelled to the canal, pipeline or penstock can be used to drive the waterwheel or turbine. The amount of power obtained by the water channelling method depends on the volume/water discharge and headwater (the effective height between the reservoir and the turbine). The water discharge (m^3/s) is determined by the flow rate of the river water (m/s) and the cross-sectional area of the river (m).

IOP Conf. Series: Earth and Environmental Science **144** (2018) 012005 doi:10.1088/1755-1315/144/1/012005

Water discharge is strongly influenced by the amount of rainfall. Rainfall is the amount of water falling on the flat surface for a certain period measured, in units of mm^3, above the horizontal surface in the absence of evaporation, run-off and infiltration. Measurements of instantaneous water discharge at the watershed location where the dam will be built are conducted to determine the smallest water discharge during the dry season through hydrological studies. Prediction of rainfall becomes very important for the initial planning of MHPP development.

In this study, the prediction of rainfall is done in the area of East Kalimantan which has 13 watersheds, which in principle has the potential for the construction of MHPP. Rainfall time series data is modeled using AR (Auto Regressive) Model based on FIS. Training is done to obtain the best FIS structure. The FIS structure of the training results is then used to predict the next two years rainfall.

2. Method

The number of rainy days in a month depends on the amount of rainfall (mm). The number of rainy days is what affects the condition of river water discharge. Rainfall or precipitation is an element of the hydrometer, which is a collection of liquid or solid particles that fall or float in the atmosphere resulting from the condensation process of water vapor (cloud) [1]. Rainfall data can be categorized as time series data with seasonal and non stationary patterns.

2.1 Fuzzy Logic

Fuzzy logic provides an inference structure that enables appropriate human reasoning capabilities. The theory of fuzzy logic provides an inference mechanism under cognitive uncertainty. The theory of fuzzy logic is based upon the notion of relative graded membership and so are the functions of mentation and cognitive processes. The utility of fuzzy sets lies in their ability to model uncertain or ambiguous data.

Fuzziness in a fuzzy set is characterized by its membership functions. It classifies the element in the set, whether it is discrete or continuous. The membership functions can also be formed by graphical representations. The graphical representations may include different shapes. There are certain restrictions regarding the shapes used. The rules formed to represent the fuzziness in an application are also fuzzy [2]. The "shape" of the membership function is an important criterion that has to be considered. There are many membership functions (MF) that can be used, one of which is Triangular MF as shown in Figure 1.

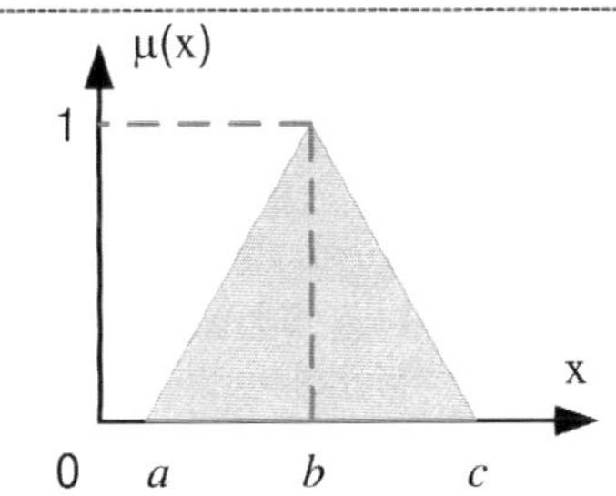

Figure 1. Triangular MF

From Figure 1, fuzzy set is declared as

$$\mu(x) = trimf\left(x,a,b,c\right) = \begin{cases} 0 & x \le a,\ x \ge c \\ \dfrac{x-a}{b-a} & a \le x \le b \\ \dfrac{c-x}{c-b} & b \le x \le c \end{cases} \tag{1}$$

2.2 FIS (Fuzzy Inference System)

FIS is a system that processes the mapping formulation of a given input to produce an output using Fuzzy Logic. FIS is shown in Figure 2.

2.2.1.Fuzzify inputs. The first step is to take the inputs and determine the degree to which they belong to each of the appropriate fuzzy sets via membership functions. The input is always a crisp numerical value limited to the universe of discourse of the input variable and the output is a fuzzy degree of membership in the qualifying linguistic set. Fuzzification of the input is performed by considering either a table lookup or a function evaluation. In this way, by the rules, each input is fuzzified over all the qualifying membership functions required.

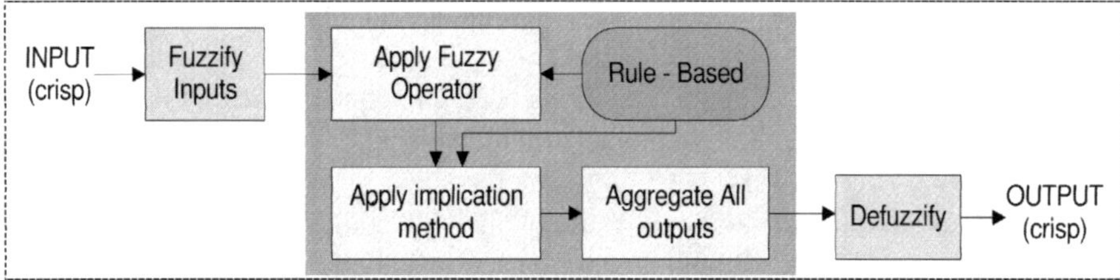

Figure 2. Fuzzy Inference System

2.2.2. Applying Fuzzy Operator. After the inputs are fuzzified, the degree to which each part of the antecedent is satisfied for each rule. If the antecedent of a given rule has more than one part, the fuzzy operator is applied to obtain one number that represents the result of the antecedent for that rule. This number is then applied to the output function. The input to the fuzzy operator is two or more membership values from fuzzified input variables. The output is a single truth value. There are two commonly used fuzzy operators, that are OR (max) and AND (min). For example, there are two inputs (A and B) and one output (C) which have linguistic variables (**Low**, **Medium**, **High**) and have the same fuzzy set as shown in Figure 3. If A(x) = 20 and B(x) = 65 then for the rule:

$$\overbrace{IF\ \boldsymbol{A}\ \text{is}\ \boldsymbol{Low}\ AND\ \boldsymbol{B}\ \text{is}\ \boldsymbol{Medium}}^{antecedent}\quad \overbrace{THEN\ \boldsymbol{C}\ \text{is}\ \boldsymbol{Medium}}^{consequent} \tag{2}$$

it is illustrated in Figure 4.

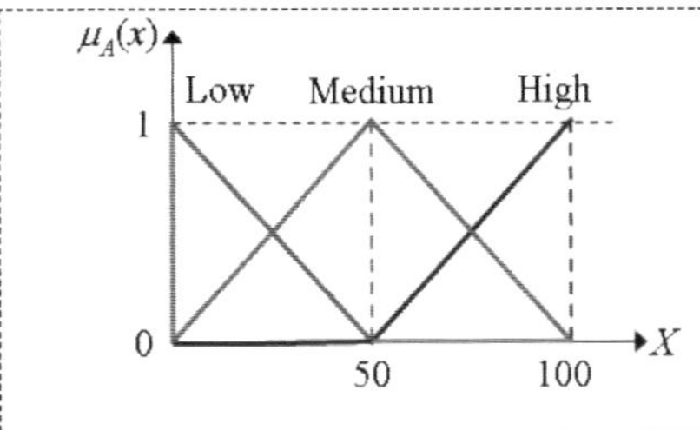

Figure 3. Fuzzy Set

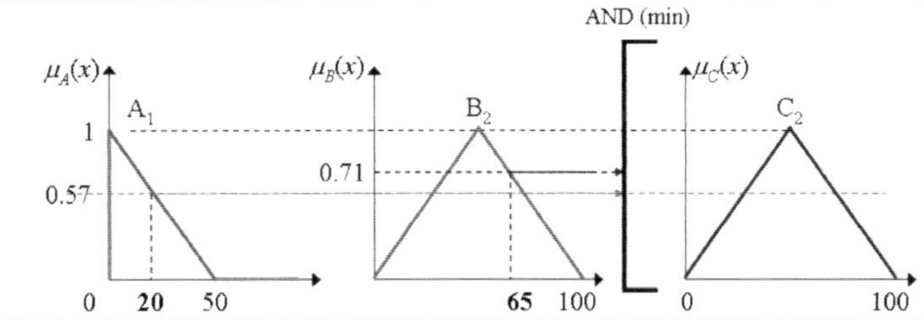

Figure 4. Applying Fuzzy Operator : AND operator

2.2.3. Applying Implication Method. Before applying the implication method, we must determine the rule's weight. Every rule has a weight (a number between 0 and 1), which is applied to the number given by the antecedent. Generally, this weight is 1 (as it is for this example) and thus has no effect at all on the implication process. After proper weighting has been assigned to each rule, the implication method is implemented. A consequent is a fuzzy set represented by a membership function, which appropriately weights the linguistic characteristics that are attributed to it. The consequent is reshaped using a function associated with the antecedent (a single number). The input for the implication process is a single number given by the antecedent, and the output is a fuzzy set. Implication is implemented for each rule. For implication operation it is used AND method as shown in Figure 5.

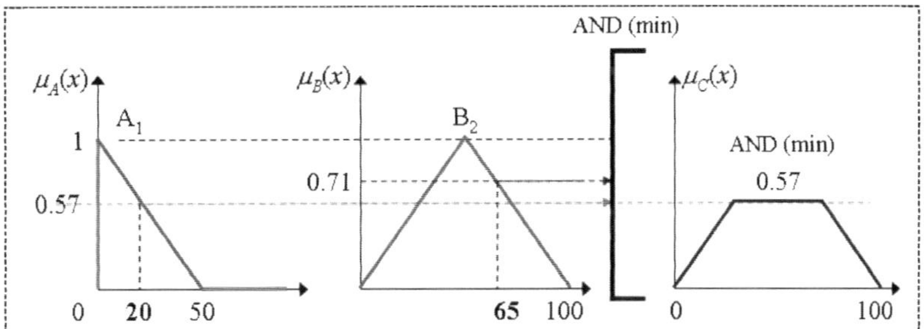

Figure 5. Applying implication method : AND method

2.2.4. Aggregate all outputs. Because decisions are based on the testing of all of the rules in a FIS, the rules must be combined in some manner in order to make a decision. Aggregation is the process by which the fuzzy sets that represent the outputs of each rule are combined into a single fuzzy set. Aggregation only occurs once for each output variable, just prior to the fifth and final step, defuzzification. The input of the aggregation process is the list of truncated output functions returned by the implication process for each rule. The output of the aggregation process is one fuzzy set for each output variable. For Fuzzy Mamdani, the aggregation stage is using OR (max) method. For example, it is shown in Figure 6.

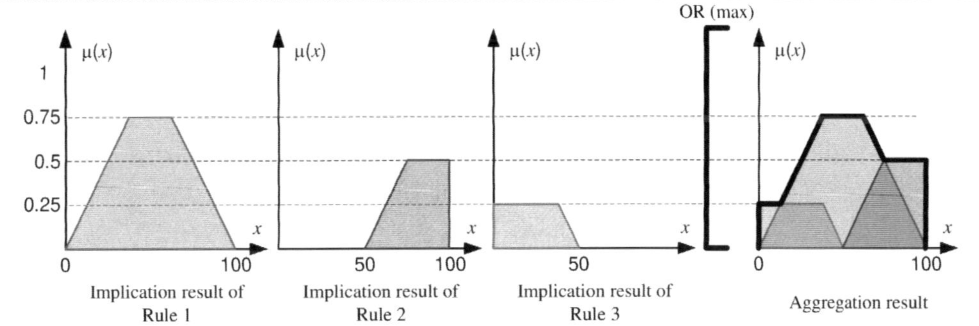

Figure 6. Example of aggregation stage using OR (max) method

2.2.5.Defuzzify. The input for the defuzzification process is a fuzzy set (the aggregate output fuzzy set) and the output is a single number. As much as fuzziness helps the rule evaluation during the intermediate steps, the final desired output for each variable is generally a single number. However, the aggregate of a fuzzy set encompasses a range of output values, and so must be defuzzified in order to resolve a single output value from the set. There are various methods of defuzzification, one of the most popular is the centroid method which returns the center of area under the curve. For a Fuzzy Set X with finite universe of discourse, the centroid method is expressed by:

$$x_c = \frac{\sum_{i=1}^{n} x_i . \mu(x_i)}{\sum_{i=1}^{n} \mu(x_i)} \tag{3}$$

where x_c is the centroid value, x_i is x value of ith point correspond to $\mu(x_i)$ from the aggregation result, $\mu(x_i)$ is the degree of membership (fuzzy number) of x_i, n is the number of $(x_i, \mu(x_i))$ pairs in the curve.

2.2.6. AR Model. AR model is declared as [4] :

$$y(t) = -a_1 y(t-1) - ... - a_n y(t-n) + e(t) \tag{4}$$

where a_1 is a constant, n is order number of system, and $e(t)$ is a white noise. Equation (4) can be decomposed into the following equation [4]:

$$\begin{aligned} y(t) &= (a_1 q^{-1} + ... + a_n q^{-n}) y(t) + e(t) \\ &= -\left(\sum_{i=1}^{n} a_i q^{-i} \right) y(t) + e(t) = A(q^{-1}) y(t) + e(t) \end{aligned} \tag{5}$$

where q^{-1} is the delay operator, and $A(q^{-1})$ is polynomial to be estimated.

2.2.7. Data Labeling. For data modeling it is needed to use AR model based on fuzzy approach. Hence, it is required labeling data. The range of the universe of discourse is divided into several intervals where each interval represents a linguistic variable. For example, the range of the universe of discourse is [0, 100] being divided into three intervals, i.e.: [0, 33], [33, 66], and [66, 100]. Each interval is labeled sequentially so that each interval corresponds to [0, 33]□[1], [33, 66]□[2], or [66,100]□[3]. Data labeling is used to build Rule-Based. For example, a fuzzy set that uses Triangular MF has a range of universe of discourse [0, 100] with 3 (three) linguistic variables (Low, Medium, High). Low is in the range [0, 50], Medium: [0, 50, 100], and High: [50, 100, 100]. The examples of data labeling are as follows:

$$\begin{matrix} A & B & C \end{matrix}$$

$$\begin{bmatrix} 65 & 25 & 60 \\ 75 & 60 & 50 \end{bmatrix} \Rightarrow \begin{bmatrix} 2 & 1 & 2 \\ 3 & 2 & 2 \end{bmatrix} \Rightarrow \begin{matrix} IF\ A\ is\ \mathbf{Medium}(2)\ AND\ B\ is\ \mathbf{Low}(1)\ THEN\ C\ is\ \mathbf{Medium}(2) \\ IF\ A\ is\ \mathbf{High}(3)\ AND\ B\ is\ \mathbf{Medium}(2)\ THEN\ C\ is\ \mathbf{Medium}(2) \end{matrix}$$

time series data data labeling Rule Based

The algorithm of prediction using AR model based on FIS is shown in Figure 7.

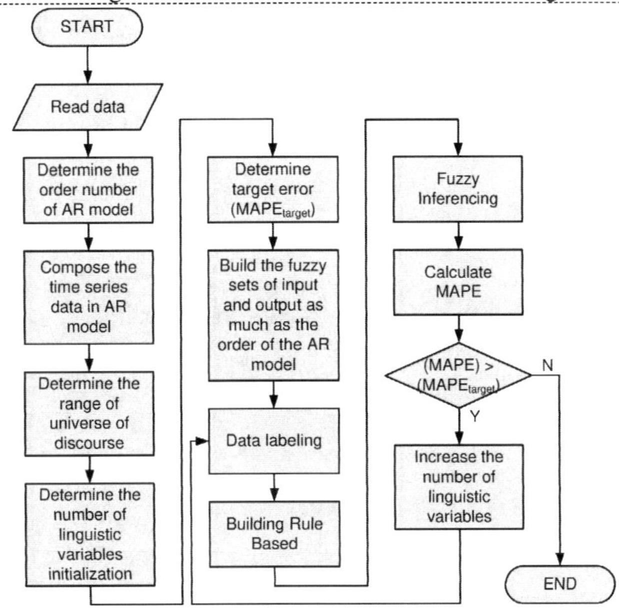

Figure 7. The algorithm of prediction using AR model based on FIS

2.2.8. Data Time Series Modeling. The monthly rainfall data are obtained from Statistics of Samarinda Municipality [1] in period 2006 – 2016 as shown in Table 1.

Table 1. Monthly rainfall data in period 2006 - 2016

Year	Jan	Feb	Mar	Apr	May	Jun	Jul	Aug	Sep	Oct	Nov	Dec
2006	228	207	215	207	307	185	24	98	108	70	191	110
2007	307	220	260	340	112	213	279	133	183	181	85	141
2008	143	194	211	259	51	205	333	149	153	208	501	350
2009	164	196	279	309	186	41	157	123	99	232	165	211
2010	148	162	157	164	223	320	259	144	202	235	207	224
2011	332	320	368	332	389	95	238	124	132	218	197	244
2012	330	206	257	371	128	172	147	140	110	116	228	220
2013	176	309	284	337	234	161	45	90	256	223	363	276
2014	273	197	319	126	190	211	50	81	82	111	300	449
2015	345	193	198	344	214	259	163	58	0	73	61	191
2016	159	99	318	369	225	202	163	99	226	175	292	357

By using 2nd order AR model then it can be stated:

$$y(t) = -a_1 y(t-1) - a_2 y(t-2) + e(t)$$

By using FIS then it can be stated:

$$y(t) = Fuzzy\left(y(t-1), y(t-2)\right) + e(t)$$

where $y(t-2)$ is rainfall data from year 2006 – 2014, $y(t-1)$ is rainfall data from year 2007 – 2015 as training input data, and $y(t)$ is rainfall data from 2008 – 2016 as training target. By training *Fuzzy* (g) such that $e(t) \to 0$ then obtaine *Fuzzy* $(g) \to y\ (t)$.

In this study, the FIS uses 5 linguistic variables for initialization and MAPE target $e(t) = 7\%$. The performance function of training result using MAPE (Mean Absolute Error) which is declared as:

$$APE_i = \frac{abs\left(y(t_i) - y_{fuzzy}(t_i)\right)}{y(t_i)} \times 100\%$$
$$MAPE = \frac{1}{N}\sum_{i=1}^{N} APE_i \tag{6}$$

where N is the number of training data, $y(t_i)$ is the ith training target, $y_{fuzzy}(t_i)$ is the ith output of FIS. Implementation of FIS is done by using Matlab programming tool.

All the training data are organized into sequential form by year then by month so that there will be data. Because using 2nd order AR model then each variable will consist of 130 data. By using 5 linguistic variables for initialization then the interval width of each data label is expressed by:

$$\text{interval} = \max(data)/(\text{Lvar}-1) \tag{7}$$

where *Lvar* is the number of lingistic variable for initialization. By using the rainfall data as shown in Table 1 so that : $\text{interval} = \max(data)/(\text{Lvar}-1) = 501/(5-1) = 125.25$. From labeling of data it is obtained the number of rule-based : $130+5=135$. The training result is shown in Figure 8, and the prediction result for next 2 years is shown in Figure 9.

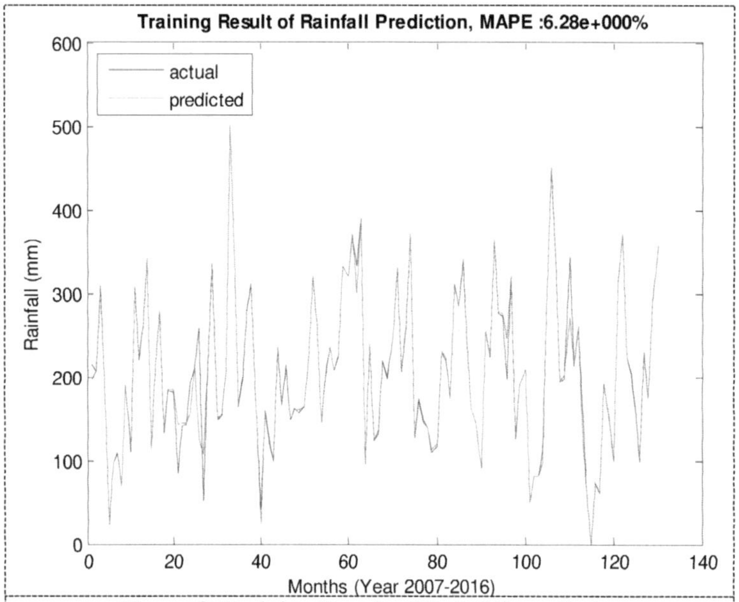

Figure 8. Training Result of Rainfall prediction

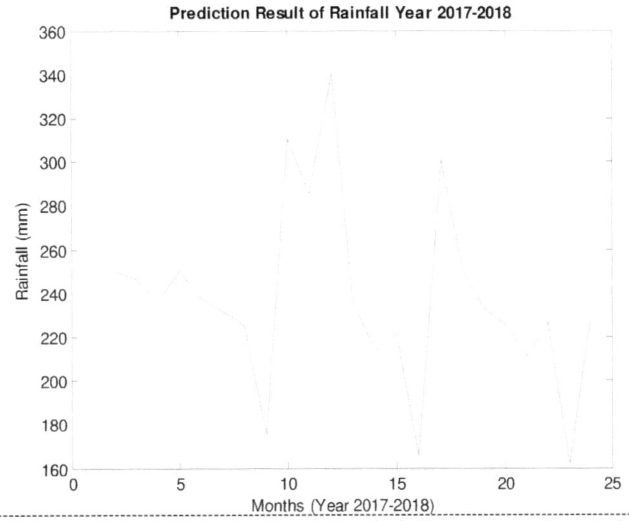

Figure 9. Prediction Result of Rainfall years 2017 - 2018

3. Result and Discussion

Summary of the training process is shown in Table 2. The training performance by MAPE is shown in Figure 10. From the training results obtained MAPE 6.28% achieved with 300 linguistic variables. The final interval width is 3.05 with the number of linguistic variables is 300. Table 2 and Figure 10 show that changes in the number of linguistic variables affect MAPE changes. Changes in the number of linguistic variables affect the change in interval width of each data label and number of rule-based. To predict the next two years (2x12 = 24 predictive data) it is done by means of each prediction result used to add rule-based in such a way that it is done until it reaches 24 predictive data.

Table 2. Summary of training process

LVar	5	25	45	65	85	105	125	145	165
Rule-Based	135	155	175	195	215	240	260	280	300
Interval Width	125	20.88	11.39	7.83	5.96	4.82	4.04	3.48	3.05
MAPE	162	47.07	26.77	21.13	11.98	8.58	7.45	7.36	6.28

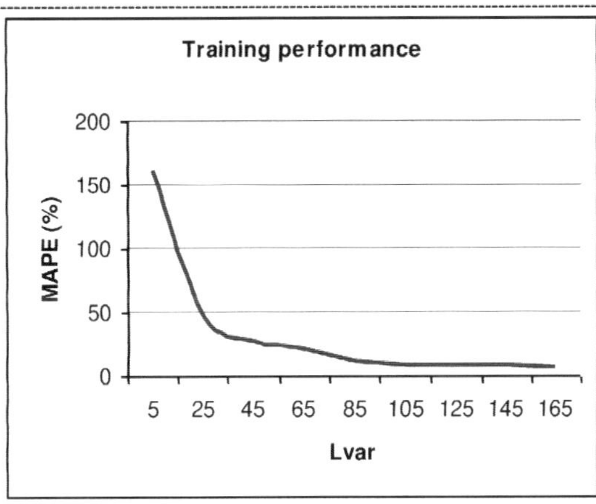

Figure 10. Training performance

4. Conclusion

The time series data prediction by using AR model based on FIS depends on the order number of AR model that determine the number of FIS inputs and with one FIS output. The interval width of each data label depends on the number of linguistic variables. The number of rule-based relies on the amount of training data already labeled and the number of linguistic variables. The change in the number of linguistic variables depends on the specified MAPE target.

Acknowledgment
The authors would like to express heartfelt thanks to The Modern Computing Research Center, Department of Information Technology, State Polytechnic of Samarinda for providing all their support.

References

[1]. Statistics of Samarinda Municipality, 2007-2017 *Samarinda Municipality in Figures 2007-2017*, Regional Development of Planning of Samarinda City, BPS-Statistics of Samarinda Municipality, ISSN : 0215-2398
[2]. S N Sivanandam 2007 *Introduction to Fuzzy Logic using Matlab*, Springer-Verlag Berlin Heidelberg
[3]. Peter J Brockwell, 2002 *Introduction to Time series and Forecasting – Second Edition*. Springer-Verlag New York, Inc. ISBN 0-387-95351-5
[4]. Ming Tao Chou, 2011 Long-term Predictive Value Interval with The Fuzzy Time Series, *Journal of Marine Science and Technology*, **19**(5) pp. 509-513
[5]. I Aydin, M Karakose, E Akin 2009 The Prediction Algorithm Based on Fuzzy Logic Using Time Series Data Mining Method, World Academy of Science, Engineering and Technology
[6]. Ashwini Tale, Arvind Singh Gusain, and Jyotirmoy Baguli 2017 Study of Load Forecasting Techniques using Fuzzy Logic *International Journal of Advanced Research in Electrical, Electronics and Instrumentation Engineering* **6**(2) ISSN (print): 2320-3765, (online): 2278-8875

Mechanism of action vasodilation *Annona muricata* L. leaves extract mediated vascular smooth muscles

S Ismail[1,2,*], N Hayati[1] and N Rahmawati[3]

[1]Laboratory of Pharmacology, Faculty of Medicine, University of Mulawarman.
Jl. Kerayan Kampus Gunung Kelua, Samarinda 75119, East Kalimantan, Indonesia
[2]Research Center on Drugs and Public Health, University of Mulawarman.
Jl. Kerayan Kampus Gunung Kelua, Samarinda 75119, East Kalimantan, Indonesia.
[3]Center for Research and Development of Medicinal Plants and Traditional Medicine.
Jl. Raya Lawu, Tawangmangu, Karanganyar, Central Java 57792, Indonesia.

*E-mail address: ismail8997@yahoo.com and ismail@fk.unmul.ac.id

Abstract. *Annona muricata* L. leaves (AML) is used as ethnomedicine by the Dayak Abai ethnicity in North Kalimantan for its already known use to reduce blood pressure. However, the mechanism of action in the vessel is still poorly understood. Aim study to prove the mechanism of action of AML in blood vessels. AML was extracted with a maceration technique using ethanol solvent. Mechanism of action test was performed with isolated rat aortic with endothelium (endo-intact) and without endothelium (endo-denuded). AML extract intervention on rats aorta with endo-intact and endo-denuded can induction vasodilatation activity. Increasing AML extract concentration can improve decrease vasodilatation activity on isolated rats aortic with endo-intact compared to endo-denuded, it means that endothelium can weaken vasodilatation activity of aorta mediated by vascular smooth muscle after the extract was given.

1. Introduction

Hypertension is still a major health problem in the world with increasing prevalence. Hypertension is also one of the most important risk factors of cardiovascular disease, which is the main cause of death worldwide. Incidence of hypertension worldwide in 2000 is approximately 26.4% with a total of 972 million people. It causes death in 7.1 million people. Most of the patients with hypertension are in developing countries and are predicted to increase to 60% or 1.56 billion people in 2025 [1,2].

Hypertension is generally treated in a long-term duration to keep the blood pressure well-controlled. Long-term medication not only can cause adverse events, but can also increase medication cost. New anti-hypertensive medicines are needed to be studied and discovered in order to obtain safer and more affordable anti-hypertensive medication. In addition, discovery of new anti-hypertensive medicines also had economic value because of the high number of patient, increased hypertension prevalence, and long duration of medication. These medicines can be obtained through ethnomedicine research of traditional herbs, which are used by certain ethnicities for hypertension medication. These herbs are the source of potential medicinal molecules [3].

Indonesia is rich in medicinal plants variety which use these plants as medicine. Many plants are used for anti-hypertensive medicine by several ethnicities in Indonesia. In 2015, Indonesian exploration research of local ethnomedicine knowledge and community-based medicinal plants, known as Medicinal Plants and Herbs Research (RISTOJA) on the Dayak Abai ethnicity, showed the usage of *Annona muricata* L. leaves or soursop for reducing blood pressure [4]. Water extract of *A. muricata* leaves (AML), which is administered intravenously can cause a hypotensive effect in normotensive rats [5]. There is no study about the mechanism of action of AML on blood vessels, which has to be studied to understand whether it is mediated by endothelium and/or the smooth muscle of the blood vessel. Study of the mechanism of action in endothelium or the smooth muscle of the blood vessel on vasodilatation activity of medicinal plant is very important. A medicine where its mechanism of action is in the endothelium will provide no benefit if given to patients with endothelial dysfunction. Therefore, this study aimed to prove the mechanism of action of AML vasodilatation on blood vessels.

2. Material and Methods

2.1. Instruments and Materials
Instruments used in this study were bioassay instrument for isolated rat aortic, which consisted of Octal Bridge Amplifier and PowerLab/16SP and program chart ver. 5.0 from AD Instrument. Isometric transducer 7003 from Ugo Basile. Materials used in this study: ethanol, DMSO, HCl, NaOH, NaCl, KCl, $CaCl_2$, $MgSO_4$, KH_2PO_4, $NaHCO_3$ was analytical grade and glucose for biochemistry from Merck. Carbogen gas (Mix 95% O_2 + 5% CO_2) from Aneka Gas. Phenylephrine and acetylcholine from Sigma Aldrich.

2.2. Animal Subject
The animal subject used in this study was a male white Wistar rat, which was obtained from the Pharmacology Laboratory, Medicine Faculty, Mulawarman University, with 4-6 month of age, 200-300 gram of weight. The study protocol approved by Ethical Committee of Health Research, Medicine Faculty of Mulawarman University with voicher number 170/KEPK-FK/VI/2016 on 13 June 2016.

2.3. Sample Collection and Identification
AML was collected in July 2016 from the place of origin of the information regarding traditional medicine in RISTOJA 2015 that was Malinau Regency, North Kalimantan. Once the AML arrived at the Pharmacology Laboratory of Medicine Faculty, the AML was sorted, washed, and dried at 60^0C for 5 days. The herbarium specimens were stored in the Pharmacology Laboratory of Medicine Faculty, Mulawarman University with voicher number ALR.2016.S001, which can be used for re-identification if needed. Identification of the plants was performed by taxonomy specialist in Mulawarman University with voicher number 56/UN17.4.3.08/LL/2016.

2.4. Extraction
The crushed simplicia of AML was macerated using ethanol ~~solvent~~. The production of the extract followed the description by Indonesia Herbal Pharmacopoeia [6]. After the dried extract was obtained, the yield was measured using percentage of weight (w/w) between the yield with powder weight of simplicia used by weighing. The extract was also tested for its water level quality using Gravimetric method by following the procedure described in Indonesia Pharmacopoeia [7]. Dried extract stored in the freezer at -20°C before further study.

2.5. Vasodilatation Activity Test
The vasodilatation activity test was performed using the isolated organ of rat aorta with endothelium (endo-intact) and without endothelium (endo-denuded) to understand whether the mechanism of action of AML extract is affected by endothelium or only by the smooth muscle of the blood vessel. The repetition was performed five times using five rats. The rats were euthanized using high dose of ketamine and their neck was dislocated after no movement was observed. Aorta was collected and

prepared by the method described by Pieper & Dondlinger [8]. It was then placed in *Kreb's-Henselheit* solution, which consisted of (mM): NaCl 118, KCl 4.7, $CaCl_2$ 1.5, $MgSO_4$ 1.1, KH_2PO_4 1.2, $NaHCO_3$ 25, and glucose 10 with pH of 7.4.

2.6. Vasodilatation Activity Test in Rat's Aorta with Endothelium-intact
Aorta was prepared and placed in a 10-ml bath organ. After being equilibrated for 60 minutes, the aorta was contracted using KCl solution to observe the contraction response (the increased tonus of aorta indicates that the contraction of the smooth muscle is good). Next, the aorta was washed using *Kreb's-Henselheit* solution for three times every 15 minutes and was tested using phenylephrine plus acetylcholine to test the endothelial integrity. After being equilibrated, the aorta was contracted using phenylephrine [10^{-6} M] until it reached flat contraction peak and acetylcholine [10^{-5} M] was given. If there is a response of relaxation, it indicates that the aorta have an endothelium-intact. Intact endothelium of the aorta, or aorta with endothelium-intact, used for this study was the one with relaxation response of more than 60% [9].

After contraction responses test and endothelial integrity test, aorta was washed three times with *Kreb's-Henselheit* solution, and then washed every 15 minutes until it returned to basal tonus and ready for vasodilatation test. In vasodilatation test, aorta was contracted with phenylephrine solution [10^{-6} M], after flat contraction was reached, we added extract solution cumulatively 0.03, 0.1, 0.3, and 3 (mg/ml) [10]. If a decrease in aorta tonus was observed, it mean that the extracts have vasodilatation activity. Negative value signifies vasodilatation activity and positive value signifies vasoconstriction activity. We also experiment on negative control (extract solvent or DMSO-ethanol 10%).

2.7. Vasodilatation Activity Test on Rats Aorta with Endothelium-denuded
Aorta endothelium were removed mechanically as. Aorta was considered to not have endothelium when vasodilatation activity become <10% after being contracted with phenylephrine [10^{-6} M] and reaching flat maximal contraction response after being added acetylcholine [10^{-5} M] [11]. Aorta was washed three times with *Kreb's-Henselheit* solution, and then washed again every 15 minutes until it returned to basal tonus and ready for vasodilatation test, similar in aorta with endothelium. We also experiment on negative control. Results were compared to aorta with endo-intact.

2.8. Data Analysis
Data presented as mean + SEM and dose-response curve. Statistic test used was *t*-test for control and extract group on aorta with endo-intact and endo-denuded, results differ significantly if $p<0.05$. Data processing software and analysis used was SigmaPlot version 12.5.

3. Result and Discussion
The yield from AML extraction is 9.24% and quality test result on the extract found that it contained water as much as 3.15%. AML extract intervention on rats aorta with endo-intact and endo-denuded can induction vasodilatation activity. Increasing AML extract concentration can improve decrease vasodilatation activity on isolated rats aortic with endo-intact compared to endo-denuded. Clear results can be seen on dose-response chart in Figure 1.

Figure 1 shows vasodilatation activity test results on isolated rats aortic with endo-intact and endo-denuded in each concentration intervention. It can be seen that control intervention on rat's aorta with endo-intact and endo-denuded can induce vasodilatation activity. Increasing control concentration can improve vasodilatation activity in isolated of rats aortic with endo-intact compared to the endo-denuded.

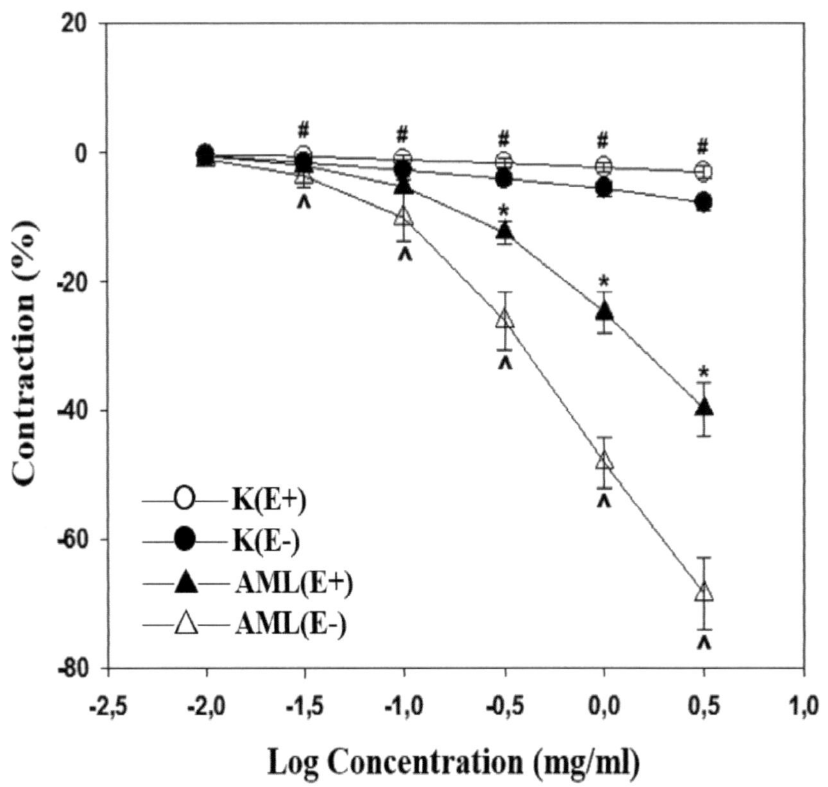

Figure 1. Vasodilatation activity on aorta in Control and AML intervention in rats aorta with endo-intact and endo-denuded

Note: n=5 rats; data in mean ± SEM; (E+)=endo-intact; (E-)=endo-denuded; K=Control (DMSO-Ethanol10%); AML= Extract ethanol AML; Statistical analysis with *t*-test. *differs significantly if compared to K(E+). ^differs significantly if compared to (E-). #differs significanty if compare to K(E+) and K(E-).

Figure 1 also shows vasodilatation activity test result on isolated rats aortic with endo-intact in each concentration intervention of AML extract. It can be seen that AML extract intervention on rats aorta with endo-intact and endo-denuded can induction vasodilatation activity. Increasing AML extract concentration can improve decrease vasodilatation activity on isolated rats aortic with endo-intact compared to endo-denuded, it means that endothelium can weaken vasodilatation activity of aorta mediated by vascular smooth muscle after the extract was given. Clear results can be seen on extract intervention at 3 mg/ml concentration.

In intervention with 3 mg/ml concentration, the extracted AML effect on aorta with endo-intact was a vascular vasodilatation activity by as much as -39.73% and -66.44% in aorta endo-denuded. This showed that vascular endothelium plays a role in weakening vasodilatation activity in aorta with endo-intact by 18.33%, meaning that intervention of AML extract at vascular endothelium can reduce vasodilatation, mediated by vascular smooth muscle. In intervention with 3 mg/ml concentration on control group, greater vasodilatation activity can be observed on aorta with endo-intact (-4.37%), compared to the one with endo-denuded (-3.20%).

Vascular tone is mediated by endothelium and smooth muscle of the vessel itself. Vascular endothelium is responsible in vasoconstriction and vasodilatation. Endothelium role in vasodilatation is mediated by muscarinic-M, histaminergic-H2, vasopressin, adrenergic-α2, serotonergic, kinin, thrombin, purinergic, ET_{B1}, and AT2 receptors which activate endothelial nitric oxide (NO), resulting in a decrease of calcium ion inside cytosol of the vascular smooth muscle, eventually causing vasodilatation. Endothelium is also responsible for vasoconstriction. Some study proved an existence of endothelium-dependent contractions factors (EDCF) that caused vascular vasoconstriction that is affected by the mediator released from endothelium [12], but what mediator was released remained a question. EDCF role in this study may be influenced by the AML extract intervention that can reduce vasodilatation activity mediated by vascular smooth muscle, but further research is needed.

Vasodilatation mechanism on blood vessel mediated by endothelium, can occur immediately on vascular smooth muscle through the effect of (1) calcium antagonist channel; (2) AT1 antagonist receptor; (3) ET antagonist receptor; (4) α1-receptor antagonist; (5) opening of potassium ion channel resulted in hyperpolarization; or in presence of (6) NO donor that caused sGC activation. All these processes cause a decrease of calcium ion in the cytosol of vascular smooth muscle, making the calcium calmodulin-ion complex to not be formed, causing the MLCK that activates myosin ATPase to not happen, therefore the myosin head will be detached from actin, causing vascular relaxation [12].

Study on rats proved that AML can reduce blood pressure, and its mechanism of action is not influenced by muscarinic, histaminergic, adrenergic or endothelium-dependent pathways. The results were not clearly explained. This confirmed our study that AML extract on aorta with endo-intact can weaken vasodilatation effect mediated by vascular smooth muscle, but mechanism of how the endothelium contributed in reducing vasodilatation effect will need further research. Hypotension effect from AML was related with alkaloids and essential oil contained in it. Alkaloids isoquinoline, coreximine, and anomurine are suspected to have weak lowering effect on blood pressure, and essential oil beta-cryophyllene shows hypotension and vasodilatation effect. Furthermore, reticuline as another alkaloid from AML was suspected to cause hypotension through voltage-dependent Ca^{2+} channel blocker and/or inhibiting the release of calcium ion from intracellular reserve, but further research needed to prove this [5]. AML also contain quercetin, it can dilate the vessels through sGC activation and cyclooxygenase-dependent pathway [13]. This might contribute to vasodilatation that happened in rat aorta with endo-denuded, hence further research on quercetin needed to be considered.

AML is already used by Dayak ethnicity at Abai in North Borneo from generation to generation as a treatment for hypertension. This herbal potion has been used in other countries, the results of the ethnobotany research has been widely applied in countries such as Bolivia [14], Nigeria [15], and Mauritius [16]. This research also confirms that the role of ethnobotany exploration is very useful in finding medicinal plants efficacy that are needed to be proved scientifically in pre-clinical and clinical studies. Our research proved the efficacy of medicinal plant of AML that has been used by Dayak Abai ethnicity to treat hypertension from generation to generation that was explored through Local Scientific Research of Community-based Ethnomedicine and Medicinal Plants in Indonesia at Abai ethnicity. Data from Local Scientific Research of Community-based Ethnomedicine and Medicinal Plants in Indonesia should have been explored by government and conserved so not to become lost as time goes on, and this intellectual property of local societies need to be thought out and protected. This abundant data is coming from Indonesia itself that is rich in medicinal plants and have more than 200 ethnics utilizing it as treatment for many kinds of diseases. If this database was available, it is the focus of the next generation researchers to scientifically prove the benefits of those medicinal plants so that they can be used in many health services and for the sake of Indonesia's prosperity.

4. Conclusion

Vasodilatation mechanism of action from ethanol extract of *A. muricata* was mediated by smooth muscle of blood vessel and endothelium that contributes to the weakening of vasodilatation activity.

Acknowledgements

Research source of funding is grant fund No. LB.02.01/VI.3/1226/2016 from the Ministry of Health of the Republic of Indonesia, Center for Research and Development of Medicinal Plants and Traditional Medicines in year 2016.

References

[1] WHO 1999 *J. Hypert.* **17**(2) 151-83
[2] Kearney P M, Whelton M, Reynold K *et al* 2005 *Lancet* **365**(9455) 217-23
[3] Babayi H, Kolo I, Okogun JI *et al* 2004 *Biokemis* **16**(2) 106-11
[4] Ismail S, Suwasono R A, Supriyoko W *et al* 2015 *Research Report on Indonesian Exploration Research of Local Ethnomedicine Knowledge and Community-based Medicinal Plants (RISTOJA) on The Dayak Abai Ethnicity in North Kalimantan*. (Tawangmangu: Center for Research and Development of Medicinal Plants and Traditional Medicine) [Indonesian]
[5] Nwokocha C R, Owu D U, Gordon A *et al* 2012 *Pharm. Biol.* **50**(11) 1436-41
[6] MoH R I 2008 *Indonesia Herbal Pharmacopoeia* (Jakarta: Ministry of Health Republic of Indonesia) [Indonesian]
[7] MoH R I 1995 *Indonesia Pharmacopoeia* (Jakarta: Ministry of Health Republic of Indonesia) [Indonesian]
[8] Pieper G M and Dondlinger L A 1997 *J. Pharmacol. Exp. Ther.* **283**(2) 684-91
[9] Zhang L, Nie P, Zhang G *et al* 2011 *J. Med. Plants* Res. **5**(11) 2187-91
[10] Lee K, Shin MS, Ham I *et al* 2015 *BMC Complement Altern. Med.* **15** 395
[11] Niazmand S, Harandizadeh F, Mahmoudabay M *et al* 2014 *Adv. Biomed. Res.* **4** 9.
[12] Boulanger CM and Vanhoutte PM 1994 *The Endothelium: A Pivotal Role in Health and Cardiovascular Disease* (Courbevoie; Servier International) p 11-20
[13] Jankowski V, Tolle M, Vanholder R *et al* 2005 *Nat. Med.* **11**(2):223–7
[14] Nawwar M, Ayoub N, Hussein S *et al* 2012 *Arch. Pharmacal. Res.* 35(5) 761–767
[15] Hajdu Z and Hohmann J 2012 *J. Ethnopharmacol.* **139**(3) 838–857
[16] Ezuruike U F and Prieto J M 2014 *J. Ethnopharmacol.* **155**(2) 857–924.

Thrombocyte counts in malaria patients at East Kalimantan

L R D Siagian[1*], V Asfirizal[2], V D L Toruan[2] and N Hasanah[3]

[1]Clinical Pathology Laboratory of Medicine Faculty Mulawarman University, Samarinda Indonesia
[2]Parasitology Laboratory of Medicine Faculty Mulawarman University, Samarinda Indonesia
[3]Histology Laboratory of Medicine Faculty Mulawarman University, Samarinda Indonesia

*Corresponding author: lonita_ui@yahoo.com

Abstract. Malaria still becoming a serious health problem in Indonesia. Beside disorders of erythrocytes, there are some data that Plasmodium caused the other blood cells like leukocyte and thrombocyte. In malaria, changes of thrombocyte is thrombocytopenia that would be a complication from malaria vivax or malaria falciparum. The aim of this study is to know the thrombocyte count of malaria patients in East Kalimantan. Design of this study is descriptic retrospective from medical record's data from 2011-2016 in 7 hospitals (AW Syahranie at Samarinda, Kanudjoso at Balikpapan, Penajam Paser Utara at Panajam, AM Parikesit at Tenggarong, Taman Husada at Bontang, Kudungga at Sangata and Abdul Rivai at Tanjung Redeb. We collected the data from June-August 2017. There are 1041 malaria patients with male and female respectively 88.2% and 11.2%. The etiology of malaria were Plasmodium falciparum, Plasmodium vivax and mixed infection (P.f and P.v) respectively 62.6%, 38% and 6.1%. We found thrombocyte count was normal, decrease and increase respectively 11%, 85% and 1.7%. The degree of thrombocytopenia in malaria patients were mild (100.000-150.000/μl) 31.8%, moderate (50.000-100.000/μL) 45.6% and severe (<50.000/μl) 33.6%. The thrombocyte count in malaria patients at East Kalimantan was thrombocytopenia with moderate degree of thrombocytopenia.

1. Introduction

Malaria still becoming a serious health problem in Indonesia. There are 12 provincies that have Annual Parasitic Index (API) more than national API, one of them is East Kalimantan. Plasmodium as an infectious agent of malaria will destroy the erythrocytes. Beside disorders of erythrocytes, there are some data that Plasmodium caused the other blood cells like leukocyte and thrombocyte. In malaria, changes of thrombocyte is decrease of thrombocyte count or thrombocytopenia that would be a complication from malaria vivax or malaria falciparum. In clinical practice, we found late diagnosis of malaria because thrombocytopenia with infectious disease have relationship with viral infection like dengue infection. Thrombocytopenia has been reported to be associated with malaria but have been not studied well. The aim of this study is to know about thrombocyte count in malaria patients at East Kalimantan.

2. Materials and Methods

This study design was descriptive retrospective from medical record's data from 2011-2016 in 7 hospitals at East Kalimantan. That hospitals are Abdul Wahab Syahranie at Samarinda, Kanudjoso

Djatiwibowo at Balikpapan, Penajam Paser Utara at Panajam, Aji Muhamad Parikesit at Tenggarong, Taman Husada at Bontang, Kudungga at Sangata and Abdul Rivai at Tanjung Redeb. We collected the data from June-August 2017. The sample are inpatients and outpatients who diagnosed malaria and have positive blood thick smear in that hospital and have thrombocyte counts in complete blood count results. The variable was sex, type of Plasmodium, thrombocyte counts and degree of thrombocytopenia. The variable of thrombocyte counts were normal, thrombocytosis if >450,000 cells/μL and thrombocytopenia if <150,000 cells/μL. We classified the thrombocytopenia into three degree. Mild thrombocytopenia if 100,000-150,000 cells/μL, moderate if 50,000-100,000 cells/μL, severe <50,000 cells/μL.

3. Results and Discussion

Our study found a total of 1041 malaria patients from 2011-2016 in 7 hospitals in East Kalimantan consists of 918 males (88.18%) and 113 (10.85%) females (table 1).

Table 1. Sex distribution of malaria patients in East Kalimantan from 2011-2016

Hospital	Total Patients (%)	Male (%)	Female (%)
Abdul Wahab Sjahranie Samarinda	152 (14.60)	139 (91.45%)	13 (8.55)
Kanujoso Djatiwibowo Balikpapan	64 (6.15)	57 (89.1)	7 (10.94)
Panajam Paser Utara Penajam	286 (27.47)	247 (86.37)	39 (13.63)
AM ParikesitTenggarong	181 (17.39)	170 (93.92)	11 (6.08)
Taman HusadaBontang	15 (1.44)	14 (93.33)	1 (6.67)
Kudungga Sangatta	46 (4.42)	38 (82.61)	8 (17.39)
Abdul Rivai Tanjung Redeb	297 (28.53)	263 (88.55)	34 (11.45)
Total	**1041 (100)**	**918 (88.18)**	**113 (10.85)**

Table 2 showed the distribution of malaria patients based on Plasmodium type in East Kalimantan from 2011-2016. We found that Plasmodium falciparum was the majority caused of our malaria patients (62.63%) followed by *P. vivax* (31.32%) and mixed infection (6.15%).

Table 2. Distribution of malaria patients based on Plasmodium type in East Kalimantan from 2011-2016

Hopitals	Plasmodium types					Total
	P.falciparum (%)	*P.vivax* (%)	*P.ovale* (%)	*P.malariae* (%)	*MixedP.f +P.v*	
Abdul Wahab Sjahranie Samarinda	88(57.89)	57(37.5)	0(0)	0(0)	11(7.24)	152
Kanujoso Djatiwibowo Balikpapan	34(53.12)	22(34.38)	0(0)	0(0)	8(1.25)	64
Panajam Paser Utara Penajam	149(52.1)	112(39.16)	0(0)	0(0)	24(8.39)	286
AM ParikesitTenggarong	151(83.42)	29(16.02)	0(0)	1(0.55)	0(0)	181
Taman Husada Bontang	2(13.33)	13(86.67)	0(0)	0(0)	0 (0)	15
Kudungga Sangatta	29(63.04)	17(36.96)	0(0)	0(0)	0(0)	46
Abdul Rivai Tanjung Redeb	199(67)	76(25.59)	0(0)	1(0.33)	21(7.07)	297
Total	**652(62.63)**	**326(31.32)**	**0(0)**	**2(0.19)**	**64(6.15)**	**1041**

Distribution of thrombocyte counts in malaria patients at East Kalimantan was showed on table 3. We found 885 patients got thrombocytopenia (89.66%), 100 patients got normal thrombocyte count (10.13%) and only 2 patients got thrombocytosis (0.20%).

Table 3. Distribution of thrombocyte counts in malaria patients at East Kalimantan from 2011-2016

Hopitals	Thrombocyte count			
	Thrombocytopenia	**Normal**	**Thrombocytosis**	**Total**
Abdul Wahab Sjahranie Samarinda	93(86.91)	13(12.15)	1(0.93)	107
Kanujoso Djatiwibowo Balikpapan	58(90.63)	6(9.37)	0(0)	64
Panajam Paser Utara Penajam	252(20.49)	30(10.6)	1(0.35)	283
AM Parikesit Tenggarong	170(93.92)	11(6.01)	0(0)	181
Taman Husada Bontang	12(85.71)	2(14.29)	0(0)	14
Kudungga Sangatta	43(95.55)	2(4.44)	0(0)	45
Abdul Rivai Tanjung Redeb	257(87.71)	36(12.29)	0(0)	293
Total	**885(89.66)**	**100(10.13)**	**2(0.20)**	**987**

Out of 885 patients with thrombocytopenia, 404 patients (45.65%) had moderate thrombocytopenia, followed with 281 patients with mild thrombocytopenia (31.75%) and 200 patients with severe thrombocytopenia (22.60%).

Table 4. Distribution of thrombocytopenia in malaria patients at East Kalimantan from 2011-2016

Hospitals	**Degree of thrombocytopenia**			
	Mild 100.000-150.000/µL (%)	**Moderate 50.000-100.000/µL (%)**	**Severe < 50.000/µL (%)**	**Total**
Abdul Wahab Sjahranie Samarinda	19(20.43)	31(33.33)	43(46.24)	93
Kanujoso Djatiwibowo Balikpapan	10(17.24)	20(34.44)	28(48.28)	58
Panajam Paser Utara Penajam	78(30.95)	124(49.2)	50(19.84)	252
AM Parikesit Tenggarong	85(50)	65(38.24)	20(11.76)	170
Taman Husada Bontang	6(50)	6(50)	0(0)	12
Kudungga Sangatta	16(37.21)	22(51.16)	5(11.63)	43
Abdul Rivai Tanjung Redeb	67(26.07)	136(52.92)	54(21.01)	257
Total	**281(31.75)**	**404(45.65)**	**200(22.60)**	**885**

We know that East Indonesia regions (Kalimantan, Sulawesi, Maluku, Irian Jaya, Nusa Tenggara) are endemic area for malaria. There are difference in number of malaria cases depend on region. Some area which known as malaria endemic area (Penajam Paser Utara, Tanjung Redeb district area) found as the most area of malaria cases in our study. In this study, *P. falciparum* is the most etiology of our malaria patients. This study found that thrombocytopenia affected in the majority of our malaria patients. We found that thrombocytopenia affected in 89.66% malaria patients, with moderate thrombocytopenia (45.65%) as the majority. Another study reported by Jadhav et al., in Bombay found that only 21.6% patients among malaria patients came with normal thrombocyte count, with mean thrombocyte count in malaria vivax 115.390/µL and malaria falciparum 100.900/µL [1]. Previous study by Morales et al., at Santos Anibal Dominicci Hospital in Venezuela between 2000-2002 found that 64.6% tertian malaria patients had decreased thrombocyte count (<150.000/mm^3), with 43% came with severe thrombocytopenia [2]. Ansari et al., found a decreased of thrombocytopenia (69.18%) among their malaria patients, and with the normal thrombocyte count in 30.81%. Between the studies, thrombocytopenia were more often found in malaria caused by *Plasmodium falciparum* or *Plasmodium vivax* [3]. Study by Chirag et al., from 781 malaria patients infected with Plasmodium falciparum, thrombocytopenia affected in 60%, while only 56% experienced with anemia [4].

Thrombocytopenia in malaria patients also reported by Kumar et al., in their study at Sasaram District Bihar, India [5]. Consistent with study by Hafeez et al., that reported 88% patients of Plasmodium vivax malaria had ëthrombocytopenia [6]. Tangvarasittichai et al., reported that thrombocytopenia was the common feature in both *P.falciparum* and *P.vivax* malaria infection, which incidence varies from 60%-80% [7].

Patel et al., in their study reported that thrombocytopenia can be used to be a highly sensitive test for diagnosed malaria disease. Also they found that severe thrombocytopenia may be more likely caused by *Plasmodium falciparum*. Even not as a criteria of severa malaria, but thrombocytopenia could be a complicated of malaria vivax and falciparum [8]. D'Acremont et al., reported that the presence of thrombocytopenia could be a laboratory predictor for diagnosis malaria among traveler to tropical countries with acute fever disease [9].

There are some suggested mechanism involved in the evidence of thrombocytopenia in malaria disease. Thrombocyte destruction mediated immune process and coagulation disorders [10], sequestration of thrombocyte in spleen [11] and oxidative stress [12].

4. Conclusion

The thrombocyte count in malaria patients at East Kalimantan was thrombocytopenia with moderate degree of thrombocytopenia. The further study was needed to know about the mechanism of thrombocytopenia in malaria.

Acknowledgments
We sincerely thank Islamic Development Bank Project for Mulawarman University on 2017 for support this study and also all our medical students of Mulawarman University for collecting the data. We thanks to the director of the hospitals where become study site of the study.

References

[1] Jadhav UM, PatkarVS and KadamNN 2004 Thrombocytopenia in malaria-correlation with type and severity of malaria *JAPI* **52**

[2] Morales, Rodriquez AJ, Sanchez E, Vargas M, Piccolo C, Colina R, Arria M and Franco-Paredes C 2005 Occurrence of thrombocytopenia in *Plasmodium vivax* malaria *Clinical Infectious Disease* Jul 1 **41**(1) pp 130-1

[3] Ansari S, Khoharo H., Abro A, Akhund IA and Qureshi F 2009Thrombocytopenia in *Plasmodium falciparumMalaria. Journal Ayub Med Coll Abbottabad* **21**(2)

[4] Chirag RC, Shubhangi DV, Himanshu RM, Varsha GY, Amul P, Vaibhav P, Dimple D and Manlik P 2010 *Plasmodium falciparum* versus *Plasmodium vivax*: Which is alLesser evil *National Journal of Community Medicine* **3**(3) ,July-September 2012.

[5] Kumar ., Kumar A, Vandan and Nag S 2014 Thrombocytopenia as a marker for diagnosis of malaria *JOSR Journal of Dental and Medical Sciences* **13**(9) Ver.III: pp 36-8

[6] Hafeez M, Lodhi FR, Akhtar Z, Ali MZ and Aijaz A 2015 Severity of thrombocytopenia in patients with Plasmodium vivax Malaria; a single center study. *J Ayub Med Coll Abbottabad* **27**(1)

[7] Tangvarasittichai O, Srikong and Tangvarasittichai S 2016 Platelet count and platelet indices used as potential markers for first malaria infection diagnosis *International Journal of Pharmaceutical and Clinical Research* **8**(10) pp 1454-58

[8] Patel U, Gandhi G, Friedman S and Niranjan S 2004 Thrombocytopenia in malaria*Journal of The National Medical Association* **96**(9)

[9] D'Acremont V, Landry P, Mueller I, Pecoud A and Genton B 2002 Clinical and Laboratory Predictors of Imported Malaria in an Outpatient Setting: an Aid to Medical decision making

in returning travelers with fever *Am J Trop Med Hyg* **66**(5) pp 481-6

[10] de Mast Q, Groot E, Lenting PJ, de Groot PG, McCall M and Saurerwein RW2007 Thrombocytopenia and release of activatedvon willebrand factor during early *Plasmodium falciparum* malaria *JID* **196**(15) pp 622-8

[11] Bhandary N, Vikram GS and Shetty H 2011 Thrombocytopenia in malaria: A Clinical Study. *Biomedical Research* **22**(4) pp 489-91

[12] Araujo CF, Lacerda M, Abdalla DS and Lima ES 2008 The role of palatelet and plasma markers of antioxidant status and oxidative stress in thrombocytopenia among patients with vivax malaria *Mem Ist Oswaldo Cruz* **103**(6) pp 517-21

IOP Conf. Series: Earth and Environmental Science **144** (2018) 012008 doi:10.1088/1755-1315/144/1/012008

The addition of pineapple flesh and pineapple peels extracts to increase the quality of used cooking oil

R Kusumawardani[1]*, N Hasanah[1] and Sukemi[1,2]

[1]Chemical Education Department, Faculty of Teacher Training and Education, Mulawarman University, Samarinda, Indonesia
[2]SMA N 2, Samarinda, Indonesia

*Corresponding author: ratna.kusuma@fkip.unmul.ac.id

Abstract. In Indonesia, reuse of cooking oil is high and common. Heating process and reuse of the cooking oil causes a change in its chemical constituents and decrease its qualities. This research aimed to investigate the addition of pineapple flesh extract (PFE) and pineapple peel extract (PPE) on the increment of the quality of oxidized (used) cooking oil. The cooking oil has been used three times. Treatment was done by mixing the used cooking oil with the extract (2:1) at 50°C. Peroxide value, FFA and iodine number of treated and untreated used cooking oils were measured by using titration method. The result showed that the treatment could increase the quality of the used cooking oils. PPE was better than PFE to increase the quality of the used cooking oil.

1. Introduction

The use of cooking oil in Indonesia is increasing and Indonesia is known as a leading global supplier of cooking oil [1-2]. It is proven by the increment of cooking oil brand produced by cooking oil industries. An increment on the number of cooking oil production is caused by the increment of society demand on the cooking oil. It is used in food processing, especially in cooking. Cooking oil is widely used for frying foods because it has a high smoke point. Moreover, cooking oil tasty taste), create a crisp texture and increase nutrition value of foods [1-3].

Heating process and reuse of the cooking oil causes a change in its chemical constituents. During heating process and contact with oxygen, the cooking oil is oxidized and produces some aldehydes, ketones, and aromatic compound that create rancid odors [3]. It also increases the number of free fatty acid (FFA) and produce trans-fatty acid which cause some diseases such as coronary heart disease and diabetes [3].

In Indonesia, reuse of cooking oil is high and common. Commonly the societies use the cooking oil 2 – 4 times [3]. Indonesian calls the used cooking oil as *minyak jelantah*. As mentioned above, reuse the cooking oil can change its chemical constituents and decrease its quality. Indonesian government establishes a standard quality of cooking oil. The cooking oil can be consumed maximum value 0.3% (w/w) of FFA, 0.01 mg O_2/g of peroxide, and 0.6 mg KOH/g of acid value (AV) [3-4].

To maintain the quality of cooking oil, antioxidant such as butylated hydroxyanisole (BHA) and butylated hydroxytoluene (BHT), can be added to inhibit the oxidation of cooking oils. The use of BHA and BHT may contribute to carcinogenicity or tumorigenicity. There are some natural antioxidants that can replace the function of BHA and BHT such as vitamin E and Carotenoid and other phenolic compounds that can be found from plants; herb, vegetable and fruits[5-7].

One of the fruits that can be used as antioxidant is pineapple. Each 100 g of pineapple consist 24.0 mg of vitamin C and 39 μg of vitamin A [8-9]. It has been known that pineapple contain some phenolic compounds e.g. ferulic acid, myricetine, quercetin and tyramin [10].

According to the reason above, this research was conducted to investigate the addition of pineapple and pineapple juice on the increment of the quality of oxidized cooking oil.

2. Materials and Methods

2.1. Sample and chemicals

Pineapple fruits were purchased from traditional market in Samarinda city, East Kalimantan, Indonesia. The pineapple fruits were cleaned from dirt and separated between its flesh and peels. Both of the flesh and peels were blended by using commercial blender. Then, the juices were squeezed and filtered. The filtrate was called as pineapple flesh extract (PFE) and pineapple peels extract (PPE) were kept at cleaned and sealed glass container.

The used cooking oil was obtained from food vendor at Jalan Perjuangan I, Samarinda, East Kalimantan, Indonesia. It has been used three times. All chemicals; acetic acid glacial, ethanol, hydrochloric acid, starch, phenolphthalein, chloroform, sulphuric acid, iodine, bromide, potassium iodide, potassium iodate, potassium hydroxide, sodium thiosulfate, were purchased from Merck.

2.2. Treatment procedure of used cooking oil

100 mL of used cooking oil was loaded into a 250mL beaker glass and followed by the addition of 50 mL PFE. The mixture was mixed using a stirring rod, heated at 50°C for 5 min using a water bath. After reaching room temperature, the mixture was filtered using Whatman filter paper. The procedure above was done also by substitute PFE with PPE. All the experiment was repeated twice.

2.3. Analysis of peroxide value (PV)

PV was determined by following the procedure of Andarwulan *et al* with some modifications [11]. Weighed sample (untreated and treated used cooking oils) was loaded into 100 mL Erlenmeyer flask. 30mL of acetate glacial – chloroform (3:2 v/v) was added into the flask followed by 0.5mL saturated potassium iodide. The Erlenmeyer flask was covered by using stopper and shaken strongly for 2 min. Then, 2 drops of starch solution was added and blackish-blue colour appeared. The mixture was titrated with 0.01 N sodium thiosulphate. The titration was stopped when the blackish-blue colour gone. Titration was also performed for blanks (distilled water). The PV was calculated by using formula:

$$PV=\frac{(V_s-V_b)xNx1000}{W} \quad (1)$$

where: V_s is $Na_2S_2O_3$ volume used in titration of sample (ml), V_b is $Na_2S_2O_3$ volume used in titration of blank (ml), N is $Na_2S_2O_3$ normality (N), and W is weight of sample (g).

2.4. Analysis of free fatty acid (FFA)

FFA was performed using a method of Andarwulan *et al* with minor modifications [11]. Weighed sample (untreated and treated sample) was loaded into weighed 100 mL Erlenmeyer flask. 25 mL of 95% ethanol (v/v) was added into the flask and the mixture was heated at 50-75°C for 5 min.using a water bath and stirred. 3 drops of phenolphthalein solution was added and titrated with 0.1 N sodium hydroxide. The titration was stopped when pink colour appeared.

$$FFA=\frac{(V \times N \times 25.6)}{W} \quad (2)$$

where: V is volume of NaOH used in the sample until the pink color persisted (mL), N is the normality of NaOH, W is weight of the sample (g).

2.5. Analysis of iodine value (IV)

Iodine value (IV) was performed using a method of Daniali *et al* with minor modifications [12]. Weighed g sample (untreated and treated sample) was loaded into 100 mL Erlenmeyer flask with stopped cock followed by the addition of 10mL of chloroform and 25 mL of iodine bromide. The mixture was rotated and left at the dark condition for 30 min after the addition of 10 mL 15% potassium iodide (w/v) and 100 mL of boiled water. The mixture was titrated with 0.1 N sodium thiosulfate until pale yellow colour appeared. Two drops of starch solution were added and the titration was continued until the blackish-blue colour gone. The IV of sample was calculated by following formula:

$$IV = \frac{(V_b - V_s) \times N \times 12.69}{W} \quad (3)$$

where: V_s is volume of sodium thiosulfate used in titration of blank (ml), V_b is the volume of sodium thiosulfate solution used in titration of sample, N is the concentration of sodium thiosulfate (N) and W is weight of sample/blank (g).

3. Result and Discussion

The used cooking oil in this research was the cooking oil that has been used 3 times. Physically, the oil was more viscous and darker than unused cooking oil and produced rancid odor. Its viscosity was thicker than that of unused cooking oil. The viscosity of treated used cooking oils with PFE and PPE were almost same as the viscosity of unused cooking oil. Its colour was also nearly the unused cooking oil (pale yellow) and their rancid odor was gone. There was pineapple odor on the treated used cooking oils, whereas the pineapple odor of treated used cooking oil with the PFE was stronger than that of treated used cooking oil with the PPE. The PV, FFA and IV of treated and untreated used cooking oil are shown in table 1.

Table 1. The quality of untreated and treated used cooking oils

Samples	Odor	Colour	PV	FFA	IV
Untreated used cooking oil	rancid	Brown	2.40 ± 0.20	0.19 ± 0.01	24.66 ± 0.03
Treated used cooking oil-PFE	Pineapple aroma	Yellow	1.60 ± 0.40	0.14 ± 0.01	27.12 ± 0.58
Treated used cooking oil-PPE	Pineapple aroma	Yellow	1.13 ± 0.12	0.13 ± 0.01	25.99 ± 0.47

Lipid oxidation is one of the chemical reactions occurred in heat processing of cooking oil and produces peroxides [12]. According to Republic of Indonesia Standard, the maximum value of PV of cooking oil is 2 Meq/kg [3]. Tabel 1 shows that untreated used cooking oil was exceeded the allowed PV and the PV of both of PFE and PPE treated used cooking oils were lower than that of maximum PV allowed. It means that the PPE and PFE could decrease the PV of used cooking oil. FFA indicates the hydrolysis reaction occurs. 0.3 % is the maximum amount of FFA in cooking oil that allowed by Republic of Indonesia Standard. Table 1 shows that the PFE and PPE could decrease the FFA of the used cooking oil and the amounts were lower than the maximum amount of FFA allowed. IV was used to determine the unsaturated fatty acid amounts. The IV is decrease during the heat and the saturation rate is increased [12]. The IV values of untreated and treated used cooking oils were low. It indicates that during the heat process, the unsaturated bonds were broken. The highest saturated fatty acid causes the cooking oil resistant to oxidation reactions.

4. Conclusion

The use of PFE and PPE to increase the quality of used cooking oil has been discussed. It can be concluded that the PFE and PPE could increase the quality of used cooking oil and the PPE was better than PFE to increase the quality of used cooking oil.

Acknowledgement

Authors gratefully thank Chemistry Laboratory of Teacher Training and Education Faculty of Mulawarman University that facilitated this research.

References

[1] Wildan F 2002, Penentuan bilangan peroksida dalam minyak nabati dengan cara titrasi, *Temu Teknis Fungsional Non Peneliti*, p 63
[2] Andarwulan N, Gitapratiwi D, Laillou A, Fitriani D, Hariyadi P, Moench-Pfanner R, and Martianto D 2014, uality of vegetable oil prior to fortification is an important criteria to achieve a health impact, *Nutrients*, **6** p 5051
[3] Ilmi IMB, Khomsan A, and Marliyati SA 2015, Kualitas minyak goreng dan produk gorengan selama penggorengan di rumah tangga Indonesia, *Jurnal Aplikasi Teknologi Pangan*, **4** p 61
[4] Suroso AS 2013, Kualitas Minyak Goreng Habis Pakai Ditinjau dari Bilangan Peroksida, Bilangan Asam dan Kadar Air, *Jurnal Kefarmasian Indonesia*, **3** p 77
[5] Taghvaei M and Jafari SM 2015, Application and stability of natural antioxidants in edible oils in order to substitute synthetic additives, *J. Food Sci. Technol.*,**52** p 1272
[6] Kahl R and Kappus H 1993, Toxicology of the synthetic antioxidants BHA and BHT in comparison with the natural antioxidant vitamin E, *Z Lebensm Unters Forsch.*, **196** p 328
[7] Saito M, Sakagami H, and Fujisawa S. 2003, Cytotoxicity and apoptosis induction by butylated hydroxyanisole (BHA) and butylated hydroxytoluene (BHT), *Aticancer Res.*, **23** p 4693
[8] Rashad MM, Mahmoud AE, Ali MM, Nooman MU, and Al-Kashef AS 2015, Antioxidant and anticancer agents produced from pineapple waste by solid state fermentation, *International Journal of Toxicological and Pharmacological Research*,**7** p 287
[9] Santoso HB 1998 *Manisan Nanas* (Jakarta: Kanisius) p 13
[10] Yapo ES, Kouakou HT, Kouakou LK, Kouadio JY, Kouamé P, and Mérillon JM 2011, Phenolic profiles of pineapple fruits (*Ananas comosus* L. Merrill) influence of theorigin of suckers, *Australian Journal of Basic and Applied Science*s, **5** p 1372
[11] Andarwulan N, Gitapratiwi D, Laillou A, Fitriani D, Hariyadi P, Moench-Pfanner R and Martianto D 2015, Quality of vegetable oil prior to fortification is an important criteria to achieve a health impact, *Nutrients*, **6** p 5051
[12] Daniali G, Jinap S, Hajeb P, Sanny M, Tan CP 2016, Acrylamide formation in vegetable oils and animal fats during heat treatment, *Food Chemistry,* **212** p 244

IOP Conf. Series: Earth and Environmental Science **144** (2018) 012009 doi:10.1088/1755-1315/144/1/012009

Water Level Prediction of Lake Cascade Mahakam Using Adaptive Neural Network Backpropagation (ANNBP)

Mislan*[1], A F O Gaffar[2], Haviluddin[3], N Puspitasari[4]

[1]Department of Physics, Faculty of Mathematics and Natural Science, Mulawarman University, Indonesia
[2]Department of Information Technology, State of Polytechnic Samarinda, Indonesia
[3,4]Department of Computer Science, Faculty of Computer Science and Information Technology, Mulawarman University, Indonesia

*Corresponding author: airmasadepan@yahoo.co.id, haviluddin@gmail.com

Abstract. A natural hazard information and flood events are indispensable as a form of prevention and improvement. One of the causes is flooding in the areas around the lake. Therefore, forecasting the surface of Lake water level to anticipate flooding is required. The purpose of this paper is implemented computational intelligence method namely Adaptive Neural Network Backpropagation (ANNBP) to forecasting the Lake Cascade Mahakam. Based on experiment, performance of ANNBP indicated that Lake water level prediction have been accurate by using mean square error (MSE) and mean absolute percentage error (MAPE). In other words, computational intelligence method can produce good accuracy. A hybrid and optimization of computational intelligence are focus in the future work.

1. Introduction

A natural hazards and flood phenomenon are part of a natural phenomenon that always exists. For example, some floods are caused by dam damage, landslides including overflowing of lake water. In general, during dry period, a greater possibility of flooding and also river becomes very low. The rising sea level will make it more difficult for the rivers in the delta to drain into the sea. Rivers do not recognize national borders. The flood is a climatological phenomenon that is affected by geology, geomorphology, relief, soil, and vegetation. The meteorology and hydrology processes are actually predictable whether caused by natural factors, humans or both. Therefore, precautions can be evaluated and implemented as well as necessary[1]. Furthermore, all these processes are expected to help communities better understand the steps to reduce risk exposure [2, 3].

Therefore, in order to produce an accurate result of a cascade lake water level forecasting, several methods were developed. Among them, statistical models have been used to create waterfall levels of lakes. This model works with the decline of the data itself (SVE), decomposition, exponential smoothing method (ES) and autoregressive integrated moving average (ARIMA). Several studies have shown that lake water levels are cascaded because the data is not linear. However, in some cases, the falling waterfall method of cascade lakes is also capable of producing good and accurate predictions [4, 5].

Along with the development of computing technology, many researchers are trying to make predictions using the computational intelligence i.e., Fuzzy, neural network, Self-organizing Maps

(SOM) [6]etc. methods in the field of hydrology.Farajzadeh et.al have presented Feed-forward Neural Network and Autocorrelation Regressive Integrated Moving Average (ARIMA) methods in order to forecast the monthly rainfall in Urmia lake basin, Iran. The results showed that these methods could be predicted the rainfall for a 6-year period (2012-2017) [7]. Cramer et al have implemented and compared six machine learning algorithms such as Genetic Programming, Support VectorRegression, Radial Basis Neural Networks, M5 Rules, M5 Model trees, and k-Nearest Neighbors. This studied was used the rainfall time series across data sets for 42 cities (20 cities around Europe and 22 cities around USA). The results showed that machine learning based intelligent system have had accuracy prediction for rainfall [1]. This paper is purpose of prediction of Lake Cascade water level in order to describe flood pattern and its probabilities. In this study, the water level prediction of Lake Cascade Mahakam using Adaptive Neural Network Back-Propagation based on AR model have been performed.

2. Method

The result of observation of a particular event in the past can be expressed in terms of historical data, called time series. Where, this time series data model can be used to generate new data values for forecasting. Meanwhile, the accuracy of forecasting depends on how well the time series data model is built [2].

In this study, the water level of Mahakam Cascade Lake is calculated from sea level as reference point 0. Then, water level data can be categorized as time series data with seasonal and non-stationary pattern. The AR model is declared in Equation 1.

$$y(t) = -a_1 y(t-1) - \cdots - a_n y(t-n) + e(t) \quad (1)$$

Where,$a_n$is a constant, n is order number of system, and $e(t)$ is a white noise. Thus, Equation 1 can be decomposed in Equation 2.

$$\begin{aligned} y(t) &= (a_1 q^{-1} + \cdots + a_n q^{-n}) y(t) + e(t) \\ &= -\left(\sum_{i=1}^{n} (a_i q^{-1} + \cdots + a_n q^{-n}) y(t) + e(t) \right) \end{aligned} \quad (2)$$

Where,q^{-1} is the delay operator, and $A(q^{-1})$ is polynomial to be estimated.

The AR models are approximated by using Feed Forward Neural Network (FFNN) which declared in Equation 3.

$$(t) = N_{ff}(A(q^{-1})y(t)) + e(t) \quad (3)$$

By training $N_{ff}(\ldots)$, such that$e(t) \to 0$ then $N_{ff}(\ldots) \to y(t)$ is obtained. In its implementation, $e(t)$ is set as small as possible. The FFNN architecture based on AR (Equation 3) is shown in Figure 1.

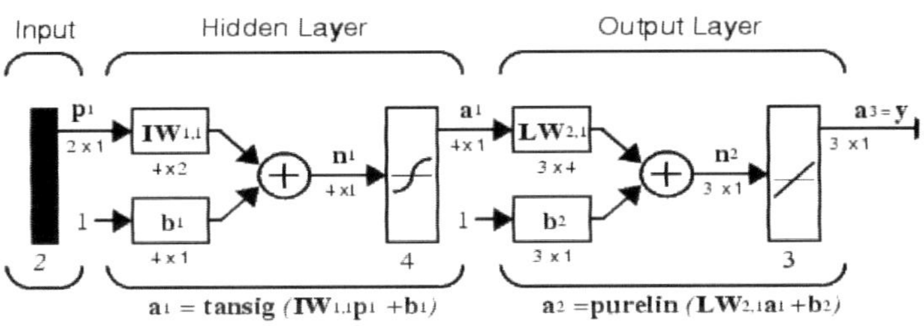

Figure 1. FFNN Architecture [2]

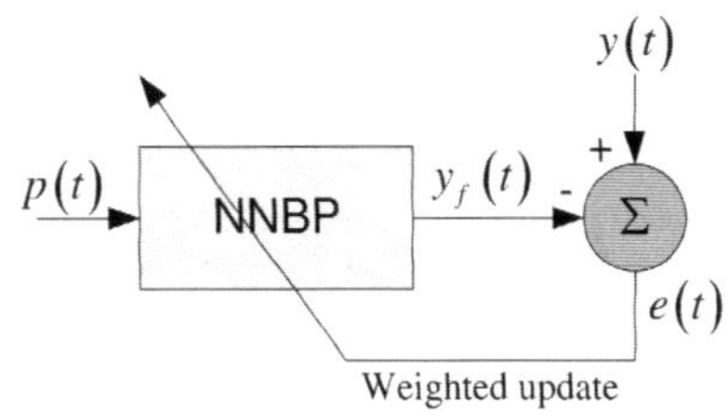

Figure 2. NNBP training

In principle, neural network backpropagation (NNBP) is doing back-propagation to fix the weight of each layer such that to achieve target error [3], as shown in Fig. 2. Under certain circumstances, for various reasons such as the amount of data, data patterns (i.e., non-linearity, non-stationary, and discontinuity), and error targets that are too small. Thus, NNBP maybe cannot reach the target error.

Furthermore, ANNBP is NNBP with adaptively weighted adjustment based on random selection both on input weights and on layer weights. The selected weights are updated by adding a very small random number which multiplied by current training error (MSE_{train}). The network weights are updated adaptively which conducted continuously to achieve the target error. The training model of ANNBP is shown in Fig. 3. Then, the algorithm as shown in Fig. 4.

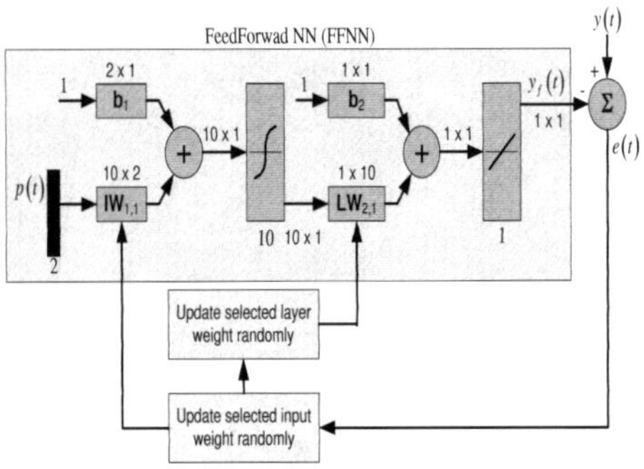

Figure 3. The training model of ANNBP

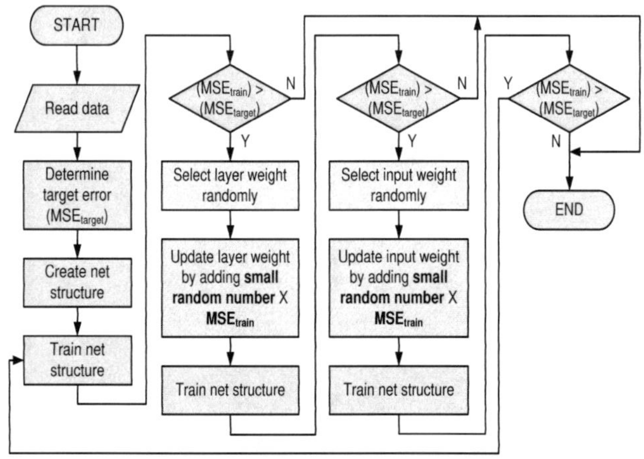

Figure 4. The algorithm of ANNBP

In this study, Lake Cascade Mahakam Water Data, East Kalimantan Province in 1989 – 2016 period or 1008 datasets has been used. The datasets consists of three types: maximum, minimum, and average water levels as shown in Tables 1, 2 and 3.

Table 1. Average Water Level of Lake Cascade Mahakam, years 1989 – 2016.

Years	Jan	Feb	Mar	Apr	May	Jun	Jul	Aug	Sep	Oct	Nov	Dec
1989	1,60	1,09	0,74	0,77	2,10	2,65	3,39	2,75	2,35	1,87	3,16	3,28
1990	4,48	2,43	1,93	2,76	2,48	0,94	-1,09	-0,98	-0,78	-0,69	0,11	-0,19
1991	0,43	0,47	1,71	2,12	4,25	4,84	3,37	0,67	0,03	-0,10	0,53	2,14
1992	2,18	0,95	1,54	1,98	2,14	2,68	1,39	1,78	1,22	1,91	2,63	4,64
1993	4,02	2,21	2,03	3,76	3,45	2,30	1,77	0,25	0,29	0,80	0,88	2,53
1994	3,42	3,31	2,74	3,92	4,89	4,88	3,80	1,70	1,16	0,93	0,97	2,75
1995	3,38	2,79	2,47	3,83	5,01	3,31	1,43	2,14	1,34	2,01	2,46	1,32
1996	2,66	4,65	4,19	3,68	4,33	2,38	1,14	-0,16	2,33	1,22	1,92	4,33
1997	3,84	2,91	1,22	-0,19	-1,32	-2,25	-1,63	-0,66	-1,62	-2,07	-2,17	-2,56
1998	-0,85	0,94	2,38	3,70	3,08	2,71	4,13	4,82	3,43	2,95	2,95	3,22
1999	3,99	2,77	3,02	4,42	4,82	4,11	2,80	1,30	0,06	0,67	1,01	0,09
2000	1,50	2,95	2,08	0,46	-0,86	-1,41	-0,28	0,53	-0,20	1,24	3,14	3,99
2001	2,87	4,22	5,43	3,85	4,68	2,19	-1,62	-1,02	-0,15	2,47	2,78	3,31
2002	4,26	4,79	4,93	5,35	5,20	4,30	2,15	-0,75	0,20	2,38	2,98	3,99
2003	2,98	1,61	0,83	0,75	0,57	1,56	1,04	0,62	0,18	0,46	-0,31	-0,21
2004	1,44	3,87	4,97	3,56	3,69	1,87	1,35	0,49	0,76	0,61	2,31	3,50
2005	2,38	2,59	2,58	5,24	5,54	4,59	2,87	1,54	1,96	2,29	3,92	5,65
2006	4,49	3,19	4,37	3,75	5,51	5,97	3,94	0,93	-0,97	-2,03	-1,60	-0,07
2007	3,19	4,09	2,50	2,96	7,67	6,38	2,57	0,88	1,50	1,08	1,91	2,57
2008	1,94	1,56	2,13	2,21	2,67	2,21	1,83	1,12	2,02	3,31	3,55	3,29
2009	3,96	1,48	2,73	4,22	2,79	-0,16	-0,48	0,62	0,66	0,32	1,95	3,50
2010	4,29	3,94	3,16	1,56	1,20	2,23	0,91	-0,03	1,08	3,36	5,69	4,96
2011	4,01	3,11	3,02	4,14	4,34	4,08	2,79	1,20	-0,04	0,77	1,01	0,12
2012	2,58	2,59	2,88	5,24	5,34	4,59	2,97	1,64	2,06	2,39	4,02	5,55
2013	2,94	2,56	2,13	2,21	2,67	2,21	1,91	1,12	2,02	3,37	3,68	3,35
2014	3,39	3,86	3,66	1,58	1,90	2,23	0,91	0,32	1,28	3,36	5,50	4,92
2015	3,92	1,49	0,51	0,60	0,57	1,38	0,93	0,49	0,35	0,27	-0,31	-0,21
2016	0,12	0,48	2,00	2,25	2,79	0,84	0,95	0,62	0,66	0,32	1,72	2,50

Table 2. Maximum Water Level of Lake Cascade Mahakam, years 1989 – 2016.

Years	Jan	Feb	Mar	Apr	May	Jun	Jul	Aug	Sep	Oct	Nov	Dec
1989	2,74	2,51	0,98	1,14	2,82	3,32	4,01	3,71	2,81	2,45	4,01	4,32
1990	4,87	3,83	2,57	3,11	2,96	1,87	-0,11	-0,53	-0,12	0,00	0,49	0,56
1991	1,23	1,98	2,12	3,12	5,56	5,57	4,95	1,99	0,47	0,37	3,13	2,80
1992	4,49	4,79	5,43	5,35	7,67	6,38	4,13	4,82	3,43	3,37	5,69	5,65
1993	5,43	2,85	2,70	4,65	4,43	2,81	2,19	1,97	0,60	1,08	2,03	2,99
1994	3,76	3,72	3,21	4,51	5,18	5,13	4,61	2,88	1,75	1,50	1,57	3,51
1995	3,70	3,09	3,12	4,79	5,47	4,26	2,31	2,45	2,13	2,97	3,17	2,04
1996	3,99	5,27	5,22	4,35	5,09	2,89	1,92	1,50	2,94	2,00	3,13	5,05
1997	4,64	3,29	2,30	0,15	-0,76	-1,85	-0,99	-0,38	-1,12	-1,62	-1,59	-2,16
1998	0,15	1,53	3,13	3,90	3,69	3,37	4,63	5,09	4,20	3,23	3,37	3,76
1999	4,46	3,46	3,73	4,93	5,08	4,76	3,32	2,02	0,49	1,31	1,52	3,54
2000	3,62	3,27	2,86	1,18	-0,26	1,97	0,21	0,80	0,23	2,11	4,21	4,79
2001	3,43	5,03	5,98	4,72	5,50	4,23	-0,34	-0,18	0,78	3,01	3,45	3,82
2002	4,79	5,09	5,13	5,75	5,44	4,79	3,49	-0,18	1,27	2,85	4,34	4,79
2003	3,42	2,45	1,08	1,26	0,80	3,62	1,35	1,26	0,86	0,80	-0,15	1,50
2004	2,84	4,82	5,30	3,96	4,59	3,37	3,08	3,13	1,00	1,21	3,65	3,71
2005	2,69	2,82	2,74	6,42	6,05	5,26	3,51	2,22	2,35	2,45	5,93	5,99
2006	5,14	4,82	5,20	4,57	6,26	6,37	6,28	1,84	0,01	-1,68	-1,01	1,59
2007	4,08	4,40	3,27	5,66	8,68	7,96	4,32	1,67	2,22	1,33	3,02	3,19
2008	2,43	1,95	2,50	2,75	3,13	2,50	2,14	1,44	2,50	4,18	4,28	4,23
2009	4,50	2,66	81,94	4,96	4,93	0,38	0,17	1,15	1,13	1,00	2,90	5,04
2010	4,64	4,73	4,04	4,01	2,00	2,62	1,62	0,43	2,13	4,30	6,90	6,35
2011	2,38	2,59	2,58	5,24	5,54	4,59	2,87	1,54	1,96	2,29	3,92	5,65
2012	4,49	3,19	4,37	3,75	5,51	5,97	3,94	0,93	-0,97	-2,03	-1,60	-0,07
2013	3,19	4,09	2,50	2,96	7,67	6,38	2,57	0,88	1,50	1,08	1,91	2,57
2014	1,94	1,56	2,13	2,21	2,67	2,21	1,83	1,12	2,02	3,31	3,55	3,29
2015	3,96	1,48	2,73	4,22	2,79	-0,16	-0,48	0,62	0,66	0,32	1,95	3,50
2016	4,29	3,94	3,16	1,56	1,20	2,23	0,91	-0,03	1,08	3,36	5,69	4,96

Table 3. Minimum Water Level of Lake Cascade Mahakam, years 1989 – 2016

Years	Jan	Feb	Mar	Apr	May	Jun	Jul	Aug	Sep	Oct	Nov	Dec
1989	0,57	0,48	0,44	0,41	1,21	1,85	2,58	1,83	1,64	1,01	2,53	2,18
1990	3,82	-0,37	1,25	2,15	1,65	-0,06	-1,52	-1,37	-1,34	-1,05	-0,39	-0,73
1991	-1,03	-1,13	1,30	1,74	2,22	4,17	2,24	-0,36	-0,50	-0,73	-0,27	1,28
1992	1,72	0,15	0,30	0,44	1,15	1,90	0,46	0,56	0,82	1,08	1,52	3,88
1993	2,07	1,42	1,36	2,78	2,52	1,83	1,31	-0,84	0,00	0,40	0,31	2,13
1994	3,05	2,80	2,32	3,27	4,53	4,67	2,95	0,85	0,71	0,56	0,56	0,94
1995	2,87	2,30	1,86	3,17	4,46	2,40	0,84	1,56	0,87	1,28	1,70	0,76
1996	0,80	3,80	3,09	3,10	3,20	1,94	-0,50	-0,52	0,79	0,76	1,04	3,23
1997	2,91	2,40	0,15	-0,70	-1,80	-2,47	-2,21	-1,04	-2,26	-2,45	-2,77	-2,80
1998	-1,39	0,27	1,61	3,25	2,35	2,15	3,48	4,30	2,60	2,68	2,53	2,53
1999	3,18	2,15	2,26	3,83	4,61	3,29	2,14	0,56	-0,29	0,16	0,32	-0,33
2000	0,71	2,34	1,30	-0,22	-1,58	-1,83	-0,93	0,18	-0,53	0,35	2,22	3,34
2001	2,35	3,51	4,15	3,38	3,83	-0,31	-2,51	-3,47	-0,75	1,03	2,14	2,86
2002	3,70	4,35	4,69	4,15	4,75	3,54	0,82	-1,65	-0,75	1,40	2,03	3,34
2003	2,50	0,71	0,52	0,32	0,31	0,80	0,21	-0,05	-0,56	-0,13	-0,53	-0,52
2004	0,26	2,86	4,04	2,76	2,60	0,12	0,16	-0,72	0,43	0,10	0,26	3,20
2005	2,03	2,39	2,33	2,68	5,30	3,55	2,29	1,21	1,33	2,08	2,48	5,17
2006	3,50	1,91	3,19	3,05	4,63	5,68	1,87	-0,12	-1,86	-2,16	-4,90	-1,34
2007	1,65	3,38	1,61	1,46	5,81	4,44	1,21	0,48	0,70	0,78	0,97	2,20
2008	-1,36	1,27	1,42	1,81	2,20	1,90	1,20	0,87	1,49	2,53	2,61	2,40
2009	2,73	0,74	0,88	2,99	0,44	-1,01	-1,04	0,19	0,24	0,00	1,04	3,04
2010	3,79	3,34	2,22	0,98	0,74	1,65	0,14	-0,44	0,11	2,20	4,32	5,12
2011	2,91	2,01	1,92	3,04	3,24	2,98	1,69	0,10	-1,15	-0,33	-0,09	-0,98
2012	1,48	1,49	1,78	4,14	4,24	3,49	1,87	0,54	0,96	1,29	2,92	4,45
2013	1,84	1,46	1,03	1,11	1,57	1,11	0,81	0,02	0,92	2,27	2,58	2,25
2014	2,29	2,76	2,56	0,48	0,80	1,13	-0,19	-0,78	0,18	2,26	4,40	3,82
2015	2,82	0,39	-0,59	-0,50	-0,53	0,28	-0,17	-0,61	-0,75	-0,83	-1,41	-1,31
2016	-0,98	-0,62	0,90	1,15	1,69	-0,27	-0,15	-0,48	-0,44	-0,78	0,62	1,40

In this study, all data is compiled into time series data. Then, the AR with 240 model has been applied using Equation 1, as stated.

$$y(t) = -a_1 y(t-1) - a_2 y(t-2) \ldots - a_{240} y(t-240) + e(t)$$

Then, by using FFNN is stated.

$$N_{ff} = \left(y(t-1), y(t-2), \ldots, y(t-240)\right) \rightarrow y(t)$$

Where,$y(t\text{-}1)$ is water level data in 1989 – 1994;$y(t\text{-}2)$ is water level data in 1990 – 1995; $y(t\text{-}240)$ is water level data in 2003 – 2008 as a training data; $y(t)$ is water level data in 2009 – 2014 as a testing data. Before the ANNBP process, the normalized datasets into the interval [-1 ... 1] has been performed using Equation 4.

$$X_n(i) = \frac{2x(X(i) - X_{min})}{X_{max} - X_{min}} - 1 \quad (4)$$

In this study, the ANNBP net structure by using five hidden neurons and error target$e(t) = 10^{-4}$ have been explored. Then, the performance of training result by using MSE (Mean Squared Error), Equation 5 and MAPE (Mean Absolute Error), Equation 6 have been implemented.

$$MSE = \frac{1}{N}\sum_{i=1}^{N}(y(t_1) - y_{ff}(t_i))^2 \quad (5)$$

$$APE_i = \frac{abs(y(t_i) - y_{ff}(t_i))}{y(t_i)} x100\%$$

$$MAPE = \frac{1}{N}\sum_{i=1}^{N} APE_i \quad (6)$$

Where, $N$is the number of training data; $y(t_i)$ is the ith training target; $y_{ff}(t_i)$ is the ith output of adaptive NNBP. In this experiment, MATLAB for ANNBP have been used.

3. Result and discussion

In this section, the water level prediction of Lake Cascade Mahakam by using Adaptive Neural Network Backpropagation (ANNBP) models is presented.

The time series is a dataset of observations ordered in time. A time series is an ordered sequence of observations and many ways are used to forecast the time series data. In this experiment, Lake Cascade Mahakam Water Datasets in 1989 – 2016 period or 1008 datasets consists of three types: maximum, minimum, and average water levels has been used.

Based on experiment, average performance training data (2009-2014) by using MSE of $8.78x10^{-4}$ and MAPE of 1.17% and performance testing data (2015-2016) MSE of 2.16 and MAPE of 3.73% of average water levels have been performed, Fig. 5, 6. Then, maximum performance training data (2009-2014) by using MSE of $1.14x10^{-1}$ and MAPE of 0.53% and performance testing data (2015-2016) MSE of 3.25 and MAPE of 18.8% of average water levels have been settled, Fig. 7, 8. Finally, minimum performance training data (2009-2014) by using MSE of $2.7x10^{-3}$ and MAPE of 0.77% and performance testing data (2015-2016) MSE of 2.25 and MAPE of 6.59% of average water levels have been settled, Fig. 9, 10. The next process, each ANNBP net structure 240-5-1 for each type of data that has been trained, then to predict the water level of 2017-2018 as shown in Fig. 11.

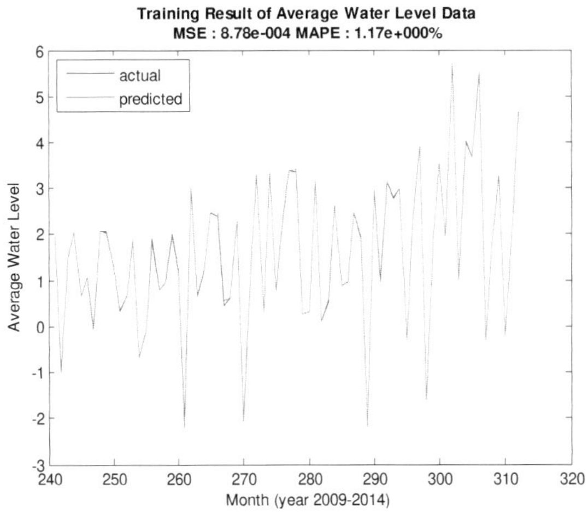

Figure 5. Training Result of Average Water Level Data (year 2009-2014)

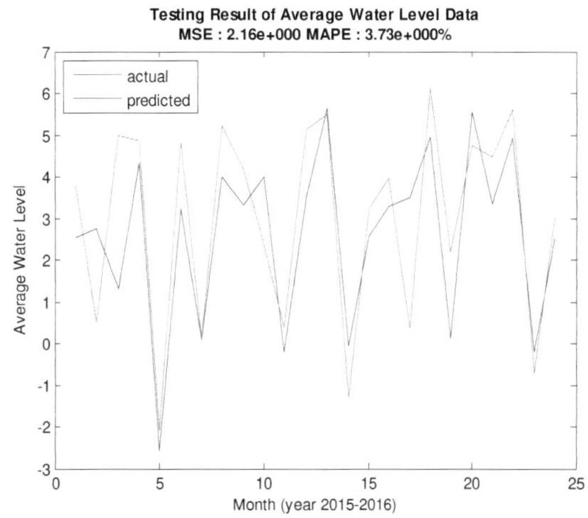

Figure 6. Testing Result of Average Water level Data (year 2015-2016)

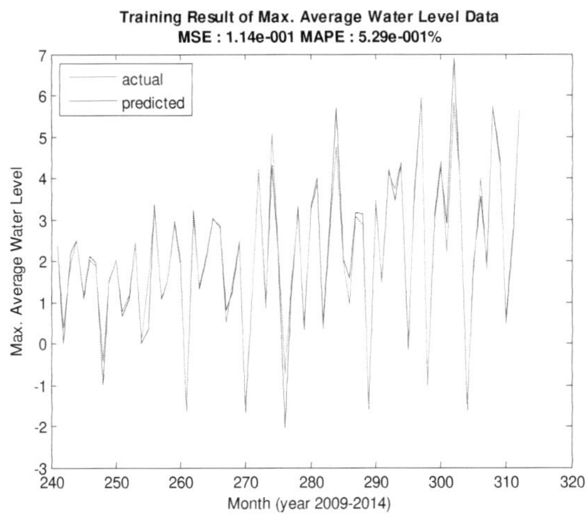

Figure 7. Training Result of Max.Water Level Data(year 2009-2014)

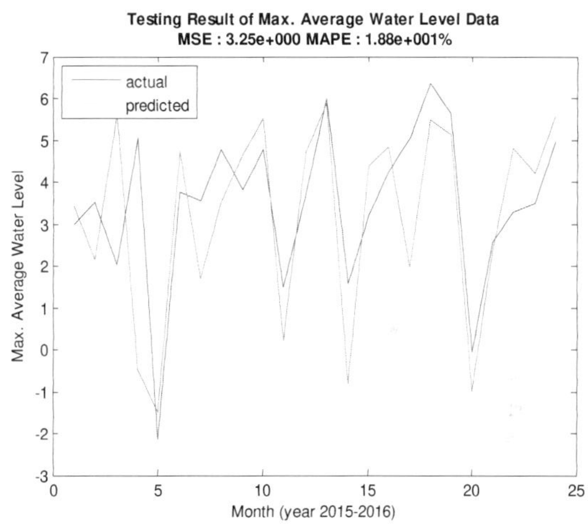

Figure 8. Testing Result of Max. Water level Data (year 2015-2016)

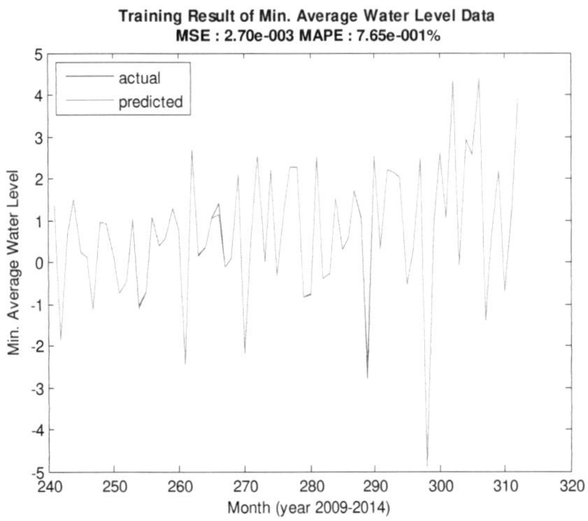

Figure 9. Training Result of Min.Water Level Data(year 2009-2014)

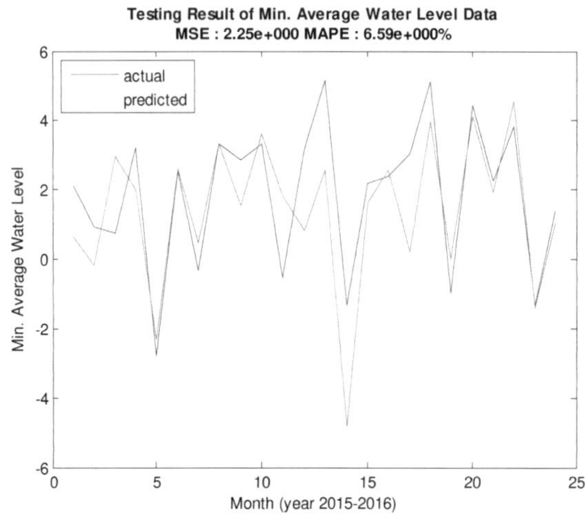

Figure 10. Testing Result of Min. Water level Data (year 2015-2016)

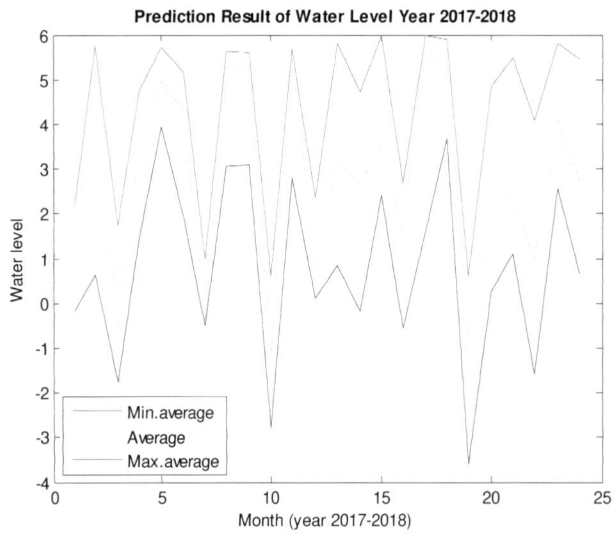

Figure 11. Prediction Result of Water level (year 2017-2018)

4. Conclusion

In this paper, ANNBP algorithm has been implemented and explored to model and predict water level of Lake Cascade Mahakam in East Kalimantan -Indonesia. After testing, the best MSE and MAPE values obtained were 9.7% with 240-5-1 NNBP architecture. In other words, this study have demonstrated that ANNBP models can be used as a predictive algorithm that provides a good predictive accuracy of water level Lake Cascade Mahakam. Future work is suggested to hybrid and optimization of a few computational intelligence methods in order to obtain more accurate prediction results.

Acknowledgments

Thanks to the Artificial Intelligence research group, Faculty of Computer Science and Information Technology (CSIT) and the IDB Project.

References

[1] Cramer S, Kampouridis M, Freitas A A, and Alexandridis A K, 2017 An extensive evaluation of seven machine learning methods for rainfall prediction in weather derivatives *Expert Systems with Applications* **85**, pp 169-81,.

[2] Mislan, Haviluddin, S Hardwinarto, Sumaryono, and M Aipassa, 2015 Rainfall Monthly Prediction Based on Artificial Neural Network: A Case Study in Tenggarong Station, East Kalimantan - Indonesia," in Int Conf. on Computer Science and Computational Intelligence (ICCSCI 2015), Jakarta, Indonesia pp. 142-51.

[3] R Islam, R Kamaruddin, S A Ahmad, S J Jan, and A R Anuar, 2016 A Review on Mechanism of Flood Disaster Management in Asia International Review of Management and Marketing,.

[4] J. Wu, J. Long, and M. Liu, 2015 Evolving RBF neural networks for rainfall prediction using hybrid particle swarm optimization and genetic algorithm Neurocomputing **148** pp 136–42,.

[5] Haviluddin and A Jawahir 2015 Comparing of ARIMA and RBFNN for short-term forecasting," International Journal of Advances on Intelligent Informatics (IJAIN) **1**(1) pp. 15-22,

[6] Haviluddin, A. Yunianta, A. H. Kridalaksana, Z. Arifin, B. Kresnapati, F. Rahman, A. F. O. Gaffar, H. Y. Irawan, M. Mulyo, and A. Pranolo, "Modelling of Network Traffic Usage Using Self-Organizing Maps Techniques," in 2016 2nd International Conference on Science in Information Technology (ICSITech) pp 334-8.

[7] J Farajzadeh, A F Fard, and S Lotfi 2014 Modeling of monthly rainfall and runoff of Urmia lake basin using “feed-forward neural network” and “time series analysis” model Water Resources and Industry **7-8** pp. 38–48

Preliminary study on biodiversity of arbuscular mycorrhizal fungi (AMF) in oil palm (*Elaeis guineensis* Jacq.) plantations in Thailand

Auliana[1*] and W Kaonongbua[1]

[1]Department of Microbiology, Faculty of Science, King Mongkut's University of Technology Thonburi, Bangkok 10140 Thailand

*Corresponding Author: wittaya.kao@mail.kmutt.ac.th

Abstract. Oil palm (*Elaeis guineensis* Jacq.) is one of the promising crop plants which has been used as raw material for producing daily products. In agricultural ecosystems, crop plants could develop a plant-fungal association with arbuscular mycorrhizal fungi (AMF). The objectives of this study were to determine the AMF biodiversity and mycorrhizal infection percentage (MIP) from field-collected soil samples of three oil palm plantations from Nong Khai, Surat Thani, and Chiang Rai provinces of Thailand. Soil characteristics (moisture content, pH, and available phosphorus) were also measured. Thirteen AMF species belonging to seven genera were identified from all soil samples, whereas *Glomus* spp. and *Acaulospora* spp. were most commonly found species. AMF biodiversity value from Chiang Rai was statistically different from other two provinces ($p < 0.05$). MIP value of soil samples from Surat Thani was statistically different as well. Furthermore, soil pH showed a positive correlation with AMF biodiversity. These results confirmed that AMF normally occurs in oil palm plantations, but at different levels of biodiversity possibly due to different environmental factors in each plantation. Nevertheless, this information could be useful for using AMF in plant growth promoter and pathogen resistance programs in order to achieve the agricultural sustainability, especially in oil palm plantations.

1. Introduction

Oil palm has become an economically important crop plant among Southeast Asian and African countries. For example, the Royal Thai government made a decision on massive land expansion for oil palm plantations during the last 20 years. Most terrestrial plants have a mutualistic association with arbuscular mycorrhizal fungi (AMF) where these fungi develop their structures within the cortical cells of plant roots and give beneficial values such as improving the nutrient content, especially for phosphorus and water absorption [1]. Obtaining the beneficial effects of AMF application could be an alternative solution for minimizing the chemical fertilizer and pesticide demand for oil palm expansion [2]. However, information on the biodiversity of AMF directly associated with oil palm is limited. In this study, we determined the biodiversity of AMF in field-collected soil samples from oil palm plantations of three provinces in Thailand (Nong Khai (NK), Surat Thani (ST) and Chiang Rai (CR)) by applying wet-sieving method and determined the mycorrhizal infection percentage (MIP) by using these soil samples as starting inocula and sorghum as host plants [3].

2. Materials and methods

2.1. Study site

Soil samples for this experiment were collected from three oil palm plantations in Thailand: 1) Rattanawapi Sub-district, Rattanawapi District, Nong Khai province; 2) Saisopa Sub-district, Phra Saeng District, Surat Thani province; and 3) Pa Tueng Sub-district, Mae Chan district, Chiang Rai province. The detailed information about these oil palm plantations is shown in table 1.

Table 1. Land use history of oil palm plantations.

Province	Coordinate	Land use before oil palm plantation	Fertilizer application	Herbicide application
NK	18°12’31.3”N 103°10’56.9”E	Rice field	• Inorganic fertilizers • Organic fertilizers (animal manure)	Yes
ST	8°35’5.9634”N 99°4’55.3116”E	Rubber plantation	• Inorganic fertilizers • Organic fertilizers (empty fruit bunch of oil palm)	Yes
CR	20°06’40.9”N 99°45’17.9”E	Rice field	• Inorganic fertilizers	No

NK = Nong Khai, ST = Surat Thani and CR = Chiang Rai.

2.2. Soil sampling

All soil samples were collected between late November 2014 and early March 2015. For each plantation, 5 samples were taken from 5 randomly placed one-square-meter plots that were separated by at least 5 m apart. Within each plot, 5 subsamples (approx. 300 cc) were randomly collected down to the soil depth of 15 cm before being mixed together to obtain a pooled sample for each plot. A total of 15 pooled soil samples of approx. 1.5 liter were collected. All soil samples were stored in plastic bags and kept at room temperature until used.

2.3. Soil characteristics analyses

Soil moisture content was calculated from each soil sample as percentage oven-dry weight of soil by drying at 105 °C for 48-72 h. Soil pH was measured in water (1:2, *wt/vl*). Available phosphorus (P) was extracted with Bray 2 solution [4].

2.4. AMF spore extraction

AMF spores obtaining from each air-dry soil sample (50 g each) were extracted by wet sieving, decanting and a 60% sucrose gradient centrifugation [3,5]. The supernatant was rinsed with tap water for 2 min in a 38-µm sieve and transferred to 9 cm petri dishes for observation. Healthy spores were observed under a stereomicroscope (Olympus SZX-ILLB200) at 40-100X magnification. Then, 5-10 spores of each spore type collected from each soil sample were mounted on slides with polyvinyl alcohol-lactic acid-glycerol (PVLG) or PVLG mixed 1:1 (*v/v*) with Melzer’s reagent and observed under a light microscope (Olympus CH2) at a magnification of up to 400X. Identification was based on the morphological descriptions provided by the International Culture Collection of Arbuscular and Vesicular-Arbuscular Mycorrhizal Fungi, INVAM: http://invam.caf.wvu.edu.

The AMF biodiversity was calculated by Shannon-Wiener (H’) index. Furthermore, it was calculated from the following equation (2), where *pi* is the proportion of individuals found in the *i*th species (*pi* is calculated from *ni*/N, [where *ni* is the number of individuals in the *i*th species and N is the total number of individuals]) [6].

$$H^{'} = -\sum pi\,(\ln pi) \tag{1}$$

2.5. *MIP test*

The MIP test for each soil sample was conducted using field-collected soil samples as starting inocula and sorghum as the host plants. Sorghum seeds were grown into the growth substrate, consisting of rhizosphere soil (15 mL) from oil palm plantations and sterilized diluents (1:10, *v/v*). Sterilized diluents consisted of a mixture of soil and sand (1:2, *v/v*) (autoclaved for 2 h at 121 °C and repeated two times after a 24-h interval). Each soil sample was conducted in triplicate for a total of 45 MIP tests. The seeds of sorghum were previously surface-sterilized with 3% $NaClO_3$ for 2 min and used at 5-8 seeds for each pot. They were maintained with watering as needed for 28 days after shoots appearance. At the end of the test, plant shoots were discarded and roots were gently detached from the growth substrate and washed clean with water.

The mycorrhizal colonization of sorghum roots was assessed on 20 root pieces (approximately 1-1.5-cm segments) from each replication. The root segments were cleared in pre-boiled KOH (10%, *w/v*; 10 min), rinsed in tap water, acidified with HCl(1%, *v/v*; 30 min), stained in pre-boiled Trypan blue (0.02%, *w/v;* 15 min) and de-stained in acetic glycerin for at least 12 h. The stained roots were mounted on slides in PVLG with coverslips, oven-dried for 24 h at 60 °C, observed under a light microscope at 400X for quantification and scoring. Colonization percentage was scored based on the presence or absence of mycorrhizal structures (mycelium, vesicles, and arbuscules) of AMF observed in the stained roots. The colonization percentage by AMF was calculated using the following formula [7]:

$$MIP = \frac{total\ number\ of\ root\ segments\ colonized}{total\ number\ of\ root\ segments\ studied}\ x\ 100 \tag{2}$$

2.6. *Statistical analyses*

All data in this experiment were reported as the mean value of three replications from each soil sample. The differences in soil samples from different plantations on the soil characteristics (soil moisture content, pH, and available P) and AMF values (AMF biodiversity and MIP) were calculated by univariate ANOVA tests. The significance of differences between the samples from different plantations in soil characteristics and AMF values were tested using Least Significant Difference test (LSD) at $P<0.05$. The relationships between soil characteristics and AMF biodiversity and between AMF values (AMF biodiversity and MIP) were calculated through Pearson correlation coefficients. Statistical procedures were conducted with the software package SPSS 16.0 for Windows.

3. Results and discussions

3.1. *Soil characteristics*

No significant difference was found among moisture contents (%) and pH(s) of soil samples from all provinces (figure 1a and 1b). However, available phosphorus values were significantly different between soil samples from the three provinces ($p < 0.05$) (figure 1c).

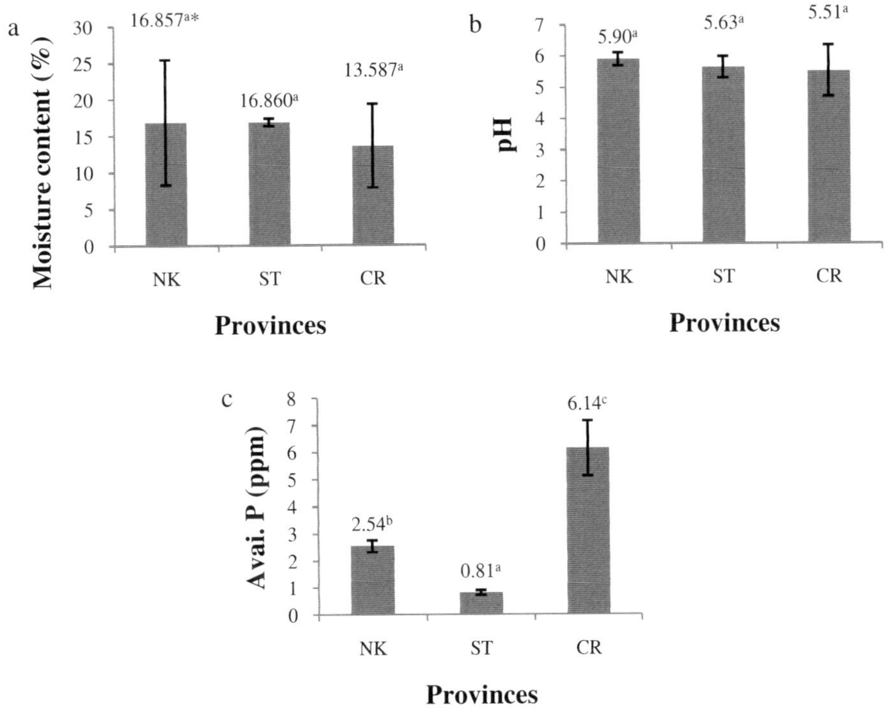

Figure 1. Mean values of (a) moisture content (%); (b) soil pH; (c) available phosphorus from all provinces. *Mean value followed by the same letters are different significantly ($p < 0.05$). NK = Nong Khai, ST = Surat Thani and CR = Chiang Rai. Standard deviation (moisture content and pH) and standard error (available P) of each province are indicated by error bars.

Soil moisture contents might vary in the range of a few percent in drier soils to excess than 100% in higher-plasticity clays [8]. In this study, the soil moisture contents were low (13.587 – 16.860%) (figure 1a). Soil pH(s) in arable land is commonly within the range of 5.0 to 6.2 [9]. Soil pH(s) in this study were within this range (figure 1b). Moreover, available P concentrations in soils might vary from less than 11 to more than 25 ppm [10]. Thus, the available P concentrations in this study were very low and ranged from 0.81 to 6.14 ppm (figure 1c).

3.2. AMF biodiversity and root colonization by AMF

We identified 13 species belonging to 7 genera from three different oil palm plantations. The mycorrhizal genera were *Gigaspora, Funneliformis, Glomus* (3 species), *Acaulospora* (3 species), *Dentiscutata, Scutellospora* and *Rhizophagus* (3 species) (figure 2). The number of species found in this study was slightly low when compared with other studies: twenty five AMF species from soil samples of two date palm plantations and its surrounding areas [11]; thirty four AMF species from rhizospheric samples of physic nut in Thailand [12]; sixteen AMF species from soil samples in dune and twenty four AMF species from soil samples in salt marsh of Mediterranean ecosystems [13].

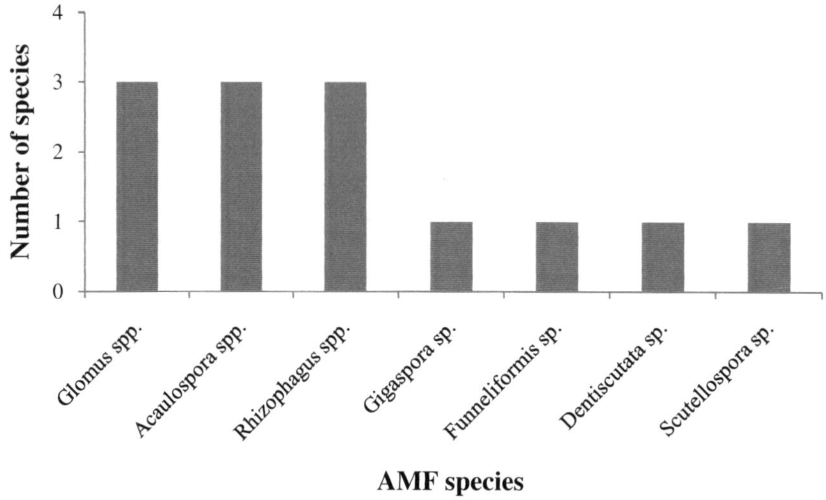

Figure 2. A number of mycorrhizal species observed from oil palm plantations, Thailand.

Glomus and *Acaulospora* were commonly found mycorrhizal genera from soil samples in this study. The remaining genera (*Rhizophagus, Gigaspora, Funneliformis, Dentiscutata,* and *Scutellospora*) were also present in this study, but in low frequency. Similar to our results, *Scutellospora, Acaulospora, Gigaspora,* and *Glomus* were mycorrhizal genera which could be found in disturbed and natural habitats in tropical Australia [14]; from the hot-dry valley of the Jinsha River, southwest China [6]; in mined soil samples [15].

AMF biodiversity of Chiang Rai was significantly different when compared to the other two provinces ($p < 0.05$) (figure 3a). These values of soil samples from three different oil palm plantations varied from low (0.81) to slightly moderate (1.33 – 1.51) (figure 3a). Other studies found AMF biodiversity in various numbers: AMF biodiversity within four shrub plants in a semi-arid Mediterranean ecosystem after revegetation was 0.78 [16]; from soil samples of physic nut in Thailand was 0.28 - 0.83 [12]; from soil samples in five host plants of rhododendron species were 1.75 - 2.40 [7]. The AMF biodiversity value could be determined from a mutual selection between mycorrhizal fungal communities and environmental factors [17]. For example, the two areas of Yungas forest have various values of AMF biodiversity where these values were significantly affected by seasonal changes. The higher values of AMF biodiversity were obtained in spring (wet season) [18]. Moreover, the highest value of AMF biodiversity from mountain grassland, Cordoba, Argentina was obtained in wet seasons (summer and spring) [19]. In this study, the relatively low value of mean AMF biodiversity values could be because these samples were collected in a dry season.

Moreover, mycorrhizal infection percentage (MIP) value of soil samples from Surat Thani was significantly different as well (figure 3b). MIP values give information about the activity of mycorrhizal community in certain ecosystems [20] and information on whether certain plants are mycorrhizal dependent or not [21]. Overall, the MIP values of soil samples from oil palm plantations were low for each province (3.00 – 9.63%) (figure 3b). Moreover, MIP values might vary within seasonal changes. The MIP values in soil samples from conventional and organic apple orchards in Brazil ranged from 43.70 – 59.70 (dry season) and 17.50 – 24.60 (wet season). In this study, we conducted MIP test in rainy period (wet season). In addition, the low number of MIP values might be due to mycorrhizal spores in soil samples have less adaptation level to different soil and climate conditions [22].

IOP Conf. Series: Earth and Environmental Science **144** (2018) 012010 doi:10.1088/1755-1315/144/1/012010

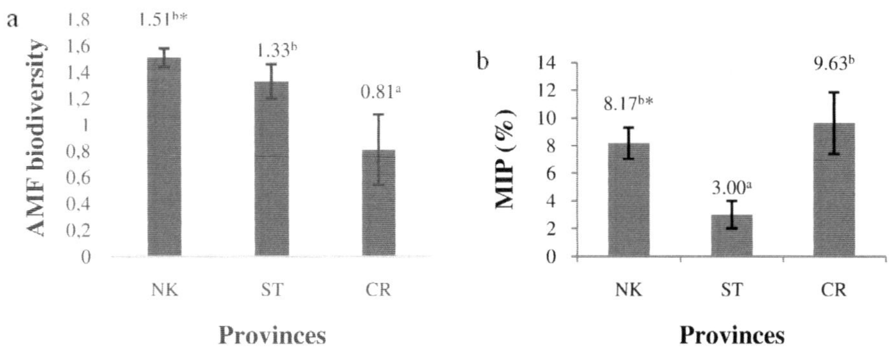

Figure 3. Mean values of (a) AMF biodiversity and (b) MIP from all provinces. *Mean value followed by the same letters are different significantly (p <0.05). NK = Nong Khai, ST = Surat Thani and CR = Chiang Rai. Standard errors of each province are indicated by error bars.

3.3. Relationships between soil characteristics, AMF biodiversity, and MIP values

There is a significant negative correlation between soil moisture content (%) and available phosphorus values from all samples ($p < 0.05$) (figure 4a). Furthermore, soil pH(s) have positive correlation with AMF biodiversity ($p < 0.05$) (figure 4b).

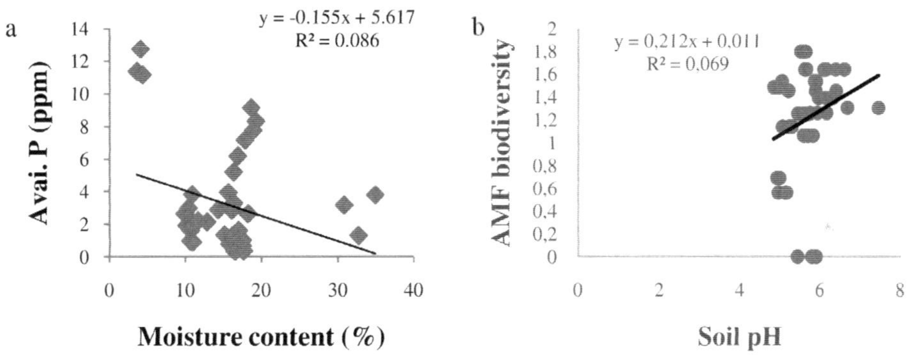

Figure 4. Correlation between (a) soil moisture contents (%) and available P and (b) soil pH and AMF biodiversity values from all samples.

Soil characteristics could be significantly affecting the arbuscular mycorrhizal fungal communities in diverse habitats [14]. Furthermore, the relationship between soil characteristics and AMF values might be affected by land use history such as fertilizer and herbicide applications in plantations. The non-significant correlation between AMF biodiversity and MIP values was comparable with several studies which showed no significant correlation between AMF biodiversity found in the samples and MIP of host plants: from rhizospheric soil samples of rhododendron species in Kumaun region of the Indian Central Himalaya [7] and from both transects in arid grassland [23].

4. Conclusion

Thirteen AMF species were found from the rhizosphere of oil palms. They were placed in seven genera: *Gigaspora, Funneliformis, Glomus* (3 species), *Acaulospora* (3 species), *Dentiscutata, Scutellospora* and *Rhizophagus* (3 species). *Glomus* spp. and *Acaulospora* spp. were the predominant species because they were found in all soil samples. This result indicates that both mycorrhizal genera (*Glomus* and *Acaulospora*) might have significant values such as improving the plant nutrient uptake and protecting oil palms from pathogens. This study also showed low to slightly moderate AMF biodiversity in the rhizosphere of oil palms from three different plantations with the highest AMF biodiversity from Nong Khai province. The root colonization of AMF was highest from Chiang Rai soil samples; however, it was not significantly different from Nong Khai soil samples. Moreover, this study found a positive correlation between AMF biodiversity and soil pH and this correlation might be affected by land use history (fertilizer and herbicide applications) in oil palm plantations. Indeed, the difference in soil characteristics may determine the difference in AMF communities. Nevertheless, the results from this study support the potential for exploiting the AMF species in plant growth promoter and pathogen resistance programs in order to achieve the sustainability, especially in oil palm plantations.

Acknowledgments

The authors wish to thank the Scholarship of East Kalimantan from the Government of Indonesia, Double Degree Program of Forest Resources and Renewable Energy Management, Faculty of Forestry, Mulawarman University and Faculty of Science, King Mongkut's University of Technology Thonburi for supporting the first author's exchange study abroad and Department of Microbiology, Faculty of Science, King Mongkut's University of Technology Thonburi for financing this experiment.

References

[1] Bever J D, Schultz P A, Pringle A and Morton J B 2001 Arbuscular mycorrhizal fungi: More diverse than meets the eye and the ecological tale of why *Bioscience* **51 (11)** pp 923-31

[2] Phosri C, Rodriguez A, Sanders I A and Jeffries P 2010 The role of mycorrhizas in more sustainable oil palm cultivation *Agriculture, Ecosystems and Environment* **135** pp 187-93 doi:10.1016/j.agee.2009.09.006

[3] Sylvia D M, Kaonongbua W and Bever J D 2007 Isolation, culture and detection of arbuscular mycorrhizal fungi *Manual of Environmental Microbiology* 3rd ed, ed C J Hurst (Washington D C: ASM Press) pp 627-37

[4] Frank K, Beegle D and Denning J 2011 Phosphorus *Recommended Chemical Soil Test Procedures for the North Central Region* ed J R Brown (North Central Regional Research Publication) No 221 pp 21-30

[5] Kaonongbua W, Morton J B and Bever J D 2010 Taxonomic revision transferring species in *Kuklospora* to *Acaulospora* (Glomeromycota) and a description of *Acaulospora colliculosa* sp. nov. from field collected spores *Mycologia* **102 (6)** 1497-509 DOI: 10.3852/10-011

[6] Dandan Z and Zhiwei Z 2007 Biodiversity of arbuscular mycorrhizal fungi in the hot-dry valley of the Jinsha River, Southwest China *Applied Soil Ecology* **37** 118-28 doi:10.1016/j.apsoil.2007.06.003

[7] Chaurasia B, Pandey A and Palni, L M S 2005 Distribution, colonization and diversity of arbuscular mycorrhizal fungi associated with Central Himalayan Rhododendrons *Forest Ecology and Management* **207** 315-24 doi:10.1016/j.foreco.2004.10.014

[8] Kalinski, M E 2006 Measurement of moisture content *Soil Mechanics Lab Manual* (USA: Wiley-John Wiley & Sons Inc) pp 7-9

[9] Faithfull N T 2002 Lime application rates for arable land *Methods in Agricultural Chemical Analysis: A Practical Handbook* (Aberystwyth: CABI Publishing) pp 226

[10] Wijanarko A and Taufiq A 2008 Penentuan kebutuhan pupuk P untuk tanaman kedelai, kacang tanah dan kacang hijau berdasarkan uji tanah di lahan kering masam ultisol *Bul. Palawija* **15** pp 1-8

[11] Al-Yahya'ei M N, Oehl F, Vallino M, Lumini E, Redecker D, Wiemken A and Bonfante P 2011 Unique arbuscular mycorrhizal fungal communities uncovered in date palm plantations and surrounding desert habitats of Southern Arabia *Mycorrhiza* **21** 195-209 doi 10.1007/s00572-010-0323-5

[12] Charoenpakdee S, Phosri C, Dell B and Lumyong S 2010 The mycorrhizal status of indigenous arbuscular mycorrhizal fungi on physic nut (*Jatropha curcas*) in Thailand *Mycosphere* **1 (2)** pp 167-81

[13] Estrada B, Beltran-Hermoso M, Palenzuela J, Iwase K, Ruiz-Lozano J M, Barea J M and Oehl F 2013 Diversity of arbuscular mycorrhizal fungi in the rhizosphere of *Asteriscus maritimus* (L.) Less., a representative plant species in arid and saline Mediterranean ecosystems *Journal of Arid Environments* **97** pp 170-5 http://dx.doi.org/10.1016/j.jaridenv.2013.05.019

[14] Brundrett M, Abbott L K, Jasper D A and Ashwath N 1995 Mycorrhizal associations in disturbed and natural habitats in tropical Australia *Int. Symp. and Workshop: Mycorrhizas for Plantation Forestry in Asia* (Kaiping Guangdong Province P R China) pp 34-40

[15] Singh A K and Jamaluddin 2011 Status and diversity of arbuscular mycorrhizal fungi and its role in natural regeneration on limestone mined spoils *Biodiversitas* **12 (2)** pp 107-11

[16] Alguacil M M, Torres M P, Torrecillas E, Diaz G and Roldan A 2011 Plant type differently promote the arbuscular mycorrhizal fungi biodiversity in the rhizosphere after revegetation of a degraded, semiarid land *Soil Biology & Biochemistry* **43** 167-73 doi:10.1016/j.soilbio.2010.09.029

[17] Yang A N, Lu L and Zhang N 2011 The diversity of arbuscular mycorrhizal fungi in the subtropical forest of Huangshan (Yellow Mountain), East-Central China *World J Microbiol Biotechnol* **27** 2351-58 DOI 10.1007/s11274-011-0702-x

[18] Becerra A G, Cabello M N and Bartoloni N J 2011 Native arbuscular mycorrhizal fungi in the Yungas Forests, Argentina *Mycologia* **103 (2)** 273-9 DOI: 10.3852/10-193

[19] Lugo M A and Cabello M N 2002 Native arbuscular mycorrhizal fungi (AMF) from mountain grassland (Cordoba, Argentina) I. seasonal variation of fungal spore diversity *Mycologia* **94 (4)** pp 579-86

[20] Sturmer S L, Sturmer R and Pasqualini D 2013 Taxonomic diversity and community structure of arbuscular mycorrhizal fungi (Phylum Glomeromycota) in three maritime sand dunes in Santa Catarina State, South Brazil *Fungal Ecology* **6** pp 27-36 http://dx.doi.org/10.1016/j.funeco.2012.10.001

[21] Tao L and Zhiwei Z 2005 Arbuscular mycorrhizas in a hot and arid ecosystem in Southwest China *Applied Soil Ecology* **29** 135-41 doi:10.1016/j.apsoil.2004.11.005

[22] Purin S, Filho O K and Sturmer S L 2006 Mycorrhizae activity and diversity in conventional and organic apple orchards from Brazil *Soil Biology & Biochemistry* **38** pp 1831-9 doi:10.1016/j.soilbio.2005.12.008

[23] Jacobson K M 1997 Moisture and substrate stability determine VA-Mycorrhizal fungal community distribution and structure in an arid grassland *Journal of Arid Environments* **35** pp 59-75

Antiplasmodial activity of flavonoids from *Macaranga tanarius* leaves

E Marliana[1,*], R Hairani[1], T S Tjahjandarie[2] and M Tanjung[2]

[1]Department of Chemistry, Faculty of Mathematics and Natural Science, Mulawarman University, Samarinda, Indonesia
[2]Natural Product Chemistry Group, Organic Chemistry Division, Department of Chemistry, Faculty of Science and Technology, Airlangga University, Surabaya, Indonesia

* Corresponding author eva_samarinda@yahoo.com

Abstract. Malaria is one of the leading causes of death in the world which is caused by *Plasmodium sp.* This parasite tends to have mutation and shows resistance towards malaria drug. Due to the emergence and spread of *Plasmodium sp.* resistance towards malaria drugs, an exploration to find new effective and selective malaria drug is essential. In this study, four flavonoids, namely nymphaeol C (**1**), solophenol D (**2**), nymphaeol A (**3**), and nymphaeol B (**4**) were isolated from ethyl acetate fraction of *Macaranga tanarius* leaves. The structures of those compounds were characterized by NMR analysis. Furthermore, antiplasmodial activity of ethyl acetate fraction and four isolated compounds (**1–4**) were evaluated by Giemsa method against *Plasmodium falciparum* strain 3D7. According to this assay, it showed the IC_{50} values were 0.30, 0.24, 0.31, 0.05, and 0.05 μg/mL, respectively. The results provide important evidence of the antiplasmodial activity of flavonoids in traditional use. In addition, it can be indicated that *Macaranga tanarius* is potential to be developed as antiplasmodial agents.

1. Introduction

Malaria is known as the world's most important parasitic disease especially when *Plasmodium sp.* is the causative agent. World Health Organization (WHO) has reported that about 300 million people were at risk of malaria with 1.5 million deaths per year caused by *Plasmodium sp.* infections [1]. Kalimantan is one of the malaria endemic areas in Indonesia, therefore, local people in this island try to find the treatment to cure this disease using some medicinal plants [2].

Macaranga is one of plants that is used for malaria treatment. This genus belongs to Euphorbiaceae family containing flavonoids and stilbenoids integrated with terpenoids. Based on the chemical structure, the combination of flavonoids with terpenoids yields many varieties of flavonoid derivatives structures which further provides an opportunity for scientist to discover new compounds and new frameworks that have potency as antiplasmodial agent.

One species of Macaranga genus that had been reported for its antiplasmodial activity is only *Macaranga triloba* which is known containing 6-prenyl-3'-methoxyeriodictyol, nymphaeol B, nymphaeol C, and 6-farnesyl eriodictyol [3]. Another species from this genus named *Macaranga tanarius* is suspected also to have antiplasmodial activity due to some compounds in *M. tanarius* also belong to flavonoids. Several isolated compounds from *M. tanarius* had been reported, such as tanariflavanone A, tanariflavanone D, nymphaeol A and nymphaeol B [3, 4].

To date, there is no study about antiplasmodial activity from *M. tanarius*, hence in this study, the isolation of bioactive compounds and the investigation of their antiplasmodial activity from *M. tanarius* leaves were performed to understand the potency of this species.

2. Materials and Methods

2.1. General information

All reagents used in this research were purchased from Merck Chemical, Co. Vacuum liquid chromatography (VLC) and radial chromatography were conducted using Silica gel 60 GF_{254} and Silica gel 60 PF_{254}. The isolation was monitored by Thin Layer Chromatography (TLC) and visualized under UV light at 254 and 356 nm with $CeSO_4$ as staining agent. Pre-coated silica gel plates (Merck Kieselgel 60 GF $_{254}$, 0.25 mm thickness) were employed for TLC analysis. ^1H and ^{13}C NMR spectra of isolated compounds were recorded with a JEOL ECS 400 spectrometer operating at 400 and 100 MHz, respectively, in $CDCl_3$ and acetone-d6 using TMS as the internal standard.

2.2. Plant material

The leaves of *Macaranga tanarius* were collected from Samboja forest, Kutai Kartanegara, East Kalimantan. The identification of this plant was taken by Herbarium of Wanariset, East Kalimantan.

2.3. Extraction and isolation

The air-dried powder of *M. tanarius* leaves (2 kg) were extracted by maceration method using methanol at room temperature for two times. The filtrate was evaporated to obtain crude extract (276 g). A part of this crude extract (150 g) was partitioned with *n*-hexane and ethyl acetate, respectively. Ethyl acetate fraction (50 g) was separated using Vacuum Liquid Chromatography (VLC) and eluted with the mixtures of *n*-hexane and ethyl acetate by increasing the polarity (9:1, 4:1, 7:3, and 1:1) to yield two major fractions, i.e. A and B. Separation of A fraction using column chromatography and eluted with the mixtures of *n*-hexane and ethyl acetate by increasing the polarity (9:1, 4:1, 7:3, and 1:1) yielded three major sub-fractions, i.e. A1, A2, and A3. Further separation of A2 and A3 gave compound **1** (65 mg) and compound **4** (25 mg). While for separation of B fraction using column chromatography and eluted with the mixtures of *n*-hexane and ethyl acetate by increasing the polarity (9:1, 4:1, 7:3, and 1:1) yielded two major sub-fractions, i.e. B1 and B2. Further separation of B1 and B2 gave compound **2** (4.8 mg) and compound **3** (12.7 mg).

2.4. Determination of antiplasmodial activity

Determination of antiplasmodial activity of ethyl acetate fraction and isolate compounds was conducted by Trager and Jensen method [5] and followed by Giemsa staining as described by Widyawaruyanti, et al [6]. In this method, sample was dissolved in 20 μL of DMSO and diluted with 180 μL of RPMI 1640 medium until obtained various kinds of concentration. A total of 50 μL of test solution was inserted into micro well plate, then added 950 μL of parasitic suspension of *Plasmodium falciparum* strain 3D7, incubated for 48 h, then centrifuged. The top of the suspension (supernatant) was removed until the concentrated suspension was obtained, then the preparation of a thin layer of blood (monolayer) stained with Giemsa 20% was then calculated for the percentage of parasitemia and the growth percentage of *P. falciparum* and its resistance, by counting the number of infected erythrocytes every 1000 erythrocytes under microscope [7]. The test was done in triplicate by varying the concentration of ethyl acetate fraction and isolated compounds. Chloroquine diphosphate was used as positive control, while solvent was used as negative control. The IC_{50} value evaluation was determined based on Probit Analysis with SPSS program.

3. Results and Discussion

3.1. Flavonoids from Macaranga tanarius leaves

There were four isolated compounds obtained from ethyl fraction of *M. tanarius* leaves, named nymphaeol C (**1**), solophenol D (**2**), nymphaeol A (**3**), and nymphaeol B (**4**). The position of protons and carbons NMR (1D NMR) of compounds **1**–**4** are presented in table 1. In addition, the Heteronuclear Multiple Bond Correlation (HMBC, 2D NMR) of those compounds are shown in figure 1.

Table 1. NMR data of compounds **1–4** in $CDCl_3$ or acetone-d6

Compounds	^{1}H NMR, 400 MHz δ_H (ppm)	^{13}C NMR, 100 MHz δ_C (ppm)
1	5.49 (1H, *dd, J*=13.2, 2.8 Hz, H-2), 3.12 (1H, *dd, J*=17.2, 13.2 Hz, H-3_{ax}), 2.73 (1H, *dd, J*=17.2, 2.8 Hz, H-3_{eq}), 5.97 (1H, *s*, H-8), 6.81 (1H, *d, J*=8.4 Hz, H-5'), 6.95 (1H, *d, J*=8.4 Hz, H-6'), 3.33 (2H, *d, J*=7.2 Hz, H-1"), 5.22 (1H, *t, J*=8.4 Hz, H-2"), 1.80 (3H, *s*, H-4"), 1.74 (3H, *s*, H-5"), 3.44 (2H, *d, J*=6.8 Hz, H-1'"), 5.15 (1H, *t, J*=6.8 Hz, H-2'"), 2.05 (2H, *t*, H-4'"), 2.07 (2H, *m*, H-5'"), 5.02 (1H, *t*, H-6'"), 1.58 (3H, *s*, H-8'"), 1.65 (3H, *s*, H-9'"), 1.75 (3H, *s*, H-10'"), 12.34 (1H, *s*, 5-OH), 5.64 (1H, *s*, 4'-OH).	76.5 (C-2), 42.8 (C-3), 197.1 (C-4), 103.1 (C-4a), 161.5 (C-5), 107.6 (C-6), 161.6 (C-7), 95.7 (C-8), 164.1 (C-8a), 128.7 (C-1'), 126.7 (C-2'), 142.8 (C-3'), 144.8 (C-4'), 113.1 (C-5'), 118.9 (C-6'), 21.4 (C-1"), 121.6 (C-2"), 135.2 (C-3"), 18.1 (C-4"), 26.1 (C-5"), 25.5 (C-1'"), 121.8 (C-2'"), 138.7 (C-3'"), 39.8 (C-4'"), 26.6 (C-5'"), 123.9 (C-6'"), 132.4 (C-7'"), 18.0 (C-8'"), 26.0 (C-9'"), 16.5 (C-10'").
2	6.25 (1H, *d, J*= 2.0 Hz, H-6), 6.38 (1H, *d, J*=2.0 Hz, H-8), 6.84 (1H, *d, J*=8.4 Hz, H-5'), 6.93 (1H, *d, J*=8,.4 Hz, H-6'), 3.48 (2H, *d, J*=6.8 Hz, H-1"), 4.98 (1H, *t*, 6.8 Hz, H-2"), 1.79 (2H, *m*, H-4"), 1.86 (2H, *m*, H-5"), 5.14 (1H, *t*, H-6"), 1.57 (3H, *s*, H-8"), 1.45 (3H, *s*, H-9"), 1.49 (3H, *s*, H-10"), 12.29 (1H, *s*, 5-OH).	147.0 (C-2), 135.3 (C-3), 176.9 (C-4), 104.7 (C-4a), 164.7 (C-5), 99.0 (C-6), 162.5 (C-7), 94.4 (C-8), 158.3 (C-8a), 129.3 (C-1'), 125.0 (C-2'), 142.7 (C-3'), 144.3 (C-4'), 113.1 (C-5'), 122.7 (C-6'), 26.6 (C-1"), 123.4 (C-2"), 131.6 (C-3"), 40.3 (C-4"), 27.3 (C-5"), 123.6 (C-6"), 130.7 (C-7"), 25.7 (C-8"), 17.6 (C-9"), 16.2 (C-10").
3	5.35 (1H, *dd, J*=12.8, 2.8 Hz, H-2), 3.11 (1H, *dd, J*=17.2, 12.8 Hz, H-3_{ax}), 2.70 (1H, *dd, J*=17.2, 2.8 Hz, H-3_{eq}), 6.01 (1H, *s*, H-8), 7.02 (1H, *s*, H-2'), 6.85 (2H, *d, J*=8.4 Hz, H-5'/H-6'), 3.24 (2H, *d, J*=6.8 Hz, H-1"), 5.24 (1H, *brt*, H-2"), 1.94 (2H, *t, J*=8.4 Hz, H-4"), 2.03 (2H, *m*, H-5"), 5.07 (1H, *t*, H-6"), 1.55 (3H, *s*, H-8"), 1.61 (3H, *s*, H-9"), 1.75 (3H, *s*, H-10"), 12.47 (1H, *s*, 5-OH).	79.9 (C-2), 43.6 (C-3), 197.2 (C-4), 102.8 (C-4a), 164.8 (C-5), 108.9 (C-6), 162.1 (C-7), 95.2 (C-8), 161.8 (C-8a), 131.5 (C-1'), 114.6 (C-2'), 146.1 (C-3'), 146.4 (C-4'), 119.1 (C-5'), 115.8 (C-6'), 21.5 (C-1"), 123.3 (C-2"), 134.9 (C-3"), 39.4 (C-4"), 27.3 (C-5"), 125.0 (C-6"), 131.5 (C-7"), 17.7 (C-8"), 25.8 (C-9"), 16.0 (C-10").
4	5.55 (1H, *dd, J*=13.2, 2.8 Hz, H-2), 3.12 (1H, *dd, J*=17.2, 13.2 Hz, H-3_{ax}), 2.73 (1H, *dd, J*=17.2, 2.8 Hz, H-3_{eq}), 5.99 (1H, *d, J*=2.4 Hz, H-6), 5.95 (1H, *d, J*=2.4 Hz H-8), 6.82 (1H, *d, J*=8.4 Hz, H-5'), 6.94 (1H, *d, J*=8,4 Hz, H-6'), 3.44 (2H, *d*, J=6.8 Hz, H-1"), 5.15 (1H, *brt, J*=6.8 Hz, H-2"), 2.05 (2H, *brt, J*=6.8 Hz, H-4"), 2.09 (2H, *brq*, H-5"), 5.01 (1H, *brt, J*=6.8 Hz, H-6"), 1.57 (3H, *s*, H-8"), 1.65 (3H, *s*, H-9"), 1.75 (3H, *s*, H-10"), 12.04 (1H, *s*, 5-OH), 5.63 (1H, *s*, 4'-OH).	76.4 (C-2), 42.4 (C-3), 196.5 (C-4), 103.0 (C-4a), 164.7 (C-5), 96.7 (C-6), 164.2 (C-7), 95.5 (C-8), 163.4 (C-8a), 128.2 (C-1'), 126.4 (C-2'), 142.5 (C-3'), 144.8 (C-4'), 113.0 (C-5'), 119.0 (C-6'), 25.4 (C-1"), 121.2 (C-2"), 139.0 (C-3"), 39.5 (C-4"), 26.2 (C-5"), 123.6 (C-6"), 132.3 (C-7"), 17.2 (C-8"), 25.7 (C-9"), 16.3 (C-10").

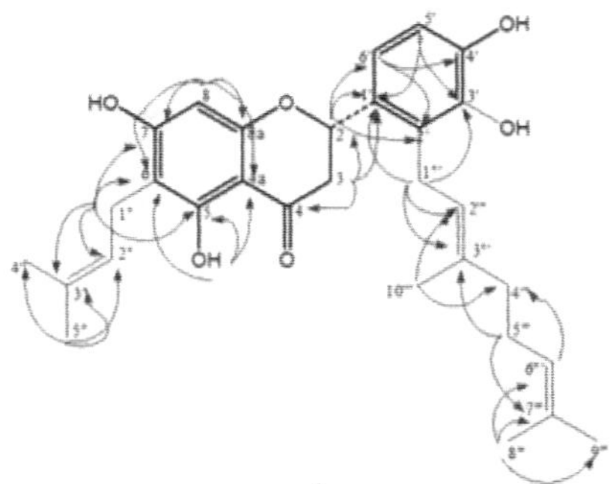

1

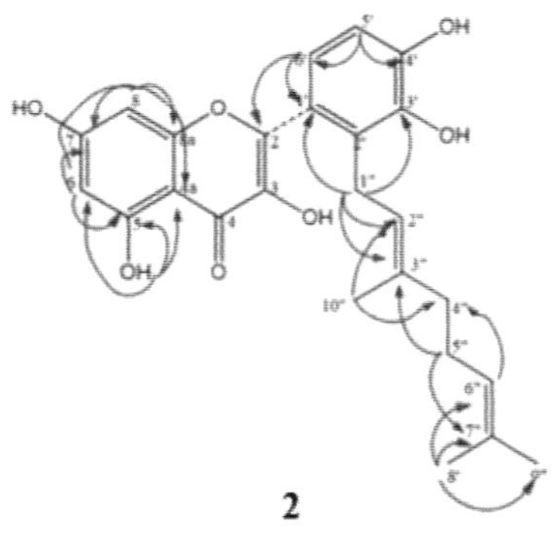

2

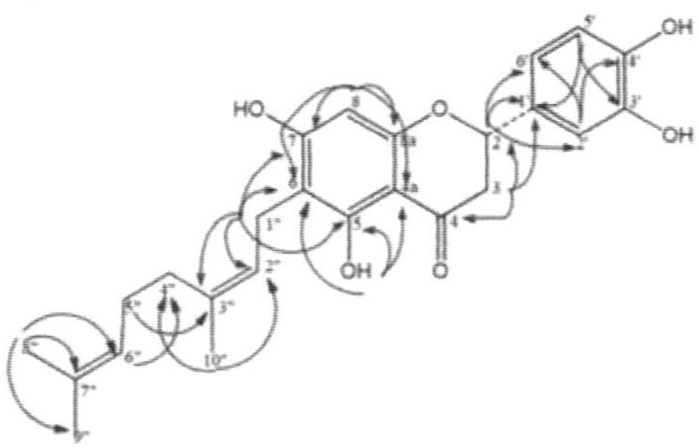

3

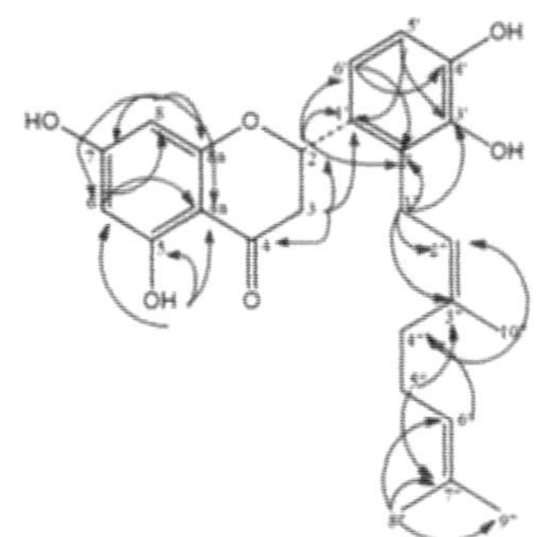

4

Figure 1. HMBC of nymphaeol C (**1**), solophenol D (**2**), nymphaeol A (**3**) and nymphaeol B (**4**)

Compound **1** was obtained as yellow oil. The UV spectra in methanol gave the maximum absorption at λ_{max} nm (log ε): 243 (4.69), 291.50 (4.67), (MeOH+NaOH): 300 (4.58), 330 (4.66, (MeOH+$AlCl_3$): 297.5 (3.94), (MeOH+$AlCl_3$+HCl): 308 (4.00) 361 (3.57). The UV spectrum profile of the isolated compound indicated that the compound belong to flavanone derivative. In addition to (MeOH+$AlCl_3$), a batochromic shift and absorption change after addition of HCl indicated that the compound can be classified as flavanone having ortho-dihydroxy [8]. ^{1}H NMR spectra analysis of compound **1** (table 1) indicated the presence of flavanone moiety which showed by three proton of doublet-doublet signals at δ_H 5.49 ppm (J=13.2, 2.8 Hz, H-2), 3.12 ppm (J=13.2, 17.2 Hz, H-3ax), and 2.73 ppm (J=17.2, 2.8 Hz, H-3eq). In addition, the spectra exhibited proton signals of five methyl singlets (δ_H 1.80, 1.74, 1.58, 1.65, 1.75 ppm) along with three vinyl triplet signals (δ_H 5.24, 5.15, 5.02 ppm) and four methylene signals (δ_H 3.33, 3.44, 2.05, 0.07 ppm), indicated that the isolated compound has one isoprenyl group and one geranyl group. Spectral analysis of ^{13}C NMR of compound **1** (table 1) showed 30 perfectly separate carbon signals comprising seven methyl carbon atoms, five methylene carbon atoms, five methyl carbon atoms, and thirteen quartz carbon atoms. The ^{13}C NMR spectra exhibited a carbonyl carbon signal (δ_C 197.1 ppm), an oxy carbon methyl carbon signal (δ_C 76.5 ppm), and five oxy acetyl carbon signals (δ_C 164.1, 161.6, 161.5, 144.8, 142.8 ppm) which are the basic skeleton of eriodictyol derivatives. Based on HMBC (figure 1), the correlation of two methyl signals with one methyl carbon signal (δ_C 121.6 ppm) and one quarterner carbon signal (δ_C 135.2 ppm) indicated the isoprenyl substituent bound at C-6. Based on 1D and 2D NMR spectral data, compound **1** was identified as 6-isoprenyl-2'-geranyleriodictyol or known as nymphaeol C according to the previous study that reported nymphaeol C from *M. triloba* having molecular formula $C_{30}H_{36}O_6$ and molecular mass m/z=492 [9].

Compound **2** was obtained as yellow solid. The UV spectra of compound **2** in methanol displayed maximum absorption at λ_{max} nm (log ε): 255 (4.30), 297.50 (3.96) and 348 sh (4.02) which is a characteristic of flavone or flavonol [7]. The maximum absorption of compound in (MeOH+NaOH) λ_{maks} nm (log ε): 270 (4.21), 325 (3.86), (MeOH+$AlCl_3$): 265 (4.60), 297 (4.14) (MeOH+$AlCl_3$+HCl): 256 (4.83), 302.50 (4.24), 362 (4.17). In addition to MeOH+$AlCl_3$ a shift in batochromic and absorption change after addition of HCl indicated that the compound belong to flavonol having ortho-dihydroxy. ^{1}H NMR spectra analysis of compound **2** (table 1) showed a pair of ortho doublet proton signals in the aromatic region (*J*=8.4 Hz) at δ_H 6.84 ppm and 6.93 ppm (1H each), corresponding to the compound pattern of nymphaeol C i.e. 2', 3', 4'-trisubstitutes in ring B. The presence of three signals of methyl singlet (δ_H 1.57, 1.45, 1.49 ppm) together with two vinyl triplet signals (δ_H 6.8, 6.8 ppm) and three methyl doublet doubles (δ_H 3.48, 1.79, 1.86 ppm), indicated that this compound has one geranyl group. ^{13}C NMR analysis (table 1) showed 25 perfectly segregated carbon signals, comprising six methyl carbon atoms, three methylene carbon atoms, three methyl carbon atoms, and 13 quartz carbon atoms. The position of geranyl substituent of compound **2** was determined by HMBC (figure 1). The correlation of proton aromatic signals at δ_H 6.93 ppm (H-6') with two quarterner carbon signals (δC 147.0, 129.3 ppm) showed δ_C 147.0 ppm bound in C-2 and δ_C 129.3 ppm in C-1'. The correlation of methylene signals in δ_H 3.48 ppm with four quaternary carbon signals (δ_C 142.7, 131.6, 129.3, 125.0 ppm) and one methyl signal (δ_C 123.4 ppm) indicated geranyl substituent bound at C-2'. Based on UV, 1D and 2D NMR data, compound **2** was identified as 2'-geranylquercetin or known as solophenol D, corresponds to an isolated compound from propolis with a positive quasi-molecular ion mass at m/z=439.1679 and molecular formula=$C_{25}H_{27}O_7$ [10].

Compound **3** was gained as yellow oil and the UV spectra in methanol showed the maximum absorption at λ_{max} nm (log ε): 291.50 (4.60), 331 sh (3.91) (MeOH+NaOH): 330 (4.64), (MeOH+$AlCl_3$): 294.50 (4.69), 348 sh (3.91), (MeOH+$AlCl_3$+HCl): 301.50 (4.59), 354.50 sh (3, 92). ^{1}H NMR spectra analysis of compound **3** (table 1) indicated the presence of flavanone moiety which showed by three signals of doublet doublet proton at δ_H 5.35 ppm (*J*=12.8, 2.8 Hz, H-2), 3.11 ppm (*J*=12.8, 17.2 Hz, H-3_{ax}), and 2.70 ppm (*J*=17.2, 2.8 Hz, H-3_{eq}). In addition, the spectra also showed three signals of methyl singlet proton (δ_H 1.55, 1.61, 1.75 ppm) to gather with two vinyl triplet signals (δ_H 5.24. 5.07 ppm) and three methylene signals (δ_H 3.24. 1.94, 2.03 ppm) indicated the presence of one geranyl group. ^{13}C NMR spectra analysis of compound **3** (table 1) showed 24 carbons signals which representative 25 carbon atoms comprising seven of methine carbon atoms, four methylene carbon atoms, three methyl carbon atoms, and ten quarterne carbon atoms. Likewise with compound **2**, this compound also has eriodictyol moiety. Geranyl position was determined by HMBC correlation (figure 1). It showed singlet signal at δ_H 12.47 ppm (5-OH) with three quarternary carbon atoms aromatic (δ_C 164.8, 108.9, 102.8 ppm) placed carbon signal δ_C 164.8 ppm at C-5, δ_C 108.9 ppm at C-6, and δ_C 102.8 ppm at C-4a. The spectra indicated one geranyl group bounded at C-6. Based on 1D and 2D NMR, compound **3** was identified as 6-geranyleriodictyol or nymphaeol A which had been isolated from *M. alnifolia* with structure formula $C_{25}H_{29}O_6$ and m/z $[M+H]^+$ =425.1948 [11].

Compound **4** was isolated as yellow oil and UV spectra in methanol showed maximum absorption at pada λ_{maks} nm (log ε): 289 (4.86), 327.50 *sh* (4.16) (MeOH+NaOH) 324.50 (4.92), (MeOH+$AlCl_3$): 308.50 (4.94) 375 *sh* (4.15), (MeOH+$AlCl_3$+HCl): 308 (4.85), 374.50 *sh* (4.12). Analysis of ^{1}H NMR spectra of compound **4** also indicated flavanone moiety i.e. the presence of three doublet doublet proton signals at δ_H 5.55 ppm (*J*=13.2, 2.8 Hz, H-2), 3.12 ppm (*J*=13.2, 17.2 Hz, H-3ax), and 2.73 ppm (*J*=17.2, 2.8 Hz, H-3eq). The isolated compound spectra exhibited proton signals of three methyl singlets (δ_H 1.57, 1.65, 1.75 ppm) along with two vinyl triplet signals (δ_H 5.15, 5.01 ppm), and three methylene signals (δ_H 3.44, 2.05, 2.09 ppm), indicated that the isolated compound has one gingival group. ^{13}C NMR spectra analysis of compound **4** showed 25 perfectly segregated carbon signals, comprising seven methyl carbon atoms, four methylene carbon atoms, three methyl carbon atoms and 11 quartz carbon atoms. This compound also has an eriodictyol skeleton. The geranyl substituent position that bounded at C-2' was defined by HMBC (figure 1). Based on 1D dan 2D NMR spectrum, compound **4** was identified as 2'-geranyleriodictyol or known as nymphaeol B which had been isolated from *M. triloba* with m/z = 424 ($C_{25}H_{28}O_6$) [9].

3.2. Antiplasmodial activity
Antiplasmodial activity test was conducted on ethyl acetate fraction and four isolated compounds of *M. tanarius* leaves using Giemsa method against *P. falciparum* strain 3D7. Levels of antiplasmodial activity were expressed in concentration inhibition (IC_{50}) value, the concentration required to inhibit 50% of parasitemia.

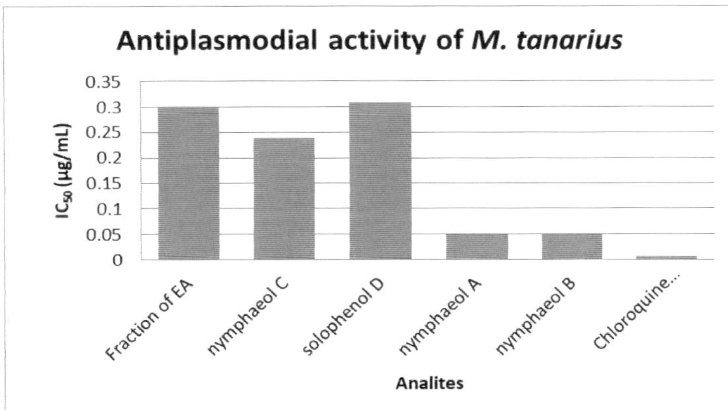

Figure 2. Antiplasmodial activity

Based on the result in figure 2, it showed that ethyl acetate fraction and four isolated compounds (**1**–**4**) are active fraction and active compounds in antiplasmodial activity with IC_{50} values of 0.30, 0.24, 0.31, 0.05, and 0.05 µg/mL, respectively. In this study, chloroquine was used as positive control with IC_{50} value of 0.006 µg/mL. If compared with chloroquine, the ethyl acetate fraction and four isolated compounds (**1**–**4**) have lower activity than chloroquine. Even though ethyl acetate fraction has higher IC_{50} value than chloroquine, the IC_{50} value indicated that this fraction can be categorized as an active fraction. Previous study stated that the extract with IC_{50} value < 25 µg/mL is categorized as an active extract in determination of antiplasmodial activity [12]. The same result also be found for isolated compounds which belong to flavonoids, but particularly, those isolated flavonoids can be categorized as active compounds in inhibiting plasmodial.

There are two main targets of the mechanism of flavonoids that capable for inhibiting the growth of plasmodium: 1) malaria parasite food vacuoles by inhibiting the process of hemoglobin degradation [13]; 2) membranes formed intra eristrositic stage malaria parasite i.e. New Permeation Pathway (NPP) by inhibiting the transport of nutrients needed by parasites [14]. The high antiplasmodial activity of the compounds is suspected from the mechanisms that act on food vacuoles. Flavonoid and its derivatives play a role in blocking the formation of hemozoin by the formation of free heme complexes with active compounds. Heme free (Fe^{3+}) is highly toxic because it can cause highly reactive oxygen species which can trigger oxidative reactions so that the parasites die. Therefore, the parasite converts into a non-toxic substance by forming a polymer from the heme residues by a co-ordination bond between Fe^{3+} heme with another heme hydroxyl group to form the *β*-hematin molecule, further forming a larger aggregate called hemozoin. This aggregate formation process of hemozoin is a process that can be used as a target for antimalarial therapy.

Pharmacokinetic effects of compounds can be reviewed from three aspects of physical properties of the compound including electronic effects, hydrophobic and molecular size [15]. The presence of terpenyl substituent can increase the lipophilic properties of compound. In this study, nymphaeol C (**1**), solophenol D (**2**), nymphaeol A (**3**), and nymphaeol B (**4**), having terpenyl substituent at C-6 and or C-2', can inhibit the growth of *P. falciparum*. The lipophilic aspect contributes to the activity of the compound in terms of its ability to bypass the semi-permeable lipid membrane of parasite. Solophenol D (**2**) which belongs to flavonol has lower antiplasmodial activity than nymphaeol C (**1**), nymphaeol A (**3**) and nymphaeol B (**4**) which belong to flavanones. The presence of double bonds on aromatic bridges and substituent of 3-OH in solophenol D (**2**) can reduce the antiplasmodial activity. Solophenol D (**2**) is more polar, so its lipophilic property is less than flavanones. Flavonol has a low solubility in fat, hence, it cause a reduced ability of compound to pass through semi permiabel lipid membrane of parasite. Furthermore, it can make the compound complicate to enter the food vacuole for inhibiting the formation of heme polymers (hemozoin) [7]. The nature of lipophilic is also associated with the mechanism of inhibition of the NPP by inhibiting the transport of nutrients needed parasites [14].

For pharmacodynamic effects of compounds can be evaluated based on the types and position of terpenyl substituent bound to flavonoids. Substitution of isoprene chain in the basic framework of flavonoids may increase the antiplasmodial activity [16]. Nymphaeol A (**3**) and nymphaeol B (**4**) exhibited the highest activity as antiplasmodial, this can be noticed that the presence of terpenyl group at C-6 or C-2' play important role in antiplasmodial activity.

4. Conclusions

According to the results, it can be concluded that flavonoids from *M. tanarius* leaves showed good activity in antiplasmodial screening. Compounds **3** and **4** which have ortho-dihydroxy in ring B and have a terpenyl substituent at C-6 or C-2' are active against *P. falciparum*. The presence of two terpenyl substituents at C-6 and C-2' (compound **1**) exhibited lower activity compare with compound **3** and **4**. Moreover, the presence of double bonds on the aromatic ring bridge in compound **2** decreases antiplasmodial activity. Particularly, it can be stated that *M. tanarius* leaves has a potency as antiplasmodial activity in the future.

Acknowledgment

Thanks are extended to the Institute of Tropical Disease Laboratory, Airlangga University, for assisting in determination of antiplasmodial activity and NMR instrument.

References

[1] Laurent D and Pietra F 2006 Antiplasmodial Marine Natural Product in the Perspective of Current Chemotherapy and Prevention of Malaria A Review *Mar Biotechnol* **8** pp 433-47

[2] Kuswantoro F 2017 Traditional anti malaria plants species of Balikpapan Botanic Garden, East Kalimantan–Indonesia *ICBS Conference Proceedings, International Conference on Biological Science* pp 78–85

[3] Zakaria I, Ahmat N, Jaafar F M *et al* 2012 Flavonoids with Antiplasmodial and Cytotoxic Activities of *Macaranga triloba Fitoterapia* **83** pp 968–72

[4] Phommart S, Sutthivaiyakit P, Chimnoi N *et al* 2005 Constituents of the Leaves of *Macaranga tanarius J Nat Prod* **68** pp 927–30

[5] Trader W and Jensen JB 1976 Human malaria parasites in continuous culture *Science* **193** pp 673–5

[6] Widyawaruyanti A, Zaini N C, and Syafruddin 2011 Antimalarial activity and Mechanism of action of flavonoid compounds isolated from *Artocarpus champeden* Spring stembark *JBP*

13(2) pp 67–77

[7] Diallo D, Maiga A, Diakite C *et al* 2004 *Malarial-5: Development of an Antimalarial Phytomedicine in Mali* ed Willcox M, Bodeker G and Rasoanaivo P (Boca Raton: CRC Press) pp 1–14

[8] Mabry T J and Markham K R 1970 *Flavonoids: The Systematic Identification of Flavonoid* (New York: Sringer-Verlag) pp 35–61

[9] Zakaria I, Ahmat N, Ahmad R and Jaafar F M 2010 Flavanones from the Flower of *Macaranga triloba World Appl Sci J* **9**(9) pp 1003–7

[10] Kumazawa S, Murase M, Momose M and Fukumoto S 2014 Analysis of antioxidant prenylflavonoids in different parts of *Macaranga tanarius*, the plant origin of Okinawan propolis *Asian Pac J Trop Med* **7**(1) pp 16–20

[11] Yoder B J 2007 *Isolation and Structure Elucidation of Cytotoxic Natural Products from The Rainforest of Madagascar and Suriname* (Faculty of The Virginia Polytechnic Institute and State University) Dissertation pp 42–7

[12] Kohler I, Siems K J, Siems K *et al* In vitro Antiplasmodial Investigation of Medicinal Plants from El Salvador *Z. Naturforsch* **57c** pp 277–81

[13] Sun S, Chen W, Cao W *et al* 2008 Research on the chelation between quercetin and Cr(III) ion by Density Functional Theory (DFT) method *J Mol Struct THEOCHEM* **860** 40-4

[14] Sherman I W 1998 *Malaria, Parasite Biology, Phatogenesis and Protection* (Washington DC: ASM press)

[15] Hansch C, Lien E J and Helmer F 1968 Structure-Activity Correlations in the Metabolism of Drugs *Arch Biochem Biophys* **128**(2) pp 319–30

[16] Widyawaruyanti A, Zaini N C and Syafruddin 2011 Mekanisme dan Aktivitas Antimalaria dari Senyawa Flavonoid yang Diisolasi dari Cempedak (*Artocarpus champeden*) *JBP* **13**(2) pp 67–77

Influence of the addition and storage time of crude extract of tea leaves (*camellia sinensis* l.) toward value of free fatty acid in crude palm oil

Erwin*, E Wahifiyah, R Hairani, and A S Panggabean
Department of Chemistry, Faculty of Mathematic and Natural Sciences, Mulawarman University Samarinda. East Kalimantan, Indonesia

*Corresponding Author : winulica@yahoo.co.id

Abstract: The purpose of this study was to determine the effect of the crude extract of tea leaves (*Camellia sinensis* L.) and storage time on the content of free fatty acid in palm oil. The dried tea leaves were macerated and concentrated by rotary evaporator. The extract obtained was added to crude palm oil with various added mass of the extract and various storage times. Phytochemical tests indicated the presence of secondary metabolites including alkaloids, triterpenoids, steroids, phenolics and flavonoids. The ANOVA test showed a decrease in free fatty acid content in crude palm oil with the addition of tea leaves extract. The LSD (Least Significant Difference) test showed the best influence on ALB of palm oil is on the total extract mass of 2 grams and the storage time of 20 days.

1. Introduction

Tea is widely consumed as a beverage, also to provide a fresh flavor and is also used as an additive to food, cosmetics and drugs, and does not cause negative effects [1]. Tea derivative products can be used for body slimming, herbal tea, preventing the formation of dental caries, ingredients of medicine for treating diabetes, hypertension and cancer [1-2].

Studies on the efficacy of the tea leaves have been carried out. Tea extract has high antioxidant activity [3-6]. Tea extract has potential as antimutagenic, anticarcinoggenic and protective agents against cardiovascular diseases [3]. Tea can be used in a precautionary measure for cancer, prostate cancer, kidney or liver disease [4]. Chemical compound content of leaves tea are polyphenols, flavonoids, flavones, flavonone, isoflavones, anthocyanins and catechins. In addition, tea also contains alkaloids (caffeine, theobromine and theofolin) of about 7-8% [1-2,7].

Caffeine contained in tea leaves has many negative effects if consumed in excess, can increase stress, accelerate bone damage, to affect heart health and stomach acid [2]. Caffeine compounds in tea can be used to lower free fatty acid content in palm oil, and can increase its economic value. Moreover, in tea leaves contain many basic metabolite compounds that are alkaline so it is expected to reduce free fatty acids in palm oil.

Based on the above description, the research to reduce free fatty acid in crude palm oil by utilizing crude extract of tea leaves (*Camellia sinensis* L.) as a base source has been done. Some important parameters such as the total amount of crude extract of tea leaves (*Camellia sinensis* L.) and the various of storage time are also studied. This research method is environmentally friendly and can increase the economic value of crude palm oil and tea.

2. Material and Methods

2.1. Material

Equipment used in this study are rotary evaporator, glass funnel, beaker glass, pipette, test tubes, press equipment, tube, centrifuge, vial, spatula, measuring flask, analytical balance, Erlenmeyer flask, magnetic stirrer, electric water bath, the stand poles, clamps, measuring cups and buret. The materials used in this study include crude palm oil, dried tea leaves, ethanol, distilled water, cotton, oxalic acid, KOH and phenolftalein.

2.2. Metodology

2.2.1. Preparation of crude palm oil. The fruit of palm oil is separated between the seeds and the flesh. The flesh is smoothed using a mortar to a pulp and filtered with a thin cotton cloth into an oil press. The extract obtained was centrifuged for 60 minutes at a speed of 6000 rpm. The oil produced is ready for use.

2.2.2.Extraction of tea leaves. The amount of 678 g of powder dried tea leaves (*Camellia sinensis* L.) was macerated with ethanol for 3 times 24 hours at room temperature. The filtrate obtained was concentrated using a rotary evaporator at a temperature of 30-40 °C and obtained crude extract of tea leaves [8-9].

2.2.3.Alkaloids test. The crude extract of tea leaves (*Camellia sinensis* L.) were dissolved in ethanol, added with a few drops of 2N H_2SO_4, while shaken. The formation of orange to red brown deposits after the addition of Dragendrorff reagent indicated the presence of alkaloids [10].

2.2.4.Triterpenoid and steroid test. The crude extract was dissolved in ethanol, put in a glass tube following by the addition of 10 drops of glacial acetic acid and 2-3 drops of concentrated sulfuric acid (Liebermann Burchard reagent). The formation of reddish orange or purple colour indicated the presence of triterpenoids, while green color indicated the presence of steroid [8].

2.2.5. Saponin test. A certain amount of crude extract was dissolved in 2 mL of distilled water then shaken strongly in a test tube. The presence of saponin was characterized by the formation of durable foam on the surface of the liquid. Foam has remained stable after the addition of a few drops of concentrated HCl [11].

2.2.6.Phenolic test. A certain amount of crude extract was dissolved in ethanol and 3 drops of 1% iron (III) chloride was added to a total of 1 mL of the extract. The presence of phenolic was indicated by the appearance of green, red, purple, blue or black colors [12]

2.2.7.Quinoline test. A total of 5 mL of extract solution was added 1N sodium hydroxide. The presence of quinone was indicated by the formation of red color [13].

2.2.8.Flavonoids test. Flavonoid test was done by Wilstater method. A certain amount of extracts is dissolved in a suitable solvent. A total of 1 mL of extract was added with a small amount of Mg powder and 5 drops of concentrated HCl. The presence of flavonoids gives the color orange, red to dark red [14].

2.2.9.Addition of crude tea leaves into crude palm oil. Amount of 5 g of crude palm oil is inserted into each of 63 vials. 1st - 21st (negative control) , 22st-42st, 43rd -63rd vial were added 0, 1, and 2 gram crude extract of tea leaves, respectively. Each vial was homogenized with a stirrer. Furthermore, the samples were moved into a dark glass bottle sealed and stored at room temperature. Determination of free fatty acids performed on the storage of varied time of 0 – 30 days.

2.2.10.Determination of Free Fatty Acid. A total of ± 2 g of oil is weighed into the Erlenmeyer flask. The oil is then added 25 mL of neutral alcohol and 2 drops of phenolphtalein indicator. The solution was titrated with a 0.1N KOH-alcoholic solution until a pink color was obtained and not dissipated for 30 s.

3. Result and Discussion

3.1.Extraction and Phytochemical Test

The extraction process in this study, from 678 g of tea leaves (*Camellia sinensis* L.) was obtained from Bogor, West Java which has been dried and macerated with ethanol of 1.9 L for several days and then concentrated with a rotary evaporator to obtain a total was 65.62 g (9.72%). The resulting total extract tested its secondary metabolite content, as shown in Table 1.

Table 1. The chemical compound content crude extract of tea leaves

Secondary metabolites	Alkaloid	Triterpenoid	Steroid	phenolic	Quinoline	Flavonoid	Saponin
result	+	+	+	+	+	+	+

Based on test results obtained information that the phytochemical extracts containing coarse alkaloids, triterpenoids, steroids, phenolic, quinone, and flavonoids.

3.2.Free Fatty Acid (FFA) Test

Determination of free fatty acids was done by using titration of acidimetry. In this study, the amount of free fatty acids in palm oil was determined for various variations in total amount of added extract of tea leaves (*Camellia sinensis* L.) and storage duration time. The determination of free fatty acid in palm oil is done by triplo respectively.

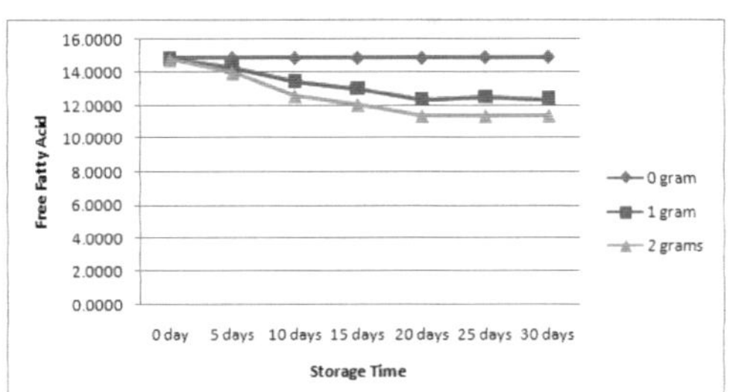

Figure 1. FFA content at various variations addition of crude extract tea leaves and duration of storage on crude palm oil

The results of the study (**Fig. 1**) showed the comparison of FFA value in crude palm oil which was not added with crude extract of tea leaves (*Camellia sinensis* L.) (as negative control) and with added crude extract of tea leaves. The value of FFA in negative control is higher than that of the added crude palm oil with crude extract of tea leaves. The increase in FFA is directly proportional to the length of storage, this is due to good hydrolysis caused by water and lipase enzymes during the treatment and storage [15-16]. The FFA value of the controlled palm oil increases with the length of storage time. This is due to the increased FFA caused by oxidation and enzyme hydrolysis reaction during processing and storage time [17,18]. In contrast to the oil that has been added to the crude extract tea leaves (*Camellia sinensis* L.), the FFA value of the oil decreases with the amount of total extract added to the oil.

It can be concluded that oil storage is good only up to 20 days. After 20 days there is an increase in the value of free fatty acids. The oil will be physically damaged in the form of a color that turns into turbid and is chemically a formed FFA. The color of the oil will become turbid, because the salt deposition of the reaction between alkaloids and free fatty acids (shown in **Fig. 2**). This is presumably because the alkaloid content of the crude extract tea leaves (*Camellia sinensis* L.) added in palm oil has reacted with free fatty acids, so that no alkaloids neutralize the free fatty acids formed as oil hydrolysis. Alkaloids are natural bases that have basic Lewis properties because they contain N atoms in heterocyclic rings that can contribute a single pair [19].

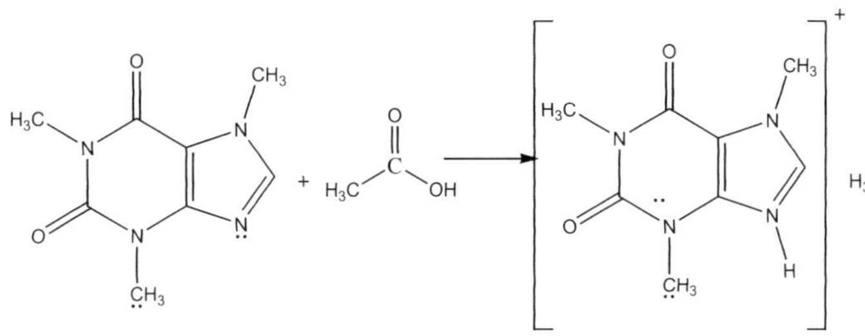

Figure 2. Reaction between caffeine with free fatty acid

To prove the hypothesis, is there any effect of addition of the crude extract tea leaves (*Camellia sinensis* L.) toward free fatty acid on palm oil done by ANOVA method. After the test is obtained $F_{count} > F_{table}$. $F_{count} = 19.9148$ and $F_{tabel} = 9.33$ with significance level of 1 %. If $F_{count} > F_{table}$. Ho is accepted and Ha is rejected, it is concluded that there is an effect of addition of the crude extract tea leaves (*Camellia sinensis* L.) and variation of storage time to free fatty acid in crude palm oil.

LSD (Least Significant Difference) test is performed to determine parameters that give the best effect to free fatty acid value. The best amount of the crude extract tea leaves (*Camellia sinensis* L.) was 2 g. The average value of FFA on the addition of 2 g of extract significantly different with LSD value, was 2.3686 > 1.0343 with significance level of 1%. This indicates that the addition of mass of the crude extract tea leaves (*Camellia sinensis* L.) 2 g can lower the larger free fatty acids. From the calculation of LSD to the storage period, it is obtained at 20 days day storage. This can be seen from the average value of FFA on 20 days of storage is significantly different to the LSD value, was 1.954 > 1.0343 with a significance level of 1%. It can be concluded that the length of storage of oil mixture with the crude extract tea leaves (*Camellia sinensis* L.) is a maximum of 20 days. After 20 days of storage, palm oil will undergo physical changes and FFA on palm oil will increase so that the quality of the palm oil is damaged.

4. Conclusion

Based on this research, the optimum mass addition of the crude extract of tea leaves was 2 g with a reduction of free fatty acids of 22.81% with the optimum length of storage time of 20 days.

References

[1] Towaha, J 2013 *Warta Penelitian dan Pengembangan Tanaman Industri*, **19** (3) pp 12-6.
[2] Siregar, N 2009 *Influencer time of soaked at tea leaf to rate of tannin beverage in coca-cola company bottling Indonesia medan.* (Medan: Universitas Sumatera Utara)
[3] Gramza A, Korczak J, Amarowicz R., 2005 *Polish Jo. of Food and Nut. Sci.* **14/55**(3) pp 219–35.
[4] Armoskaite V, Ramanauskiene K, Maruska A, Razukas A, Dagilyte A, Baranauskas A, and Briedis A 2011 *J. of Med. Plants Res*, **5**(5) pp 811-6.
[5] Yashin A, Yashin Y, and Nemzer B 2011 *Am. J. Biomed. Sci*, **3**(4), pp 322-35
[6] Dian-Nashiela F, Noriham A, Nooraain H, and Azizah, A. H. 2015. *Int. Food Res. J.,* **22**(3), pp 1189-94 .
[7] Rohdiana D and Shabri 2012 *Jurnal Penelitian Teh dan Kina* **15**(2) pp 81-8.
[8] Erwin, Sulistyaningsih S and Kusuma I W 2015 *J. Pharm. Bio. Sci.*, **6** (1) pp 598 – 604.
[9] Erwin, Anggeraini D and Suryani 2016 *Der Pharmacia Lettre* **8**(1) pp 233-6
[10] Seervidya N and Mehrotra S 2003 *J. of AOAC Int* **86**(6) pp 1124-7.
[11] Kokate A 1999 *Phytotherapy* **78** pp 126-9.
[12] Raaman N 2006 *Phytochemical technique* NIPA Pitam Pura New Delhi pp 19-24.
[13] Ciulei I 1984 *Methodology for Analysis* of *Vegetable Drugs, Chemical Industries Branch. Division-Industrial Operation UNIDO* Bucharest-Rumania pp 11-23.
[14] Rao U S M, Abdurrazak M, Mohd K S 2016 *Malays. J. of Anal. Sci,*, **20**(5) pp 1181 –90.
[15] Gunawan T M M A and Rahayu A 2003 *JSKA* **VI**(3) pp 1-6.
[16] Basyuni M, Amri N, Putri L A P, Syahputra I, Arifiyanto D 2017 *Indo. J. Chem* 2017 **17**(2) pp 182 – 90.
[17] Sastrohamidjojo H 1996 *Sintesis Bahan Alam* (Yogyakarta:UGM Press).
[18] Rinaldi A, Alimuddin, Panggabean A S 2015 *Molekul* **10**(2) pp 112–20.
[19] Sanjiani N, Suaniti N, dan Rustini N 2015 *Jurnal Kimia* **9**(2) pp 259-66.

IOP Conf. Series: Earth and Environmental Science **144** (2018) 012013 doi:10.1088/1755-1315/144/1/012013

Relation analysis of physical activity and food insurance with obesity on the participant the new me program in vico indonesia badak field

D E Lufitasari*, D L Setyowati

Occupational Health and Safety
Public Health Faculty, Mulawarman University, Gunung Kelua Campus. East Kalimantan, Indonesia.

*Corresponding Author: delufitasari29@gmail.com

Abstract: Obesity is often defined as an abnormal condition because of the serious excess of fat in the adipose tissue that interferes with health. Obesity tendency is more common in individuals who have a lifestyle with mild activity levels and consume high-calorie foods as well as low in micro-substances. The purpose of this study is to determine the relationship between physical activity and food intake with the incidence of obesity in the employee/contractor participants program The New Me. The design of this research is case control study, using random sampling with lottery system. Data collection using standard questionnaire Beacke, SQFFQ and DASS42 Item and data analysis using Chi-square test (α=10%). The results showed that there was no relationship between physical activity (p=0,230,OR=0.457, CI90%=0.310-0,602) with obesity. Food intake using two comparison parameters iscurrent weight (p=0.45,OR=3.036, CI90%=0.624-12.806) and ideal weight (p=1,000, OR=1.238, CI90%=0.310-2.429), each of which has nothing to do with the incidence of obesity. In this study it can be concluded that there is no relationship between physical activity and food intake. Physical activity is a propective factor and food intake is a risk factor to obesity occurrence at employee/contractor participant program The New Me in the VICO Indonesia Badak Field Muara Badak, East Kalimantan.

1. Introduction

In 2008, some 2.8 million adults died because obesity, about 300 million people who are clinically classified as obese, who are the main contributors of degenerative diseases such as diabetes, heart disease and cancer. Obesity is a condition in which body fat accretion is based on the value of the body mass index (BMI) [1]. Determination of obesity using BMI, weight calculation results (in kilograms) divided by the square of height (in meters). [2]

According Riskesdas in 2013, the highest prevalence of non-communicable diseases that will affect work productivity and the labor force population group work are obese (26.6%), hypertension (25.8%), lung disease chronic obstructive (3.8%), diabetes mellitus (2.1%), coronary heart disease (1.5%), cancer (1.4%) and stroke (1.21%). And the highest prevalence of obesity has grouped the population in the productive age. [3]

The main cause of obesity is excess energy intake are not in accordance with the long-term energy expenditure. [4]Obesity tendency is more common in individuals who have a lifestyle with mild activity levels and consume high calorie foods as well as low in micro. Risk factors that can lead to overweight and obesity in the workplace include work with little activity or movement as part of sedentary life style, workplace design that is not formed for free movement, limited opportunities for physical activity during work hours and options unhealthy food in the canteen. [2]

According to the Ministry of Health RI 2011, stated that the importance of applying diverse food consumption patterns, balanced nutrition and adequate physical activity and regular in the workplace in relation to the optimal productivity of workers was still not given attention.[5]

According to the Ministry of Health RI 2011, obesity is a major risk factor for degenerative diseases, heart disease and blood vessels, diabetes and cancer. These include the effects of lifestyle changes, which are the effects of globalization and industrialization including the development of fast food that is generally not balanced nutrition that is high in fat and salt, and low fiber content. Mobility is very high, making people tend to consume fast food without considering the nutritional content [5]

VICO Indonesia Badak Square, Muara Badak East Kalimantan is one of Indonesia's natural gas oil companies where most of its employee / contractor work together with the buffet method provided by the contracting company.The food menu in this company contains a lot of fat, low fiber, taste in foods that are quite salty, desserts of snack containing flour and not often available ice cream at lunch hour.The purpose of this study was to determine the relationship between physical activity and food intake with the incidence of obesity in New Me employees / program contractors.

2. Method

This research is an analytic survey method with case-control approach. This research was conducted in the field of VICO Indonesia LLC Badak - East Kalimantan. The research population is the employee/contractor who follow The New Me program as many as 45 people. The sample in this study used a 1: 1 ratio of 19 cases (obese people): 19 controls (people who were overweight) and the total sample was 38 people. Sampling of case and or control was conducted random sampling technique with lottery method. The instrument of this study was used: Questioner for Physical Activity, Depression Anxiety Questionnaire (DASS) 42 and Quantitative Food Frequency Questionnaire. By Chi-square test ($\alpha = 10\%$).

3. Result and Discussion

3.1.Individual Characteristics

The frequency distribution of respondents based on individual characteristics can be seen in table 1.

Table 1. Distribution of Respondents Frequency Based on Individual Characteristics

No	Characteristics	N	%
1	Age		
	> 30 years	29	76,3
	≤ 30 years	9	23,7
2	Gender		
	Male	32	78,9
	Female	6	21,1
3	Educational status		
	Collage	34	89,5
	High School	4	10,5
4	Marital Status		
	Married	27	78,9
	Single	11	21,1
5	Work Stress		
	Stress	13	34,2

Not Stress	25	65,8
Total	38	100

3.2. Univariate Distribution

Univariate frequency distribution of physical activity and food intake is presented in table 2.

Table 2. Distribution of Univariate Frequency of Physical Activity and Food Intake

No	Individual Characteristics	n	%
1	Physical Activity		
	Light Activity	3	7,9
	Medium Activity	35	92,1
2	The Need For Ideal Body Weight Energy Intake		
	Excess Energy Intake	20	52,6
	Energy Intake Is Not Excessive	18	47,4
3	The Need For Energy Intake Of Current Weight		
	Excess Energy Intake	7	18,4
	Energy Intake Is Not Excessive	31	81,6
	Total	38	100

Table 2 shows 38 employees / contractors as new program participants as many as 3 respondents who have light category (7.9%). And food intake using two comparison parameters, namely ideal body weight and weight now. Each of 38 employee / contractor as participants of the new me program with ideal body weight ratio, there are 20 respondents who have Excess Intake Energy (52,6%), and from comparison of current weight intake, there are 31 respondents who have Intake Energy Not Excessive (81,6%)

3.3. Bivariate Distribution

The bivariate frequency distribution of physical activity and food intake is shown in table 3.

Table 3. Distribution of Univariate Frequency of Physical Activity and Food Intake

Variable	The Incidence of Obesity				Total		OR (90% Cl)	P Value
	Obesity		*Overweight*					
	N	%	N	%	n	%		
Physic Activity								
Light	3	15,8	0	0	3	7,9	0,457	0,230
Medium	16	84,2	19	100	20	92,1	(90%Cl; 0,310-,602)	
The Need For Energy Intake of Current Weight								
Excess Energy Intake	5	26,3	2	10,5	7	7,9	3,036	0,405
Energy Intake Isn't Excessive	14	68,4	17	89,5	31	92,1	(90%Cl; 1,248-17,158	
The Need For Energy Intake Of Ideal BodyWeight								
Excess Energy Intake	10	52,6	10	52,6	20	52,6	1,000	1,000
Energy Intake Isn't Excessive	9	47,4	9	47,4	18	47,4	(90%Cl; 0,336-4,047)	
Total	19	100	19	100	38	100		

The results obtained p value is 0,230 then there is no relationship between physical activities with the incidence of obesity of employees/contractors. OR = 0,457 indicates that physical activity is a propective factor. That is, respondents who have medium activity has a prospective level of 0.405 times greater than those who have mild activity are obese.

Regular activity is useful for regulating weight and strengthening the heart and blood vessel system. Most respondents (92,2%) had mild activity and had an obesity probability of about 0.2-fold for non-obesity. Respondents in this research work well in it are miners who work in the room and field. This tends to cause less chance for exercise even though the company has provided relatively comprehensive sports facilities. Some respondents do sports activities such as jogging and badminton. The respondents do this sport as a hobby rather than a necessity, which can be concluded that the respondent did not do it regularly.

According to Ministry of Health 2014, the study shows a lot of decrease in risk of premature death if people do about 2.5 hours (150 minutes) aerobic exercise that is being intensified (moderate-intensity aerobic physical activity) every week. [6] Based on the work schedule, there are workers who go home every afternoon as well as workers with a work schedule of 5 days and 14 days. This certainly affects physical activity, especially sports habits performed. For the respondents who go home every day, the time to exercise becomes an obstacle because it is tired during the day work plus using a bicycle vehicle either to the workplace and back home. While for those who stay on location when finished working prefer to take a break compared to doing sports activities. Respondents who are on the site are required to follow aerobic gymnastics every Wednesday organized by the Health Department in collaboration with the Sport Hall.

In this study, there is a similar phenomenon with Rissanen research, where some respondents while still having obesity nutritional status, some of these respondents do regular exercise. However, when the nutritional status of some respondents is in the line of overweight or ideal, some of these respondents relax the sport activities. And impact in the nutritional status of respondents back on the status of obesity. Research conducted by Rissanen which shows that low physical activity is a factor responsible for the occurrence of obesity. For example, obesity does not occur in active athletes whereas athletes who stop doing exercise more often gain weight and obesity.[7]

The result of this research is p value equal to 0,405 according to comparison with current weight energy intake so there is no relation between food intake with obesity of employee / contractor. The value of OR = 3.036 corresponds to the ratio of current heavy energy intake, which indicates that food intake is a risk factor for obesity. That is, respondents who have excessive energy intake have a chance, 3 times the occurrence of obesity from those who do not have excessive energy intake for obesity.

Then, the result p value of 1,000 in comparison with the ideal weight energy intake so there is no relationship between food intake with the incidence of obesity of employees / contractors. OR = 1000, indicating that energy intake is not a risk factor for obesity. This means that respondents who have excessive intake at least 1-fold greater than those who do not have excessive energy intake.

According to comparison of dietary intake of body weight now some respondents (7.9%) have excessive energy intake and have obesity chances about 3 times. compared with ≤ enough energy intake. Meanwhile, according to the ideal food intake weight ratio of most respondents (52.6%) have excessive energy intake and have no chance of obesity. Then from the calculation of the ideal diet intake is to maintain a weight that means intake of food to maintain weight between normal weight and ideal body weight.

The high proportion of respondents with fatter intake indicates that the food menu provided by the catering is a diet high in fat, especially animal fat. The results of observations on the menu served by the catering is always available standard menu for breakfast, lunch and dinner that contains a lot of fat. The details of each set menu consists of 2 types of animal side dishes (meat, fish or chicken), 1 type vegetable side dish of tofu / tempe and complementary foods such as vegetables, boiled potatoes, soups, crackers (prawns, onions, melinjo crackers), pickles and various sauce. For the breakfast menu always provided omelets, sausages, pancakes, sandwiches, boiled potatoes, rice, crackers, boiled eggs, toast, bread, cereal, milk and assorted jams. As dessert provided 6 types of food such as ice cream, pudding, cake and other pastries and 3 kinds of fresh fruit and various kinds of fruit preparations such as salad and salad. In addition, midnight menu is provided for employees/contractors who work at night and low cholesterol menu as an option for low-cholesterol diet. According to Hu, the intake of energy derived from fats and carbohydrates is associated with energy spending that is more influential on weight.

Behavioral factors (current consumption with weight ratio of body weight now) consuming excess fat intake has an obesity chance of about 5 times compared to ≤ enough energy intake. [8]

The provision of food in this company is done in the form of a buffet (all you can eat) so it can affect a person to take food in an uncontrolled amount such as excessive energy consumption. Especially if the food is served food that is preferred, dense energy and delicious so the total energy intake increasingly not well controlled. VICO Indonesia employees / contractors also have a habit of eating fast food that is sold at the company's stalls. According to Tee, states that fast food times are usually done during the afternoon or after work hours. In addition, there is also a change in family eating habits. The presence of family tendencies for food outdoors, and an increase in fast food consumption, the use of more vegetable oil and sweet drinks[9].

4. Conclusion

It can be concluded that there is no relationship between physical activity and food intake of obesity. The relationship of physical activity with obesity is a protective factor. Meanwhile, the relationship of food intake by the ratio of current body weight energy intake with the incidence of obesity is a risk factor and the relationship of food intake in accordance with the ratio of ideal body energy consumption with the incidence of obesity is not a risk factor for obesity because of the calculation of food intake of body weight is now to reduce weight which means food intake to lower excess body weight to ideal body weight. on the employee / contractor of The New Me program at VICO Indonesia Badak Field Muara Badak, East Kalimantan.

Acknowledgment

Thanks to VICO Indonesia Badak Field Muara Badak, East Kalimantan, especially to the Health Department and the employees /contractor participant of The New Me program which has given opportunity and trust to the researcher to finish the research in this company.

References

[1] WHO. (2010). *Obesity: Preventing And Managing Global Epidemic. Report of A WHO Consultation Technical Report Series 894.* Geneva: WHO.

[2] National Institute for Health (NIH). (2006). *Obesity: the prevention, identification, assesment and management of overweight and obesity in adults and children.* National Institute for Health and Clinical Excellence National Collaborating Centre for Primery Care.

[3] KementrianKesehatan RI. (2013). *Riset Kesehatan Dasar (Riskesdas) 2013.* Jakarta: Badan Penelitian dan Pengembangan Kesehatan.

[4] RiccardiG ,Aggett P , Brighenti F , Delzenne N , Frayn K , Nieuwenhuizen A , Pannemans D , Thels S , Tuijtelaars S , Vessby B. (2004). *PASSCLAIM - body weight regulation. Insulin sensitivity and diabetes risk.*Eur J. Nutr. 43 (2): 117-146.

[5] KementrianKesehatan RI. (2011). *Strategi Nasional Penerapan Pola Konsumsi Makanandan Aktivitas Fisik Untuk Mencegah Penyakit Tidak Menular.* Jakarta: Direktorat Jendral Bina GizidanKesehatanIbudanAnak.

[6] Kementrian Kesehatan RI. (2014). *Pedoman Gizi Seimbang (PGS) (Pedoman Teknis Bagi Petugas Salam Memberikan Penyuluhan Gizi Seimbang).* Jakarta: Direktorat Jenderal Bina GizidanKesehatanIbudanAnak.

[7] Rissanen AM. (1991). *Determinant of Weight Gain and Overweight in Adults Finns.*Eur J ClinNutr. 45:419-430.

[8] Hu , FB , Manson JE , Willet WC. (2001). *Types of dietary fat and risk of coronary heart diseases a critical review.* J Am Coll Nutr. 2001:20:5-19.

[9] Tee ES. (2002). *Priority Nutritional Concerns In Asia.* Food an Nutrition Bulletin 23(4):345-348.

Amoebic meningoencephalitis in Samarinda East Kalimantan

L R D Siagian[1,2*], V M LToruan[1,3], Y O Hutahaen[4], C Gunawan[5]

[1]Departement of Parasitology, Faculty of Medicine, Mulawarman University, Samarinda, Indonesia
[2] Department of Clinical Pathology, Faculty of Medicine, Mulawarman University, Samarinda, Indonesia
[3] Department of Dermato venereology, Faculty of Medicine, Mulawarman University, Samarinda, Indonesia
[4]Department of Neurology, Faculty of Medicine, Mulawarman University, Samarinda, Indonesia
[5]Departmen of Internal Medicine Faculty of Medicine, Mulawarman University, Samarinda, Indonesia

*Coresponding Author :lonita_ui@yahoo.com

Abstract. Meningoencephalitis is a disease of central nervous system caused by s microorganism such as bacteria, Rickettsiae, Spirochaeta, virus, fungus, protozoa and helminth. Free living amoeba protozoa has been reported as an etiology agent of menigoencephalitis. Amoebic meningoencephalitis is a very rare condition and often fatal because of late diagnosis. We reported a 24 years old woman, hospitalized at Abdul Wahab Syahranie's Hospital, Samarinda East Kalimantan with decreased of consciousness and fever. The brain MRI showed focal lession suggested as a granuloma. ed Microorganisms like cyst and trophozoit with pseudopodia amoeboic form were identified from the cerebrospinal fluid. Based on these findings the patient was diagnosed with amoebic meningoencephalitis infection and got intravenous metronidazole. The patient died on the twenty third days of hospitalization.

1. Introduction

Meningoencephalitis could be caused by microorganism such as bacteria, Rickettsiae, Spirochaeta, virus, fungus, protozoa and helminth. Some of free living amoeba protozoa like Acanthamoeba, Neigleria and Balamuthia genus have been reported as an etiology of central nervous system infection in human. Amoebic meningoencephalitis is very rare but often fatal because of late diagnosis. In this patient, the formation resembles protozoa found in cerebrospinal fluid (CSF), suggesting an amoebic protozoa. They found as the formation of cysts, and trophozoit with the presence of pseudopodia. We reported a fatal case with amoebic meningoencephalitis.

2. Case Report

A 24-year-old woman came to the Emergency Unit of AW. Sjahranie Hospital Samarinda East Kalimantan on February 23, 2016 with a major complaint of decreased consciousness since 4 hours before admission. The patient had fever since one week before admission. Approximately one month before admission, the patient had severe headache and was hospitalized for two weeks. On physical examination there was no meningeal excitatory sign nor focal neurologic deficits. Magnetic

Resonance Imaging (MRI) angiography was performed on January 29, 2016. The results did not indicate the existence of focal lesions of the parenchyma and blood vessels. The patient discharged from the hospital with a diagnosis of migraine headaches and treated with Depakote 250 mg two times daily, and the headache did not improve. On physical examination, the Glasgow Coma Scale (GCS) score was E1V1M5 , diameter of pupil 2.5 mm bilaterally, normal light reflex bilaterally, blood pressure 120/80 mmHg, pulse frequency 96 times per minute, respiratory rate 20 times per minute, temperature of 38,4°C , Babinski and clonus positive on the left side. Laboratory examination on 23 February 2016 was o Hb 12.8 gr/dL, leucocytes 10.780 / µL, platelet 284.000 / µL, blood glucose 121 mg / dL, albumin 4,2 g / dL, ureum 24,0 mg/dL, creatinine 0.5 mg/dL, sodium 133 mEq/L, potassium 4.0 meq L, chloride 100 mEq/L, magnesium 2.3 mEq/L, procalcitonin <0.05 ng/mL, non reactive HBsAg , non reactive HIV antibodies, negative IgM dengue, negative IgG, negative NS1 dengue. Radiological examination of chest X-ray showed normal. Head CT scan on February 23, 2016 showed a lateral ventricular dilatation that impressed an obstructive hydrocephalus. Patients are diagnosed with Hydrocephalus as a result of bacterial meningoencephalitis. The patient was treated with intravenous Ceftriaxone two grams twice daily intravenous and Dexamethasone 10 grams four times daily. On February 25, 2016, the installation of Extra Ventricular Drainage (EVD) was performed and found an intracranial pressure increased. The analysis of cerebrospinal fluid (CSF) obtained clear color, cloudy, cell count 7 cells/mL with MN cells 85% and PMN 15%, protein 170 mg/L, glucose 10 mg dL, None test (+), Pandy test (+) . There was no growth of germs in the CSF culture. TORCH immunological examination on February 26, 2016 showed Rubella IgG positive (186 IU/mL) and CMV IgG positive (205.2 IU/mL) and CMV IgG avidity was 72.2%. The patient was diagnosed temporarily with Fungal and TB Meningoensefalitis and the presence of reinfection of the Cytomegalovirus virus. The patient was given additional injection therapy of Flukonazole 450 mg once daily and Gancyclovir 250 mg twice daily and combination of antituberculosis drugs (Rifampicin 450 mg, INH 300 mg, Etambutol 750 mg, Pyazinamide 1000 mg. Provision of Amphotericin B Injection 0.5 g/kg body weight day but the drug was not available. On February 29, 2016 sputum culture examination was performed and *Acinetobacter baumanii* was found, in addition nasal swab culture obtained *Staphylococcus haemolitycus*. On March 1, 2016, analysis of CSF showed clear and slightly cloudy fluid, 150 leucocytes/mL (of with MN 85% and PMN 15%), Pandy test (+), protein 4,232 mg/L and glucose 71 mg/dL .

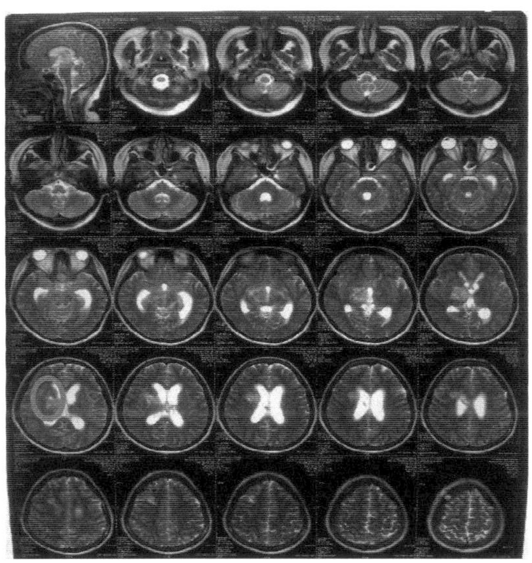

Figure 1.MRI results on the 8th day of an impressive treatment of a granuloma

No growth of germs was found from CSF culture. In sputum culture examination, *Pseudomonas aeruginosa* was obtained and in left and right axillary swap cultures, *Klebsiella penumeonia* was obtained. On March 3, 2016, patient's GCS score was E1M5Vx (tracheostomy was performed). Microbiological examination of CSF obtained negative acid fast bacilli and examination of real time PCR showed positive DNA *Mycobacterium tuberculosa complex*. The patient was given additional therapy of 1 gram Streptomycin once daily. In addition to the contrasted MRI results in the presence of focal lesions in T1 and T2 in the distal regions of the ganglia dextra basal which suggest a granuloma (figure 1).

Laboratory examination on March 4, 2016 was showed Hb 11.3 gr/dL, leucocytes 19,440/µL, 261,000 µL platelets, sodium 128 mEq/L, potassium 4.6 mEq/L, chloride 97 mEq/L, and Procalcitonin 0.79 ng mL., SGOT 112 mIU/mL, SGPT 256 mIU/mL. In the direct preparation of CSF with ink parker, a formation of microorganisms was found, did not absorb the color, three times erythrocytes size, with cyst (figure 2) , with the existence of pseudopodia and found amoeboid formation. It was something like amoeba (figure 3 and figure 4) On March 7, 2016, patient was given additional treatment of Metronidazole 500 mg injection three times daily. Patient was diagnosed with amoebic Meningoencephalitis.

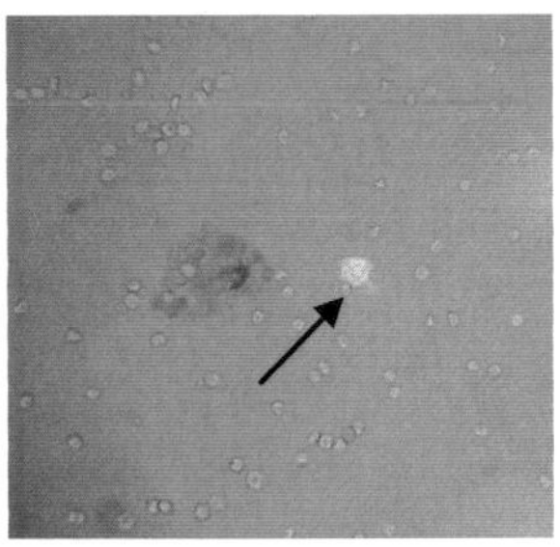

Figure 2. Amoebic cyst

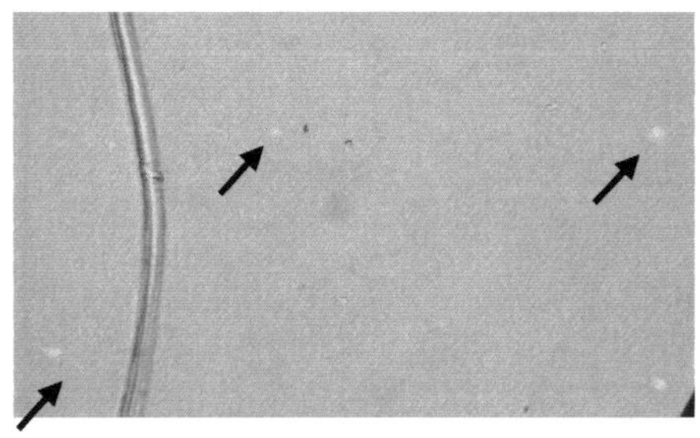

Figure 3. (a) Trofozoit with pseudopodia. (b,c) amoebic cysts

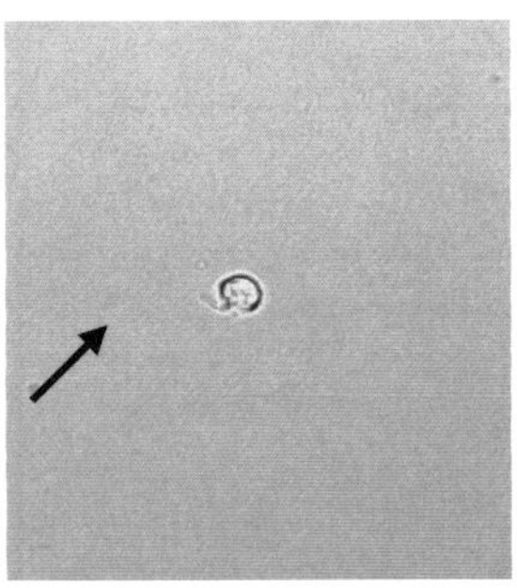

Figure 4 . Amoeba with pseudopodia

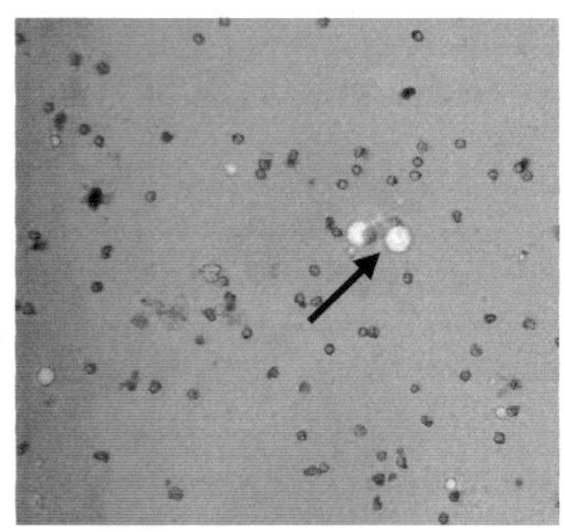

Figure 5 . Cysts with nucleous

3. Discussion

Meningoencephalitis can be established clinically through anamnesis and neurological physical examination. The presence of fever, headache, behavioral changes, meningeal excitatory signs and focal neurologic deficits suggest an inflammatory process in the meninges region accompanied by focal lesions in the brain parenchyma. Acute meningoencephalitis can be caused by various microorganisms such as gram-positive bacteria, gram-negative bacteria, anaerobic bacteria and *Mycobacterium tuberculosis*, Rickettsiae, Spirochaeta, viruses, fungi, protozoa and helminth. In the initial phase of the course of the patient's illness, the presence of meningeal excitatory signs or focal neurologic deficits was not found, only headache and fever was exist. Symptoms of meningeal excitatory and even global focal neurologic deficits was found in the advanced phase of the disease course. The patient had decreased of consciousness approximately 1.5 months from the first time complaining of headaches. Etiology of meningoensefalitis was established through examination of CSF and microbiological analysis and contrast head imaging. Since the patient was admitted to the hospital, several times the CSF was examined but until day 18 showed no bacterial growth but showed a significant increase in protein levels since admission. The results indicated the possibility of infection caused by tuberculosis, fungi or parasite that is with the increase of protein content, decrease of glucose level and cell count is dominated by MN (85%) than PMN (35%). Meningoencephalitis can also be caused by frequent fungal infections due to *Cryptococcus neoformans*. In this patient, we performed a microbiological examination for Cryptococcus with direct preparation using ink parker but no Cryptococcus was found. According to the reference, Cryptococcus identification uses Indian ink. Immunologic examination results showed positive CMV IgG antibody and continued with an avidity test, indicating the possibility of reinfection or past infection. CMV infection may manifest as interstitial pneumonia, hepatitis, Guillain-Barre Syndrome, retinitis, myocarditis, colitis, thrombocytopenia and haemolytic anemia, skin eruption, and meningoencephalitis.

Meningoencephalitis due to CMV generally occurs in immunocompromised patients. In AIDS patients, CMV infection is the most common viral opportunistic infection (21-44%). Meningoencephalitis due to CMV is characterized by sensory impairment and motor weakness such as radiculopathy, severe headache, photopobia, lethargy. CSF shows a slight increase in lymphocytes. The exact diagnosis is based on the discovery of CMV DNA by PCR examination. However, in this patient there is no CMV DNA testing due to equipment limitations. Antiviral drugs that are effective in the treatment of CMV infections are ganciclovir, valganciclovir, foscarnet, cidofovir, maribavir[1] Ganciclovir has been given in this patient. Based on results of brain fluid analysis indicating the possibility of tuberculosis or fungal infection, the patient was given antituberculose therapy but did not give any improvement. *Mycobacterium tuberculosa complex* positive DNA test results can be interpreted as an infection of other Mycobaterium species, namely *M.bovis*, *M. bovis Bacillus Calmette-Guerin*, *M. africanum*, *M. canetti*, *M.microti* [2]. The 11th day of treatment, from the microscopic examination of the direct supply of cerebrospinal fluid found a formation that resembles protozoa. There are several types of protozoan infections that can infect central nervous system, such as amoeboeasis , cerebral malaria, Toxoplasmosis, Tripanosimiasis, Microsporodiasis, Leismanisasis[2]. In this patient, the formation resembles protozoa found in CSF, suggesting an amoebic formation as it is found in the formation of cysts, amoeboid formations and the presence of pseudopodia. Some types of free living amoeba from the genus Acanthamoeba, Balamuthia and Neigleria have been known to infect the central nervous system of humans [2,3]. It could be manifest as opportunistic infections mainly in immunocompromised hosts or as non opportunistics [4] Some Acanthamoeba species cause chronic infections ie. granulomatous amoebic encephalitis (GAE) particularly in immunocompromised individuals. Balamuthia can cause GAE in immunocompromised and immunocompetent individuals. *Neigleria fowlerii* has been reported to infect children and immunocompetent young adults and cause acute and fulminant illness called Primary Amoebic Meningoencephalitis (PAM) [3].

Free living amoeba can live in soil and water such as lake water, swimming pool water or water for domestic. A study in Malaysia detected Acanthamoeba naegleria species from some swimming pool of four sites in Petaling Jaya and Kuala lumpur [5]. PAM cases caused by Neigleria fowlerii that contaminate domestic water used for bathing have been reported. Infection can occur when amoeba enters through the nose, attaches to the olfactory mucosa, and migrates through the cribriform plate along the olfactory nerve. When entering the olfactory bulb in the brain, this amoeba will divide rapidly and death occurs within 7-10 days. *N.fowlerii* may result in fulminant hemorrhagic brain necrosis [6]. Symptoms of PAM infection are characterized by severe and sudden headache, fever, nausea and vomiting and rhinitis. The symptomps will next followed by tense neck, diplopia, diminished olfactory sensations, confusion and convulsions, which develop rapidly into coma and result in death. Diagnosis with Head CT scan or MRI, indicating lesion but not specific. This condition is often indistinguishable from meningoencephalitis due to bacteria, fungi or viral[4]. Microscopic examination with direct preparations of CSF is the choice because the presence of moving amoebae can be observed with a light microscope. Amoeba can be misinterpreted with macrophages, but the *N. Fowlerii* nucleus contains a large nucleus located centrally that distinguishes it from host cells [6] . Acanthamoeba is the most commonly isolated amoeba from the environment. Amoeba is often present in domestic water reservoirs, hospital water, air and soil. It has two morphological forms, trophozoites and cysts, in which the trophozoite form is an infective form. Some Acanthamoeba species cause GAE which is a progressive chronic infection and rarely in the central nervous system. Symptoms of GAE include headache, slight fever, seizures, hemiparese, personality disorders, nausea, strained neck, decreased consciousness and coma. The clinical symptoms of GAE are not specific so often previously diagnosed as bacterial leptominingitis, tuberculosis meningitis, viral ensevalitis, toxoplamosis, fungal infections, neurosistiserkosis or brain tumors. On brain imaging studies, hydrocephalus found with meningens thickening, multifocal lesions and edema [2,6] The definitive diagnosis of GAE is confirmed by the discovery of amoebae in tissue or amoebic isolation. Microscopic examination with a light microscope and electron can detect Acanthamoeba from brain fluids or brain tissue. The shape of trofozoid can be distinguished from macrophage cells in the presence of a nucleus structure with a large central nucleolus that forms a halo image. Biopsy examination of brain tissue by HE staining will show a picture of brain granuloma, necrosis with multinucleated cells, inflammatory infiltrate and amoeba [2,6]. *Acanthamoeba castellanii, A. polyphaga,* and *Balamuthia mandrillaris* are the species that cause brain infections. These amoeba can be found on the ground and have been reported to be found in water. Balauthia-induced encephalitis presents symptoms that are not specific, chronic in the form of headache, nausea, vomiting, fever, myalgia, seizures, weight loss, hemiparese, speech disorders and are usually associated with previous lesions of skin granuloma. Balamuthia is often identified as Acanthamoeba because both cause GAE. *B.madrillaris* is not visible in the fluid reservoir of the brain but can only be isolated from the biopsy of brain tissue [2,6].

This patient showed a course of sub-chronic disease. A contrast head imaging examination results in an impressive granuloma. With the discovery of a cyst formation, trophozoite and the presence of protozoan pseudopods in CSF, it may be suspected of an Acanthamoeba protozoal infection in the central nervous system. This patient had been given antifungal, antiviral, antituberculosis, and metronidazole but the patient's condition has not shown any improvement since hospitalization. Survival rate is determined by the speed of diagnosis and immediate management. Administration of Amphotericin B, Fluconazole IV followed by oral Rifampicin was reported to show improvement in protozoan meningoencephalitis cases [6] . In this patient, Amphotericin could not be administered because of inavailability of the drug. This condition could be contributed to the patient's worsening, since it had been proven that only amphotericin B has in vitro and in vivo amoebicidal effect [7]. On the 19th day of treatment, this patient had nosocomial pneumonia in which sputum culture was obtained Pseudomonas aeruginosa. *Pseudomonas aeruginosa* is one of the major causes of nosocomial infections and is currently feared for resistance to various antibiotics, including the carbapenem (MDR /XDR P. aeruginosa) group [6,8] . This patient suffered from a condition of acquired or secondary immunodeficiency resulting in infection of various microorganisms such as bacteria, fungi, viruses, parasites. Acquired immunodeficiency occurs through two major pathogens: biological complications of a disease (infection, malnutrition, cancer) or complications of therapy against other diseases (immunosuppressive therapy). Immunodeficiency occurs due to impairment of cellular mediated immunity (CMI), ie. reduced number of lymphocytes or decreased lymphocyte activation (CD4 and CD8 effector cells) that play a role in the adaptive immune response to intracellular microorganisms [9,10] .

4. Conclussion

A case of a 24-year-old woman with meningoencephalitis is probably caused by amoeba. Clinical symptoms are not specific early in the course of the disease, making the diagnosis difficult to enforce. In the advanced phase, after being given anti-amuba therapy, the patient did not show clinical improvement. Anti-amoebic therapies such as Amphotericin B and fluconazole were required. Amoebic meningoencephalitis is a rare case and the diagnosis is difficult to established. For that reason, consider the development of immunological examination methods to help detecting the presence of amoeba in brain fluids since amoebic morphological examination is highly subjective and needs special skills.

Acknowledgements

We thanks very much to the director of the Abdul wahab syahranie hospital and very greatfull to all patient's family.We hope, this case report could give new information about amoebic meningoencephalitis.

References

[1] Crumpacker CS, Zhang JL 2010 Cytomegalovirus, In: Mandell LG, Bennett JE, Dolin R (Eds). Mandell, Douglas, and Bennett's Principle and Practice of Infectious Diseases. 7th edition.

[2] Frotingham R, Harold G H, Kenneth W H, 1994 Extensive DNA eequence conservation throughout the *Mycobacterium tuberculosis* complex. *J of Clinical Microbiology* July vol **32** 7:1639-1643

[3] Chimelli L, 2011 A *morphological approach to the diagnosis of protozoal infections of the central nervous system* SAGE Hindawi Access to Research Pathology Research International Vol. **2011**, article ID 290853:1-15

[4] Trabelsi H, Dendana F, Sellami A, Sellami H, Cheikhrouhou F, Neji S, Makni F, Ayadi A 2012. Review. Pathogenic free living amoebae: Epidemiology and clinical review. Elsevier Masson. *Pathologie Biologie* 60:399-404

[5] Init I. Lau YL Arin Fadzlun, A, Foead.AI, Neilson RS, Nissapatorn, V 2010 Detection of free living amoebae, Acanthamoeba and Naegleria, in swimming pools, Malaysia. *Tropical Biomedicine* **27**(3): 566-577

[6] Vivesvara GS 2010 Free living Amebae as Opportunistic Agents of Human Diseases Ashdin *Publishing Journal of Neuroparasitology* **1**(2010) Article ID N100802 :1-13

[7] Castillo MM, Zuniga RC, Velazquez DC, Debnath A, Luna JS, Shibayama M 2016 Naegleria fowleri after 50 years: is it a neglected pathogen? *Journal of medical microbiology* **65**: 885-896.

[8] Bruno A da Rocha Herbert BT, 2009 *Diagnotic of Infectious by Pathogecnic Free Living Amoeba Hindawi Publishing Corporation Interdiciplinary Perpective Infectious Disease*, Vol **2009** id 250416:1-14

[9] Pier G B, Ramphal L, 2010 *Pseudomonas aeruginosa* In Mandell L G, Bennett J E, Dolin R, (Eds) Mandell D, Bennett's, *Principle and Practice of Infectious Diseases 7th edition Churchill Livingstone Philadephia* pp 2835-2860

[10] Abbas A K, Lichtman A H, Pillai S, 2007 *Cellular and Molecular Immunology 6th edition Saunder Elsevier Philadelphia* pp 463-488

Inhibition of fish bacteria pathogen in tilapia using a concoction three of Borneo plant extracts

EH Hardi[1], G Saptiani[1], IW Kusuma[2], W Suwinarti[2], A Sudaryono[3]

[1] Department of Aquaculture, Faculty of Fisheries and Marine Science, Mulawarman University. Jl. GunungTabur, GunungKelua, Samarinda Ulu, Samarinda-75123, East Kalimantan, Indonesia. Tel./Fax.: +62-541-749159.
[2] Faculty of Forestry, Mulawarman University, Samarinda-75123, East Kalimantan, Indonesia.
[3] Department of Aquaculture, Faculty of Fisheries and Marine Science, Diponegoro University

*Corresponding author: estieriyadi2011@gmail.com

Abstract. This study was conducted to evaluate the antibacterial activity of concoction *Solanum ferox*, *Boesenbergia pandurata* and *Zingimber zerumbet* extract (SF, BP, and ZZ) to inhibit pathogenic bacteria in tilapia with the each concentrations 600 ppm BP, 900 ppm SF and 200 ppm ZZ. Antibacterial activity was measured by testing the concoction of three plants extract against single isolate *Aeromonas hydrophila* and *Pseudomonas* sp. and combined both bacteria (10^5 colony-forming units per milliliter). In this research, oxytetracycline was used as a control. Clear zone inhibition was observed at 6, 12, 18 and 24 hours after incubation at 30 °C. The results showed that the different concoction of BP: SF: ZZ have inhibitory zones against both single and joint isolate bacteria. The ratio of3:3:4 and 1:8:1 had higher antibacterial activity towards *Pseudomonas* sp. and 1:1:3 ratios both inhibit joint bacteria. The ZI% higher of concoction extracts against *A.hydrophila* is 1:1:8; 1:3:1; 3:4:3. The ZI% concoction extracts against *Pseudomonas* sp. ware 3:3:4 and 1:8:1 ratio. While the two bacteria combined, just 1:1:3 ratio had higher Z%. The conclusion is that a concoction of SF:BP:ZZ is effective to inhibit the growth of *A. hydrophila* and *Pseudomonas* sp., even its antibacterial ability is similar to the effectiveness of antibiotic oxytetracycline.

1. Introduction

Aeromonas hydrophila and *Pseudomonas* sp. are two common gram-negative bacteria found to infect freshwater fish [1,2,3,4,5]. Fish infected with *A. hydrophila* bacteria will appear to have ulcers in the external organs while the infected fish *Pseudomonas* sp. will experience a watery internal organ and a ruptured gall bladder [6, 7]. It reinforced by the statement of [8], *A. hydrophila* is a primary pathogen, it produces an ulcerative disease under stress conditions or an association with other pathogens, such as *Aphanomyces* spp. [9].

Research on the use of plant extracts to control the disease has increased showed by many articles that explain the benefits of plant extracts as a natural antibacterial on fish farming. The sole extract from *B. pandurata*, *S. ferox*, and *Z. zerumbet* which contain levamisole, flavonoid, steroid, carbohydrate, can inhibit the growth of pathogenic bacteria such as *Aeromonas hydrophila* and *Pseudomonas* sp. that infecting tilapia both in vitro and in vivo [10, 11, 12, 13], plant extract from

Ocimum sanctum [14, 15], *Azadirachtin* [16], *Viscum albums*, *Urticadioica* and *Zingiberofficinale* [17], *Radix astragalin* seu Hedysari and *Radix angelicaesinensis* [18, 19], *Astragalus radix* and *Scutellariae radix* [20]. Seven Chinese plant extracts of *Aloe vera* Mill. (Aloaceae), *Angelica s*pecies (Umbelliferae), *Astragalus membranaceous* (Leguminosae), *Ganoderma lucidum* (Fr.) Karst. (Ganodermataceae), *Panax ginseng* C.A Mey. (Araliaceae), *Scutellaria species* (Lamiaceae) and *Zingiber officinale*Rosc. (Zingiberaceae) has an antimicrobial ability [21].

The combined use of multiple extracts for disease control has a higher antibacterial ability compared to the use of a single extract. *Curcuma longa*, *Ocimum sanctum*, and *Azadirachta indica* extract combined with a 1:1:1 had higher antibacterial activity ratio in Goldfish *Carassius auratus* were challenged with *A. hydrophila* infection [24]. Further trials on the treatment using a combination of the three extracts can improve survival and assist the healing process of wounds caused by bacterial infection of *A. hydrophila* in carp (*Carassius auratus*) [22]. In this paper, we will discuss the antibacterial activity of the combined *S. ferox, B. pandurata* and *Z. zerumbet* extracts against single bacteria and a combination of *A. hydrophila* and *Pseudomonas* sp.

2. Method

This research was conducted from January to March 2017 at Aquatic Microbiology Laboratory, Faculty of Fisheries and Marine Sciences and Wood Chemical Laboratory, Faculty of Forestry Mulawarman University, East Kalimantan.

2.1. Bacteria test

The pathogen bacteria used in this study were *A. hydrophila* (EA-01) and *Pseudomonas* sp. (EP-01) from the Laboratory of Aquatic Microbiology Faculty of Fisheries and Marine Sciences Mulawarman University. The bacteria was grown in BHI (Brain Heart Infusion Broth, DIFCO®) and BHIA (Brain Heart Infusion Agar, DIFCO®) solid medium for 24 hours at 30 °C, the density used was each of the bacteria was 10^5 CFU mL-1. The test bacteria used in the antibacterial test was single bacteria *A. hydropila*, *Pseudomonas* sp. and the combined both bacteria with a ratio 1:1.

2.2. Preparation of B. pandurata, Z. zerumbet and S. ferox Extract

Spices obtained from traditional markets in Samarinda, East Kalimantan Indonesia and the extraction process using ethanol solution by following the basic procedures [23, 10]. The extraction process was carried out at the Wood Chemical Laboratory at the Faculty of Forestry, Faculty of Fisheries, and Marine Sciences at the University of Mulawarman. The air-dried plant samples were mashed using a blender then dried at room temperature for 48 hours. A total of 100 g of dried samples were soaked in 100 mL of ethanol in an Erlenmeyer at room temperature. The extract solution was filtered with sterileWhatman no. 42 filtration paper and filtered sample was centrifuged for 24 h at 50 rpm to obtain a crude extract.

The extract concoction was prepared by mixing 0.6 g *B. pandurata*, 0.9 g *S. ferox* and 0.2 g *Z. zerumbet* ware weighed, soaked in 1 L of sterile distilled water [23, 25]. In this study will use a combination of *S. ferox* extract, *B. pandurata* and *Z. zerumbet* (SF: BP: ZZ) with a ratio of 1: 1: 8; 8: 1: 1; 1: 8: 1; 1: 1: 3; 3: 1: 1; 1: 3: 1; 3: 3: 4; 4: 3: 3; 3: 4: 3; 2: 2: 1; 1: 2: 2; 2: 1: 2.

Table 1. Combined extract of *S. ferox*, *B. pandurata* and *Z. zerumbet* used in the study

Extract concoction	*S. ferox* (mL)	*B. pandurata* (mL)	*Z. zerumbet* (mL)
1:1:8	10	10	80
1:1:3	20	20	60
3:3:4	30	30	40
2:2:1	40	40	20
8:1:1	80	10	10

Extract concoction	*S. ferox* (mL)	*B. pandurata* (mL)	*Z. zerumbet* (mL)
3:1:1	60	20	20
4:3:3	40	30	30
1:2:2	20	40	40
1:8:1	10	80	10
1:3:1	20	60	20
3:4:3	30	40	30
2:1:2	40	20	40

2.3. *Antimicrobial activity of concoction plant extracts*
The In vitro antibacterial test was performed by following diameter clear zone method [26]. It began with growing each test bacteria at BHIA at room temperature of 0.5 mL for 30 min. It was followed by dripping 25 μl of each extract on a 6 mm sterile antibacterial paper, and placed on the inoculated BHIA media with each of the test bacteria and incubated for 24 hours at 30 °C [27]. Observation of the inhibitory zone diameter was performed at 6, 12, 18 and 24 incubation periods. Sterile aquadest (distilled water) is used as a negative control and commercial antibiotic Oxytetracycline (OT) is used as a positive control.

The measured result was a clear zone or zona inhibition (ZI); the ZI% of the extract used was compared with the antibiotic OT as the standard. The percentage of ZI was calculated using the [28].

$$ZI\% = 100\%X\left(\frac{extract\ ZI - \ ZI\ water}{antibiotic\ ZI - ZI\ water}\right) \quad (1)$$

2.4. *Minimum inhibitory concentration (MIC)*
The minimum inhibitory concentration (MIC) of the extracts against single and combined bacteria was carried out using TSA media following the Mandal et al (2002)method. The test was performed by adding each of the combined extracts of 0.5 mL or 0.5 mL of oxytetracycline antibiotics into TSB medium containing bacterial cultures each (10^3 CFU / mL). The cultures were then incubated for 24 h at 30 °C [30, 31, 25].

Furthermore, 1 mL of the suspense was diluted with 9 ml sterile aquadest and the dilution was inoculated into TSA medium and culture for 24 hours at 30 °C, then the total number of growing bacteria was calculated based on [30, 31, 29, 10]. The percentage of MIC is calculated using the following formula.

$$MIC\% = 100\%X\left(\frac{extract\ CFU - aquadest\ CFU}{antibiotic\ CFU - aquadest\ CFU}\right) \quad (2)$$

3. Result and Discussion

3.1. *Inhibition zone*
Figure 1 shows the combined effectiveness of the three extracts of *A. hydrophila* bacteria. The inhibition zone formed between 8 - 20.33 mm, in general the 6th hour after the incubation period of the inhibition zone begins to form and continues to increase until 24 hours. There are 7 compounds extracts that have a sensitive inhibitory zone (> 14 mm) against *A. hydrophila* bacteria, which is a combination of extract with a ratio of 1: 1: 8; 2: 2: 1; 3: 1: 1; 4: 3: 3; 1: 2: 2; 1: 8: 1; and 1: 3: 1. Combined extract with a ratio of 1: 2: 2 is the best ratio with 20.33 mm inhibition zone at 24 hours incubation compared to other ratios.

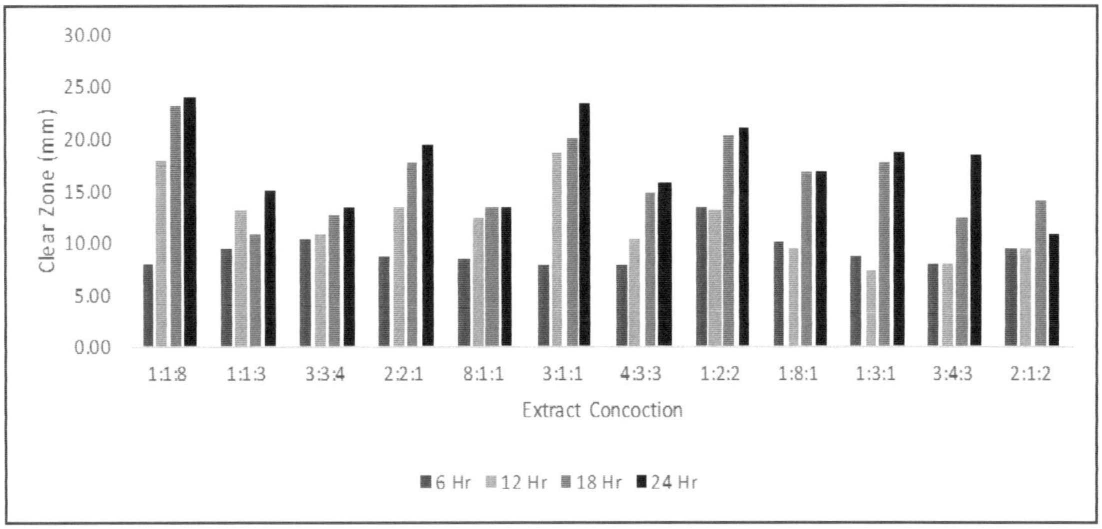

Figure 1. Antibacterial activity of concoction of *S. ferox*, *B. pandurata* and *Z. zerumbet* against *Aeromonas hydrophila*

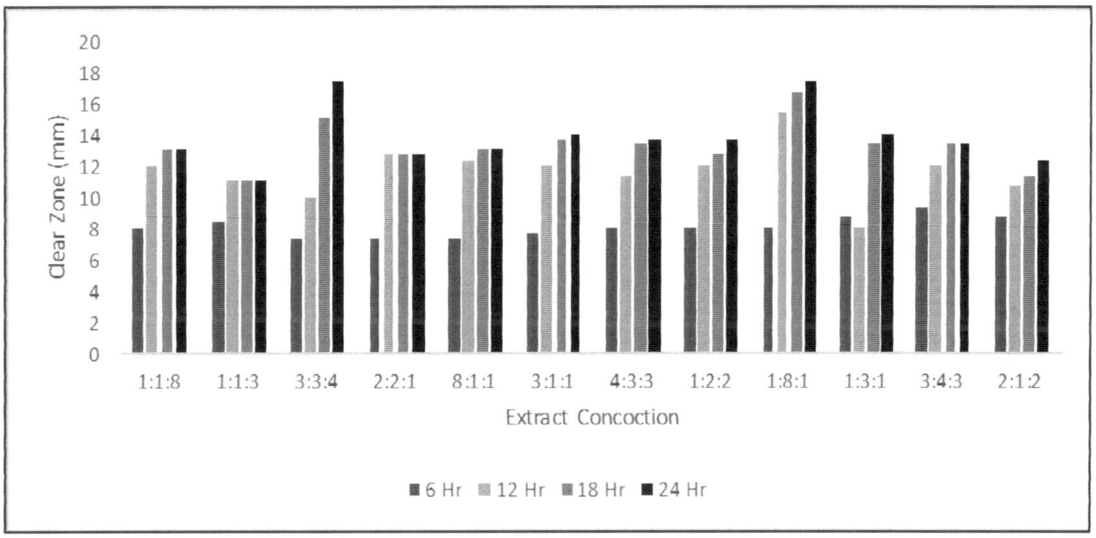

Figure 2. Antibacterial activity of *S. ferox, B. pandurata* and *Z. zerumbet* extracts against *Pseudomonas* sp.

Pseudomonas sp. is a pathogenic bacterium that is also inhibitory assay using a combination of all three extracts. Figure 2 shows the inhibitory zone from the overall combination of extracts to the bacteria *Pseudomonas* sp. The clearly zone is generally smaller than the zone of inhibition against *A. hydrophila* bacteria that is ranged from 8 - 17.33 mm. There are only 3 combined extracts that have a sensitive inhibitory zone (> 14 mm) against *Pseudomonas* sp. namely the combination of extract with a ratio of 3: 3: 4; 3: 1: 1; and 1: 8: 1. Combined extract with 3: 3: 4 and 1: 8: 1 ratio is the best ratio with 17.33 mm inhibition zone at 24 h incubation time compared to other ratio

Both bacteria *A. hydrophila* and bacteria *Pseudomonas* sp. always found to infect tilapia simultaneously and the pathogenicity rate of bacterial compounds is generally higher than that of single infections [35]. The trials of the combined trials of the extracts were carried out on a combination of both in vitro bacteria. The results show the average inhibition zone formed from the overall combination of extracts to the two bacteria can be seen in Figure 3. What formed is ranged from 8 - 18.67 mm. There are 8 combinations of extracts that have a sensitive inhibitory zone (> 14 mm) against both bacteria (*A. hydrophila* and *Pseudomonas* sp.), Which are combined with a ratio of 1: 1: 3; 3: 3: 4; 2: 2: 1; 8: 1: 1; 4: 3: 3; 1: 3: 1; 2: 1: 1; and 3: 4: 3. Combined extract with a ratio of 1: 1: 3 is the best ratio with 20.67 mm inhibition zone at 24 h incubation compared with other ratios.

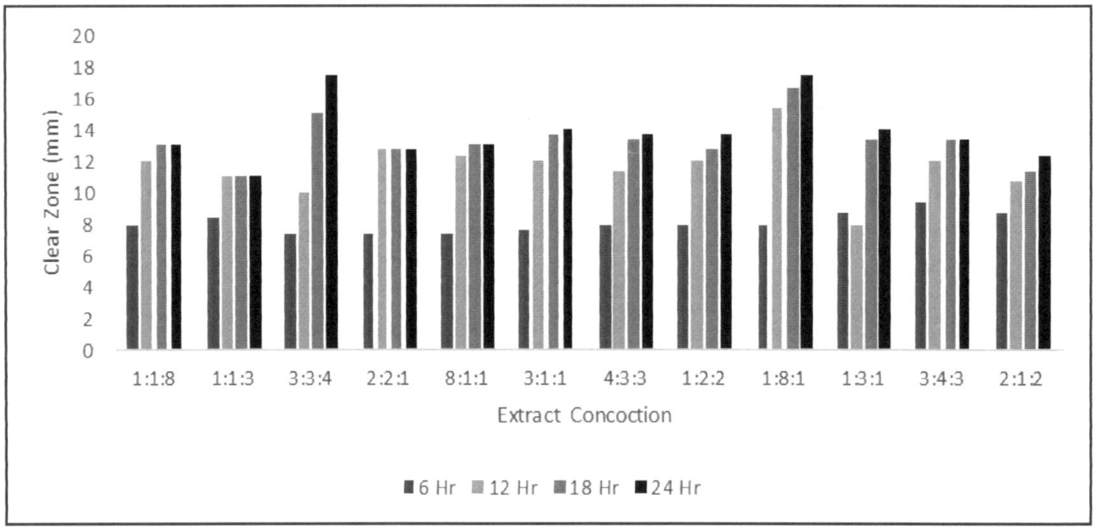

Figure 3. Antibacterial activity of concoction *S. ferox* extract, *B. pandurata* and *Z. zerumbet* against *Aeromonas hydrophila* and *Pseudomonas* sp. combination.

The ethanol extracts of the various comparison ratios have different inhibitory zones when compared with the antibiotic oxytetracycline, only a few ratios that have a ZI% value of percentage. Generally, a ZI% value is formed after 12 h of incubation time and begins to increase with increasing length of incubation. Zone inhibition (ZI%) against *A. hydrophila* bacteria ranged from 5.6 to 26.32%; 19.5% against *Pseudomonas* sp. and 5.7-42.5% against both bacteria. The overall average ZI% formed can be seen in Table 2.

Table 2. Percentage of Zone Inhibition (ZI%) concoction extract of *S. ferox*, *B. pandurata* and *Z. zerumbet* against *Aeromonas hydrophila*, *Pseudomonas* sp. and combination of both bacteria based on time (hours/h) of incubation

Concoction Extract (SF:BP: ZZ)	Bacterial test											
	A. hydrophila				*Pseudomonas* sp.				Combined Bacteria			
	6 hr	12 hr	18 hr	24 hr	6 hr	12 hr	18 hr	24 hr	6 hr	12 hr	18 hr	24 hr
1:1:8	0	1	21.05	26.32	0	0	0	0	0	0	0	0
1:1:3	0	1	21.05	26.32	0	0	0	0	0	0	26.4	42.5
3:3:4	0	1	21.05	26.32	0	0	3.45	19.5	0	0	0	5.7
2:2:1	0	0	0	0	0	0	0	0	0	0	0	28.7
8:1:1	0	0	0	0	0	0	0	0	0	17.2	28.7	0
3:1:1	0	0	0	0	0	0	0	0	0	0	0	0
4:3:3	0	0	0	0	0	0	0	0	0	0	26.4	35.6
1:2:2	0	0	0	0	0	0	0	0	0	0	0	0
1:8:1	0	0	0	0	0	5.7	14.9	19.5	0	0	0	0

1:3:1	0	0	1	5.26	0	0	0	0	0	0	0	0
3:4:3	0	0	0	1	0	0	0	0	0	0	5.7	12.6
2:1:2	0	0	0	1	0	0	0	0	0	0	0	26.4

The ratio of SF: BP: ZZ of 3: 3: 4 is a ratio that has a good ZI% against single bacteria *A. hydrophila, Pseudomonas* sp. as well as the combination of both bacteria, compared with other ratios. Successive amount of the ratio of the three combined extracts to *A. hydrophila* bacteria, *Pseudomonas* sp. and both bacteria are as follows 4, 2, 7 ratio. The largest ZI% score was 35.6% in a 4: 3: 3 ratio against both bacteria.The value of combined MICs SF: BP: The highest ZZ is a 1: 1: 8 ratio to *A. hydrophila*; 1: 8: 1 against *Pseudomonas* sp. and 1: 1: 3 against the combined bacteria.

Table 3. MIC% concoction of *S. ferox*, *B. pandurata* and *Z. zerumbet* against *A. hydrophila*, *Pseudomonas* sp. and a combination of both bacteria at the 24 h incubation period.

Concoction extract (SF:BP:ZZ)	Bacteria Test		
	A.hydrophila	*Pseudomonas* sp.	Combined Bacteria
1:1:8	94%	57%	56%
1:1:3	70%	45%	85%
3:3:4	63%	64%	67%
2:2:1	80%	52%	67%
8:1:1	63%	57%	45%
3:1:1	92%	59%	54%
4:3:3	70%	55%	67%
1:2:2	80%	55%	45%
1:8:1	77%	67%	45%
1:3:1	80%	60%	65%
3:4:3	79%	55%	64%
2:1:2	55%	45%	70%

The three-plant extracts namely *S. ferox*, *B. pandura* and *Z. zerumbet* have antibacterial ability against bacteria *A. hydrohila* and *Pseudomonas* sp. If the extract of *B. pandurata* and *Z. zerumbet* effectively inhibited the growth of *A. hydrophila* bacteria both in vitro and in vivo, *S. ferox* extract was more effective in inhibiting *Pseudomonas* sp. also in vitro [10, 11] and in vivo [12, 13]. The concoction of the three extracts effectively inhibited both bacteria in vitro, and the results showed higher compared with a single extract. The concoction of the extract to inhibit pathogenic bacteria proved to be better than that of a single extract such as the results of [25] study that combined the *O. support*, *T. avicennoides*, and balsam apple (*Momordica balsamina*) (290.6-750.0 lg / mL) against some bacteria: *A. hydrophila*, S. *dysenteriae*, *Shigella flexneri, Shigella sonnei, Shigella bodyii,* and *E. coli.* Similarly, the results of the [24] study combined three extract *Curcuma longa, Ocimum sanctum*, and *Azadirachta indica* with a 1: 1: 1 ratio effectively suppressing *A. hydrophila* bacteria compared to a single extract.

Phytochemical results show that the extract of *S. ferox* contains alkaloids and carbohydrates; *B. pandurata* contains alkaloids, flavonoids, and carbohydrates while *Z. zerumbet* contains alkaloid flavonoids, steroids and carbohydrates [10]. Crude extracts are potential to inhibit bacterial growth than fractionation of components that would normally affect the loss of antibacterial cytokines [34], so the use of crude extracts for field applications would be more effective.

Differences of the material contained resulted in differences in antibacterial activity against bacteria, when the three extracts were combined, showed an increase in antibacterial ability due to synergism between the three extracts. The components contained in the extracts increase each other's ingredients. However, further research is needed for in vivo testing whether the incorporation of the three extracts can be useful in enhancing fish immunity or not. The control of *A. hydrophila* bacteria, *Pseudomonas* sp. both single and combined in fish in vivo can be performed.

4. Conclusion

In this study showed that the combined extract of *S. ferox, B. pandurata*, and *Z. zerumbet* with the ratio 1: 1: 8; 1: 8: 1; and 1: 1: 3has the potential to inhibit the growth of *A. hydrophila* bacteria, *Pseudomonas* sp. or both.

Acknowledgments

Authors are grateful to the RISTEK DIKTI with the project in 2016 and 2017, Mulawarman University, which made this work possible. The authors are grateful to the Department of Fisheries and Marine Science for the facilities made available.

References

[1] Newman S G 1983 *Aeromonas hydrophila*: a review with emphasis on its role in fish disease. *Collection Foundation Marcel Merieux* (France: Lyon).
[2] Cipriano R C, Bullock G L, Pyle S W 1984 *Fish Disease Leaflet* **68**:87-97.
[3] Choudhury D, Pal A K, Sahu N P, Kumar S, Das S S, Mukherjee S C 2005 *Fish and Shellfish Immunology* **19**:281–291.
[4] Kumari J, Sahoo P K 2005 *Journal of Fish Diseases* **29**:95–101.
[5] Yavuzcan H 1998 *The Israeli journal of aquaculture* **50**(2):82-85.
[6] Hardi E H, Pebrianto C A, Hidayanti T, Handayani R T 2014 *Jurnal Kedokteran Hewan* **8**(2):130- 134. [Indonesia]
[7] Hardi EH, Pebrianto C A 2012 *Jurnal Ilmu Perikanan Tropis* **16**(2):35-39. [Indonesia]
[8] Ansary A, Haneef R M, Torres J L, Yadav M 1992 *Journal of Fish Diseases* **15**:191–196.
[9] Lilley J H, Hart D, Richards R H, Roberts R J, Cerenius L, Soderhall K 1997 *Veterinarian Record* **140**:653–654.
[10] Hardi E H, Kusuma I W, Suwinarti W, Agustina, Abbas I, Nugroho R A 2016 *AACL Bioflux* **9**(3):638-646.
[11] Hardi E H, Kusuma I W,Suwinarti W, Agustina, Nugroho RA 2016 *Nusantara Bioscience* **8**(1):18- 21.
[12] Hardi E H, Saptiani G, Kusuma I W,Suwinarti W, Nugroho R A 2017*AACL Bioflux* **10**(2):182:190.
[13] Hardi E H, Kusuma I W, Suwinarti W, Saptiani G, Sumoharjo, Lusiastuti A M 2017*Nusantara Bioscience* **9**(2):220- 228.
[14] Logambal SM, Venkatalakshmi S, Michael R D 2000 *Hydrobiologia* **430**:113–20.
[15] Venkatalakshmi S, Michael R D 2001 *Journal Aquatic Tropics* **16**:1–10.
[16] Logambal SM, Michael R D 2001 *Journal Aquatice Tropics* **16**:339–47.
[17] Dugenci S K, Arda N, Candan A 2003*Journal Ethnopharmacol***88**:99-106
[18] Jian J, Wu Z 2003 *Aquaculture* **218**:1–9.
[19] Jian J, Wu Z 2004 *Fish Shellfish Immunol* **16**:185–91.
[20] Yin G, Jeney G, Racz T, Xu P, Jun X, Jeney Z 2006 *Aquaculture* **253**:39–47.
[21] Tan B K, Vanitha J 2004 *Current Medicinal Chemistry* **11**:1423-1430.
[22] Harikrishnan R, Balasundaram C, Heo M S 2010 *Fish and Shellfish Immunology* **28** :354-361.
[23] Limsuwan S, Voravuthikunchai S P 2008 *FEMS Immunology Med Microbiol* **53**: 429–436.
[24] Harikrishnan R, Balasundaram C 2008 *Journal of Aquatic Animal Health* **20**:165–176.
[25] Iwalokun BA, Gbenle GO, Adewole TA, Akinsinde K A 2001 *Journal of Health, Population,*

and Nutrition **19**:331–335.
[26] Dülger B, Gonuz A 2004 *Pharma Biology* **42**: 301-304.
[27] Bradshaw L J 1992 Laboratory of microbiology, 4th edition. Saunders College, Philadelphia.
[28] Singh R, Chandra R, Bose M,Luthra P M 2002 *Current Science* **83**:737–740.
[29] Mandal S, Mandal M, Pal NK, Halder P K 2002 *Indian Journal of Experimental Biology* **40**:614–616.
[30] Ilori MO, Sheteolu AO, Omonigbehin EA, Adeneye A A 1996 *Journal of Diarrhoeal Diseases Research* **14**:283–285.
[31] Nagamura CV, Nakamura TU, Bando E, Fernandes A, Melo N, Cortez DAG, Filho B P 1999 *Memorias do Instituto Oswaldo Cruz* **94**:675– 678.
[32] Mandal S, Mandal M, Pal NK, Halder PK 2002 *Indian Journal of Experimental Biology* **40**:614–616.
[33] Harikrishnan R, Balasundaram C 2008 *Journal of Aquatic Animal Health* **20**:165–176.
[34] Babu B, Jisha VK, Salitha CV, Mohan S, Valsa A K 2002 *Indian Journal of Microbiology* **42**:361–363.
[35] Rauhun 2014 *SKRIPSI* Fakultas Perikanan dan Ilmu Kelautan Universitas Mulawarman Samarinda [Indonesia].

Prospect development of local beef cattle from South Kalimantan as supporting to food sovereignty in Indonesia

S N Rahmatullah[1*], A Sulaiman[2], Askalani[3] and N M K Azizi[4]

[1]Mulawarman University, Kampus Unmul Gunung Kelua, Pasir Balengkong Street, Samarinda, East Kalimantan, 75119
[2]Lambung Mangkurat University, A. Yani Street No. Km 36, Banjarbaru, South Kalimantan
[3]Agency of Integrated Agricultural Research and Development, South Kalimantan
[4]Center of Research and Development Local Government, South Kalimantan

E-mail :suryanr@faperta.unmul.ac.id;surya_pato@yahoo.co.id

Abstract. Development production of Indonesia local cattle didn't balance increasing of consumption livestock in Indonesian so that's makes imports are still high both local beef production. Indonesian local cattle is one of cattle the largest contributed the national meat production of ruminant particular group of cattle so that the livestock has the potential to be developed as a profitable business if doing of farmer. One of advantages is potential areas in South Kalimantan mainly swamp lands was very large. It's one effort in national promoting food sovereignty programme and the provinces especially about introduction of Indonesian local cattle in the swamp area are also important. Availability of region in South Kalimantan spacious and great forage production, then prospects for cattle productivity there is huge so that's can even increase income for farmers. The programme more serious support from the governance of South Kalimantan province to develop Indonesian local cattle in swamp area and in that's a real policy for encouraging farmers to maintain Indonesian local cattle to make sustainable food in Indonesia.

1. Introduction

Efforts to realize national food sovereignty especially animal protein products of livestock are continuously improved by implementing programs to reduce national dependence on imported meat by seeking productivity and livestock population increase especially in Indonesia local cattle that had not been optimally developed in the region[1]. Increased productivity and population of Indonesian local cattle has been intensified over the past few years with various programmes such as Estrus Synchronization and Artificial Insemination (2013-2016) and Mandatory Cow Pregnant Breeding Cattle (SIWAB) implemented since 2017 until now. These programs aim to increase the population utilizing by Indonesian local cattle.

Increasing the potential for Indonesia local cattle is helping to meet the demand for beef in the domestic market is quite high and always increases from year after year. Indonesian fresh meat per capita consumption of 2011-2015 reached grow than average of 2.08 kgs per capita with 10.28%. The level of consumption of 2017 only increased slightly 0.6 percent, compared with 2016. But when compared with some neighboring countries, Indonesia is still very far behind example Singapore have per capita consumption of meat is 55 kgs per year while Philippine reaches 7 kgs per year[2].The development population of Indonesian local cattle such as Kalimantan especially in South Kalimantan areas is very large. A few factors that support the development of Indonesian local cattle in South Kalimantan include potential nature and human resources the establishment of cattle production systems based on the potential for regional areas, and there is a technology, economic, social and partnership developed in collaboration between Government, Academician, Farmer and Businessmen sector.

The origin Indonesian cattle is a native Indonesian cattle that has always originated from the domestication of wild animals in Indonesia, not from imports cattle. One example of the original Indonesia cattle among others Bali cattle, Madura cattle, Pesisir cattle, and Katingancattle. While the sense of Indonesia local cattle is a cattle that has its own distinct characteristics with the same cattle from imported cattle due to the breeding process and selection conducted by Indonesian farmers.Indonesian local cattle have characteristics small cattle with a size that is almost the same weight as some other small cattle in Africa and India. These data also show that body weight variation on various age levels in Bali cattle is considerable, so the chance of development through selection will still be effective. The results of Talib *et al.*[3] shows that one of Indonesia local cattle such asOngole cross-breed cattle has an average body weight of 250-300 kg with the age of 1.5 - 2 years that's same result research with Hastuti[4]. Male carcass cattle weights produced 50.46% [5]while the productivity in cattle associated distance between from 15.74 months and service per conception 1.67 [6]. Indonesian local cattle had some advantages such as having good adaptation level in tropical, good palatability to low-quality feed, relatively resistant to internal and external parasite, good productivity and carcasses ranging from 49 to 55% [7,8,9,10,11]. This showed the superior performance of Indonesian local cattle but the most important thing is that this cattle is the Indonesiagermplasm that became the treasure of the Indonesia[12].

South Kalimantan with its capital city of Banjarmasin one of province in Kalimantan island have a slope of soil with 4 classification class shows that as much as 43.05% at South Kalimantan Province has a slope of 0-2% land. The of South Kalimantan area according to the height class is divided into 6 classes of elevation indicates the area of South Kalimantan mostly located in the class height of 25-100 m above of 31.29% from sea surface. Land in the area of South Kalimantan Province is mostly forest with details of dense forest (780,319 Ha), forest of grove (377,774 Ha) , and swamp forest (90,060 Ha), forest type (352,840 Ha) of soil covering 870,314 Ha, In the form of grass (50,119 Ha) and for others (83.014). While the use for rice fields 413,107 Ha, plantations 437,037 ha and for the township of 57,903 Ha and for Tegalan (48,612 Ha). Government data inSouth Kalimantan province showed swampland area reached 235,677 hectares and that extent swampland that has been planted with food crops reaches 78,544 hectares and other potential land that can be utilized for food in swamps reach 90,000-hectare Lebak (Livestock office Service of South Kalimantan[13].

This article aims to examine the potentials and strategies of developing Indonesian local beef cattle in South Kalimantan that are related to the development of meat productivity as an animal protein food in Indonesia. The information presented is expected to be an input in formulating models of development and institutional of beef cattle business that efficiently and effectively to realize Indonesian food sovereignty especially meat production.

2. Materials and methods

In this research using literature study method based on data onto potential of South Kalimantan beef cattle period 2008-2016 from Livestock Service of South Kalimantan Province and with direct observation and discussion with breeder and livestock policy at Livestock Service Office of South Kalimantan Province. Activities lasted for 6 (six) months from March of August 2017. The duration of this activity is related to data collection and discussion activities with the respondents. The data sources of this activity consist of Government Agency's Performance Accountability Report. Livestock Service Office of South Kalimantan Province 2008-2016, Livestock Statistics Data of South Kalimantan Year 2008-2016, Location Location Quotient (LQ) Cattle Year 2008-2016, Government

Policy South Kalimantan Province, questionnaire and discussion with respondents (farmer, academician, researcher and businessmen).

3. Results and discussion

3.1. Potency of cattle in South Kalimantan

South Kalimantan as one of the areas in tropical climate has variation in potential and condition of region that can rely on agriculture sector as supporting of regional economy because contribution to agriculture sector to regional economy is one of the biggest for regional income. The contribution to the agriculture sector (food crops, horticulture, livestock, and plantation) to the total area of South Kalimantan is around 22.77%, of which 2-3% comes from the livestock sub-sector. The achievement of livestock population of South Kalimantan over the last five years especially in view of the development from 2013 to 2015 has experienced significant growth.

South Kalimantan Province continues to play an active role to encourage and develop the livestock sector in this area. Various policies are done maximally to create a conducive business for both breeders, as well as for investors who are willing to invest their capital in South Kalimantan. Some of the programme implemented include the intensification of natural mating with provision of bulls and enhancement of artificial insemination, agribusiness development, strengthening the institutions of artificial insemination center, developing cattle in plantation and agriculture for the provision of organic fertilizer and bioenergy, development of cattle breeding, cultivation development cattle and Development of feed and territorial. The pattern of development of beef cattle farming is in according with the opinion of Dwiyanto [14]; Talib and Noor[15]; Mayulu *et al.* [16]; Priyanto [17]; Matondang and Rusdiana [18] stated that efforts to increase the productivity of beef cattle with several programme such as the utilization of local resources, intensive breeding, with a business development base focused on the downstream industry, the utilization of integrated feeding potential for food crops, plantations and forestry and has led to semi commercialventures as well as the addition of technological innovations towards the commercialization of livestock product development.

3.2. Development Beef Cattle in South Kalimantan

The development of Indonesian local beef cattle in South Kalimantan has considerable potential. This is greatly supported by the existence of geographical conditions for which there are extensive forage for pastures, the potential for swamp areas and integration between plantations, agriculture, and cattle. Based on the potential for human resources, natural resources and the level of its utilization and attention to livestock commodities and marketing situation, in general, business prospects and opportunities for the development of livestock sector in South Kalimantan are very likely to be developed especially beef cattle that came from Indonesia local cattle. Some of Indonesia local cattle well-developed in South Kalimantan included Bali cattle, Ongole cross-breed cattle (PO), Sumba Ongole cross-breed (SO), Madura cattle and Simmental Bali cross-breed (Simba)

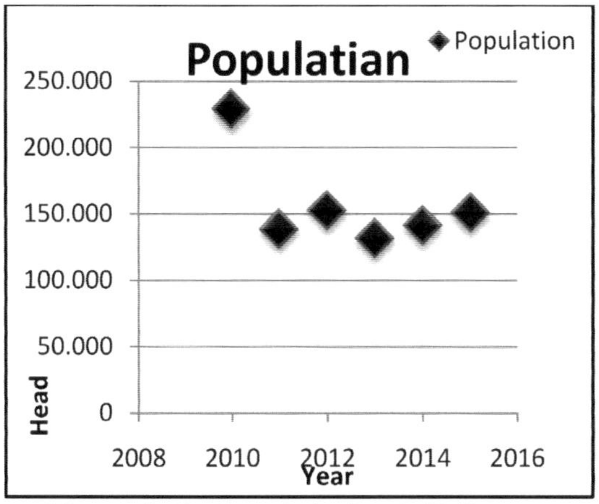

Figure 1. Beef cattle in South Kalimantan (Livestock Office Service of South Kalimantan, 2015).

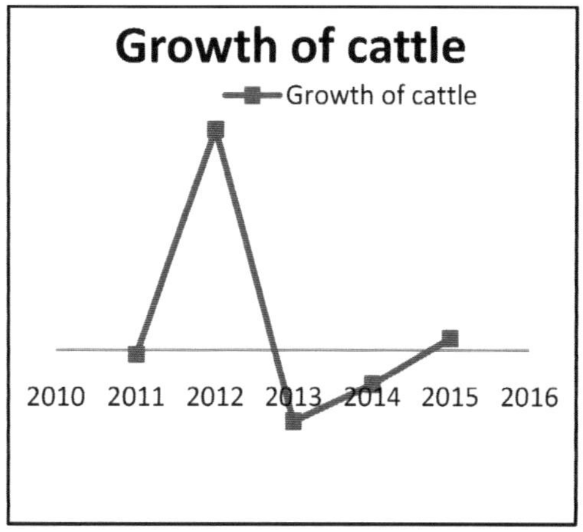

Figure 2. Graphic of growth addition beef cattle population in South Kalimantan.

Carrying capacity of potential land for cattle especially Bali cattle can still be developed up to 675,000 Livestock Units or equivalent to 877,500 Bali cattle while the population of beef cattle in South Kalimantan are 150,875 or 17.19%, so it's still open the potential to increase the number of beef cattle population in South Kalimantan. Based on this capacity, South Kalimantan are still open to increasing the population of 726,625 heads [19].

Land and climate suitability in South Kalimantan area are suitable to be developed Sundaicus Bos breed cattle, Bos in discus and Cross-breed, not covered by buffalo, goat and livestock engages and dairy cattle. According to Mukson *et al.*,[20]opportunities for cattle development in the outer islands of Java are very large, one of them in Kalimantan, it is related to land conditions and topography and the availability of forage for abundant cattle make the potential as a center cattle in Indonesia

The development of cattle in South Kalimantan is aimed at increasing the local cattle population of Indonesia by utilizing the natural potential that is owned by considering environmental sustainability. The potential for the land is sufficient to accommodate 877,500 heads of cattle for cattle fed and the utilization of local feed waste from agricultural waste (corn and rice straw), plantation (palm oil) was expected to contribute in the provision of feed for 300,000 head of cattle per day. South Kalimantan seeks to develop the development of beef cattle breeding based on the development of areas such as cattle fattening development areas, the development of integrated areas of cattle with plantations and the development of breeding or breeding areas. According to Saputra *et al.*[21], efforts to develop beef cattle based on the development of the area are able to support the increase of beef cattle productivity in Aceh Province. Until now, based on data into from governance, development of local beef cattle in South Kalimantan makes this area as one supplier of beef cattle and meat to neighboring areas such as Central Kalimantan and East Kalimantanprovince[22].

3.3. Productivity of Beef Cattle in Swamp Land South Kalimantan

The productivity of Indonesia local n beef cattle in South Kalimantan swamp land is relatively positive about population growth until 2015, 150,875 heads in South Kalimantan. In addition reports on farmers as well as direct observation in the field, the weight of bull of Bali cattle kept in swamps at 4 months, can reach 260 kgs (at the beginning of maintenance around 190 kg, so ADG (average daily gain), ranges from 0.58 kg/day). This does not close the possibility could be more because if this is done with good management then the weight of the cattle can reach 280 kgs or more.

The reproductive factor, the local cow of Indonesian cattle has no difficulty in running the breeding program because, besides the existence of natural mating, AI (artificial insemination) has also developed in South Kalimantan. However, the attention on ranchers in the swamp area of South Kalimantan is the lack of supervision of recording. So, we often encounter breeders that use the estimation system in the form of mating or weight of cattle to be sold. According to Anggraeny *et al.* [23], the performance of Indonesian local cattle reproduction in South Kalimantan is relatively better especially Calving Interval (CI) compared to CI value from Brahmancross breeder in East Java. In terms of the productivity of Indonesian local beef cattle development there is no difference between the maintenance in the swamp and dry land, the problem is the system of sales, because it is related to the watery swamp conditions, the existing transportation like a boat is very limited, so sometimes to overcome it, farmers will jointly sell cattleor sell in the form of carcasses because they themselves do the slaughtering.

The Government of South Kalimantan Province has also undertaken several programs as part of the development of Indonesian local cattle in swamp areas with the aim of making South Kalimantan as a supplier of cattle seed for other areas in South Kalimantanself-area and Central Kalimantan to East Kalimantan. Model of breeding and fattening beef cattle has grown in South Kalimantan with the growth centers of livestock business such as Bali development reproduction center of Barito Kuala regency, cattle breeding center covers Tanah Laut and Barito Kuala regency, while the center of beef cattle development includes Banjar, Tanah Laut, Balangan, Tapin and Hulu Sungai TengahRegency[24].

3.4. Management Pattern of Beef Cattle in Swamp Land

Management of local beef cattle raising by farmers in South Kalimantan based on observations made consists of 3 systems namely extensive, semi-intensive and intensive. Differences between this system based on the location of the breeder are located. The following will showed some maintenance management patterns that farmers can do in the swamp area.

3.4.1. Extensive system. This extensive system is a local beef cattle raising system that a small proportion of South Kalimantan people still applied. This maintenance method is one of the many maintenance system applied for the people farmers in Indonesia[25]. Usually, farmers that used this system are a breeder who is in the area of pasture grazing area such as in the Tanah Laut, Tanah Bumbu and in the Hulu Sungai Tengah regency. In this system, the cattle are allowed to feed themselves but at night the cattle is stacked (but very rarely the ranchers do the warehouse) and in the morning the cattle is issued and left in the pasture. The application of this extensive system is widely used in the swamp area in Hulu Sungai Tengah and Hulu Sungai Utara regency. Farmer systems in the swamp area usually using this system during the dry season.

3.4.2. Semi-intensive system. This semi-intensive system is used by some cattle ranchers in South Kalimantan. This pattern of maintenance example in the morning, the cattle is brought to a place that has a pasture, while in the afternoon, the cattle is brought back to the cage [26]; Susanti *et al.* [27]. In swampy areas, this pattern is applied by farmers from Barito Kuala regency as most of these areas have swamps pasture. The potential for swamp land with forage conditions derived from swamp plants can still be converted by Bali cattle, Ongole cross-breed cattle and buffalo for feed[28].

3.4.3. Intensive system. This intensive system is mostly applied by farmers in South Kalimantan, both on dry land and in wet areas (however in swamp areas only a small percentage example in unlogged swamp areas). Farmer are looking for breeding with forage for cattle while cattle are left in cages. The privileges of intensive systems farmer can focus on caring for cattle as well as health and safety awake [29]. However, not all farmers of this system are using good management such as cage management, recording until breeding and mating arrangement. The farmer of this system is a breeder specializing in development systems rarely using this system in breeding programme. Farmers in the swamp area using this system are located in the Wanaraya area (Barito Kuala regency).In swamp, areas the use of this system is very rare. So it needs to be explained and given further information about the breeder.

This is related to the swamps that cause difficulties breeders in building the cage. Some of the forage feeds used by farmers to be given to Bali cattle in swamplands vary considerably from padi *hyiang* (likely rice plant but grasses), *kumpai minyak* (grassesgroup), *parupuk laki* and *bini*(regional term for grasses swamp).

4. Conclusions

The development of beef cattle production especially Indonesian local cattle in South Kalimantan has a very big prospect to be improved and can be used as part of sustainable economic growth to improve the welfare of farmers in South Kalimantan. Utilization of the potential for geography, topography and natural conditions of South Kalimantan accompanied by government programme support to make the swamp areas in South Kalimantan into cattle development centers so that the concept of cattle development in the region can be realized. The potential for South Kalimantan as the center of beef cattle production in Kalimantan can be realized with the coordination of farmers, government, academia, researchers and businessmen sector aimed at creating breeding and development of cattle agribusiness patterned as business partners for the establishment of a better community economy. The development of the areas into a breeding center, purification cattle by utilizing the condition of South Kalimantan province which mostly has wetland and swamp land is expected to increase the production of beef cattle with based Indonesian local cattle in Kalimantan so as to make the economy in South Kalimantan into an economy based sustainable economic system.

Acknowledgments

Authors would like to thank the collaboration of Mulawarman University, Lambung Mangkurat University and Local Government of South Kalimantan Province which has facilitated the implementation of this research as well as supporting data and information in participation in the writing of this scientific paper.

References

[1] Suryana and M Yasin 2014 Prospect of integrated palm oil-cattle development in South Kalimantan. *J. Litbang Pertanian* **34(1)** 9-18.

[2] Directorate General of Livestock and Veterinary, Agricultural Ministry, Indonesia. 2015. Report of accountability working of office service in 2015. Agricultural Ministry. Jakarta.

[3] Talib, C, A Bamualim, and A.Pohan 1998 Problem of Bali cattle development in maintenance in grazing area. *Prosiding of National Conference Livestock and Veterinary*. Bogor, 1-2 December 1998:248-254.

[4] Hastuti, M 2002. Performance of Ongole male cross breed give supplementation withbioplus. Undergraduate thesis. Gadjah Mada University. Yogyakarta.

[5] Iswanto, Y 2003 Production of carcass and non-carcass of Ongole cross breed give fed king grass and concentrate with probiotic addition. Undergraduate thesis. Gadjah Mada University. Yogyakarta.

[6] Hidayat, N 2003 Estimation of natural increase of beef cattle in Majalengka Regency, East Java. Undergraduate thesis. Gadjah Mada University. Yogyakarta.

[7] Talib, C 2002 Bali cattle in germ source areas and opportunities and development chance. *Wartazoa* **12(3)** 100-107.

[8] Astuti, M 2004 Potential and diversity of genetic resources of Ongole cross breed.*Wartazoa* **14(4)** 30-39.

[9] Abdullah, M.A.N, R.R. Noor, H. Martojo, D.D. Solihin, and E. Handiwirawan 2006 The phenotypic variability of Aceh cattle in Nanggroe Aceh Darussalam. *J. Indon.Trop. Anim. Agric* **32(1)**11-21.

[10] Rahmatullah, S N, Jakarta, Ronny R. Noor. 2016. Identification of growth hormone gene variation in exon region at Indonesian local cattle based on PCR-sscp method. *Biodiversitas* **17(2)** 492-497.

[11] Sengkey, N.M, A.H.S. Salendu, E. Wantasen, P.O.V. Waleleng 2017 Potential development of beef cattle in Subdistrict West Tampaso. *Zootek* **37(2)** 350 – 359.

[12] Chamdi, A N. 2005. The characteristics of genetic resource of Bali cattle (bos-bibos banteng) and the alternative of its conservation method. *Biodiversitas* **6(1)** 70-75.

[13] Livestock Office Service of South Kalimantan. 2016. Report of accountability working of office service in 2016. South Kalimantan Province. Banjarbaru.

[14] Dwiyanto, K. 2008. Utilization of local resources and technology innovation in supporting the development of beef cattle in Indonesia.*J.Pengembangan Inovasi Pertanian* **1(3)** 173-188.

[15] Talib, C and Y.G. Noor 2008 Beef cattle production in supporting Indonesia food safety and security. *Prosiding of National Conference Livestock and Veterinary*. 2008: 44-50.

[16] Mayulu, H, Sunarso, I. Sutrisno, and Sunarso 2010 Policy on the development of beef cattle in Indonesia. *J. Litbang Pertanian* **29(1)** 34-41.

[17] Priyanto, D 2011 The strategy of beef cattle business in support of beef and buffalo self-sufficiency programme in 2014.*J. Litbang Pertanian* **30(3)** 108-116.

[18] Matondang, R H and S. Rusdiana 2013 Strategic steps in achieving beef self-sufficiency in 2014. *J. Litbang Pertanian* **32(3)** 131-139.

[19] Biyatmoko, D 2015 Efforts to improve availability forage feed to livestock and carrying capacity through planting three strata forage system. *Ziraa'ah* **40(3)**184-191.

[20] Mukson, W. Roessali and H. Setiyawan 2014 Analysis development regional of cattle beef in support meat self-sufficiency in central Java. *J.Peternakan Indonesia* **16(1)**26-32.

[21] Saputra, H, A. Daryanto, D.S. Hendrawan 2009 Strategy of development of beef cattle with insight of agribusiness in Aceh Province.*J. Manajemen dan Agribisnis* **6(2)** 152-162.

[22] Hamdan, A, E.S. Rohaeni, A. Subhan 2010 Characteristics of swamp buffalo as germ plasm in South Kalimantan. *Prosiding of National Conference Buffalo*2010: 57-64.

[23] Anggraeny, Y N, Mariyono, P.W. Prihandini 2010 Reproductive performance of brahman cross in three provinces in Indonesia: case study in East Java, Central Java, and South Kalimantan. *Prosiding of National Conference Livestock and Veterinary*2010: 73-79.

[24] Livestock Office Service of South Kalimantan 2015Dynamic of Livestock Population in South Kalimantan.South Kalimantan Province. Banjarbaru.

[25] Elly, F H, B.M. Sinaga, S.U. Kuntjoro, N. Kusnadi 2008 Development of cattle business through integration of cattle-plants in North Sulawesi.*J.Litbang Pertanian* **27(2)** 63 – 68.

[26] Prasetya, A 2011 Management of beef cattle in BPTP Rambatan, West Sumatera. Thesis. Bogor Agricultural University 1-65.

[27] Susanti, A E, N. Ngadiyono, Sumadi 2015 Output estimation of beef cattle in tidal lowland Banyuasin regency South Sumatera Province. *J. Lahan Suboptimal* **4(2)** 99-109.

[28] Rohaeni, ES 2014 Analysis potency of areas for development beef cattle in Tanah Laut Regency, South Kalimantan. *Prosiding of National Conference Innovation and Technology Agriculture Specific Location*2014:493-501.

[29] Nurdin, A S, A. Fariani, Sriati 2014 Development of ruminants based on availability of forage and labor in Palembang city of South Sumatera. *J. Peternakan Sriwijaya* **3(2)**1 -11.

IOP Conf. Series: Earth and Environmental Science **144** (2018) 012017 doi:10.1088/1755-1315/144/1/012017

Evaluation of soil and leaves nutrient on the growth of cultivated tabatbarito (*Ficusdeltoidea* jack.) in Makroman Village, Sambutan District of East Kalimantan, Indonesia

H Manurung[12*], W Kustiawan[3], IW Kusuma[3] and Marjenah[3]

[1]Biology Department, Faculty of Mathematics and Natural Sciences, Mulawarman University, Samarinda, East Kalimantan, Indonesia
[2]Graduate Student of Forestry Science, Faculty of Forestry, Mulawarman University, Samarinda, East Kalimantan, Indonesia
[3]Faculty of Forestry, Mulawarman University, Samarinda, East Kalimantan, Indonesia

E-mail: hetty_manroe@fmipa.unmul.ac.id

Abstract. This study aimed to evaluate the soil and leaves nutrient status on the growth of cultivated tabatbarito (*Ficusdeltoidea* Jack) in various level ages. The field experiment was conducted during December 2015 to November 2016 at Makroman Village, Samarinda-East Kalimantan. On 6, 9, and 12 months old after planting (MAP) the data was collected to evaluate the plant height, leaf number, branch number, biomass, soil and leaves nutrient concentrations. The results showed that the average pH of soil was 3.92±0.06, categorized as a very acid. The concentration of soil nutrients were: nitrogen (1.13±0.31 %), phosphorus (0.01±0.01 ppm), potassium (297.60±50.11 ppm), calcium (2.97±1.79 cmol(+)Kg^{-1}), and magnesium (3.69±2.30 cmol(+)Kg^{-1}). The leaf nutrient concentration was 1.74±1.42 % (N), 0.25±0.19 % (P), 1.86±0.15 % (Ca), 1.88±0.29 % (Mg). The soil nutrients concentration (N, P, Mg) and the leaf nutrient (N, P, K, Ca, Mg) has a correlates with plant height increment, branch number increment, and biomass increment. The results indicated that the N, P, K, Ca, Mg played an important role in the growth of *F. deltoidea* and this nutrient should be considered well when this plant will be cultivated as a source of the medicinal plant on a large scale.

1. Introduction

Tabatbarito (*Ficus deltoidea* Jack) belong to the Moraceae family, it is a small shrub and epiphytic[1]and found along shrubland beaches and hilly forest[2]. In Kalimantan of Indonesia, these plantsare very valuable and very expensive in traditional markets because it is widely used as a traditional medicine[3]. The leaves of *F. deltoidea*are boiled and the decoction is taken as a medicine after the mother gave birth. It is believed that it helps to contract the uterine and vaginal muscles, improve blood circulation and regain body strength as well as for treating disorders related to the menstrual cycle. It is also taken as medicinal tea for general health and in treating pneumonia, diabetes, hypertension, diarrhea, and gout[4]. In the recent year, *F.deltoidea* has started to be cultivated to supply the increasing market needs. Although tabatbarito is an epiphytic plant, as long as soil nutrients are well available, this plant can be planted on the ground. Evaluation of growth performance and yield potential of selected emascotek (*Ficusdeltoidea*) accessions on bris soils have reported by Musa[5].

For proper growth and optimal yield of plants, soil nutrients must be available to plants in the correct quantity, proportion and in a usable form at the right time. The Soil is a major source of nutrients needed by plants for growth. The three main nutrients are nitrogen (N), phosphorus (P) and potassium (K). Other important nutrients are calcium (Ca), magnesium (Mg) and sulfur (S). Plants also need small quantities of iron, manganese, zinc, copper, boron, and molybdenum, known as trace elements because only traces are needed by the plant. The role these nutrients play in plant growth is complex, and this document provides only a brief outline[6].

Nitrogen is a key element in plant growth. It is found in all plant cells, in plant proteins and hormones, and in chlorophyll. Atmospheric nitrogen is a source of soil nitrogen. Some plants such as legumes fix atmospheric nitrogen in their roots; otherwise, fertilizer factories use nitrogen from the air to make ammonium sulfate, ammonium nitrate, and urea. When applied to soil, nitrogen is converted to a mineral form, nitrate, so that plants can take it up [7]. The study by Sheikh and Ishak[8]on the cultivation of tabat baritoshowed that the nitrogen application improved growth, chlorophyll contents and antioxidant activity of leaf. Information about total flavonoid and antioxidant activity in leaves and stem extract of wild and cultivated tabatbarito also has been reported by Manurung et al.[9].

Phosphorus helps transfer energy from sunlight to plants, stimulates early root and plant growth, and hastens maturity[10]. The soil solution potassium is readily taken up by the plant's root system and is then replaced by the potassium on the exchange sites it increases vigour and disease resistance in plants, helps form and move starches, sugars, and oils in plants, and can improve fruit quality[11]. Calcium is essential for root health, the growth of new roots and root hairs, and the development of leaves. Magnesium is a key component of chlorophyll, the green coloring material of plants, and is vital for photosynthesis (the conversion of the sun's energy to food for the plant). Sulfur is a constituent of amino acids in plant proteins and is involved in energy-producing processes in plants. The effects of calcium and magnesium on plant growth, biomass partitioning, and fruit yield of winter greenhouse tomato has been reported by Hao and Papadopoulos[12]. In general, the growth rates such as plant height, leaf area, diameter, leaf number, and biomass of plants affected by nitrogen, phosphorus, potassium, calcium, and magnesium that found in the soil as nutrient elements[6].

Keeping in view the increasing future demand of tabatbarito plant, its cultivation was needed to be optimized. This study aimed to evaluate the soil and leaves nutrient on the growth of cultivated tabatbarito (*Ficusdeltoidea* Jack.) in Makroman Village, Sambutan District of East Kalimantan. By getting information about soil and leaf nutrients of this plant, expected to support if this plant will be cultivated as a source of medicinal plants in large scale in Kalimantan region.

2. Materials and method

2.1. Field/experimental procedures

A field study was conducted to evaluate soil and leaves nutrient on the growth of cultivated tabatbarito (*Ficusdeltoidea* Jack.) in Makroman Village, Sambutan District of East Kalimantan, Indonesia(Coordinates: 0°21'18"-1°09'16" South and 116°15'36"-117°24'16" East). Tabatbarito seedlings from 16-week-old stem cuttings are used as plant material. They were planted in the field area (20m x 50m) that consisted 10 rows, each row consisted 50 plants. Single row planting with 100 cm between plants in the row and 200 cm between rows were used (this is the first method that used in the cultivation process of tabat barito in Kalimantan). Plants were maintained until 12 months old after planting (MAP). On 6, 9, and 12 MAP, measurements were taken on plant height, leaves number, branches number, biomass, and concentration of soil and leaf nutrient. Soil and leaf nutrient analysis were conducted in the Soil Science Laboratory at the Faculty of Forestry Universitas Mulawarman, Samarinda, East Kalimantan,Indonesia.

2.2. Soil sampling and soil nutrient evaluation

Composited soil samples were taken 500 g of each row, put into a plastic bag and analysed for pH and concentration nutrients elements (N, P, K, Ca, Mg) in the laboratory. After 6, 9, and 12 MAP, the measurements of pH, N, P, K, Ca, Mg element was evaluated. The soil sample was taken from the depth of 0-30 cm and the results were calculated after being dried in the oven with the temperature of 150°C until constant weight was reached. The composite sample was air-dried and its pH, total Nitrogen (Kjehdal), available phosphor (Bray), available potassium, calcium, and magnesium were analysed [18].

2.3. Evaluation of leaf nutrients concentration

The total N concentrate was measured using Kjeldahl method(extraction, distillation, and titration). To measure the element of P and K, the plant components were extracted using High-Pressure Digestion method at the temperature of 180°C for 10 hours with HNO as a reductant. The phosphor was measured by using calorimetric technique and using nitrate-molybdate-vanadate acid as a coloring agent and was measured by using spectrophotometer at the wavelength of 470 nm. Potassium (K), calcium (Ca), magnesium (Mg) was measured by using atomic absorption spectrophotometer (AAS) at the wavelength of 766.5 nm, 489.5 nm, and 245.2 nm[18].

3. Data analysis

The data of soil and leaf nutrient concentration were analysedand described in a tabular form. Correlation between soil N, P, K, Ca, and Mg concentration (X) with leaf N, P, K, Ca, and Mg concentration and growth of *F. deltoidea* (Y) was tested by using regression analysis.

4. Results and discussion

4.1. Soil condition and nutrient element

The results of soil condition and nutrient element analysis welllisted in table 1. Data on the soil condition showed that the soil pH average of 3.92 ± 0.06 that categorized as very acid. The pH scale goes from 0 to 14 with pH 7 as the neutral point. When the amount of hydrogen ion in the soil increases, the soil pH decreases, causing the soil tobecome more acidic. From pH 7 to 0, the soil is increasingly more acidic, and from pH 7 to 14, the soil is increasingly more alkaline or basic[7].

Results of soil nutrient element analysis showed the concentration of N range from 0.93 to 1.49 with mean 1.13±0.31 with a very high status. The concentration of nitrogen in the soil increased when the ages of plant increased. One of the key soil nutrients is nitrogen (N). Plants can take up N in the ammonium (NH_4^+) or nitrate (NO_3-) form. At pH's near neutral (pH 7), the microbial conversion of NH_4^+ to nitrate (nitrification) is rapid, and crops generally take up nitrate. In acid soils (pH < 6), nitrification is slow, and plants with the ability to take up NH_4^+ may have an advantage[13]. Soils high in organic matter such as chocolate soils are generally higher in nitrogen than podzolic soils. Nitrate is easily leached out of the soil by heavy rain, resulting in soil acidification[14].

Table 1.Soil condition and nutrient concentration of cultivated tabatbarito (*F.deltoidea*) on different plant ages.

Plant Ages (MAP)	pH (H_2O)	N (%)	P (ppm)	K (ppm)	Ca (cmol (+)Kg^{-1})	Mg (cmol(+)Kg^{-1})
6	3,94	0,93	0,01	355,05	4,65	2,19
9	3,97	0,97	0,01	274,85	3,17	2,53
12	3,86	1,49	0,001	262,90	1,08	6,34
Mean	3.92±0.06	1.13±0.31	0.01±0.01	297.60±50.11	2.97±1.79	3.69±2.30

The concentration of P has the status extremely low with the average 0.01 ± 0.01 ppm. The concentration of K nutrient status is medium (262.90) to high (355.05) with the average 297.60 that categorized as high refers to McCauley[7].The concentration of Ca element is very low (1.08 cmol(+)Kg^{-1}) to medium 4.65 (cmol(+)Kg^{-1}) with mean 2.97 ± 1.79 (cmol(+)Kg^{-1}) were categorized as low. Calcium is essential for root health, the growth of new roots and root hairs, and the development of leaves. It is generally in short supply in the acid soils. Mg nutrient categorized as high element with average 3.69 ± 2.30 (cmol(+)Kg^{-1}). Magnesium is a key component of chlorophyll, the green coloring material of plants, and is vital for photosynthesis (the conversion of the sun's energy to food for the plant). Deficiencies occur mainly on sandy acid soils in high rainfall areas, especially if used for intensive horticulture or dairying. Heavy applications of potassium in fertilizers can also produce magnesium deficiency. The three main nutrients that needed by plants for growth are nitrogen (N), phosphorus (P) and potassium (K). Other important nutrients are calcium, magnesium, and sulphur[14].In general, the growth rates such as plant height, leaf area, diameter, leaf number, and biomass of plants affected by nitrogen, phosphorus, potassium, calcium, and magnesium that found in the soil as nutrient elements.The study by Fukunawa et al.[15] reported that the soils in East Kalimantan were less fertile

4.2. Leaf nutrient concentration

The results of leaves nutrient analysis (N, P, K, Ca, and Mg) welllisted in table 2. It showed that the cultivated *F. deltoidea* leaves on 6, 9, 12 MAP had the varied concentration of different N, P, K, Ca, and Mg. The concentration of N leaves during growth proses of *F. deltoidea* ranged from 0.11 (%) to 2.67 (%), with average 1.74 ± 1.42 (%). The concentration P ranged from 0.47 to 0. 14 with the average 0.25 ± 0.19. The concentration of K ranged from 1.69 (%) to 1.98 (%) with the average 1.86 ± 0.15 (%). The concentration of Ca ranged from 2.02(%) to 4.83(%), with the average 3.82 ± 1.56 (%). Concentration of Mg ranged from 1.59 (%) to 2.16 (%) with the average 1.88 ± 0.29 (%). The present study showed that concentration of N, P, K, Ca decreased until the age of 12 MAP, whereas the concentration of Mg decreased until the age 9 MAP and then increased until 12 MAP. The lower N availability could be ascribed to the loss of N from soil ecosystems and the insufficient recovery of soil N pools under poor vegetation regrowth due to intensive land use in this area[16].Styger et al.[17]recommended that upland agricultural intensification and diversification based on improved soil fertility through optimized organic and inorganic inputs and fire-less land management that encourages the re-establishment of nutrient stocks. The negative correlation was observed from the soil nutrient K and growth of diameter and high, the positive correlation was observed from the P and K content in the leaf of plant and growth of diameter, height, and volume increment of*Macaranga gigantean*[18].

Table 2. Leaf nutrient concentration of *F. deltoidea* that cultivated in Makroman village Samarinda at 6, 9, and 12-month-old after planting (MAP)

Plant ages (MAP)	N (%)	P (%)	K (%)	Ca (%)	Mg (%)
6	2,67	0,16	1,98	4,83	1,88
9	2,45	0,13	1,92	4,61	1,59
12	0,11	0,47	1,69	2,02	2,16
Mean	1.74±1.42	0.25±0.19	1.86±0.15	3.82±1.56	1.88±0.29

4.3. Plant Growth and correlation with soil and leaf N, P, K, Ca and Mg concentration

The growth result of cultivated *F deltoidea* on 6, 9, and 12-month-old after planting were showed in table 3. The highest plant height, leaves number, branches number, and biomass found on 12 MAP as compared to 6 and 9 MAP. The highest growth increment found on leaves number followed by biomass, plant height, and branch number.

Table 3. The plant growth and plant growth increment of *F. deltoidea* on different ages.

Plant ages (MAP)	6	9	12	Mean
Plant Height (cm)	28,38	39,02	54,14	40.51±12.94
Leaves number	58,60	154,68	381,68	198.32±165.90
Branches number	9,02	22,23	34,83	22.03±12.91
Biomass ($gr/plant^{-1)}$	12.35	55.60	93.30	53.75±40.51
Plant Height increment ($cm\ month^{-1}$)	4.73	4.33	4.51	4.53±0.20
Leaves number increment	9.77	17.19	31.81	19.59±11.21
Branches number increment	1.50	2.47	2.90	2.29±0.72
Biomass increment ($gr/plant^{-1}\ month^{-1)}$	2.06	6.17	7.78	5.34±2.95

The soil nutrient: nitrogen, phosphorus, and magnesium concentration correlate with height increment, branch number increment, and biomass increment; the potassium correlate with branch number, plant height increment, and leaf number increment; calcium concentration correlate only with plant height. Relationships between soil N nutrient concentration and growth (plant height increment, branch number increment, and biomass increment) of cultivated *F. deltoidea* describe in figure 1. All of the leaf nutrient (N, P, K, Ca and Mg) had a correlate with plant height increment, branch number increment, and biomass increment. The highest average of plant height increment, leaf number increment, branch number increment, biomass increment found on 12 MAP. The growth increased rapidly in the first years of cultivation. In general, nitrogen (N), potassium, calcium, magnesium, and sulfur are more available within soil pH 6.5 to 8, while boron (B), copper, iron, manganese, nickel (Ni), and zinc are more available within soil pH 5 to 7. Phosphorus is most available within soil pH 5.5 to 7.5 At pH less than 5.5, high concentrations of H, aluminum and manganese in soil solution can reach toxic levels and limit crop production[7]. In the tropical rainforests of East Kalimantan servings of nutrient base class (K, Ca and Mg) more accumulates on vegetation than in the soil, while the portion of nutrient N largely accumulates in the soil [19], and in forest plantation reported that plant forests of *Eucalyptus deglupta* and *Paraseriantesfalcataria* at the age of 5-10 years in East Kalimantan also accumulate nutrient element of K in the largest amount, followed by calcium, nitrogen, and magnesium. In the other parts of natural forests in East Kalimantan show that 70 to 94% of base elements exist in the standing biomass [20].Sheikh and Ishak reported that the nitrogen application improved the growth of *F.deltoidea* [8].

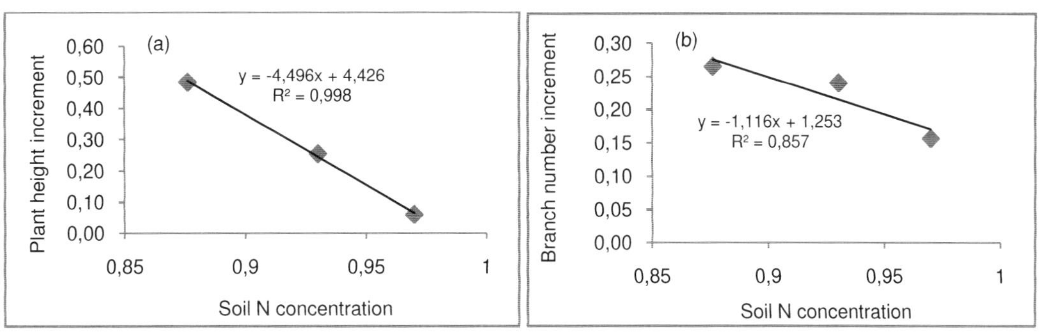

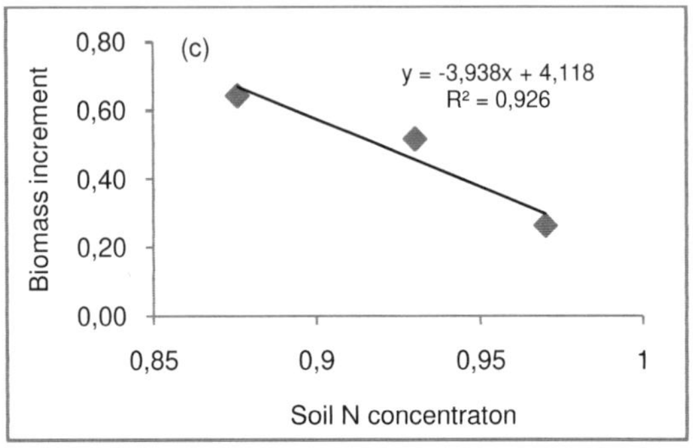

Figure 1.Relationships between soil N concentration and growth of *F. deltoidea* at different plant ages. (a) plant height increment Vs Soil N concentration (b) branch number increment Vs soil N concentration, and (c) biomass increment Vs soil N concentration.

5. Conclusions

The soil pH in the growth of *Ficus deltoidea* in Makroman Village, Sambutan District of East Kalimantan categorized as a very acid. The concentration of soil nutrients, nitrogen categorized as a very high, phosphorus: extremely low, potassium: high, calcium:very low, and magnesium: high. The soil nutrients concentration such as nitrogen, phosphorus, and magnesium have a correlate with plant height increment, branch number increment, and biomass increment; potassium correlates with branch number, plant height increment, and leaf number increment; calcium concentration correlates only with plant height. All of the leaf nutrient (N, P, K, Ca and Mg) concentration had a correlate with plant height increment, branch number increment, and biomass increment.

Acknowledgements

The author thanks to Directorate General of Higher Education Ministry of Research Technology and Higher Education, Republic Indonesia for financially this study as part of the Doctoral Programme at Universitas Mulawarman.

References

[1] Brickell C and Zuk JD 1997 *The American Horticultural Society A-Z. Encyclopedia of Garden Plants* (New York: DK Publishing Inc.)p 1104

[2] De Padua LS, Bunyapraphatsara N and Lemmens RHMJ 1999 *Plants Resources of South East Asia (Prosea) Medicinal and Poisonous Plants*.Prosea Bogor **12(1)** p 283

[3] Sulaiman MR, Hussain MK, Zakaria ZA, Somchit MN, Moin S, Mohamad AS and Israf DA 2008 Evaluation of the antinociceptive activity of *Ficusdeltoidea* aqueous extract*Fitoterapia***79**557-561

[4] Musa A and Ogbadoyi EO 2012 Influence of applied nitrogen fertilizer on the bioaccumulation of micronutrients, antinutrients and toxic substances in *Telfairiaoccidentalis* (fluted pumpkin). *Int J Pl.An and Env. Sci.* **34** 12-6

[5] Musa Y 2006 Evaluation of growth performance and yield potential of selected emascotek (*Ficusdeltoidea*) accessions on bris soils *J. Trop. Agric. And Fd. Sci.* **34** 229-235

[6] Brady NC and Weil RR 2016 *The Nature and Properties of Soil*15th ed(New York: Pearson Education Inc) p139

[7] McCauley A, Jones C and Olson-Rutz K 2017 *Soil pH and Organic Matter.* Nutrient Management Module No **8** (4449-8) Montana State University p 16

[8] Sheikh S and Ishak CF 2016 Effect of nitrogen fertilization on antioxidant activity of Mas cotek (*Ficusdeltoidea* Jack) *J. Med. Plants. Stud.***4**208-214

[9] Manurung H, Kustiawan W, Kusuma IW and Marjenah 2017 Total flavonoid content and antioxidant activity in leaves and stems extract of cultivated and wild tabatbarito

(*Ficusdeltoidea* Jack). AIP Conference Proc.**1813**ed RA Nugroho (American Institute of Physics) pp 020007 1-6 http://dx.doi.org/10.1063/1.4975945

[10] Abdolzadeh A, Wang X, Veneklaas EJ and Lambers H 2010. Effects of phosphorus supply on growth, phosphate concentration and cluster-root formation in three *Lupinus species*. *Annals of Botany***105** 365-374

[11] Prajapati K and Modi HA 2012 The importance of potassium in plant growth-A Review.*Indian Journal of Plant Sciences* **1** 177-186

[12] Hao X and Papadopoulos AP 2004 Effects of calcium and magnesium on plant growth, biomass partitioning, and fruit yield of winter greenhouse tomato.*HortScience* **39** 512-5

[13] Hardjowigeno S 2003 *Ilmu Tanah*(Jakarta Indonesia: AkademikaPressindo) p 286

[14] Departments of Primary Industries 2017 Plant nutrient in the soil NSW Government. Wollongbar Agricultural Institute, for CaLM and NSW Agriculture Wollongbar Agricultural Institute, for CaLM and NSW Agriculture, North Coast region, under the National Landcare Program. Retrieved from https://www.dpi.nsw.gov.au/agriculture/soils/improvement/plant-nutrients

[15] Funakawa S, Yonebayashi K, Kaewkhongkha T and Makhrawie 2010 *Soil Ecological Study on Shifting Cultivation in Southeast Asia*(Thailand: Faculty of Natural Resources, Prince of Songkla University) p 12 Retrieved from http://natres.psu.ac.th/Link/SoilCongress/bdd/symp27/2062-t.pdf

[16] Effendi MBW, Tanaka S, Joseph JK, Logie S, Brangking U, Jonathan LAT, Arifin A, Yoshinori M and Sakurai K 2009 Vegetation conditions and soil fertility of fallow lands under intensified shifting cultivation systems in Sarawak, Malaysia. *Tropics* **18** 115-126.

[17] Styger E, Rakotondramasy HM, Pfeffer MJ, Fernandes ECM and Bates DM 2007 Influence of slash-and-burn farming practices on fallow succession and land degradation in the rainforest region of Madagascar.*Agric. Ecosyst. Environ.***119** 257-269

[18] Susanto D, Ruchiyat D, Sutisna M and Amirta R 2016 Soil and leaf nutrient status on growth of *Macarangagigantea* in secondary forest after shifting cultivation in East Kalimantan, Indonesia. *Biodiversitas***17**409-416

[19] Ruhiyat D 1993 *Dynamics of Nutrient in Natural and Plantation Forest: Cycle of Forest Biogeochemistry*. Proc. Workshop to construction of tropical rain forest with vision of environment to increase productivity 1-3 Marc 1993 Samarinda, Indonesia.

[20] Meckensen J, Ruhiyat D and Folster H 2001 Volume-based nutrient content of *Acacia mangium*, *Eucalyptus deglupta* and *Paraserianthesfalcataria* in industrial tree plantations in East Kalimantan, Indonesia *J.Trop. For. Sci.***13** 512-526

Seed Germination and Cuttings Growth of *Piper Aduncum*

D Susanto[1*], Sudrajat[1], W Suwinarti[2] and R Amirta[2]

[1]Department of Biology, Faculty of Mathematics and Natural Sciences, Mulawarman University, Samarinda 75123, East Kalimantan, Indonesia.
[2]Faculty of Forestry, Mulawarman University, Indonesia, Samarinda, Indonesia

*Corresponding Author: susantodwiki@yahoo.com

Abstract. *Sirih hutan (Piper aduncum* L*)* is one of group shurbs tropical species, has potential to be developed as raw material of biomass based electricity. The aim of this research was to know seed germination and cuttings growth of *P. aduncum* plant as the first step in cultivation of this plant. Observation of flowers and fruits were done in secondary forest, while seed germination and growth of shoot cuttings were done in the laboratory. The results showed that *P. aduncum* seeds can be germinated in a relatively short time of 17 to 25 days with a fairly high germination percentage of 90 ± 8.16% and germination rate of 4.7 ± 0.34%. The growth of seedlings at 2 months old was 4.78 ± 0.42 cm, plant height 3.97 ± 0.27 cm, and relative growth rate 0.33 ± 0.14%. The treatment of synthetic growth regulator had significant effect on shoot growth and root number on the plant stem cuttings. Preparation of seedlings ready to plant in a generative and vegetative for cultivation of these plants in the experimental plot.

1. Introduction

Piper aduncum L (Sirih hutan) is classified as an invasive plant, naturally spreading from Fiji, the Solomon Islands and the United States. It spreads also to Southeast Asia, Melanesia and Papua New Guinea [1, 2]. *Piper aduncum* is usually called sirih hutan (Indonesia), while in Sunda area usually is called Gedebong or Seuseureuhan, which can be found at altitude 1 - 2.000 dpl (above sea level).

In Indonesia, this plant was first introduced in Bogor Botanical Gardens, West Java in 1860. Then spread widely in the island of Java, Sumatra, Seram, Papua and other islands. In Kalimantan, *P. aduncum* was first collected in 1952 [3]. According [4] , the spread of invasive *P. aduncum* plants on the island of Borneo, particularly in East Kalimantan occurred through a network of logging roads stretching from the southern part of Kutai Barat Regency to the north in Malinau District. Many grow on the older (old) logging roads in the south and there are no plants in the younger (new) streets in the north. The minimum spread rate is estimated to be five to seven kilometers per year, where a river with a width of 30 m cannot prevent the spread of this plant. On the previous report [5],this plant is found in secondary forest of shifting cultivation fields with a fallow age of 0.5 to 3 years, but not found in the fallow period of 6 to 10 years.

Forest betel plant extracts have the potential to be *Aedes aegypti* mosquito larvae of dengue virus spreader [6], drawing the growth of *Fusarium solani,* causing fusariosis disease in *Piper nigrum* [7] .

The above ground biomass from *P. aduncum* reached 48 Mg DM ha^{-1} at 23 months and three quarters of the biomass consisted of wood. The growth rate of *P. aduncum* averaged 69 kg DM ha^{-1} per day and increased with higher rainfall [3]. Rapid growth and large biomass production from these plants could potentially be developed as a source of renewable raw materials for biomass-based power generation, especially for the inland areas of East Kalimantan. On the pastreport [8],the woody shrub species of *P. aduncum* produce an energy potential of 2.19 Mwh per ton of dry biomass, compatible and can be used as a biomass feedstock for sustainable electricity production. Recently, information on germination and cultivation of this plant has not been widely known and never been cultivated. Based on the above description, our study focused on the generative and vegetative reproduction of *P. aduncum* plant as a preliminary study for the provision of ready-to-plant seedlings in the cultivation of this plant.

2. Materials and Methods

2.1. Study area
Flower and fruit development research were conducted in forest research Faculty of Forestry Mulawarman University, East Kalimantan, Indonesia (Botanical Garden of Mulawarman University of Samarinda), located between the coordinates of 0 ° 25'10 "- 0 ° 25'24" South Latitude and 117 ° 14'00 "- 117 ° 14'14" East Longitude. Seed germination research from fruit and growth of shoot / stem cuttings were done at Physiology Laboratory of Faculty of Mathematic and Natural Sciences,Mulawarman University in Samarinda, East Kalimantan, Indonesia.

2.2. Observation of flowers and fruit of P. aduncum
Observations were made in the secondary forest of the flowering and fruiting *P. aduncum* plants. Then performed fruit harvesting and morphological observation and fruit anatomy that has been mature physiologically. Seed extraction was done to separate the seeds from the other pieces of fruit [5].

2.3. Fruit and Seed
The harvesting mature fruits were conducted in April - May 2017, the fruit was collected in bulk (mixed from several trees in the same place). The harvesting mature fruits were done with a hooked pole and under a tree laying plastic to accommodate falling fruits. Samples taken as 1 kg of fruit peeled and seeds extracted. [9] .

2.4. Seed Extraction
The seeds separate from the other pieces of fruit such as skin, meat, wings, stems then extraction of seeds. In dry extraction, the fruits were done in the open place for 3 - 4 days until the fruit was broken, and the seeds were easy to remove from the fruit. Performed sifting to separate the seeds from the skin of the fruit [9] . While on the wet extraction, the fruit was immersed in a burlap sack for one week (every day watered with moisture) until the outer skin of the fruit becomes soft, then dried until the fruit rupture and seeds were easy to remove from the fruit. Performed sifting to separate the seeds from the skin of the fruit [5].

2.5. Seed germination
The research was done by using completely randomized design with two treatments which were 1) the way of extraction of seeds consist of: wet extraction and dry extraction, 2) germination media. Seeds were sown evenly on germination medium of 20 seeds with 3 replicates for each treatment and observed for 60 days. The parameters observed were germination time of first seed germination (GTFS), germination time of the last germinated seed (GTLS), average germination time (MGT), germination percentage (G), and germination rate (GR).

2.6. The growth of seedlings
The seeds germinated in the laboratory were transferred into a 30 cm high polybag and 15 cm in diameter, filled with a topsoil soil topsoil medium. Tillers were kept in a greenhouse with a shade of

paranet 50% for 1 month. High data, leaf number, wet weight and dry weight and relative growth rate (RGR) were measured at the end of the study.

2.7. Growth of shoot / stem cuttings

Shoot cuttings were done by planting shoots on some planting medium: M0 = top soil, M1 = soil top soil + manure (1: 1), M2 = soil top soil + manure (2: 1), M3 = soil top soil + fertilizer cage (3: 1), and second treatment with synthetic growth regulator; R0 = control, R1 = 50 ppm, R2 = 100 ppm, R3 = 150 ppm and R4 = 200 ppm. Shoots cuttings were maintained in plastic hoods to keep moisture in the greenhouse with a 50% paranet shade for 2 months, then measured for growth and rooting.

2.8. Data analysis

Data obtained from field observations were analyzed descriptively, while germination data and seedbed were analyzed by analysis of variance (Anova) and advanced test with Duncan's Multiple Range Test (DMRT) at 95% levels.

3. RESULTS AND DISCUSSION

3.1. Morphology of flowers and fruit

Observations of the flowers and *P. aduncum* fruit of the forest showed that these plants flower and bear fruit continuously throughout the year. Compound interest grows continuously on the leaf axillary at the end of the branch and the fruit was increasingly ripe often with the increasing of flowers from the end of the branch. The morphology of flowers, fruits and seeds is presented in Figure 1, long ang diameter seed of *P. aduncum* in Table 1.

a b

Figure 1. a. Plant of *Piper aduncum*, b. Branch and fruits

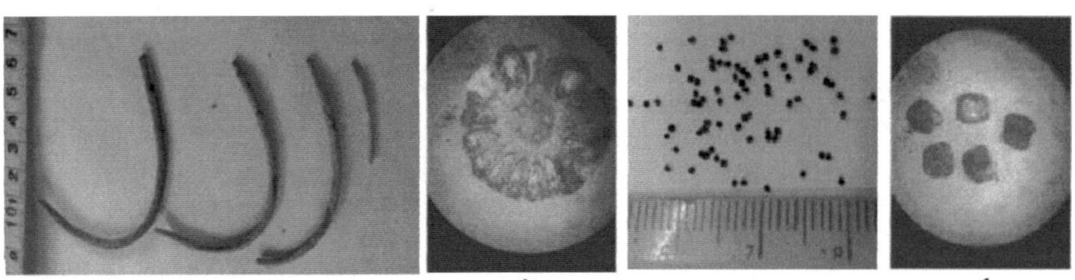

a b c d

Figure 2. Fruits and seeds of *Piper aduncum* Description: a. flowers, young fruit and ripe fruit physiological, b. cross section of fruit, c. seeds, d. microscope photo seeds (4 x10 magnification)

Table 1. Mean size of fruit and seed of *P. aduncum*

Organ	Long (cm)	Diameter (cm)
Fruits	11.83±1.07	0.35±0.04
Seeds	< 0.1	<0. 1

Fruit was round and long, the measurement of ripe fruit obtained the average length of 11.83 ± 1.07 cm, diameter 0.35 ± 0.44 cm with a small seed size of less than 1 mm.

3.2. Seed germination

The method of seed extract and planting media was influenced and statistically significant (Anova, $P \leq 0.05$) on the seed germination percentage. The percentage of germination and rate of seed germination of wet extraction was greater than dry extraction seeds. The highest percentage of germination on the combination of wet extraction with compost media was 65%while the germination rate (MGT), early germination (GTFS) and germination end (GTLS) seeds from wet extraction were faster than dry extraction seeds (Table 4). The percentage percentage of germination of forest is presented in the table and the figure below.

Table 2. Effect of seed extract method on percentage of seed germination of *Piper aduncum* seed.

Extraction treatment	G(%)	GTFS (days)	GTLS (days)	GR (%)
Weet extraction	90±8.16	17	25	4.7±0.34
Dry exraction	0	0	0	0

G, mean germination (%) ; GTFS (days), germination time of the first seed; GTLS (days), germination time of the last seed. GR' germination rate (%)

Table 3. The growth of seedlings of tropical shrubs 2 month

No	Species	∑Leaves	Stem hight	RGR
	P. aduncum	4.78±0.42	3.97±0.27	0.33±0.14

RGR= relative growth rate

3.3. The growth of stem cuttings

3.3.1. Application of growth regulators

Tropical bush-type plants that have the potential as seeds of energy until now have not been cultivated. In this study carried out vegetative propagation experiments of several types of tropical shrubs, including stem cuttings with the application of plant hormones to stimulate root and shoot growth. This is important in the preparation of ready-to-plant seedlings when these types of plants will be cultivated. The results of stem cuttings measurements with IAA (Indole asetic acid) and NAA (Naptaline asetic acid) hormone applications are presented in the table below.

Table 4. Effect of plant hormone treatment on *P. aduncum* cuttings growth

Treatment	∑ shoots	∑ leaves	Hight	∑ Roots
K0	1.00	5.33	26.33^{a}	11.33^{a}
K1	1.00	6.33	26.67^{a}	22.67^{b}
K2	1.33	9.67	45.33^{b}	34.67^{c}

3.3.2. Application of commercial growth regulators.

In this study used commercially plant growth regulators on the market and commonly used by farmers in vegetative propagation such as grafts or stem cuttings or shoots. The content of these growth regulators were IBA, NAD, MNAA, MNAD and Oysters. The result of analysis of variance using anova showed that media had significant effect on high growth of shoot cuttings (0.010 <0.05), while commercial growth regulator did not significantly affected on the growth of shoot cuttings (0.319> 0.05). The interaction between medium and commercial growth regulator did not significantly affect the growth of plant height (0.434 <0.05). The average growth rate of *P.aduncumt* is presented in the Table 5.

Table 5. Influence of planting medium and commercial growth regulator on the average height of *P. aduncum* at 2 months.

Treatment	M0	M1	M2	M3	Mean
R0	16.40±14.29	20.73±1.16	16.33±14.44	22.50±3.28	18.99±9.23
R1	26.83±1.75	24.83±6.64	22.00±1.90	0.00±0.00	18.42±11.65
R2	22.76±3.57	8.83±15.29	7.50±12.99	7.00±12.12	11.52±12.19
R3	21.96±19.05	17.40±15.51	13.86±12.03	8.00±13.86	15.31±14.12
R4	26.30±9.89	8.80±15.24	6.33±10.97	5.16±8.95	11.65±13.28
Mean	22.85±10.5[b]	16.12±12.3[ab]	13.21±11.3[a]	8.53±11.0[a]	

Further test results with DMRT showed that giving planting medium decreased the growth of plant height cuttings. The highest average at M0 was 22.85 ± 10.50 and the lowest on M3 namely 8.53±11.0. M1 is not significantly different from M2 but is different from M0 and M3.The result of analysis of variance showed that media and commercial growth regulator have significant effect on the growth of leaf buds leaf cuttings (0.043 <0.05), (0.045 <0.05). The interaction between medium and commercial growth regulator did not significantly influence on the growth of shoot leaves leaf number (0.075 <0.05). The average number of piper plant leaves is presented in the Table 6.

Table 6. Influence of planting medium and commercial growth regulator to the average number of leaves of plant of *P. aduncum* age 2 months.

Treatment	M0	M1	M2	M3	Mean
R0	4.00±3.46	5.67±0.58	3.33±3.05	5.67±0.58	4.67±2.27[b]
R1	2.33±0.58	4.67±2.52	6.00±1.00	0.00±0.00	3.25±2.67[ab]
R2	3.67±0.58	3.00±5.19	1.33±2.31	0.67±1.15	2.17±2.79[a]
R3	5.33±0.58	3.00±3.60	1.67±1.53	1.67±2.89	2.91±2.61[ab]
R4	5.00±2.00	1.33±2.31	0.67±1.15	1.00±1.73	2.00±2.41[a]
Mean	4.07±1.91[b]	3.53±3.14[ab]	2.60±2.58[ab]	1.80±2.48[a]	

Further test results with DMRT showed that the planting medium decreased the growth of the number of leaves of plant cuttings. The highest average at M0 is 4.07 ± 1.91 and the lowest on M3 is 1.80 ± 2.48. M0 was not significantly different from M1 and M2 but significantly different from M3. Administration of commercial growth regulator also decreased the growth of the number of leaves of plant cuttings. The highest average at M0 was 4.67 ± 2.27 and the lowest at R4 was 2.00 ± 2.41. R0 was not significantly different from R1 and R3 but significantly different from R2 and R4.

The result of analysis of variance showed that media and commercial growth regulator have significant effect on growth of root root cuttings (0.00 <0.05), (0.036 <0.05). The interaction between media and commercial growth regulator did not significantly affect the growth of leaf cuttings (0.213> 0.05). The average height of pipe plant is presented in the Table 7.

Further test results with DMRT showed that the planting medium decreased the growth of the number of plant root cuttings. The highest average at M0 was 21.07 ± 10.30 and the lowest at M2 was 3.40 ± 4.67. M0 was significantly different from M1, M2 and M3. Application of commercial growth regulator also decreased the growth of the number of leaves of plant cuttings. The highest average at R0 was 13.33 ± 10.08 and the lowest at R2 is 5.42 ± 7.84. R0 was not significantly different from R1 but significantly different from R2, R3 and R4.

Table 7. Influence of planting medium and commercial growth regulator to the average number of Roots of *P. aduncum* age 2 months.

Treatment	M0	M1	M2	M3	Mean
R0	19.00±16.46	8.33±6.50	8.00±7.55	18.00±5.29	13.33±10.1^{b}
R1	28.67±5.13	5.33±7.50	5.67±5.51	0.00±0.00	9.92±12.33^{ab}
R2	16.67±5.67	3.67±6.35	1.00±1.73	0.33±0.58	5.42±7.84^{a}
R3	16.67±14.47	4.67±8.08	2.00±2.64	2.00±3.46	6.33±9.67^{a}
R4	24.33±6.03	1.00±1.73	0.33±0.58	0.33±0.58	6.50±11.09^{a}
Mean	21.07±10.30^{b}	4.60±5.97^{a}	3.40±4.67^{a}	4.13±7.60^{a}	

4. Discussion

Field observations showed that *P aduncum* flowering and fruitful continuously. The flowers grow on the leaf's armpit and the physiological ripe fruit had a length of 11.83 ± 1.07 cm, diameter 0.35 ± 0.44 cm. Small seeds with dimensions less than 1 mm. Reported [10] that *P. aduncum* flowers consist of compound interest, jar form, one or two pairs, protective leaves 0.5-1.25 mm. Fruit in the form of buni fruit, was short stalk, long grain 12-14 cm.

In this study, the wet extraction treatment yielded high seed germination which was 90 ± 8.16%, while the extraction of dry beans had no seeds that germinated. On the report [11] stated that the percentage of germination of *P. aduncum* reached 49-68% under different thermal regimes in the presence of light, without prior permanence hydrated in darkness for a month (no) and after permanence (yes). *P. aducum* seeds are orthodox that can be stored for a long time [12] . According to author name [13] the growth of *P.aduncum* seedlings is very slow.

In this study, the treatment of planting media and plant growth regulators were remarkable on the growth of cuttings of *P. aduncum* crops. The best planting medium was topsoil soil, the application of manure decreases the growth of leaf number, plant height and number of roots. Likewise, in commercial application of commercial growth agents, the best growth in control (without the application of growth regulators). This is in contrast to those reported by author name [14] that the growth regulator treatment enhanced the growth of cuttings of Agarwood crops (*Aquilaria malacensis* Lamk.). On the report [13] reported that for multiplication of plants in large quantities, cuttings are a more efficient method, namely by the exposure of the hormones IAA and NAA with the media from the sand.

5. Conclusion

P. aduncum seeds can germinate in a relatively short time of 17 to 25 days with a fairly high germination percentage of 90 ± 8.16% and germination rate of 4.7 ± 0.34%. The growth of seedlings at 2 months old is 4.78 ± 0.42 cm, plant height 3.97 ± 0.27 cm, and relative growth rate 0.33 ± 0.14%. The treatment of synthetic growth regulator has significant effect on shoot growth and root number on plant stem cuttings. Preparation of seedlings is ready to plant in a generative and vegetative for cultivation of these plants in the experimental plot.

Acknowledgements

This work was financially supported by the Grant of Mulawarman University Research of Excellent provided by the Directorate General of Research and Development, the Ministry of Research , Technology, and Higher Education of Indonesia. We acknowledge the Forest Education , Faculty of Forestry, Mulawarman University for providing permission to conduct the field work.

References

[1] Tan HTW, Ibrahim AB and Tan KX 2008 Nature in Singapore I p 55-59.
[2] Hartemink AE 2000 International Soil Reference and Information Centre (ISRIC) Netherland
[3] Hartemink AE 2010 J Trop Forest Sci **22**(2): 202-213.
[4] Padmanaba M and Sheil D 2014 J Trop Corcervation Sci **7**(1): 34-44.
[5] Susanto D, Ruhiyat D, Sutisna Mand Amirta R 2016 J **Biodiversitas 17**(1): 192-199.
[6] Sudrajat, Susanto Dand Rahmat A 2010 J BioProspek **7**(1): 90-94
[7] Benchimol RL, Sutton, Basthos CN, and Dia-Filho MB 2003 XXXVI Congresso Brasileiro de Fitopatologia, manejo integradodedo nças de plantasu berlândia/MG- Center Convention
[8] Amirta R, Yuliansyah, Angi EM, Ananto BR, Setiyono B, Haqiqi MT, Septiana HA, Lodong Mand Oktavianto RN 2016 J Nusantara Biosci **8**(1): 22-31
[9] Suita E and Nurhasybi 2009 J Info Benih **13** (1): 170-175
[10] Syamsuhidayat SS and Hutapea JR 1991 Departemen Kesehatan RI Jakarta
[11] Dousseau S, De Alvarenga AA, Alves E and Da Silva AJ 2011 J Revista Brasil Bot **34**(3):297-305
[12] Da Silva TL, Gomes HT and Scherwinski-Pereira JE 2017 DOI 10.1080
[13] Mendoza-C, Celis A, Pach ME 2010 International Symposium on Medicinal and Aromatic Plants IMAPS 2010 and History of Mayan Ethnopharmacology IMAPS
[14] Yusnita E, Puspitasari P and Susanto D 2017 Int J Sci & Tech Research **6** (07): 73-77

Floyd-warshall algorithm to determine the shortest path based on android

Ramadiani*[1], D Bukhori[1], Azainil[2] and N Dengen[1]
[1]Faculty of Computer Science and Information Technology Mulawarman University, Samarinda, East Kalimantan, Indonesia
[2]Faculty of Teacher Training and Education Mulawarman University, Samarinda, East Kalimantan, Indonesia

**Corresponding Author*: ilkom.ramadiani@gmail.com

Abstract. The development of technology has made all areas of life easier now, one of which is the ease of obtaining geographic information. The use of geographic information may vary according to need, for example, the digital map learning, navigation systems, observations area, and much more. With the support of adequate infrastructure, almost no one will ever get lost to a destination even to foreign places or that have never been visited before. The reasons why many institutions and business entities use technology to improve services to consumers and to streamline the production process undertaken and so forth. Speaking of the efficient, there are many elements related to efficiency in navigation systems, and one of them is the efficiency in terms of distance. The shortest distance determination algorithm required in this research is used Floyd-Warshall Algorithm. Floyd-Warshall algorithm is the algorithm to find the fastest path and the shortest distance between 2 nodes, while the program is intended to find the path of more than 2 nodes.

1. Introduction

A donation program is one of the ideas developed and run by social institutions with the aim of collecting as much aid as possible to be distributed to refugees and victims of natural disasters. In order to be immediately channeled, the collection of donations must be carried out with a fast process that requires the mechanism as effectively and efficiently as possible. We need a medium that adapts to current trends, and one solution is to make an application based android donation pickup using the application of digital maps. Particularly in this research, the shortest route selection algorithm named Floyd-Warshall Algorithm is applied to a digital map of donation pickup program, which is expected to streamline the distance that must be passed by donation pickup institution to donor location.

1.1. Geographic Information System (GIS)

Technological advancements have given people more access to more difficult information. One of the ease that can be felt is access to a spatial and geographic area or geographic information that currently has been digitized or commonly called digital mapping.

Long ago, the use of maps is very important in many areas of life. For example, visiting a new place in the city or out of town is a very reasonable thing. The problem is sometimes some people difficult to recall the location of a place or direction of the route to the place. So, some people take a long time to rediscover [1].In the event of a disaster, the inhabitants of the hospital are mostly running around with no direction or guidance. There are several empty areas that can be used as assembly points, ie in the north, west, and south. However, this area of utilization has not been maximized because most hospital residents are unaware of the existence of the assembly point. With the evacuation map, expected during a disaster, residents can easily follow the direction of the evacuation arrow to the designated place. The simulation model will also be performed to evaluate the direction of the grooves in the evacuation map being applied [2].

For this reason, the importance of developing Geographic Information System (GIS) capable of storing or documenting spatial information in digital form, so that in various practice the use of the current map can be done more easily. Features not previously included in the mapping can be manifested such as navigation, GPS, more specific place/location search, and so on.

1.2. Android Operating System

Android is a popular and licensed free mobile platform (open source) that has taken over the mobile phone world on a large scale. Most of the smartphone industry chose to use Android as its operating system, so quickly android became the trend center among smartphone users. The developers of programs and applications are competing to create various android apps, so the development of Android in the world into a wave of great changes that have even changed the lifestyle of the crowd [3].

As an open platform, everyone is free to develop application and customization of the operating system from Android without worrying about licensing issues as Google releases Android codes under the Apache license, a software license and an open standard mobile device. Android Studio is one of the IDE for Android development which is a spin-off version of Intellij Java IDE. Android Studio is the latest solution from Google Inc. in the middle so crowded android development needs of late. Android Studio has a new unified build system called Gradle that is fully integrated to support the flexibility of the Android development process [4].

2. Method

Floyd-Warshall's algorithm is one of the variants of dynamic programming, a method that solves problems by looking at the solution to be obtained as an interrelated decision. This means that solutions are formed from solutions that come from the previous stage and there is the possibility of more than one solution [5].

The problem solving that is the focus of the discussion in this study using the Floyd-Warshall algorithm is the shortest path selection between 2 or more marker positions or location markers in a digital map. The Floyd-Warshall algorithm compares each route variation with the distance from each point to the destination point and accumulates the distance between the nodes (intersection points) that are passed. The line that connects between nodes has a value in the magnitude of the number indicating the distance weight. The Floyd-Warshall algorithm will take each line as a calculation material in determining the shortest route. The solution of the shortest route selection problem using the Floyd-Warshall Algorithm will be described in following steps [6-7]:

Step 1: Delete all loop paths

Delete all loop paths in the node set. Each path that starts and ends on the same node is a loop as in node A. A loop path is not needed in route selection because it does not connect more than one node.

• Step 2: Erase the parallel path

Remove parallel paths or paths more than one between two nodes and leave only one path with the smallest weights. Removing the parallel path and making it a single path will streamline route selection.

• Step 3: create a matrix with the Distance Table and Sequence Table

To find the shortest path between two nodes, there are several requirements in the form of a matrix with two tables called Distance Table (D) and Sequence Table (S). Distance Table (D) contains the value of the distance between two nodes, and the Sequence Table (S) contains the name of the node that serves to indicate which nodes are passed and included in the route. If a routing graph has an

number of ns then table D and table S will have n rows and n columns (table n x n). To do the problem solving used iteration method or looping k as much as n times.
K = iteration
n = number of iterations
D_k = Table D on the k-iteration
S_k = Table S on the k-iteration
d_{ij} = distance between node i and node j
C_{ij}= cell holding d value between node i and j

There is a 0 iterations condition or k = 0 ie the state before any change of value or initial condition of the table based on the distance weight on the original route graph. Then in Table D_0 inserted a value of the weighted distance between every two nodes in accordance with the route graph. In cells that have the same row names and column names, a zero value (0) indicates the absence of a distance weight. Whereas between 2 nodes that do not have a direct connecting line (inserted value) is inserted value infinity (∞).

For Table S_0 we include values in the name of the node in which the node must be passed or is in the middle of node i and node j called pit stop node. However, on the 0th iteration, the value in table S_0 is still adjusted to table D_0 which uses the direct line so it does not have a pit stop node. The value entered into table S_0 is the value of the original node. Especially in cells that have value ∞ or 0 in table D_0 then adjust to the condition value entered in table S_0 is strip (-).

• Step 4: Iterate

The iteration is done as many times in sequence and each subsequent matrix of the input value is heavily dependent on the previous matrix. In iteration 1 and so on begin to enter value pit stop node to find the combination of routes along with the distance variation. More detailed information regarding the use of pit stop nodes on each iteration can be seen in Table 1.

Table 1. PitstopNodeat Each Iterations

Iterations Number	Table	*Pitstop (p)*
0	D_0, S_0	-
1	D_1, S_1	*A*
2	D_2, S_2	*B*
3	D_3, S_3	*C*
4	D_4, S_4	*D*
N	D_5, S_5	p_n

The purpose of using a pit stop node is to compare the distance (d_{ij}) from node i to node j. The smallest value of the result of the comparison is what will be the value in the table in the next iteration. Mathematically filling the value in the 1st iteration table and so on using the formula as follows:

$$d_{ij} = min\left\{\left(d_{ij}\right)^{k-1}, \left(d_{ip}\right)^{k-1} + \left(d_{pj}\right)^{k-1}\right\} \quad (1)$$

The 1st iteration begins by creating D_1andS_1tables and entering values based on the comparison results using the above formula. Similar to Table D_0 and S_0, in cells with the initial node and the same destination node (C_{ij} | i = j) it loads zero (0) and strip (-) values. Next, each cell of the row and column A in Tables D_1 and S_1 is filled with the jo from Table D_0 and S_0. This is because the 1st iteration uses node A as a pit stop (p) in which the accumulated distance will not gain an addition so that the value will be equal to d_{ip} + d_{kp}. As for the remaining cells are filled with values in accordance with the formula, namely by comparing the value of the dip with d_{ip} + $d_{kp}$and find the minimum value.

In the Floyd-Warshall Algorithm, if a numeric value is confronted with an infinite value (∞) then the minimum value of a number is determined because it is assumed that the infinity value (∞) is always more likely. The same is done for other remaining cells, all looking for the least possible distance if the node travel route is always through the pit stop A (p = A) node. Any change of value is found in table D_1, in table S_1 is also given a change of value by filling the value according to the pit stop node in the iteration. The iteration is stopped at the nth iteration as the limit of the number of iterations to be performed in accordance with the number of nodes present. Back to step making of matrix which explains that Table D is Distance Table (distance) and Table S is Sequence Table (node name), then Table D and Table S which is requirement to find the shortest path between nodes is Table Dn and Table Sn.

• Step 5: check the shortest path using Table D and Table S
In the last step, it is about how to use Table D and Table S to find the shortest path between nodes. Each value in the rows and columns connected in Table D is the value of the shortest distance between the two connected nodes so that it is obtained from table D the value of the dij. After knowing the shortest distance, the next thing to know is which line is used. For that, we will switch to Table S with respect to the value of C_{ij}in it. With regard to the shortest distance (d), it will be known the route of the node to be passed.

3. Result and Discussion

The use of Floyd-Warshall algorithm in this research will use case study of donation picking from donors located in Samarinda City area of East Kalimantan.

3.1. Sample of Reseach

Used 20 respondents as the research sample. This number is an estimate of the number of requests for picking up daily donations.

Table 2. The Nodes of Research Sample

Node	Nama	Latitude	Longitude
0	Kantor KNRP Kaltim	-0.5046652	117.1082298
1	M. Sabransyah	-0.478543	117.151212
2	Richa Rosmita	-0.4722673	117.1508941
3	E. Mutmayanti	-0.4512754	117.1675302
4	Lasiyo	-0.4979957	117.1207656
5	MegaPutri	-0.4536098	117.1585432
6	Suhartanto	-0.4623663	117.1164232
7	HadiMulyadi	-0.4772829	117.1333146
8	Irvana	-0.458934	117.1593112
9	Windy	-0.4618985	117.1839463
10	Thamrin	-0.4527096	117.1332746
11	Siti Fatimah	-0.4944913	117.1545706
12	M. Ubaid	-0.4743301	117.1802164
13	Aditya H.	-0.4778609	117.127978
14	AyuFadhilah	-0.4631318	117.1499509
15	TopanYudha	-0.473763	117.163867
16	Firza	-0.5011503	117.1417186
17	Asmiliyani	-0.47919	117.124319
18	AzrinaNindia	-0.489814	117.107383
19	Robby A.	-0.4895625	117.1407127
20	Nur Wahidah	-0.464537	117.153216

3.2. Route Graph

After the location or coordinates of the donors obtained then will form a routing graph that shows the connection between the nodes (the position of each donor). With Google Maps API and Google service, each node symbolized with a marker will appear in the fragment folder. Each marker has a weighted distance to another marker in a different location. Visually the graph of the formed route can be seen in Figure 1.

Table 3. DistanceTablefrom Node 0 to 20

D	0	1	2	3	4	5	6	7	8	9	10	11	12	13	14	15	16	17	18	19	20
0	99999	5588	5948	8855	1578	7952	4765	4118	7609	9664	6385	5279	8686	3690	6531	7073	3748	3338	1644	3982	6690
1	5588	99999	694	3519	4014	2875	4265	1997	2348	4082	3485	1802	3262	2587	1709	1504	2713	2994	5035	1688	1564
2	5948	694	99999	2969	4397	2231	3990	2033	1746	3853	2919	2491	3271	2624	1015	1453	3353	3055	5217	2223	892
3	8855	3519	2969	99999	7333	1033	5819	4772	1246	2172	3816	4991	2914	5293	2355	2519	6218	5715	7936	5180	2165
4	1578	4014	4397	7333	99999	6463	3969	2682	6088	8086	5197	3782	7116	2366	5041	5495	2358	2116	1742	2408	5170
5	7952	2875	2231	1033	6463	99999	4787	3839	594	2972	2814	4542	3327	4332	1422	2305	5580	4744	6961	4443	1345
6	4765	4265	3990	5819	3969	4787	99999	2501	4789	7516	2158	5536	7223	2142	3733	5429	5130	2057	3197	4043	4102
7	4118	1997	2033	4772	2682	3839	2501	99999	3534	5887	2717	3036	5231	597	2424	3423	2800	1023	3201	1588	2625
8	7609	2348	1746	1246	6088	594	4789	3534	99999	2761	2978	3966	2883	4067	1140	1716	5062	4493	6713	3969	918
9	9664	4082	3853	2172	8086	2972	7516	5887	2761	99999	5731	4866	1435	6475	3786	2591	6397	6907	9064	5702	3433
10	6385	3485	2919	3816	5197	2814	2158	2717	2978	5731	99999	5192	5746	2842	2184	4125	5438	3093	5013	4158	2576
11	5279	1802	2491	4991	3782	4542	5536	3036	3966	4866	5192	99999	3622	3484	3505	2514	1608	3768	5278	1636	3315
12	8686	3262	3271	2914	7116	3327	7223	5231	2883	1435	5746	3622	99999	5828	3589	1821	5211	6245	8286	4708	3194
13	3690	2587	2624	5293	2366	4332	2142	597	4067	6475	2842	3484	5828	99999	2938	4020	2995	433	2646	1919	3172
14	6531	1709	1015	2355	5041	1422	3733	2424	1140	3786	2184	3505	3589	2938	99999	1944	4302	3360	5581	3098	395
15	7073	1504	1453	2519	5495	2305	5429	3423	1716	2591	4125	2514	1821	4020	1944	99999	3905	4443	6533	3113	1564
16	3748	2713	3353	6218	2358	5580	5130	2800	5062	6397	5438	1608	5211	2995	4302	3905	99999	3106	4022	1286	4245
17	3338	2994	3055	5715	2116	4744	2057	1023	4493	6907	3093	3768	6245	433	3360	4443	3106	99999	2221	2155	3601
18	1644	5035	5217	7936	1742	6961	3197	3201	6713	9064	5013	5278	8286	2646	5581	6533	4022	2221	99999	3710	5817
19	3982	1688	2223	5180	2408	4443	4043	1588	3969	5702	4158	1636	4708	1919	3098	3113	1286	2155	3710	99999	3097
20	6690	1564	892	2165	5170	1345	4102	2625	918	3433	2576	3315	3194	3172	395	1564	4245	3601	5817	3097	99999

With the use of Google Maps API then the creation of Sequence tables becomes very simple because basically, the weight of the distance drawn from the line between the marker is already the shortest distance selection.

3.3 Route Selection Algorithm

In order for the purpose of research and the making of this donation pickup program can be achieved, it takes a method or a path or route determination algorithm. Basically, the Floyd-Warshall algorithm is the algorithm for finding the fastest path and the shortest distance between 2 nodes, while the donation pickup program is intended to find paths between more than 2 nodes [6][7].

Theoretically, determining the route can be done by finding the shortest distance from one node to another node in the sequence. Because the actor who runs the pick-up system is KNRP Kaltim, the initial node of the route starts from the KNRP office in East Kalimantan. In table D, the location or coordinates of the KNRP East Kalimantan office marker are located at node 0. The problem is on the determination of which next node will be the second, third, and so on markers. For that, it takes a special algorithm as follows:

a. Anested loop is performed to check the smallest value in each row of Table D.
   $K_1, K_2, K_3, K_4, \ldots K_n$

IOP Conf. Series: Earth and Environmental Science **144** (2018) 012019 doi:10.1088/1755-1315/144/1/012019

b. For each c where i = j changed its value to a high value, in this study used the number 99999 meters or 99 km, assuming there is no distance between 2 nodes in the city of Samarinda (Area 712km^2) that exceeds 99 km.C | i = j, C_{ij} = 99999
c. Perform a value comparison of each cell within each row of Table D with the formula: $d_{ij} = min \{C_{i0}, C_{i1}, C_{i2}, C_{i3}, \ldots C_{in}\}$.
d. Each value of the column with the smallest distance in the previous iteration is replaced with a value of 99999. C | j=(j)$^{k-1}$, $C_{ij} = 99999$.

The creation of distance tables will be based on the weighted distance information between the nodes of the route graph formed using the Google Maps API in Figure 1and Figure 2.

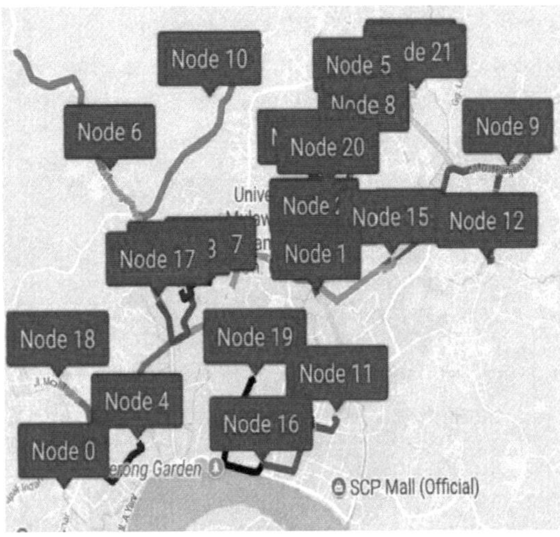

Figure 1. The Route Graph Between the Nodes

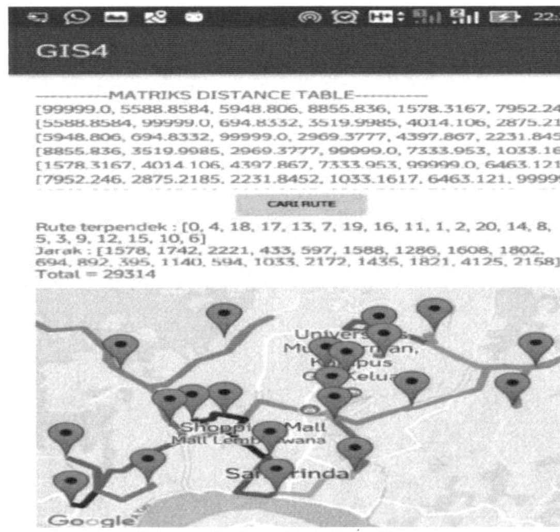

Figure 2. Application Output

3.4 Testing Program

To test whether the program has been created and all the output generated from the program is appropriate, then required comparative data. Using the formula contained in the route selection algorithm, we get:

Table 4. Testing Program

Iteration	Distance value between node i and j	
K_1	$d_{0j} = min \{ 99999, 5588, 5948, 8855, 1578, 7952, 4765, 4118, 7609, 9664, 6385, 5279, 8686, 3690, 6531, 7073, 3748, 3338, 1644, 3982, 6690 \} = \mathbf{1578}$	j=4
K_2	C_{40} = 99999 $d_{4j} = min \{ 99999,4014,4397,7333,99999,6463,3969,2682,6088,8086,5197,3782,7116,2366,5041,5495,2358,2116,1742,2408,5170 \} = \mathbf{1742}$	j=18
K_3	$C_{18\,0}$ = 999999 $C_{18\,4}$ = 999999 $d_{18\,j} = min \{ 99999,5035,5217,7936,99999,6961,3197,3201,6713,9064,5013,5278,8286, 2646,5581,6533,4022,2221,99999,3710, 5817 \} = \mathbf{2221}$	j=17
$K_4, K_5, \ldots K_{20}$		

In the same way, we get the d value from the next iteration up to the last iteration. Manual calculations of the overall iteration performed result in the selection of routes with the shortest weights in sequence as follows:

Table 5. Manual Calculations

Nodes i to j				Distance
0	→	4	=	1578 meters
4	→	18	=	1742 meters
18	→	17	=	2221 meters
17	→	13	=	433 meters
13	→	7	=	597 meters
7	→	19	=	1588 meters
19	→	16	=	1286 meters
16	→	11	=	1608 meters
11	→	1	=	1802 meters
1	→	2	=	694 meters
2	→	20	=	892 meters
20	→	14	=	395 meters
14	→	8	=	1140 meters
8	→	5	=	594 meters
5	→	3	=	1033 meters
3	→	9	=	2172 meters
9	→	12	=	1435 meters
12	→	15	=	1821 meters
15	→	10	=	4125 meters

From the manual data calculation formed a route pattern in the sequence: 0-4-18-17-13-7-19-16-11-1-2-20-14-8-5-3-9-12-15-10-6. This pattern shows the route that must be taken and the order of donor locations that must be visited starting from node 0 to node 6. As for the total distance from node 0 to node 6 is: 0 ➔ 6 = 1578 + 1742 + 2221 + 433 + 1802 + 694 + 892 + 395 + 1140 + 594 + 1033 + 2172 + 1435 + 1821 + 4125 + 2158 = 29314 meters

After the manual testing that has been done, it is found that conformity with the output of the donation pickup program as shown in figure 2 and the program making in this research has resulted in a donation pickup route to 20 donor markers scattered throughout the area in Samarinda with a total distance of 29,314 meters or about 29 km follows the line path of the route to be skipped shown in the map fragment view.

4. Conclusion

The design and implementation carried out to get some conclusions that can be submitted as a result of research, among others; A digital map from Google Maps can be used in making a donation pickup application. The donor location can be poured into a marker form on Google Maps so that the purpose of picking up donations becomes clearer and from the existing 20 donor nodes or markers can be found the best route that has the shortest distance using the Floyd-Warshall algorithm and the route-specific algorithm.

Acknowledgments

Authors wishing to acknowledge for Rector and the Dean of Computer Science and Information Technology Faculty, Mulawarman University through Vice Rector for Academic who has provided funding the research stimulant.

References

[1] Hati, GM, Suprayogi, Andri, S and Bandi 2013Aplikasi Penanda Lokasi Peta Digital Berbasis Mobile GIS pada Smartphone Android *Jurnal Geodesi Undip* **2(4)**pp 26-40

[2] Iftadi, Irwan, Jauhari, Wakhid A &Nugroho, Beny 2011Perancangan Peta Evakuasi Menggunakan Algoritma*Floyd-Warshall* untuk Penentuan Lintasan Terpendek *Performa* **10(2)** pp 95-104

[3] Annuzi, J., Darcey, L., and Conder, S. (n.d.) *Introduction to android application development.*

[4] Grant, K. and Haseman, C. 2014 *Beginning Android programming*. [S.l.]: PeachPit.

[5] Novandi, Raden AD. 2017 Perbandingan Algoritma Dijkstra dan Algoritma *Floyd-Warshall* dalam Penentuan Lintasan Terpendek (Single Pair Shortest Path). *Makalah IF2251 Strategi Algoritmik, 2,* pp. 1-5

[6] Gutenschwager, Radtke, V¨olker, and Zeller 2012 The Shortest Path Comparison of Different Approaches and Implementations for the Automatic Routing of Vehicles. Proc., of the 2012 Winter Simulation Conference, pp 3312-3323.

[7] Kairanbay M, Hajar M J. 2013 A Review and Evaluations of Shortest Path Algorithms, *International Journal of Science & Technology Research (IJSTR)* **2(6)** June, ISSN 2277-8616 pp 96-104

Identification of Active Compounds in the Root of Merung (*Coptosapelta tomentosa* Valeton K. Heyne)

Fitriyana[1]

[1]Chemical Engineering Department, State Polytechnic of Samarinda, Jl. Dr. Cipto Mangunkusumo Kampus Gunung Lipan Samarinda 75131

E-mail : fitriyana79@gmail.com

Abstract. The roots of Merung (*Coptosapelta tomentosa* Valeton K. Heyne) are a group of shrubs usually found on the margins of secondary dryland forest. Empirically, local people have been using the roots of Merung for medical treatment. However, some researches show that the plant extract is used as a poisonous material applied on the tip of the arrow (dart). Based on the online literature study, there are less than 5 articles that provide information about the active compound of this root extract. This study aimed to give additional information more deeply about the content of active compound of Merung root extract in three fractions, *n*-hexane (nonpolar), ethyl acetate (semi polar) and methanol (polar). The extract was then analysed using Gas Chromatography-Mass Spectrometry (GC-MS). GC-MS analysis of root extract in *n*-hexane showed there were 56 compounds, with the main compound being decanoic acid, methyl ester (peak 5, 10.13%), 11-Octadecenoic acid, methyl ester (peak 15, 10.43%) and 1H-Pyrazole, 3- (4-chlorophenyl) -4, 5-dihydro-1-phenyl (peak 43, 11.25%). Extracts in ethyl acetate fraction obtained 81 compounds. The largest component is Benzoic acid (peak 19, 22.40%), whereas in methanol there are 38 compounds, of which the main component is 2-Furancarboxaldehyde, 5-(hydroxyl methyl) (peak 29, 30.46%).

1. Introduction

Merung (*Coptosapelta tomentosa* Valeton K. Heyne) are classified as shrubs and normally found at the edge of the secondary forest. The present of Merung in the wild is very rare but can be abundant in certain areas. This plant is the original plant of Asia continent both tropical and sub-tropical spread from Myanmar to Indo-China, Thailand, Peninsular Malaysia, Sumatra, Bangka, Java and Kalimantan.

Traditionally, people uses Merung root for healing treatments and other uses. Root extract is usually given to children infected with parasitic worms [1], anti-parasites [2], postpartum drugs [3], anti-malaria [4] and others. The abundance of the use of roots as a traditional medicine makes the starting point of some research on the content of active chemical compounds in it. Active compounds are studied as drug candidates or lead compounds to optimize for potentially more potent compounds minimal toxicity.

Several basic studies have reported that the active ingredients contained in Merung Roots include: 1-hydroxy-2-hydroxy methyl anthraquinone [2], anthraquinones (1 and 3-5) and one naphthoquinone (2) [5], flavonoids, tannins, saponins, terpenoids, alkaloids and phenolic [6][7][8]. However, most of the compounds produced are still general so there is still a gap of research that needs to be filled in order for the merit and health benefits of Merung for health can be known more fully.

This is the basis for the necessity of this research is conducted to enrich information content of active compound of Merung root in order to be utilized more broadly, especially in the field of pharmaceutical. In this research will look more deeply information of active chemical compound content of crude extract of Merung root in three fractions, namely *n*-hexane (nonpolar), ethyl acetate (semi polar) and methanol (polar). The extracts of the three fractions were then analyzed using Gas Chromatography-Mass Spectrometry (GC-MS).

2. Materials and methods

The materials used were the root of Merung from Mekar Baru Village, Busang District, East Kutai Regency, methanol, *n*-hexane and ethyl acetate, all solvents with grade pure analysis (pro analyst). This research is experimental. The obtained data was the result of GC-MS showing chemical components at the root of Merung.

2.1. Place and Time of Research

This research was conducted in research laboratory of Chemical Engineering Department, State Polytechnic of Samarinda and Organic Chemistry Laboratory of Gajah Mada University Yogyakarta in April to August 2017.

2.2. Tools

The tools used for maceration roots and fractionation roots were digital scales, measuring cups, beaker, filter paper, rotary evaporator, separating funnel and glass funnel. While the tools used for the analysis of chemical components was Gas Chromatography-Mass Spectroscopy (70 eV ionization energy, GCMS-QP2010S SHIMADZU). The column used was Rxi 5MS with size 30 m×0.25 mm, thick layer of column 0.25 μm and carrier gas that was Helium.

2.3. Methods

The research will be conducted through the following stages:

2.3.1. Preparation and extraction. The roots were cleaned and chopped into a coarse powder. Rough powder was dried and stored in tightly sealed containers. A total of 100 g of rough powder of Merung root was extracted by maceration technique using methanol for 3×24 hours. Every 24 h, the filtrate was filtered and evaporated using a rotary evaporator at 50°C obtained fixed weight.

2.3.2. Partition / fractionation. The methanol extract was then stratified partitioned from *n*-hexane and ethyl acetate. Liquid-liquid fractionation of each solvent was carried out three times.

2.3.3. Analysis. The chemical contents of *n*-hexane, ethyl acetate and methanol extracts of Merung roots were analysed using Gas Chromatography-Mass Spectroscopy (GC-MS) (ionization energy 70 eV). Individual liquid extract to GC tool. The injector temperature is set to 300 °C. The column temperature is gradient: the initial temperature is 70 °C for 5 min, raised by 5 °C/min to 300 °C, 300 °C for 19 min. Qualitative identification of chemical components is done by comparing the retention time and the resulting mass spectrum with the reference components present in the library.

3. Results

The results of the root component analysis of Merung of each fraction using GC-MS can be seen in Figures 1, 2 and 3. The relative amounts of each component can be determined based on the percentage of the peak region relative to the total peaks.

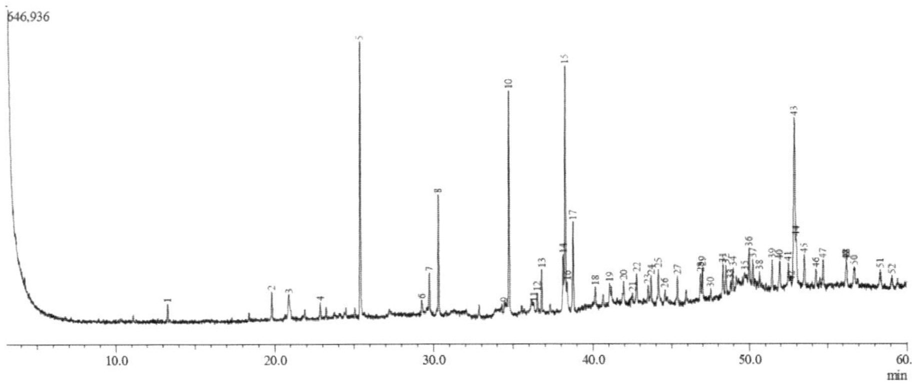

Figure 1.GC-MS chromatogram of extracts of Merung roots in *n*-hexane solvent

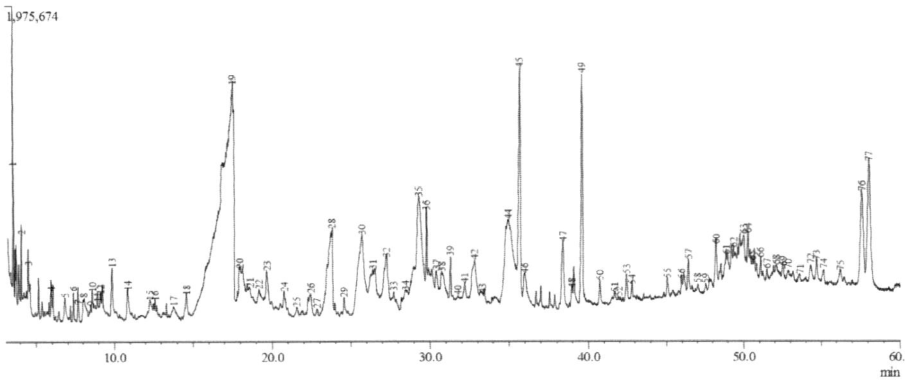

Figure 2. GC-MS chromatogram of extracts of Merung roots in ethyl acetate solvent

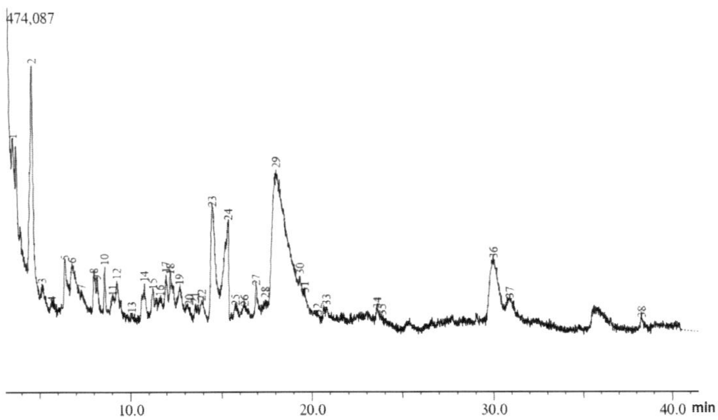

Figure 3.GC-MS chromatogram of extracts of Merung roots in methanol solvent

Based on the results of chromatogram shows there are 56 chemical compounds contained in roots extract *n*-hexane, 81 compounds in ethyl acetate extract and 38 compounds in root-methanol extract. Each peak analysed its molecular weight by a mass spectrometer. The results of the analysis showed that peak number 11, 11 and 10 of the main compounds of the roots in *n*-hexane, ethyl acetate and methanol were sequentially. The main constituent compounds of root Merung extract in each fraction can be seen in Tables 1, 2 and 3.

Based on the results of the matching of spectrograph data from each of the main peaks with the library research it can be known chemical compounds.

Table 1. Characterization of *n*-hexane soluble compounds

Peak	Compounds	Retention Time (min)	Area (%)	Acuracy (%)
2	Decanoic acid, methyl ester	19.811	0.99	91
5	Decanoic acid, methyl ester	25.373	10.13	91
8	Tetradecanoic acid, methyl ester	30.298	4.20	91
10	Eicosanoic acid, methyl ester	34.727	8.39	91
13	Eicosanoic acid, methyl ester	36.788	1.52	91
15	11-Octadecenoic acid, methyl ester	38.262	10.43	92
17	Octadecanoic acid, methyl ester	38.751	3.29	92
24	Hexadecane	43.696	1.14	91
27	Dodecane, 2,5-dimethyl	45.358	1.12	91
29	Tridecane, 2,5-dimethyl	46.951	1.11	91
43	1H-Pyrazole, 3-(4-chlorophenyl)-4,5-dihydro-1-phenyl	52.833	11.25	65

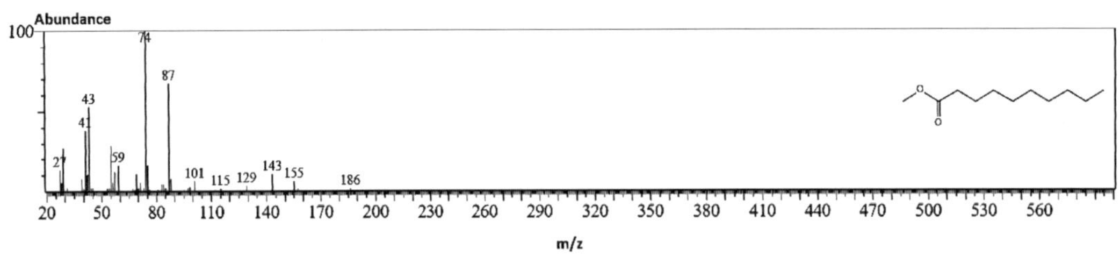

Figure 4. The mass spectrum at peak 5

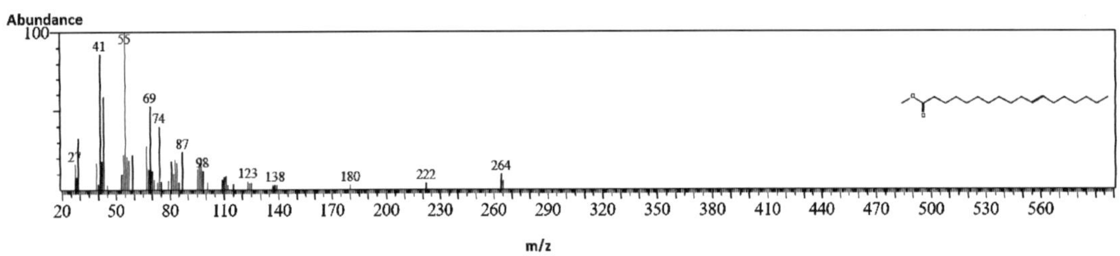

Figure 5. The mass spectrum at peak 15

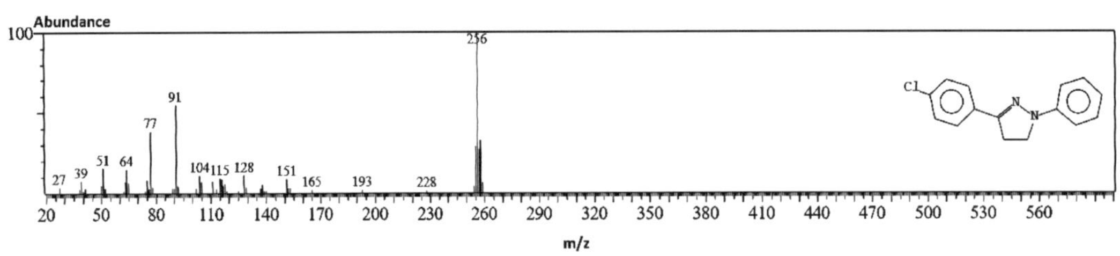

Figure 6. The mass spectrum at peak 43

Table 2. Characterization of ethyl acetate soluble compounds

Peak	Compounds	Retention Time (min)	Area (%)	Acuracy (%)
1	Toluene	3.561	1.37	94
2	Acetic acid, butyl ester	4.070	0.48	95
3	2-Furancarboxaldehyde (CAS) Furfural	4.519	0.87	95
13	2,5-Furandione, dihydro- (CAS) Succinic anhydride	9.851	0.74	95
19	Benzoic acid	17.537	22.40	90
28	Cinnamic Acid	23.762	3.19	91
31	Benzoic acid, 4-hydroxy	26.402	2.20	90
39	Tetradecanoic acid	31.275	0.59	93
44	13-Oxabicyclo [10.1.0] tridecane (CAS) Epoxycyclododecane	34.953	4.70	80
45	Hexadecanoic acid (CAS) Palmitic acid	35.700	3.16	92
49	Octadecanoic acid (Stearic acid)	39.615	2.77	93

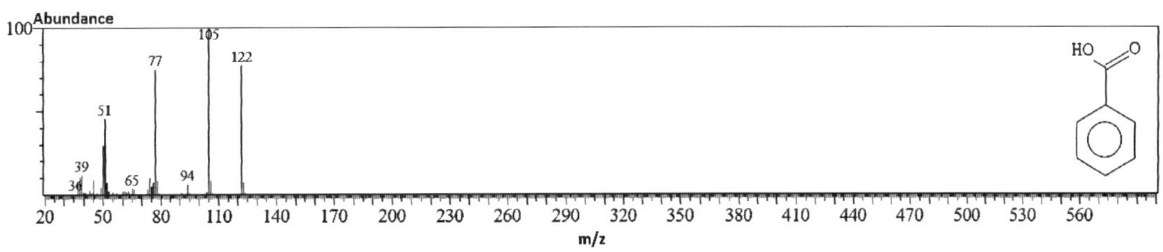

Figure 7. The mass spectrum at peak 19

Table 3. Characterization of methanol soluble compounds

Peak	Compounds	Retention Time (min)	Area (%)	Acuracy (%)
1	2-Propanone, 1-hydroxy (Acetol)	3.488	2.08	79
2	2-Furancarboxaldehyde (Furfural)	4.543	8.51	94
5	Ethanone, 1-(2-furanyl)	6.408	2.36	77
6	2-Furanmethanol (Furfuryl alcohol)	6.795	3.48	79
18	2,5-DIMETHYL-4-HYDROXY-3(2H)-FURANONE	12.208	2.44	84
19	1,4-Cyclohexanediol (Quinitol)	12.694	2.14	77
23	4H-Pyran-4-one, 2,3-dihydro-3,5-dihydroxy-6-methyl	14.496	7.01	88
24	Benzoic acid (Retardex)	15.367	6.60	93
29	2-Furancarboxaldehyde, 5-(hydroxymethyl)	18.031	30.46	87
36	Cyclopentaneacetaldehyde, 2-formyl-3-methyl-.alpha.-methylene- (dolichodial)	29.988	6.79	78

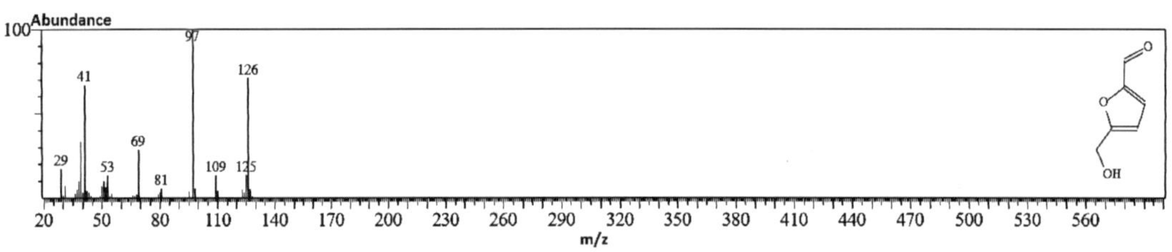

Figure 8. The mass spectrum at peak 29

4. Conclusion

Gas Chromatography-Mass Spectrometry analysis of root extract in n-hexane showed there were 56 compounds, with the main compound being decanoic acid, methyl ester (peak 5, 10.13%), 11-Octadecenoic acid, methyl ester (peak 15, 10.43%) and 1H-Pyrazole, 3- (4-chlorophenyl) -4,5-dihydro-1-phenyl (peak 43, 11.25%). There are 81 compounds in ethyl acetate fraction. The largest component is Benzoic acid (peak 19, 22.40%), whereas in methanol there are 38 compounds, of which the main component is 2-Furancarboxaldehyde, 5-(hydroxyl methyl) (peak 29, 30.46%).

5. Acknowledgement

The authors would like to thank the Head of Research and Service Center for the Society of State Polytechnic of Samarinda for this research grant, Head of Research Laboratory of Chemical Engineering Department, State Polytechnic of Samarinda for permission of laboratory use and Organic Chemistry Laboratory Team of Gajah Mada University Yogyakarta for gas chromatography analysis.

References

[1] Lin, K 2005 Ethnobotanical study of medicinal plants used by the Jah Hut peoples in Malaysia. *Indian Journal of Medical Sciences* **59(4)** 156.

[2] Hounkong, K., Sawangjaroen, N., Kongyen, W., Rukachaisirikul, V., Voravuthikunchai, S. P., & Phongpaichit, S 2014 Anti-intestinal protozoan activities of 1-hydroxy-2-hydroxymethylanthraquinone from Coptosapeltaflavescens. *Asian Pacific Journal of Tropical Disease*. **4**(6) 457–462. https://doi.org/10.1016/S2222-**1808(14)** 60606-X

[3] Grape, V H., Turreira Garcia, N., Schmidt, L H., Phourin, C. & Srisanga, P 2016 Postpartum phytomedicine and its future in maternal healthcare in Prey Lang, Cambodia. *Cambodian Journal of Natural History*. **2016(2)** 119.

[4] Arnida, Rahmawaty Sahi, E. & Sutomo 2017 Aktivitas Anti Plasmodium in vitro dan Identifikasi Golongan Senyawa dari Ekstrak Etanol Batang Manuran (*Coptosapelta tomentosa* Valeton ex K.Heyne) asal Kalimantan Selatan. *Jurnal Ilmiah Ibnu Sina*. **2(2)** 270–278.

[5] Kongyen W, Rukachaisirikul V, Phongpaichit S, Sawangjaroen N, Songsing P, Madardam H. 2014 Anthraquinone and naphthoquinone derivatives from the roots of Coptosapelta flavescens, *Natural Product Communications* **9(2)** 219-20

[6] AchmadFauzi, A. A., Muh., A., & Herman, H. 2015 Aktivitas antioksidan akar Merung (*Coptosapelta tomentosa* Valeton K.Heyne) terhadap radikal bebas DPPH (1,1 difenil-2-pikril hidrazil). Retrieved from http://documentslide.com/documents/jurnal-antioksidan-akar-merung.html#

[7] Ayudia, L 2016 Aktivitas Penghambatan Polimerisasi HEM Fraksi n -Butanol Akar Manuran (*Coptosapelta tomentosa* Valeton ex K. Heyne) Asal Kotabaru Kalimantan Selatan. Retrieved November 27, 2017, from https://www.scribd.com/document/335400174/Artikel-Ilmiah-Farmasi

[8] Supriningrum, R., Sapri, Ali Pranamala, V 2016 Uji Toksisitas Akut Ekstrak Etanol Akar KB (*Coptosapeltato mentosa* Valeton ex K.Heyne) denganMetode Brine Shrimp Lethality Test (BSLT). *Jurnal ilmiah manuntung*, **2(2)** 161–165.

Chemical compositions and antimicrobial potential of *Actinodaphne macrophylla* leaves oils from East Kalimantan

A S Putri, F F Purba, I W Kusuma and H Kuspradini*

Faculty of Forestry, Mulawarman University, Jl Ki Hajar Dewantara Kampus Gunung Kelua, Samarinda, East Kalimantan, 75116, Indonesia

*Corresponding author: hkuspradini@fahutan.unmul.ac.id

Abstract. Essential oils producing plants comprises about 160-200 species, one of which belongs to Lauraceae family. *Actinodaphne macrophylla* is a plant of the Lauraceae family and widely spread on Kalimantan island. For humans, essential oils are used in cosmetics industry, food industry, and pharmaceutical industry. This research aimed to analyze the characteristics of essential oil and potential of antimicrobial activity from *A. macrophylla* leaves oils. Essential oils were obtained by steam distillation method. Antimicrobial activity was assayed using agar diffusion method which compared with two synthetic standards including chlorhexidine and chloramphenicol. Four microorganisms were used in this study were *Candida albicans*, *Staphylococcus aureus*, *Streptococcus mutans*, and *Streptococcus sobrinus*. The obtained oil was determined for its characteristics including the yield, refractive index, and chemical components. The attained components were analyzed using GC-MS. The results of this study showed that essential oils of *A. macrophylla* leaves contained 0.1051% of yield, clearless, and refractive index was 1.425. Based on GC-MS analysis result, it showed chemical components including spathulenol, 2-monopalmitin, (+)-sabinene, copaen, camphene, and β-pinene. This plant potentially can inhibit the growth of *S. aureus*, *C. albicans*, *S. sobrinus*, and *S. mutans* with inhibition zones of 17.22, 20.89, 22.34 and 22.89 mm, respectively.

1. Introduction

Family of Lauraceae contains approximately 45 genera and 2250 species. The genus Actinodaphne belongs to the family Lauraceae. This genus distributes mostly in tropical and subtropical Asia and is an important component of tropical forest. Indonesia, Malaysia, East Asia, and North America (a few) are where the genus mainly distributes, and there are about 70 species of evergreen tree and shurbs there [1]. It is locally known as wuru (Indonesia) or medang kuning and medang kunyit (Malaysia) [2].

Several parts of this genus are pharmaceutically important. The decoction of *A. angustifolia* leaf is used to treat the kidney stones. The seed oil obtained from *A. hookeri* is used in rheumatic pain treatment. The *A. obovata* bark is used to treat fracture. The *A. lancifolia* is very useful for treating arthritis, edema, overexertion, and stomacheache [3,4,5].

The genus Actinodaphne has been reported to contain isoquinoline alkaloids compounds (aporphines, oxoaporphines) [6], lactones [7], lignans [8], and phenolic amides [9]. The alkaloids are important for pharmacologists as antitumor, antibacterial, and antifungal [10]. In some areas, the plant of the genus Actinodaphne is used as an ornamental plant.

Research on the essential oil of family Lauraceae showed that there are several classes of secondary metabolite compounds such alkaloids, phenyl propanoid, flavonoids, 2-pyrron derivatives, and benzyl-esters [11]. However, there are variations in the components of essential oil derived from different growing sites even within the same species [3].

Infection disseases is one of main health problems in the develop countries include Indonesia. Some infection diseases are invading skin tissue, mouth ducts, respiratory and digestion. In the mouth and respiratory canal can occur infection of tooth tissue (dental caries) and lung. *Streptococcus sobrinus* and *S. mutans* are found in the high quantities by the form of biofilm in patients with tooth decay. The growth and development of *Streptococcus sobrinus* and *S. mutans* occur after both of them paste on the tooth and decay it [12]. *S. mutans* is also associated with the severity of periodontal disease in adults [13]. *Staphylococcus aureus* and *Eschericia coli* are a major cause of various humans infections. The first causes skin and soft tissues infections, surgical site infections, and bone and joint infections [14]. *Candida* species are major human fungal pathogens that cause both mucosal and deep tissue infections [15]. *Candida* are known to be aetiological agents of human infection, however, more than 90% of invasive infection are caused by *Candida albicans*, *C. glabrata, C. parapsilosis, C. tropicalis* and *C. krusei* [16].

There are no literature reports on the chemical composition and biological activities of the essential oils from the leaves of *Actinodaphne macrophylla* grown in East Kalimantan. Therefore, this present study used steam distillation to extract the leaf oil of this species and analyzed the chemical components of oil using a GC-MS, also its antimicrobial activity by agar well diffusion.

2. Materials and methods

2.1. Plant Material

Fresh leaves of *Actinodaphne macrophylla*, Lauraceae, were collected in 2015 from Education Forest of Forestry Faculty, Mulawarman University, Samarinda, East Kalimantan. The voucher specimen was deposited in the Dendrology Laboratorium of Mulawarman University. Before extraction, the leaves were air dried at room temperature protected from the light for 24 hours.

2.2 Isolation of essential oils

The essential oil of *A. macrophylla* leaves (10 kg) were extracted using steam distillation method [17]. After extraction, the volume of oil obtained was measured and stored in vial bottle sealed with cap, covered with alumunium foil to protect the contents from light, and kept in the dark storage until used. The volume of leaf oil was presented in the percentage of yield.

2.3 GC-MS analysis

The essential oil was analyzed by GC-MS (Shimadzu-QP-5050A) in the Hasanuddin University, Makassar, with the following setting: column: HP-5 MS, 60 m x 250 µm ID x 0.25 µm film thickness; temperature program: from 70°C to 290°C (40 minutes) at 15°C.minute^{-1}; and injection temperature: 290°C. The injection port temperature was 290°C, and the detector temperature was 250°C. The injection mode was split (50:1), and the inlet pressure was 18.03 psi. The carrier gas was helium with a flow rate of 1 ml. minute^{-1}. The mass spectrometer conditions were as follows: ionization voltage: 70 eV; MS source temperature at 250°C; MS quadruple temperature at 150°C; interface temperature at 290°C; and electron ionization mass spectra were acquired over the mass range of 40-800 m/z. The compositions were reported as a relative percentage of the total peak area.

2.4 Antimicrobial activity

The in vitro antimicrobial activity of the oil against human oral pathogen were evaluated by agar diffusion method using Nutrient agar for bacteria and fungi [17]. The microbial included three bacterias are *Streptococcus mutans*, *Streptococcus sobrinus*, and *Staphylococcus aureus*, and one fungus is *Candida albicans*. The final concentration used in this test were 100% (pure oil), 10%, and 1% which dilluted in 40% ethanol. Two kinds of antibiotic standard as a positive control were used in this study, Chlorhexidine for bacteria and Chloramphenicol for fungi. The 40% ethanol was served as a negative control. After incubation at 32°C for 18-24 hours, the antimicrobial activity values were determined. The diameter of inhibition zones and activity index were measured [18] and calculated [19] in this study. All experiments were performed in triplicate.

3. Results and discussions

Steam distillation of *A. macrophylla* leaves produced a yellowpale-colored oil (figure 1). The oil from *A. macrophylla* leaves obtained 0.11% of a yield (Table 1). The refractive index was measured by hand refractometer was 1.425.

Table 1. Characteristics of *A. macrophylla* leaves oil

Name of plant	*Actinodaphne macrophylla*
Family	Lauraceae
Yield	0.11%
Color	Yellow pale
Refractive index	1.425

Figure 1. The essential oil from leaves of *A. Macrophylla*

Thirty five compunds have been evaluated by GC-MS. The chemical components were identified comparing by retention time, molecular formula, molecular weight, and percentage of area (Table 2). The leaf oil of *A. macrophylla* contained sesquiterpenes were predominant (49.3%), followed by monoterpenes (25.05%), aliphatic and mixed compounds (19.02%), and benzene (0.60%). Among the sesquiterpenes, copaen (7.01%), spathulenol (11.58%), and caryophyllen oxide (5.82%) were the major compounds, and of the monoterpenes, camphene (4.23%), (+)-sabinene (8.49%), β-pinene (4.73%) were the chief components. Among the aliphatic and mixed compounds, 2-monopalmitin was the main component. This is the first study on the chemical characterization from the leaf oil of *A. macrophylla* grown in East Kalimantan.

Table 2. Chemical compositions of *A. macrophylla* leaves oil

No	RT[a] (minutes)	Name of Compounds	MF[b]	MW[c] (g/mol)	Area (%)
1	3.062	Camphene	$C_{10}H_{16}$	136.24	4.23
2	3.193	(+)-Sabinene	$C_{10}H_{16}$	136.24	8.49
3	3.252	β-Pinene	$C_{10}H_{16}$	136.24	4.73
4	3.542	meta-Cymene	$C_{10}H_{14}$	134.22	2.29
5	3.581	D-Limonene	$C_{10}H_{16}$	136.24	1.39
6	3.616	Eucalyptol (1,8-Cineole)	$C_{10}H_{18}O$	154.25	0.85
7	4.010	(+)-Linalool	$C_{10}H_{18}O$	154.25	1.97
8	4.409	(+)-Camphor	$C_{10}H_{16}O$	152.24	1.74
9	4.600	4-Terpineol	$C_{10}H_{18}O$	154.25	1.62
10	5.223	Acetic acid, 1,7,7-trimethyl-bicyclo[2.2.1]hept-2-yl ester	$C_{12}H_{19}ClO_2$	230.73	0.60
11	5.608	α-Cubebene	$C_{15}H_{24}$	204.36	2.82
12	5.811	Copaene	$C_{15}H_{24}$	204.36	7.01

No	RT[a] (minutes)	Name of Compounds	MF[b]	MW[c] (g/mol)	Area (%)
13	5.871 6.472	γ-Muurolen	$C_{15}H_{24}$	204.36	3.04
14	6.629	(-)-δ-Cadinol	$C_{15}H_{26}O$	222.37	1.89
15	6.629	β-Cadinene	$C_{15}H_{24}$	204.36	1.04
16	6.752	(-)-Calamenene	$C_{15}H_{22}$	202.34	1.36
17	6.941	(-)-Spathulenol	$C_{15}H_{24}O$	220.35	1.04
18	6.988	Longipinocarveol, trans-	$C_{15}H_{24}O$	220.35	0.78
19	7.141	Spathulenol	$C_{15}H_{24}O$	220.35	10.54
20	7.187	Caryophyllene oxide	$C_{15}H_{24}O$	220.36	5.82
21	7.246	Epiglobulol	$C_{15}H_{26}O$	222.37	1.15
22	7.325	Humulene oxide	$C_{15}H_{24}O$	220.35	3.10
23	7.385	Cubenol	$C_{15}H_{26}O$	222.37	3.73
24	7.455	(-)-δ –Cadinol	$C_{15}H_{26}O$	222.37	2.07
25	7.507 7.546	Cadala-1(10),3,8-triene	$C_{15}H_{22}$	202.34	3.24
26	7.609	Longiverbenone	$C_{15}H_{22}O$	218.34	1.56
27	7.859	2-Cyclohexen-1-one, 4-butylidene-3-methyl-5-propyl-	$C_{10}H_{16}O$	152.24	1.09
28	7.923	Acetamide, N-tricyclo[4.3.1.1(3,8)]undec-3-yl-	$C_{12}H_{19}NO$	193.29	0.79
29	8.085	Acetamide, N-(3,4-dichlorophenyl)-N- methoxy-	$C10H11Cl_2NO_2$	248.11	0.90
30	8.148	22,23-dibromostigmasterol acetate	$C_{31}H_{48}Br_2O_3$	628.53	0.77
31	8.222	Boronia butenal	$C_{14}H_{22}O$	206.32	1.46
32	8.342	Platambin	$C_{15}H_{26}O_2$	238.37	1.06
33	8.442	Stigmasterol	$C_{29}H_{48}O$	412.70	3.87
34	8.583	Dihydro-β-ionone	$C_{13}H_{22}O$	194.32	1.05
35	9.440	2-Monopalmitin	$C_{19}H_{38}O_4$	330.50	11.23

[a]RT was Retention Time
[b] MF was Molecular Formula
[c]MW was Molecular Weight

From the results presented above, the leaf oil contituents of *A. macrophylla* were primarily sesquiterpenoids. According to the study of Salleh and Ahmad [20], they have been evaluated the chemical compositions of the *A. macrophylla* leaf oil were collected from Simpan Bangi Forest, Selangor, Malaysia, and extracted by hydrodistillation, their main components were germacrene B (16.8%), globulol (16.10%), t-muurolol (7.2%), δ-cadinene (6.7%), cis-cadin-4-en-10-ol (5.6%), and epi-cubenol (4.1%). It is proved that the chemical differences may depend on the procedure of extraction, the season, and the distinct habitat in which the plant was collected.

The essential oil was screened for their possible antimicrobial activity by using agar diffusion. The results are given in Table 3, indicated that a moderate to strong against all four microbes. The most sensitive microorganisms were *S. mutans,* followed by *S. sobrinus, C. albicans,* and *S. aureus* with the diameter of inhibition zones were 20.89, 22.34, 20.89, and 17.22 mm, respectively. The best inhibition zone on the 1% of oil was against *C. albicans* with the diameter 11.89 mm. While On the 10% of oil, the best inhibition was towards *S. aureus* (15.22 mm).

Table 3. Antimicrobial activity of the leaves essential oil of *A. macrophylla*

Sample Conc.	*C. albicans* ZI[a]	*C. albicans* AI[b]	*S. aureus* ZI	*S. aureus* AI	*S. mutans* ZI	*S. mutans* AI	*S. sobrinus* ZI	*S. sobrinus* AI
10 μg/well[c]	31.67 ± 1.53[d]	1.00	15.22 ± 0.19	1.00	15.22 ± 0.39	1.00	16.33 ± 0.39	1.00
100%	20.89 ± 0.58	0.66	17.22 ± 0.86	1.13	22.89 ± 1.51	1.50	22.34 ± 0.77	1.37
10%	12.78 ± 0.39	0.40	15.22 ± 0.86	1.00	11.67 ± 1.22	0.77	11.89 ± 0.58	0.73
1%	11.89 ± 0.19	0.38	10.78 ± 0.39	0.71	0.00 ± 0.00	0.00	9.44 ± 0.77	0.58

[a]ZI was Zone of Inhibition
[b] AI was Activity Index
[c]Concentration of positive control
[d]Chlorhexidine used as positive control against *C. albicans*

The inhibition zones of essential oils against microorganism mainly depend on their chemical composition and the quantity of the major single compounds, concentration of the oil, as well as kind and amount of bacteria [21, 22]. Essential oils primarily destabilize the cellular architecture, leading to the breakdown of membrane integrity and increased permeability, which disrupts many cellular activities. The essential oils pass through the cell wall and cytoplasmic membrane, which may disrupt the arrangement of dissimilar fatty acids, phospholipids bilayers, and polysaccharides molecules. All these events may be responsible for the coagulation of inner cellular components in the cytoplasm and break down of the bonds between the lipid and protein layers [23].

Sokovic et al [24] reported that β-pinene, D-limonene, 1,8-cineol, linalool, and camphor on the concentration of 1.0 μg/ml could inhibited the growth of human pathogen bacterias, especially *Staphylococcus aureus*. The methanolic bark extract of *A. macrophylla* have a good antioxidant activity. Its leaves and bark extract also exhibited moderate the lipoxigenase enzimes inhibitory effect. It has been reported to the antityrosinase, acetylcholinesterase and anti-inflammatory activities of the leaves and bark of *A. macrophylla* [25]. The root bark extracts of *A. lanata* Meissner were potential as antibacterial properties against *Streptococcus pygenes, S. aureus, Bacillus subtilis, Klebsiella penumonia, E. coli,* and *Pseudomonas aeruginosa* [26].

This is the first report describing the effect of *A. macrophylla* leaves oil from East Kalimantan on the human oral pathogen.

4. Conclusion

In conclusion, the results of this study indicated that the essential oil of the *Actinodaphne macrophylla* leaves showed a good antimicrobial activity against all microorganisms. The oil examined in this report are worthy of further study due to their antimicrobial activity. It could be evaluated as one of the alternative natural sources in antimicrobial industry.

Acknowledgements

The authors gratefully acknowledged to Ministry of Research, Technology and Higher Education, Republic of Indonesia, and also members of Laboratory of Forest Products Chemistry.

References

[1] Purkayastha S K 1985 *Indian woods. Their Identification, Properties and Uses Vol V. Oleaceae to Santalaceae* (Delhi: The controller of publications) 165
[2] Burkill I H 1935 *A Dictionary of the Economic Products of the Malay Peninsula* (Kuala Lumpur Malaysia: Ministry of Agriculture Malaysia) 42-43
[3] Palazzo M C, Agius B R, Wright BS *et al* 2009 Chemical Compositions and Cytotoxic Activities of Leaf Essential Oils of Four Lauraceae Tree Species from Monteverde, Costa Rica *Article Records of Natural Producs* **3**(1) 32-37
[4] Sharma M 2014 Comparative wood anatomy of *Actinodaphne* species *International Journal of Plant, Animal and Environmental Sciences* **4** 165-169
[5] Kim M R, Jung H J, Min B S *et al* 2002 Constituents from the Stems of *Actinodaphne lancifolia Phytochemistry* **59** 861-865

[6] Uprety H, Bhakuni D S and Dhar M M 1972 Aporphine alkaloids of *Litsea sebifera*, *L. wightiana* and *Actinodaphne obovata* *Phytochemistry* **11** 3057-3059
[7] Tanaka H, Nakamura T, Ichino K *et al* 1989 Two lactonic compounds, lancifolide and isolancifolide from *Actinodaphne lancifolia* *Phytochemistry* **28** 626-628
[8] Tanaka H, Nakamura T, Ichino K *et al* 1989 A lignan from *Actinodaphne longifolia* *Phytochemistry* **28** 952-954
[9] Tanaka H, Nakamura T, Ichino K *et al* 1989 A phenolic amide from *Actinodaphne longifolia* *Phytochemistry* **28** 2516-2517
[10] Rachmatiah T, Mukhtar M R, Nafiah M A *et al* 2009 (+)-*N*-(2-Hydroxypropyl)lindcarpine: A New Cytotoxic Aporphine Isolated from *Actinodaphne pruinosa* Nees *Journal Molecules* **14** 2850-2856
[11] Guenther E 2006 *Essential Oils 1st Edition* Translate by S Ketaren (Jakarta: UI-Press)
[12] Tamura H and Kato H 2007 *Streptococcus sobrinus* Gene for Hypothetical Protein Partial Cds *NCBI Sequence View* v2.0
[13] Contardo M S, Diaz N, Lobos O *et al* 2011 Oral colonization by *Streptococcus mutans* and its association with the severity of periodontal disease in adults *Rev. Clin. Periodoncia. Implantol. Rehabil. Oral* **4**(1) 9-12
[14] Abulreesh H H and Organji S R 2011 The prevalence of multidrug-resistant staphylococci in food and the environment of Makkah, Saudi Arabia *Res. J. Microbiol.* **6**(6) 510-523
[15] Sardi J C O, Scorzoni L, Bernardi T *et al* 2013 *Candida* species: current epidemiology, pathogenicity, biofilm formation, natural antifungal products and new therapeutic options *Journal of Medical Microbiology* **62** 10-24 DOI: 10.109/jmm.0.045054-0
[16] Pfaller M A, Diekema D J, Procop GW *et al* 2007 Multicenter comparison of the VITEK 2 antifungal susceptibility test with the CLSI broth microdillution reference method for testing amphotericin B, flucytosine, and voriconazole against *Candida* spp *J. Clin. Microbiol.* **45** 3522-3528
[17] Kuspradini H, Putri A S, Sukaton E *et al* 2016 Bioactivity of essential oils from leaves of *Dryobalanops lanceolata*, *Cinnamomum burmannii*, *Cananga odorata*, and *Scorodocarpus borneensis* *Agriculture and Agricultural Science Procedia* **9** 411-418 DOI: 10.1016/j.aaspro.2016.02.157
[18] Kusuma I W, Murdiyanto, Arung E T *et al* 2014 Antimicrobial and Antioxidant Properties of Medicinal Plants Used by The Bentian Tribe from Indonesia *Food Science and Human Wellness* **3** 191-196 DOI: 10.1016/j.fshw.2014.12.004
[19] Dharajiya D, Patel P, Patel M *et al* 2014 In vitro antimicrobial activity and qualitative phytochemical analysis of *Withania somnifera* (L.) Dunal extracts *Int. J. Pharm. Sci. Rev. Res.* **27** 349-54
[20] Salleh W M N H W and Ahmad F 2016 Antioxidant and Anti-inflammatory Activities of Essential Oils of *Actinodaphne macrophylla* and *A. pruinosa* (Lauraceae) *Natural Product Communications* **11**(6) 853-855 ISSN 1555-9475
[21] Baydar H, Sagdic O, Ozkan G *et al* 2004 Antibacterial Activity and Composition of Essential Oils froms Origanum, Thymbra and Satureja Spesies with Commercial Importance in Turkey *Food Con.* **15** 169-172
[22] Nazzaro F, Fratianni F, De Martino L, *et al* 2013 Effect of Essential Oils on Pathogenic Bacteria *Pharmaceuticals* **6**(12) 1451–1474
[23] Swamy M K, Akhtar M S, Sinnah U R 2016 Antimicrobial Properties of Plant Essential Oils against Human Pathogens and Their Mode of Action: An Updated Review *Evidence-Based Complementary and Alternative Medicine* http://dx.doi.org/10.1155/2016/3012462
[24] Sokovič M, Glamočlija J, Marin PD *et al* 2010 Antibaterial Effects of the Essential Oils of Commonly Consumed Medical Herbs Using an In Vitro Model *Molecules* **15** 7532-7546
[25] Salleh W M N H W and Ahmad F 2016 Preliminary investigations of in vitro antioxidant, antityrosinase, acetylcholinesterase and anti-inflammatory activities of *Actinodapne* species

Marmara Pharmaceutical Journal **20** 137-143 DOI: 10.12991/mpj.20162048388
[26] Vimal S and Kumar S R 2014 Antibacterial activity of root bark extract of *Actinodaphne lanata* Meissner *International Journal of Research in Pure and Applied Microbiology* **4**(2) 43-45 ISSN 2277-3843

Forecasting hotspots in East Kutai, Kutai Kartanegara, and West Kutai as early warning information

S Wahyuningsih[1*], R Goejantoro[1], N A Rizki[1]

[1]Study Program of Statistics, Department of Mathematics, Faculty of Mathematics and Natural Sciences, MulawarmanUniversity, Samarinda, Indonesia

Corresponding author: swahyuningsih@fmipa.unmul.ac.id

Abstract. The aims of this research are to model hotspots and forecast hotspot 2017 in East Kutai, KutaiKartanegara and West Kutai. The methods which used in this research were Holt exponential smoothing, Holt's additive dump trend method, Holt-Winters' additive method, additive decomposition method, multiplicative decomposition method, Loess decomposition method and Box-Jenkins method. For smoothing techniques, additive decomposition is better than Holt's exponential smoothing. The hotspots model using Box- Jenkins method were Autoregressive Moving Average ARIMA(1,1,0), ARIMA(0,2,1), and ARIMA(0,1,0). Comparing the results from all methods which were used in this research, and based on Root of Mean Squared Error (RMSE), show that Loess decomposition method is the best times series model, because it has the least RMSE. Thus the Loess decomposition model used to forecast the number of hotspot. The forecasting result indicatethat hotspots pattern tend to increase at the end of 2017 in KutaiKartanegara and West Kutai, but stationary in East Kutai.

1. Introduction

Hotspots areone cause of forest fires. According to BNPB, Indonesian National Board for Disaster Management, East Kalimantan Province has the most hotspots in Kalimantan [1]. Forest fires caused by hotspots are a serious problem.One solution to minimize the problem of forest fires is to provide information as early as possible about the number of hotspots for the future. Temporal analysis is one of the method that can be used to determine the level of potential vulnerability of the region.

There are several preliminary studies related to forecasting hotspots. Detection of fire, spreading of fire, and smoke in forest fires can be used a genetic algorithm [2]. Elman Recurrent Neural Network can be used to forecast air temperature as forest fire factor [3]. Correlation analysis, linear regression and nonlinear regression method can be used as seasonal forecasting of fire over Kalimantan [4]. Prediction and verification forest fire can be used smoothing method and multiple regression [5]. Elman recurrent neural network were used to predict hotspots in Riau Province [6], and Adaptive Neuro Fuzzy Inference System (ANFIS) have used for classification model for forest fire hotspot occurences prediction [7].

The aims of this research are to model hotspots and forecast hotspot 2017 in East Kutai, KutaiKartanegara and West Kutai. The methods which used in this research were Holt's exponential smoothing, Holt's additive dump trend method, Holt-Winters' additive method, additive decomposition method, multiplicative decomposition method, Loess decomposition method and Box Jenkins method.The results of this study are expected to contribute to the government, to immediately be able to provide early information about the number of hotspots in East Kutai, KutaiKartanegera, and West Kutai. Thus forest fires can be prevented.

2. Hotspots

Hotspots by definition can be interpreted as areas that have relatively higher surface temperatures than the surrounding area based on certain temperature thresholds observed by remote sensing satellites. The typology is a point and is calculated as the sum rather than an area. Hotspots are the result of fire/land fire detection at a certain pixel size (eg 1 km x 1 km) which is likely to burn when the satellite passes in relatively cloud-free conditions.

How remote sensing satellites monitor completely the fires of land / forest in a region is illustrated in Figure 1. It is explained that in case of land / forest fires in a location it can be detected by satellites in a hotspot (left), two fires still within a 500 m radius can be detected only one hotspot point (center), otherwise a very large fire incident can be detected as 4 or more hotspots. This illustration indicate that the hotspot point is not the same as the number of land and forest fire incidents in the field.

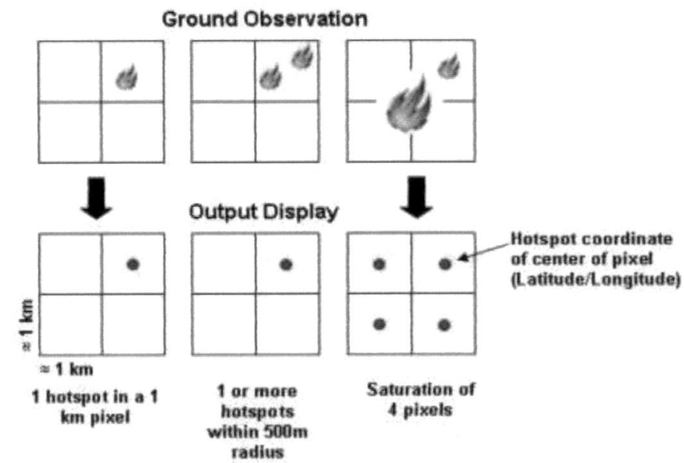

Figure 1. Illustration of land fires using remote sensing satellite data.

3. Methodology

The variable of this study is the number of hotspots of East Kalimantan province (2014-2016) by regency/city taken from LAPAN website (http://modis-catalog.lapan.go.id/monitoring) [8]. The population in this study is the number of hotspots in East Kalimantan, with the sample is the number of hotspots in East Kalimantan 2014-2016. Software used were R 3.1.3 and QGIS 2.14.18. The methods used were the exponential smoothing method, the decomposition method, and the Box-Jenkins method, which is Holt's saturation, Holt's additive damped trend, Holt-Winters' additive, additive decomposition, multiplicative decomposition, Loess decomposition, and Autoregressive Integrated Moving Average (ARIMA).

4. Results and discussion

The plot of hotspots in the province of East Kalimantan from 2014 to 2016 can be seen in Figure 2. The total number of hotspots of 2014 -2016 by regency/city can be seen in Figure 3. Based on Figure 2 and 3 it was found that the areas to watch out for are East Kutai, KutaiKartanegara, and West Kutai, as they had higher number of hotspots than other regions.

IOP Conf. Series: Earth and Environmental Science **144** (2018) 012022 doi:10.1088/1755-1315/144/1/012022

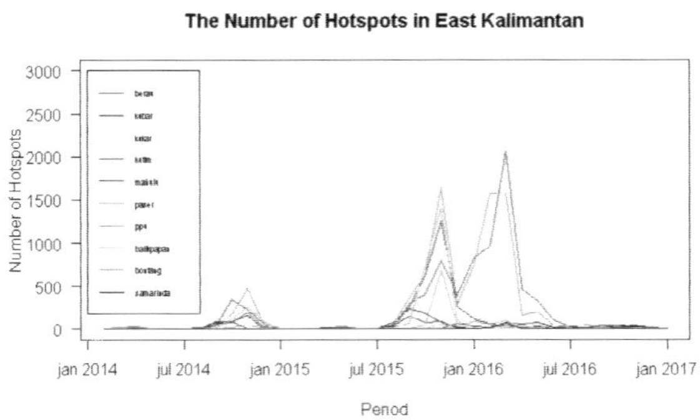

Figure 2. Hotspots distribution in East Kalimantan 2014-2016.

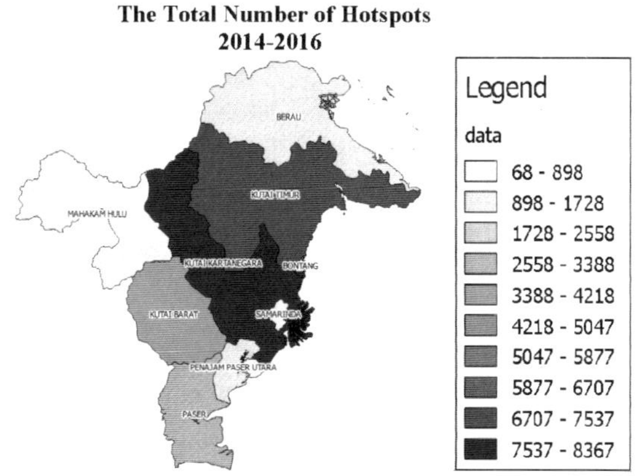

Figure 3. Spasial distribution of total hotspots 2014-2016 in East Kalimantan.

4.1. *Exponential Smoothing Method*

The exponential smoothing method used is Holt's linear trend method, additive damped trend, and Holt-Winters' seasonal method. The RMSE values for each model are shown in Table 1.

Table 1. Fitting Model by Exponential Smoothing

Region	Model	Parameter	RMSE
East Kutai	Holt's linear trend method	$\alpha = 0.6406, \beta = 1\times10^{-4}$	333.5140
	Additive damped trend	$\alpha = 0.6386, \beta = 1\times10^{-4}, \phi = 0.9343$	333.3407
	Holt-Winters seasonal method	$\alpha = 0.6444, \beta = 5\times10^{-4}, \gamma = 8\times10^{-4}$	288.8492
Kutai Kartanegara	Holt's linear trend method	$\alpha = 0.5614, \beta = 1\times10^{-4}$	392.1447
	Additive damped trend	$\alpha = 0.5561, \beta = 1\times10^{-4}, \phi = 0.9144$	391.6077
	Holt's-Winters seasonal method	$\alpha = 0.6317, \beta = 0.0150, \gamma = 0.0122$	323.3836
West Kutai	Holt's linear trend method	$\alpha = 0.8640, \beta = 1\times10^{-4}$	217.0291
	Additive damped trend	$\alpha = 0.8598, \beta = 1\times10^{-4}, \phi = 0.8754$	216.8285
	Holt's-Winters seasonal method	$\alpha = 0.6827, \beta = 1\times10^{-4}, \gamma = 1\times10^{-4}$	172.9971

In Table 1 it can be seen that the RMSE value for the Holt-Winters seasonal method is the smallest for the three regions, so for the best exponential smoothing model for the forecasting model is the Holt-Winters seasonal method.

4.2. *Decomposition Method*

Decomposition method used was additive decomposition, multiplicative decomposition, and Loess decomposition. The RMSE value for each model, can be seen in Table 2.The RMSE value for Loess decomposition is the smallest for the three regions, so for the decomposition model most suitable for the hotspots forecasting model is the Loess decomposition model.

Table 2. Fitting Model by Decomposition Method

Data	Model	RMSE
East Kutai	Additive decomposition	151.3462
	Multiplicative decomposition	292.7014
	Loess decomposition	1.18×10^{-15}
KutaiKartanegara	Additive decomposition	191.177
	Multiplicative decomposition	333.8793
	Loess decomposition	9.47×10^{-15}
West Kutai	Additive decomposition	121.4384
	Multiplicative decomposition	208.8232
	Loess decomposition	3.55×10^{-15}

4.3. Autoregressive Integrated Moving Average (ARIMA)
The first step in analyzing using ARIMA model is checking stationary, model identification, and parameter estimation.This step is done by looking at the time series plot to check the stationary, then proceeded to see the Autocorrelation Functions (ACF) plot and Partial Autocorrelation Functions (PACF) plot. Both of these functions are used to determine the possibility of some ARIMA models being used. Time series plots in Figure 2 show thatthe number of hotspots was non-stationary. To determine the suitable ARIMA model, it is necessary differencing the real data. Figure 4 shows time series plot from hotspots data that have differencing, it can be seen that it is quite stationary. ACF plot for each region shows that ACF plot of East Kutaiis cut off in lag 1 and its PACF is cut off in lag 1 as well. ACF plot and PACF plot of KutaiKartanegara is cut off in lag 2. ACF plot and PACF plot of West Kutai is dies down.

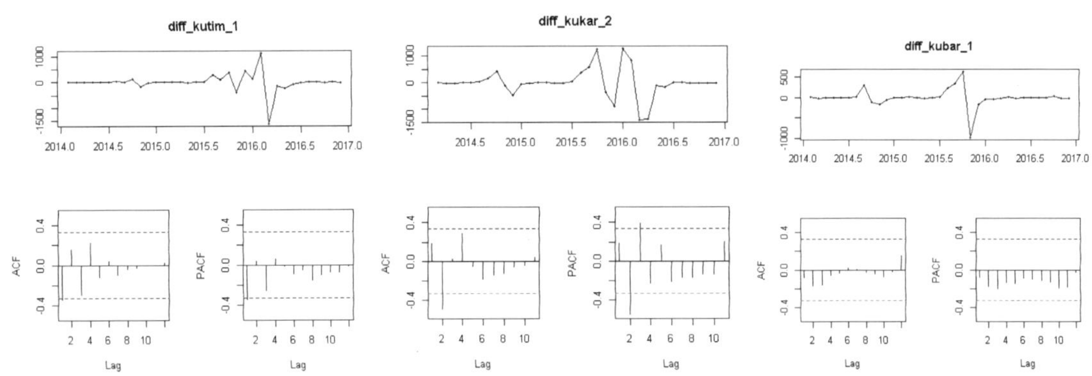

(a) (b) (c)

Figure 4. Time series plot, ACF plot, and PACF plot (a) East Kutai, (b) KutaiKartanegara, (c) KutaiBarat.

Table 3 shows the estimated parameter values of the significant ARIMA models for the number of hotspots in the East Kutai, Kutai Kartanegara, and West Kutai. Based on Table 3, the best models for East Kutai, Kutai Kartanegara and Kutai Barat are ARIMA (1,1,0), ARIMA (0,2,1), and ARIMA (0,1,0), respectively. These models have the smallest RMSE values in each region.

Table 3. The Significant ARIMA Models for The Number of Hotspots

Region	Model	RMSE	Parameter	Coefficient	t
East Kutai	ARIMA(1,1,1)	328.8194	ϕ_1	-0.7753	-3.326
			θ_1	0.5361	1.6795
	ARIMA(1,1,0)	332.7136	ϕ_1	-0.3375	-2.1662
	ARIMA(0,1,1)	333.4766	θ_1	-0.3485	-2.116
	ARIMA(0,1,0)	354.7523			
KutaiKartanegara	ARIMA(3,2,2)	364.6174	ϕ_1	-0.9547	-3.3891
			ϕ_2	-0.5233	-2.5062
			ϕ_3	-0.3868	-2.5995
			θ_1	-0.3227	-1.0225
			θ_2	-0.6773	-2.2728
	ARIMA(3,2,1)	375.2607	ϕ_1	-0.3175	-1.8721
			ϕ_2	-0.3452	-2.1191
			ϕ_3	-0.1505	-0.9205
			θ_1	-1	-6.5963
	ARIMA(3,2,0)	415.4831	ϕ_1	-0.9361	-6.7008
			ϕ_2	-0.816	-5.3368
			ϕ_3	-0.523	-3.9383
	ARIMA(2,2,2)	379.0022	ϕ_1	-0.0897	-0.2595
			ϕ_2	-0.2728	-1.5447
			θ_1	-1.1984	-3.2681
			θ_2	0.1984	0.5751
	ARIMA(2,2,1)	381.898	ϕ_1	-0.2614	-1.6156
			ϕ_2	-0.2975	-1.8841
			θ_1	-1	-9.0662
	ARIMA(2,2,0)	503.5177	ϕ_1	-0.6798	-4.532
			ϕ_2	-0.4394	-3.0075
	ARIMA(1,2,2)	387.2013	ϕ_1	0.2974	0.6897
			θ_1	-1.6228	-3.9745
			θ_2	0.6228	1.5803
	ARIMA(1,2,1)	404.9793	ϕ_1	-0.1882	-1.129
			θ_1	-1	-11.5473
	ARIMA(1,2,0)	568.4219	ϕ_1	-0.464	-3.1458
	ARIMA(0,2,2)	392.6058	θ_1	-1.3696	-6.2855

			θ_2	0.3697	1.9175
	ARIMA(0,2,1)	414.6641	θ_1	-1	-12.4069
	ARIMA(0,2,0)	646.8521			
West kutai	ARIMA(0,1,0)	217.853			

4.4. Selection of The Best Model

The comparison for the best model chosen for each method can be seen in Table 4.The model with Loess decomposition has the smallest RMSE value, making it the best model for predicting the number of hotspots in East Kutai, KutaiKartanegara, and West Kutai.

Table 4. The Comparison of RMSE Values for Each Model

Data	Model	RMSE
East Kutai	Holt	333.51
	Loess	1.18E-15
	ARIMA(1,1,0)	332.71
KutaiKartanegara	Holt	392.14
	Loess	9.47E-15
	ARIMA(0,2,1)	414.66
West Kutai	Holt	217.03
	Loess	3.55E-15
	ARIMA(0,1,0)	217.85

Figure 5 indicates that there is an increase in the number of hotspots in the second half of 2017 for KutaiKartanegara and West Kutai. The forecasting result for East Kutai is stationary (not high at the end of 2017).

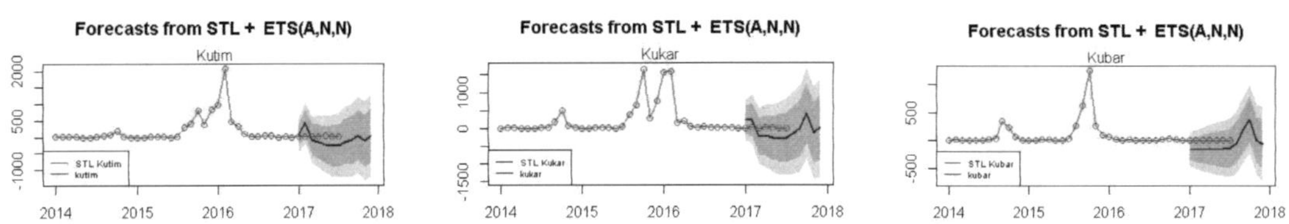

(a) (b) (c)

Figure 5. Forecasting hotspots using Loess decomposition (a) East Kutai, (b) KutaiKartanegara, (c) West Kutai.

5. Conclusions

The best model to forecast the number of hotspots is to use Loess decomposition method. The results of forecasting the number of hotspots in 2017 indicate that there is an increase in the number of hotspots in the second half of 2017 for KutaiKartanegara and West Kutai. The forecasting result for East Kutai is stationary (not high at the end of 2017).These results can be used as input for the forest fire information system.

Acknowledgement
This research was supported by BOPTN Faculty of Mathematics and Natural Sciences Mulawarman University Samarinda

References

[1] BNPB 2015 [http://geospatial.bnpb.go.id/pantauanbencana/data/datakbhutanall.php]

[2] Fithrony 2009 *Detection of Fire, Spreading of Fire,and Smoke in Forest Fire*(D4 Information Technology Department ITS)

[3] Maulida A 2011*Utilization of Elman Recurrent Neural Network to Forecast Air Temperature as Forest Fire Factor* repository.ipb.ac.id/jspui/bitsream/123456789/53985/1/ G11amal.pdf

[4] Spessa A C, Feld R D,Pappenberger F, Langner A,Englhart S,Weber U,Stockdale D,Siegert F,Kaiser J W, and Moore J 2014 Seasonal Forecasting of Fire Over Kalimantan Indonesia*Nat. Hazards Earth Syst.Sci***15**4290442 p5070-5111

[5] Halide H2016*Wild Land Fire : Prediction and Verification*(Menara Intan)

[6] Amaranggana TT 2016 *The Temporal Prediction for the Emergence of Hotpots in Riau Province using Elman Recurrent Neural Network* repository/ipb.ac.id/handle/123456789/ 8554/1/G16tta.pdf

[7] Wijayanto AK, San O, Kartika N D, and Herdiyani Y 2017Classification Model For Forest Fire Hotspot Occurences Prediction*IPB iopscience.iop.org/article/10.1088/1755-1315/54/1/012059*

[8] Lembaga Penerbangan dan Antariksa Nasional2016 *Hotspot Information of Forest / Land Fire*. (Jakarta: Center for Remote Sensing Utilization of Deputy for Remote Sensing LAPAN)

Membrane stabilization activity as anti-inflammatory mechanisms of *Vernonia amygdalina* leaves extracts

MK Nuryanto[1], S Paramita[1], A Iskandar[2]

[1]Laboratory of Community Medicine, Faculty of Medicine, Mulawarman University. Jl. Kerayan Kampus Gunung Kelua, Samarinda 75119, East Kalimantan, Indonesia.
[2]Laboratory of Microbiology, Faculty of Medicine, Mulawarman University. Jl. Kerayan Kampus Gunung Kelua, Samarinda 75119, East Kalimantan, Indonesia.

E-mail: khairul.nuryanto@gmail.com

Abstract. Inflammation is a normal process in the human body as a response to injury from healing process. Meanwhile, chronic inflammation will cause new health problems to patients. Anti-inflammatory agents generally used for those conditions, have several side effects to patients. The aim of this research was to find alternative anti-inflammatory agents, especially from natural sources.*Vernonia amygdalina* knew locally as "daun afrika" belong to family *Apiaceae* is one of those potential natural sources for alternative anti-inflammatory agents. This plant is known astraditionalmedicine from East Kalimantanfor health problems caused by the muscle stiffness and used as material in this research. The experimental method of anti-inflammatory measurement using membrane stabilization activity for *V. amygdalina* leaves extracts. The results showed that significant differences of EC_{50}($p<0.05$)achieved between indomethacin as the positive control (26.39 ± 2.91 µg/mL) with *V. amygdalina* leaves extracts for concentration 1% (131.81 ± 2.95 µg/mL) and 10% (62.54 ± 2.05 µg/mL). EC_{50} of *V.amygdalina* leaves extracts showed the potential anti-inflammatory activities. It could be concluded that *V. amygdalina* leaves extracts to have anti-inflammatory activities, which could be further developed as a new natural source of the anti-inflammatory agents.

1. Introduction

Inflammation is a physiological process in the body in response to a lesion on the body, such as onwork-related fatigue. Acute inflammation can be triggered by various stimuli and characterized by a rapid response of hosts to the site of infection or trauma tissue, i.e. delivery of leukocytes and plasma proteins such as antibodies, to the site of inflammation. Chronic inflammation may progress after acute inflammatory processes, which last for several weeks, months and even years [1].

During an acute and chronic inflammatory process, a number of chemical mediators released. A large number of inflammatory mediators released via arachidonic acid pathways, including prostaglandins, as a result of the breakdown of arachidonic acid by cyclooxygenase enzymes. Although this process is a physiological process in the body, if this process is excessive, it will appear adverse impact on patients. Anti-inflammatory drugs used to overcome this process, with a number of side effects related to the use of these drugs. Some of the most prominent side effects of using anti-inflammatory drugs are side effects on the gastrointestinal system that increase the risk of gastric ulcers; as well as the cardiovascular system that increases the risk of blood vessel occlusion due to blood clots. This is what makes incessant efforts to find alternative anti-inflammatory drugs, especially those derived from natural materials [2].

Indonesia is a country famous for its biodiversity including rich medicinal plants, with ethnic and cultural diversity. Each ethnic group has the cultural repertoire. Each ethnic community has local wisdom, including the use of plants for traditional medicine. Knowledge of the use of medicinal plants by indigenous ethnicity is very important for the traditional andmodern medicine because many plant extracts for modern medicine are found through local knowledge approach. Utilization of medicinal plant data from ethnobotany research is an effective way of finding new chemicals used for treatment [3].

One of the medicinal plants suspected of having potential anti-inflammatory activity is *Vernonia amygdalina* Delile. These shrubs included in the *Apiaceae* family. This plant is commonly named as *daun afrika* in Indonesia or bitter leaf in English. A preliminary observation of the use of *V. amygdalina* is that these plants are commonly consumed by wild chimpanzees that look sick and at some later look healthy and do normally again [4]. The traditional use of this drug is for anti-worms, anti-malaria, and constipation [5]. Based on the above, this research intends to see the mechanism of action of anti-inflammatory extract of *V. amygdalina*leaves by using membrane stabilization test.

2. Material and methods

2.1. Medicinal plant sampling

The study conducted in September 2017. Sampling place of medicinal plants conducted in Kutai Kartanegara Regency, East Kalimantan. Processing of medicinal plant samples to the membrane stabilization testconducted at the Research Laboratory, Faculty of Medicine Mulawarman University.The study protocol approved by the Research Ethics Committee on Faculty of Medicine Mulawarman University.

2.2. Medicinal plant extraction

The crushed simplicia was then macerated with ethanol solvent. In this study used absolute ethanol solvent. For extract preparation, one part of the dry powder of the simplicia put into the macerator (weighed in the maceration bottle before insertion of simplicia and recalculated after insertion of simplicia), added 10 parts of the solvent (measuring volume before inserting). Then soaked for 6 hours while stirring occasionally (using an orbital shaker at 20 rpm for 10 minutes) at room temperature, then leave for 18 hours. The filtrate was separated by filtration using Whatman filter paper. The separation process repeated twice with the same type and number of solvents. All thefiltratewascollected, then evaporated withrotavapor vacuum at50°C until the viscous extract obtained. The resulting extract was then further dried by being placed in the desiccator containing blue silica gel inside an oven at50°C temperature. After obtained dry extract, the calculated yield obtained, the percentage of weight (b/b) between yield with powder weight of simplicia used by weighing. Dry extracts stored in refrigerators -20°C before the further test [6].

2.3. Membrane stabilization test

Fresh blood samples were taken with the addition of anticoagulants. The blood samples were centrifuged at room temperature. Supernatants (plasma and leukocytes) are carefully removed, while red blood cells washed with normal saline. The washing and centrifugation process wererepeated up to 5 times until the supernatant is clear. Then the erythrocyte suspension ready for the membrane stabilization test. This test using erythrocyte suspension with indomethacin as the positive control. The mixture consists of a hyposaline solution of sodium chloride, sodium phosphate buffer, erythrocyte suspension, test material (positive control standard and plant extract) and the last mixture added to normal saline. The test material was not added to the blood control, while the control of the drug was not added to the erythrocyte suspension. The solution mixture incubated in a water bath, followed by centrifugation at room temperature. The discharged hemoglobin is read in the spectrophotometer at a wavelength of 560 nm. Percentage of membrane stability calculated by the following formula:

Membrane Stability = [100 – (Absorbance Treatment – Absorbance Control)] x 100% [7][8].

2.4. Data analysis

Membrane stabilization activity was tabulated in mean ± SD, then calculated the EC_{50} value. Differences to controls analyzed with the ANOVA test and said significant if $p < 0.05$.

3. Results and discussion

Based on Figure 1, the lowest absorbance of *V. amygdalina* 1% was found in dose of 100 mg/ml with its absorbance is 0.383 ± 0.001. Based on Figure 2, the lowest absorbance of *V. amygdalina* 10% was found in dose of 100 mg/mL with its absorbance is 0.190 ± 0.010. Thehalf maximal effective concentration(EC_{50}) value of membrane stabilization activity of *V. amygdalina* 1% is 131.81 ± 2.95 mg/mL and 10% 62.54 ± 2.05 mg/mL when compared with indomethacin 26.39 ± 2.99 mg/mL, showsinFigure 3.

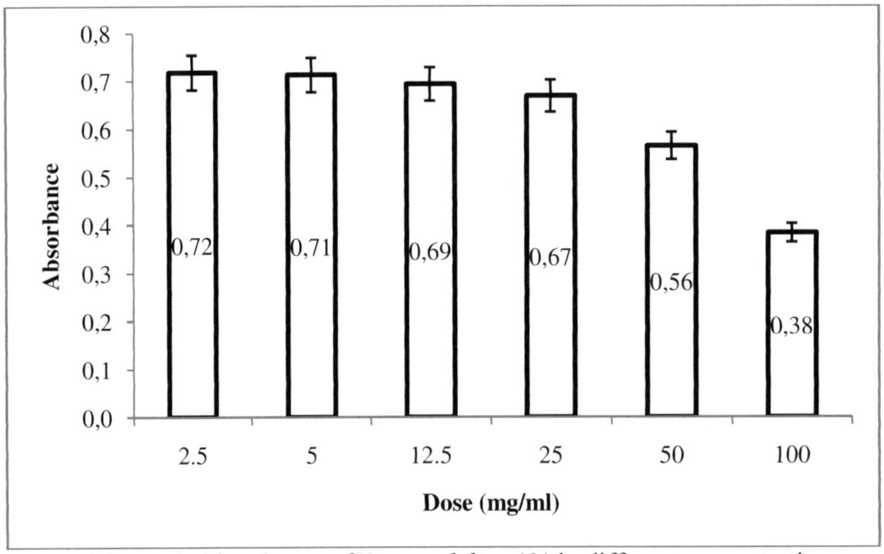

Figure 1. Absorbance of*V. amygdalina* 1% in different concentration

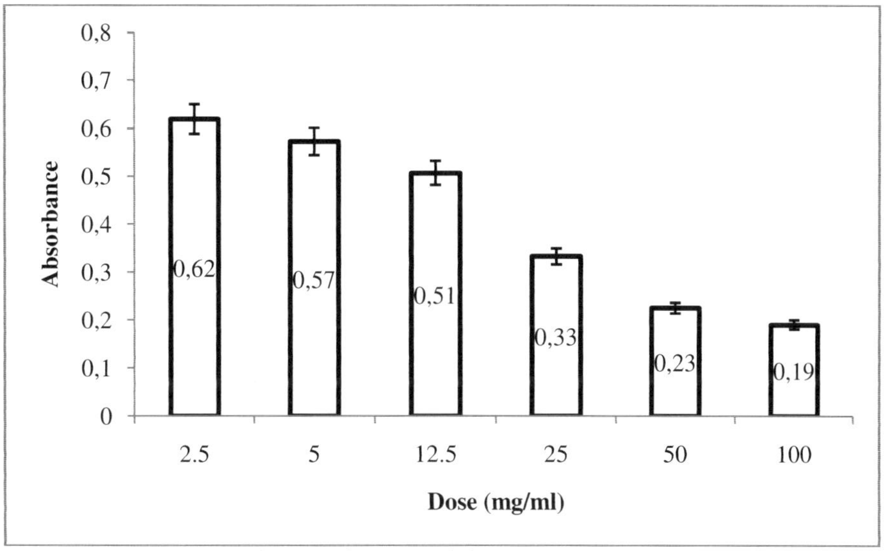

Figure 2. Absorbance of*V. amygdalina* 10% in different concentration

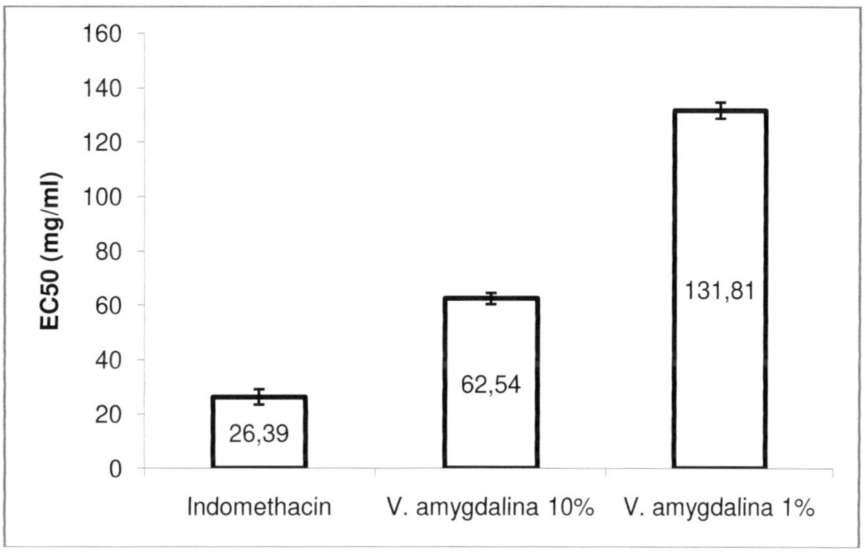

Figure 3. Differences of EC_{50} results in membrane stabilization test between *V. amygdalina*1%, 10%, and indomethacin as positive control (ANOVA p=0.000)

The smaller of EC_{50}value the better the anti-inflammatory activity as membrane stabilization. The results showed that indomethacin as a positive control had better anti-inflammatory activity than *V. amygdalina* on membrane stabilization activity. This suggests that membrane stabilization activity is a possible mechanism of action for the anti-inflammatory activity of *V. amygdalina*.

This plant synonym is*Bracheilema paniculatum* R.Br., *Cacalia amygdalina* Kuntze, *Cheliusia abyssinica* Sch.Bip.ex A.Rich., *Decaneurum amygdalinum* DC., *Vernonia adenosticta* Fenzl exWalp., *Vernonia eritreana* Klatt, *Vernonia giorgii* De Wild., *Vernonia randii* S.Moore, *Vernonia* Benth., and *Vernonia weisseana* Muschl.This plant can grow as high as 2-5 m in the tropics, has a hard stem with a blackish color, and leaves are oval-shaped green with a distinctive odor and bitter taste. The active components of *V. amygdalina* are saponins, alkaloids, terpenes, steroids, coumarins, flavonoids, phenolic acids, lignans, xanthones, anthraquinones, edotid and sesquiterpenes [9]. The biological activity of *V. amygdalina* that showas antibacterial, antifungal, antiparasitic, antimalarial, antiviral, pesticides, antimutagenic, anticancer, anticoagulant, antioxidant, antidiabetic and hypolipidemic effects [10].

There have been studies that have demonstrated central and peripheral analgesic effects of *V. amygdalina*extract, by testing of experimental animals with formalin test and induced writhing assay [11]. Other studies have also found that *V. amygdalina* extract is capable of inhibiting rat paw edema induced by carrageenan and xylene-induced rat ear edema. The study showed that *V. amygdalina* extract had in vivo anti-inflammatory activity [12].

Throughout the researcher's knowledge, this is the first study to investigate the *in vitro*anti-inflammatory activity of *V. amygdalina* leaves extracts, using membrane stability test. The mechanism of *V. amygdalina* leaves extracts as an anti-inflammatory agent suspected through the flavonoid content. An important mechanism for anti-inflammatory activity is inhibition of enzymes that produce eicosanoid, including A2 phospholipase, cyclooxygenase, and lipoxygenase, thereby lowering prostanoid and leukotriene concentrations. Other mechanisms included inhibition of histamine release, phosphodiesterase, protein kinase and transcriptase activation [13].

4. Conclusion

The results of this study show the anti-inflammatory activity of *V. amygdalina* through membrane stabilization mechanism. Further research needed to allow *V. amygdalina* developed as a new anti-inflammatory agent based on medicinal plants.

Acknowledgement

This study was funded by the research grant from Faculty of Medicine Mulawarman University year of 2017.

References

[1] Souza TRCL, Marques GS, Vieira ACQ M and Freitas JC R 2012 State of the Art of Anti-inflammatory Drugs*Pharmacotherapy* ed F Badria (Croatia: Intech) pp 116-40
[2] Pountos I, Georgouli T, Howard Band Giannoudis P V 2011 *Int. J. Interferon Cytokine Mediat. Res.***3**19-27
[3] Ministry of Health Republic of Indonesia 2016 *Proposal Guide, and Advanced Research Protocols: Research on Medicinal and Herbal Plants* (Tawangmangu: Indonesian Center for Research and Development of Medicinal and Traditional Medicines)
[4] Ijeh I I and Ejike CEC C 2011 *J. Med. Plants Res.***5(7)** 1051-61
[5] Zakaria Y, Azlan NZ, Hassan NF N and Muhammad H 2016 *J. Med. Plants Stud.***4(3)** 1-5
[6] Ministry of Health Republic of Indonesia 2008*Indonesian Herbal Pharmacopoeia*(Jakarta: Ministry of Health Republic of Indonesia)
[7] Oyedapo OO, Akinpelu BA, Akinwunmi KF, Adeyinka M O and Sipeolu F O 2010 *Int. J. Plant Physiol. Biochem.***2(4)** 46-51
[8] Omale J and Okafor P N 2008 *Afr. J. Biotechnol.***7(17)** 3129-33
[9] Farombi E O and Owoeye O 2011*Int. J. Environ. Res. Publ. Health***8** 2533-55
[10] Yeap SK, Ho WY, Beh BK, Liang WS, Ky H, Yousr AH N and Alitheen N B 2010 *J. Med. Plant Res.***4(25)** 2787-812.
[11] Njan AA, Adzu B, Agaba AG, Byarugaba D, Díaz-Llera S and Bangsberg D R 2008 *J. Med. Food***11**(3) 574-81
[12] Iroanya O, Okpuzor J and Mbagwu H 2010 *Int. J. Pharmacol.***6** 31-6
[13] Rathee P, Chaudhary H, Rathee S, Rathee D, Kumar V and Kohli K 2009*Inflamm. Allergy Drug Targets***8** 229-35

IOP Conf. Series: Earth and Environmental Science **144** (2018) 012024 doi:10.1088/1755-1315/144/1/012024

Antioxidant and antibacterial activity of *Litsea garciae*

I Wulandari, I W Kusuma and H Kuspradini*

Faculty of Forestry, Mulawarman University, Samarinda, East Kalimantan, Indonesia

*Corresponding Author: hkuspradini@fahutan.unmul.ac.id)

Abstract. *Litsea garciae* is an evergreen tree growing up to 26 meters tall and useful tropical plant species. This plant have medicinal uses. The lightly burned bark can be used to cure caterpillar sting. The purpose of this study was to analyze the characteristics of secondary metabolite compounds, total phenolic and flavonoid content, antioxidant and antibacterial activity from leaf, bark and branch of *L.itsea garciae* plant extracts. Antioxidant activity was determined by using free radical method (1,1-diphenyl 2-picrylhydrazyl). Antibacterial activity against *Propionibacterium acnes* was assayed by 2,3,5-triphenyl tetrazolium chloride method. The sample extracts were obtained using a successive maceration method with n-hexane, ethyl acetate, and ethanol solvent. The result of phytochemical analysis on *Litsea garciae* extract positive contained several secondary metabolite compounds. Among the three sample extracts, the highest of total phenol content present in all three parts of ethanol extract with a value of 0.9-1.0 μg/mg GAE. The highest total flavonoid content was 10.1 μg/mg CE. The highest antioxidant activity was found in ethyl acetate stem extract (86% ± 0.00) at 100 ppm concentration, with IC_{50} at 41.54 ppm. The present work showed that *L. garciae* ethanol extract has potential to inhibit the growth of *P. acnes* bacteria.

1. Introduction

Lauraceae (Medang-medangan) is a large plant and can grow not only in the tropics but can also grow in subtropics. In addition to containing essential oils, Lauraceae has been known to also contain several classes of other secondary metabolite compounds such as alkaloids, phenylpropanoids, flavonoids, benzyl-esters, and alkane-alkene derivatives [1]. *Litsea,* an important genus of the Lauraceae family, is often found in areas such as tropical and subtropical Asia, Australia, and from North America to South American subtropical [2]. Studies of literature related to biological activity indicate that the secondary metabolite compounds contained in the Lauraceae plants indicate the presence of insecticide and cytotoxic activities. Bioactive compounds that have insecticidal activity, among others, from the alkaloids, terpenoids, and flavonoids [3]. The solvents of ethanol and other alcohols are usually used to extract the secondary metabolite compounds of the plant. This solvent is able to increase the permeability of the cell wall and penetrate into the cell to extract the secondary metabolites more and more widely than the extraction with the n-hexane solvent. [4].

Due to rapid increase of antibiotic resistance in our region, it is interesting to determine whether their traditional uses are supported by actual pharmacological effects or merely based on folklore. This study was conducted to determine phytochemicals components, antioxidant and antibacterial activity of *L. garciae* from three different extract.

2. Methods

2.1 Plant material and sample

The branch, bark, and leaf of *Litsea garcia* were collected from Samarinda, East Kalimantan. Research was conducted in Forest Products Chemistry Laboratory of Mulawarman University, Samarinda.

2.2Maceration

The sample (50 grams) was extracted for 1x24 hours, using successive maceration methods with three solvents which has different polarity (n-hexane, ethyl acetate, and 96% ethanol). The sample were filtered to continue the next process.

2.3 Phytochemical screening

Phytochemical analysis was performed on *L. garciae* extract, for the determination of chemical elements including alkaloids, flavonoids, steroids, terpenoids, tannins, saponins, carbohydrates, coumarin, and carotenoids. The procedure for analyzing each of the bioactive compounds is described by Harborne[5]Kokate[6] Senthilmurugan, [7]

2.4 Determination of Total Phenolic Content (TPC)

Total phenolic content was determined using Folin-Ciocalteu reagent according to the method described by Velioglu et al [8]. Extract (100 µL) was mixed with 0.25 mL of Folin-ciocalteu reagent and 2.5 ml distilled water and then 1.25 ml sodium carbonate (60g/L). After 60 min at room temperature, absorbance was measured at 760 nm using spectrophotometer. Standards of gallic acid (Sigma-Aldrich) in the concentration range 10 to 100 µg/mL were run with the test samples, from which a standard curve was plotted. Result was expressed as mg gallic acid equivalents (GAE)/g of dried sample.

2.5 Determination of Total Flavonoid Content (TFC)

Total flavonoid content was determined by using colorimetric method described by Dewanto et al [9] with slight modification. Briefly, 0.1 µL of extract was mixed with 2.5 mL of distilled water in a test tube followed 1.5 mL of a 2% $AlCl_3 \cdot 6H_2O$ (Sigma-Aldrich) solution was added and allowed to stand for another 20 min. The mixture was mixed well with vortex, and the absorbance measured immediately at 420 nm using spectrophotometer. Standards of rutin (Sigma-Aldrich) in the concentration range 10 to 100 µg/mL were run with the test samples, from which a standard curve was plotted. Results were expressed as mg rutin equivalents (RE)/g of dried sample.

2.6 DPPH free radical scavenging Assay

The antioxidant potential of *L. garciae* extract assessed by measuring the scavenging 1,1-diphenyl-2-picrylhydrazyl (DPPH) free radical. The DPPH assay was performed as described by Shimizu et al [10]. The samples were mixed with 33 µl of 467 µl 96% ethanol and filled up with 500 µl DPPH solutions, to a final volume 1 ml. The absorbance of the resulting solution and the blank were recorded after 20 minute at room temperature. Ascorbic acid was used as a positive control. Scavenging of DPPH free radical was recorded spectrophotometrically at 517 nm and the free-radical scavenging activity was calculated as follows:

$$\%\,Scavenging\ Effect = \left(1 - \frac{Abs_{sample}}{Abs_{control}}\right) x100\% \quad (1)$$

2.7 Determination Antibacterial activity

Minimum Inhibitory concentration MIC was determined in the plant extracts that showed some efficacy against the tested isolates, extracts were tested against the isolates for their inhibitory activity, using a common broth microdilution method in 96 multiwell microtiter plates, in duplicate, as reported by [11] with slight modification. For susceptibility testing, 50 μl of Mueller-Hinton broth was distributed from the second to the twelfth test wells extracts from each plant part were dissolved in 96 ethanol to reach a final concentration 1.250 μg/ml. Inoculated plates were incubated at 37°C for 24 h. One hour before the end of incubation 50 μl of a 0.01% solution of 2, 3, 5- triphenyl tetrazolium chloride (TTC) was added to the wells and the plate was incubated for another hour. Since the colorless tetrazolium salt is reduced to a red colored product by biologically active organisms, the inhibition of growth can be detected when the solution in the well remains clear after incubation with TTC. The lowest concentration of each extract showing no visible growth was recorded as the minimum inhibitory concentration (MIC). Inoculated and uninoculated wells of plant extract-free broth was included (the first controls of the adequacy of the broth to support the growth of the organism; the second is a check of sterility). Chloramphenicol was used as positive controls.

3. Results and discussion

3.1 Phytochemical screening

Identification of secondary metabolite content is an important first step in the search for new bioactive compounds from natural ingredients that can be a for new drug synthesis or prototype of drugs.

Table 1. Phytochemical screened of *L. garciae*

Solvent	Part	Alk	Flav	Sap	Tan	Triter	Ste	Car	Cou	Caro
	Branch	+	-	-	+	-	-	+	+	-
n-hexane	Bark	+	+	-	-	-	+	+	+	-
	Leaf	+	+	-	+	-	-	+	+	-
	Branch	+	+	-	+	+	-	+	+	-
Ethilasetat	Bark	+	-	-	-	+	-	+	+	+
	Leaf	+	-	-	+	-	+	+	-	+
Ethanol	Branch	+	+	-	+	+	-	+	+	-
	Bark	+	+	-	+	+	-	+	+	+
	Leaf	+	+	-	-	+	-	+	+	-

Remarks :(+) positive contains, (-) negative contains
Alk; alkaloids, Flav; flavonoids, Sap; saponins, Tan; tannins, Triter; triterpenoid, Ste; steroid, Car; carbohydrate, Cou; Coumarin, Caro; Carotenoids.

Test results show that all extracts formed by using different solvents from branch, bark, Leaf from *L.garciae* extract contain phytochemicals such as alkaloids, flavonoids, terpenoids, steroids well as carotenoid. The test results showed that all the extracts formed by using different solvents from *L.garciae* plants contain phytochemicals such as alkaloids, flavonoids, alkaloids, steroids, and tannins. However, saponins were found to be absent in all of the extract samples from the three solvents.

These results indicates that the different solvents, as the differences in polarity, could selectively extract different secondary metabolites[12]. Containing flavonoid have an effect on arachidonic acid metabolism, thus could have anti-inflammatory, anti-allergic, antithrombotic or vasoprotective effects. The presence of alkaloids in this plant has further confirmed its medicinal use as antibacterial agent [13].

3.2 Phenol and Flavonoid content

Below is a table of test results of total phenol and flavonoid by method Colorimetric $AlCl_3$ method test [10,11] and Folin-Ciocalteu method test [8,9].

Table 2. Total Phenol and Flavonoid content *L.garciae*

Solvent	Part	Phenolic µg GAE/mg extract	Flavonoid µg GAE/mg extract
n-hexane	Branch	30±0.002	190±0.004
	Bark	40±0.002	170±0.003
	Leaf	30±0.001	110±0.002
ethyl	Branch	30±0.002	140±0.005
acetate?	Bark	40±0.004	160±0.005
	Leaf	40±0.004	210±0.004
	Branch	100±0.001	1010±0.002
Ethanol	Bark	90±0.009	800±0.001
	Leaf	100±0.001	240±0.001

Overall, the highest total phenol contained in *Litsea garciae* was obtained from ethanol-soluble extract, averaging 100 µgGAE/ mg extract. The soluble extract of n-hexane and ethyl acetate has a total phenol content of almost the same that ranges between 300-400 µg GAE/ mg extract.

In the total flavonoid test, the stem sample part of *Litsea garciae* extract has the total content of Flavonoid reaches 1010 µgGA / mg extract. In previous research, methanol, ethanol, acetone, ethyl acetate have been used to extract phenols [14].The recovery of phenols from plant materials is influenced by solubility of the phenolic compounds in the solvent used for the extraction process. Otherwise, solvent polarity will play a key role in increasing phenolic solubility [15]. Total flavonoid, as one of the most diverse and widespread groups of natural compounds, are probably the most natural phenolics [16]. Both flavonoid and phenolic compounds are known to have multiple biological effects, including antioxidant and anti-inflammatory properties [17].

3.3 Antioxidants

In simple terms, antioxidants are expressed as compounds capable of inhibiting or preventing oxidation.The results of the antioxidant testing of *L. garciae* plant parts of the three solvents and concentrations can be seen in the table.

Table 3. Antioxidant activity of *L.garciae*

Solvent	Part	Antioxidant (ppm)				
		100	50	25	12.5	IC_{50}(ppm)
n-hexane	Branch	25±0.004	23±0.003	20±0.001	16±0.001	nd
	Bark	49±0.004	27±0.001	17±0.006	4±0.007	nd
	Leaf	19±0.004	15±0.033	4±0.005	3±0.004	nd
Ethyl acetate	Branch	86±0.001	67±0.0010	34±0.003	27±0.008	41.54
	Bark	61±0.0013	52±0.003	52±0.002	23±0.007	55.29
	Leaf	37±0.007	23±0.002	5±0.0012	0±0.007	nd
Ethanol	Branch	62±0.002	66±0.005	57±0.006	4±0.003	52.52
	Bark	60±0.001	58±0,002	52±0.001	46±0.002	19.26
	Leaf	72±0.005	56±0.002	33±0.003	24±0.003	53.76

nd = not detected

The antioxidant activity test was performed by capture free radical of DPPH (1,1-diphenyl diphenyl-2-picrylhydrazyl) by using UV / VIS 1200 spectrophotometer. This method was chosen because it is a simple, quick, and easy method for screening the radical trapping activity of some compounds, but this method proves to be accurate and practical [18]. The antioxidant inhibitory value resulting from the extract sample testing has decreased value at each concentration from 100ppm-12.5ppm.The strong active of *L.garciae* were obtained in bark ethanol extract with IC50 at 19.26 ppm. Differences in the polarity may explain differences in extraction yield and antioxidant activity [19].

3.4 Antibacterial assay

Antibacterial testing using bacteria *Propionibacterium acnes*. Antibacterial activity was performed using Elkhair et al [9] method that has been modified. The concentrations used were 1250, 625 and 312.5 ppm.

Table 4. Minimal Inhibitory Concentration (MIC) of *L.garciae*

Solvent	Part	MIC (ppm)
n-hexane	Branch	625
	Bark	312,5
	Leaf	625
Ethyl acetate	Branch	625
	Bark	1250
	Leaf	312,5
Ethanol	Branch	625
	Bark	312,5
	Leaf	625

In the bark sample using *n*-hexane and ethanol solvent, inhibition occurred at the smallest concentration 312.5ppm. The antimicrobial agents are expressed its potency as minimum inhibitory concentration (MIC) in this method. The method was carried out in a broth dilution test, in which a specific amount of bacteria was added to the serial dilution of antimicrobial agents in broth wells.

After incubation, bacterial growth was indicated by turbidity and its lack was indicated as growth inhibited by the antimicrobial agent. Among the extractions assayed, the n-hexane and ethanol bark and ethyl acetate leaves extracts of *L.garciae*.showed least activity.The have potency as bacterial growth inhibitor. As seen in table 4, all the sample could inhibit the bacteria growth at small concentration range (312.5-625 ppm). Except the stem part ethyl acetate extracts from stem.
The investigated plants that did not show strong antibacterial activity; however, do not mean absence of bioactive constituents nor is that the plant inactive. Active compound(s) may be present in insufficient quantities in the crude extracts to show activity with the dose levels employed [20].The results can be due to the method of plant extraction, the type of microbes used in study, and variation of plant part [21].

4. Conclusion

Based on the results of research conducted, it is known that *L. garciae* extract contains secondary metabolite alkaloids, flavonoids, tannins, carbohydrates, Coumarin which can inhibit free radicals and growth of *P. acnes*.bacteria.

Acknowledgment
The authors gratefully acknowledged to Ministry of Research, Technology and Higher Education, Republic of Indonesia, and also members of Laboratory of Forest Products Chemistry

References

[1] Guenther E 2006 *Essential Oils 1st Edition* translate by S Ketaren (Jakarta: UI Press) [Indonesia]
[2] Tanaka H, Yatsuhashi S, Yasuda T, Sato M, Sakai E and Xiao C 2009 A new amide from the leaves and twigs of *Litseaauriculata.J.Nat. Med.***63** p 331-334
[3] Duke SO 1990 Natural Pesticides From Plants Edunewcrop*Proceedings***VI** p 511
[4] Cannell JPR 1998 *Natural Products Isolation* (Humana Press)
[5] Harborne JB 1987 Phytochemical Method: Modern method Guidelines of plants analysis (Bandung: ITB) [Indonesia]
[6] Kokate CK 2001 *Pharmacognosy 16th Edn*(Mumbai India: NialiPrakasham)
[7] Senthilmurugan GB, Vasanthe and Suresh K 2013 Screening and Antibacterial Activity Analysis of some Important Medicinal Plants*InternationalJournal of Innovation and Applied Studies***2**(2) p146-152
[8] Velioglu YS, Mazza G, Gao L, and Oomah BD 1998 Antioxidant activity and total phenolics in selected fruits, vegetables, and grain products. *J.Agr. Food Chem.***46:** p 4113-4117
[9] Dewanto V, Wu X, Adom KK and Liu RH Thermal 2002 processing enhances the nutritional value of tomatoes by increasing total antioxidant activity. *J. Agr. Food Chem.* **50**: p 3010-3014.
[10] Shimizu K, Kondo R, Sakai K, Takeda N, Nagahata T and Oniki T 2001 novel vitamin E Devirate with 4-substituted Resorcinol Moiety has both antioxidant and Tyrosinase Inhibitory Properties*Lipids***36**: p 21 – 25
[11] Elkhair EA, Fadda H and Mohsen UA 2010 Antibacterial Activity andPhytochemical Analysis of Some Medicinal Plants from Gaza Strip-Palestine*Journal of Al Azhar University*
[12] Rebey IB, Bourgou S, Debez IBS, Karoui IJ, Sellami IH and Msaada K 2012 Effects of extraction solvents and provenances on phenolic contents and antioxidant activities of cumin (*Cuminumcyminum L.)* seeds. Food and Bioprocess Technology **5**: p 2827–2836.
[13] Evans WE 2002 Treas and Evans Pharmacognosy. 15th ed. WB Saunders Co. Ltd, Philadelphia. Freidewald WT, RI Levy and DS Frederickson 1972. Estimation of the concentration of low-density lipoprotein cholesterol in plasma without use of the preparative ultracentrifuge. Clin.

Chem.**18** : p 499-502.

[14] Tan ES, A Aminah H M Khalid, M Y Maskat and M A Ghani 2012 Antioxidant properties of three banana cultivars (*Musa acuminata* '*Berangan*, '*Mas*' and '*Raja*') extracts Sains Malays **41** : p 319-324.

[15] Naczk M andShahidi F J 2006 Pham Biomed Anal **41**: p 1523-1542.

[16] Shimoi K, Masuda S, Shen B and Kinae N 1996 Mutat.Res**350** : p 153-161.

[17] Amarowicz R 2007 Eur. Lipid Sci. Technol**109**: p 549–551

[18] Febriani K 2012 Antioxidant Activity of extract and fraction of *Cocculusorbiculatus*(L) DC using DPPH method and Identification of Chemical components group from active fraction (Depok: FMIPA UniversitasIndonesia) [Indonesia]

[19] Naczk M, F Shahidi 2006 Phenolics in cereals fruits and vegetables: occurrence, extraction, and analysis. *Pharm. Biomed. Anal***41 :**p 1523-1542.

[20] Parekh J, and Chanda S 2007 Antibacterial and phytochemical studies on twelve species of Indian medicinal plants *J African of Biomedical Research* **10**: p 175 – 181.

[21] Ncube N S, Afolayan A J, and Okoh AI 2008: Assessment techniques of antimicrobial properties of natural compounds of plant origin current methods and future trends *J. African of Biotechnology***12** : p 1797-1806.

Antibacterial and glucosyltransferase enzyme inhibitory activity of *helmyntostachyszelanica*

H Kuspradini1*, AS Putri1 and T Mitsunaga

Faculty of Forestry, Mulawarman University, Samarinda, East Kalimantan, Indonesia

**Corresponding author :* hkuspradini@fahutan.unmul.ac.id

Abstract. *Helminthostachyszeylanica* is a terrestrial, herbaceous, fern-like plant of southeastern Asia and Australia, commonly known as tunjuk-langit. This kind of plant have a medicinal properties such as treatment of malaria, dysentery and can be eaten with betel in the treatment of whooping cough. To evaluate the scientific basis for the use of the plant, the antimicrobial activities of extracts of the stem and leaves were evaluated. The bacteria used in this study is *Streptococcus sobrinus,* a species of gram-positive, that may be associated with human dental caries. The dried powdered plant parts were extracted using methanol and 50% aqueous extract and screened for their antibacterial effects of *Streptococcus sobrinus* using the 96 well-plate microdilution broth method. The inhibitory activities of its related enzyme were also determined. The plant extracts showed variable antibacterial and Glucosyltransferase enzyme inhibitory activity while some extracts could not cause any inhibition. It was shown that 50% ethanolics of *Helminthostachyzeylanica*stem have a potency as anti dental caries agents.

1. Introduction

Plant species are sources of well known and medically useful secondary products. The most important secondary plant metabolites are phenols and flavonoids [1,2].They have distinctive biological activity as natural antibacterials which exceed many other synthetic ones [3].

One of the most common infectious disease in the world is dental caries. It is well known that mutans streptococci are the major etiological agents in dental caries. *S. sobrinus* species cause important dental caries diseases because it produces acids,and extra cellular glucans and fructans from sucrose, which are critical factors in the expression of virulence by these microorganisms [4].Therefore the use of antimicrobial agents to control the colonization and accumulation of these cariogenic bacteria on the tooth surface is a logical approach to prevent this common oral disease.*Helminthostachyszeylanica* Linn commonly called "tunjuklangit". This is a pteridophyte and have high medicinal value and used in various parts of the world. The whole plant parts like roots and leaves use as medicine. Whole parts of this plant are used as various medicinal purposes not only Indonesia. In India, the rhizome of *Helminthostachyszeylanica* is chewed with areca to treat whooping cough or used for curing impotency.[5]. In Malaysia, the rhizome used as antidiarrheal agent [6].

Several studies reported that *Helminthostachyszeylanica* contained flavonoid such as ugonstilbene, ugonin and quercetin glucosides [7,8,9,10,11]. However, a detailed study on the antimicrobial, especially on the oral pathogen causing dental caries of this extract has not been conducted yet. Therefore, this research aims were screened for in vitro evaluation of the antimicrobial of plant species against *Streptococcus sobrinus*. The related enzyme glucosyltransferase inhibitory were also assessed.

2. Materials and Methods

2.1. Plant material

Fresh leaves of *C. sumatrana* were collected from East Kalimantan and were provided by the Department of Forest Product Technology, Mulawarman University. A voucher specimen is kept in our laboratory for future reference.

2.2. Preparation of Extracts

Plant material were extracted with methanol and 50% ethanol. After evaporating the extracts to dryness, the extracts were dissolved with 40% ethanol for measuring the polyphenol content, antimicrobial of *Streptococcus sobrinus* and glucosyltransferase (GTase) inhibitory activity.

2.3. Total Phenolic (TP) and Total Flavonoid (TF) Content

Total polyphenol content was determined using the Folin-Ciocalteu method[12], adapted to a microscale. In a 1.5-ml eppendroff tube, 0.79 ml distilled water, 0.01 ml sample extract appropriately diluted, and 0.05 ml Folin-Ciocalteu reagent was added and mixed. After exactly 1 min, 0.15 ml of sodium carbonate (20g/100ml) was added, and the mixture was mixed and allowed to stand at room temperature, for 120 min. The absorbance was read at 750 nm, and total polyphenol concentration was calculated from a calibration curve, using gallic acid as standard (50-800ug/ml), and expressed as mg/g gallic acid equivalent.

The content was determined as described by other authors [13]based on the total content of quercetin, adapted to a microscale. Briefly, 200 ul of aliquots were mixed with 200 ul of methanol containing 5% anhydrous aluminium chloride ($AlCl_3$) and completed to 1 ml with 40% ethanol. After 30 minutes, the absorbances were read at 420 nm against blank containing 40% ethanol (600 ul) and 5% $AlCl_3$ (200 ul). The percentages of flavanoids were calculated from standard curve of quercetin prepared in methanol, and expressed as mg/g quercetin equivalent

2.4. Antimicrobial assay

Themicroorganism that used in this study was *Streptococcus sobrinus* 6715, which are proven cariogenic pathogens. Since mutans streptococci are involved in caries formation, we choose *S. sobrinus* as the test microorganism in this study. *S. sobrinus* 6715 were cultured on Todd Hewwit Broth Agar containing 1% sucrose[14]. The medium was autoclaved at 121°C for 15 minutes and incubated at 37°C for 24 hours without shaking. The crude extracts were tested for bacterial inhibitory in sterile 96-well plates. Fifty micro litters of microbial inoculums were added into the wells containing sample and medium to achieve a final volume of 200µl. The concentration of test extract was prepared with the range of 0 - 450µg/ml using two-fold dilution method. Solvent and medium controls were included on each test plate. In order to dissolve the sample extracts 40% ethanol was used in this study, which showed no significantly inhibitory effect on *S. sobrinus* growth.

Furthermore, the bacterial inhibitory effects of isolated compounds were also measured. Triplicate samples were taken for each test concentration. After incubation for 12 hours at 37°C, *S. sobrinus* growth was estimated spectrophotometrically at 590 nm using micro plate reader. Percent growth was calculated with concentration tested [15].

2.5. Inhibition assay for glucosyltransferase enzyme activity

Preparation of GTase and assay for GTase-inhibitory activity followed the methods as previously described [16] :

2.5.1. Preparation of GTase. *Streptococcus sobrinus* 6715 was grown for 16 hr at 37 °C in 4L of Todd Hewwit (TH) broth. After centrifugation of the liquid medium at 5000 rpm for 10 min, the cell were collected and then extracted with 75 ml of 8M urea at 20 °C for 1 hr with stirring. The crude enzyme solution containing urea was dialyzed against 10 mM potassium phosphate buffer (pH 6) until the urea was removed entirely. One ml of the crude enzyme solution was pipette into microtube and stored in a freezer at -80 °C.

2.5.2. Assay for GTase inhibitory activity. Insoluble glucan synthesized by GTase was measured tubidimetrically. GTase was incubated in 300μl of 0.1 M phosphate buffer (pH 6.0) containing 1% sucrose, 0.1% sodium azide, 0.5% dextran T-10, and in the presence or absence of sample at 37°C for 3 hr. The volume of the crude GTase solution used in the presence in the assay was determined by that giving an absorbance of 1.0 at 590 nm. Inhibition rate is expressed by the following equation: Inhibition rate (%) = 100 x (Ac – As)/Ac. (Ac and As represent absorbance obtained in the control and in the sample dose, respectively.).

3. Result and Discussion

The results for yield, total phenolic and total flavonoid content of *Helminthostachyszeylanica* are presented in Table 1.

Table 1. Yield, Total Phenolic, Total Flavonoids of *Helminthostachyszeylanica*

Part	Solvent	Yield (%)	Phenol (mg GAE/g)	Flavonoid (mg QE/g)
Stem	MeOH	26.62	45.11	31.05
	50% EtOH	21.00	63.44	39.00
Leaves	MeOH	28.21	55.7	36.77
	50% EtOH	20.17	30.34	0.77

Table1 show salist of the extraction yields, total phenolic concent (TPC) and total flavonoid content (TFC) obtained from the two solvent extractions of *Helminthostachyszeylanica* leaves.Themethanol extract of *Helminthostachyszeylanica* gave the greatest yield, whereas the 50% EtOH extract had the lower yield. According to some researchers, aqueous methanol and ethanol have been proven as effective solvents to extract phenolic compounds from different plants [17, 18]. Our results is similar to the report[19, 20]where methanol solvent was most effective in extracting phenolic components from oat bran and young ginger. The phenolics compounds often associated with other biomolecules (polysaccharides, proteins, terpenes, chlorophyll, inorganic compounds etc) and a solvent must be found that it is suitable for extracting them. However, ethanol/water or acetone /water were better solvents compared to ethanol or acetone [21, 22]. These authors also showed that the methanolic extract was better for flavonoid extraction such ascatechin, epicatechinand epigallocatechin. Based on the results of TPC and TFC ,the best extracting solvent for stem was 50% ethanol, while methanol was for leaves. According to the our results, it seems that the yield and efficiency of the phenolics extraction depends on the type and kind of the solvent as well on the flavonoids. For total phenolics and flavonoids extraction from stem parts, 50% ethanol was more efficient compared to methanol, while from leaves parts methanol was more efficient compare to 50% ethanol.

Table 2. Antimicrobial and Glucosyltransferase Inhibitory Activity of *Helminthostachyszeylanica*

	Antimicrobial (%)					GTase inhibition (%)			
Concentration (μg/ml)	0	56	112.5	225	450	0	112.5	225	450
Stem-MeOH	0	3.84	6.59	5.50	35.39	0	36.43	46.88	41.61
Stem-50% EtOH	0	3.23	6.81	11.62	11.77	0	28.39	41.25	47.88
Leaves-MeOH	0	0.00	0.00	0.00	5.38	0	37.19	36.46	44.21
Leaves-50% EtOH	0	8.60	11.19	12.36	12.35	0	69.35	75.42	83.01

Table 2 shows the antimicrobial of *Helminthostachyszeylanica*. The antimicrobial activity of *S. sobrinus* from the plants studied here showed that a direct extraction with methanol is effective to extract the antimicrobial component(s) from stem part of *Helminthostachyszeylanica*. Their effects on the growth of *S. sobrinus* were most likely due to the release of chemicals from the crude extracts into the medium when they were mixed. A number of studies have confirmed antibacterial mainly attributable to its content in polyphenol, even though the correlation was not perfectly matched in this study. The different reactions of *S. sobrinus* to the different extracts indicated that each solvent extracted different chemical components of plants.The differences in the antibacterial activity could be because of the presence of specific antibacterial compounds, in addition to the phenolic compounds [23].Fifty percent ethanol extract of *Helminthostachyszeylanica* crude extracts showed the remarkable value in inhibiting the GTase enzyme at concentration 450 ppm. The mechanism of the toxicity of polyphenols against microbes may be related to inhibition of hydrolytic enzymes (proteases) or other interactions that inactivate microbial adhesins, cell envelope transport proteins and non-specific interactions with carbohydrates [24].

4. Conclusion

The result of the present study suggests that *Helminthostachyszeylanica* can prevent dental caries, since it demonstrated antimicrobial and GTase inhibitory activity against *S. sobrinus*. Therefore, the above active extracts deserve further studies in order to identify and characterize their active components.

References

[1] Hartmann T 2007 From waste products to ecochemicals: fifty years research of plant secondary metabolism. *Phytochemistry***68**pp2831-2846

[2] Jenke-Kodama H, Müller R, Dittmann E 2008 Evolutionary mechanisms underlying secondary metabolite diversity. *Prog Drug Res***65**pp 121-140

[3] Suhaj M 2006 Spice antioxidants isolation and their antiradical activity: A review. *J. Food Composi-tion and Analysis***19** p 531-6

[4] Hamada S, Slade HD 1980 Biology, immunology and carciogenicity of *Streptococcus mutans. Microbiology Review***44**pp 331-84

[5] Kamini S 2007 ImportanceofFernsinHumanMedicine Ethnobotanical Leaflets **11** p231-3

[6] Jalil J, Bidin AA, and Chye TS 1986 Phytochemical Study of Ophyoglosaceae Family Species, *Proc. Malays Biochemistry Society Conference* Fakultas Sains Hayati Universitas Kebangsaan Malaysia Bangi Malaysia**12**, 160164.

[7] Murakami T, Hagiwara M, Tanaka K, Chen CM. 1973 Chemicsheuntersuchungenuber die inhaltsstoffe von *Helminthostachyszeylanica* (L). *Chem Pharm Bull***21** p 1851-22

[8] Chen CC, Huang YL, Yeh PY, Ou JC 2003 Cyclized geranyl stilbenes from the rhizomes of *Helminthostacyszeylanica*. *Planta Med***69**(10) p 964-73

[9] Fitrya, Anwar L, dan E MUgonin J2010 flavonoid from tunjuklangit (Helmynthostchyszeylanica) root extract. *Indones J Chem Univ Gadjah mada*, **10**pp226-31

[10] Yamauchi K, Mitsunaga Tand Batubara I2013 Novel quercetin glucosides from *Helminthostachyszeylanica* root and acceleratory activity of melanin biosynthesis*J Nat Med***67** 369. https://doi.org/10.1007/s11418-012-0672-9.

[11] Hsieh HL, Yang SH, Lee TH et al. 2016 Evaluation of Anti-Inflammatory Effects of *Helminthostachyszeylanica* Extracts via Inhibiting Bradykinin-Induced MMP-9 Expression in Brain Astrocytes *MolNeurobiol***53** 5995. https://doi.org/10.1007/s12035-015-9511-9.

[12] Kaur Sand Poonam M 2014 Study of Total Phenolic and Flavonoid Content,AntioxidantActivityandAntimicrobialPropertiesofMedicinalPlants. *J MicrobiolExp***1**(1): 00005

[13] John B, Sulaiman CT, Satheesh Gand V R K Reddy 2014 Total Phenolics and Flavonoids in Selected Medicinal Plants from Kerala*Int J Pharm Pharm Sci***6**(1), p 406-2

[14] Szynol A, de Soet JJ, Sieben-van Tuyl E, Bos JW, Frenken LG 2004 Bactericidal effects of a fusion protein of Llama heavychain antibodies coupled to glucose oxidase on oral bacteria antimicrobial agents and chemotherapy. *Antimicrob Agents Chemother* **48** p 3390–5

[15] Cai L and Wu CD1996 Compounds from *Syzigium aromaticum* possessing growth inhibitory activity against oral pathogens*J of Natural products***59**(10)p 987-990

[16] Kuspradini H, MitsunagaT and Ohashi H2009 Antimicrobial activity against *Streptococcus sobrinus* and glucosyltransferase inhibitory activity of taxifolin and some flavanonol rhamnosides from kempas (*Koompassia malaccensis*) extractsJ *Wood Sci***55** p 308. https://doi.org/10.1007/s10086-009-1026-4

[17] Siddhuvaj,P.andBecker,K.2003Antioxidantproperties of various solvent extractsof totalphenolic constituents from three different agroclimatic origins of drumstick tree (*Moringa oleifera* Lam) leaves. *JournalofAgriculturalandFoodChemistry***54** p2144-2155

[18] Sultana,B., Anwar,F.and Ashraf,M.2009.Effectof extractionsolvent/techniqueontheantioxidantactivity of selected medicinal plant extracts.*Molecules***14** pp 2167-2180.

[19] SunT,HoC(2005).Antioxidantactivitiesofbuckwheatextracts.*FoodChem.***90** p743–749.

[20] Ghasemzadeh,A.,Jaafar,H.Z.E.andRahmat,A.2011 Antioxidant activities, total phenolics and flavonoids content in two varieties of Malaysia young ginger (*Zingiber officinale* Roscoe). *Molecules* **15** p 4324-4333.

[21] KallithrakaS,Garcia-VigueraC,BridleP,BakkerJ1996Surveyofsolvents for the extraction of grape seed phenolics. *Phytochem. Anal.* **6**p265–267

[22] YilmazY,ToledoR(2006).Oxygenradicalabsorbancecapacitiesof grape/wine industry by products and effect of solvent type on extractionofgrapeseedpolyphenols.*J.FoodCompost.Anal.*,**19** p 41-48.

[23] Ouerghemmiab I, Rebeya IB, Rahalia FZ, Bourgoua S, Pistellic L, Ksouria R, Marzouka B, Tounsia MS 2017 Antioxidant and antimicrobial phenolic compounds from extracts of cultivated and wild-grown Tunisian *Rutachalepensis**Journal of Food and Drug Analysis*25 p 350-359

[24] PylaR , TJ Kim, JL Silva, YS Jung 2010 Enhanced antimicrobial activity of starch-based film impregnated with thermally processed tannic acid, a strong antioxidant*Int. J. Food Microbiol.* **28** pp 154-160

Potential of antioxidant and toxicity of some medical plants used by sub-ethnic communities of Bahau in East Kalimantan

P Rohim, E T Arung and I W Kusuma*

Faculty of Forestry, Mulawarman University, Samarinda, East Kalimantan, Indonesia. 75119.

*Corresponding author : kusuma_iw@yahoo.com

Abstract. Thepurpose of this research is to assay the potential antioxidant and toxicity of several plants from Bahau, a sub-ethnic in East Kalimantan in regard to their utilization as traditional medicines. This research includes phytochemical analysis, DPPH radical and superoxide radical scavenging activity as well as toxicity assay using *Artemiasalina* shrimp larvae. The results of the extraction showed the highest yield was 2,91% obtained from avung tanaq(*Ficus uncinata*), while the lowest is 1.14% obtained from tevoqsalah(*Saccharum sp.*) species. The result of phytochemicals showed that all plants contain alkaloid and carbohydrate.While carotenoids, saponins, triterpenoids and steroids were absence in all plant extracts. The DPPH radical scavenging activity test showed that the lowest IC_{50} value of kayog kue(*Dictamnus albus*)by 23.96 μg/mL. The superoxide radical scavenging activity assay showed IC_{50} values of all extract samples were >100 μg/mL. The toxicity assay showed that LC_{50} values of all samples of extract tested were >1000 μg/mL. The present research suggested good potential activity of some plants from Bahau ethnic and further research oriented to wide uses of the plants as herbal products is needed.

1. Introduction

Antioxidant compounds are widely found in plants, both in flowers, leaves and fruits. Plants containing bioactive compounds such as flavonoids, alkaloids, and terpenoids are potential raw materials that can be used as natural antioxidants [1]. Exploration of medicinal plants as natural antioxidants is needed, especially in medicinal plants commonly used by rural communities. Indonesia has enormous potential in supplying raw materials of medicinal plants comprising 30,000 plant species from a total of 40,000 species of plants in the world, of which 940 are medicinal plants, accounting for 90% of the total number of medicinal plants in Asia [2]. Bahau is a local tribe in East Kalimantan, Indonesia. As a Dayak sub-tribe, the bahau community still uses medicinal plants as an alternative health care for some people Types of diseases such as scabies, wounds, sore eyes, bone breakouts, arthritis, pregnant and postpartum care, diabetes, fever and others [3]. The purpose of this research is to assume the potential antioxidant and toxicity of several plants from Bahau, a sub-ethnic in East Kalimantan in terms of their utilization as traditional medicines.

2. Materials and Methods

2.1. Plant Materials
The bark and roots of medicinal plants were collected from the village of Matalibaq, East Kalimantan, Indonesia. The collected plants were: *Dictamnusalbus, Ficusuncinata, Saccharumsp,* and*Saprosmalatifolia*. Plants were identified by, Raharjo Ari Suwasono, from the Laboratory of Forest Dendrology and Ecology,Mulawarman University and confirmed by reference. The voucher specimen was deposited at the Forest Products Chemistry Laboratory, at same university.

2.2. Preparation of Extracts
Plant samples ca 100 g were extracted with 96% ethanol at room temperature by continuously shaking with a shaker for 48 hours. Afterwards the extract solution was filtered by filter paper to get crude extract using a rotary evaporator at 40 ◦C. The gummy extracts then put in vacuum oven to near dryness to produce plant extracts.

2.3. Phytochemical analysis
Phytochemical analysis of the plant extracts was conducted by a standard procedurePhytochemical analysis of the plant extracts was conducted by a standard procedure which included testing of alkaloid, flavonoid, carbohydrate, carotenoid, coumarin, tannin, saponin, triterpenoid and steroid [4,5,6].

2.4. Antioxidant assay

*2.4.1. DPPH radical inhibition (1,1-diphenyl-2-picrylhydrazyl).*Free radical scavenging activity against DPPH was done with refence to the method of [7]. Thirty three microliters of extract was diluted mixed by 467 μL and 500 μL DPPH. The final sample concentrations were 50, 25, 12.5, and 6.25 μg/mL. The solution was incubated for 20 min and absorbance was measured at 517 nm using a UV-VIS spectrophotometer. Ascorbic acid was used as a positive control.

2.4.2. Superoxide radical inhibition. The antioxidant activityby means of radical scavenging activity was evaluated by superoxide radical scavenging assay with reference to[8]. 500 μL extract diluted was mixed by 500 μL NBT, 500 NADH μL and 50 μL PMS. The final sample concentrations were 50, 25, 12.5, 6.25, 3.125, and 1.56 μg/mL. The solution was incubated for 10 min and absorbance was measured at 560 nm using a UV-VIS spectrophotometer. Ascorbic acid was used as a positive control.

2.5. Toxicity assay
Toxicity activity was evluated by means of a brine shrimp lethality test using [9]. Plant extracts were tested at concentrations of 125, 250, 500, and 1000 μg/mL. After incubation 48 hours in room temperature (25-29 °C), nauplii (larvae) collected by pipette and used for testing. Survivors were counted after 24 hours.

The value of LC_{50} was obtained from the line corresponding to the line plotted concentration versus the lethal percentage.

3. Results and Discussion

3.1. Plant extracts
Four medicinal plants were extracted with ethanol at room temperature to give crude plant extract (Table 1).

Table 1. Yield of MedicinalPlantExtracts in EthanolSolvents.

No	Local Name	Scientific Name	Part used	Yield %
1	AvungTanaq	*F.uncinata*	Root	2.91
2	KayogKue	*D.albus*	Root	1.42
3	KayuJeluh	*S. latifolia*	Stem Bark	2.60
4	Tevoq Salah	*Saccharum sp.*	Root	1.14

3.2. Phytochemical analysis

The results on phytochemical screening of plant extracts were presented in Table 2. Phytochemical screening showed in all extracts containing alkaloids and carbohydrates, whereas flavonoids were only positive in the *F. uncinata* and *Saccharum* sp. Samples, the carotenoids were only positive in the *Saccharum* sp. latifolia, Tannin was only positive in the samples of *F. uncinata* and *S. latifolia.* In the analysis of saponin metabolite compounds, triterpenoids and steroids showed negative results in all extracts.

Tabel 2. Phytochemical analysis of Four Medicinal Plants

No	Local Name	Scientific Name	Part used	Phytochemicals							
				Alk	Flav	Carbo	Caro	Coum	Tan	Sapo	Tri and Ste
1	AvungTanaq	*F.uncinata*	Root	+	+	+	-	+	+	-	-
2	KayogKue	*D.albus*	Root	+	-	+	-	+	-	-	-
3	KayuJeluh	*S.latifolia*	Stem Bark	+	-	+	-	-	+	-	-
4	Tevoq Salah	*Saccharum*sp.	Root	+	+	+	+	+	-	-	-

Remarks:Alk= alkaloid; Fla= flavonoid;Carbo= Carbohydrate Car= carotene; Cou= coumarin; Tan= tannin; Sap= saponin; Tri and ste= triterpenoid or steroid.

The results of phytochemical analysis showed that the samples tested have the potential to be developed as traditional medicine, because they have secondary metabolite compounds that support for the cure of disease. Alkaloid is group of compound having some significant roles of biological activity such as ephedrine for asthma, analgesics for morphine, and anticancer effects of vinblastine [10]. Flavonoidsare known to have anticancer agents, antiviral, antioxidant, antibacterial and anti-inflammatory [11]. Coumarin also showed to play important roles in many useful applications include, antioxidants [12], analgesic [13], anti-cancer [14], anti HIV [15], anti-inflammatory [16], antibiotics [17], anticoagulants [18] and anti tumor [19].

3.3. Antioxidant activity

3.3.1 Inhibition DPPH radical activity

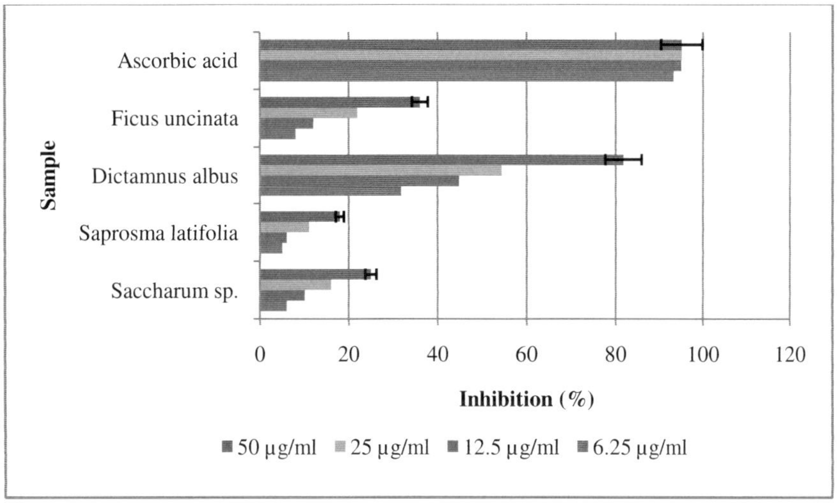

Figure 1. Inhibition of DPPH Radical Percentage Graph

Free radical inhibitory activity was evaluated by the DPPH radical scavenging mechanism. All samples tested showed an increase in inhibition by increasing concentration as presented in Figure 1. The best DPPH radical inhibition activity was demonstrated by a root sample of *D. albus* that caused inhibition of 82% at concentration of 50 µg/mL and was comparable to ascorbic acid with 95%inhibiton. The IC_{50} value presented in Table 3 shows the sample of *D. albus* with the lowest IC_{50} value of 23.96 µg/mL. The result showed that *D. albus*possesses a very strong antioxidant as confirmed by [20].

Tabel 3. Results of DPPH radical inhibition activity assay

No	Local Name	Scientific Name	Part used	IC_{50} (µg/mL)
1	AvungTanaq	*F.uncinata*	Root	49.36
2	KayogKue	*D.albus*	Root	23.96
3	KayuJeluh	*S.latifolia*	Stem Bark	142.69
4	Tevoq Salah	*Saccharum sp.*	Root	99.52
	Ascorbic acid	-	-	3.12

3.3.2. Inhibition of Superoxide Radical Activity

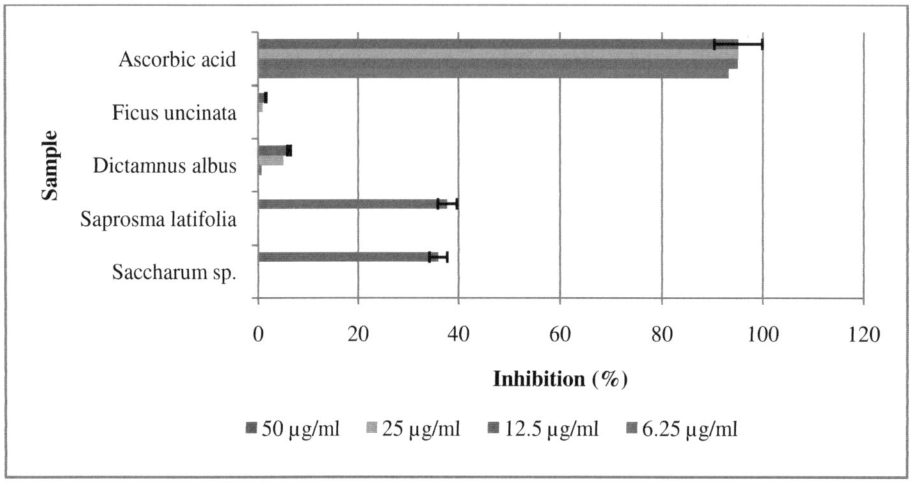

Figure 2. Radical Scavenging Activity of Plant Samples againts-Superoxide Radical

The mechanism of the assay is based on the reduction of nitro blue tetrazolium (NBT) in the presence of NADH (……) and phenazine methosulfate (PMS) under aerobic conditions. All samples tested showed low inhibitory activity as presented in Figure 2. *Saccharum sp.* and*S. latifolia* inhibited superoxide radical by 36% and 37.81% at concentration 50 µg/mL. The result also showed that the IC_{50} values of the extractswere > 100 µg/ml.

Tabel 4. Superoxide Radical Scavenging Activity of The Plants Extracts

No	Local Name	Scientific Name	Part of Plant	IC_{50} (µg/mL)
1	Avung Tanaq	***F. uncinata***	Root	>100
2	Kayog Kue	***D. albus***	Root	>100
3	Kayu Jeluh	***S. latifolia***	Stem Bark	>100
4	Tevoq Salah	***Saccharum sp.***	Root	>100
	Ascorbic acid	-	-	26.24

3.4. Toxicity
Toxicity test by using brine shrimp lethality test as presented in Table 5 showed that all samples had very low toxicity effects with values of LC_{50}> 1000 µg/mL. A low LC_{50} value will exhibit a high toxicity effect, while higher LC_{50}indicates the low toxicity. According to [21] and [22], extracts having LC_{50} values greater than 1000 µg/mL did not show significant toxicity to shrimp brines. This indicates that all of these extracts may not be toxic to humans.

Tabel 5. Toxicity of Plant Extractsagaints Brine Shrimp

No	Local Name	Scientific Name	Part used	ToxicityLC_{50} (µg/ml)
1	Avung Tanaq	***F. uncinata***	Root	>1000
2	Kayog Kue	***D. albus***	Root	>1000
3	Kayu Jeluh	***S. latifolia***	Stem Bark	>1000
4	Tevoq Salah	***Saccharum sp.***	Root	>1000

4. Conclusion

The presentresult indicated that four medicinal plants include the *Ficusuncinata*, *Dictamnusalbus*,*Saprosmalatifolia* and *Saccharumsp*. collected from the Bahau tribe in Matalibaqhave antioxidant potential. The result might be a scientific basis for the development of medicinal plants as natural antioxidant agents.

Acknowledgments

This work was supported by Research and Development of Health, Ministry of Health Republic of Indonesia as advanced research grant funding RISTOJA 2016 (Contract No. LB.02.01/2/1515/F.032/2016).

References

[1] Purwanto D Bahri S Ridhay A 2017*Antioxidant Activity Test of Purnajiwa (Kapsia arborea Blume) Fruit Extract With Various Solvents* Tadulako University KOVALEN**3** (1) : 24-32
[2] APFORGEN 2010*Medicinal Plants Indonesia* (in Indonesia) Asia Pasific Forest Genetic Resources Programme News Letter Edition **2** pp 1-4
[3] Yusro F Mariani Y Diba F 2014*Inventory of Medicinal Plants for Fever used by Four Dayak Sub-Ethnict in West Kalimantan* Indonesia Kuroshio sci **8** pp 33-8
[4] Kokate C K Purohit A P Dan Gokhale S B 2001*Carbohydrate and Derived Products Drugs Containing Glycosides Drugs Containine Tannins Lipids and Protein Alkaloids*Text Book of Pharmacognosy edition **7** pp 133-166 167-254 255-269 272-310 428-523
[5] Harborne JB 1998*Phytochemical Methods A Guide to Modern Techniques of Plant Analysis* 3rd (India: Springer) pp 5-32
[6] Senthilmurugan G B Vasanthe K Suresh 2013*Screening and Antibacterial Activity Analysis of some Important Medicinal Plants* (India: Innovative Space of Scientific Research Journals) **2**(2) pp 14652
[7] Shimizu K Kondo R Sakai K Takeda N Nagahata T Oniki T Novel 2001*Vitamin E derivative with 4-substituted resorcinol moiety has both antioxidant and tyrosinase inhibitory properties :Lipids* (Japan: Springer) **36** (12) pp 1321–6
[8] Liu F VEC Ooi Chang S T 1997*Free radical scavenging activity of mushroom polysaccharide extracts* Life Sci **60** pp 763–71
[9] Meyer H N 1982*Brine Shrimp Lethality Test* (Amsterdam: Plant Research) Med **45** Hipokrates verlag Gmbrl pp 31-34
[10] Lu J J Bao J L Chen X P Huang M and Wang Y T 2012*Review Article Alkaloids Isolated From Natural Herbs as the Anticancer Agents* Evidence-Based Complementary and Alternative Medicine Article ID 485042 12 Pages
[11] ZhenhuaY ZhangW Feng FZhang Y and KangW 2014*α-Glucosidase inhibitors isolated from medicinal plants*(Beijing: Sciencedirect) **3**(3-4) pp 136–74
[12] Vazquez-Rodriguez SFigueroa-Guinez R Matos M J Santana L Uriarte E Lapier M Maya J D Olea-Azar C 2013*Synthesis of Coumarin-chalcone Hybrids and Evaluation of Their Antioxidant and Trypanocidal Properties* Med Chem Commun **4** p 993
[13] Khode S Maddi V Aragade P Palkar M Ronad P K Mamledesai S Thippeswarny A Satyanarayana D 2009*Synthesis and Pharmacological Evaluation of a Novel Series of 5-(Substituted)aryl-3-(3-Coumarinyl)-1-Phenyl-2-Pyrazolines as Novel Anti-inflammatory and Analgesic Agents* Eur J Med Chem **44** p 1682
[14] Wu X Q Huang C Jia Y M Song B A Li J Liu X H 2014 *Novel Coumarin–dihydropyrazole Thio-ethanone Derivatives : Design Synthesis and Anticancer Activity* Eur J Med Chem **74** p 717
[15] Bhavsar D Trivedi J Parekh S Savant M Thakrar SBavishi A Radadiya A Vala H Lunagariya J Parmar M 2011*Synthesis and in Vitro Anti-HIV Activity of N-1,3-benzo[d]thiazol-2-yl-2-*

(2-oxo-2H-chromen-4-yl) Acetamide Derivatives using MTT Method Bioorg Med Chem Lett **21** p 3443

[16] Timonen J M Nieminen R M Sareila O Goulas A Moilanen L J Haukka M Vainiotalo P Moilanen E Aulaskari P H 2011*Synthesis and Anti-inflammatory Effect of a Series of Novel 7-hydroxycoumarin Derivatives* Eur J Med Chem **46** p 3845

[17] Chimenti F Bizzarri B Bolasco A Secci D Chimenti P Carradori S Granese A Rivanera D Lilli D Scaltrito M M 2006*Synthesis and in Vitro Selective anti-Helicobacter Pylori Activity of N-substituted-2-oxo-2H-1-benzopyran-3-carboxamides* Fur J Med Chem **41** p 208

[18] Van S R M Wadelius M Kamali F Daly A K Manolopoulos V G De Boer A Barallon R Verhoef T I Kirchheiner J Haschke-Becher E 2009*Genotype-guided dosing of Coumarin Derivatives : The European Pharmacogenetics of Anticoagulant Therapy (EU-PACT) Trial Design* Pharmacogenomics **10** p 1687

[19] Gouda M A Berghot M A Baz E A Hamama W S 2012*Synthesis Antitumor and Antioxidant Evaluation of Some New Thiazole and Thiophene Derivatives Incorporated Coumarin Moiety* Med Chem Res **21** p 1062

[20] Molyneux P 2004*The use of The Stable Free Radical Diphenylpicryl hydrazyl (DPPH) For Estimating Antioxidant Activity* Songklangkarin J Sci Technol **26** (2) pp 211-9

[21] Mbwambo Z H Moshi M J Masimba P J Kapingu M C Nondo RS O 2007Antimicrobial activity and brine shrimp toxicity of extracts of*Terminalia brownii* roots and stem BMC Compl Altern Med **7**:9 doi 10.1186/1472-6882-7-9

[22] Moshi M J Innocent E Magadula J J Otieno D F Weisheit A MbabaziP K Weisheit A Nondo R S O 2010 Brine shrimp toxicity of someplants used as traditional medicines in Kagera Region north westernTanzania Tanzania J Health Res **12** (1) pp 63-7

IOP Conf. Series: Earth and Environmental Science **144** (2018) 012027 doi:10.1088/1755-1315/144/1/012027

Antidiabetic potential of *Caesalpinia sumatrana*, a medicinal herbs traditionally used by local tribe in East Kalimantan

D A Wicaksono, E Rosamah and I W Kusuma*

Faculty of Forestry, Mulawarman University, Jl. Ki Hajar Dewantara, Samarinda, East Kalimantan, Indonesia. 75119.

*Corresponding author: kusuma_iw@yahoo.com

Abstract. The aims of the research was to analyze the content of phytochemicals, to examine the antioxidant and antidiabeticpotentials of *n*-hexane, chloroform, ethyl acetate, and ethanol extracts of *Caesalpinia sumatrana*. Method to measure antioxidant capacity of sample involves the use of the free radical, 1,1-diphenyl-2-picrylhydrazyl (DPPH) which is widely used to test the ability of compounds to act as free radical. Analysis the potential of antidiabeticactivity of the extracts was determined by α-glucosidase and α-amylase inhibitory assay. Of all extracts obtained by successive maceration, ethanol maceration gave the highest extract by 2.63% of extract on the dry weigh basis. The result of phytochemicals showed that all extracts contain alkaloid and flavonoid. The highest antioxidant activity was 82.32% with IC_{50} value of 5.00 µg/ml obtained by ethanol extract. The results of enzyme inhibitory assay of α-glucosidase showed that ethanol extract of *C. sumatrana*had IC_{50} value 17.16 µg/mLto inhibit α-glucosidase activity and IC_{50} value 16.78 µg/mLfor α-amylase. The present result displayed potential of the plant to be developed as natural antidiabetic and antioxidant agents.

1. Introduction

Indonesia has approximately 30,000 variety of plants, and 7,000 of them are herbs. This needs to be proved, to be developed and to be used to increase the level of healthiness [1].*Tetracera* sp., *Tristaniopsis whiteana*, *Bauhinia purpurea*, *Syzygium* sp., *Parkia roxburghii*, *Cananga odorata*, *Alpinia galangal*, *Passiflora foetida*, *Cinnamomum burmanii*, and *Ligodium circinatum*are several types of plants used by some local tribes in Kalimantan as natural antidiabetic [2]. Bahau ethnic in Matalibaq village in East Kalimantan uses *Caesalpinia sumatrana*, as herbs for diabetic. These herbs need to be validated scientifically. Diabetes is one of the largest global health emergencies of the 21st century. In 2015 there are 153.2 million adults estimated to have diabetes in the Western Pacific and are expected to continue rising in 2040 to 214,8 million people with diabetes.Indonesia is in position 7 of 10 countries with the largest diabetics in the world with an estimated 10 million people with diabetes[3]. The aims of the research are to analyze the content of phytochemicals, to examine the activity of antioxidant, and to analyze antidiabetic potential of *n*-hexane, chloroform, ethyl acetate, and ethanol extracts.

2. Materials and Methods

2.1. Plant material

Fresh leaves of *C. Sumatrana* were collected from Matalibaq village, Mahakam Ulu district in East Kalimantanand were authenticated by laboratory of Forest Dendrology and Ecology, Faculty of Forestry, Mulawarman University. A voucher specimen is kept in our laboratory for future reference.

2.2. Preparation of Extracts
Plant material was extracted successively with *n*-hexane, chloroform, ethyl acetate, and ethanol in the increasing order of their polarity using rotary evaporator to obtain the extracts. The yield of the extract in each solvent was recorded.

2.3. Phytochemical analysis
The *n*-hexane, chloroform, ethyl acetate, and ethanol extracts of following plants was subjected tophytochemical tests for the detection presence of alkaloid, flavonoid, coumarin, carotenoid, tannin, saponin, steroid as secondary metabolites using standard procedures [4].

2.4. DPPH radical scavenging capacity
The DPPH (1,1-diphenyl-2-picrylhydrazyl) radical scavenging method was used to evaluate the antioxidant property. Ascorbic acid is used as a comparator of antioxidant activity. The antioxidant activity of each sample was expressed in terms of IC_{50} and was calculated from the graph after plotting inhibition percentage against extract concentration DPPH assay was carried out of the standard protocol [5]. 1000 µL of 0.1 mM DPPH solution was mixed with 1.5 mg of various concentrations (6.25 to 50 µg/mL) of leaf extract. The mixture was shaken vigorously and incubated at room temperature for 20 min in the dark. DPPH free radical scavenging was measured by a spectrophotometer at 517 nm . DPPH solution with no extract and ethanol is used as a control. The experiment was done in three repetitions. Percentage of DPPH free radical inhibition was calculated by the formula:

$$Inhibition\ (\%) = \frac{Acontrol - Asample}{Acontrol} \times 100 \tag{1}$$

Where $A_{control}$ is the absorbance of the control (Ascorbic acid) and A_{sample} is the absorbance of reaction mixture samples (in the presence of sample).

2.5. Inhibition assay for a-amylase activity
α-Amylase inhibitory activity was assayed according to standard protocol with a slight modification [6]. Amylase activity was determined using soluble starch (1%) as a substrate in 0.05 M sodium phosphate buffer (pH 6.8) containing 1 mM $CaCl_2$. Briefly, 100 µL of a-amylase solution (20 U/mL) was mixed with 250µL of sodium phosphate buffer (pH 6.8) and 50 µL of the extracts in 75% aqueous ethanol (6.25 to 50µg/mL). After incubation at 37°C for 10 min, 75µL of the starch solution (1%) was added and the mixture re-incubated at 37°C for 20 min. The reaction was terminated by adding 50 µL of dinitrosalicylic acid solution and boiled for 10 min in a boiling water bath. The total volume was made up to 2.0 mL of distilled water. The absorbance was measured at 540 nm against a control containing a buffer solution in place of the enzyme solution. Quercetin was used as a positive control. The assay was performed in triplicate. Inhibition of α-amylase was calculated by the formula:

$$Inhibition\ (\%) = \frac{Acontrol - Asample}{Acontrol} \times 100 \tag{2}$$

Where Acontrol is the absorbance of the control (quercetin) and A sample is the absorbance of reaction mixture samples (in the presence of sample).

2.6. Inhibition assay for a-glucosidase activity
α-Glucosidase inhibitory activity was assayed according to standard protocol with a slight modification [7]. α-Glucosidase (0.075 unit) was premixed with the extract at various concentrations (12.5 to 100 µg/mL), and 3 mM p-nitrophenyl α-D-glucopyranoside as a substrate in phosphate buffer was added to the mixture to start theiraction. The reaction was incubated at 37°C for 30 min and stopped by adding 2 ml of 0.1 M Na_2CO_3. α-Glucosidase activity was determined by measuring release of *p*-nitrophenol from p-nitrophenyl α-D-glucopyranoside at 400 nm. Quercetin was used as a positive control. The assay was performed in triplicate. Inhibition of α-α-glucosidase was calculated by the formula:

$$Inhibition\ (\%) = \frac{Acontrol - Asample}{Acontrol} \times 100 \tag{3}$$

Where Acontrol is the absorbance of the control (quercetin) and Asample is the absorbance of reaction mixture samples (in the presence of sample).

3. Result

The results for phytochemical screening of the plant extracts are presented in Table 1. Phytochemical screening revealed that the all extracts contained alkaloid and flavonoid, while coumarin was only absence in *n*-hexane extracts, carotenoid were presence in *n*-hexane and ethyl acetate extracts. Tannin, saponin, and steroid were absence in all extracts. Flavonoids have been reported to possess anticancer, antiviral, antioxidant, antibacterial, and anti-inflammatory agents. Recent studies have determined that flavonoid compounds can be very effective in inhibiting α-glucosidase activity [8].

Table 1. Phytochemical analysis of*Caesalpinia sumatrana*extracts

Phytochemicals	Plant extracts			
	n-Hexane	Chloroform	Ethyl acetate	Ethanol
Alkaloid	+	+	+	+
Flavonoid	+	+	+	+
Coumarin	-	+	+	+
Carotenoid	+	-	+	-
Tannin	-	-	-	-
Saponin	-	-	-	-
Steroid	-	-	-	-

+ sign indicates presence and - sign indicates absence.

The results of antioxidant activity showed the radical-scavenging activity of *C. sumatrana* extracts was estimated by comparing the percentage inhibition of formation of DPPH radicals with that of ascorbic acid. Ethyl and ethanol extracts showed moderate antioxidant activity when compared with ascorbic acid. The results of antioxidant activity are presented in Figure 1 and Table 2.

Table 2. Antioxidant analysis of*Caesalpinia sumatrana*extracts

Samples	Extracts	IC_{50} (μg/mL)
Caesalpinia sumatrana	*n*-Hexane	75.28
	Chloroform	66.79
	Ethyl Acetate	12.08
	Ethanol	75.28
Ascorbic acid*		3.12

*Ascorbic acid was used as a positive control

The DPPH radical scavenging activity of *C. sumatrana* extracts increased with increasing the concentration. The ethyl and ethanol extracts appeared to be more active than other extracts. The activities of these plant extracts were 82–89% at 50 μg/mL, which was comparable to that of the positive control ascorbic acid (95% activity). The IC_{50} value obtained was as 5.00 and 3.12 μg/mL for ethanol and ascorbic acid extract respectively. It means ethanol extract of plant at higher concentration captured more free radicals formed by DPPH resulting into decrease in absorbance and increase in IC_{50} value.

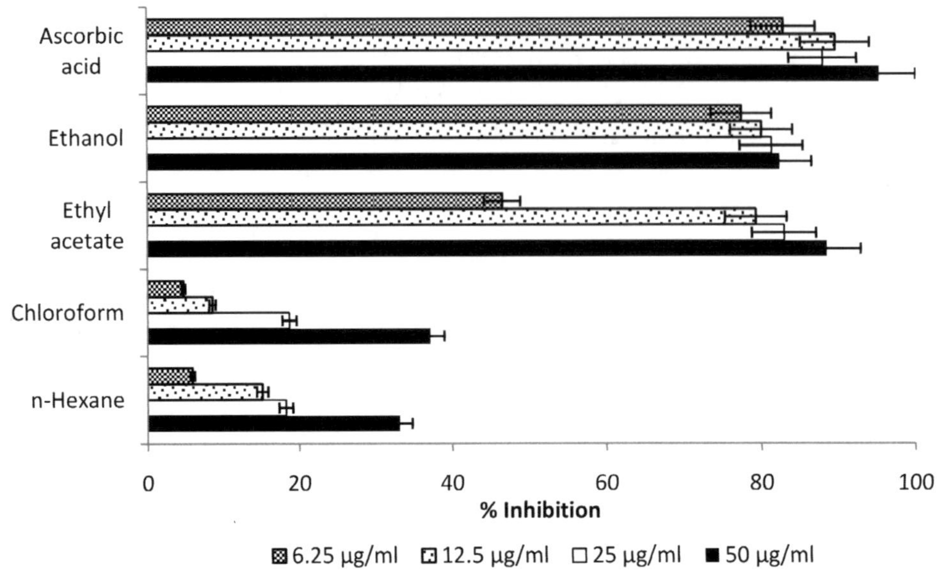

Figure 1. DPPH (1,1-diphenyl-2-picrylhydrazyl) scavenging activity of plant extracts

The inhibitory activity of the *n*-hexane, chloroform, ethyl acetate, and ethanol extract of *C. sumatrana* against yeast α-glucosidase is shown in table 3. The ethanol extracts inhibited yeast α-glucosidase activity by 95.58, 88.76, 75.10 and 64.42% at concentrations of 100, 50, 25 and 12.5 µg/mL, respectively. Quercetin, an α-glucosidase inhibitor used as positive control, inhibited enzyme activity by 98.85, 86.26, 82.06 and 62.90% at concentrations of 100, 50, 25 and 12.5 µg/L respectively. The ethanol extract had an IC_{50} value of 17.16 µg/mL.

Table 3. Inhibitory activity of the extract against α-glucosidase

Samples	Extracts	IC_{50} (µg/mL)
Caesalpinia sumatrana	*n*-Hexane	45.88
	Chloroform	44.36
	Ethyl Acetate	31.74
	Ethanol	17.16
Quercetin*		2.00

* Quercetin was used as a positive control

For α-amylase inhibition, ethanol extracts inhibited α-amylase activity by 95.58, 88.76, 75.10 and 64.42 % at concentrations of 50, 25, 12.5 and 6.25 µg/mL, respectively. Ethanol extracts had the highest inhibition activity with IC_{50} values 16.78µg/mL moderate than quercetin was used as a positive control 4.04 µg/ml. Some flavonoids, such as rutin, myricetin, kaempferol, and quercetin, have been previously reported to inhibit α-glucosidase and α-amylase. These flavonoids exhibit both hypoglycemic and antioxidant effects in diabetic animals [9]. The results of our experiments seemed to agree with this finding, probably due to the presence of flavonoids in phytochemicals analysis.

Table 4. Inhibitory activity of the extract against α-amylase

Samples	Extracts	IC_{50} (µg/mL)
Caesalpinia sumatrana	*n*-Hexane	18.22
	Chloroform	28.50
	Ethyl Acetate	27.31
	Ethanol	16.78
Quercetin*		4.04

* Quercetin was used as a positive control

4. Conclusions

In the light of the results, phytochemicals revealed that alkaloid and flavonoid were present in the extracts and our study indicates that ethanol extracts of*C. sumatrana* leaves have highest α-glucosidase and α-amylase inhibitory and scavenging activities against DPPH. The present result displayed potential of the plant to be developed as natural antidiabetic and antioxidant agents.

Acknowledgments

This work was supported by Laboratory of Forest Chemistry Product, Faculty of Forestry, Mulawarman University and National Institute of Health Research and Development, Ministry of Health Republic of Indonesia as advanced research grant funding RISTOJA 2016 (Contract No. LB.02.01/2/1515/F.032/2016). The kindness of Bahau people in Matalibaq is gratefully acknowledged.

References

[1] Lestari, P. 2016. *Study of typical plants of North Sumatra are nutritious medicine*(in Indonesia). (Indonesia: Journal of Farmanesia)Vol. 1 No. 1, p 11-21.

[2] Widuri, S.A., Noorcahyati, P. 2013. *ModernPotential of Some Types of Plants Efficacious Antidiabetes by Ethnic Kalimantan as Secondary Metabolite Source for The Development Of Modern Drugs*(in Indonesia).(Indonesia:Journal of BPTK) p 1-13

[3] Cho, N.H., Whiting, D., Forouhi, N., et al. 2015. *International Diabetes Federation Diabetes Atlas.*Seventh Edition. (Belgium:IDF Diabetes Atlas) 13-17.

[4] Senthilmurugan, G., B. Vasanthe, K, Suresh. 2013. *Screening and Antibacterial Activity Analysis of some Important Medicinal Plants.* International Journal of Innovation and Applied Studies.(India: Innovative Space of Scientific Research Journals) Vol. 2 No. 2. 146-152.

[5] Shimizu K, Kondo R, Sakai K, Takeda N, Nagahata T, Oniki T. Novel.2001.*Vitamin E derivative with 4-substituted resorcinol moiety has both antioxidant and tyrosinase inhibitory properties.* Lipids. (Japan: Springer)Volume 36, Issue 12, pp 1321–1326

[6] Wang, H., Du, Y. J., & Song, H. C. 2010. *a-Glucosidase and a-amylase inhibitory activities of guava leaves.* Food Chemistry 123 (China: Elsevier), p 6–13.

[7] Kim YM, YK Jeong, MH Wang, WY Lee, HI Rhee. 2005. *Inhibitory effect of pine extract on alpha-glucosidase activity and postprandial hyperglycemia.* Nutrition. (Korea: Elsevier) Vol 21, p756-761.

[8] Zhenhua, Y., Zhang, W., Feng, F., Zhang, Y., Kang, W. 2014. *α-Glucosidase inhibitors isolated from medicinal plants.* Food Science and Human Wellness. (Beijing: Sciencedirect). Volume 3, Issues 3-4, p 136–174.

[9] Kamalakkannan, N., and Prince, P. S. M. 2006. *Rutin improves the antioxidant status in streptozotocin-induced diabetic rat tissues.* Molecular and cellular biochemistry. (India: Springer), p 211–219.

Microbial diversity on sedimentated rice fields due to coal mining activities in Tenggarong Seberang subdistrict of Kutai Kartanegara

Rosfiansyah[1], Sopialena [1]
[1] Agroecotechnology Department of Agriculture Faculty, Mulawarman University Samarinda, East Kalimantan, Indonesia

E-mail: rosfi.plantprotection@gmail.com, sopialena88@gmail.com

Abstract. The results showed that on upland rice fields with sediment found five genus of fungus with number of colonies 4.0 x 10^3 cfu/g to 9.3 x 10^4 cfu/g; three bacterial families with number of colonies 7,1 x 10^4 cfu/g to 2,8 x 10^5 cfu/g; and five genera of nematodes with the amount of 2.6 x 10^2 /kg of soil to 1.1 x10^3 /kg of soil.In unpolished upland rice fields were found four genus of fungus with colonies of 2.4 x 10^3 cfu/g to 8.4 x 10^4 cfu/g, three bacterial families with number of colonies 1.2 x 10^5 cfu/g to 2.7 x 10^5 cfu/g and four genera of nematodes with the amount of 9.6 x 10^2 /kg of soil to 1.1 x10^3 /kg of soil. The most common microbes are *Aspergillus, Cladosporium, Penicillium, Phytium* and *Trichoderma* (fungi), Achromobacteraceae, Brevibacteriaceae, Micrococcaceae (Bacteria), as well as *Dorylaimus, Hemicycliophora, Mononchus, Meloidogyne, Paratrichodorus, Radopholus, Rotylenchulus, Rhabditis, Seinura* and *Trichodorus* (Nematodes). Fungi, bacteria and nematodes have a good role in the process of soil decomposition. The results of soil chemical analysis showed that soil fertility is lower in upland rice fields with sediments compare to those without sediment.
Keywords: bacterial, fungus, nematodes, soil

1. Introduction

Environmental problems will lead to wider aspects in East Kalimantan, such as the decrease of soil productivity, soil compaction, sedimentation, the movement of soil or avalanche, disturbance to the security and health of the population, disruption of flora and fauna, as well as the microclimate change [1]. Sediment deposition caused by erosion that occurs during mining activities and after mining can lead to disruption of plant growth due to reduced oxygen. Factors that cause erosion by water in the form of rainfall, topography, soil type, land use and cover crop. Sedimentary sediments on land usually consisting of clay, dust, sand and large sand. The presence of sediment deposition usually does not affect soil fertility, it can even increase soil fertility in the land because it can bring nutrients and microbes from eroded land if the land has a good soil fertility rate.

The presence of microbes in the soil plays an important role in biogeochemical cycles and is highly responsive for recycling of organic compounds. Soil microbes affect ecosystem conditions in the soil by their contribution in the provision of plant nutrients, plant growth, soil structure and soil fertility [2-4].

The soil fertility criterion is determined by a combination of three interacting factors, ie physical, biological and chemical factors. Physical characteristics and soil chemistry can be understood more perfectly than their biological characteristics. Therefore, more is known physical status and soil chemistry, and little information about the biological status of the soil. There is little difficulty in determining the biological status of the soil because the substance is alive, dynamic and can change in space and time.

The dynamic nature of the soil biological status provides great opportunities for its management. The soil biological status can provide early warning of soil degradation, making it possible to implement more sustainable land management practices [5]. The biological aspects of the soil are complex and require better understanding because there is not much information about the amount and diversity of soil microbes, and how their activity levels in maintaining fertile and productive soils [6].

Microbes are organisms that have a very narrow niche that is very vulnerable to environmental changes. These vulnerabilities spur mutated microbes to survive in new environmental conditions [7]. The research aims to find out the diversity of microbes as well as the level of soil fertility in upland rice fields sedimentation due to coal mining activities in Tenggarong Seberang Subdistrict of Kutai Kartanegara.

2. Material and Methods

2.1.The methods

This research is a qualitative descriptive research. The sampling is aimed to identify microbial and chemical analysis. The soil samples were collected from six different locations in field near the operational activity area of PT MSJ mining company in Bukit Pariaman Village, Tenggarong Seberang Subdistrict, Kutai Kartanegara. They were the areas which were three upland rice fields with sediments *(UTM 50 M0522356 9976041, UTM 50 M0519390 9971851 (S2), UTM 50 M0519289 9971079 (S3))*, and three upland rice fields without sediment *(UTM 50 M0522371 9976059 (U1), UTM 50 M0519413 9971802 (U2), UTM 50 M0519392 9971058 (U3))*. These samples then were analyzed in the Laboratory of Plant Pests and Diseases Science and Soil Science Laboratory of the Faculty of Agriculture, Mulawarman University, Samarinda, East Kalimantan, Indonesia.

2.2.Soil Microbial Analysis

For microbial analysis, the sample was taken from the rhizosphere area of the sample plants to identify the total population of bacteria, fungi, and nematodes. 1 kg of each soil sample was taken from the depth of 0-30 cm and mixed into one, then from that mixture, 1 kg was taken to represent a sample. Furthermore, the samples are packed in a plastic bag and labeled corresponding to its coordinate. Then, the sample was picked up to the laboratory in cool box, to maintain the existing microbes well preserved. This process then will be followed by the determination of the total population of microbes on the standard procedure for microbial analysis. In relation to be method used to calculate the microbes population, this study applied plate count method with a colony counter. A total of 1 g of soil was added by 9 mL of distilled water and made serial dilution to 10^{-6}. Dilution is used to define the population of each of the different parameters. The total population of bacteria was calculated using the dilution with a ratio of 10^{-5} and 10^{-6} under the media of *nutrient agar* (NA), within 2-days incubation period. While for the total fungi, this study used a ratio of dilution at 10^{-3} and 10^{-4} and using the media of *Potato Dextrose Agar* (PDA) within three days period of incubation. On the other hand, to determine the total population of nematodes, using a stereo microscope, a direct counting from a total of 50 g of soil were taken and extracted according to *Baermannfunnel method* within 2-3 days period of incubation.

2.3.Soil Chemical and Physical Analysis

The soil chemical analysis comprises more aspects such as pH, organic C, C/N ratio, total-N, P-available, K-available, and CEC. The analysis was conducted in the laboratory. The soil pH level was usually measured in water suspension. The physical analysis was focused on the soil texture such as sand, silt, and clay

3. Results And Discussion

3.1.Microbial Biomass

Research shows that the presence of sedimentation has little effect on microbial development on upland rice fields, compared with land without sediment. Although on the third location without sediment the number of microbes and the population was slightly higher than the third location with sediment (Table 1).

In the first field, the diversity of unsedimentary field fungi has three genera (Aspergillus, Phytium, and Trichoderma) better than sedimentary fields with two genera (Aspergillus and Phytium), but sedimentary field has a slightly better number of fungal colonies. The diversity of nematodes in unsedimentary land has only one genus (Seinura) equal to the sedimentary field but differentiated genus (Rhabditis), but the nematode density on unsedimentary field is better than the sedimentary field. Diversity of bacteria in unsedimentary field is two families (Micrococcacea and Achromobacteraceae), whereas in the sediment field there is only one bacterial family (Micrococcacea) but has better colony than unsedimentary field.

In the second field, the diversity of fungi on unsedimentary field has three genera (Aspergillus, Penicillium, and Trichoderma) equal to the field with sediment containing three genera (Phytium, Cladosporium, Penicillium) and there are two different genera, besides the number of fungi colonies is better than sedimentary field. The diversity of nematodes in the unsedimentary field has only two genera (Rotylenchulus, Rhabditis), whereas the sedimentary field has three genera (Dorylaimus, Rhabditis, Hemicycliophora) and more nematodes. The diversity of bacteria and the number of colonies in the unsedimentary field there is one family (Micrococcacea) is the same as the field with the sediment has one bacterial family (Micrococcacea).

In the third field, the diversity of fungi on unsedimentary field has three genera (Aspergillus, Penicillium, and Trichoderma) lower than the sedimentary field having four genera (Aspergillus, Penicillium, Trichoderma and Cladosporium) and there is addition of the Cladosporium genus. Diversity nematodes on unsedimentary land have seven genera (Paratrichodorus, Meloidogyne, Trichodorus, Radopholus, Mononchus, Rhabditis, and Seinura) more than the sedimentary field which has two genera (Mononchus and Paratrichodorus). Diversity of bacteria in unsedimentary field there is one family (Micrococcacea and Achromobacteraceae), whereas in sedimentary field there is only 1 bacterial family.

Sedimentation that occurs due to erosion will bring soil particles together with microbes present in the soil particles. In addition, sediment deposits that occur will cover the pores of the soil so that microbes in sediment-covered soil cannot develop because there is no oxygen intake. So that the microorganisms in the sedimented soil are the microbes carried on the sediment.

Soil fertility can be detected from the number of microbial populations living in it. The high number of microbes is a sign of the high level of soil fertility because microbes require organic materials for growth. Organic matter is subsequently overhauled by microbes to become nutrients available to plants and in the soil. The soil fertility provides nutrients that allow well-adapted species within a community to grow rapidly and dominate [8]. In addition, soil microbial diversity plays a key role in maintaining multifunctional ecosystems because the quality and quantity of soil organic matter is strongly influenced by the structural dynamics and functions of soil microbes that are strongly supportive in the decomposition process of litter and mineralization of organic matter [9-10].

Microbes are involved in activities related to plant growth, by encouraging nutrient cycling and stress reduction, play a beneficial role in nutrient-deprived lands [11]. Soil microbial biomass in soil fractions is responsible for the energy and nutrient cycles, as well as the transformation of organic matter [12,13]. In this regard, Soil microorganisms play a role in the mineralization of nutrients through the decomposition of organic matter. Microbial biomass is a small but important nutrient reservoir (C, N, P, and S) and many of these nutrient transformations occur in microbial biomass [14]. Soil microbial biomass also contributes to soil structure and soil stabilization [10], microbial as an indicator compared to superior organisms are more sensitive to different ecological factors such as plant diversity, soil organic matter content, humidity, and climate change, therefore microorganisms are widely used as an indicator of soil quality[15].

Table 1. The inventory of microbial diversity in upland rice fields with sediment and unsediments

Sample	Fungi		Nematode		Bacteria	
	cfu/g	Genus	Population /kg of soil	Genus	cfu/ g	Characteristic/family
Location 1						
Sediment	9,3 x 10^4	*Aspergillus* sp. *Phytium* sp.	2,6 x 10^2	*Rhabditis* sp.	2,8 x 10^5	Colonies of beige, thick, serrated, rough, coccus, gram (+) *Micrococcaceae*
Un-sediment	1,1 x 10^4	*Aspergillus* sp. *Phytium* sp. *Trichoderma* sp.	9,6 x 10^2	. *Seinura* sp.	1,8 x 10^5	Colonies of beige, thick, serrated, rough, coccus, gram (+) *Micrococcaceae* Colonies white, smooth, serrated, basil, gram (-) : *Achromobacteraceae*
Location 2						
Sediment	4,0 x 10^3	*Phytium* sp. *Cladosporium* sp. *Penicillium* sp.	1,1 x 10^3	*Dorylaimus*sp. *Rhabditis* sp. *Hemicycliophora* sp.	9,4 x 10^4	Colonies of beige, thick, serrated, rough, coccus, gram (+) *Micrococcaceae*
Un-sediment	8,4 x 10^4	*Aspergillus* sp. *Penicillium* sp. *Trichoderma* sp.	1,0 x 10^3	*Rotylenchulus* sp *Rhabditis* sp.	9,4 x 10^4	Colonies of beige, thick, serrated, rough, coccus, gram (+): *Micrococcaceae*
Location 3						
Sediment	5,0 x 10^3	*Aspergillus* sp. *Penicillium* sp. *Trichoderma* sp. *Cladosporium* sp.	5,8 x 10^2	*Mononchus* sp. *Paratrichodorus* sp.	7,1 x 10^4	Colonies beige, thick, serrated, smooth, small basil, gram (-): *Brevibacteriaceae* Colonies White, smooth, serrated, basil, gram (-): *Achromobacteraceae* Colonies of beige, thick, serrated, rough, coccus, gram (+): *Micrococcaceae*
Un-sediment	2,4 x 10^3	*Aspergillus* sp. *Penicillium* sp. *Trichoderma* sp.	1,1 x 10^3	*Paratrichodorus* sp. *Meloidogyne* sp. *Trichodorus* sp. *Radopholus* sp. *Mononchus* sp. *Rhabditis* sp. *Seinura* sp.	1,2 x 10^5	Colonies beige, thick, serrated, smooth, small basil, gram (-): *Brevibacteriaceae* Colonies of beige, thick, serrated, rough, coccus, gram (+): *Micrococcaceae*

Microorganisms form bothsaprophytic or symbiotic relationship with plants,where the habit of symbionts may be parasitesor mutualistic. Among the favorable saprophytics specific bacterial microbial rhizosphere, i.e.called plant growth that promotes rhizobacteria(PGPR), and fungi, such as pathogen antagonists Trichoderma.

Aspergillus, Cladosporium, Penicillium, Phytium and Trichoderma in soils play a greater role in soil decomposition [16].Micrococcus and Achromobacter are two of the many PGPR found in rooting in the soil [17-18]. The role of PGPR for plants is to act as biofertilization in the soil, stimulate root growth, rhizoremediation, and control stress in plants, while PGPR in biological control mechanisms play a role in reducing the impact of diseases on plants due to plant capability in antibiosis, systemic resistance induction, and competition for nutrition and recesses [19].

Meloidogyne, Radopholus, Rotylenchulus more acts as parasitic nematodes in plants that are more commonly found in East Kalimantan [20] than other parasitic nematodes such as Hemicycliophora, Paratrichodorus andTrichodorus [21].Seinura in addition to its role as decomposer also plays a role in controlling the number of soil microorganisms as a predator in other organisms. Rhabditis plays a role in soil decomposition and is a fungus eater on the ground[22-23].Dorylaimus is an omnivorous nematode capable of feeding on different food sources depending on environmental conditions and food availability [22, 24]. Nematodes play a role in the decay of organic matter so that it has a role such as earthworms, which play a direct role in the decomposition process even innematodes are often considered saprophagous. In fact, such nematodes feed on the bacteria that abound in decaying organic matter and not on the organic matter itself[22-25].

3.2.Chemical Analysis of Soil Fertility

Based on the analysis of soil chemical properties on the first location, it was found that the soil observed at the sediment site had high cation exchange capacity (CEC), high base saturation (BS), slightly acid pH, high organic carbon (first location of sediment) and very high (first location without sediment), low N total element, low available P (first location is seded) and very high (first location without sediment) and K available very high. Based on the soil fertility criterion[26] that the soil fertility level at the first location with sediment studied was moderate. From the analysis of soil chemical properties tend to decrease soil fertility quality due to sediment, such as very low N content available, resulting in very low soil fertility.

Table 2. Results of Chemical Analysis in upland rice fields with sediment and unsediments

Sample	pH	C Orga-nik	N Total	C/N Ratio	P_2O_5 Bray 1	K_2O Mor-gan	KT K	Kej Basa
		%			ppm P	ppm K		%
Location 1								
Sediment	6,95	4,92	0,19	28	4,6	88,8	24,3	94,9
Unsediment	6,6	5,35	0,18	27	29,9	103,8	25,1	77,3
Location 2								
Sediment	7,05	4,77	0,16	30	2,5	77,8	13,5	100,0
Unsediment	6,78	3,89	0,21	18	5,3	90,3	13,5	89,6
Location 3								
Sediment	6,10	4,02	0,24	17	4,5	92,8	17,5	77,7
Unsediment	5,95	3,98	0,22	18	4,6	100,3	18,6	78,0

In the second location, based on the analysis of soil chemical properties, it was found that the soil observed at the sediment site had low cation exchange capacity (CEC), very high alkaline saturation (KB), slightly acid pH, high organic carbon, low total N element (location second sediment) and medium (second location without sediment), P is very low (second location with sediment) and low (second location without sediment) and K available is very high.

In the third location, the level of soil fertility in the studied site was low. The analysis of soil chemical properties has a tendency to decrease the chemical quality due to sediment, such as low N content available, resulting in very low soil fertility. Based on the analysis of soil chemical properties it is known that the soil observed at the sediment site has medium cation exchange capacity (CEC) moderate, base saturation (BS) is very high, soil pH is slightly acidic, organic carbon is high, element of moderate total N, P is low and K available very high.

The organic chemistry of organic matter affects the decomposition rate of organic matter on the soil, the quality of organic matter is determined by the ratio of C / N, N total, P total, lignin and polyphenols[27-28]. In the decomposition occurs the physical and chemical processes of organic matter into other chemical compounds with the help of microbes [9-13].

CEC is very important to know because able to give criteria of soil fertility, the higher the CEC then the higher also long-term soil fertility [30]. CEC in a high soil indicates the soil has a large enough negative charge so that the land is capable absorbing more potassium, to be subsequently released when the potassium content in the solution is reduced[31].

3.3.Physical Analysis of Soil Fertility

The result of analysis of soil physical characteristic that is texture on upland rice field and un-sediment field (Table 3). Sand is one part of a fraction that has a particle size of 2μm-50μm and belongs to a class group with a coarse texture [31]. The percentage of fine and coarse sand on the sedimentary ground in the first location of 24.0% and 9.8%, the second sites 26.1% and 4.2% and the third sites 15.1% and 3.2%. On land without sediment in the first location 1.8% and 1.5%, second location 20.8% and 0.7% and third location 11.5% and 1.9%. From the data, it was found that the fine sand content of the sediments on site 2 was higher in fine sand content than the first and third locations, but the first location of the sand content was higher than the second and third locations. In addition, if sediments are compared to land without sediment, it is seen that the sediment area has a finer sand content than roughly sediments. The sediment transport mechanism is based on the depth of the sediment flow carrying the flow of water with a base sedimentary charge in the form of coarse particles such as sand moving and can be rolled or jumped from high to low.

Dust is one of the soil materials that have smaller particle sizes of the sand is 50μ-2μ [31]. Dust is included in the form of rinse load fine particles carried by the flood stream down to a calm/stagnant stream or lower ground. The results showed that the dust content of the sampling sites in the sediments field was higher with 44.5%, 55.5%, and 63.1% respectively of the sediments, ie 41.9%, 49, 2% and 59, 4%.

Table 3. Results of physical Analysis in upland rice fields unsediments

Sample	Particle Distribution				Texture
	Clay	Silt	Sand		
			Fine	coarse	
	%				
Location 1					
Sediment	24,3	41,9	24,0	9,8	L

Unsediment	52,2	44,5	1,8	1,5	SiC
Location 2					
Sediment	20,5	49,2	26,1	4,2	L
Unsediment	23,5	55,0	20,8	0,7	SiL
Location 3					
Sediment	22,3	59,4	15,1	3,2	SiL
Unsediment	23,5	63,1	11,5	1,9	SiL

The clay has a particle size of less than 2μ, with this measure making the clay the smallest of all soil textures [31]. The results showed that the clay content at the sampling sites in the land without sediment was higher at 52.5%, 23.5%, and 23.5% respectively of the sediments i.e. 24.3%, 20.5% and 22.3 %.

The many sediments containing clay fractions have a slow rate of water exchange, which is due to the very small size of the clay particles. As a result, the 0_2 exchange brought by water entering into the sediment becomes obstructed. However, at the site of seeded in the early days of flooding and puddles then muddy will be more dangerous because of the influence of clay coating and cover the pores of the soil so that no oxygen enters the soil.

Severe quality deterioration is the effect of sediment on soil physical properties. The accumulation of sediment causes oxygen cannot enter the root zone. This effect is similar to a prolonged flood. This means that the existing sediments inhibit the entry of oxygen to the root zone during the sediment is wet and during that time the roots of plants lack oxygen. Oxygen will re-enter the root zone after the sediment is dry and by that time the new flood effect ends.

3.4.Land improvement with sediment

Dedicated land can be repaired and used if channel normalization and sediment deposition ponds are made so that water does not pass through the land and disturb the land. The damaged land can only be used if the mud is dry and fertility improvement is done by using organic fertilizer or addition of soil microorganisms.The mixture of topsoil and fertilizer has successfully increased the number of fungal genus genera, the number of nematode genus and they can improve the soil fertility in post-mining areas, can increase C, N and P, as well as the growth of plant, regarding the pH soil (which is closed to neutral) [1]. The increase in microbial activity depends on the annual increase in C and N in cultivation of plants. Soil microbial activity is not only able to release nutrients for plants but also able to stimulate the process of mineralization and mobilization of pollutants and xenobiotic, so that microbial activity is important in the biogeochemical cycle[10, 12, 32]. Therefore, microbial activity depends on nutritional conditions, temperature and availability of water in the soil.

4. Conclusion

Microbial life in the soil plays an important role in controlling the stability of soil ecosystems. Soil sedimentation and soil management may affect the structure of microbial communities in the soil. By understanding the soil environmental conditions well from the physical, chemical and biological aspects, so the management of the land to maintain the fertility of the land can be done appropriately.Fungi, bacteria, and nematodes have a good role in the process of soil decomposition. The results of soil chemical and physical analysis showed that soil fertility is lower in upland rice fields with sediments compare to those without sediment.

Acknowledgement

The authors would like to thank the environmental agency of Kutai Kertanegara district of East Kalimantan province which has supported this research.

References

[1] Sopialena, Rosfiansyah, and Sila S 2017 The benefit of top soil and fertilizer mixture to improve the ex-coalmining land.*Nusantara Bioscience* **9** pp 36

[2] Koorem K, Gazol A, Öpik M, Moora M, Saks Ü, Uibopuu A, Sõber V, and Zobel M 2014 Soil nutrient content influences the abundance of soil microbes but not plant biomass at the small-scale. *PLoS ONE* **9** pp 1

[3] Lehman R M, Cambardella C A, Stott D E, Acosta-Martinez V, Manter D K, Buyer J S, Maul J E, Smith J L, and Collins H P 2015 Understanding and enhancing soil biological health:The solution for reversing soil degradation *Sustainability*7 pp 988

[4] Jacoby R, Peukert M, Succurro A, Koprivova A, and Kopriva S 2017The role of soil microorganisms in plant mineral nutrition—current knowledge and future directions. Front *Plant Sci* **8** pp 1617

[5] Trivedi P, Delgado-Baquerizo M, Anderson I C, and Singh B K 2017 Response of soil properties and microbial communities to agriculture: implications for primary productivity and soil health indicators *Plant Sci* **7**990

[6] Fitter A H, Gilligan C A, Hollingworth K, Kleczkowski A, Twyman K M, Pitchford J W and The members of the Nerc Soil Biodiversity Programme 2005 Biodiversity and ecosystem function in soil.*British Ecological Society* **19** p 369

[7] Metting F B 1996 Biodiversity and application of microalgae. *JInd Microbiol* **17** p 477

[8] Pold G and DeAngelis K M 2013 Up against the wall: the effects of climate warming on soil microbial diversity and the potential for feedbacks to the carbon cycle*Diversity* **5** p 409

[9] Delgado-Baquerizo1 M, Maestre F T, Reich P B, Jeffries T C, Gaitan J J, Encinar D, Berdugo M, Campbell C Dand Singh B K 2015 Microbial diversity drives multifunctionality in terrestrial ecosystems. *Nat Commun*7 p 10541

[10] Schulz S, Brankatschk R, Dumig A, Kogel-Knabner I, Schloter M and Zeyer J 2013 The role of microorganisms at different stages of ecosystem development for soil formation. *Biogeosciences* **10** p 3983

[11] Zolla G, Bakker M G, Badri D V, Chaparro J M, Sheflin A M, Manter D K and Vivanco J 2013*Understanding root-microbiome interactions*, ed F.J. de Bruijn (Wiley Blackwell, pp:745-754)

[12] Azcón-Aguilar C and Barea J M 2015 Nutrient cycling in the mycorrhizosphere. *J. Soil Sci. Plant Nutr.***25** p 372

[13] Araújo A S F and MeloW J 2010 Soil microbial biomass in organic farming system*Ciência Rural***40** p 2419

[14] Balota E L, Machineski O, and Matos M A 2012 Soil microbial biomass under different tillage and levels of applied pig slurry *R. Bras. Eng. Agríc. Ambiental* **16** p 487

[15] Martinez-Salgado M M, Gutiérrez-Romero V, Jannsens M and Ortega-Blu R. 2010*Biological soil quality indicators: a review(Current Research, Technology and Microbiology and Microbial Biotechnology)*vol 1ed Mendez Vilas A (Spain: Formatex Research Center) pp 319–328 Available Online At http://www.formatex.info/ microbiology2/319-328.pdf

[16] Sharma T S and Singh R 2015 Isolation, identification and screening of cellulose degrading microfungi from decomposed *E.crassipes*. *Inter. J. Innov. Scie Res*. **4** p 509

[17] Anandaraj M, R. Dinesh 2008 *Use of microbes for spices production* ed V.A. Parthasarathy, K. Kandiannan, V. Srinivasan (New Delhi: New India Publishing Agency) pp 101-132

[18] DawwamGE, Elbeltagy A, Emara H M, Abbas I H and Hassan M M 2013 Beneficial effect of plant growth promoting bacteria isolated from the roots of potato plant. *AOAS* **58** p 195

[19] Vejan P, Abdullah R, Khadiran T, Ismail S and Boyce A N 2016 Role of plant growth promoting rhizobacteria in agricultural sustainability—A Review *Molecules* **21** p 573

[20] Suyadi and Rosfiansyah 2017 The role of plant-parasitic nematodes on productivity reduction of banana and tomato in East Kalimantan. *Asian J Agric* **1** p 40

[21] Zeng Y, Ye W, Martin S B, Martin M and Lane Tredway 2012 Diversity and occurrence of plant-parasitic nematodes associated with golf course turfgrasses in North and South Carolina, USA. *JNematol* **44** p 337
[22] Vinciguerra MT 1979 Role of nematodes in the biological processes of the soil *Bull. Zool* **46** p 363
[23] Hosen M J, Hossain M, Martinez J G, Sumaya N H, Mei Y, Manandhar S, dos Santos G and Moens T 2011 Effect of bacterivorous and predatory nematodes on macroalgal detritus decomposition *Proceedings of the Pakistan Academy of Sciences* **48** p 137
[24] Elele K, Nzeako S O and Asor J E 2017 Nematode occurrence and distribution in an organically managed *Solanum melongena* L field. *Scientia Agriculturae* **20** p 71
[25] Thakur M V, Reich P B, Fisichelli N A, Cesarz A S S, Dobies T, Rich R L, Hobbie S E and Eisenhauer N 2014 Nematode community shifts in response to experimental warming and canopy conditions are associated with plant community changes in the temperate-boreal forest ecotone *Oecologia* **175** p 713
[26] Soil Research Center 1983*Assessment criteria for soil chemical analysis properties* (Bogor: Central Agricultural Research and Development, Agriculture Ministry of Indonesia)
[27] Swift M J and Woomer P 1993 *Organic matter and sustainability of agricultural systems: Definition and measurement*, ed Mulongoy, K. and Merckx, R. (Chichester: John Wiley and Sons Ltd.)
[28] Nurhidayati, Arisoesilaningsih E, Suprayogo D and Hairiah K 2012 Earthworm population density in sugarcane cropping system applied with various quality of organic matter. *J. Tropical Life Science* **2** p 103
[29] Syukur A 2005 Effect of organic matter giving on soil properties and caisim growth in coastal sandy soil. *Soil Sciences and Environments* **5** p 30
[30] Subiksa IG M, Adiningsih S, Sudarsono and Sabiham S 2004 Effect of amelioration and K fertilization on parameters of potassium Q-1 in acid mineral soils. *Soil and Climate* **22** p 40.
[31] Hardjowigeno S 2007 *Soil Science*. (Jakarta: Akademika Pressindo)
[32] Chirinda N, Olesen JE, and Porter JR. 2008. Effects of organic matter input on soil microbial properties and crop yields in conventional and organic cropping systems. 16 the IFOAM Organic World Congress, Modena, Italy, June 16-20 2008. Available online At http://orgprints.org/view/projects/conference.htm

Antioxidant activity of an invasive plant, *Melastoma malabathricum* and its potential as herbal tea product

N M Sari*, H Kuspradini, R Amirta and I W Kusuma

Faculty of Forestry, Mulawarman University, Jl. Ki Hajar Dewantara, Kampus Gn. Kelua Samarinda, 75119, East Kalimantan, Indonesia

nurmaulidsr@gmail.com

Abstract. East Kalimantan possesses abundant biodiversity of tropical medicinal plant. *Melastoma malabathricum* (known locally as karamunting, senduduk) is an invasive plant along with other species in the family of Melastomataceae with traditional medicinal purposes. This research explored the potential of Karamunting (*M. malabathricum*) plant for its antioxidant activity and the potential as a material for herbal tea product. The plant was macerated to yield ethanolic extract, and at the same time plant powder was packed into tea bags and extracted with hot water to obtain the infused water. Antioxidant activity was evaluated by DPPH radical scavenging assay. The results showed that the ethanol extracts of plant samples displayed ability to inhibit DPPH free radical formation by 82% at 50 ppm concentration. Evaluation of the tea water extract showed that the highest inhibition obtained by leaves powder by 90% and fruit 88% at 1 minutes immersion time. This finding suggest that leaves and fruit of *M. malabathricum* plants display potential as herbal tea material having antioxidant activity if the safety aspect can be assured.

1. Introduction

Indonesia is a tropical country with abundant natural resources. Various studies on medicinal plants are closely related to the local people have been widely published. Therefore, medicinal plants are believed to have various medicinal properties. Nowadays, traditional medicine show an increase developed populations. According to the World Health Organization [1], traditional medicine are the sum total of the knowledge and practice, also experiences to different cultures based on their beliefs through generations. Today, Asian population especially Indonesia prefers traditional plants as their health medicine. Beside the traditional plants being easy to obtain, it also had no side effect and safe to be consumed.

Melastomataceae plants are originally growth in the tropic and subtropic regions, there are more than 4000 species in the world. *Melastoma* genus in Southest Asian region, were found around 22 species, 2 subspecies and 3 varieties [2].

M. malabathricum Linn is a shrub and an invasive plant commonly found in previously cleared land that belongs to the family *Melastomataceae*. Locally known in Indonesia as karamunting or senduduk, have been used as traditional medicinal purposed. Traditionally, the leaves can be used to heal wounds by chewing and applied to the wound. The leaves part is used to treat some disease such as haemorrhoid, dysentery, diarrhea and also prevent the scars [3,4]. *M. malabathricum* leaves, scientifically have been reported to exert antibacterial [5], anti inflammatory and antipyretic [6,7], antioxidant and cytotoxic [8], antidiarrheal and antiulcer [9] also anticancer [10] activities.

Antioxidant are widely used to prevent oxidation in natural food system. Natural food system was using antioxidant to reduce oxidation. Antioxidant of plant are safe and natural when used in small scale based on natural source from local people. However while the plants used in large scale to reduce oxidation of the food system, the safety effects of the plants are unknown. The tea extracts are highly recommendly used as natural antioxidant. Many studies of antioxidant properties of green tea and black tea showed the highest antioxidant activity of the extracts [11].

Natural antioxidant was apply as food additives, bioactive nutraceuticals and bio-pharmaceuticals on daily activities. Related to the extraction, characterization and utilization of natural antioxidant insentively performed to find potential candidates in againts the aging process [12].

The present study aims to explore the potential of karamunting plant extract for its antioxidant activity and the potential of the plant as material for herbal tea product.

Figure 1. Morphology of *Melastoma malabathricum* Plant (Photo source: personal doc.)

2. Methods

2.1. Plant materials and chemical

The leaves and fruits of *M. malabathricum* Linn. were collected from Lempake, Samarinda, East Kalimantan. Voucher specimen was kept in our laboratory for future reference. DPPH (*1,1 diphenyl-2-picrylhydrazyl*) was purchased from Tokyo Kasei Kogyo (Tokyo, Japan). DMSO (*dimethyl sulfoxide*), sulfuric acid, hydrochloric acid, acetic anhydride, and potassium iodide were purchased from Merck (Darmstadt, Germany). Ascorbic acid, 1-naphtol and bismuth (III) nitrate were obtained from Sigma (St. Louis, MO, USA).

2.2. Preparation of extracts

Plant materials were washed thoroughly with water to remove the extraporeneous matter and prepared by drying for 3 days under the shade and milled with a blender. The leaves and fruit part of the samples were extracts with ethanol at room temperature with continuous shaking on a shaker (7400 Tubingen, Edmun Buchler, Germany) for 48 hours. Following filtration of the suspension through Whatman paper No. 2 (Maidstone, UK), the crude extracts were evaporated in a rotary evaporator at 40°C and put in a vacuum oven near dryness to yield the plant extract.

2.3. Preliminary phytochemical screening

The ethanol extracts of *M. malabathricum* was subjected to preliminary screening such as alkaloids, flavonoids, tannin, steroids, triterpenoids, carbohydrates and saponins [13,14].

2.4. Antioxidant activity

2.4.1. Ethanol extracts
The leaves and fruit part of the samples were dissolved in DMSO and used at 3 times dilution for the actual experiment. The DPPH radical scavenging method was performed as previously described by Shimizu et al [15] with slight modification. UV absorption was measured on a Shimadzu UV-VIS 1240 Spectrophotometer (Shimadzu Corp., Kyoto, Japan).

2.4.2. Tea extracts
The powdered of *M. malabathricum* leaves and fruit were weighed 2 gr and packed into tea bags. The tea extracts were obtained from infusion process for 1 minute, 3 minute and 5 minute of immersion time. The DPPH radical scavenging method was performed as previously described by Shimizu et al [15] with slight modification. UV absorption was measured on a Shimadzu UV-VIS 1240 Spectrophotometer (Shimadzu Corp., Kyoto, Japan).

3. Results and Discussion

3.1. Plant extracts
Leaves and fruit of *M. malabathricum* plants were macerated by ethanol at room temperature (Table 1). The ethanol maceration yielded 6.82-7.76% extracts on the basis of sample dry weight. The result showed that leaves part gave higher extract than the fruit.

Table 1. The yield extract of *M. malabathricum* Linn

Plant Part	Yield (%)
Leaves	6.82
Fruit	7.76

*Yield was calculated on the basis of sample dry weight

M. malabathricum fruit has purple color and solid that produce an extractive content more concentrated and caused the fruit extract having the highest yield.

3.2. Phytochemical screening
The leaves and fruit part of *M. malabathricum* plants showed the presence of secondary metabolites as shown as the table 2. The leaves part of plant contains steroids while the fruit contains triterpenoids.

Table 2. Phytochemical screening of *M. malabathricum* Linn

No	Plant Part	Compounds						
		alkaloids	flavonoids	saponin	tannin	triterpenoids	steroids	carbohydrates
1	Leaves	+	+	+	+	-	+	+
2	Fruit	+	+	+	+	+	-	+

*(+): Contained positive; (-) Contained negative

According to the research, tannin and flavonoids compound are known having antibacterial, antitumor, antiviral, also have strong antioxidant, anticancer and antimicrobial activities [16]. In a previous study, methanol extract of *M. Malabathricum* contained terpenoids, phenolic compound, tannin, flavonoids, triterpenes and saponin [17]. The result of our research were similar to those obtained in the literature.

3.3. Antioxidant activity

The DPPH radical scavenging of ethanol extracts was using Ascorbic acid as a positive control and DMSO as a negative control. Radical DPPH has been used to determination free radical scavenging activities of pure antioxidant compounds such as fruit and plant extracts and also food materials [18]. Radical scavenging activity was expressed as percentage according to the following formula:

$$\%\ \text{DPPH radical scavenging activity} = \frac{\Delta\text{control} - \Delta\text{sample}}{\Delta\text{control}} \text{ x } 100 \qquad (1)$$

Antioxidant assay of *M. malabathricum* leaves and fruit of ethanol extracts displayed ability to inhibit DPPH free radical formation by 82 % at 50 ppm concentration on the leaves part while the fruit part by 77% at 25 ppm concentration (Fig. 2).

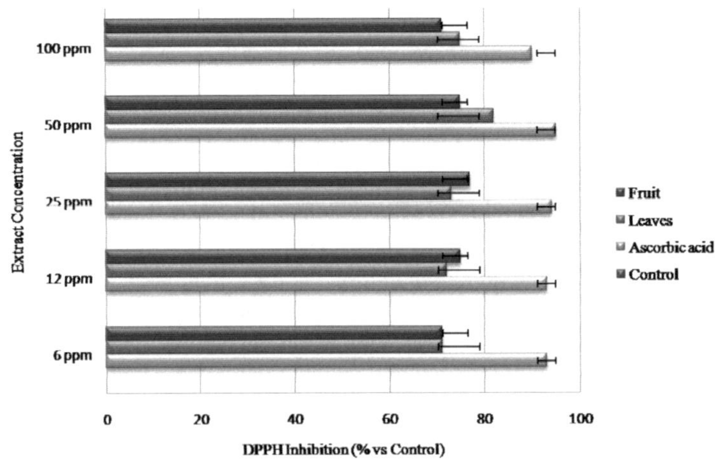

Figure 2. Antioxidant activity of ethanol extracts against DPPH

The DPPH radical scavenging of tea extracts was using a commercial Green tea (Tea A) and Black tea (Tea B) as a positive control and distilled water as a negative control. Tea extracts evaluation of *M. malabathricum* showed the leaves powder had the highest inhibition by 90% and the fruit powder by 88% at 1 minute immersion time (Fig. 3).

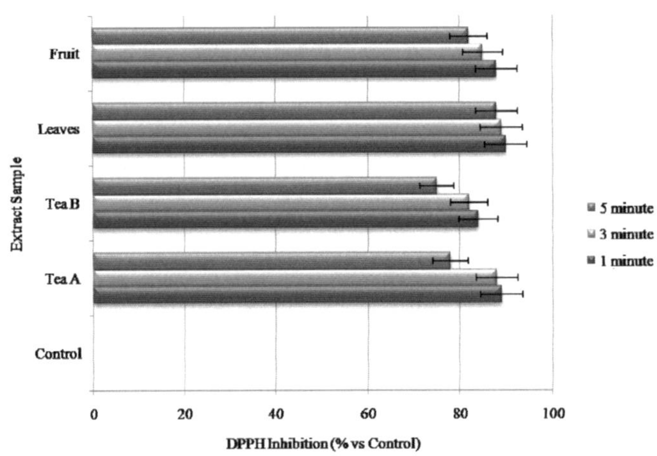

Figure 3. Antioxidant activity of tea extracts against DPPH

The results of antioxidant activity of *M. malabathricum* ethanol extracts showed that the lower concentration of sample, percentage of antioxidant activities were higher. This caused by the higher the concentration of extracts, the colour of sample more concentrated and the absorption of DPPH reduction are less than maximum.

Based on literature study of medicinal plants, *M. malabathricum* plant used to treat toothache disease, diarrhea and dysentery. *M. malabathricum* also contain ellagic acid, anthocyanin and nobotannin β [19]. Flavonoids, phenolics compounds, alkaloids, tannin, amino acids and volatile constituents were investigated as main active ingredients have a potential antioxidant activities [20]. In this research, the leaves and fruit part of *M. Malabathricum* presence of flavonoids and tannins known having potent as antioxidant properties.

The results of antioxidant activity of *M. Malabathricum* tea extracts showed that the longer time of immersion time, the lower percentage of antioxidants produced. This can be proven by the results of antioxidant *M. malabathricum* tea extracts that show the highest percentage of antioxidant at 1 minute immersion time. Therefore, the investigated of these plants confirmed have a potential effect on treating digestive system disorders.

4. Conclusions

The present study has proved that the extracts of leaves and fruit of *M. malabathricum* plants possess antioxidant properties. The results display that leaves and fruit of the plant having antioxidant activity as potential herbal tea material.

References

[1] World Health Organization (WHO) 2005 "*National Policy on Traditional Medicine and Regulation of Herbal Medicines*" (Geneva:WHO).

[2] Nurfariza AR, Nur AMA, Hadji A, Mohammad SBAR 2014 Cytotoxicity screening of *Melastoma malabathricum* extracts on human breast cancer cell lines *in vitro*. *Asian Pac J Trop Biomed* **4**(7) pp 545-8. doi: 10.12980/APJTB.4.2014C658 .

[3] Z.A. Zakaria, E.S. Jaios, M.H. Omar, S.Abd. Rahman, S.S.A. Hamid, S.M. Ching, L.K. Teh, M.Z. Salleh, S. Deny and M. Taher 2016Antinociception of petroleum ether fraction derived from crude methanol extract of *Melastoma malabathricum* leaves and its possible mechanisms of action in animal models. *BMC Complementary and Alternative Medicine* **16** p 488. doi: 10.1186/s12906-016-1478-1.

[4] Joffry S M, Yob N J, Rofiee M S, Affendi M M, Suhaili Z, Othman F, et al. 2012 *Melastoma malabathricum* (L.) Smith ethnomedicinal uses, chemical constituents and pharmacological properties: a review. *Evid Based Complement Alternat Med.* doi:10.1155/2012/258434.

[5] Sunilson JAJ, Anandarajagopal K, Kumari AVAG, Mohan S 2009 Antidiarrhoeal activity of leaves of *Melastoma malabathricum* (L). *Indian J Pharmaceutic Sci.* **71** pp 691–5. https://dx.doi.org/10.4103%2F0250-474X.59556.

[6] Zakaria ZA, Raden MNRNS, Hanan Kumar G 2008 Antinociceptive, anti-inflammatory and antipyretic properties *of Melastoma malabathricum* (L) leaves aqueous extract in experimental animals. *Can J Physiol Pharm* **84** pp 1291 –9. https://doi.org/10.1139/y06-083 .

[7] Zakaria ZA, Rofiee MS, Mohamed AM, Teh LK, Salleh MZ 2011 In vitro antiproliferative and antioxidant activities and total phenolic contents of the extracts of *Melastoma malabathricum* leaves. *J Acupunct Meridian Stud*; **4**(4) pp 248-56. https://doi.org/10.1016/j.jams.2011.09.016.

[8] Nazlina I, Norha S, Noor Zarina AW, Ahmad IB 2008 Cytotoxicity and antiviral activity of *Melastoma malabathricum* (L) extracts. *Malays J Appl Biol.* **37** pp 53–5. doi: 10.12980/APJTB.4.2014C658.

[9] Hussain F, Abdulla MA, Noor SM, Ismail S, Ali HM 2008 Gastroprotective effects of *Melastoma malabathricum* (L) aqueous leaf extract against ethanol-induced gastric ulcer in rats. *Am J Biochem Biotechnol.* **4** pp 438–41. doi: 10.3844/ajbbsp.2008.438.441.

[10] Sirat HM, Susanti D, Ahmad F, Takayama H, Kitajima M. 2010 Amides, triterpene and flavonoids from the leaves of *Melastoma malabathricum* L. *J Nat Med*; **64**(4) pp 492-5. https://doi.org/10.1007/s11418-010-0431-8.

[11] Elham RN, Alireza SM, Mohammad G, Mehdi K 2015 Evaluation of antioxidant interactions in combined extracts of green tea (*Camellia sinesis*), rosemary (*Rosmarinus officinalis*) and oak fruit (*Quercus branti*). *J Food Sci Technol* **52**(7) pp 4565-71. doi: 10.1007/s13197-014-1497-1.

[12] Irawan WK, Murdiyanto, Enos T A, Syafrizal, Yong-ung Kim 2014 Antimicrobial and antioxidant properties of medicinal plants used by the Bentian tribe from Indonesia. *Food Science and Human Wellness* **3** pp191-196. http://dx.doi.org/10.1016/j.fshw.2014.12.004 .

[13] Kokate CK 2000 *Practical Pharmacognosy* (Valllabh Prakashan: India).

[14] J.B. Harborne 1998 P*hytochemical Methods: A guide to modern techniques of plant analysis*, 3rd ed (London : Chapman& Hall, an imprint of Thomson Science)

[15] Shimizu K, Kondo R, Sakai K, et al. 2001 Novel vitamin E derivative with 4- substituted resorcinol moiety has both antioxidant and tyrosinase inhibitory properties, *Lipids* **36** pp 1321–6. https://doi.org/10.1007/s11745-001-0847-9.

[16] Turgut T, Duygu T 2017 In vitro anti-urease, antioxidant activities and phythochemical composition of *Geranium purpureum*. Food Measure **11** pp 2102-9. doi: 10.1007/s11694-017-9594-2.

[17] Deepak Kumar Jha, Likun Panda, Sudha Ramaiah, Anand Anbarasu 2014 Evaluation on Comparison of Radical Scavenging Properties of Solvent Extracts from *Justicia adhatoda* Leaf Using DPPH Assay. *Appl Biochem Biotechnol* 174 pp 2413-25. doi: 10.1007/s12010-014-1164-z.

[18] Vikas Kumar, Danish Ahmed, Pushpraj S Gupta, Firoz Anwar and Mohd Mujeeb 2013Anti-diabetic, anti-oxidant and anti-hyperlipidemic activities of *Melastoma malabathricum* Linn. Leaves in streptozotocin induced diabetic rats. *BMC Complementary and Alternative Medicine*, **13** pp 222. doi: 10.1186/1472-6882-13-222.

[19] Kornkanok Tangjitman, Chalobol Wongsawad, Kaweesin Kamwong, Treetip Sukkho, Chusie Trisonthi 2015 Ethnomedicinal plants used for digestive system disorders by the Karen of northern Thailand. *Journal of Ethnobiology and Ethnomedicine* **11** p 27. doi: 10.1186/s13002-015-001.

[20] Dereje B, Minaleshewa A and Mirtachew TA 2016 Determination of some selected secondary metabolites and their invitro antioxidant activity in commercially available Ethiopian tea (*Camellia sinensis*). *SpringerPlus* **5** p 412. doi: 10.1186/s40064-016-2056-1.

Phytochemicals and antidiabetic activity of *Eusideroxylon zwageri* stem bark collected from East Kalimantan, Indonesia

I W Kusuma[1*], Rahmini[1], R Ramadhan[1], N Rahmawati[2], R A Suwasono[1], and NM Sari[1]

[1]Faculty of Forestry, Mulawarman University, Jl. Ki Hajar Dewantara, Samarinda, East Kalimantan, Indonesia. 75119
[2]Research and Development Agency of Medicinal Plants and Traditional Medicine, Ministry of Health, Jalan Raya Lawu, No. 11, Tawangmangu, Indonesia. 57792

*Corresponding author: kusuma_iw@yahoo.com

Abstract. *Eusideroxylon zwagery* (Lauraceae), a tropical tree species known as ulin or borneo iron wood and traditionally used for the treatment of diabetes in the Ethnic of Kutai. Plant extract was prepared by maceration using ethanol. The plant extract was evaluated its DPPH and superoxide radicals scavenging activity, the inhibition on α-glucosidase and α-amylase activity as antidiabetic potential and the analysis of the total phenolic, total flavonoids and proanthocyanidin contents. The ethanolic extract of the stem bark was 8.62% on the dry weight basis. The IC_{50} values of antioxidant activity of the extract in DPPH and superoxide radical scavenging mechanisms were 44.90 μg/ml and 30.47 μg/ml. In antidiabetic assay, the *E. zwageri* stem bark extract showed IC_{50} value 58.45μg/ml in α-glucosidase inhibition, and 9.04 μg/ml in α-amylase inhibition. Quercetin, an antidiabetic activity-having flavonoid, displayed IC_{50} values 2.00 μg/ml and 4.04 μg/ml in α-glucosidase and α-amylase inhibitory assays. In phytochemical assay, the extract had 31.28 GAE/g extract (mg), 30.48 CE/g extract (mg) and 183.3 PE/g extract (mg) for the total phenolic, total flavonoid and total proanthocyanidin contents. The limited reports of *E. zwageri* indicated the needs to search the active compounds from plant as potential antidiabetic agents by considering plant conservation status.

1. Introduction

Diabetes mellitus (DM) is one of the chronic diseases associated with metabolic disorders of the body. There are two main categories of diabetes mellitus, type 1 and type 2. Type 1 diabetes, or diabetes with insulin dependence, is a condition in which the body has a low ability or even not able to secrete insulin to prevent metabolic disorders. Type 2 diabetes, or diabetes without insulin dependence, is a condition in which the body experiences insulin deficiency or insulin resistance to regulate blood sugar levels for optimal needs for the body. The impaired metabolism followed by the release of excessive free radicals will cause imbalance generally lead to oxidative stress which is the main factor associated with the severity and death in diabetes. Diabetes mellitus is a major cause of disability, figuring among the top ten killers worldwide [1,2].

An approach to diabetes therapy is through the inhibition mechanism of enzymes that play an important role in the absorption of glucose in the human metabolism, such as α-glucosidase and α-amylase enzymes. Some studies have illustrated the potential for inhibition of the enzyme α-glucosidase by plants and also its effect on blood sugar levels [3]. Inhibition of the α-glucosidase enzyme also shows a slowing in the absorption of carbohydrates [4].

Therefore, α-glucosidase and α-amylase inhibitors are potential for use in the treatment of type 2 DM and other diseases associated with hyperglycemia. Plants are a potential source of natural medicine as there are currently a number of commercial drugs derived from plants. The search for natural ingredients derived from plants for enzyme inhibitors that play a role in glucose absorption has been the focus of intensive research [5].

Indonesia is one of the countries having a richness of plant diversity including medicinal plants. The richness is also supported by the potential of local wisdom that grows and develops in various ethnic groups residing in Indonesia. The objectives of this study were to evaluate the antidiabetic activity of the ethanol extract of *Eusideroxylon zwageri* stem bark through the inhibition mechanism of α-glucosidase and α-amylase enzymes and to investigate the DPPH and superoxide free radical scavenging activities.

2. Materials and Methods

2.1. General

Plant sample was collected from Kutai Kartanegara (Kutai ethnic) districts in July 2016. The sample was identified by taxonomist at the Laboratory of Dendrology, Faculty of Forestry Mulawarman University. Plant specimens were stored in the same laboratory.

Chemicals and biological enzymes used include ethanol, *n*-hexane, chloroform, ethyl acetate, 1,1-diphenyl-2-picryl hydrazil (DPPH), sodium hydrogen phosphate, disodium hydrogen phosphate, β-nicotinamide adenine dinucleotide, nitroblue tetrazolium, phenazine metosulfate, α-glucosidase enzyme, α-amylase enzyme, nitrophenyl α-glucopyranoside, dinitrosulfanilic acid, dimethyl sulfoxide, acarbose, ascorbic acid, starch , gallic acid, Folin-Ciocalteau reagents, sodium carbonate, aluminum chloride, ascorbic acid, quercetin, pelargonidin and other reagents were purchased from chemical companies.

2.2. Preparation of Extracts

The process of extraction was conducted by using maceration method. Approximately 50 g of *E. zwageri* stem bark were soaked in ethanol for 2 x 24 hours. The maceration was repeated 2 times according to the response of the plant material to the solvent. The sample solution was filtered by Whatman filter paper no. 2. The resulting solution of the filter was concentrated with a rotary evaporator at the temperature of 40° C. The condensed extract obtained was then stored in a vacuum oven at 35 ° C until a solid extract was obtained.

2.3. Antioxidant activity

2.3.1.DPPH radical scavenging capacity. The antioxidant activity assay used in this study was based on Shimizu et al method [6]. The test was performed using a UV-Vis spectrophotometer at room temperature (25 ° C) with a wavelength of 517 nm and using DPPH solution (1,1-diphenyl-2-picrylhydrazyl). A total of 33 µl of plant extracts (1.5-200 µg/ml) in DMSO were added to the cuvette, 467 µl of ethanol and 500 µl of DPPH 60 µM solution in ethanol were added. Samples were incubated for 20 minutes in a room in the darkness at room temperature. Antioxidant activity was determined by decolourization of DPPH at 517 nm wavelength using UV spectrophotometer. Ascorbic acid is used as a positive control. Antioxidant activity is determined by percentage reduction of DPPH using equation:

$$\text{Scavenging activity (\%)} = \frac{A\ \text{DPPH (t)} - A\ \text{sampel (t)}}{A\ \text{DPPH (t)}} \text{ x } 100 \tag{1}$$

Where A DPPH is the absorbance of DPPH and A sample is the absorbance of sample extract. All tests were run in triplicates (n=3) and average values were calculated.

2.3.2. Superoxide anion radical scavenging activity. The measurement of the superoxide anion radical scavenging activity of plant extracts was carried out according to the method as reported by Liu et al. [7]. Superoxide radicals were formed on the PMS-NADH system with NADH oxidation and tested by NBT reduction. In this experiment, superoxide radicals were formed in 3 ml Tris HCl buffer containing NBT, NADH and extract solutions. The reaction was initiated by addition of the PMS solution into the mixture. The mixture was subsequently incubated at 25 ° C for 5 min and the absorption on the spectrophotometer was measured at 560 nm. Decreased absorption values indicate an increase in superoxide anion radical scavenging activity. Percentage of activity was calculated based on the ratio of absorption value on the extract sample to the absorption of the control. Ascorbic acid was used as a positive control. The activity was presented in the form of IC_{50} value (inhibitory concentration, concentration causing 50% inhibition of radical formation) to that of IC_{50} value of ascorbic acid.

2.4. Enzyme Inhibitory Assay
2.4.1. α -Glucosidase inhibitory activity assay. α-Glucosidase inhibitory activity was evaluated according to the method by Kim et al. [8]. The test solution contained 150 µl of 0.1 M sodium phosphate buffer, 0.1 UN α-glucosidase and extract concentration (1.5 - 200 µg/ml). The test mixture was pre-incubated at 37°C for 10 min. Subsequently, 50 ml of 2 mM *p*-nitrophenyl α-D-glucopyranoside in 0.1 M sodium phosphate buffer was incubated at 37°C for 25 min. The reaction was stopped by adding 50 ml of 0.1M sodium carbonate (Na_2CO_3). The yellow colour formed was measured by a spectrophotometer at a wavelength of 405 nm. Mixtures containing α-glucosidase enzymes but without extracts were used as controls with 100% enzyme activity, and quercetin was used as a positive control.

$$\%\ \text{inhibition} = \frac{Absorbance\ of\ control - Absorbance\ of\ extract}{Absorbance\ of\ control} \ x\ 100 \qquad (2)$$

The inhibitory activity was presented in the form of IC_{50} (inhibitory concentration, concentration causing 50% inhibition of enzyme) to then be compared to the IC_{50} value of quercetin.

2.4.2. A-Amylase inhibitory activity assay. The test sample (1.5 – 200 µg/mL) in test tube was added to 500 µl 0.02 M sodium phosphate buffer (pH 6.9) containing 0.5 mg/ml of the α-amylase enzyme solution and pre-incubated at 25°C for 10 min. After pre-incubation, 500 µl 1% starch solution in 0.02 M sodium phosphate buffer (pH 6.9) was added to each test tube. The subsequent mixture was incubated at 25°C for 10 min. The reaction was terminated by adding 1 ml of the 3.5-acid dinitrosalisilic dye reagent. The reaction tube was incubated in boiling water for 5 min and then cooled to room temperature. The reaction mixture was then added with 10 ml of distilled water and measured by UV spectrophotometer at 540 nm wavelength [9]. Quercetin which is an α-amylase inhibitor is used as a positive control.

2.5. Total phenolic content analysis
Total phenolic content analysis was carried out with reference to the modified method of Yangthong et al. [10]. The total phenol content analysis was performed using a UV spectrophotometer at room temperature (25°C) at 765 nm. A calibration curve was made using gallic acid. The sample was put into a reaction tube of 0.3 ml, with 1.5 ml of Folin-Ciocalteu reagent and 1.2 ml of Na_2CO_3. Gallic acid was used as a positive control. Samples were incubated for 60 minutes in a room in the darkness at room temperature. After 60 min incubation, absorbance of the test sample was measured by using a UV / VIS spectrophotometer at 720 nm. The total concentration of phenol was calculated by the

standard calibration curve calibration expressed in mg/g gallic acid equivalent (GAE). The total content of phenol was calculated using the formula:

$$C = c.v/m \tag{3}$$

where C is the total phenolic content (mg GAE/g), c is extract concentration (μg/ml), v is volume and m is extract weight (μg).

2.6. Total flavonoid analysis

The method used refers to Sharma et al. [11] with modification, especially in terms of sample weight used. Total flavonoid test method was performed using UV spectrophotometer at room temperature (25°C) and with maximum absorbance at 500 nm. The standard catechin calibration curve was made with serial catechin concentration (0-10 μg/ml). The sample was put into a 0.1 ml reaction tube, then added with distilled water of 0.7 ml, 0.1 ml $NaNO_2$ (5%) solution, $AlCl_3$ (10%) solution 0.1 m and 0.5 ml of NaOH (1 M), after which the solution was incubated for 10 minutes in a room in the darkness at room temperature. After incubation for 10 min, the absorbance of the test sample was measured by using a UV spectrophotometer at a wavelength of 500 nm. The total flavonoid content was calculated by a standard catechin curve expressed in mg CE/g Catechin equivalent (CE). The total flavonoid content was calculated using the formula (3) above.

2.7. Total proanthocyanidin content analysis

The analysis of proantocyanidin in plant extracts was carried out with reference to Porter et al. [12]. Six ml of a *n*-butanol/HCl solution (950 ml of *n*-butanol and 50 ml of conc. HCl), 1 ml of extract solution, 0.2 ml of ferric ammonium sulfate in 2 mol/l HCl were added and mixed. The mixture vial was further heated in a water bath at a boiling temperature of 50 minutes. Then the vial was cooled and the solution was transferred to a volumetric flask and added with a solution of *n*-butanol/HCl up to 25 ml. The absorption was measured on a spectrophotometer at a wavelength of 550 nm. Pelargonidin was used as a standard. The results were calculated based on the absorption unit at 550 nm per 1 mg extract.

3. Results and Discussions

α-Glucosidase is one of the key enzymes that play a role in the absorption of carbohydrates in the body. On the other hand, α-amylase is an endoglucanases enzyme that hydrolyses the 1,4-glucosidic binders in the starch. These enzymes have been known to be therapeutic targets for controlling hyperglycaemia that may potentially lead to diabetes [13,14].

E. zwageri stem bark was extracted and evaluated its antidiabetic and radical scavenging activities. Maceration done with ethanol gave ethanolic extract 8.62% as presented in Table 1.

Table 1. Yield of *E. zwageri* Stem Bark Extract and Radical Scavenging Activities

Samples	Scientific name	Part used	Extract yield (%)	DPPH IC_{50} (μg/ml)	Superoxide IC_{50} (μg/ml)
Ulin	*E. zwageri*	Stem bark	8.62	44.90	30.47
Ascorbic acid				3.10	20.64

The results on DPPH free radical scavenging activity assay showed that the *E. zwageri* extract scavenged the radicals with IC_{50} 44.90 μg/ml. The value was much higher than the IC_{50} of ascorbic acid, the reference antioxidant that showed 3.10 μg/ml. In superoxide radical scavenging assay, the plant extract showed better activity than against DPPH. The IC_{50} of plant extract to scavenge the radicals was 30.47 μg/ml. In the assay, ascorbic acid showed the IC_{50} 20.64 μg/ml. The assay of radical scavenging activity showed that the *E. zwageri* stem bark extract was more active to scavenge the superoxide radicals rather than DPPH.

Furthermore, the inhibitory activity against α-glucosidase and α-amylase enzymes of the *E. zwageri* stem bark extract was presented in Table 2. The results showed that the extract inhibited the α-glucosidase activity with IC_{50} 58.45 μg/ml. The value was higher than the IC_{50} of the control, quercetin, showing 2.00 μg/ml. In α-amylase inhibitory assay, the extract showed IC_{50} 9.04 μg/ml that was close to the IC_{50} of quercetin, showing 4.04 μg/ml. In relation to a scientific validation of ethnopharmacological information of *E. zwageri* stem bark as natural antidiabetic agen, the results indicated that the activity of *E. zwageri* is more effective as α-amylase inhibitor rather than the α-glucosidase.

Table 2. Antidiabetic Activity of *E. zwageri* Stem Bark Extract by Means of α-Glucosidase and α-Amylase Inhibitory Assays.

Samples	Scientific name	Part used	α-Glucosidase IC_{50} (μg/ml)	α-Amylase IC_{50} (μg/ml)
Ulin	*E. zwageri*	Stem bark	58.45	9.04
Quercetin			2.00	4.04

Results of total phenolic content, total flavonoid content and total proantocyanidin content analyses from the stem bark extract of *E. zwageri* were shown in Table 3. The results showed that the extract contains 31.28 mg GAE/g extract, 30.48 mg CE/g extract, and 183.30 mg PE/g extract represented the total phenolic, total flavonoid and total proanthocyanidin contents, respectively. The results suggested the potential contribution of phenolic, flavonoid and proanthocyanidin compounds in scavenging and inhibition of the extracts to radicals and key enzyme of diabetes. Several studies have shown that there are differences in the ability of phenolic compounds to inhibit the activity of α-glucosidase and α-amylase enzymes. In relation with that, *Orthosiphon stamineus* and sugar molasses have been reported to possess a good inhibitory activity on α-amylase. Although they are also good inhibitors of the α-glucosidase enzyme, the sequence of activity against these two enzymes is different [15,16].

Table 3. Total Phenolic, Total Flavonoid and Total Proanthocyanidin of *E. zwageri* Stem Bark

Samples	Scientific name	Part used	Total Phenolic (mg GAE/g extract	Total Flavonoid (mg CE/g extract)	Total Proanthocyanidin (mg PE/g extract)
Ulin	*E. zwageri*	Stem bark	31.28	30.48	183.30

The stem bark of *E. zwageri* is used by the Kutai ethnic in Indonesia for treating the diabetes in traditional manner. To the best of our knowledge, reports on the biological active compounds from *E. zwageri* are very limited. Some neolignans have been isolated from the heartwood of the plant and reported to be the responsible compounds for the durability of the wood [17]. Furthermore some lignans have also been isolated from the plants as reported by Yoosu et al. [18]. However, no extensive reports on the biological activity and related biological active compounds have been reported from the stem bark of *E. zwageri*. Isolation and identification of the active compounds from the plant stem bark are required to answer the mechanism of the bioactivity. Regarding the sustainability of plant source, the efforts toward the bioproduction of the active compounds of *E. zwageri* should be put in more priority rather than the extensive uses of the plant materials as the main source.

Currently, efforts to find alternative sources for the treatment of type 2 diabetes are in increase, especially those from medicinal plants having high antidiabetic activity and with a lower risk of side effects than commercial drugs currently available. In line with this, the research has shown that natural material-based compounds and natural-based products have the potential to have excellent ability to inhibit the activity of key enzymes that cause diabetes and with low side effects [19-21]. Therefore, the sustainable uses of plant sources as natural antidiabetic offer good solution for diabetes treatment.

In addition, regarding the information where medicinal plants were collected, efforts to strengthen the benefit sharing aspect to the community where samples of plants are collected need to be continued

4. Conclusions

An ethnobotanically-selected medicinal plant, *E. zwagery* has been investigated its antidiabetic and antioxidant activities. The results informed that the stem bark ethanolic extract of the plant showed good antidiabetic activity only for α-amylase inhibition and weak activity to α-glucosidase. The extract was more active to scavenge the superoxide radical than DPPH. Further investigation is needed to find the responsible compound in the plant and to explore possibility of bioproduction for the active compounds.

Acknowledgments

The research was funded by the Health Research and Development Agency, Ministry of Health, Indonesia, through Advanced RISTOJA 2016 under the contract number LB.02.01/2/1515/F.032/2016. Acknowledgments were conveyed to the Kutai community in Sebulu Kampung, East Kalimantan who have supported the research by sharing ethnopharmacological information and plant specimens. The same remarks were presented to the Laboratory of Forest Dendrology and Ecology, Faculty of Forestry, Mulawarman University who has assisted the identification of plant species.

References

[1] Dieye AM, Sarr A, Diop SN, N'Diaye M, Sy GY, Diarra M, Gaffary LR, Sy AN and Faye B 2008 *Fundamental Clin. Pharmacol.* **22** 211
[2] American Diabetes Association (ADA) 2012 *Diabetes Care* **35** S64
[3] Kumar S., Narwal S, Kumar V and Prakash O 2011 *Pharmacogn.* Rev. **5** 19
[4] Heacock PM, Hertzler SR, Williams JA and Wolf BW 2005 *J. Am. Diet Assoc.***105** 65
[5] Yao Y, Cheng X, Wang L, Wang S and Ren G 2011 *Int. J. Mol. Sci.* **12** 6445
[6] Shimizu K, Kondo R, Sakai K, Takeda N, Nagahata T and Oniki T 2001 *Lipids* **36** 1321
[7] Liu F, Ooi VEC and Chang ST. 1997 *Life Sci.* **60** 763
[8] Kim YM, Jeong YK, Wang MH and Lee WY 2005 *Nutrition* **21** 756
[9] Kwon GJ, Choi DS and Wang MH 2007 *J. Korean Food Sci. Technol.* **39** 569.
[10] Yangthong M, Hutadilok-Towatana N and Phromkunthong W 2009 *Plant. Foods Hum. Nutr.* **64** 218
[11] Sharma K, Ko EY, Assefa AD, Ha SY, Nile SH, Lee ET and Park SW 2015 *J Food and Drug Anal.***23** 243
[12] Porter LJ, Hirstich LN and Chan BG 1986 *Phytochem.* **25** 223
[13] Yilmazer-Musa M, Griffith AM, Michels AJ, Schneider E and Frei B 2012 *J. Agric. Food Chem.* **60** 8924
[14] Meng Y, Su1 A, Yuan S, Zhao H, Tan S, Hu C, Deng H and Guo Y 2016 *Plant Foods Hum. Nutr.* **71** 444
[15] Mohamed EAH, Siddiqui MJA, Ang LF, Sadikun A, Chan SH, Tan SC, Asmawi MZ and Yam MF 2012 *BMC Complement. Altern. Med.* **12** 176
[16] Kong F, Yu S, Zeng F and Wu X 2016 *Sugar Tech* **18** 333
[17] Syafii W, Yoshimoto T and Samejima M 1985 *Mokuzai Gakkaishi* **31** 952

[18] Yoosu S, Namseok C and Terazawa M 2009 *Chemistry of Natural Compounds* **45** 3.
[19] Harbilas D, Martineau LC, Harris CS, Adeyiwola-Spoor DC, Saleem A, Lambert J, Caves D, Johns T, Prentki M, Cuerrier A, Arnason JT, Bennett SA and Haddad PS 2009 *Can. J. Physiol. Pharmacol.* **87** 479
[20] Petrovska BB 2012 *Pharmacogn. Rev.* **6** 1.
[21] Mahomoodally MF 2013 *Evid. Based Complement. Alternat. Med.* 617459.

Somatic embryogenesis of East Kalimantan local upland rice varieties

Nurhasanah[*1], **Ramitha**[1], **B Supriyanto**[1] **and W Sunaryo**[1]

[1]Department of Agroecotechnology, Faculty of Agriculture, Mulawarman University. Jl. Pasir Balengkong No.1 Kampus Gunung Kelua, Samarinda, East Kalimantan-Indonesia 75119, Tel./Fax. +62-541-749159/738341

[*]Email: nurhasanah_2710@yahoo.com

Abstract. Somatic embryogenesis is the formation, growth and development of embryos from somatic cells. Somatic embryo induction is one of the in vitro plant propagation techniques that is very important for plant developmental purposes. Four local upland rice varieties of East Kalimantan, Mayas Pancing, Gedagai, Siam and Serai, were used in this study. A total of 200 explants (mature rice grains) for each varieties were inoculated on MS solid medium supplemented with 1 mg L-1 2,4 Dichlorophenoxy acetic acid (2,4-D) and 0.5 mg L-1 6-Benzylaminopurine (BAP). The results showed that response of each variety differed to embryosomatic induction, indicated by callus induction rate and callus quality, in terms of callus color and structure. The fastest callus formation was sobserved in Gedagai variety (8 days) while Mayas Pancing (13 days) was the latest one. The rate of callus induction varied from 60 to 98.5 %, and Serai variety has the highest callus induction rate. The highest friable callus structure was found in Siam variety (89.1%) and the lowest was in Gedagai (62.5%). Callus color was dominated by the yellowish-white (transparent) on all varieties tested. Most of the callus was potential as embryogenic callus characterized from the nodular and globular of friable callus structure and its yellowish-white color.

1. Introduction

Research on in-vitro culture has been rapidly developed in higher-plant cells and tissues. Widespread application of tissue culture technologies to crop improvement have been reported included in cereal species such as rice. Organogenesis and embryogenesis are types of cells differentiation in the regeneration of in vitro derived plants. Organogenesis is characterized by the unipolar bud primordial with subsequent development of shoot and roots. On the other hand, the bipolar structure, having both shoot and root apical meristem in a single cell, is the character of embryogenesis pathway [1-2].

Somatic embryogenesis is a somatic cell that develop to form new plant through specific embryonic developmental stages without gamete fusion [1, 3]. Embryogenesis begins from a single cell then proliferates with differentiating organs and tissues. Embryogenesis is the most popular plant regeneration due to its prospect for specific plant developmental purpose, such as genetic transformation, cryopreservation, synthetic or artificial seeds, protoplast production for somatic hybridization, and somaclonal variation.

Various factors critically influence somatic embryogenesis, including plant genotype, explant source, media composition, and plant growth regulator [4]. The use of growth regulating hormones should consider the type and concentration of the substances in accordance with the plant genotype as well as physiological conditions of explant, since each genotype and plant tissue has specific response to the provision of growth regulators. The successful induction of somatic embryos and subsequent recovery of viable plants is not routine or efficient for the majority of species. Highly genotypic variation resulting in specific morphogenetic response is a major limitation in rice embryogenesis. Therefore, optimization of embryogenesis protocol for desired genotype is essential.

Somatic embryogenesis using East Kalimantan local upland rice variety has not been reported. The local rice variety carry several important traits based on previous studies [5-6], and potential to be developed as new superior rice variety in plant breeding program including plant biotechnology approach. Therefore, the application of somatic embryogenesis in East Kalimantan local upland rice variety can be utilized for genetic quality improvement through genetic manipulation, genetic engineering and increasing genetic diversity of rice through somaclonal variation.

2. Materials and methods

2.1. Plant materials

Four local upland rice varieties from East Kalimantan, Serai Gunung, Gedagai, Mayas Pancing, and Siam were used in this study. These varieties were included into superior local rice varieties having good grain quality in terms of taste characteristics. The mother plants were grown in the screen house and maintained according to general upland rice cultivation procedure.

2.2. Explants and Media Preparation

Mature embryos were used as explants source in this study. The embryos is derived from a physiologically mature rice grain obtained from the ready harvested plants. The grains were selected with the following criteria, free from pests and diseases, and the grains size were uniform.

A total of 200 explants were used for each variety. The selected rice seeds were cleansed using detergent and rinsed with distilled water. The lemma and palea of the seeds were removed manually. The caryopsis then were surface sterilized in 70% alcohol solution for 1-2 minutes, then with 30% of sodium hypochlorite for 15 minutes and followed with 10% of sodium hypochlorite for 5 minutes. Explants finally were washed with sterile distilled water twice for 3-5 minutes and drained on sterile filter paper. The sterilization steps were conducted in LAFC (Laminar Air Flow Cabinet).

2.3. Inoculation and Somatic Embryos Induction

Somatic embryos were induced by inoculating the sterilized seeds explant on the callus induction medium. The callus induction medium was agar solidified MS [7] media containing sucrose (30 gL-1) and supplemented with 1 mgL-1 2,4-D and 0.5 mgL-1 BAP. Inoculation was conducted by lying the explants horizontally on the media. The cultures then were incubated in dark condition (25+2 °C) for somatic embryos induction.

3. Results and discussion

3.1. Callus Formation

The rate of callus formation varied in different plant genotypes (Figure 1). The fastest callus formation was observed in Gedagai variety which needs only 8 days for callus formation, and Mayas Pancing was the latest one. Variation was also observed in the callogenesis response among the four varieties (Tabel 1). The percentage of explants forming callus varied from 60-98.5%. Among the genotypes, Serai variety produced the highest explants forming callus. On the other hand, Gedagai which was the fastest in forming callus has the lowest ability of explants forming callus.

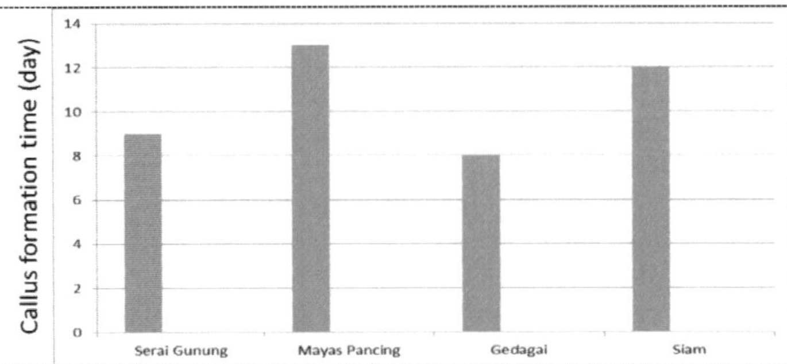

Figure 1. Callus formation time of four local upland rice varieties of East Kalimantan.

Table 1. Number of explants forming callus in four local upland rice varieties of East Kalimantan[a].

Variety	Explants forming callus (Mean ± SD[b])	Percentage (%)
Serai	197±0.31	98.5
Gedagai	120±1.09	60.0
Mayas Pancing	138±0.98	69.0
Siam	169±0.88	84.5

[a] Data was observed from 200 explants.
[b] SD=Standard Deviation.

Each variety has different ability of callus induction in somatic embryogenesis as observed in this study. This indicates that the ability of embryogenesis is influenced by the genotype of the plant. Significant genotypic differences in callus induction using mature embryo explant in rice somatic embryogenesis have been reported earlier [8-9]. Genetic variation in each genotypes is resulted by the different levels of endogenous phytohormones influencing callus induction and proliferation [10]. Therefore, somatic embryogenesis is not only a routine techniques using standard use of media to promote the induction and development stages of embryo, since the successful induction of somatic embryos is not always efficient for all plant species even within species.

The initiation of callus in mature embryos with endosperm-supported explant developed from scutellar tissue layer, the tissue between endosperm and zygotic embryo (Fig. 2A). Then the callus continued to grow as a result of the proliferation of callusing cells (Fig. 2B), forming a larger mass of cells. Calli induced from scutellar tissue of mature seed are the excellent source of cells for in vitro regeneration and for the production of transgenic rice [11-12]. Therefore, when using endosperm without zygotic embryo as an explant source in rice somatic embryogenesis, the removing process of zygotic embryo should be as thoroughly as possible that it does not take away the scutellar layer together.

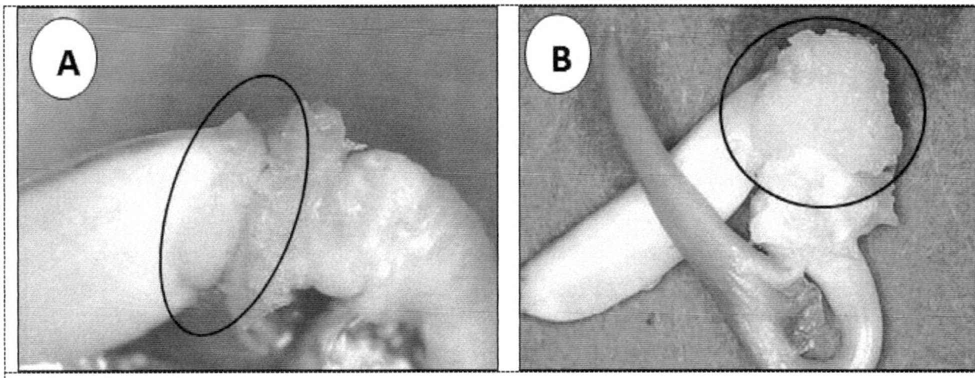

Figure 2. Initial stage of callus development in mature embryo with endosperm supported explant; A. Early formation of callus in the scutellum tissue; B. Callus enlarged forming a larger mass of the cells.

3.2. Callus Structure

Callus structure observation was carried out to evaluate whether the cell structure was compact or not. The compact callus structure is the composite callus consisting of closely packed cells. The non-compact callus structure is the crumbling one or friable (do not converge between the callus composite cells) (Figure 3). The callus structure observation of the four local upland rice varieties of East Kalimantan is presented in Table 2. The callus structure was dominated by the non-compact callus structure. The highest non-compact callus structure was found in Serai variety in which around eighty percent of the calli are friable, while the highest compact callus structure was observed in Gedagai variety.

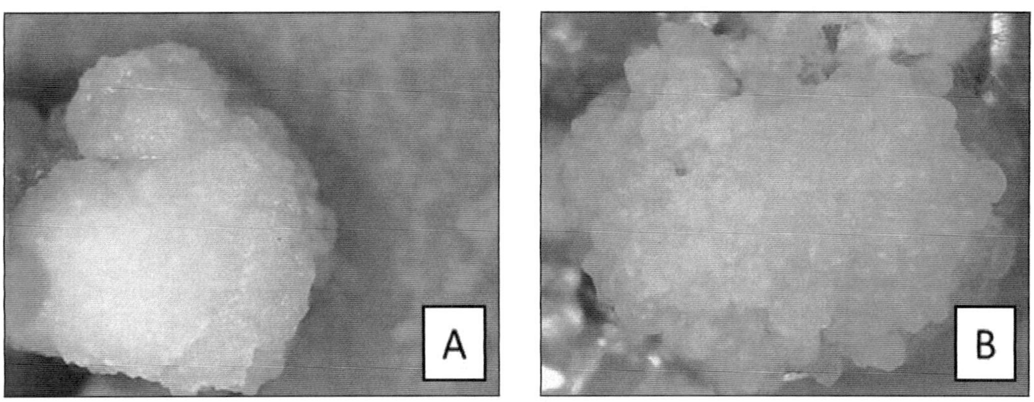

Figure 3. Differences of non-compact and compact callus structure. A. Compact callus structure; B. non-compact callus structure.

The quality of a callus can be observed from the callus structure. The good quality callus is assumed to have a non-compact structure (friable). The friable callus structure is embryogenic potential. The embryoid development was abundant in this structure and plants were easily regenerated [13]. There was a striking difference between friable and compact callus from the microscopical sections. Vascular bundles which were prominent in compact callus, were not present in friable callus [14].

Table 2. Callus structure of East Kalimantan local upland rice variety.

Variety	Compact (Mean ± SD)	Percentage (%)	Non-compact (Mean ± SD)	Percentage (%)
Serai	34±0.86	17.3	163±0.89	82.7
Gedagai	45±0.69	37.5	75±1.04	62.5
Mayas Pancing	40±0,61	29.0	98±1.05	71.0
Siam	17±0.48	10.1	152± .75	89.1

It was observed that the friable callus was initially a compact form of invisible nodules of the embryo (Figure 2), after two weeks the compact callus began to form nodules of the embryonic candidate (3B), depending on the variety and combination and concentrations of growth regulator substances. The next stage, is the nodules formation having a sphere of smooth and thin walls on the surface of the cell which is called as globular phase (Figure 3B). The smooth and glossy surface of the callus is when the somatic embryo of the globular phase begin, whereas in the compact callus structure, the surface of the callus does not form nodules (Figure 3A).

3.3. *Callus Color*

Callus growth indicator on in vitro culture can be observed from callus color. Callus color depicts the visual appearance of the callus, it indicates whether a callus is healthy and good growing or degenerate. Visual observation of the callus color of the local upland rice somatic embryos is presented in Table 3. There were yellowish-white or creamish, white and brown callus color. The callus color is dominated by the yellowish-white color (transparent), ranging from 77.5 to 87.6 % depending on the plant genotype. The highest number of yellowish colored callus was found in Serai variety, while Gedagai variety produced the lowest one.

Besides callus structure, callus color also describes the regeneration ability of the cells in forming buds and roots [4]. The yellowish-white color is one of the embryogenic callus characteristic. Nodular and globular callus with transparent color (pale yellow) (Figure 3B), usually have a higher ability to form buds than the compact and browning colored callus [15].

Table 3. Callus color of East Kalimantan local upland rice varieties.

Variety	Yellowish-white (Mean ± SD)	White (Mean ± SD)	Brown (Mean ± SD)
Serai	165 ± 0,83	13 ± 1,01	19 ±0,46
Gedagai	93 ±1,04	13 ± 0,95	15 ± 0,38
Mayas Pancing	112 ± 0,94	14 ± 0,28	10 ± 0,0
Siam	148 ± 0,77	10 ± 0,0	13 ± 0,29

It was observed that the brown callus color present in low percentage in this study. The brown color of the callus indicates the degeneration or cells growth declining, and even may lead to death [16]. It is a type of non-embryogenic callus [13], and has low ability of regeneration. Brown callus occurs frequently in compact callus structure, indicated by lignification of the browning cells [17]. The brown callus was initially white, then the color changes to brown.

3.4. *Embryosomatic Developmental Stage*

Somatic embryos are bipolar propagules, derived from somatic tissue but the development, morphology and physiology are similar to sexually derived zygotic embryos [1]. The development of embryo occurs through several phases. There is a slight difference between the somatic embryo developmental stage of dicot and monocot plants. In dicotyledonous plants, the stages are globular, heart, and torpedo [18], whereas in monocots, the stages are globular, scutellar, and coleoftillar [15, 19]. In this study we captured the somatic embryos developmental stages of East Kalimantan local upland rice variety (Figure 4) representing the globular, scutellar, and coleoftillar stages. Based on morphological observations on the callus, the most frequently callus observed is the globular phase (Figure 4A). The scutellar and coleoftillar developmental phase was rarely found (Figures 4B and 4C).

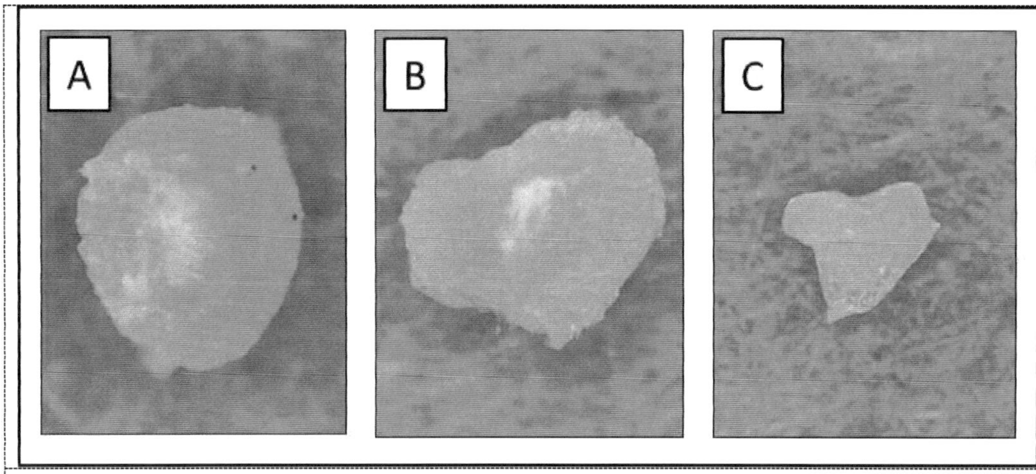

Figure 4. Visual observation of rice somatic embryo developmental stage. A. Globular; B. Scutellar; C. Coleoftillar phase.

The success of somatic embryogenesis occurs when the callus or cell is embryogenic, characterized by bipolar structures, having both basal and apical meristem initiating root and shoot development simultaneously [2]. In monocots, apical and basal cell lineages are usually incorporated into a pear-shaped proembryo and are difficult to distinguish from each other [20], which might be described as the scutellar and colleoptillar developmental of stage (Figure 4B and 4C).

Acknowledgements

Authors acknowledge technical staff and member of Plant Biotechnology Laboratory, Faculty of Agriculture, Mulawarman University. This study was financed by IDB 4in1 project 2017 (399/UN17.11/PL/2017) to which the authors are highly indebted.

References

[1] von Arnold S, Sabala I, Bozhkov P, Dyachok J and Filonova L 2002 *Plant Cell Tiss. Org. Cult.* **69** 233-249
[2] ten Hove C A, Lu K J and Weijers D 2015 *Development* **142** 420–430
[3] Merkle S A, Parrott W A and Flinn B S 1995 Morphogenic Aspects of Somatic Embryogenesis *In Vitro Embryogenesis in Plants Current Plant Science and Biotechnology in Agriculture* Ed T A Thorpe (Dordrecht: Springer)
[4] Khan U W, Ahmed R, Shahzadi I and Shah M M 2015 *Sarhad J. of Agric.* **31** 199-209
[5] Nurhasanah, Sadaruddin and Sunaryo W 2016 *Biodiversitas* **17** 401-408
[6] Nurhasanah, Sadaruddin and Sunaryo W 2017 *Biodiversitas* **18** 1165-1172

[7] Murashige T and Skoog F 1962 *Physiol. Planta.* **15** 473-497
[8] Pawar B, Kale, Bahurupe J, Jadhav A, Kale A and Pawar S 2015 *Rice Sci.* **22** 283-289
[9] Verma D, Joshi R, Shukla A and Kumar P 2011 *Ind. J. Exp. Biol.* **49** 958-963
[10] Deo P C, Harding R M, Taylor T, Yagi M A and Becker D K 2009 *Plant Cell Tiss. Org. Cult.* **99** 61-71
[11] Wani S H, Sanghera G S and Gosal S S 2011 *New Biotechnol* **28** 418-422
[12] Joyia F A and Khan M S 2013 *Int. J. Agric. Biol.* **15** 27-33
[13] Armstrong C L and Green C E 1985 *Planta* **164** 2007-2014
[14] Fransz P F and Schel J H N 1991 *Can. J. Bot.* **69** 26-33
[15] de Alcantara G B, Dibax R, de Oliveira R A, Filho J C B and Daros E 2014 *Acta Sci. Agron* **36** 63-72
[16] He Y, Guo X, Lu R, Niu B, Pasapula V, Hou P, Cai F, Xu Y and Chen F 2009 *Plant Cell Tiss. Org. Cult.* **98** 11–17
[17] Laukkanen H, Rautiainen L, Taulavuori E and Hohtola A 2000 *Tree Physiol.* **20** 467–475
[18] Sadeq M A, Pathak M R, Salih A A, Abido M and Abahussain A 2014 *American J. of Plant Sci.* **5** 2342-2353
[19] Gray D J and Purohit A 1991 *Critical rev. in plant sci.* **101** 33-61
[20] Zhao P, Begcy K, Dresselhaus T, Sun M-X 2017 *Plant Physiol.* **173** 130–142

IOP Conf. Series: Earth and Environmental Science **144** (2018) 012032 doi:10.1088/1755-1315/144/1/012032

Mapping urban green open space in Bontang city using QGIS and cloud computing

F Agus[*1], Ramadiani[1], W Silalahi[1], A Armanda[1] and Kusnandar[2]

[1]GIS and Environment Modeling Research Group, Computer Science Department, Faculty of Computer Science and Information Technology-CSIT, Mulawarman University Samarinda, Kalimantan Timur Province, Indonesia

[2]Informatics Department, STMIK Widya Cipta Dharma Samarinda, Kalimantan Timur Province, Indonesia

[*]Corresponding Author: fahrulagus@unmul.ac.id

Abstract. Digital mapping techniques are available freely and openly so that map-based application development is easier, faster and cheaper. A rapid development of Cloud Computing Geographic Information System makes this system can help the needs of the community for the provision of geospatial information online. The presence of urban Green Open Space (GOS) provide great benefits as an oxygen supplier, carbon-binding agent and can contribute to providing comfort and beauty of city life. This study aims to propose a platform application of GIS Cloud Computing (CC) of Bontang City GOS mapping. The GIS-CC platform uses the basic map available that's free and open source. The research used survey method to collect GOS data obtained from Bontang City Government, while application developing works Quantum GIS-CC. The result section describes the existence of GOS Bontang City and the design of GOS mapping application.

1. Introduction

Geographic Information Systems (GIS) have been developed and applied to various areas of life. The existence of SIG also helps the community to provide various information. In the field of health, GIS can assist and provide useful information such as health information mapping or public health spatial database [1]. GIS functions for crime mapping and geographic crime analysis solution for operational, tactical, investigative, and strategic policing and crime reduction purposes [2]. In elections to the region or president, GIS is linked to the economy and elections can help illustrate the distribution of voters and their choices [3]. There are many more GIS applications that have been developed for various purposes such as mapping of public facilities [4], handicraft industry center [5], or GIS for urban green space management [6-7].

Mapping techniques are also undergoing rapid development. Supported by increasingly sophisticated cartography techniques, making the GIS development process easier. Digital map creation is much different from 5-10 years ago. Implementation of spatial data infrastructure for a government can be done by applying free open source GIS software on the state of on limited financial budget [8]. Outside there has been a lot of digital maps available for free and can be accessed freely with still include the source, such as the surface map of the earth by google maps (https://maps.google.com/), or a digital highway map that is the open street map (OSM: http://www.openstreetmap.org).

There are two approaches to making GIS developed into web-based, proprietary and open source approaches. The proprietary approach means that users are willing to pay for licenses as software used for development. For this section, the available software is ArcGIS (www.arcgis.com) made by ESRI. While the 2nd approach using open source alias free, namely Quantum GIS for desktop versions [9-10] and QGIS Cloud (http://qgiscloud.com) for the development of web GIS.

Green Open Space (GOS) in the city area is at the heart of urban life. Its existence can provide great benefits as an oxygen supply, as a carbon-binding agent, sustains beauty and comfort and reduces the impacts of climate change [11]. The problems that often occur in aspects of GOS management are the low quantity, overlapping of land ownership as the location of GOS and inappropriate location selection and less attention to the public interest.

Bontang City is the second highest city in the Human Development Index in East Kalimantan Province. The city has a rapid development so as to improve the overall human welfare. The area of green space owned by Bontang City has exceeded the target of regulation Public Works Minister at 30 percent of the total area. GOS area of this city reached 33.9 percent of the city area outside the protected forest, which is 9,206.00 hectares. Targets promoted by the Bontang City Government according to Local Regulation RTRW Bontang city that is equal to 38.37 percent or 3,532.34 hectares. As an industrial city, the city also has the same problem in terms of GOS management, which is still targeting many new locations of green open space [12].

This study aims to build a model of geospatial application software for the management of GOS location of Bontang City using Cloud Computing based GIS.

2. Materials and Method

The study area. The study was conducted in the administrative area of Bontang City. The city is geographically located $0^0$01' North Latitude - $0^0$12' North Latitude and $117^0$23' East Longitude - 117^0 38' East Longitude, with an area of 14,780 Hectares or 497.57 Km^2. Bontang City administration area consisting of 3 (three) districts:

a. North Bontang District;
b. Bontang Selatan District; and
c. District Bontang Barat.

The boundaries of the planning area of RTRW Kota Bontang include:

a. Regency of East Kutai in the north;
b. East bordering Makassar Strait;
c. Regency of Kutai Kartanegara in the south; and
d. Regency of East Kutai in the west.

Residents of Bontang City currently number 118,000 souls with the potential of natural resources include seaweed, fishery and plantation and natural gas. Tourism potential developed by the city government include Pupuk Kaltim Lake, Wetland Beach and Kaba Bay. Administrative Map of Bontang City as described in Figure 1 (Source: http://www.bpkp.go.id/kaltim/konten/1174/Profil-Pemkot-Bontang).

RTRW Planning Bontang City divides the area into 3 Sections of City Region (BWK), namely BWK I consists of 8 (eight) urban villages include: Bontang Kuala Village, Village of Gunung Elai, Bontang Baru Village, Api-Api Sub-District, Central Berbas Village, Tanjung Laut Sub-District, Tanjung Laut Indah Sub-District; BWK II consists of 6 (six) villages covering: Satimpo Village, Gunung Telihan Village, Canaan Village, Belimbing Village, Loktuan Village, Guntung Village and BWK III covering 1 (one) urban village of Bontang Lestari. Planning of RTRW Bontang City 2012-2032 is described with the map as in Figure 2 (Source: Bontang City RTRW Regulation No. 11 The year 2012.

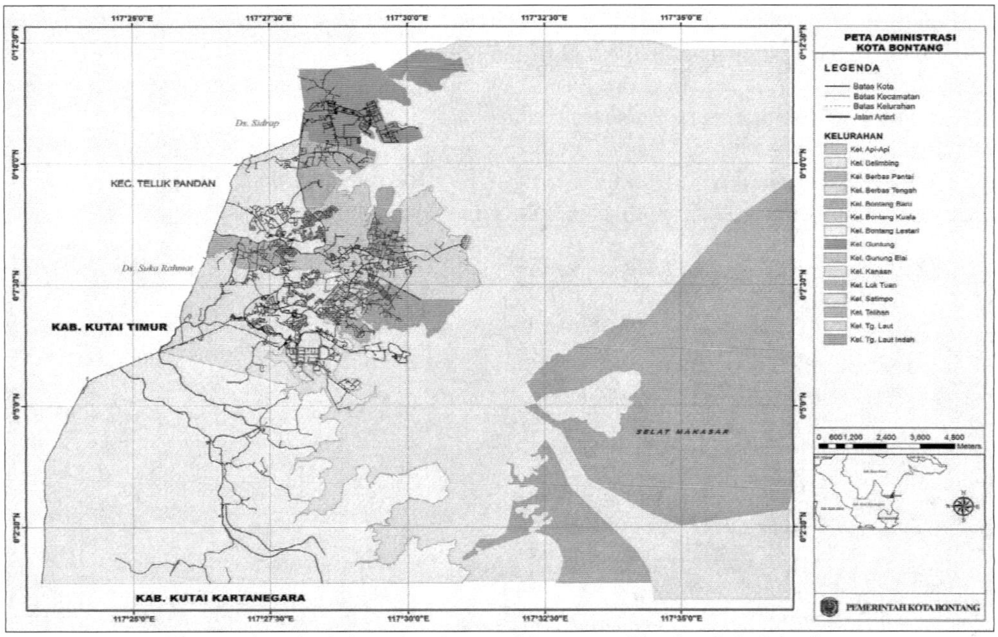

Figure 1. Administrative Map of Bontang City by Sub-District

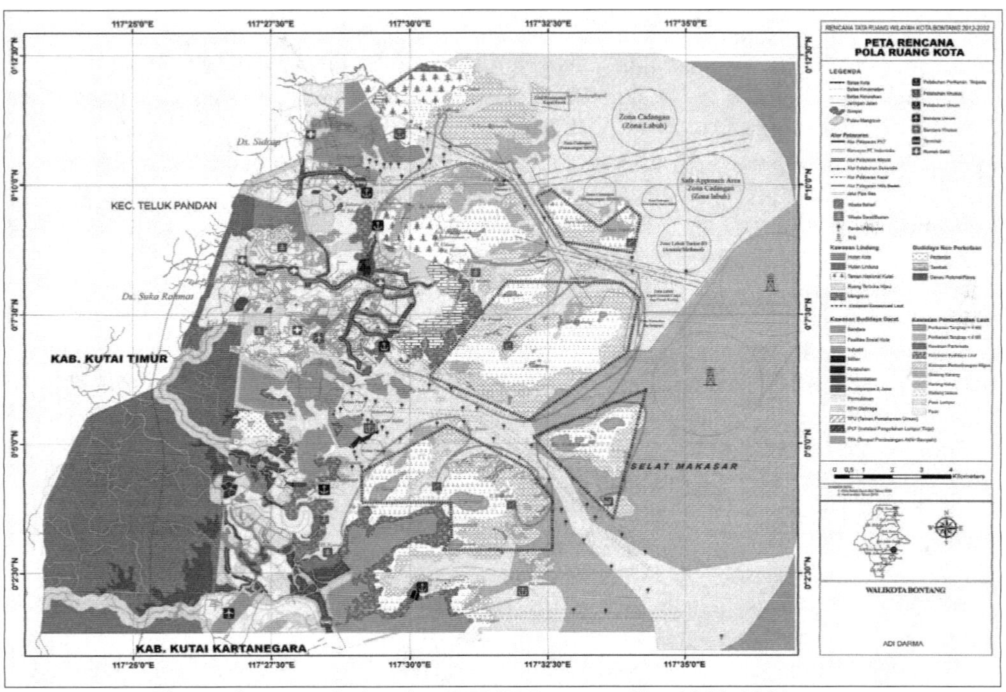

Figure 2. Map of Planning RTRW Bontang City 2012-2032

2.1. Field data collection

This stage is to collect data and materials in the form of an in-depth survey on the policy and implementation of GOS management of Bontang City. Data collection was conducted by two institutions directly related to the management of Green Open Space in Bontang City, namely the Office of Spatial Planning and City and the Office of the Environment.

2.2. Data analysis

At this stage data analysis has been collected in the previous stage. Primary data were obtained from interviews for the design of software products, as well as secondary data and other supporting data collected and used as complementary research data.

This stage performed the process of analyzing spatial data and attributes. It used Quantum GIS Desktop 2.0 Dufour software with Ubuntu 14.04 LTS 32 Bit operating system (Figure 3a). Minimum hardware requirements for processing this spatial data as described in Figure 3b.

Figure 3a. Quantum GIS Desktop 2.0 Dufour Version

Figure 3b. Minimal System Requirement QGIS Desktop Processing

2.3. Process Integration using Cloud-Computing GIS

At this stage, processed of integration between Quantum GIS Desktop applications with Cloud Computing applications such as QGIS-Cloud or GIS-Cloud and embedded coding process with WebGIS platform development script or Mobile GIS platform will be developed. This study is limited to the scope of application of Cloud Computing GIS to generate WebGIS platforms.

3. Result and Discussion

3.1. Distribution of GOS Bontang City

The result of the in-depth survey on the existence of GOS Bontang City resulted in 14 locations of green space spread in the three sub-districts. South Bontang region is the area that has the widest GOS, followed by North Bontang and last Bontang Barat. The distribution of GOS locations is described in the following table.

Table 1. Distribution of GOS Bontang City

No.	GOS Name	GOS Address	Wide(m^2)	Ownership Status
1.	Bontang Lestari	Mohammad Roem Street, Bontang Lestari, South Bontang	80,000.00	Government
2.	Guntung	Tari Jepen Street, Guntung, Nort Bontang	31,000.00	Government
3.	Taman Tanjung Laut	Jl. Jendral Soedirman, Tanjung Laut, South Bontang	12,083.00	Government
4..	Hutan Kota Wana Khatulistiwa	Urip Sumoharjo Street, Bontang Lestari, South Bontang	12,083.00	Privat
5.	Belimbing	Arief Rahman Hakim Street, Belimbing, West Bontang	12,000.00	Government
6.	Loktuan	Kapal Selam 3 Street, Loktuan, Nort Bontang	9,019.00	Government
7.	Hutan Kota DSM	Bontang Lestari, South Bontang	-	Privat
8.	Taman Basket HOP VI	Ternate Street, Gn Elai, Nort Bontang	-	Privat

No.	GOS Name	GOS Address	Wide(m^2)	Ownership Status
9.	Taman Nasional Kutai	RA Kartini Street, Gn Elai, Nort Bontang	-	Government
10.	Taman Wisata Graha Mangrove	Cut Nyak Dien Street, Bontang Baru, Nort Bontang	-	Privat
11.	Taman Cibodas	Kembang Sepatu Street, Belimbing, West Bontang	-	Privat
12.	Taman Palembang	Palembang Street, Telihan, West Bontang	-	Government
13.	Taman Adipura	Basalt Street, Bontang Kuala, Nort Bontang	-	Government
14	Lapangan Lang-Lang	Aip II KS Tubun Street, Api-api, Nort Bontang		Government

Based on the data in the table, done pointing to create GOS Location layer on Quantum GIS application.

3.2. Data Layer Processing
Data processing GOS location Bontang City has 4 main layers, namely the Administration Boundaries layer, Boundaries of Protected Forest layer, Coastal Area layer and GOS Location layer. In addition to the four layers, data processing also uses maps from openstreetmap.org in the form of OpenCycleMap (OCM) Landscape layer and OpenStreetMap layer and Google Maps of Google Satellite Layer. All of these maps are spatial data developed free and open source by mentioning the original source.

After the Dufour QGIS application loads successfully on the computer, open all the shapefiles containing pre-prepared data layers. Next, do the process of installing the plugin for the open layer map. After the install process is successful, do the loading process for each open layer as the background screen. Do not forget that the background screen is placed on the first screen. The map visualization results in Quantum GIS with the background of each layer as described in Figures 4a, 4b, and 4c. Each data layer can be done editing process. The editing process is done to give the map more exclusive, elegant and dynamic. Such as giving a coloring style in accordance with the color of the map object displayed or setting the transparency value of each layer.

3.3. Processing with Cloud Computing
The next is GIS data processing using Cloud Computing (CC) system. The CC system is a popular terminology of the present era. CC has explained the leap of products and technology services that are very powerful in the field of information technology. CC can be applied to be a solution to the challenges of GIS applications in the future [12]. GIS-CC can also be applied to build spatial data on local government [13]. such as policy and implementation of one data one map of East Kalimantan Provincial Government (http://onedataonemap.kaltimprov.go.id/geoportal/).

The QGIS-Cloud operation is very easy and fast. First, the user must register an account at https://qgiscloud.com/en/account/sign_in. Next, do the install process plugin on Quantum GIS desktop. If the install is successful then QGIS-CLoud will appear on the left side of QGIS desktop, as shown in Figure 4a-c. Next, do the login process user. The last step of this process is uploading local data layer. If the upload process is successful then the entire map layer on the desktop will be converted into a web platform.

In this case, GOS processing of Bontang City uses GIS-Cloud (giscloud.com) as GIS Cloud Computing platform. This platform is a proprietary application, but the researchers used trial versions for the purposes of this study. The results of the GIS-Cloud processing as described in Figure 5a, as well as the downloaded maps on GIS Cloud Computing, are shown in Figure 5b.

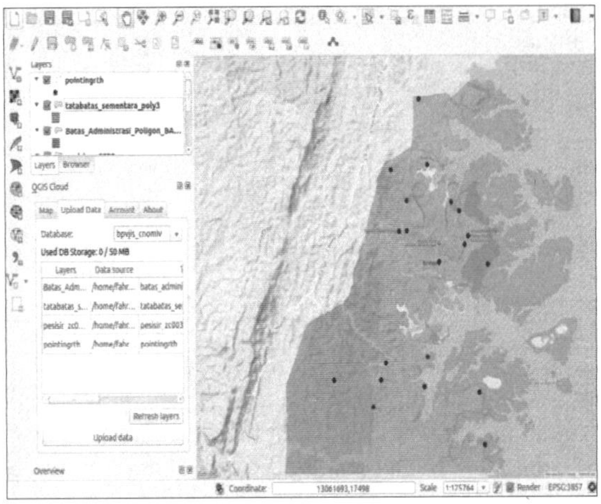

Figure 4a. Data layer processing on Quantum GIS Desktop and QGIS-Cloud with OpenCycleMap layer as background

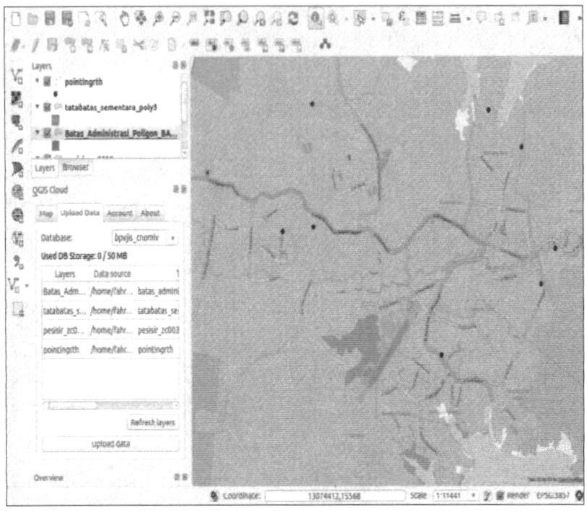

Figure 4b. Data layer processing on Quantum GIS Desktop and QGIS-Cloud with OpenStreetMap (OSM) layer as background

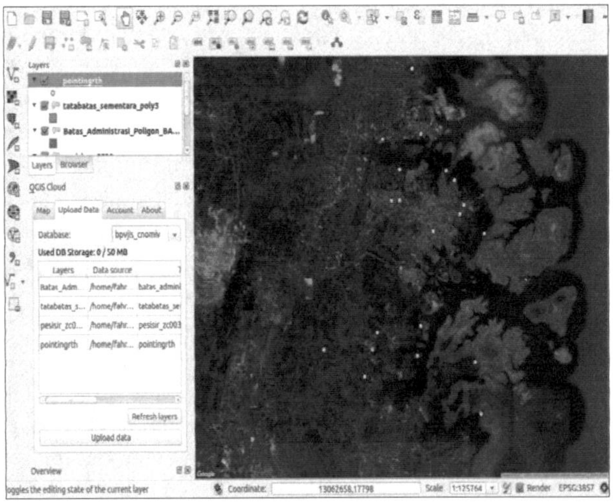

Figure 4c. Data layer processing on Quantum GIS Desktop and QGIS-Cloud with Google Satellite layer from Google Maps as background

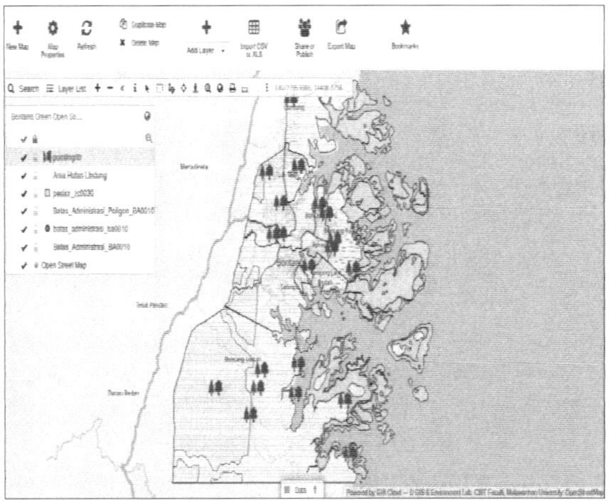

Figure 5a. Processing data layer on GIS Cloud Computing using GIS-Cloud

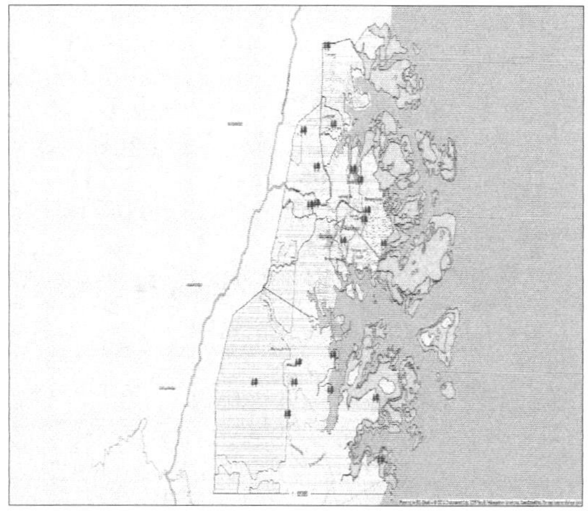

Figure 5b. Downloaded maps from Cloud Computing using GIS -Cloud

The GIS-Cloud result is a web-platform application with the address http://viewer.giscloud.com/map/783971/bontang-green-open-source-giscloud. The results of this map can be shared and accessed openly with the terms and conditions that the user has registered on the web giscloud.com. It can be seen in Figure 5a locations that are GOS areas in Bontang City. There are as many as 14 GOS locations scattered in the Bontang City area which is marked with a picture of a green tree.

The successful implementation of the architecture 14 GOS locations scattered in the Bontang City area can be provided through the determination of successful implementation factors. Factors that determine the success of the implementation of this system, among others: 1. Involvement, support, and management commitment. Strong and consistent management commitment and direct involvement will greatly help accelerate implementation; 2. Must be assigned the implementing person in charge so that he/she can act fully in implementing the architectural plan with the responsibility and authority given. 3. The quality of available human resources competent with information technology. 4. Develop SOP (Standard Operations Procedure) 5. The existence of special training on Architecture Planning. So that each unit can master the concept and how to use it. 6. Ability to evaluate the need for new technology [15].

4. Conclusion

This research has successfully implemented Cloud Computing GIS platform to build a web-based application of GOS location of Bontang City. The CC-based GIS platform makes the WebGIS application development process much easier and faster without the need to set up a domain name and hosting server.

Acknowledgments

Authors wishing to acknowledge for Rector of Mulawarman University through Vice Rector for Academic who has provided funding through PNBP 2017 research stimulant.

References

[1] Cromley EK, McLafferty S. Chapter 3: Spatial Databases for Public Health. In: GIS and Public Health. 2012.
[2] Chainey S, Ratcliffe J. GIS and Crime Mapping. GIS and Crime Mapping. 2013.
[3] Parker HD. The Economy, the Election, and GIS. GIS World. 1993;
[4] Kaswidjanti W, Santosa B, Wicaksono RS. Sistem Informasi Geografis Pemetaan Fasilitas Umum. Seminar. ISSN: 1979-2328; 2008 (semnasIF).

[5] Sari WNM. Membangun Aplikasi Sistem Informasi Geografis berbasis Dekstop untuk Industri Kerajinan Bnatul Yogyakarta. 2010

[6] Wicaksono SA, Agus F, Arifin Z. Penentuan Calon Ruang Terbuka Hijau Kota pada Kawasan Padat Penduduk Menggunakan ArcGIS. 2016 **1**(1) pp109–12.

[7] Fahrul A, Sumaryono, Lambang S, Afif R 2014 1st International Conference on Information Technology, Computer, and Electrical Engineering: Green Technology and Its Application for a Better Future, ICITACEE 2014 – Proceedings (2015) pp. 260-265

[8] Steiniger S, Hunter AJS. Free and open source GIS software for building a spatial data infrastructure. Geospatial Free Open Source Softw. 2012

[9] Dempsey C. Getting Started With QGIS: Open Source GIS. GIS Lounge. 2012

[10] QGIS Development Team. Welcome to the QGIS project! Qgis. 2016.

[11] Computing C, Govindarajulu D. Urban green space planning for climate adaptation in Indian cities. Urban Clim [Internet]. 2014 **10**(P1) pp 35–41. Available from: http://dx.doi.org/10.1016/j.uclim.2014.09.006

[12] Rina Nurhayati. Laporan Kinerja Instansi Pemerintah Dinas Tata Ruang, Pemerintah Kota Bontang 2016; http://ppid.bontangkota.go.id/wp-content/uploads/2016/10/LAKIB-TATA-RUANG-2016.pdf

[13] Bhat MA, Ahmad B. Cloud Computing: A solution to Geographical Information Systems (GIS). Int J. 2011;

[14] Bediroglu S, Yildirim V,1- Nisanci R. Building spatial cloud-based local government services. Proc Inst Civ Eng - Munic Eng. 2016;

[15] Ramadiani, Dyna Marisa Khairina, Afifah Khaerani Aziz, Perancangan e-Government Kelurahan Sempaja Timur Menggunakan Enterprise Architecture Planning. Prosiding Seminar Ilmu Komputer dan Teknologi Informasi Vol. 2, No. 1, Maret. e-ISSN 2540-7902 dan p-ISSN 2541-366X, 2017 pp 301-308

IOP Conf. Series: Earth and Environmental Science **144** (2018) 012033 doi:10.1088/1755-1315/144/1/012033

Restoration of mangrove forest landscape in Babulu Laut village, sub district of Babulu, Penajam Paser Utara district

W K Febrina[1,2*], Marjenah[1*] and Sumaryono[1*]

[1] Faculty of Forestry, Mulawarman University, Samarinda, East Kalimantan, Indonesia
[2] Regional Secretariat of East Kalimantan Province, Samarinda, East Kalimantan, Indonesia

*Corresponding Author: wekaef15@gmail.com, marjenah_umar@yahoo.com, msumaryono@gmail.com

Abstract. The reforestation activities of mangrove forest carried out in various regions have not been well known as the success and influence of landscape in rehabilitation area. Utilization of existing land along the coastal Babulu Laut Village has reduced the area of mangrove forest from day to day. Due to the use of land by the community without considering the conservation aspect causes the loss of mangrove forest. This study aims to determine the final condition of the success rate of forest and land rehabilitation, land cover and the benefits of mangrove forest restoration for the surrounding people. The research method used is the preparation and orientation of research location, data input, codefication, data processing, the field verification and analysis of data. The results of the execution of the inventory mangrove in 22 research location in the Babulu Laut Village, Babulu Subdistrict, Penajam Paser Utara District of 125 ha of plant a whole is kind of *Rhizophora sp*, where the intensity of sampling 2% with the growing plants of 65.92 % or 2,175 stem/ha then success rate of Mangrove Forest Rehabilitation at Babulu Laut Village Babulu Subdistrict is medium level (55-75%).

1. Introduction

Realizing the function of mangrove forests for the life of the community and the global ecosystem is requiring rehabilitation efforts of mangrove areas. The rehabilitation efforts are carried out to achieve the function, namely the function of mangrove forest as an economic zone and environmental function where mangrove forest is a buffer zone of life in the coastal area. Thus, the rehabilitation efforts are demonstrated in all mangrove forest areas, both in areas are controlled by the community (whole mangrove in the tambak area) or in areas are not controlled by the community (rehabilitation of pure mangrove) [1]

Research on landscape cover in the restoration of mangrove forest restoration area located in Babulu Laut Village within Babulu Subdistrict, the district of Penajam Paser Utara is one effort to restore the ecological carrying capacity of mangrove forest based on its function, with the rehabilitation of mangrove forest. With the hope of mangrove forest that formed can contribute to the ecological conservation of coastal areas in Babulu Laut Village within Babulu Subdistrict, the district of Penajam Paser Utara and also may provide benefits to the surrounding people.

The above description indicates that the successful rehabilitation of mangrove forest did inlcude the interference of the surrounding people. Therefore the researcher hopes this research can provide information to the government and the community in the coastal village of Babulu Laut in Babulu Subdistrict. It will be showing that the restoration of mangrove forest is very useful for the environment and resources nature so that the preservation of mangrove can be maintained.

Based on technical plan data of mangrove forest rehabilitation and beach seagrass, Babulu Laut Village in Babulu Subdistrict categorized as Priority I with minimum planting 3,300 stems/ha. Therefore, this study aims to determine the final condition of the success rate of forest and land rehabilitation in Babulu Laut Village, Babulu Subdistrict, to know the extent of land cover from the rehabilitation of forest and mangrove forest land in Babulu Laut Village, Babulu Subdistrict so that the achievement of mangrove forest landscape is sustainable.

2. Methods

This research was conducted in Babulu Laut Village, Babulu Subdistrict, Penajam Paser Utara District. Northside adjoining Labangka Village and Babulu Village of West Sea, South of Maresi Village in the south, Babulu Darat Village adjoining Babulu Darat Village and East of Makassar Strait Sea.

The research materials are East Kalimantan Province Administration Map, Peta Rupa Bumi (Topographic Map) scale 1: 50.000 Year 2013 Geospatial Information Agency, Map of Appointment of Forest and Fishery Area of East Kalimantan Province (attachment to Minister of Forestry Decree No. 718 / Menhut-II / 2014 dated August 29 2014), Map of Mangrove Forest Rehabilitation Block Area BPDAS Mahakam Berau Year 2013 Scale 1: 15.000 Babulu Subdistrict North Penajam Paser, Landsat Image Map 8 OLI (LC08_L1TP_117061_20140724_20170421_01_T1 coverage July 24, 2014 and LC08_L1TP_117061 _20160611_20170324_01_T1 coverage June 11, 2016).

The tools needed to create and support the overlay process of the map are Computers, ArcGIS Software 10.5, printer, GPS Garmin Map Csx 60, phiband, Suunto brand compass, photo camera, stationery.

Research activities include: Preparation and orientation of research sites, data input, codefication, data processing, field verification and data analysis, as follows:

2.1. Preparation and Orientation of Research Sites

Activities undertaken in the preparation of the study is to explore the information areas related to the restoration of mangrove forest from 2014 to 2016 that have been completed. It will be used to identify and inventory research areas and data processing programs (image processing).

2.2. Input Data

Landsat 8 OLI image (path / row: 121/65) downloaded from the United States Geological Survey website via http://earthexplorer.usgs.gov/ is LC08_L1TP_117061_20140724_20170421_01_T1 coverage July 24, 2014 and LC08_L1TP_117061_20160611_20170324_01_T1 coverage June 11, 2016 at research site Babulu Laut Village Babulu Subdistrict Penajam Paser Utara District.

2.3. Codefication

The first step is to make a link between the map with tabular data, then the process of giving the code for each object on the map. The code should be the same as the one on the tabular data. The next process is the projection of the position coordinate format. All digital maps are projected into a geographical or UTM projection, in order to have a universal coordinate system [2].

2.4. Data Processing

Data processing includes the following process:

2.4.1. Primary Data and Secondary Data. Primary data is data collected through measurement or direct ground checking to research variable in research area. Meanwhile secondary data is data collected from related government institution, literature study, and interview result. Primary data includes classifying the land cover of mangrove forest rehabilitation activities in 2014 and 2016, overlaying between mangrove forest rehabilitation maps and landsat 8 OLI coverage maps of 2014 and 2016 downloaded year, composite band process combining bands in Citra Landsat with combination of 4, 3, 2 channels of natural color and vegetation analysis 6,5,4 with Red, Green, Blue (RGB) filter obtained by False Color Composite (FCC) to facilitate the separation of the appearance of natural vegetation, surface water, and open land.

2.4.2. Delineation of Object Research. Delineation object research aims to explain the boundaries of coastal areas in Babulu Subdistrict to obtain information in the form of supporting data. Determination of delineation is to limit the results of land cover inventory and mangrove identification.

2.5. *Field Verification*

Field verification parameters under study are as follows:

2.5.1. Mangrove Forest Rehabilitation Plant. Observation of mangrove forest rehabilitation plants growing. Observation aims to identify the physical condition of plants as the basis for the management of Forest and Land Rehabilitation (RHL) further and according to the design of activities. Plant observation includes: measurement of plant area; number and type of plants; and the calculation of healthy plant growth percentage. The measurement of cultivation area is carried out on the realization of cultivation areasshown in the total area planted in units of Ha and compared to the plan of plant area according to the design (Regulation of Directorate General of River Basin Management and Social Forestry Number : P.1 / V-SET / 2013).

2.5.2. Making of Research Fields. Conducted by sampling technique method Systematic Sampling with Random Start, the first measuring plot is made randomly and the next measuring plots are made systematically. Sampling intensity (IS) is adjusted to the field condition IS 2%, each plot represents 0,2 ha (20 meter x 100 meter), as shown in Figure 1 below.

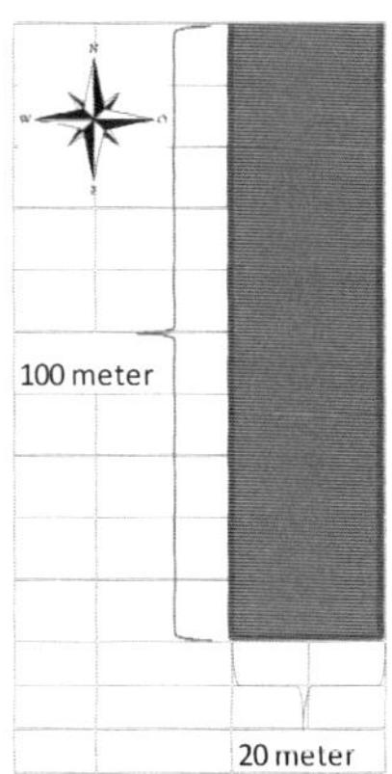

Figure 1. Sample Field of Research

2.6. *Analysis of Data*
Some of the data analyzed are as follows:
2.6.1. Final Condition of Plant Success Rate. The result of mangrove forest inventory and percent of plant growth is calculated by comparing the number of plants present in a plot with the number of crops that should be present in the measuring plot, as the following formula :

$$T = \left(\frac{\sum hi}{\sum ni}\right) \times 100\% = \frac{(h1+h2+\cdots.+hn)}{(n1+n2+\cdots+nn)} \times 100\% \quad (1)$$

(Regulation of Directorate General of River Basin Management and Social Forestry Number : P.1/V-SET/2013)
where: T = Percent (%) grows healthy plants
hi = Number of healthy plants contained in plot to i
ni = Number of plants that should be on plot to i.
Which results can be presented as Table 2.

2.6.2. Conversion Score In Percentage Form. The total score of RHL success evaluation is then converted into percentage form, and given the predicate according to the percentage value. How to convert a score into a percentage is by comparing the scored vector score with the maximum vector score score that can be achieved, then multiplied by one hundred percent. Percentage of success rate has been obtained then given the predicate of success rate. The percentage class and predicate RHL success as presented in Table 1 below.

Table 1. Class of Percentage and Predicate of RHL Success

Number	Percentage of Class	Remarks
1	91 % - 100 %	Very good
2	76 % - 90 %	Good
3	55 % - 75 %	Moderate
4	< 55 %	Less

2.6.3. Land Cover Area. In this study used the Map of Rehabilitation of Mangrove Forest Activities in 2013 and Citra LC08_L1TP_117061_20140724_ 20170421_01_T1 coverage July 24, 2014 and LC08_L1TP_117061_ 20160611_20170324_01_T1 coverage June 11, 2016 with Band 4, 3, 2 that is natural color and vegetation analysis 6,5,4 with filter Red, Green, Blue (RGB) so that the appearance of closure and land use is clearly visible using unsupervised classification because the data obtained will be more accurate using field data. The calculation results of changes in mangrove forest cover can be presented as Table 3.

3. Results and Discussion

Research Sites in Babulu Laut Village in Babulu Subdistrict, the Penajam Paser Utara District, located at position between 116 ° 29 '46.572 "BT, -1 ° 33' 38.9592" LS and 116 ° 32 '54.42 "BT, -1 ° 29' 3.7644" LS, with the boundaries of the northern part of Labangka Village, south of Maruat Village, west of Sebakung Jaya Village and the Eastern Babulu Laut Strait of Makassar, as Figure 2.

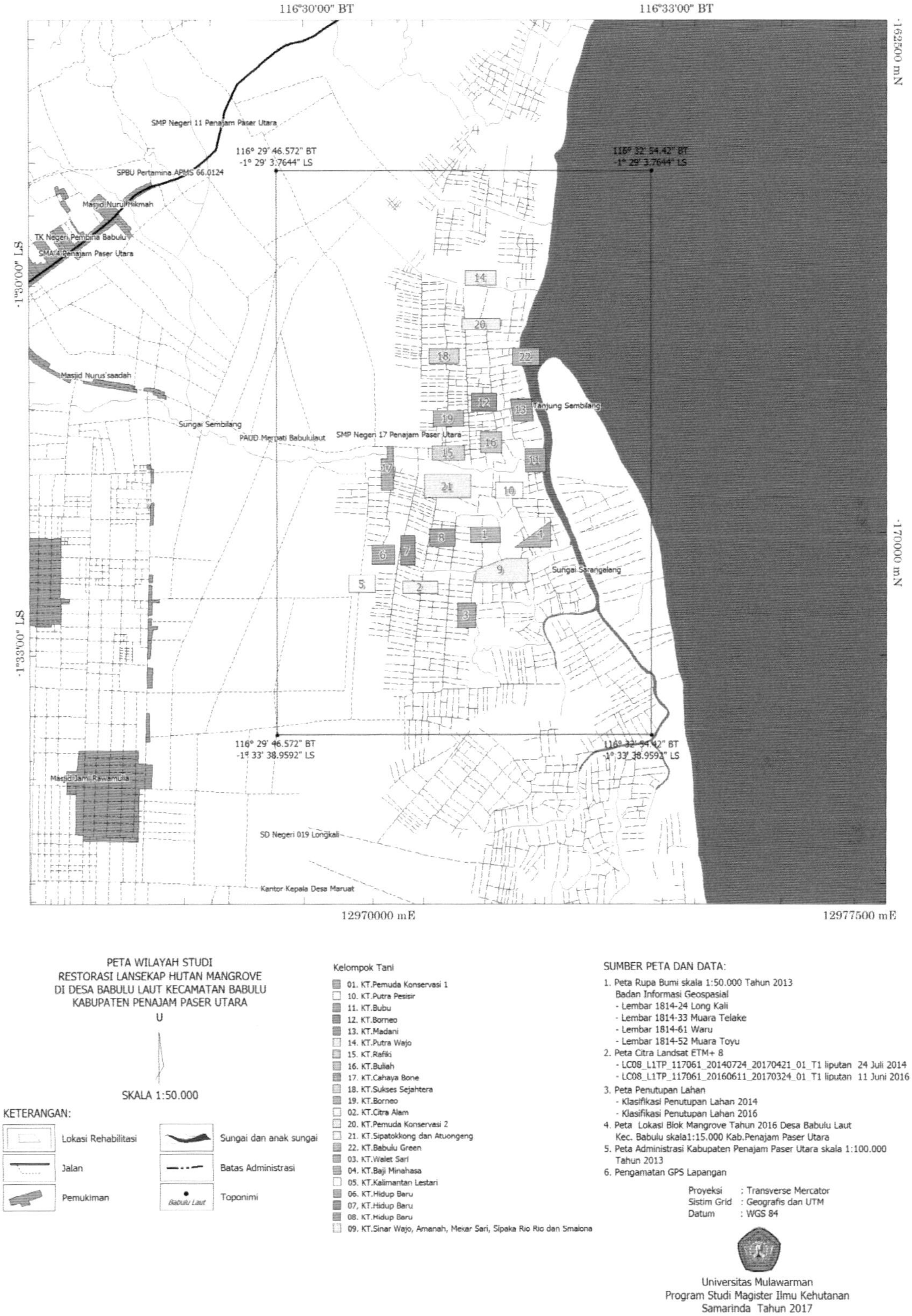

Figure 2. Study Area Map

Rehabilitation planting of mangrove forest implemented by Mahakam Berau Protected Forest Watershed Management Center in 2013 by involving community and farmer group, list of farmer group name and area of research location is as follows:

Table 2. List of Farmers Group and Area of Research

Locations	Names of Farmer's Group	Wide (Ha)
1	Pemuda Konservasi 1	10.78
2	Citra Alam	10.47
3	Walet Sari	10.69
4	Baji Minasa	10.29
5	Kalimantan Lestari	10.58
6	Hidup Baru	10.33
7	Hidup Baru	10.25
8	Hidup Baru	10.31
9	Sinar Wajo, Amanah, Mekar Sari, Sipaka Rio Rio, and SMA Lona	25.19
10	Putra Pesisir	10.46
11	Bubu	10.39
12	Bomeo	10.47
13	Madani	10.86
14	Putra Wajo	10.24
15	Rafiki	10.88
16	Buliah	10.06
17	Cahaya Bone	10.77
18	Sukses Sejahtera	10.25
19	Bomeo	10.93
20	Pemuda Konservasi 2	10.17
21	Sipatokkong and Atuongeng	25.45
22	Babulu Green	10.34
	Total	**260.16**

3.1. Plant Assessment of Mangrove Forest

The result plant assessment of Mangrove at 22 locations of research in Babulu Laut Village Babulu Subdistrict of Penajam Paser Utara District with area of 5,2 Ha with total plant species is type of *Rhizophora sp* end stand at each location with 2% Sampling Intensity, with plant growth percentage 65.92%.

Table 3. Percentage of Plant Growth

Site	Area (ha)	Area IS 2% (ha)	Trees (stem)	Σ Average (stem/ha)	Percentage of Growth (%)
1	10.78	0.22	662	3,071	93.05
2	10.47	0.21	554	2,646	80.17
3	10.69	0.21	351	1,642	49.75
4	10.29	0.21	504	2,449	74.21
5	10.58	0.21	428	2,023	61.29
6	10.33	0.21	320	1,549	46.94
7	10.25	0.21	376	1,834	55.58
8	10.31	0.21	414	2,008	60.84
9	25.19	0.50	1537	3,051	92.45
10	10.46	0.21	372	1,778	53.88
11	10.39	0.21	300	1,444	43.75
12	10.47	0.21	388	1,853	56.15
13	10.86	0.22	576	2,652	80.36
14	10.24	0.20	516	2,520	76.35
15	10.88	0.22	502	2,307	69.91
16	10.06	0.20	394	1,958	59.34
17	10.77	0.22	545	2,530	76.67

Site	Area (ha)	Area IS 2% (ha)	Trees (stem)	Σ Average (stem/ha)	Percentage of Growth (%)
18	10.25	0.21	470	2,293	69.48
19	10.93	0.22	488	2,232	67.65
20	10.17	0.20	454	2,232	67.64
21	25.45	0.51	967	1,900	57.57
22	10.34	0.21	398	1,925	58.32
Average					**65.97**

As comparative data of plant assessment conducted by BPDASHL Mahakam Berau, on Assessment of Mangrove Forest Rehabilitation (P-0) of 125 Ha in 2013 0f 76.97%, and Evaluation of RHL Activities (P-1) covering 100 Ha in 2014 of 66.37% and research results Year 2016 amounted to 65.92%, then based on the criteria set by the Director General for Management of Watershed and Social Forestry Number: P.1 / V-SET / 2013, the success rate of Mangrove Forest Rehabilitation in Babulu Laut Village Babulu Sub-district North Penajam Paser District is moderate (55-75 %) and declared successful.

3.2. Classification of Land cover

The land cover classification of 2014 and 2016 underwent a land cover change with a 206,96 ha classified land cover by 2016, and an open field classification of 57,17 Ha, as Table 4 and Figure 3 and 4.

Table 4. Land Cover Year 2014 and Year 2016

Site	Area	Year of 2014		Year of 2016	
		Land Cover	Open Field	Land Cover	Open Field
1	10.78	5.19	5.58	9.5	1.27
2	10.47	4.65	5.82	8.94	1.53
3	10.69	3.66	7.02	7.33	3.35
4	10.29	5.62	4.67	9.36	0.93
5	10.58	8.68	1.89	9.08	1.5
6	10.33	2.52	7.8	9.02	1.3
7	10.25	4.63	5.62	7.97	2.29
8	10.31	3.31	7.00	8.09	2.23
9	25.19	20.37	4.82	16.75	8.45
10	10.46	6.00	4.46	8.24	2.21
11	10.39	7.05	3.35	7.42	2.97
12	10.47	7.03	3.44	9.74	0.72
13	10.86	4.92	5.93	8.24	2.61
14	10.24	6.06	4.17	9.37	0.87
15	10.88	6.78	4.11	9.26	1.63
16	10.06	5.71	4.35	6.12	3.94
17	10.77	2.87	7.9	9.57	1.19
18	10.25	1.02	9.23	7.22	3.03
19	10.93	1.17	9.76	9.71	1.22
20	10.17	1.31	8.86	5.53	4.64
21	25.45	15.25	10.2	23.57	1.88
22	10.34	4.18	6.16	2.93	7.41
Total	**260.16**	**127.98**	**132.14**	**202.96**	**57.17**

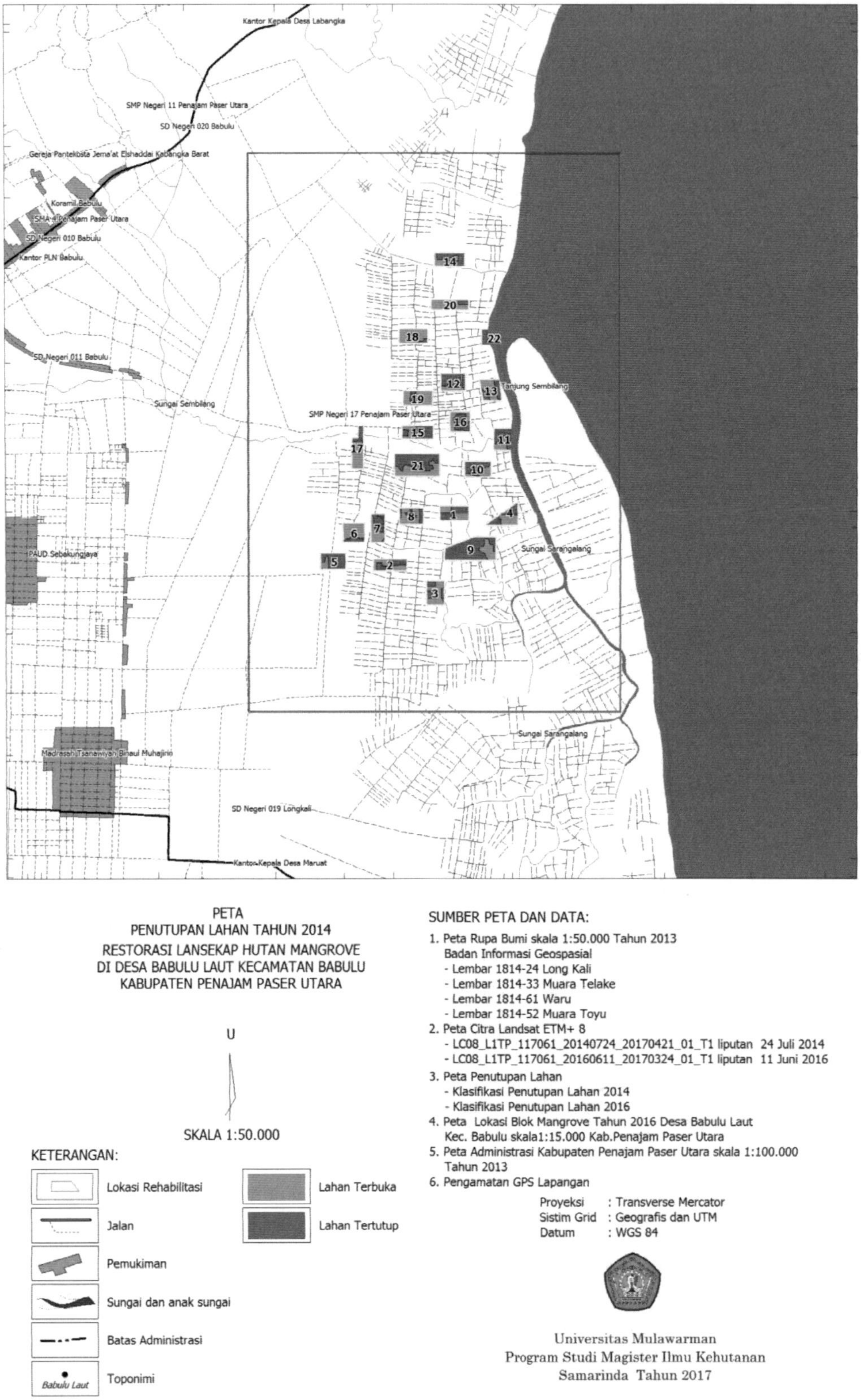

Figure 3. Map of Land Cover Classification Year 2014

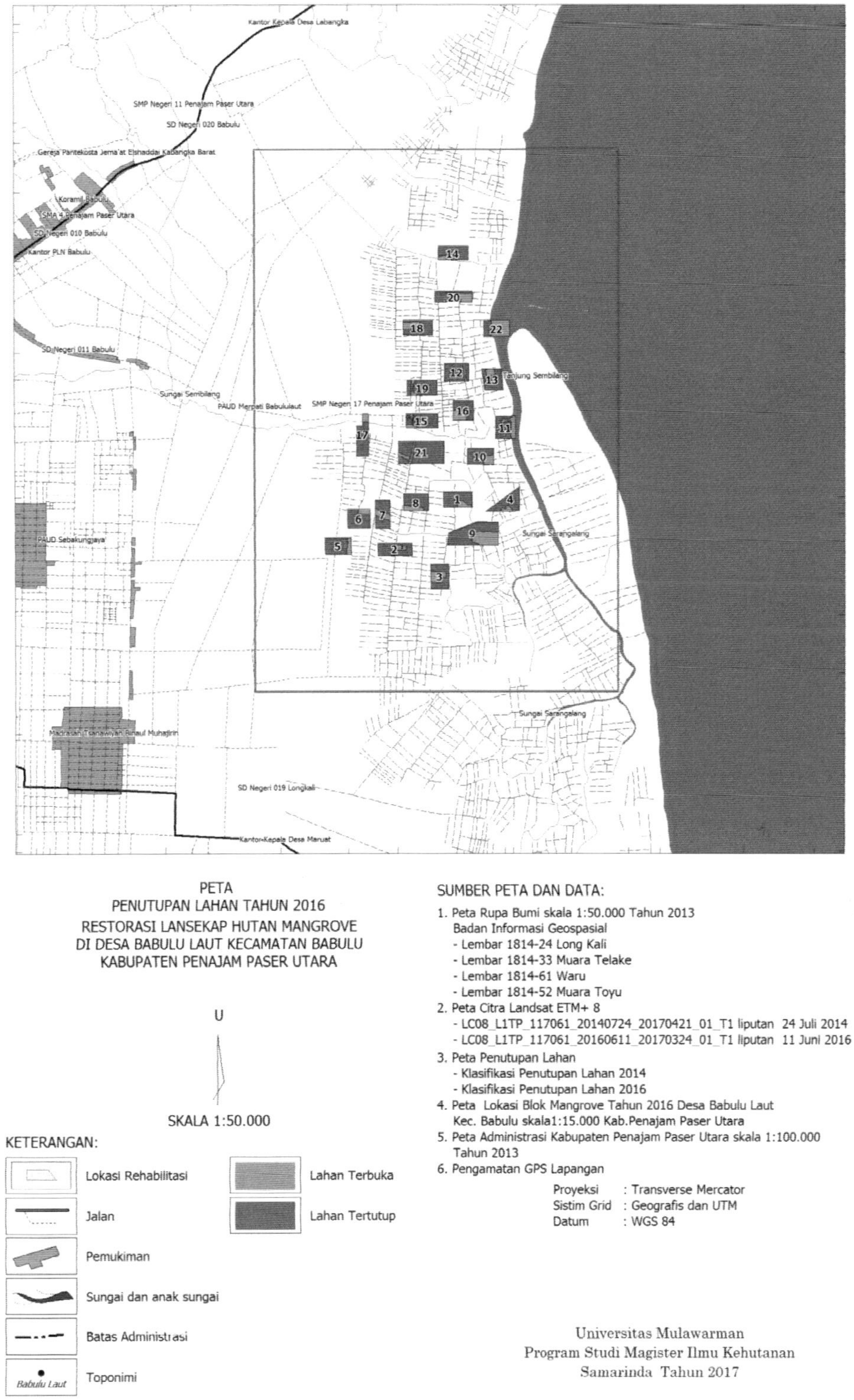

Figure 4. Map of Land Cover Classification Year 2016

IOP Conf. Series: Earth and Environmental Science **144** (2018) 012033 doi:10.1088/1755-1315/144/1/012033

From the land cover classification results in 2014 and 2016, the additional land cover area of 74,98 ha and reduction of land cover area of 74,97 ha. Thus there is a change of land cover from 2014 to 2016.

3.3. Land cover change

The overlay results of 22 (twenty two) mangrove forest rehabilitation locations and the land cover classification results of 2014 and 2016 and Landsat ETM + 8 B654 in 2014 and 2016 obtained the result of land cover change in the research location obtained result of changes in land cover area of 56.74% , as Table 5 and Figures 5, 6 below:

Table 5. Open Coverage Changes in 2014 and 2016

Site	Area	Open Field (Ha)		Percentage of Changes Land Cover (%)
		2014	2016	
1	10.78	5.58	1.27	77.24
2	10.47	5.82	1.53	73.71
3	10.69	7.02	3.35	52.28
4	10.29	4.67	0.93	80.09
5	10.58	1.89	1.5	20.63
6	10.33	7.80	1.3	83.33
7	10.25	5.62	2.29	59.25
8	10.31	7.00	2.23	68.14
9	25.19	4.82	8.45	-75.31
10	10.46	4.46	2.21	50.45
11	10.39	3.35	2.97	11.34
12	10.47	3.44	0.72	79.07
13	10.86	5.93	2.61	55.99
14	10.24	4.17	0.87	79.14
15	10.88	4.11	1.63	60.34
16	10.06	4.35	3.94	9.43
17	10.77	7.90	1.19	84.94
18	10.25	9.23	3.03	67.17
19	10.93	9.76	1.22	87.50
20	10.17	8.86	4.64	47.63
21	25.45	10.20	1.88	81.57
22	10.34	6.16	7.41	-20.29
Total	**260.16**	**132.14**	**57.17**	**56.74**

Table 5 shows that location 9 under the coordination of Sinar Wajo Farmer Group, Amanah, Mekar Sari, Sipaka Rio Rio & Sma Lona experienced a change in land cover reduction by 75.31%. After observation in the field this is because the distance of location 9 too close to the beach so that the mangrove easily flooded water and low success rate of planting. And plot 22 under the coordination of Sipatokkong & Atuongeng Farmer Group experienced a change in land cover reduction by 20.29%, this is due to the lack of seriousness of the farmer group in maintaining the mangrove plants that have been planted.

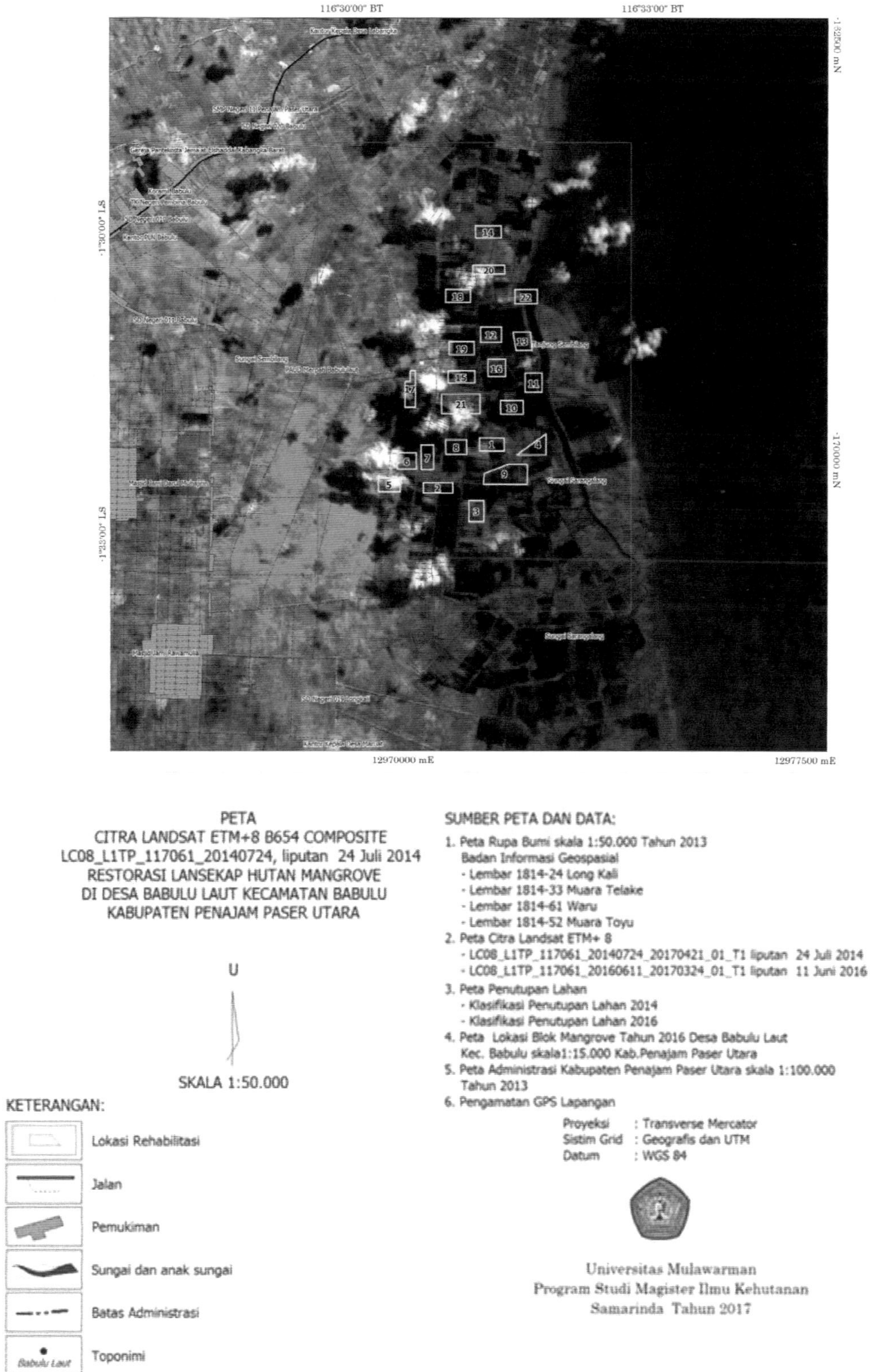

Figure 5. Map of Citra Landsat ETM+8 B654 Composite Year 2014

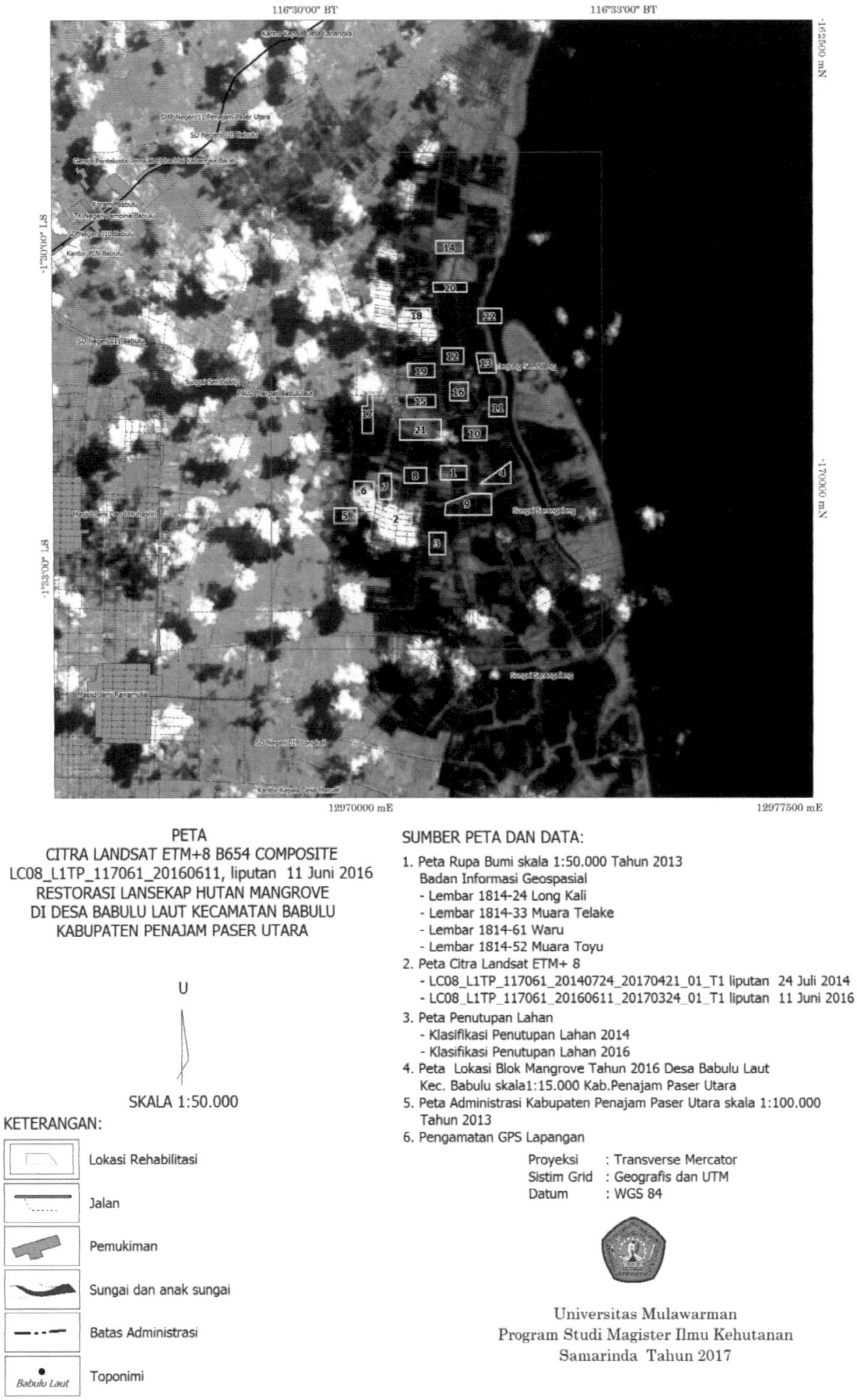

Figure 6. Map of Citra Landsat ETM+8 B654 Composite Year 2016

3.4. Changes in Mangrove Forest Landscape

The landscape change in mangrove forest cover from 2014 open land area of 132,14 ha to an area of 57,17 ha in 2016 with a percentage of 56.74% as shown in Figure 7 below :

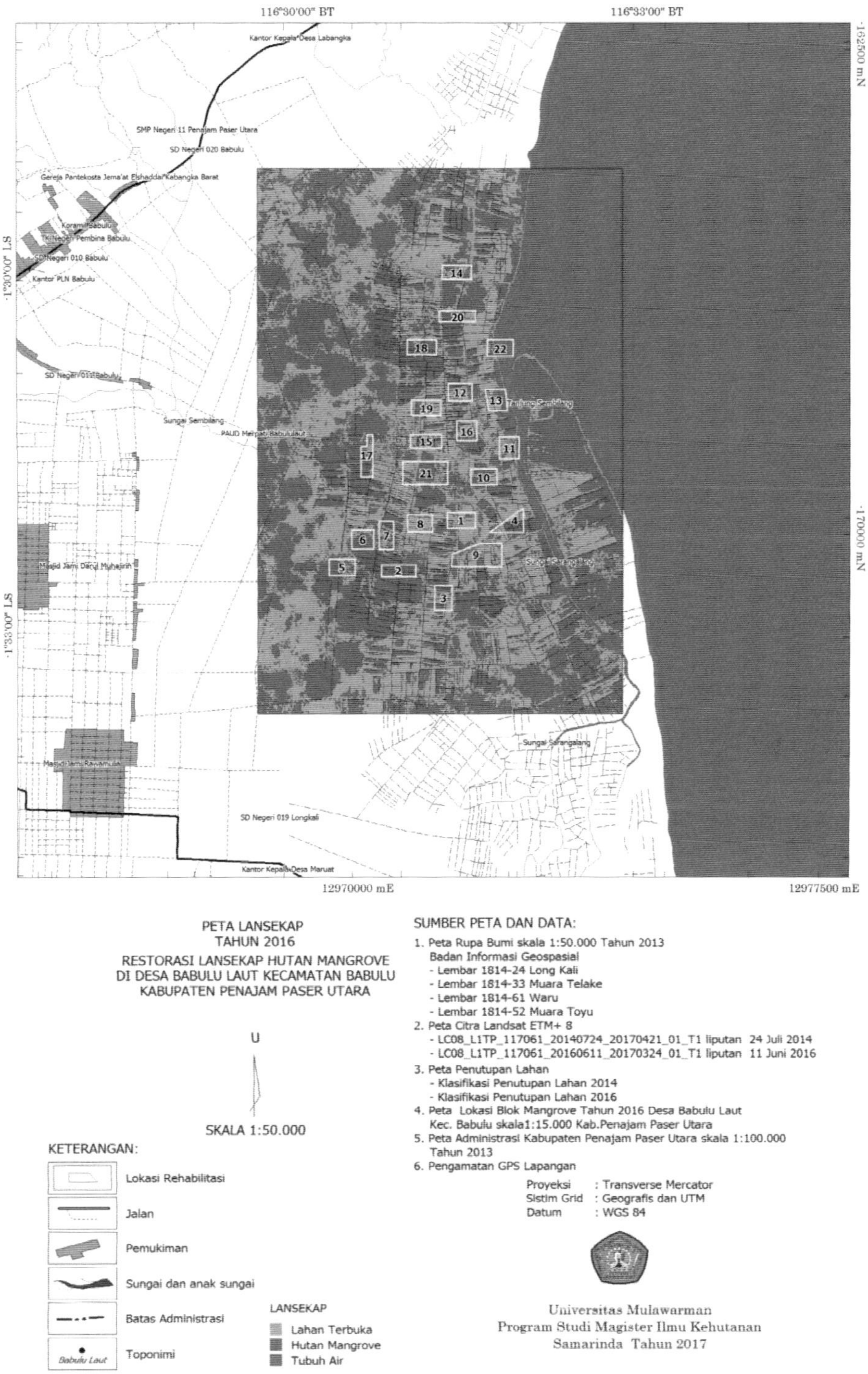

Figure 7. Mangrove Forest Landscape Map 2016

The type of plant that becomes the dominant structural element in the landscape is the type of tree. Trees in the landscape have an important functional and aesthetic role. Aesthetically, trees can function as a complement, unifying, enforcing, marking and framing the environment [3].

In his research [4] stated that the economic value of the mangrove area emerging as a result of its ecological role and its harvest products is often ignored so that the area is widely converted into agricultural areas, fishponds, salt ponds, forestry, and infrastructure [4]. So with the restoration of mangrove forest landscape in Babulu Laut Village Babulu Subdistrict North Penajam Paser Regency, can create a fertile ecosystem and become habitat for some kind of biota of economic value, of course public welfare can be guaranteed.

Forest landscape restoration approaches can therefore play a positive role in the conservation and sustainable management of this globally important forest type. In addition, our research suggests that forest restoration can be cost-effective, at least in some situations, in that the increase in benefits provided to people can outweigh the costs incurred. This is consistent with recent reviews highlighting the ecological feasibility restoration [5] and its potential benefits for both biodiversity and provision of ecosystem services [6].

Acknowledgments
The researcher presents his sincere appreciation goes to Dr. Rudianto Amirta, S.Hut, MP as the Dean of Faculty of Forestry, Mulawarman University, my first advisor, Prof. Dr. Ir. Hj. Marjenah, MP, second advisor Dr. Ir. H. M. Sumaryono, M.Sc. who has helped him patiently finishing this undergraduate thesis by giving suggestion, guidance, and correction until the completion of this article.

References

[1] Alviya I M, Salminah V B, Arifanti R, Maryani E, Syahadat 2015 Stakeholders' of Perception of Forest Landscape Management of Tulang Bawang Watershed *J. of Social Research and Economic Forestry* http://ejournal.fordamof.org/latihan/index.php/JPSE/article/view/191 pp 171-184
[2] Putra A D 2012 Study of Zoning Mapping Based on Geospatial Information Forest Education and Research Sebulu, Sub District Sebulu Kutai Kartanegara District pp 44-45
[3] Booth N K 1983 Basic Elements of Landscape Architecture Design Waveland Press Inc Illnois p 314
[4] Ronnback P 1999 The ecological basis for economic value of seafood production supported by mangrove ecosystems *Ecological Economics* **29** pp 235 - 252
[5] Jones H P, Schmitz O J 2009 Rapid recovery of damaged ecosystems *PLoS ONE* **4**:e5653doi: 10.1371/journal.pone.0005653
[6] González-Espinosa M, Ramírez-Marcial N, Galindo-Jaimes L, Camacho-Cruz A, Golicher D, Cayuela L, Rey-Benayas J M 2009 Tendencias y proyecciones del uso del suelo y ladiversidad florística en Los Altos de Chiapas, México *Investigación Ambiental* **1** pp 40–53.

Production of high-calorie energy briquettes from bark waste, plastic and oil

W Suwinarti*, **R Amirta and Yuliansyah**

Industrial Biotechnology Laboratory, Faculty of Forestry, Mulawarman University Jl. Ki Hajar Dewantara Samarinda, Indonesia

*Corresponding author: w_suwinarti@yahoo.com

Abstract. Bark is the waste generated from the utilization of plantation timber, while plastics and oil waste are produced from daily human activity. These waste has the potential to be used as energy briquettes raw materials, especially for fuel in power plants. It would be worth very strategic for the environment and the welfare of society, considering that at this time we are not yet fully capable of well managing all three waste types. On the other hands most of the power plants that operate today still use diesel and coal as fuel. Therefore, the best composition of mixing bark, plastic and oil will be studied as well as its influence on the physical and chemical quality of the briquettes produced. The results show that the addition of the oil waste (70%) and used plastic (30%) as additive give effect to the performance of the briquette formation with the highest calorific value of 33.56 MJ/kg.

1. Introduction

Since fossil fuel sources across the globe are fast depleting, renewable and non-conventional energy sources are urgent to be explored to ensure the world is not threatened by a vacuum of energy. Due to the erratic change of energy prices and unfavourable forecast of world economy, considerable efforts have been devoted to substitute raw fossil fuels with various other sources for the production of energy. The various factors, such as ever increasing diesel consumption, large outflow of foreign exchange, and concern for environment, contribute to the search for a suitable environmental friendly alternative to fossil fuel [1, 2].

Biomass is one of the renewable and potentially sustainable energy sources and has many possible applications varying from heat generation to the production of advanced secondary energy carriers. It has almost zero or very low net CO_2 emission since carbon and energy are fixed during the biomass growth. There are different types of technologies for converting biomass to electricity or to a secondary fuel such as thermal conversion, chemical conversion and bio-chemical conversion [3, 4]. Biomass is generally defined as any hydrocarbon material, mainly consists of carbon, hydrogen, oxygen, and nitrogen. It has been recognized as a major world renewable energy source to supplement declining fossil fuel resources. However, the use of first generation biofuels, which have been mostly taken from food and oil crops, such as rapeseed oil, sugar cane, sugar beet, maize, vegetable oils, and animal fats, has generated a lot of controversies, primarily because of their impact on global food markets and food security. Hence, it drives the need to base non-food biomass resource biofuels [5, 6].

As one of the countries with abundant biomass reserves and also experiencing energy and fuel crises, the Government of Indonesia has set a target of diversifying energy sources to replace coal, gas and petroleum with renewable energy sources, including those with developing biofuels and bioenergy derived from the use of various types of biomass, which can be obtained from agriculture, plantation, forestry and others [7]. Plantation forest were developed as part of efforts to provide sufficient raw materials for the existence of the pulp and paper industries that are currently expanding rapidly in Indonesia. The development of the pulp and paper industry that generally uses wood raw materials, in practice, many also leave environmental problems, especially those derived from bark as a disturbing waste and usually only burned or used as an artificial soil matrix.

On the other hand, the plastics have become one of the most important and indispensable materials in our contemporary world. These plastics are not presently biodegradable and are extremely troublesome components for land filling. The waste plastics are known for creating a very serious environmental challenge because of their huge quantities and the disposal problems caused by them. To avoid the impact of the plastic in the environment, the recycling of plastics constitutes a valid alternative [2].

Another waste material that also needs to be taken seriously and observed is the vehicle oil waste. The amount of oil waste from motor vehicles, industrial machinery and power plants from time to time increases as the number of vehicles grows in the midst of society. Generally, used oil is only collected in order to facilitate the transportation and recycling, although there are many people who immediately throw it into the environment. Therefore, serious attention needs to be given to prevent the recurrence of this incident, especially by seeking alternative use of this waste oil into raw materials or additives from derivative products that we may produce from the processing.

Responding to this waste problem, a strategic thought was needed to process and exploit the potential of these three waste types into an energy product that can be utilized as fuel for power generation, since to date most of the area in Kalimantan is also facing the lack of electricity supply. This study aims to know the potential utilization of mixture of acacia bark waste, plastics and oil as raw materials in the production of energy briquette; the optimum composition of these mixture that can be applied to obtain energy briquette products with high calorific value (high quality); and the quality of energy briquette that can be produced from the composition of this mixture.

2. Material and Methods

2.1. Material

Acacia mangium bark waste was collected from chip mill industry of PT. Sarana Bina Semesta Alam (SBSA), located in the Village of Tanjung Karas, Muara Kaman, Kutai Kertanegara and also PT. Kutai Chip Mill (KCM) in Balikpapan, East Kalimantan. Prior to use, the ingredients have been previously dried under the sun for several days to reach a water content of 12 ~ 14%, before further made into chips and converted into powder.

2.2. Methods

Energy briquettes are made from a mixture of acacia bark waste, plastics and used oil. All ingredients based on each ratio are mixed in the mixer at low speed, heated and then printed into briquette using briquette press, to obtain energy briquettes with final dimensions: length ± 20 cm, width ± 10 cm and thickness ± 5 cm. In this study, the ratio of ingredients has been developed by placing the use of acacia bark waste as the main factor of the constituents (50 - 60%), while the plastics and additives of used oils are minor factors, each of which will be attempted to be developed by varying amounts of mixing in the range of 5-20% for plastics and 30-40% for used oil, where the respective portion of the material will be calculated from the weight of the waste bark waste used (w/w). The mixed compositions that will be developed in this study can be seen in table 1.

Tabel 1. The composition of a mixture of bark, plastic and oil

Sample Number	The Mixture		
	Plastic (g)	**Oil (g)**	**Bark (g)**
1	40	60	100
2	25	70	100
3	30	70	100
4	35	70	100
5	40	70	100
6	20	80	100
7	20	60	120
8	10	70	120

Acacia bark waste, due to its abundant amount, is expected to become the main component of the forming and filling of this biomass energy briquette product. While plastic waste is used as an adhesive substitute to strengthen the structure and increase the calorific value. Used waste oil is deliberately added to increase the calorific value of this super briquette product, since waste oil has a high enough calorific value of ± 10,000 Kcal/kg and also to facilitate the initiation process of combustion of energy briquette product [8-11].

To know the characteristic and quality of raw material and energy briquette product, especially to evaluate the best composition that can be developed further, there will be a series of test which includes the analysis of physical properties and also the chemical content of the product. The complete series of tests is carried out with reference to the methods contained in table 2.

Tabel 2. Parameters and methods of briquette testing [12]

Number	Parameter	Methods
1	Density	ASTM D 4784
2	Moisture	ASTM D 3173-11
3	Ash Content	ASTM D 3174-12
4	Volatile Matter	ASTM D 3175-11
5	Fixed Carbon	ASTM D 3172-07a
6	Total Sulphur	ASTM D 4239-12
7	Calorific Value	ASTM D 5865-12

3. Results and Discussion

The result showed that increasing the addition of plastics in the mixed briquette can increase the density value and lower water content as seen in table 3. This may occur because the function of adding plastic to this mixture can increase the adhesiveness and compactness of this briquette texture. These results are also in line with the results of research which states that the materials are pressed under the different pressure caused the material amount transported into do compacting matrix. If the

IOP Conf. Series: Earth and Environmental Science **144** (2018) 012034 doi:10.1088/1755-1315/144/1/012034

matrix capacity is large and thus also the material amount is larger, then the pressing machine produces a higher pressure, the final briquettes are longer and therefore their density and strength are also higher [13, 14].

Tabel 3. The physical properties of briquettes

Mixture Composition (plastic:oil:bark)	Density (kg/m^3)	Moisture Content (%)
40:60:100	0.79	3.58
25:70:100	0.68	5.45
30:70:100	0.70	3.43
35:70:100	0.70	3.12
40:70:100	0.79	3.66
20:80:100	0.67	4.93
20:60:120	0.64	6.49
10:70:120	0.56	6.05

The ash content, volatile matter and fixed carbon contained in the briquette samples were also determined and the results as presented in table 4. The mixture of briquettes with large amount of bark waste has higher ash content and fixed carbon than a small one, while the volatile matter is lower than other briquettes. The results of this study are also in line with previous research that convey the characteristics of briquettes produced from several types of biomass [15, 16].

Tabel 4. The chemical properties of briquettes

Mixture Composition (plastic:oil:bark)	Ash (%)	Volatile Matter (%)	Fixed Carbon (%)	Sulphur content (%)
40:60:100	2.41	85.15	8.86	0.13
25:70:100	2.80	80.37	11.38	0.17
30:70:100	2.19	86.26	8.12	0.16
35:70:100	2.85	84.62	9.41	0.16
40:70:100	2.57	85.04	8.73	0.20
20:80:100	2.51	82.32	10.24	0.18
20:60:120	4.06	75.69	13.76	0.13
10:70:120	3.42	77.25	13.28	0.17

Table 4 also shows that the higher the addition of oil, the higher the sulphur content. On the other hand, adding plastics to the mixture does not seem to affect the quality of the briquettes due to their very small amounts and plastic functions only as adhesives.

Figure 1 presents the calorific value generated by the briquettes mixture. The highest calorific value of 33.56 MJ/kg was obtained by a mixture of 30:70:100 (plastics: oil: bark). These results indicate that the optimal mixture is not always produced by the amount of ingredients most widely added but the best mixed combinations that can produce the high quality of briquette. Compared to calorific value obtained in the other study of different acacia species [14], these results indicate a higher calorific value. It can be caused by the addition of plastic and oil on the production of this briquette.

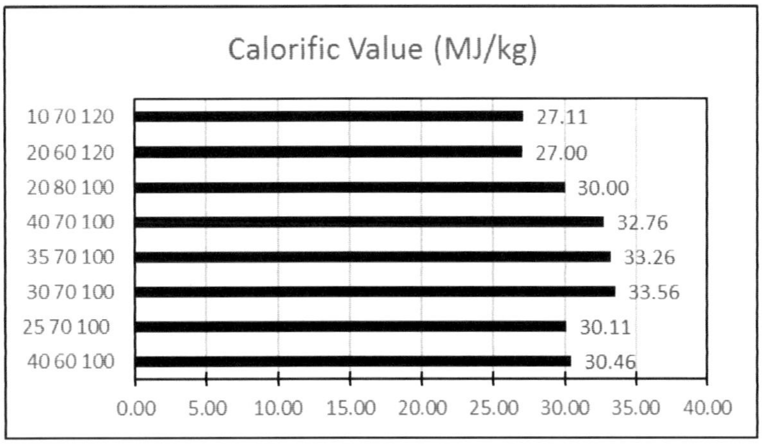

Figure 1. The calorific value of briquettes

The relationship among ash content, fixed carbon and volatile matter with the calorific value is depicted in figures 2-3.

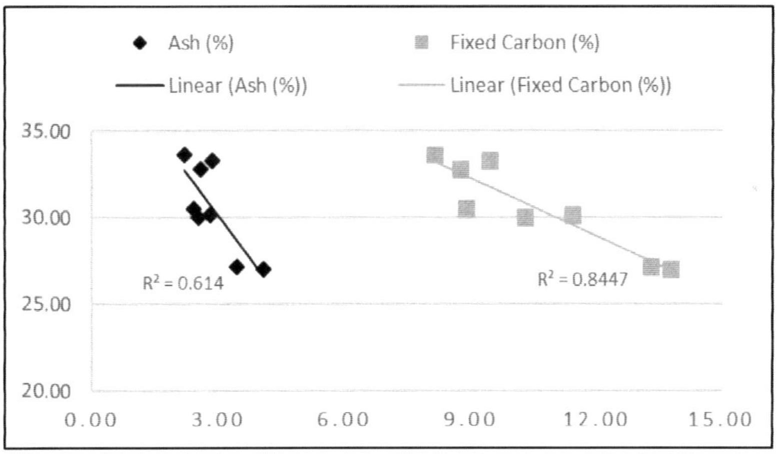

Figure 2. The relationship among ash, fixed carbon and calorific value

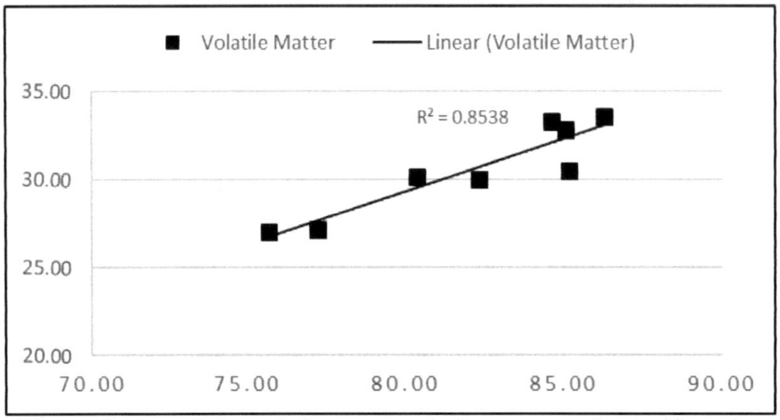

Figure 3. The relationship between volatile matter and calorific value

The relationship between ash content and calorific value is indicated by a smaller R value compared to the relationship between fixed carbon and calorific value of 0.61401 and 0.84473, respectively, whereas the relationship between volatile material and calorific value gives a higher R value of 0.85382.

The highest calorific value of briquette obtained in this study shows that it can exceed the calorific value of coal used as conventional fuel during this time and wood (dry) [16] as shown in table 5. According to European standard EN 14961-3 (EN*plus* A2) for wood briquette, the calorific value of the briquettes produced in this study may be stated to meet this standard whose calorific value is ≥ 15.3.

Tabel 5. The comparison of briquettes quality

Mixture Composition	Calorific value (MJ/kg)			
	Sample	Coal[a]	Wood (dry)[a]	EN 14961-3 (EN*plus* A2)[b]
40 60 100	30.46	15.0-27.0	14.4-17.4	≥ 15.3
25 70 100	30.11			
30 70 100	33.56			
35 70 100	33.26			
40 70 100	32.76			
20 80 100	30.00			
20 60 120	27.00			
10 70 120	27.11			

[a] [17]
[b] [18]

Although the testing of briquette quality in this study has shown good results, the most optimum briquette research study still needs to be continued to reduce the oil addition so that the sulfur released can still be below the standard determined threshold.

4. Conclusions

Bark waste, plastic and oil very potential to be used as raw material for energy briquette production. The mixture of these three waste types produced a high-calorie energy briquettes. The highest calorific value of 33.56 MJ/kg is obtained by briquette mixture of 30:70:100 (plastics:oil:bark). Overall, the calorific value of the briquettes produced in this study may be stated to meet the standard.

Acknowledgment

This work was financially supported by the Grant of Mulawarman University Research of Excellent Program, provided by the Directorate General of Higher Education, the Ministry of Higher Education, Research and Technology of Indonesia. Special thanks also addressed to Sarana Bina Semesta Alam and Kutai Chip Mill Company for providing the samples.

References

[1] Kumar P S, Bharathikumar M, Prabhakaran C, Vijayan S and Ramakrishnan K 2017 Conversion of waste plastics into low-emissive hydrocarbon fuels through catalytic depolymerization in a new laboratory scale batch reactor. *Int J Energy Environ Eng* **8**:167–173.

[2] Siddiqui M N and Redhwi H H 2009 Catalytic coprocessing of waste plastics and petroleum residue into liquid fuel oils *J Anal Appl Pyrol* **86**:141–147

[3] Arena U 2012 Process and technological aspects of municipal solid waste gasification. A review *Waste Manag.* **32**:625–639

[4] Tilahun M, Sahu O, Kotha M and Sahu H 2015 Cogeneration of renewable energy from biomass (utilization of municipal solid waste as electricity production: gasification method) *Mater Renew Sustain Energy* **4**:4

[5] Brennan L and Owende P 2010 Biofuels from microalgae—a review of technologies for production, processing, and extractions of bio- fuels and co-products *Renew. Sustain. Energy Rev.* **14**(2):557–577

[6] Tabatabaei M, Tohidfar M, Jouzani G S, Safarnejad M and Pazouki M 2011 Biodiesel production from genetically engineered microalgae: future of bioenergy in Iran. *Renew. Sustain. Energy Rev.* **15**(4):1918–27

[7] Rahmat M H 2017 *General Plan of National Energy* http://setkab.go.id/ruen-rencana-umum-energi-nasional Accessed October 25th, 2017

[8] Brandy S, Tam K, Leung G and Salam C 2009 Zero waste biodiesel: Using glycerin and biomass to create renewable energy *UCR Undergraduate Research Journal* 5-11.

[9] Chaiyaomporn K and Chavalparit O 2010 Fuel pellets production from biodiesel waste *Environment Asia* **3**(1):103-110

[10] Melero J A, Van Grieken R, Morales G and Paniagua M 2007 Acidic mesoporous silica for the acetylation of glycerol: Synthesis of bioadditives to petrol fuel *Energy Fuels* **21**(3):1782–91

[11] Sanchez E A, Pasache M B and Garcia M E 2014 Development of briquette from waste wood (sawdust) for use in low-income households in Piura, Peru. *Proceedings of the World Congress on Engineering 2014*. Vol II, London

[12] ASTM International 2017 *Annual Book of ASTM Standard-ASTM Volume 05.06 Gaseous Fuels; Coal and Coke; Catalysts; Bioenergy and Industrial Chemicals from Biomass* https://www.astm.org Accessed March 25th, 2017

[13] Plistil L, Brozek M, Malatak J, Roy A and Hutla P 2005 Mechanical characteristic of standard fuel briquettes on Biomass basis *Res. Agr. Eng.* **52**(2):66-72

[14] Mitchual S J, Mensah K F and Nicholas D A 2013 Effect of species, particle size and compacting pressure on relaxed density and compressive strength of fuel briquettes *International Journal of Energy and Environmental Engineering* **4**:30

[15] Munjeri K, Ziuku S, Maganga H, Siachingoma B and Ndlovu S 2016. On the potential of water hyacinth as a biomass briquette for heating applications *Int J Energy Environ Eng* **7**:37–43
[16] Tiwari C 2011 Producing fuel briquette from sugarcane waste *EWB-UK national Research and Education Journal* 220-550
[17] The Engineering Tool Box 2017 *Higher calorific values for some common fuels—coke, oil, wood, hydrogen and others* www.engineering toolbox.com Accessed November 10th, 2017
[18] European Standard EN 1496-3 2013 *ENplus Briquettes Manual for the Certification of Wood Briquettes For the End-user Market*. Deutsches Pelletinstitut. Berlin.

IOP Conf. Series: Earth and Environmental Science **144** (2018) 012035 doi:10.1088/1755-1315/144/1/012035

Albumin profile of snakehead fish (*Channastriata*) from East Kalimantan, Indonesia

A N Asikin[1*] and I Kusumaningrum[1]

[1] Department of Fishery Product Technology, Faculty of Fishery and Marine Science, Mulawarman University

Corresponding author: asikin63@yahoo.com

Abstract. This study aimed to determine the properties of albumin of snakehead fish (*Channa*strata) by various method extraction. The extraction of snakehead fish albumin was done using water (W), NaCl 0,9% (N), HCl 0.1 M (H). This research used three groups weight of snakehead that were 300-600 g (small; S), 600-900 g (medium; M) and 900-1200 g (large; L). Raw materials (snakehead fish) obtained from Middle Mahakam area, East Kalimantan, Indonesia. The parameters of this research were yield, proximate, albumin, and colour. The data were analyzed by using completely randomized design which consist two factors of treatments (solvent and weight of snakehead) and three replications.

1. Introduction

Snakehead fish (*Channa strata*) is one of the local fish in East Kalimantan, especially in Middle Mahakam area. The local name of snakehead fish is known as haruan, ruan or gabus. This fish has a high economic value due to its high content of albumin [1]. Albumin is kind of globular protein dissolving in water, salt, acidand could be coagulated by heat [2, 3].

Albumin is very important for human body needed, especially for the healing of wounds [4]. Protein from snakehead fish enhances dermal wound healing and reduces post-operative pain and discomfort. In addition, the amino acid composition of snakehead fish was reported to play a role in the process of wound healing [5].

The quality of albumin depend on the method was used during extraction. In addition, the weight of fish have the difference nutrition content also effect to albumin quality. The objective of this research was to evaluate the properties of albumin extracted by various solvent and the difference of fish weight from Middle Mahakam area, East Kalimantan. The properties of albumin in this study included yield, proximate composition, amino acid composition and lightness of snakehead protein extracted.

2. Method

2.1. Sample preparation

Raw materials used in this research were snakehead fish obtained from Middle Mahakam area, East Kalimantan, Indonesia. The other materials were used in this research include NaCl,HCl, clean water and distilled water. The fish was remove the scales, gills, and stomach content. Fish was washed until there was no blood and mucus and then was filleted. The fillet was cut into small pieces and was blend using solvent appropriated the experimental design. The ratio solvent and flesh was 1:1 and then was heated using waterbath at the temperature of 60°C for 15 minutes. After that, the samples were filtered to separate filtrate and substrate. Then, the filtrate was dried by oven at the temperature of 55°C for about 48 hours. The dry extracted was grinded and sieved using 60 mesh. The albumin powder was analyzed consist of yield, chemical composition, lightness, amino acid and albumin content.

2.2. *Statistical Analysis*

Data analysis was applied using completely randomized design consist of nine treatments with three replications using analysis of variance following by Tukey test at $\alpha = 0.05$. The sample code used in this study were SW (small fish with water solvent), MW (medium fish with water solvent), LW (large fish with water solvent), SN (small fish with NaCl solvent), MN (medium fish with NaCl solvent), LN (large fish with NaCl solvent), SH (small fish with HCl solvent), MH (medium fish with HCl solvent) and LH (large fish with NaCl solvent).

3. Result and discussion

3.1. *Yield and chemical composition*

Table 1 shows the chemical composition of albumin extract. The yield of samples ranged between 2.31% and 3.58%., average being 2.89%. Snake head fish protein concentrate which was made using distilled water solvent had high water content among the other concentrated produced from other solvents, thereby affecting the mass of the product. The amount of yield of the products can be known from the moisture content of the product. The higher the moisture content of a product, the higher the yield produced, because high water levels resulted in increased product mass [6].

Proximate analysis showed that the lowest moisture and ash content obtained from treatment LW that were 9.63% and 16.91%, respectively. Treatment LW also gave the highest protein and the lowest fat content that were 67.67% and 0.75% respectively. While the lightness, all treatment of the the samples ranged from 74.57 to 83.21, average being 80.31.

Table 1. Chemical composition of snakehead albumin extract.

Samples	Yield (%)	Moisture (%)	Ash (%)	Protein (%)	Fat (%)	Lightness
SW	2.56	12.16	16.91	63.79	1.74	80.00
SN	3.29	12.53	23.01	56.92	1.86	81.04
SH	2.93	12.18	21.93	59.24	1.20	81.10
MW	2.31	11.04	18.29	65.22	1.29	79.11
MN	3.35	11.19	22.69	60.21	1.42	82.55
MH	2.50	11.64	22.34	63.33	2.16	83.21
LW	2.89	9.63	18.02	67.67	0.75	74.57
LN	3.58	12.94	21.55	57.38	1.61	80.57
LH	2.64	12.48	21.83	63.59	1.44	80.68

High or low value of protein content can also be influenced by the amount of moisture loss (dehydration) of the material. Protein content will be even greater if the amount of water loss increases. The protein content depends on the amount of ingredients added and is largely influenced by moisture content [7]. Moreover, protein content of the fish is affected by moisture, where there is an inverse relationship between protein and water content, in which the higher the protein content increases as the water content decreases [8].

Low fat content is one of theindicator of the quality of good fish protein concentrate, because the high fat content also affected the process of rancidity in product [9]. Rancidity occurs when components of taste and smell of volatile formed as a result of oxidative damage of unsaturated fats and oils. These components caused the smell and taste of unwanted fat and oil and products containing fats and oils [8]. Low level of fat in the snakehead fish protein concentrate is also influenced by the process of meat cutting which can help releasing fat and water because it can increase the surface area in contact with hot material upon heating. Downsize will facilitate the penetration of moisture and hot air in the cells containing fat. The smaller the surface area formed as a result of cuts, the greater the contact material with the heat, causing the composition of fatty damaged impact on the levels of fat [10].

The result showed that the highest yield was obtained using salt solvent. While the lowest yield was generated by using water. The acid solvent resulted the extract more lightness. While the water solvent gave lower L value.

3.2. Amino acid composition
The profile of amino acid of snakehead albumin obtained is shown in Table 2. The table shows that the highest of total amino acid was generated by water solvent on medium size (fish weight 600-900 g). As a whole, the highest components of amino acid are aspartic acid, glutamic acid, lysine and leucine. Leucine is one of the components absolutely needed for children growth as a brain function trigger, prevents muscular degradation and play a role in the wound healing process [11].

Table 2. Amino acid profile of snakehead albumin extract (the values represented in percentage; %).

Amino Acid	Solvent								
	Water			NaCl 0.9%			HCl 0.1 M		
	S	M	L	S	M	L	S	M	L
Aspartic acid	5.51	7.44	7.43	5.74	6.17	6.43	5.26	5.42	5.52
Glutamic acid	4.60	6.86	6.51	5.50	5.66	5.87	4.41	4.53	4.92
Serine	1.58	2.16	2.36	1.48	1.80	2.11	1.47	1.40	1.78
Histidine	0.67	1.37	1.35	0.07	1.06	1.09	0.74	0.73	0.85
Glycine	3.66	3.31	3.09	2.57	2.74	2.75	3.69	3.75	3.72
Threonine	1.56	2.19	2.22	1.11	2.07	1.94	1.64	1.60	1.77
Arginine	1.64	2.83	2.77	2.15	2.30	2.43	1.52	1.52	1.80
Alanine	3.43	4.22	4.05	3.58	3.54	3.65	3.40	3.41	3.64
Tyrosine	0.94	1.39	1.67	1.46	1.48	1.27	0.79	0.64	0.80
Methionine	0.90	1.10	1.37	1.01	1.42	0.92	0.62	0.59	0.74
Valine	1.85	3.22	3.06	2.11	2.50	2.59	1.76	1.83	1.84
Phenylalanine	3.36	4.10	4.11	2.70	3.57	3.62	3.22	3.37	3.29
I-leucine	1.93	3.21	2.92	2.01	2.65	2.56	1.86	2.08	1.92
Leucine	3.38	4.98	5.00	4.24	4.10	4.28	3.14	3.28	3.27
Lysine	4.35	5.89	5.52	3.67	5.32	4.75	4.84	4.69	4.37
Amino Acid Total	39.37	54.25	53.43	39.41	46.39	46.23	38.36	38.83	40.21

3.3. Albumin content
Result of soluble protein (albumin) content of snakehead fish range from 13.95 % to 19.61 %. Figure 1 shows that albumin or dissolved protein in the snakehead fish could be soluble amongst all of solvent. Solubility of albumin in water shows that the albumin is polar.In addition, to dissolve a compound that was polar, the polar solvent must be used.In this case, this research used distilled water, 0.1 M of HCl and 0.9 % of NaCl were the polar solvent [13].

Level increase of soluble protein after treatment was due to hydrolysis of the soluble protein in the fish. Hydrolysis of proteins is the process of the breakup or breakdown of peptide bonds of proteins into simpler molecules with solvents assistance. Hydrolysis of peptide bonds will cause some changes in the protein, which increases the solubility due to the increase in the ion content of the amine (NH_3^+) ions and carboxyl (COO^-), thus, results in lower molecular weight proteins or polypeptides as well as unraveling the structure of globular proteins [13].

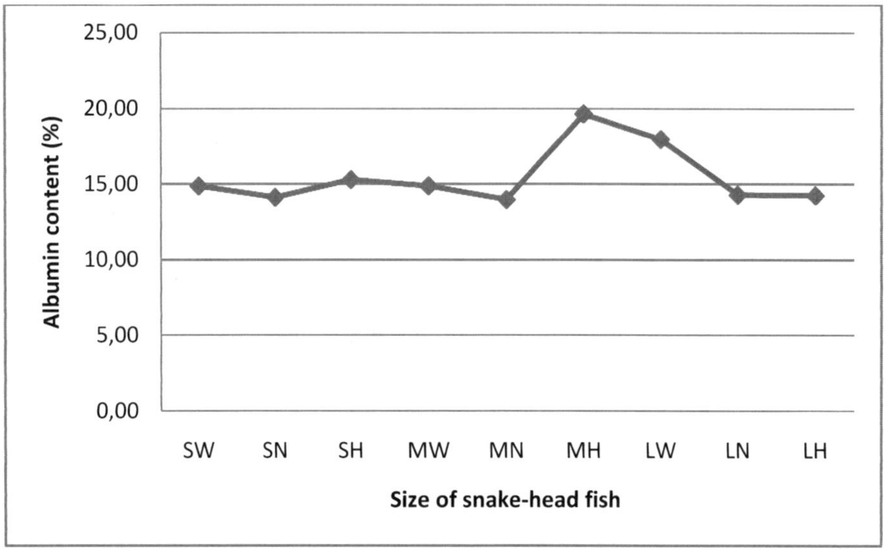

Figure 1. Albumin content of snake-head fish extract on various solvent(SW:small fish with water solvent, MW: medium fish with water solvent, LW: large fish with water solvent, SN: small fish with NaCl solvent, MN: medium fish with NaCl solvent, LN: large fish with NaCl solvent, SH: small fish with HCl solvent, MH: medium fish with HCl solvent and LH: large fish with NaCl solvent).

4. Conclusion

As a conclusion, salt solvent was the best treatment to resulted albumin extract based on yield and lightness. While the highest albumin content obtained from acid solvent (HCl 0,1 M) at medium size of fish (600-900 g).

Acknowledgment

The study was financially supported by The Ministry of Research, Technology and Higher Education Republic of Indonesia through the competitive grants program 2017 No: 350/UN17.41/KL/2017.

References

[1] Mustafa A M,Aris W and Yohanes K. 2012 Albumin and zinc content of Snakehead Fish (*Channastriata*) extract and its role in health *IEESE International Journal of Science and Technology (IJSTE)***1(2)** pp 1-8.
[2] Ramadhoni A R, Afriyanto E and Pratama RI 2016 Extraction of snakehead fish (*Ophiocephalusstriatus*) into fish protein concentrate as albumin source using various solvent*JurnalTeknologi* **78(4)** pp 1-6
[3] Winarno FG 2004 *Kimia Pangan dan Gizi*[Food Chemistry and Nutrient](Jakarta: PT. Gramedia Pustaka Utama) [Bahasa Indonesia]
[4] Suprayitno E 2014Profil albumin fish cork (*Ophiocephalusstriatus*) of different ecosystems*Int.J.Curr.Res.Aca.Rev* **2(12)** pp 201-8
[5] Gam L H, Leow C Y and Baie S 2006 Proteomic analysis of snakehead fish (*Channastriata*) muscle tissue *Malaysian Journal of Biochemistry and Molecular Biology* **14** pp 25-32
[6] Apriliyanti, T 2010 *Physicochemical Properties and Sensory Studies of Purple Sweet Potato Flour (Ipomoea batatasBlackie) with Drying Variation Process* Thesis (UniversitasSebelasMaret)

[7] Sebranek J G 2009 Functional Properties of Muscle Proteins: Implications for Processed Meat Product Characteristics*Proc. Reciprocal Meat Conference* **62** ed Johnson K and T Powell (Arkansas, USA) pp 1–7
[8] Buckle K A, Edwards R A, Fleet G Hand Wootton M1987 *Food Science* Translated by Purnomo H and Adiono(Jakarta: University of Indonesia)
[9] Asfar M, Tawali A B, Abdullah N et al., 2014 Extraction of albumin of snakehead (*ChannaStriatus*) in producing the fish protein concentrate (FPC) *Int Journal of Scientific & technology Research* **3(4)** pp 85-8
[10] Irianto H E and Giyatmi S 2009 *Teknologi Pengolahan Hasil Perikanan* [Fishery Products Processing Technology](Jakarta: Penerbit Universitas Terbuka) [Bahasa Indonesia].
[11] Firlianty, Suprayitno E, Nursyam H, Ardoko and Mustafa A 2013 Chemical composition and amino acid profile of *channidae* collected from central Kalimantan, Indonesia*IEESE International Journal of Science and Technology* **2(4)** pp 25-9
[12] Yuniarti D W, Sulistyati T D and SuprayitnoE 2013 Effect of Temperature on the Vacuum Drying to the Quality of Snakehead Fish Albumin Powder *THPI Student Journal*1(1) pp 1–9
[13] Nielsen H, Engelbrecht J, BrunakS and Heijne G von 1997 Identification of Prokaryotic and Eukaryotic Signal Peptides and Prediction of Their Cleavage Sites *Journal of Protein Engineering* **10** pp 1–6

Extraction optimization and characterization of gelatine from fish dry skin of Spanish mackerel (*Scomberromorus commersoni*)

I Kusumaningrum[1*], Y Pranoto[2] and S Hadiwiyoto[2]

[1]Department of Fishery Product Technology, Mulawarman University, East Kalimantan-75119, Indonesia
[2]Department of Food and Agricultural Product Technology, Gadjah Mada University, Yogyakarta-55281, Indonesia

*Corresponding author: iinklaten81@gmail.com

Abstract. This work was to optimized gelatin extraction from dry skin of Spanish mackerel (*Scomberromorus commersoni*) using Response Surface Methodology (RSM). The aim of this study was to determine the optimal condition of temperature and time for extraction process and properties of the gelatin extracted from dry mackerel skin. The optimal condition for extraction was 59.71°C for 4.25 hours. Results showed that predicted yield by RSM was 13.69% and predicted gel strength was 291.93 Bloom, whereas the actual experiment for yield and gel strength were 13.03% and 291.33 Bloom, respectively. The gelatin extracted from dried skin were analyzed for their proximate composition, yield, gel strength, viscosity, color, and amino acid composition. The results of dried skin gelatin properties compared to the commercial gelatin. Gelatin extracted from the dried skin gave content lower moisture, ash and protein content but higher fat compared to commercial gelatin. This study also shows that the gelatin extracted from the dried skin gave higher gel strength and pH but the lower amino acid composition compared to commercial gelatin.

1. Introduction

The use of gelatin is wide enough in many applications, but there are still some obstacles for consumers to consume these products. In addition to the controversy and concerns of many people because of the taboo for Hindus to eat beef, a ban on the Jewish people to consume any products derived from pigs and Islam forbid pork, as well as the emergence of a security risk. This happens when the raw material of gelatin made from pig and the cow disease Bovine Spongiform Encephalopathy (BSE), foot and mouth disease (foot and mouth disease) and that can be transmitted through gelatin [1-2]. It is therefore very important to find the source of gelatin other than cows and pigs to meet the needs of gelatin in the country and to overcome various obstacles.

Raw material of gelatin can be derived from fish, especially the skin and a small portion of fish bones. Utilization of fish skin is also in line with efforts to reduce waste in the fish processing industry. Fish processing waste that is currently underutilized in the fishery industry and gelatin from fish skins would be a good way to add value to this waste product [3]. Gelatin derived from fish tend to be more secure and abundant raw materials. In addition, the utilization of fish skin as a raw material for gelatin can improve the value added to the waste treatment [2, 4].

Dried fish skin gelatin as a source of raw materials has been reported in recent studies, e.g. on cobia skin [5] and seawater fish skin [6]. They studied the effect of the two methods of Cobia skin preservation, i.e. drying and freezing on the properties of gelatin produced, in terms of gelatin yield, proximate composition, gel strength, viscosity, color and amino acid composition. While [6] investigated the gelatins properties extracted from fresh and sun-dried skins of seawater fish.

One type of fish skin that could potentially be used is the skin of Spanish mackerel fish that are found throughout the year in the waters of Indonesia. Spanish mackerel is one result of capture fisheries in marine waters, especially in East Kalimantan. The volume of marine capture fisheries production of mackerel in Indonesia were 132 705 tonnes in 2011, while the volume production of mackerel from East Kalimantan were 1 793 tonnes [7].

The high production of fish fillet dumped as the waste of processing. One of the processing industries used Spanish mackerel were the manufacturing industries of "amplang" crackers. The processing waste of these industries generally just thrown away. The skin of Spanish mackerel contains a large amount of collagen. It is suggested that the skin can be useful sources of gelatin.

Yield and quality of gelatine are highly dependent on the method of extraction was done. The quality was determined by the gel strength of gelatin which is the main characteristic of gelatin. While the quantity, the yield is an indicator of production efficiency. The method used for the extraction of gelatin from fish skin, in general, was the acid extraction. Collagen derived from fish skin collagen, which is a type of easily extracted by weak acid treatment, followed by heating process for dissolving collagen. Heating time and temperature on the extraction process is a thing that must be controlled to obtain the yield of gelatin with a better quality [8].

Optimization of temperature and extraction time was necessary for this research. Response surface methodology (RSM) was a method used in the optimization between the effects of temperature and extraction time on yield and gel strength as one of the determinants of the quality of gelatin. Therefore, the objectives of this study were to determine the optimum conditions for extracting collagen from the dried skin of Spanish mackerel using RSM and to compare its physicochemical and rheological properties with the commercial gelatin.

2. Experimental details

2.1. Materials

Raw materials to be used was the dried skin of Spanish mackerel obtained from some of the fish processing industry in Samarinda, East Kalimantan. The fresh skins were kept in a cool box and washed to remove any contaminants. Fish skin was cleaned from the remaining flesh was scraped with a spoon and also removed the bones and then washed with running water. The cleaned fish skin was dried under the sun for about 2-3 days until dried.

2.2. Pre-treatment

Dried fish skin was rehydrated first by soaked in the water (dried fish skin: water = 1: 4) during 4 hours. After that, the skins were dipped in warm water (60-70°C). Once cooled, cut into small sizes ± 0.5 x 0.5 cm. Skins that has been cut into small pieces and then were treated with 0.05 M acetic acid with a ratio of fish skin: acetic acid = 1:4 (w / v) at room temperature for 10 hours. Then, the skins were washed to remove excess chemicals until neutral condition.

2.3. Gelatin extraction

The extraction of gelatin was carried out as described inMuyonga procedure [9]. First, the pre-treated fish skins were transferred to beakers and filled with distilled water at 1:3 (w/v) ratio. The extractions were carried out on control temperatures and time in accordance with the conditions that have been designed. The gelatin liquor was then filtered through four-layered cheesecloth to obtain gelatin filtrate. The filtrate was put on the trays and dried in a cabinet drier at 55°C for 48 h to obtain gelatin sheets. The sheets were ground to result in gelatin granules. The gelatin granules were packed in a plastic and kept in a refrigerator temperature until used for analysis. The yield and gel strength was used as a parameter to determine the optimum conditions for extraction.

2.4. Yields and proximate composition analysis
Yield was expressed in a percentage (%), calculated by weighing the resulted gelatin granules divided by the weight of fish skin in dry basis (DB), after considering the moisture content. The proximate compositions of gelatin granules were analyzed in terms of moisture content, crude protein, crude lipid and ash content following the AOAC method [10].

2.5. Determination of gel strength
The gel strength was carried out according to [11], determined by using Universal Testing Machine INSTRON Model 4510 (INSTRON Co, Canton, Mass, USA), with load cell 5kN, cross-head speed 1 mm/s, equipped with a 1.27-cm diameter of fat-faced cylindrical Teflon plunger. Gel sample was prepared by dissolving gelatin in distilled water to make concentration 6.67% and then kept at 9-10°C for 17-18 h for maturation. Gel strength was expressed inBloom, obtained from maximum force (g) when the plunger had penetrated 4 mm into the gelatin gel.

2.6. Determination of viscosity
The viscosity of gelatin solution was expressed in centipoises (CPS), measured by using a rotary Viscometer (BOHLIN Instruments Ltd., Gloucestershire, UK). Gelatin solution (6.67%) was prepared by dissolving gelatin granules in distilled water at 60°C, and the viscosity measurement was conducted at a temperature of 40-50°C.

2.7. Determination of color
To determine the color, the colorimeter (Minolta Cr-300 series, Japan) was used. L* (lightness), a* (redness/greenness) and b* (yellowness/blueness) value were measured.

2.8. Amino acid analysis
Gelatin samples were hydrolyzed in HCl 6N at 110°C for 24 h, and the hydrolyzates were analyzed by an amino acid analyzer.

2.9. Experimental design and analysis of data
Optimum conditions for the extraction of dried fish skin gelatin were determined using Response Surface Methodology (RSM). A central composite design (CCD) was used to determine the effects of the independent variables on the yield and gel strength of gelatin extraction. The design consists of a 2^k factorial, where k is a factor that was observed (X) at least two factors with the addition of axial points (α) for two factors is 1.414 (α = 2k / 4) [12]. The two factors were hot water extraction temperature in °C (X1) and extraction time in hours (X2). The experiment was conducted at 13 experimental points. The experimental design is shown in Table 1. The experimental data from different treatments was analyzed using Minitab statistical software (Version 15.1). The fitting was done to 2nd order model for each response. This model can be expressed by the following equation:

$$Y_i = \beta_0 + \sum_{i=1}^{k} \beta_i X_i + \beta_{ii} X_i^2 + \sum\sum_{i<j} \beta_{ij} X_i X_j \quad (1)$$

where Y is dependent variable (yield and gel strength), β_0 is the constant, β_i, β_{ii}, and β_{ij} are regression coefficients and Xi, Xj are levels of the independent variables.

3. Results and discussion

3.1. Optimization experimental design

The experimental results using RSM with CCD model are presented in Table 1. Two independent variables used in this research that the extraction temperature (X1) and extraction time (X2) with five levels for each variable. While the dependent variable is the yield (%) and gel strength (Bloom).

Table1. Central composite design for optimizing the extraction condition and responses of the dependent variables of gelatin to changes in the independent variables.

Run no.	Variable levels		Response	
	X1	X2	Y1	Y2
1	-1	-1	9.66	208.67
2	1	-1	7.20	105.00
3	-1	1	9.84	169.63
4	1	1	10.45	103.32
5	-1.414	0	5.62	73.46
6	1.414	0	8.08	80.62
7	0	-1.414	10.06	200.95
8	0	1.414	11.91	218.62
9	0	0	13.00	308.76
10	0	0	14.00	264.96
11	0	0	14.74	306.69
12	0	0	13.15	290.17
13	0	0	13.20	292.44

X1 (extraction temperature, °C), X2 (extraction time, h), Y1 (yield, %), Y2 (gel strength, Bloom).

3.2. Conditions for optimum responses

For the optimization of gelatin extraction, the two independent variables were of extraction temperature (X1) and extraction time (X2) was set in range while the dependent variables were fixed at maximum. The optimal solutions of extraction temperature were (59.7 °C, X1) and extraction time were (4.25 h, X2). The predicted values of yield and gel strength were 13.69% and 292 Bloom, while the experimental values of extraction yield and gel strength were 13.03% and 291 Bloom, respectively.

The yield obtained based on the weight of dried skin with moisture content of fish include in its calculations. In preliminary studies, the results obtained for the water content of Spanish mackerel dried skin was 13.24% (WB). The results indicated the yield of the resulting optimization of the dried skins extraction was 13.03% (DB) that obtained at 59.71°C for 4.25 hour treatments.

High and low yields also affected by the loss of proteins that occur during immersion in acid solution (swelling process) and during the extraction process with a high temperature or too low. The low yield is also associated with the loss of extracting collagen through leaching during the series of washing steps or due to incomplete hydrolysis of the collagen [13]. The yield of fish gelatin is lower than mammalian gelatin, which ranges from 6-19%. Wide range of yield due to differences in collagen content in the raw material [9].

a.

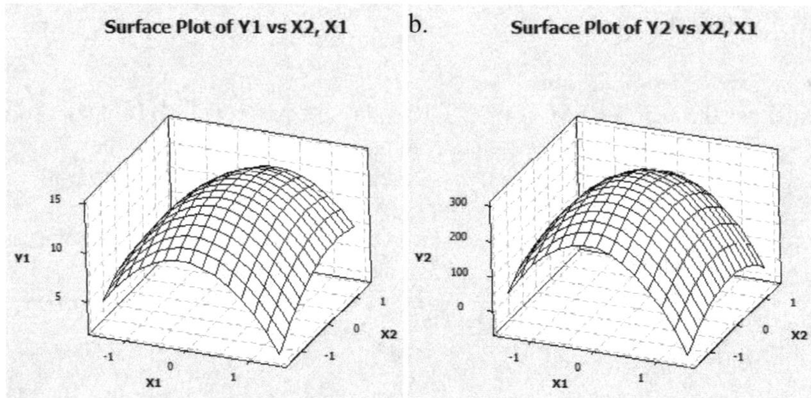

Figure 1. Surface and contour plot optimization of mackerel fish skin gelatin (a. based on yield response; b. based on gel strength response)

The three dimensions plot (Figure 1) shows that the yield of fish skin gelatin increased on the extraction temperature beyond 60 °C and decreases at temperatures above 70 °C and 55 °C below. So also with longer extraction time will affect the yield value. The results showed that the extraction time for 4 to 6 hours will increase the value of the yield of gelatin. Based on the response of gel strength (Figure 2.), the extraction of gelatin increased as temperatures between 50 °C to 70 °C. While based on the length of time of extraction, gel strength will increase with an extraction time of about 4 hours. The longer the extraction time, the resulting gel strength will be lower.

3.3. Physicochemical properties of gelatin

The physicochemical properties are shown in Table 2. The moisture content of Spanish mackerel dried skin gelatin was lower than the commercial gelatin. Indonesian National Standard [14] regulates the maximum level of gelatin moisture is 16%, while the maximum moisture content according to [15] is 14%. Therefore, the moisture content of these gelatin had fulfilled the standard regulations.

Table 2. The chemical compositions of Spanish mackerel dried skins gelatin

Properties	**Gelatin**	
	Spanish mackerel dried skins	**Bovine commercial**
Moisture (% DB)	10.81 ± 0.08	13.24 ± 0.03
Ash (% DB)	1.44 ± 0.03	1.51 ± 0.01
Protein (% DB)	89.93 ± 0.15	92.57 ± 0.02
Fat (% DB)	1.78 ± 0.01	0.11 ± 0.01
Gel strength (Bloom)	291.33 ± 6.98	216.98 ± 21.58
Viscosity (CPS)	8.07 ± 0.17	8.23 ± 0.13
Colour :		
L	52.15 ± 0.95	64.32 ± 0.19
a	4.90 ± 0.26	1.74 ± 0.16
b	21.77 ± 1.01	17.49 ± 0.14

Data presented as means standard deviation of triplicate determinations

Ash content of Spanish mackerel dried skin gelatin showed a low value when compared with the ash content of commercial gelatin. Indonesian National Standard [14] regulates the maximum level of gelatin moisture is below 3.25%. Gelatin with ash content of less than 0.5% indicates a high-quality gelatin, while according to the FAO recommended levels of ash in the food less than 3%.

The low ash content is important in determining the quality of gelatin. Ash content of gelatin may be contributed by the residual of chemicals after processing, come from raw material, or also the possibility of mixing with the other ingredients [13]. Ash content also showed the efficiency of demineralization of raw material pre-treatment [16-17].

The protein content of the commercial gelatin gave higher protein content (92.57%) compared to Spanish mackerel dried skins gelatin(89.93%). SNI does not provide a minimum limit for protein content. High levels of protein showed a good extraction process. Levels of protein in gelatin is also influenced by the process of soaking the skin. The soaking process resulted in the termination reaction of hydrogen bonds and opening of the coil structure of collagen, therefore, contributes to the amount of protein extracted.

Fat content of Spanish mackerel dried skin gelatin is higher than the commercial gelatin. SNI does not provide rules about fat content. Commercial gelatine showed the fat content is 0.11%. Generally, commercial gelatin contains almost no fat (<0.5%). The differences in fat content might be due to the differences in fat content in differences raw materials.

Gel strength is one of the most important functional properties of gelatin. The gel strengths of commercial gelatins are expressed using Bloom values and the value is the weight in grams that is required for a specified plunger to depress the surface of a standard. The average molecular weight of gelatin is largely responsible for its gelling behavior. After overnight maturation (18 hour, 7-8°C), gelatin extracted from Spanish mackerel dried skin showed higher gel strength (291 Bloom) than that of commercial bovine (217 Bloom).

The gelling strength of commercial gelatin ranges from 100 to 300, but gelatin with Bloom values of 250–260 are the most desirable. Gel strength is one physical properties and most important function of gelatin and it is governed by molecular weight, as well as by complex interactions determined by the amino acid composition. The large range of Bloom values found for the various gelatin arises from differences in proline and hydroxyproline content in collagens of different species and is also associated with the temperature of the habitat of the animals. Hydrophobic amino acids (Ala, Val, Leu, Ile, Pro, Phe, and Met) could also contribute to the high Bloom value of tilapia fish gelatin. This study shows that the gel strength of Spanish mackerel dried skins was higher than the commercial bovine gelatin [18].

Viscosity is the resistance to the flow of a fluid which is defined as the ratio between the shear stress on the rate of shear rate. The more viscous fluid, the greater the force required for fluid to flow with a certain rate[19]. Viscosity is an important physical property of the gelatine after the gel strength [20]. The viscosity values (CP) for mackerel gelatine were lower than commercial gelatin. SNI does not give the terms of the viscosity of gelatin. The viscosity characteristics displayed by a given gelatin grade are primarily related to the molecular weight distribution of the gelatin molecules [21]. Higher molecular weight subunits give an increase in viscosity. The longer the chain of amino acids, the high molecular weight of gelatin, so that the higher viscosity [6].

Colour of L shows the lightness, a shows redness and b shows yellowness. Gelatin extracted from mackerel skin was darker compared to commercial gelatin. In general, satisfactory gelatin should have little color. The color of gelatin depends on the raw material. However, it does not influence other functional properties. The light color is preferred and considered as a positive attribute since it is easier to incorporate these light colored gelatin into any food system without imparting any strong color attribute to the product [13].

3.4. Amino acids composition

The amino acids composition of the gelatin from mackerel skin gelatin compared to commercial gelatin are shown in Table 3. The amino acids (proline and hydroxyproline) and glycine contents of the Spanish mackerel dried skins gelatin were 8.27% and 6.10%, respectively. The content of amino acid and glycine are important for gel strength. The mammalian gelatine have a high composition of these amino acids, especially proline and hydroxyproline, which are related to gelling property a low content of amino acids indicates poor gelling power [22]. Although proline is important, hydroxyproline is believed to play a singular role in the stabilization of the triple-stranded collagen helix due to its hydrogen bonding ability through its hydroxyl group. Besides the amino acids, lysine also stabilizes gelatin structure by forming crosslinking structure between chains [13].

Table 3. Amino acids composition of mackerel gelatin and commercial bovine gelatin

Amino acids	**% amino acids**	
	Spanish mackerel dried skins	**Commercial Bovine**
Aspartic acid (Asp)	5.21	5.48
Glutamic acid (Glu)	10.86	9.87
Serine (Ser)	1.86	2.32
Glycine (Gly)	6.10	8.92
Histidine (His)	2.08	3.38
Arginine (Arg)	2.79	3.74
Threonine (Thr)	1.06	1.98
Alanine (Ala)	1.29	2.12
Tyrosine (Tyr)	1.12	2.58
Valine (Val)	2.03	2.23
Methionine (Met)	2.01	2.61
Cystine (Cys)	1.71	2.35
Isoleucine (Ile)	1.13	1.76
Leucine (Leu)	1.07	1.78
Phenylalanine (Phe)	1.72	3.37
Lysine (Lys)	5.89	4.49
Hydroxyproline (Hyp)	2.89	2.68
Proline (Pro)	5.38	10.49

4. Conclusion

This study suggests that Spanish mackerel dried skins can be used in gelatin production. The optimal solution for multiple responses extraction yield and gel strength showed that an extraction temperature of 59.71°C and extraction time of 4.25 hour gave the best result. This study shows that the gelatin gel strength extracted from Spanish mackerel dried skin was higher than that of commercial gelatin. However, the gelatin extracted from Spanish mackerel dried skin gave lower viscosity, moisture, ash, protein and amino acid content compared to commercial gelatin.

Acknowledgments
We are grateful for Directorate General of The Higher Education Ministry of Education and Culture Scholarship of Indonesia to Author 1. We sincerely thank Professor Yudi Pranoto and Suwedo Hadiwiyoto, M.Phill for valuable discussions and encouragement.

References

[1] Aewsiri T, Benjakul S, Visessanguan W Eun J, Wierenga P A and Gruppen H 2009 Antioxidative Activity and Emulsifying Properties of Cuttlefish Skin Gelatin Modified by Oxidised Phenolic Compounds *Food Chemistry* **117** pp 160-8
[2] Mohtar NF, Perera C and Quek S Y 2010 Optimization of Gelatin Extraction from Hoki *(Macruronus novaezelandiae)* Skins and Measurement of Gel Strength and SDS-PAGE *Journal of Food Chemistry* **122** pp 307-13

[3] Mohtar N F, Perera C O and Quek S Y 2011 Utilization of Gelatin from NZ Hoki *(Macruronus novaezelandiae)* Fish Skins *International Food Research Journal*. **18(3)** pp 1111-15
[4] Binsi P K, Shamasundar B A, Dileep A O, Badii F and Howel N K 2009 Rheological and Functional Properties of Gelatin from The Skin of Bigeye Snapper *(Priachantus hamrur)* Fish: Influence of Gelatin on The Gel-Forming Ability of Fish Mince *Journal of Food Hydrocolloids* **23** pp 132-45
[5] Amiza M A and Aishah S 2011 Effect of drying and freezing of Cobia (*Rachycentron canadum*) skin on its gelatine properties *International Food Research Journal* **18(4)** pp 1335-41
[6] Pranoto Y, Marseno D W and Rahmawati H 2011 Characteristics of Gelatins Extracted from Fresh and Sun-dried Seawater Fish Skin in Indonesia. *International Food Research Journal*. **18(4)** pp 1335-41
[7] Ministry of Marine Affairs and Fisheries 2012 *Capture fisheries statistics of Indonesia, 2011* (Jakarta: Directorate General of Capture Fisheries)
[8] Kasankala L M, Xue Y, Weilong Y, Hong S D and He Q2007 Optimization of Gelatin Extraction from Grass Carp *(Catenopharyngodon idella)* Fish Skin by Response Surface Methodology *Journal of Bioresource Technology* **98** pp 3338-43
[9] Muyonga J H, Cole C G B and Duodu K G 2004 Extraction and Physico-chemical Characterization of Nile Perch *(Lates niloticus)* Skin and Bone Gelatin. *Food Hydrocolloids* **18** pp 581-92
[10] Association of Official Analytical Chemists International (AOAC) 2006 Official Methods of Analysis
[11] Gómez-Guillén M C, Turnay J, Fernandez-Diaz M D, Imo N, Lizarbe M A and Montero P 2002 Structural and Physical Properties of Gelatin Extracted from Different Marine Species: A Comparative Study *Journal of Food Hydrocolloids* **16** pp 25-34
[12] Montgomery D C 2001 *Design and analysis of experiments* (New York: John Wiley dan Sons, Inc.) pp 427-510
[13] Jamilah B and Harvinder K G 2002 Properties of Gelatins from Skins of Fish Black Tilapia (*Oreochromis mosambicus*) and Red Tilapia *(Oreochromis nilotica) Journal of Food Chemistry***77** pp 81-84
[14] National Standard of Indonesia (SNI) 1995 Quality and Method of Gelatin Procedure
[15] British Standard Institution (BSI) 1975 Methods for sampling and testing gelatin (physical and chemical methods)
[16] Balti R, Jridi M, Sila A, Soussi N and Arroume N N 2011 Extraction and Functional Properties of Gelatin from The Skin of Cuttlefish *(Sepia officinalis)* Using Smooth Hound Crude Acid Protease-Aided Process *Journal of Food Hydrocolloids* **25** pp 943-50
[17] Jonjareonrak A, Rawdkuen S, Chaijan M, Benjakul S, Osako K and Tanaka M2010 Chemical Composition and Characterisation of Skin Gelatin from Farmed Giant Catfish *(Pangasianodon gigas) LWT-Food Science and Technology* **43** pp 161-5
[18] Badii F and Howell N K 2006Fish Gelatin: Structure, Gelling Properties and Interaction with Egg Albumen Proteins *Journal of Food Hydrocolloid* **20** pp 630-40
[19] Astawan M, Purwiyatno H and Ani M 2002 Analysis of Rheological Properties of Swordfish Skin Gelatin *Journal of Food Technology and Industry, Indonesia* XIII**(1)** pp 38-46
[20] Tabarestani H S, Maghsoudlou Y, Motamedzadegan A and Mahoonak A R S 2010 Optimization of Physic-chemical Properties of Gelatin Extracted from Fish Skin of Rainbow Trout *(Onchorhynchus mykiss) Journal of Bioresource Technology* **101** pp 6207-14
[21] Karima A and Bhat R 2009 Fish Gelatin: Properties, Challenges, and Prospects as An Alternative to Mammalian Gelatins *Journal of Food Hydrocolloids* **23** pp 563-76
[22] Ward A G and Courts 1977 *The Science and Technology of Gelatin* (London: Academic Press)

IOP Conf. Series: Earth and Environmental Science **144** (2018) 012037 doi:10.1088/1755-1315/144/1/012037

Catalytic hydrotreating of bio-oil derived from *Chlorococcum* sp.

R R D J N Subagyono[1], M Marshall[2], W R Jackson[2] andA L Chaffee[2]

[1]Chemistry Department, Faculty of Mathematics and Natural Sciences, Mulawarman University, Samarinda, East Kalimantan, Indonesia 75242
[2]School of Chemistry, Monash University, Clayton, Victoria 3800, Australia

*Corresponding author: dirgarini_julia@yahoo.co.id

Abstract. Hydrotreating of bio-oil derived from *Chlorococcum* sp. in the absence of NiMo/Al-SBA-15 catalysts has been investigated. The silica alumina supports decreased product yields (DCM solubles and hydrocarbon gases) due to severe coking. Incorporation of NiMo increased the product yields to up to 65 wt.% and increasedthe proportion of products in the kerosene boiling point range (C_{12}-C_{17}).The superiority of the high acidity supports persisted after incorporation of NiMo.

1. Introduction

Issues related to limited availability of fossil fuels, an increase in oil price and emissions generated from fossil fuels have created an increased interest in renewable fuels. Bio-oil production from microalgae has therefore attracted an increasing proportion of the research in this area, particularly due to the many advantages of this biomass type, including high productivity and the ability to grow on marginal land [1]. It is essential to deploy a suitable upgrading method to improve the quality of the bio-crude derived from microalgae so that the upgraded oil will be environmentally and industrially safe and also useful as commercial fuel.

Mesoporous alumina-silicates, such as Al-SBA-15, have been widely used as catalysts[2,3] or supports [4-6]in many catalytic reactions. Compared to other mesoporous silicas, e.g. MCM-41 and HMS, SBA-15 has many potential advantages, such as higher surface area, larger pore volume, thicker pore walls and higher thermal stability [7]. Introduction of aluminium into the silica framework, especially by direct synthesis, increased the catalytic activity in the Fischer Tropsch synthesis due to the presence of acid sites [4]. While metals supported on Al-SBA-15 materials have been used in many catalytic reactions, there is a dearth of information on the use of this type of catalyst for hydrotreating of bio-oil derived from microalgae.

In this work, we report hydrotreating of bio-oil derived from single strain microalgae, *Chlorococcum* sp.,catalysed by NiMo/Al-SBA-15 with different Si/Al ratios, investigating the effect of H_2 pressure and catalyst support acidity on product yields and product composition.

2. Materials and method

2.1. Materials

Biocrude oil was derived from *Chlorococcum* sp., a marine microalga provided by Biomax Australia. The preparation and the production of the bio-crude oil are described in our previous report[8]. The chemicals used in the study were the non-ionic block copolymer surfactant EO_{20}-PO_{70}-EO_{20} Pluronic P123, poly(ethylene oxide)-block-poly(propylene oxide)-block-poly(ethylene oxide) and aluminium isopropoxide (Aldrich), hydrochloric acid (32%) (Ajax), tetraethyl orthosilicate (TEOS 98%) and ammonium fluoride (NH_4F 98%) (Sigma-Aldrich), $NH_3.H_2O$ (Univar), nickel acetate (Sigma-Aldrich),ammonium heptamolybdate (Merck),N_2, H_2 and 10% NH_3/90% He (Air Liquide), liquid chromatography grade dichloromethane (DCM) and hexane (C_6H_{14}) (Merck). All chemicals were used without further purification.

2.2. Preparation of catalysts

Al-SBA-15 with different Si/Al molar ratios (10, 20) were synthesised using a modified hydrolysis-controlled approach as described in Li *et al.*[9].Al-SBA-15 with different Si/Al ratios was doped with nickel (Ni) and molybdenum (Mo). The concentrations of Ni and Mo were chosen to be 2.4 and 10%, respectively. Ni and Mo were impregnated into Al-SBA-15 using the wet impregnation method. The resulting products were denoted NiMo/Al-SBA-15. The NiMo/Al-SBA-15 catalysts were characterised with Nitrogen adsorption/desorption analyser, ammonia-temperature programmed desorption, Raman spectroscopy, and inductively coupled plasma-atomic emission spectroscopy, and X-Ray diffraction.

2.3. Hydrotreating

Hydrotreating was carried out in 27 mL stainless steel autoclaves fitted with glass liners, to minimize any catalytic effect of the autoclave walls. For catalytic hydrotreating, ~1.0 g of bio-oil, 10 % catalyst (by weight of the bio-oil) and 3, 6, or 9 MPa (cold) H_2 were charged into the autoclave. The autoclaves were placed in a pre-heated fluidized sand bath, continuously agitated for 15 minutes at 425 ^{o}C, removed, allowed to cool and finally weighed.

The gas, solid and liquid products were worked up as described in Subagyono *et al.* [8]. During work up, a fraction of the DCM solubles was taken for further analyses. The lighter fraction (DCM soluble, hexane soluble) was defined to be oil and the heavier and more polar fraction (DCM soluble, hexane insoluble) was defined to be asphaltene. The product yield was defined as the total yield of oil, asphaltene, hydrocarbon (if any, and sulphide) gases, and CO_2. The reproducibility of the reactions is indicated by error bars given on figures; error bars were calculated from duplicate experiments. It is important to note that the large variation in the oil yields was due to the fact that the oil yield is obtained by difference, not by direct weighing.DCM soluble fractions were analyzed by gas chromatography–mass spectrometry (GC–MS) on a HP6890 instrument in splitless mode as described in Subagyono *et al.*[8].

3. Results and discussion

3.1. Characterisation of catalysts

The wide-angle XRD patterns of NiMo/Al-SBA-15 (Figure 1a) showed low intensity peaks due to orthorhombic MoO_3 at 2θ values of 23, 27 and 33°. The low intensity peaks observed indicated that Ni and Mo were mainly amorphous and hence, probably evenly dispersed[10]. The small-angle XRD patterns of Al-SBA-15 showed a high intensity peak and two well defined smaller peaks, corresponding to the (100), (110), and (200) planes, between 2θ values of 0.6-1.8^{o} (Figure 1c). These three peaks similar to the SBA-15 structure with p6mm hexagonal symmetry [7]. After incorporation of Ni and Mo, the three peaks were still clearly resolved, indicating that the average structural order of the silica-alumina supports was retained (results not shown). The NH_3-TPD profiles of Al-SBA-15 (Si/Al=10) (Figure 1b) showed only one broad band peaking at 240-280 ^{o}C, indicating the presence of weak to medium strength acid sites, probably due to the tetrahedrally coordinated aluminium. In the Raman spectrum of Al-SBA-15 (Si/Al=10) (Figure 1d), peaks observed at 289, 335, and 666 cm^{-1}were due to crystalline MoO_3 [10]. A peak at 377 $cm^{-1}$was due to distorted surface polymolybdenum [11]. The presence of large clusters of MoO_3 molecules was indicated by the appearance of the most prominent peak at 819 cm^{-1} [10, 11]. Furthermore, a medium intensity peak at 994 cm^{-1} was due to asymmetric vibrations of bridging Mo-O-Mo in polymolybdates [12].

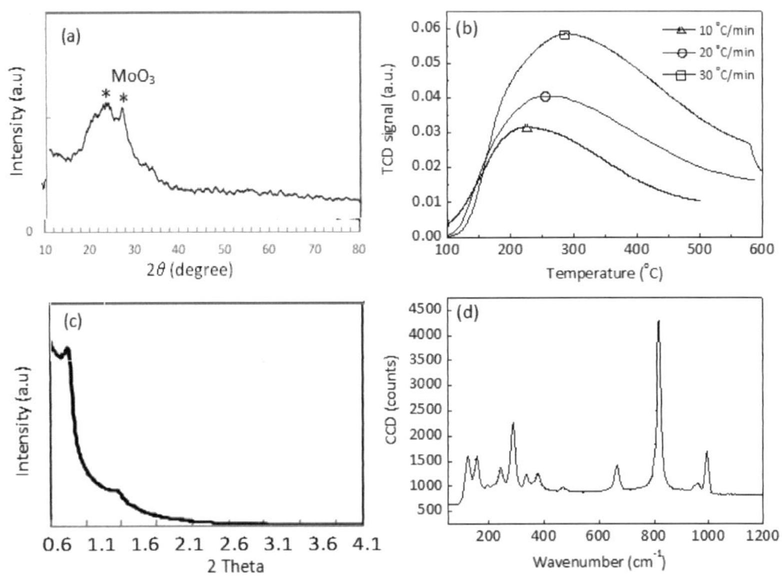

Figure 1. (a)Wide angle XRD spectrum of Al-SBA-15 (Si/Al=10) (* = trace of orthorhombic MoO_3 (PDF card 00-005-0508);(b) NH_3-TPD profiles of Al-SBA-15 (Si/Al=10) obtained at different heating rates (c) low angle XRD spectrum of NiMo/Al-SBA-15 (Si/Al=20); d) average Raman spectrum of NiMo-Al/SBA-15, Si/Al =10

The BET surface area, pore volume and pore diameter (Table 1) (NiMo)/Al-SBA-15 were calculated from the N_2 adsorption isotherms.

Table 1. Textural and physical properties of the (NiMo)-Al/SBA-15

Si/Al ratio	Measured Si/Al ratio[a]	Measured %[a]*		surface area[c] (m^2/g)	Pore volume[d] (cm^3/g)	Pore diameter[d] (nm)
		Ni	Mo			
10	10.6	-	-	332	1.34	14.3
NiMo/10		2.5	9.2	230	0.92	14.3
20	21.0	-	-	449	1.45	11.3
NiMo/20		2.3	8.9	315	1.04	11.2

[a]determined by ICP-AES
[b] determined by EDS
[c] calculated by the BET method
[d]calculated by the BJH method from desorption branch of the nitrogen isotherm
* the calculated/theoretical percentage of Ni= 2.5 and Mo=10.0

The BET surface area decreased with increasing aluminium content. The silica-alumina supports had a larger pore diameter (10-14 nm) and slightly higher pore volume (1.3-1.5 cm^3 g^{-1}) than those reported for Al-SBA-15 prepared by Li et al. and Shujie et al. [2, 9]. The larger pore diameter possibly resulted from the prolonged calcination (20 hours) necessary to completely remove the organic template and the high pore volume from formation of secondary mesoporous channels when aluminium was introduced into the silica framework by the direct-synthesis method [13]. The BET surface areas for the Al-SBA-15 materials were lower than those reported by other workers [2, 9]. This may be due to the larger pore volume of the Al-SBA-15 and not be due to the presence of non-framework aluminium in the pores [14]. After incorporation of Ni and Mo, the surface area and pore volume decreased, suggesting that Ni and Mo were incorporated inside the pore (and not just deposited on the particle surface). The ICP-AES results showed that Ni and Mo were successfully loaded into the Al-SBA-15 materials (Table 1).

3.2. Hydrotreating of bio-oil

Incorporation of NiMo into the supports decreased coking and product yields while generally increased oil, asphaltene and gas yields (Figure 2). The product yields and oil yields of reactions with NiMo/Al-SBA-15 with Si/Al=10 and 20 exceeded those of the non-catalysed reactions (results not shown). The results imply that Ni and especially Mo were catalytically active in hydrogenation reactions and therefore in reducing coke production. In the absence of catalyst, the major reactions were thermal cracking of high molecular weight components, but in the presence of catalyst, thermal and catalytic cracking both occurred.

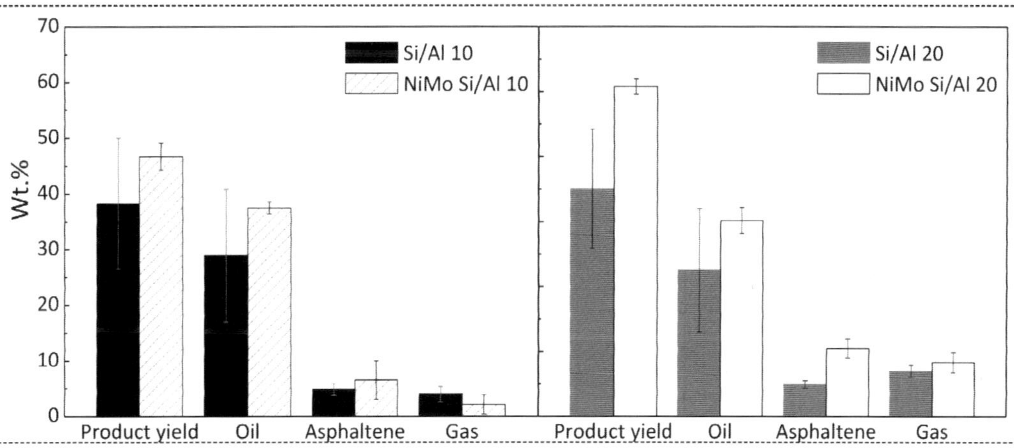

Figure 2. Product yields and oil, asphaltene and gas yields for hydrotreating of *Chlorococcum* sp. under 3 MPa cold H_2 for 15 minutes at 425 °C in the presence of Al/SBA-15 (with or without NiMo)

In the presence of NiMo/Al-SBA-15, product yields and asphaltene yields increased with increasing pressure (Figure 3). The less acidic catalyst (Si/Al=20), however, reached its best performance at a H_2 pressure (cold) of 6 MPa; the yields fell when the H_2 pressure was increased to 9 MPa (cold) H_2.

The chromatograms of DCM solubles produced in the presence NiMo catalyst under different H_2 pressures were generally similar (Figure 4). The peaks due to heavy molecules, observed in the chromatogram of the bio-crude oil, disappeared when 9 MPa H_2 was introduced to the system. Peaks due to pentadecane, heptadecane and phytane were the most prominent in the chromatograms but peaks due to other hydrocarbons were also observed. The appearance of other hydrocarbon peaks and the higher product yields in the presence of NiMo catalyst than in the presence of only the supports suggested that NiMo helped to stabilise reaction products against repolymerisation, to increase decarboxylation of fatty acids or to hydrogenate high boiling point saturated compounds.

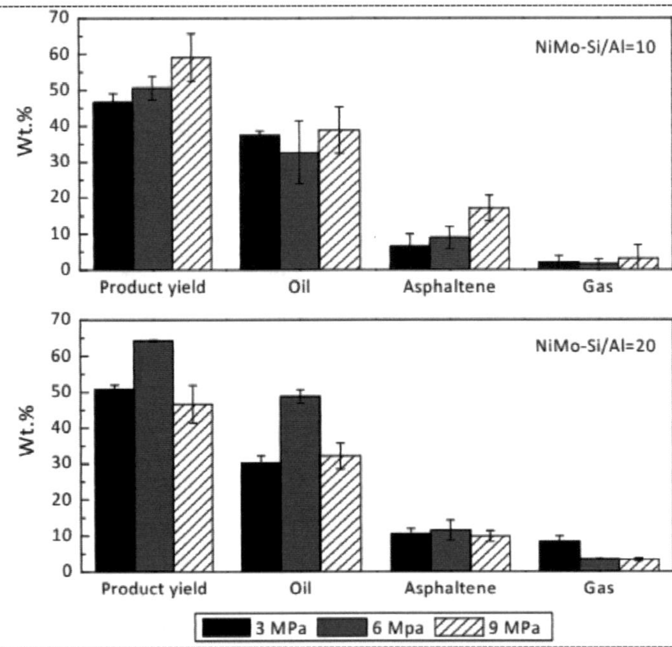

Figure 3. Product yields and oil, asphaltene and gas yields for hydrotreating of *Chlorococcum* sp. at 425 °C for 15 minutes as a function of H_2 pressure in the presence of NiMo-Al/SBA-15 catalysts

The degree of cracking, decarboxylation and thermal decomposition of the bio-oil was examined in the presence and in the absence of catalyst under different conditions. The total ion chromatograms were divided into three regions, 1. C_6-C_{11} (gasoline type, boiling point 70-200 °C); 2.C_{12}-C_{17} (kerosene type, boiling point 200-300 °C); and 3.>C_{17} (heavy oil, boiling point 300-470 °C), and the total peak area of all compounds (not only hydrocarbons) in the three ranges were calculated. The estimated weight (EW) percent of each fraction in the DCM solubles (excluding hydrocarbon gas and CO_2) was then calculated (Figure 5).

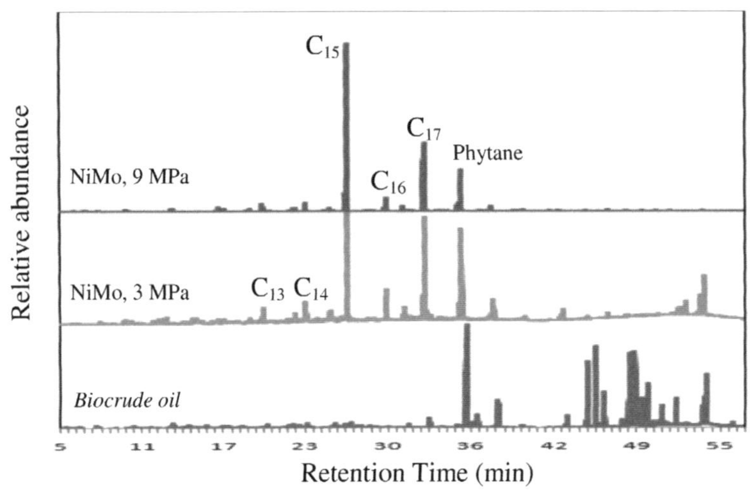

Figure 4. Chromatograms of bio-oil and DCM solubles produced by hydrotreating of *Chlorococcum* sp. for 15 minutes at 425 °C in the presence of NiMo-Al/SBA-15 (Si/Al=10)

The EW values of C_{12}-C_{17} fractions for the supports were generally only slightly higher than or similar to, the EW values found in the absence of supports. Thus, the increase in proportion of the C_{12}-C_{17} in the bio-oil produced in the presence of supports may be due to repolymerisation and coking of the high boiling materials in the presence of supports, thus reducing the proportion of high boiling point material in the products, rather than to more cracking of the long chain hydrocarbons or decarboxylation of fatty acids. Recalling that incorporation of the supports alone tended to induce coking (Figure 2) and reduced product yield; the silica alumina supports themselves are perhaps responsible for coke formation due to polymerisation of dehydrogenated species including high boiling point materials. After incorporation of the NiMo catalyst, the EW of the kerosene range rose and those of the heavy oil fell. Furthermore, the increase in the percentage of the kerosene range was greatest when the hydrogen pressure was increased to 9 MPa, suggesting more effective hydrogenation.

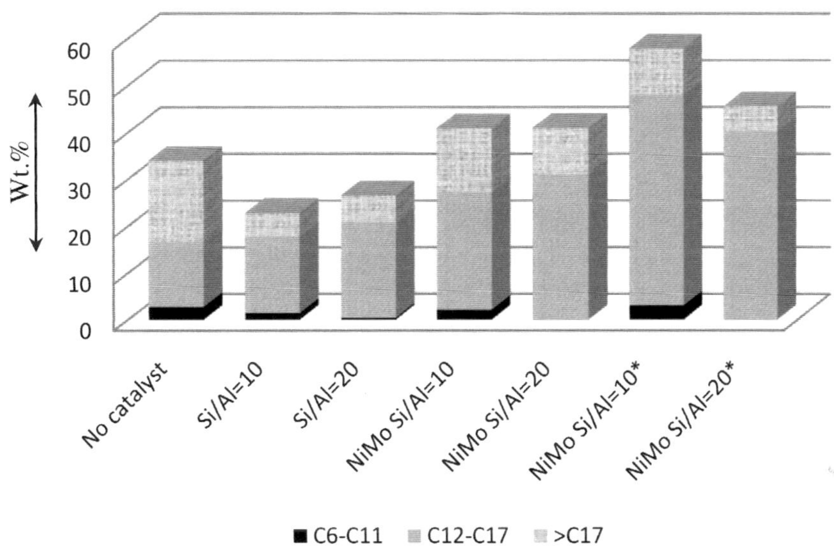

Figure 5. Estimated weight percent (%) of DCM soluble fractions from hydrotreatings of the bio-oil derived from *Chlorococcum* sp. at 425 °C for 15 minutes under 3 MPa H_2 and 9 MPa H_2 (*) in the presence of NiMo-Al/SBA-15 catalysts

4. Conclusions

The Al-SBA-15 supports did not have any upgrading effect but higher acidity NiMo/Al-SBA-15 increased the overall yield of DCM solubles up to 65 wt.% db and greatly increased the proportion of product in the lower boiling point kerosene range, by stabilizing products against repolymerization and by promoting cracking and decarboxylation of fatty acids. Thus the catalyst synthesized as described is useful. However, the extent of coking of the upgraded oil remained too high, so that improvements in the upgrading approach would be required to obtain a practically useful fuel.

Acknowledgments

We thank the Defence Science and Technology Organization (DSTO), Australia, Boeing Australia and the Victorian State Government Department of Economic Development, Jobs, Transport and Resources, through its Energy Technology Innovation Scheme (ETIS), for financial support. We also thank Biomass Australia for providing the algal samples. We also acknowledge Dr. Anthony R. Auxilio for assistance in conducting the ammonia TPD analyses.

References

[1] Jena U, Das K C and Kastner J R 2011 Effect of operating conditions of thermochemical liquefaction on biocrude production from *Spirulina platensisBioresour Technol* **102** 6221-9

[2] Wu S, Huang J, Wu T, Song K, Wang H, Xing L, Xu H, Xu L, Guan J and Kan Q 2006 Synthesis, Characterization, and Catalytic Performance of Mesoporous Al-SBA-15 for Tert-butylation of Phenol *Chin J Catal* **27** 9-14

[3] Faye J, Takahashi A, Capron M l, Fongarland P, Dumeignil F and Fujitania T 2013 Al-modified mesoporous silica for efficient conversion of methanol to dimethyl ether *RSC Advances***3** 5895-902

[4] Chen S, Li J, Zhang Y, Zhao Y and Hong J 2013 Effect of tetrahedral aluminum on the catalytic performance of Al-SBA-15 supported Ru catalysts in Fischer-Tropsch synthesis *Catal Sci Techno* **3** 1063-8

[5] Klimova T, Reyes J, Gutie´rrez O and Lizama L 2008 Novel bifunctional NiMo/Al-SBA-15 catalysts for deep hydrodesulfurization: Effect of support Si/Al ratio *Appl Catal Gen* **355** 159-71

[6] Sundaramurthy V, Eswaramoorthi I, Dalai A K and Adjaye J 2008 Hydrotreating of gas oil on SBA-15 supported NiMo catalysts *Micropor Mesopor Mat* **111** 560-8

[7] Zhao D, Huo Q, Feng J, Chmelka B F and Stucky G D 1998 Nonionic Triblock and Star Diblock Copolymer and Oligomeric Surfactant Syntheses of Highly Ordered, Hydrothermally Stable, Mesoporous Silica Structures *J Am Chem Soc* **120** 6024-36

[8] Subagyono D J N, Marshall M, Jackson W R, Chow M and Chaffee A L 2014 Reactions with CO/H_2O of two marine algae and comparison with reactions under H_2 and N_2*Energ Fuel* **28** 3143-56

[9] Li Y, Zhang W, Zhang L, Yang Q, Wei Z, Feng Z and Li C 2004 Direct synthesis of Al-SBA-15 mesoporous materials via hydrolysis-controlled approach *J Phys Chem B* **108** 9739-44

[10] Soni K K, Chandra Mouli K, Dalai A K and Adjaye J Effect of Ti loading on the HDS and HDN activity of KLGO on NiMo/TiSBA-15 catalysts *Micropor Mesopor Mat* **152** 224-34

[11] Ferdous D, Bakhshi N N, Dalai A K and Adjaye J 2007 Synthesis, characterization and performance of NiMo catalysts supported on titania modified alumina for the hydroprocessing of different gas oils derived from Athabasca bitumen *Appl Catal B Environ* **72** 118-28

[12] Montesinos-Castellanos A and Zepeda T A 2008 High hydrogenation performance of the mesoporous NiMo/Al(Ti, Zr)–HMS catalysts *Micropor Mesopor Mat* **113** 146-62

[13] Dragoi B, Dumitriu E, Guimon C and Auroux A 2009 Acidic and adsorptive properties of SBA-15 modified by aluminum incorporation *Micropor Mesopor Mat* **121** 7-17

[14] Mokaya R 1997 Post-synthesis grafting of Al onto MCM-41 *Chem Comm* 2185-6

Synthesis of N-oleyl O-sulfate chitosan from methyl oleate with O-sulfate chitosan as edible film material

Daniel[1,*], O Sihaloho[1], C Saleh[1] and A R Magdaleni[2]

[1]Department of Chemistry, Faculty of Mathematics and Natural Sciences, Mulawarman University, Samarinda, East Kalimantan, Indonesia
[2]Faculty of Medicine, Mulawarman University, Samarinda, East Kalimantan, Indonesia

*Corresponding author: daniel_trg08@yahoo.com

Abstract. The research on the synthesis of N-oleyl O-sulfate chitosan through sulfonation reaction on chitosan with ammonium sulfate and followed by amidation reaction using methyl oleate has been done. In this study, chitosan was chemically modified into N-oleyl O-sulfate chitosan as an edible film making material. N-oleyl O-sulfate chitosan was synthesized by reaction between methyl oleate and O-sulfate chitosan. Wherein the depleted chitosan of O-sulfate chitosan into O-sulfatechitosan was obtained by reaction of sulfonation between ammonium sulfate and chitosan aldimine. While chitosan aldimine was obtained through reaction between chitosan with acetaldehyde. The structure of N-oleyl O-sulfate chitosan was characterized by FT-IR analysis which showed vibration uptake of C-H sp^3 group, S=O group, and carbonyl group C=O of the ester. The resulting of N-oleyl O-sulfate chitosan yielded a percentage of 93.52%. Hydrophilic-Lipophilic Balance (HLB) test results gave a value of 6.68. In the toxicity test results of N-oleyl O-sulfate chitosan obtained LC_{50} value of 3738.4732 ppm. In WVTR (Water Vapor Transmission Rate) test results for chitosan film was 407.625 gram/m^2/24 hours and N-oleylO-sulfate chitosan film was 201.125 gram/m^2/24 hours.

1. Introduction

Edible film is a thin layer serving as a packaging or food coating that can be eaten together with packaged products [1]. Moreover, Sanford and Hutchings (1987)[2]statedthat it is not only to extend the shelf life, edible films can also be used as carriers of food components, including vitamins, minerals, antioxidants, antimicrobials, preservatives, and ingredients to improve the taste and color of packaged products. In addition, the materials used to make edible films are relatively inexpensive, easily biologically modified (biodegradable), and simple manufacturing technology. The example of the use of edible films is in wrapping candy, sausage, fruit, and dried soup [3].

Indonesia as a maritime country certainly leaves waste in the form of shrimp shells. The main element of shrimp shells is chitin, a natural polysaccharide that has many uses such as chelating, emulsifying, and adsorbent. The properties of chitin are non-toxic and easily degradable. It encourages the modification of chitin with the aim of optimizing the usefulness as well as expanding the field of chitin application. One of the derivatives of chitin is widely developed because of its wide application is chitosan. Chitosan is a polysaccharide resulting from de-acetylation of chitin. Due to its biodegradable and non-toxic properties, chitosan are recommended to be used in environmentally friendly industry [4,5].

As aforementioned, the research on the production of packaging materials or food coatings that is environmentally friendly called edible film was conducted. Thus, by utilizing the chitosan characteristics as emulsifier to be made into N-oleyl O-sulfate chitosan by reacting sulfate chitosan with methyl oleate to form nontoxic and easily degradable product[6]. The results were characterized by using Fourier Transform-Infra Red (FT-IR) spectroscopy, Water Vapor Transmission Rate (WVTR) test, Hydrophilic-Lipophilic Balance (HLB) Test, and toxicity test by Brine Shrimp Lethality Test) BSLT method[7].

2. Materials and Methods

2.1. Instruments

The instruments used in this research were three-neck flask, burette (50 mL ± 0.05 mL), analytical balance, rotary evaporator, chemical glass 250 mL, drop pipette, hot plate and magnetic stirrer, ball condenser, volume pipette, measuring flask 250 mL, glass funnel, separating funnel, dropping funnel, measuring glass 10 and 100 mL, laboratory spatula, pipe $CaCl_2$, distilled water (*aquadest*) bottle, magnetic stirrer, stainless steel cylinder, thermometer, glass plate, desiccator, box, vial bottle, and infrared spectrophotometer.

2.2. Materials

The materials used in this research were Chitosan, alcohol 96%, sodium bicarbonate (pa), acetic acid (pa), oleic acid (pa), acetaldehyde (pa), sodium hydroxide (pellet), potassium hydroxide (pellet), anhydrous sodium sulfate, phenolphthalein, hydrochloric acid (pa), methanol, distilled water (*aquadest*), filter paper, toluene, oxalic acid (pa), H_2SO_4 (pa), *n*-hexane, and shrimp larvae.

2.3. Procedures

2.3.1. Synthesis of Chitosan Aldimine. A total of 3.0057 grams of chitosan were dissolved in 150 mL of 1% acetic acid, inserted into a three-neck flask connected with a water-flushed condenser. After soluble, 2 mL of acetaldehyde was added and then refluxed at room temperature for 6 hours while dripping 3 mL of glacial acetic acid through a dropping funnel.

2.3.2. Synthesis of O-Sulfate Chitosan. The chitosan aldimine obtained from the previous stage was added with 0.1 M ammonium sulfate with a ratio of 2:5 ($W_{kitosan}/V_{ammonium\ sulfate}$) and then put in shaker for 4 hours [8]. The compound was filtered with Whatman 42 filter paper, the precipitate was washed with distilled water (aquadest) until neutral and dried for 4 hours in an oven at 60 °C.

2.3.3. Releasing Protective Group. A total of 3 grams of O-sulfate chitosan aldimine was added with sodium bicarbonate, 1:1, with toluene as solvent and sodium methoxide as catalyst at 60-70 °C for 3 hours. The result was washed with distilled water (aquadest) and filtered. Last, the precipitate was dried at 55 °C.

2.3.4. Synthesis of Methyl Oleate. A total of 28.2 grams of oleate acid (p.a) were inserted into a flat bottom flask. Then, 14 mL of methanol and 50 ml of toluene (p.a) were added and stirred. Then slowly dripped 2 mL of $H_2SO_{4(c)}$ while cooled at 0 °C and refluxed at 80 °C for 5 hours. Then, it was put in the rotary evaporation. The residue was extracted with 200 mL n-hexane (technical) and washed with distilled water (aquadest). The top layer was taken and anhydrous Na_2SO_4 was added and allowed to stand for 30 minutes and filtered. The result of filtrate was put in rotary evaporation with temperature of 40-50 °C.

2.3.5. Synthesis of N-Oleyl O-Sulfate Chitosan. A total of 1 gram of sulfate chitosan was put into a three-neck flask. Then, 3 mL of methyl oleate was added and stirred at 60-70 °C ± 3 hours. It was filtered. Then, the precipitate was dried at room temperature. The results obtained were analyzed by FT-IR spectrophotometer.

2.3.6. Synthesis of Edible Film. A total of 1 gram of N-oleyl O-sulfate chitosan was dissolved in 15 mL 1% acetic acid. It was stirred for 30 minutes at 40 °C. Then the mixture was poured into a glass plate whose sides are duct tape. The stainless steel cylinder was moved down to form a thin layer on the glass plate. It was dried at 25 °C for 24 hours.

2.3.7. Analysis of the Chemical Properties of N-Oleyl O-Sulfate Chitosan

2.3.7.1 Determination of Hydrophilic-Lipophilic Balance (HLB) Value. The analysis was done toward N-Oleyl O-Sulfate Chitosan. HLB was determined by using the formula as follow:

$$HLB = 20\left(1 - \frac{S}{A}\right) \quad (1)$$

Note: S=Saponification value
A=Acid number

2.3.7.2 DeterminationofWater Vapor Transmission Rate (WVTR) Value. The water vapor transmission rate of the film was measured by using gravimetric method, ASTM E96/E96M-10. The vapor absorbent material (silica gel) was placed in a vial bottle. Then the sample was placed in the mouth of the vial bottle covering the bottle. Later the end of the bottle was given glues so that no air enters.

The vial bottle was weighed with a precision of 0.0001 grams. It was placed in a closed container. It was given a small puddle of water. The bottle was weighed at the same hour (for 24 hours) and determined by the weight gain, the WVTR value is calculated by the formula:

$$WVTR = \frac{slope \times 100 \times 100\ m^2}{sample\ \ wide} \quad (2)$$
$$= g/\,m^2\ /24\ hours\ \ (RH25^o\ C)$$

2.3.7.3 Toxicity Test by BSLT Method. Ten shrimp larvae in 100 μL of seawater were inserted into the test vial bottle, then 100 μL of sample solution was added. For each concentration, three repetitions are performed. The control was done with 100 μL blank solution then added seawater up to 100 μL. The observation was done after 24 hours by counting the number of alive and dead shrimp larvae, and then calculated its molarity like the following equation:

$$\%\ Mortality = \frac{Number\ \ of\ dead}{Number\ \ of\ dead + Number\ \ of\ alive} \times 100\% \quad (3)$$

3. Results and Discussion

3.1. O-sulfate chitosan

Chitosan used in this research is chitosan for synthesis. The FT-IR spectra used to produce absorption peaks in the numbers: 3448.72 cm^{-1}, 2924.09 cm^{-1}, 2846.93 cm^{-1}, 2337.72 cm^{-1}, 1635.64 cm^{-1}, 1427.32 cm^{-1}, 1381.03 cm^{-1}, 1319.31 cm^{-1}, 1257.59 cm^{-1}, 1080.14 cm^{-1}, 894.97 cm^{-1} (Figure 1).

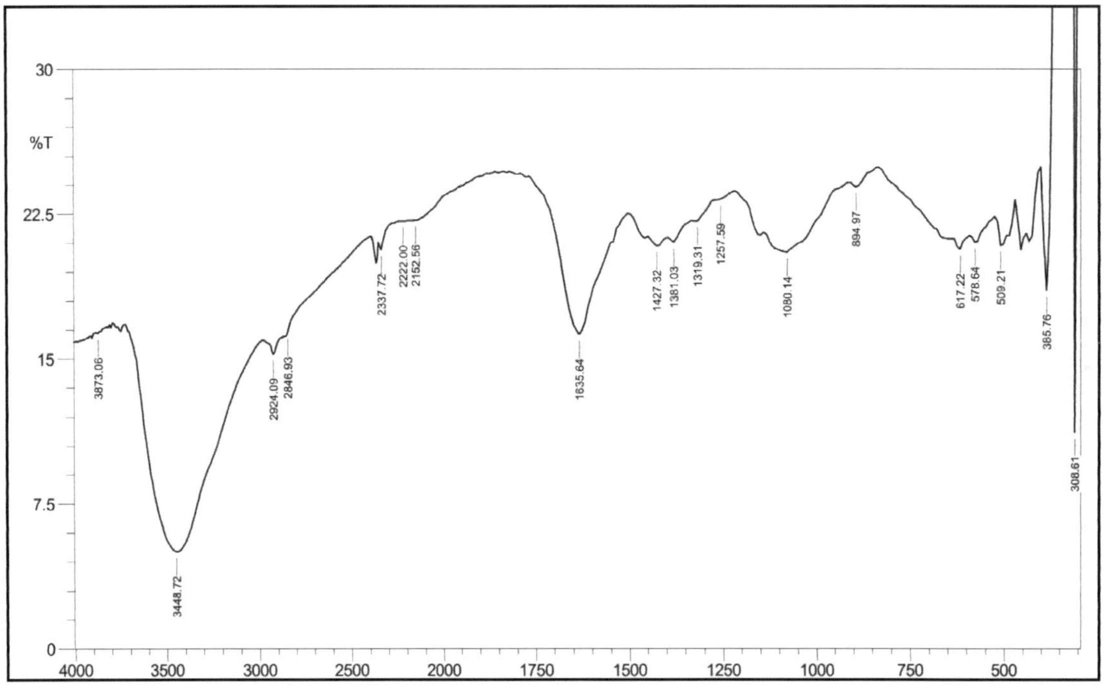

Figure 1. FT-IR spectra of chitosan

In this research, chitosan was first reacted with 1% acetic acid. The function of 1% acetic acid is as solvent for dissolving chitosan because chitosan is generally soluble in organic acids. One of them is acetic acid. A stirred situation was reacted with glacial acetic acid through a dropping funnel. The glacial acetic acid serves as a giver of acidic atmosphere in chitosan solution that was yellow. In a stirred state, an acetaldehyde solution was added to react with the chitosan solution.

In the preparation of chitosan sulfate, chitosan aldimine and ammonium sulfate served as a substitute that undergoes sulfonation reaction. $H\text{-}SO_3^+$ from ammonium sulfate substituted the CH_2O-H bonds in the O-sulfate chitosan to form the O-Sulfate chitosan aldimine with the ammonium (NH_3) and water (H_2O) as the byproducts. Ammonium and water were filtered using filter paper. It was washed with distilled water (*aquades*) to neutralize the chitosan aldimine that had been formed [9]. The conversion of chitosan aldimine to O-sulfate chitosan aldimine was essentially an electrostatic binding of sulfate ions in the chitosan reactive group (carbonyl group). This conversion was done to increase the adsorption capacity of chitosan due to the presence of sulfate ions, then the carbonyl group in chitosan will be more cationic. In addition, the presence of sulfate groups was expected to also bind the cationic groups. Sulfate ions were used because of their electrons. The reactivity of the O- sulfate chitosan is more stable than the chitosan because the carbonyl group has undergone permanent protonation with the attachment of sulfate ion. Then the deprotection against Aldine groups was done to bind to the N group on chitosan by using sodium bicarbonate. Sodium bicarbonate was added to O-sulfate chitosan aldimine to release the protective group with toluene as its medium and sodium methoxide as a catalyst.

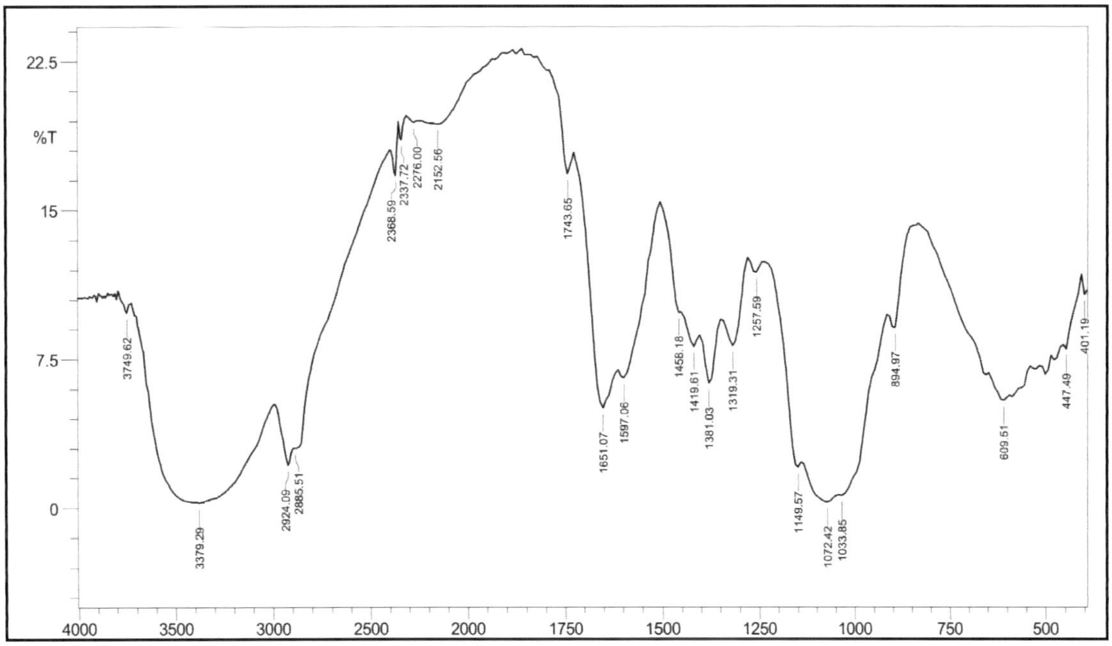

Figure 2. FT-IR spectra of O-sulfate chitosan

The FT-IR spectra of O-sulfate chitosan in figure 2 showed that the absorption peak in the wavenumber region 3379.29 cm^{-1} is the vibration of the O-H group. A stretching vibration of C-H sp^3 group was indicated on wavenumber 2924.09 cm^{-1} and 2885.51 cm^{-1}. It was supported by absorption peak at wavenumber 1419.61 cm^{-1} that is a bending vibration of C-H from CH_2 and spectrum 1319.31 cm^{-1} that is a typical absorption of C-H for CH_3.

The absorption peak at wavenumber 1651.07 cm^{-1} was a strain of the carbonyl group (C-O). It was also supported by the absorption peak at wavenumber 1072.42 cm^{-1} which was the vibration of the S = O group. It was the typical peak of O-sulfate chitosan [10].

The FT-IR spectra (Figure 2) showed the deprotection of protective group through the absorption peak at the wavenumber region 3379.29 cm^{-1} showing the overlapping vibration of O-H toward the N-H group of chitosan. The absorption peak in 2924.09 cm^{-1} and 2885.51 cm^{-1} showed the stretching vibration in C-H sp^3 and the absorption peak at the wavenumber 1597.06 cm^{-1} had weakened indicating deprecated imine.

3.2. *Methyl oleate*

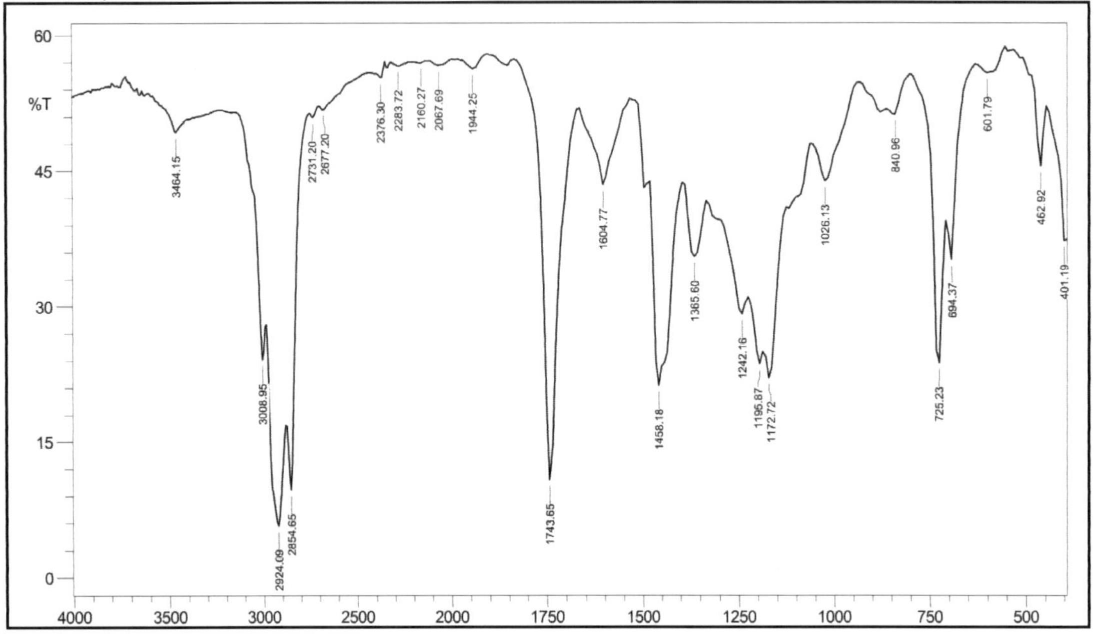

Figure 3. FT-IRspectraof methyl oleate

The FT-IR spectra in figure 3 showed the absorption peak at the wavenumber region 2924.09 cm^{-1} and 2854.65 cm^{-1}. According to Silvester, et al [10], it was the stretching vibration of C-H sp^3 supported by the absorption of bending vibration C-H sp^3 in wavenumber region 1458.18 cm^{-1}. The absorption peak at the wavenumber 1743.65 cm^{-1} was the absorption of the C=O carbonyl group of esters. It was formed and supported by the vibration of ester C-O-C in wavenumber region 1172.72 cm^{-1}. The spectra showed the absorption peak at wavenumber region 1604.77 cm^{-1} was the vibration of C=C alkene. Vibration absorption in wavenumber region 725.23 cm^{-1} was the rocking vibration of (CH_2) n at $(CH_2)_{14}$. In the FT-IR spectra of methyl oleate, the formed compound contained the C=O and C-O-C groups which were the characteristic of the ester. It did not contain the -OH group so that the compound was a methyl oleate compound.

3.3. *N-oleyl chitosan O-sulfate*

The FT-IR spectra in figure 4 showed the vibration peaks at wavenumber 2926.01 cm^{-1} and 2854.65 cm^{-1}. It was the stretching vibration of sp^3 supported by the absorption of bending vibration of C-H sp^3 at wavenumber 1450.47 cm^{-1}.The absorption peaks at wavenumber 1093.64 cm^{-1} and 1020.34 cm^{-1} were the vibration of the S=O group which was the typical peak of O-sulfate chitosan. The absorption peak at the wavenumber 1741.72 cm^{-1} was the absorption of the C=O carbonyl group of esters. It was formed and supported by the vibration of C-O-C ester in the wavenumber region 1170.79 cm^{-1}.The spectra showed the absorption peak at wavenumber region 1651.07 cm^{-1} was the vibration of C=C alkene. The absorption of stretching vibration at wavenumber region 3008.95 cm^{-1} supported by the bending vibration peak at wavenumber 725.23 cm^{-1} was the rocking vibration of $(CH_2)_n$ at $(CH_2)_{14}$. The FT-IR spectrum of N-oleyl chitosan O-sulfate contained a $(CH_2)_n$ group which was the characteristic of oleate. Thus, it can be concluded that the oleate in methyl oleate had entered and reacted with O-sulfate chitosan to form N-oleyl chitosan O-sulfate [10].

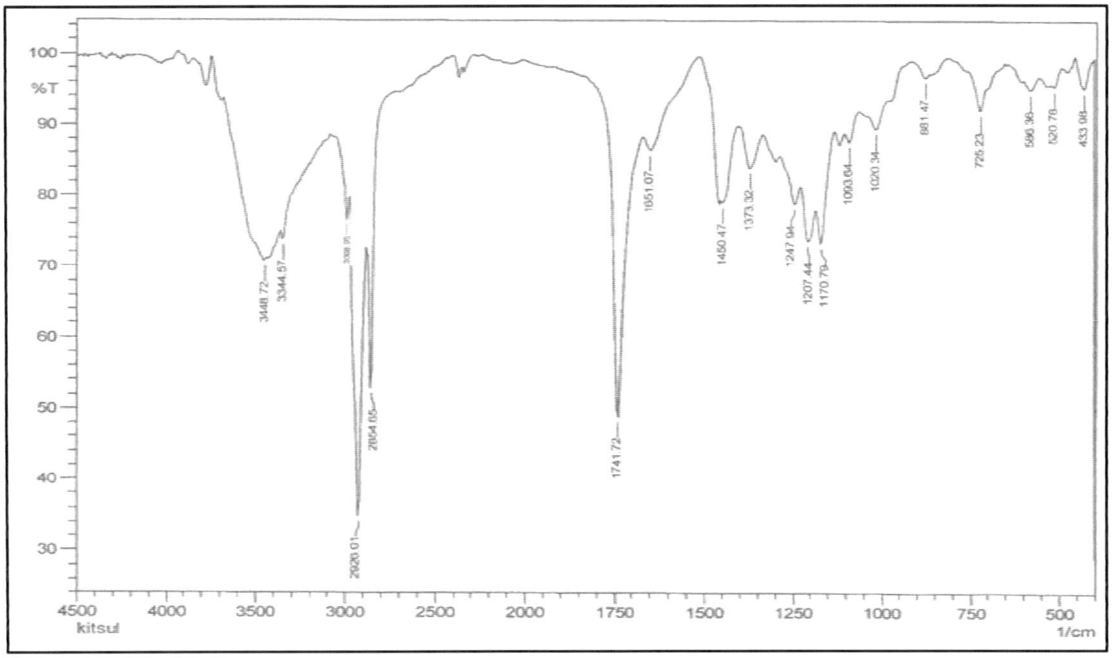

Figure 4. FT-IR spectra of N-oleyl chitosan O-sulfate

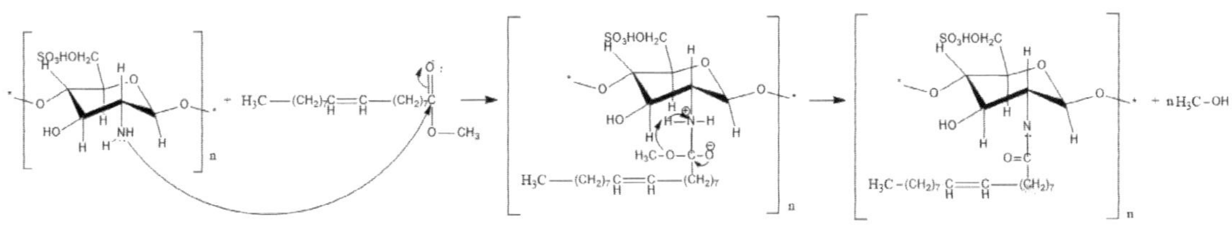

Figure 5. Mechanism reaction of N-oleylO-sulfate chitosan formation

3.4. *Edible film*

The produced edible film was transparent color, not rigid and elastic. Chitosan used as a base material gave transparent color to edible film. Then the oleate on the edible film provides elastic and non-rigid properties. The function of the use of oleate in the edible film was as plasticizer which was an organic material added to reduce the stiffness of the edible film. It had more flexible properties than the film without plasticizer.

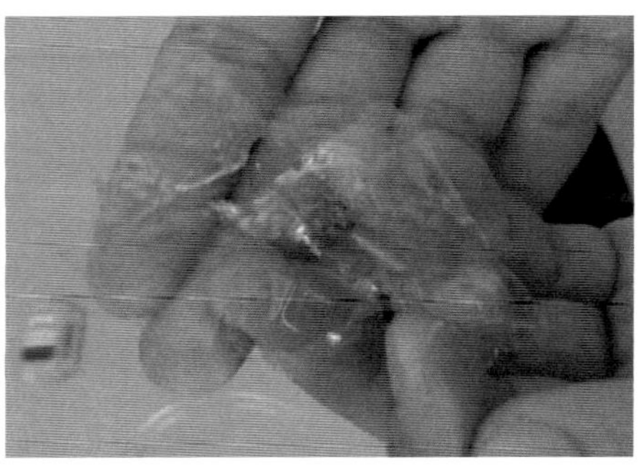

Figure 6. *Edible film*

3.5. Hydrophilic-lipophilic balance (HLB) and saponification values

Table 1. Determination of acid number

Sample	Sample Mass (Gram)	Volume of Titration (mL)	Average volume of Titration (mL)	Acid number
N-oleyl O-sulfate	0.03	0.75		
chitosan	0.03	0.75	0.75	14.02
	0.03	0.75		

Table 2. Determination of saponification value

Sample	Sample Mass(Gram)	Volume of Titration (mL)	Average volume of Titration (mL)	Saponification value
		20.75		
Blank		20.75	20.75	
		20.75		
N-oleyl O-sulfate chitosan	0.03	19.50	19.75	18.7

The HLB test gave the data of acid number and the saponification value from N-oleylO-sulfate chitosan. The acid number was 14.02 and the saponification number was 18.7. Thus, HLB value obtained from N-oleyl O-sulfate chitosan was 6.68.

The HLB value of N-oleylO-sulfate chitosan was 6.68. If it was viewed by a scale showing the surfactant function of the HLB value was at position 4-7 indicating that the compound can be used as a W/O emulsion.

3.6. Water vaportransmission rate(WVTR) test

Table 3. Determination of water vapor transmission rate

Sample of Film	Silica gel Mass + Vial bottle (gram)			Time used (hour)	Sample area (m^2)
	G1	G2	G		
Chitosan	11.7426	12.3948	0.6522	24	16.00
N-Oleyl O-Sulfate Chitosan	9.2521	9.5739	0.3218	24	16.00

The WVTR was used to determine the permeability value of a material toward water vapor. Water vapor permeability is a measure of a material as it can be passed by water vapor.

$$WVTR = \frac{24 \times 10}{A} \times \frac{W}{t} (gram/m^2/24\ hours)$$

Water vapor movement in a food product can occur when the difference of water activity (aw) happen. The difference of the humidity of the food product with its surrounding environment can be controlled by regulating the aw on the food, or by wrapping it with an edible film layer which had good barrier properties against the water vapor transmission and can avoid the loss of water vapor effectively. Water vapor transmission was strongly influenced by the temperature, thickness, type and the concentration of the plasticizer and the nature of the material to form edible film.

The data showed that the permeability of water vapor from the chitosan film was relatively high if it was compared to N-oleyl O-sulfate chitosan film. It was because of the hydrophilic properties possessed by the chitosan. The chitosan polymer of it had a large hydrogen bond so that the water vapor permeability value was high. However, the chemical polymer component of N-oleyl O-sulfate chitosan had non-polar lipid that had lower water vapor permeability that made the film can retain the water.

3.7. Toxicity test by BSLT (Brine Shrimp Lethality Test)

The observation of bioactivity potential was done based on lethal concentration 50% (LC_{50}) value. It was a value that showed the concentration of toxic substances that can cause the death of organisms up to 50%. Chemical compounds belonged to potentially active or toxic when having LC_{50} less than 1000 ppm, otherwise, it is not toxic if it has LC_{50} value more than 1000 ppm.

Based on the molarity test, by using SAS Probit Analysis, of shrimp larvae from N-oleyl O-sulfate chitosan, the LC_{50} value was 3738.4732 ppm. The value indicates that N-oleyl O-sulfate chitosan is non-toxic. It happened because the sulfate used in the synthesis of N-oleyl O-sulfate chitosan was not a concentrated sulfuric acid, but ammonium sulfate which tends to be lower level of toxicity, in addition of the ammonium sulfate used had a low concentration of 0.1 M.

4. Conclusions

Based on the results, it can be concluded that N-oleylO-sulfate chitosan can be synthesized by sulfonation reaction in chitosan and ammonium sulfate and followed by amidation reaction using methyl oleate.N-oleyl O-sulfate chitosan can reduce the rate of water vapor transmission in producing edible film through Water Vapor Transmission Rate (WVTR) analysis. In addition, N-oleyl O-sulfate chitosan can be used safely in producing edible film. It can be seen through the toxicity test by BSLT Method that shows that the value of LC_{50} was 3738.4732 ppm.

Acknowledgments

Thanks are extended to the Ministry of Research, Technology, and Higher Education Indonesia for funding this research. Authors would like also to thank to Head of Laboratory of Organic Chemistry, Faculty of Mathematics and Natural Sciences, Mulawarman University, for facility that had been given to authors.

References

[1] Joseph C S, Prashanth K V H, Rastogi N H *et al* 2011Optimum Blend of Chitosan and Poly-(Ɛ-caprolactone) for Fabrication of Films for Food Packaging Application *Journal of Food Bioprocces Technology* **4**(7) pp 1179-85

[2] Sanford P A and Hutchings G P 1987 *Chitosan-A Natural, Cationic Biopolymer: Comercial Applicationsin Industrial Polisaccharides: GeneticEngineering, Structure/Property and Applications*, ed Yalpani M (Amsterdam: ElsevierScience Publisher) p 363

[3] Susanto T and Sucipta N 1994 *Teknologi Pengemasan Bahan Makanan*(Blitar: CV Family)

[4] Purnamasari A P 2013 *Modifikasi Kitosan Sulfat Sebagai Katalis Heterogen Pada Produksi Biodiesel dari Minyak Kelapa Sawit*(Departement of Chemistry, Faculty of Science and Technology, Airlangga University, Surabaya) Thesis

[5] Supeni G and Irawan S 2012 *Pengaruh Penggunaan Kitosan Terhadap Sifat Barrier Edible Film Tapioka Termodifikasi* (Jakarta: Balai Besar Kimia dan Kemasan)

[6] Vieville C, Mouloungui Z and Gaset A 1993 Etherification of Oleic Acid by Methanol Catalyzed by P-Toluenesulfonic Acid and The Cation-exchange Resin K2411 and K1481 I Supercritical Carbondioxide *Industrial Engineering Chemical Research* **32** pp 2065-68

[7] Saputro A N C and Mahardiani L 2011 Sintesis Senyawa Turunan Kitosan, Chitosan Modified Carboxymethyl (Cs-Mcm) dan Aplikasinya Sebagai Agen Perbaikan Mutu Kertas Daur Ulang *Jurnal EKOSAINS* **III**(1) pp 47-53

[8] Mahatmanti FW 2001 *Studi Adsorpsi Ion Logam Seng (II) dan Timbal (II) pada Kitosan dan Kitosan-Sulfat Dari Cangkang Udang Windu (Peanus monodon)* (UGM, Yogyakarta)Thesis

[9] Usman M R M, Jain B U, Pawar S R, Shaikh T Y, Potda M B, Pati G H 2013 Many Potential Appliction of Chitosan as a Pharmaceutical Excipient, Int, *J.A.P.S BMS*2(1) pp 16-20
[10] Silverstein R M, Bassler G C and Morril T C 1991 *Spectrometric Identification of Organic Compounds 4th Ed* (New York: John Willey & Son Inc)

Fabrication of dye-sensitized solar cell using chlorophylls pigment from sargassum

M A Ridwan*[1], E Noor[2], M S Rusli[2], Akhiruddin[2]

[1] Departement of Agroindustrial Technology, Pangkep State Polytechnic of Agriculture Kabupaten Pangkep 90761 South Sulawesi, Indonesia
[2] Departement of Agroindustrial Technology, Faculty of Agricultural Technology, Bogor Agriculture University, Bogor 16680 West Java, Indonesia

*Corresponding Author: erlizanoor@yahoo.com

Abstract Dye-sensitized solar cell (DSSC) is a new generation of the solar cell. Its development in the dye-sensitized system is varied. Natural dyes have been the choice in developing DSSC. This study used a dye-sensitized chlorophyll pigment from Sargassum sp. as a dye-sensitized solar cell. This study aims to obtain chlorophyll pigment extract to be used as a dye in DSSC and to obtain the best energy conversion efficiency from DSSC. The chlorophyll pigments were extracted using APHA method (2012), and the TiO_2 coating method used was doctor blade method. The two fabricated cells have an area of 1 cm^2 immersed with chlorophyll dye for 30 hours. Then these cells were tested using direct sun radiation. The concentration value of chlorophyll in acetone solution was 61.176 mg/L. The efficiency value obtained was 1.50% with VOC of 241 mV, ISC 2.9 x 10^{-4} mA and fill factor 0.432.

1. Introduction

The organic solar cell is an alternative device to convert solar energy into environment-friendly electric energy. Commercial solar cells are generally made of inorganic materials such as silicon (Si) and germanium (Ge). Types of silicone-based and germanium-based solar cells have an energy efficiency of about 14%-17% with cell working time of 25 years, but the high cost of raw materials and expensive production cost make the selling prices of solar cells in the market are relatively high [1].

Therefore, researchers start to make a new breakthrough by using organic materials. The use of raw materials and easy developing technique are believed to result in the inexpensive solar panel. DSSC is a dye-sensitized solar cell developed as an alternative concept to the conventional solar cell. Dye-sensitized solar cell (DSSC) is a photoelectrochemical solar cell primarily consisting of photoelectrode, electrolyte, and opposite electrode. Titanium dioxide (TiO_2) has been a favorite semiconductor in various studies. Unfortunately, due to great band gap (3 to 3.2 eV), TiO_2 absorb only the ultraviolet part of sunlight making it having slow conversion efficiency [17]. Many studies were conducted to study the potential of DSSC as solar cell recently [2]. Dye extract or pigment of plants used as photosynthesizer include the chlorophyll extract [2] or anthocyanin [3].

Chlorophyll is the main pigment effective as photosynthesizer in photosynthesis process of plants, with maximum absorption at 670 nm, so that chlorophyll is an interesting component as the visible part of photosynthesizer [2]. There are some types of chlorophyll found as photosynthesis product, but the commonly found types in higher plants are chlorophyll a and b [4].

Chlorophyll has chemical formula as shown in Figure 1. Both types of the chlorophyll, chlorophyll a and b, have light absorption at two different wavelengths, namely at 400 nm-490 nm and 620 nm-680 nm.

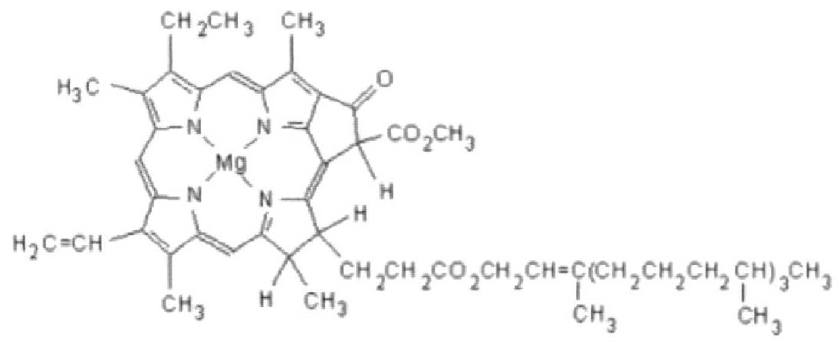

Figure 1. Molecular formula of chlorophylls [5]

This paper discusses the relationship between chlorophyll pigment extract content and absorption spectrum and sensitizer properties for DSSC dye.

2. Method

2.1 Chlorophyll Pigment Extract from Sargassum

Pigment extraction referred to APHA method [10]. One hundred grams of Sargassum powder with a size of 20 mesh was extracted using maceration method in technical acetone solution: distilled water (9:1)(v/v) for 24 hours. The ratio of alga and acetone solvent required in the extraction was 1:4 (b/v). Pigment-acetone solution was filtered, and the pigment filtrate was concentrated with rotary vacuum evaporator (Yamato) at a temperature of ±50 ^{o} C until free of solvent.

Absorption wavelengths were characterized by using UV-Vis Hewlett Packard spectrophotometer (200-800 nm). 3 mL of each chlorophyll solution was used to analyze the color absorption wavelength with a spectrophotometer. The extracted pigment concentration was calculated by measuring the extract absorbance at a specific wavelength. The following is the formula to calculate chlorophyll concentration according to Wintermans and De Mots[6].

2.2 Dye-Sensitized Solar Cell Prototype and TiO_2 Plate Voltage Current Value

Photoexcitation in the dye-sensitized solar cell does not occur in semiconductor electrode but occurs at light-absorbing dye or on the interface between dye-sensitized semiconductor and electrolyte. Injection of electrons from dye to TiO_2requires more reductive dye excitation than the conduction band of TiO_2. Electrons in the conduction band of semiconductor were collected in counter electrode and flow through the external circuit. Electrons then flow through the electrode to opposite electrode. The oxidized dye will require electron transfer from the electrolyte to bring it back to ground state. Redox reaction in the electrolyte was triggered by the presence of electrons coming into the electrolyte solution. Therefore, this cell is also called photoelectrochemical cell [7].

The thin layer was developed according to modified ICE method [16], with colloid development and coating on TCO. The colloid was developed by mixing 0.5 g and 1 mL of acetic acid with pH 3-4 and Triton X-100. The mixture was then scraped slowly until the colloid was formed. Triton X-100 functions as a surfactant to improve the adhesion in the substrate. Next, colloid was coated onto TCO which had been pre-washed with soap and immersed in ethanol. This was achieved by covering the

TCO sides with scotch tape leaving a blank cross-section (1x1 cm2). After that, colloid was applied and distributed evenly on TCO glass with the glass bar. Then, the layer in TCO was air-dried for 2 minutes and scotch tape was removed gently. Layer on TCO was heated with hotplate until the layer adhered perfectly and then put into a container prefilled with silica gel.

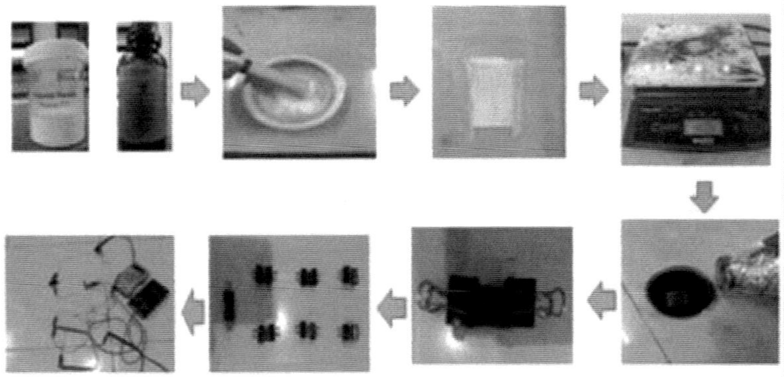

Figure 2. Prototype DSS Solar Cell in Chlorophyll Pigment

Figure 2 shows the solar cell that has been immersed in chlorophyll pigment solution. The light brown cross-section measuring 1x1 cm2 is the thin layer of TiO_2 that have absorbed chlorophyll pigment. The left and right side of the solar cell are clipped with Binder Clips No. 107 to avoid the movement of the glass. The solar cell was then tested for current and voltage. Characterization of current and voltage was conducted by testing the output current and voltage from the sample by utilizing the sunlight energy at intensity of 0.085 mW/cm^2. The curve obtained from the measurement results indicated a slope. This was in accordance with a study by Chou et al [9].

2.3 Experimental Setup

2.3.1 Immersion of TiO_2 layer in dye-sensitizer Solutions. Immersion of TiO_2 layer was achieved by varying the time: 5 hours, 15 hours, 30 hours, and 45 hours. TiO_2 layer on TCO glass was immersed in chlorophyll nano pigment solution in dark room for the pigment solution to completely absorbed. After immersion, the layered conductive glass was dripped and drained, and then the current and voltage of the solar cell kit were characterized.

2.3.2 Characterization of current-voltage. The light source used in this study was sunlight at 12.00-13.00 p.m. The current and voltage were measured using voltmeter and amperemeter which were combined parallel.The potentiometer was used as a dummy load to determine the current value variatively. The measurement was performed by rotating the potentiometer so that the resistance value changed. This resistance value change will be readable in voltmeter and amperemeter. The values were documented when a value change occurred on amperemeter. The values were then plotted in the graphic form. The observed parameters included open circuit voltage (Voc), namely the voltage when the applied external load is high. Voc was obtained when there is a value change in resistance by a potentiometer, the readable current was zero. Whereas, short circuit current (Isc) was obtained when the readable voltage value was zero. Maximum voltage value (Vmax) and maximum current (Imax) were obtained by observing the highest values from multiplication between voltage and current of each data [9].

3. Result and discussion

3.1 UV-Vis absorption spectra

Absorption is a quantity that expresses the ability of a material to absorb light. Organic compounds are able to absorb light due to their valence electrons that can be excited to higher energy level. One of the organic compounds is chlorophyll from *Sargassum Sp.*

The important role in the dye-sensitized solar cell is the ability of the dye to absorb sunlight and convert it to electric energy. Generally, the dye used is ruthenium complex. But in its development, organic dye begins to be developed because it is cheaper than ruthenium complex, having high absorption coefficient due to $\delta - \delta$intramolecular transition, and no worry about the limited resources, because it doesn't contain noble metal like ruthenium [15].

Test against chlorophyll extract dye using UV-Vis spectrophotometer to determine the absorbance of chlorophyll pigment. The principle of spectrophotometry is the light absorption by molecules. Molecules can absorb radiation from UV-Vis region because it contains electron [14]. The results of UV-Vis absorption of *Sargassum Sp* chlorophyll are shown in Figure 3.

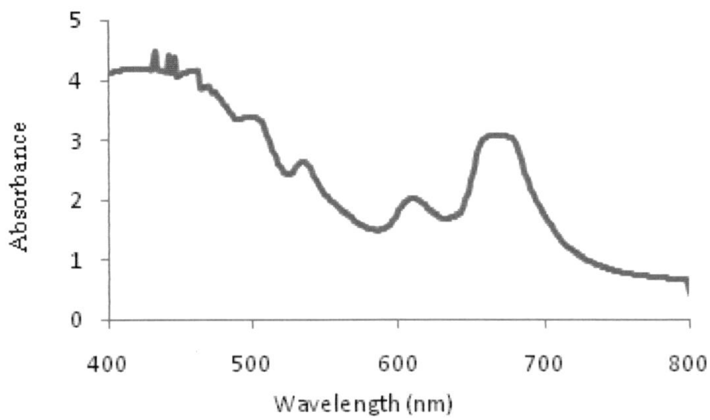

Figure 3. Absorption spectra of chlorophylls in solution

According to Figure 3, chlorophyll concentration in the sample in acetone solution is 61.176 mg/L, as calculated by the formula of Wintermans and de Mots [8].

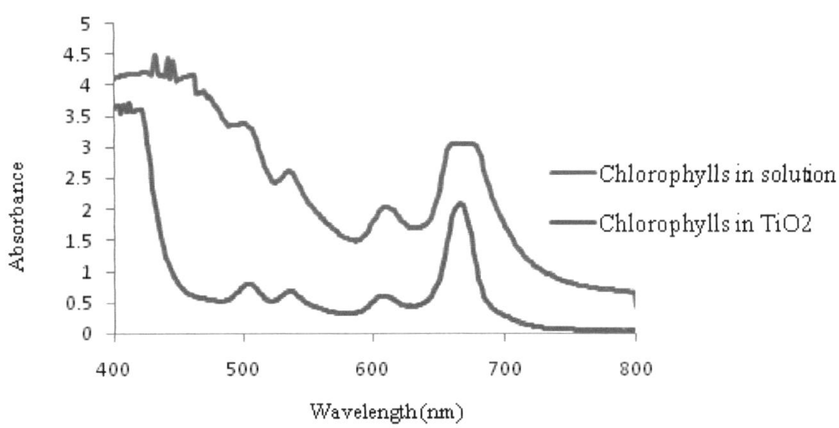

Figure 4. Absorption spectra of chlorophylls in solution and TiO_2

From figure 4 we can see the absorbance value in chlorophyll that had been applied in TiO_2 decreases. This is due to not all chlorophyll in chlorophyll pigment extract solution adhere to TiO_2. In the application of solar cell, this DSSC uses TiO_2 with nanometer size. TiO_2 in nanometer size has an advantage namely can hold more dye because there are many cavities its layers. Bigger sizes have only a few cavities so they can hold only few dye and the efficiency is smaller because produce few electron-hole pair. DSSC performance depends more on dye sensitizer. Dye absorption spectrum and adherence to semiconductor TiO_2 surface is an important factor in determining the efficiency of DSSC [11].

3.2 Photovoltaic performance of DSSC

Photovoltaic test for DSSC using chlorophyll pigment extract as sensitizer was performed by measuring the characteristic I-V curves under radiation with sunlight at 1.00 – 12.00. For example, characteristic I-V curves of DSC sensitized with black rice extract is shown in Figure 5. According to figure 5, the short-circuit current (I_{SC}) value, open-circuit voltage (V_{OC}) of DSSC can be obtained directly as shown in Table 1, respectively. Maximum power (P_{max}) of DSSC was confirmed based on the formula: P_{max} = (I x V). The FF value (fill factor) was calculated based on the formula FF = $(I \times V)_{max}/(I_{SC} \times V_{OC})$. Photoelectric parameters of DSSC sensitized with chlorophyll pigment extract from seaweed of sargassum type was measured and calculated, the results are shown in Table 1.

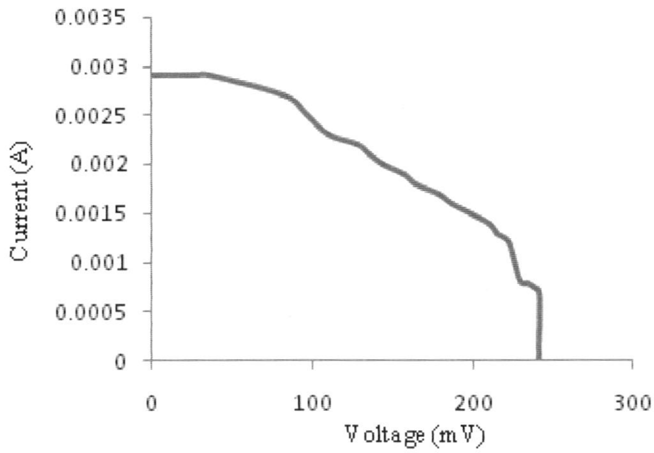

Figure 5. I–V characteristics of DSCs sensitized

From the characteristic I-V curves in Figure 5, we can calculate several parameter values in solar cell including the open voltage (Voc), short-current density, maximum voltage (Vmax), maximum current density (Imax), maximum power (Pmax), fill factor (FF) and efficiency (η) that can be seen from Table 1.

Table 1. Photoelectrochemical parameters of DSCs sensitized with chlorophylls at the working area of 1.0 cm^2

Dye	Parameter						
	Voc (mV)	Isc (mA)	Vmax (mV)	Imax (mA)	P max (mW)	FF	η (%)
Chlorophylls	241	0.0029	178	0.0017	0.3003	0.432	1.50

Table 1 shows that fill factor produced is 0.43 with the efficiency of 1.5%. The obtained results were higher compared to previous studies (0.4%) with a fill factor of 0.336). It is worth to mention that this value was higher than those reported by Wang et. Al.[12].

Experimental data obtained with photoelectrochemical studies were reported in Figure 7, where shown the I-V curves marked with the following parameters: V_{oc} 0,36, J_{sc} 0,8 $mA/cm^2$and efficiency of 0.178%. The more important point for sargassum-based DSSC is the very good efficiency (1.50%). It is worth to mention that this value was higher than reported by Wang et. al. [13], obtained by using pure sargassum as dye sensitizer. Moreover, by considering the increased photoelectrochemical obtained by these researchers with the increased dye concentration [13], the increased performance (according to current efficiency) was also expected for DSSC which was based on higher dye content of sargassum. Finally, the comparison of the obtained results with the results in the literature showed that for green alga [13]

4. Conclusions

This study was conducted to extract chlorophyll pigment from sargassum, by using simple purification method that can avoid complex procedures used by Wang et. Al. However, the use of chlorophyll pigment as a dye for DSSC cells can be an interesting alternative solution for cheap cell production. In addition, the use of chlorophyll extracted from these species will have an advantage in ecological sustainability.

The results of chlorophyll extract de in acetone solution by using UV-Vis spectrophotometer at a wavelength of 200 to 668 nm and obtained chlorophyll concentration of 61.176 mg/L. Whereas the measurement of chlorophyll absorbance in TiO2 plat indicated smaller absorption properties compared to absorption properties in chlorophyll extract solution. This was due to the fact that not all chlorophyll in dye solution adheres to TiO2 plate.

Chlorophyll extract obtained from *Sargassum sp* was used as a sensitizer in DSSC and produced higher fill factor and better current density when compared to data available in the literature. The DSSC made with sargassum sp extract indicates the efficiency of about 1.5%, respectively for J_{sc}, V_{oc}, fill factor. Increased cell efficiency based on sargassum extract was caused by the optimum procedure used for DSSC assembly ad better sensitization properties from chlorophyll compared to other chlorophyll types.

Acknowledgments

The authors would like to express gratitude to the Ministry of Research, Technology and Higher Education of the Republic of Indonesia for the financial support of this work under the research grant National Strategic Research (STRANAS) for the academic year 2015-2016.

References

[1] M. Fahlman, W.R. Salaneck, Surface, and interfaces in polymer-based electronic, surface science, 500 (2002), 904-922

[2] Y. Amao, Y.Yamada, K. Aoki, Journal of Photochemistry and Photobiology A: Chemistry 164(2004), 47-51.

[3] K. Wongcharee, V Meeyoo, S Chavadej, Solar Energy Materials and Solar Cells 91 (2007), 566-571

[4] R. Sastrawan, Photovoltaic Modules of Dye Solar Cells, Dissertation University of Freiburg (2006).

[5] Nurhayati dan Veinardi S. 2011. Isolation of chlorophyll a from spinach leaves and modification center ion with a Zn^{2+} study on its optical stability. *Journal of Mathematics and Science*, 16(2): 65-31.

[6] Wintermans JFGM, De Mots A. (1965). Spectrophotometric Characteristics of Chlorophylls A and B and Their Phenophytins in Ethanol. *Biochimica et Biophysica Acta (BBA) Biophysics including Photosynthesis*, *109*(2), 448-453.

[7] Wei D, Amaratunga G. 2007. Photoelectrochemical cell and its application in optoelectronics. *International Journals of Electrochemical Science* 2: 897-912.

[8] Wintermans JFGM, De Mots A. (1965). Spectrophotometric Characteristics of Chlorophylls A and B and Their Phenophytins in Ethanol. *Biochimica et Biophysica Acta (BBA)-Biophysics including Photosynthesis*, *109*(2), 448-453.

[9] Chou TP, Zhang Q, Cao G. 2007. Effects of dye loading conditions on the energy conversion efficiency of ZnO and TiO2 dye-sensitized solar cells. *The Journal of Physical Chemistry C* 111.50: 18804-18811.

[10] [APHA] American Public Health Association. 2012. *Standard Methods for the Examinator of Water and Wastewater 22nd Edition*. Washington DC: APHA Publisher.

[11] Khwanchit W, Meeyoo V, Chavadej S. 2006. Dye-sensitized solar cell using natural dyes extracted from Rosella and blue pea flowers. *Sol Energy Mater Sol Cell.* 91: 566 –571.

[12] Wang XF, Kitao O, Hosono E, Zhou H, Sasaki SI, Tamiaki H. 2010. TiO2-and ZnO-Based Solar Cells Using A Chlorophyll A Derivative Sensitizer for Light-Harvesting and Energy Conversion. *Journal of Photochemistry and Photobiology A: Chemistry*, *210*(2), 145-152.

[13] Desai U.V, Xu Chengkun, Wu Jiamin, dan Gao Di, 2012, Solid-state-Dye Sensitized Solar Cell Based on Ordered ZnO nanowire arrays, *Nanotechnology*, (23).205-401

[14] Hardjono S. 1991. Spektroskopi, Yogyakarta: Liberty.

[15] Kathiravan A, Chandramohan M, Renganathan R, Sekar S. 2009. PhotoinducedElectron Transfer from Phycoerythrin to Colloidal Metal SemiconductorNanoparticles. *SpectrochimicaActa Part A: Molecular and biomolecular spectroscopy*, *72*(3), 496-501.

[16] ICE (Institute for Chemical Education). 2008. Nanocrystalline solar cell kit recreating photosynthesis. Madison : University of Wisconsin-Madison.

[17] Gratzel M. 2005. Solar energy conversion by dye-sensitzed photovoltaic cells. *Inorganic Chemistry 2005;4*. 4;6 841 – 51.

Trial production of fuel pellet from *Acacia mangium* bark waste biomass

R Amirta[1]*, T Anwar[1], Sudrajat[2], Yuliansyah[1], W Suwinarti[1]

[1]Faculty of Forestry, Mulawarman University. Jl. Ki Hajar Dewantara, Samarinda,East Kalimantan, Indonesia

[2]Faculty of Mathematics and Natural Sciences, Mulawarman University. Jl. Barong Tongkok, Samarinda,East Kalimantan, Indonesia.

E-mail: r_amirta@yahoo.com /rudiantoamirta@fahutan.unmul.ac.id

Abstract. Fuel pellet is one of the innovation products that can be produced from various sources of biomass such as agricultural residues, forestry and also wood industries including wood bark. Herein this paper, the potential fuel pellet production using *Acacia mangium* bark that abundant wasted from chip mill industry was studied. Fuel pellet was produced using a modified animal feed pellet press machine equipped with rotating roller-cylinders. The international standards quality of fuel pellet such as ONORM (Austria), SS (Sweden), DIN (Germany), EN (European) and ITEBE (Italy) were used to evaluate the optimum composition of feedstock and additive used. Theresults showed the quality offuel pellet produced were good compared to commercial sawdust pellet. Mixed of Acacia bark (dust) with 10% of tapioca and 20% of glycerol (w/w) was increased the stable form of pellet and the highest heating value to reached 4,383 Kcal/kg (calorific value). Blending of Acacia bark with tapioca and glycerol was positively improved its physical, chemical and combustion properties to met the international standards requirement for export market. Based on this finding, production of fuel pellet from Acacia bark waste biomass was promising to be developed as an alternative substitution of fossil energy in the future.

1. Introduction

Indonesia and other tropical countries such as Brazil and Zaire have a great potential for creating energy from biomass as a substitute for petroleum. Taking into account the large size of Indonesia's forests and palm oil plantations, coupled with the possibility of using timber and agricultural residues, Indonesia has the ideal conditions for this change. Wood chips, oil palm and other plant biomass provide low cost residues that are renewable and underused, environmentally friendly and potentially capable of generating heat, steam and electric power [1,2,3]. Thus, they may constitute an alternative fuel for producing energy. In term of that, one of the potential sources of energy feedstock is *Acacia mangium* plantation and also its chip mill industry that produced the huge amounts of residues.

A. mangium is one of the major fast-growing species used in plantation forestry programs throughout Asia and the Pacific countries, including Indonesia. Due to its rapid growth and tolerance of very poor soils,*A. mangium* is playing an increasingly important role in efforts to sustain commercial supply of tree products while reducing pressure on natural forest ecosystems. As a Clean Development Mechanism (CDM) for climate change mitigation, *A. mangium* plantations are expected to provide efficient and profitable contributions to the sawlog and pulp markets [4]. The increase of

Acacia plantation in Indonesia is mainly driven by growing of global demand for pulp and paper. With the growth of pulp and paper production in Indonesia, the amounts of residues generated also shows a corresponding increase, mainly in formation of bark residues and sawdust.

Bark residue was produced from debarking process at harvesting areas and also chip mill industry in various types of material such as wood chunk, small log end and also sawdust (Figure 1a). The abundant of Acacia bark waste biomass was stimulated enviromental problems such as contamination of groundwater sources through leaching of extractive materials with high tannin content, trigering and causing land fires around industrial sites and also requires a large buildup location and consumes land preparation around the industry (Figure 1b).

Due to those environmental problems and also opportunity to utilize and convert the waste material into the valuable renewable energy, herein this paper potential fuel pellet production using *A. mangium* bark that abundant wasted from chip mill industry was studied with the special emphasizes to point out the quality of fuel pellet produced, to find an optimum composition among wood bark waste, tapioca and glycerol used and to know the highest energy properties, including calorific value potentially generated. Comparison between the quality of fuel pellet produced with the global standard of wood pellet quality was also discussed. Fuel pellet has been choosen due to some advantegous factors that belongs of this product such as pelletization reduces moisture content, increases energy content (MJ/kg), enhances combustion efficiency, and produces greater homogeneity of composition as compared to raw biomass [5,6,7].

2. Materials and Methods

2.1. Acacia wood bark waste biomass

Waste of *A. mangium*wood bark biomass was collected from chip wood industry of PT. Sarana Bina Semesta Alam (SBSA), located at district Muara Kaman, Kutai Kertanegara, East Kalimantan, Indonesia. Before used, Acacia bark waste biomass was chipped, converted into the dust form and air-dried until approximately 12% of moisture content.

2.2. Lab-scale trial production of fuel pellet

A. mangium bark waste fuel pellets of 8mm in diameter were prepared in a lab scale process with a modified animal feed pellet press machine equipped with rotating roller-cylinders (Figure 2). The production trials of Acacia bark waste fuel pellets were carried out by using 400g of Acacia bark and blending with 5-25% (w/w) tapioca in the presence and absence of various amounts of glycerol (5-20%, w/w). At the beginning, Acacia bark waste used as the main raw material in the absence of glycerol, however when the calorific value of the pellet was not good enough and satisfy, glycerol will be gradually added to mixture at the minimum percentage until quality of pellet increased. The temperature applied for the process was about 30-35°C, without any specific pressure used.

2.3. Standard quality of fuel pellet (quality analysis)

The international standards of pellet such as ONORM (Austria), SS (Sweden), DIN (Germany), EN (European) and ITEBE (Italy) were used to evaluate the quality of fuel pellet produced from *A. mangium* bark waste biomass.

3. Results and Discussion

3.1. Physical properties of fuel pellet

The production trials were carried out using mix of Acacia bark waste biomass (dust), and tapioca in the absence and presence of biodiesel by product, glycerol as additive. The aims of the trial productions are to know the optimum condition that can be applied to produce fuel pellet from waste of Acacia bark. The properties of raw materials used were determined as shown in Table 1.

Table 1. Chemical properties of *A. mangium* bark waste biomass and tapioca

Chemical Properties	Bark Waste Biomass	Tapioca
Moisture content (%)	12.71	12.65
Ash content (%)	5.58	0.10
Volatile matter (%)	84.58	87.33
Fixed carbon (%)	3.39	0.04
Lignin content (%)	35.91	-
Cellulose content (%)	35.73	-

Figure1.(a) Abundant Acacia wood bark and sawdust waste biomass produced daily in chip mill industry; (b) contamination of groundwater sources and vulnerable fire around industrial sites

Herein this study, we found the low technology developed from a modified common animal feed pellet press machine equipped with rotating roller-cylinders was effective to be used to made fuel pellet from Acacia bark waste biomass, even other additive materials such as tapioca and glycerol required to be blended to form the highest quality of fuel pellet (Figure 2.). Additives play a major role in wood pellet characteristics and are a subject of major interest as they act as binding agents for the biomass raw material[8].We found in the less tapioca used, they were easily broken into the small pieces form during pelletizing or producing a low physical quality of fuel pellet. This fact was simillar with the situation that reported on empty fruit bunches used as feedstock for the pellet production [9,10]. Tapioca was used to bind the small surface of bark dust material. This natural binder was applied to increase physical properties of feedstock such as low density and less surface area that hinder the smooth flow ability of biomass and also to substitute natural function of lignin as binder. Lignin as natural binder on briquette and pellet production was common used and reported previously [11].

Furthermore, we also observed that in the presence of suitable amounts of natural binder tapioca, the density of fuel pellets were good enough to reached 710-770 kg/cm3 to meet the Sweden (SS-187120) and Austria (ONORM) standars quality of pellet products (Table 2) (Figure 3).

Figure2. Modified animal feed pellet press machine equipped with rotating roller-cylinders used on trial production of fuel pellet from *A. mangium* bark waste biomass

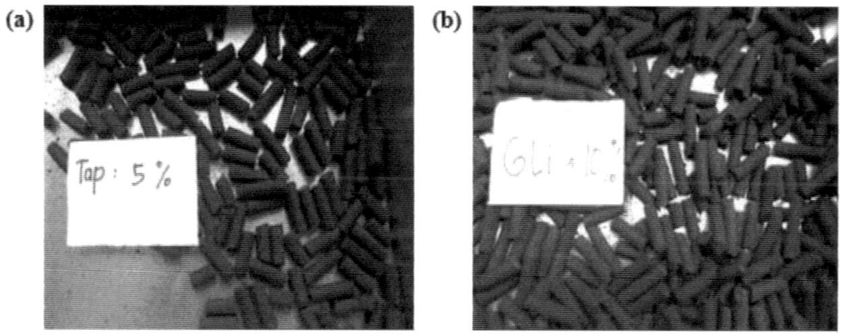

Figure 3. Long size and stable formation of Acacia bark waste fuel pellet in the presence of tapioca and glycerol as binder and additive; (a) 5% tapioca and (b) in the presence of 10% glycerol

Table 2. Physical properties of *A. mangium* bark waste biomass fuel pellet in various amounts of tapioca and glycerol

Compositions	**Physical Properties of Acacia Bark Fuel Pellet**			**Global Standard Quality (kg/m^3)**	
	Diameter (mm)	**Length (mm)**	**Density (kg/m^3)**	**SS-187120**	**ONORM**
Wood bark waste biomass (dust) + various amounts of tapioca in the absence of glycerol					
Tapioca 5%	7.47 ± 0.01	28.88 ± 0.10	770 ± 0.05	≥ 600	770 ± 0.05
Tapioca 10%	7.48 ± 0.03	29.39 ± 0.04	710 ± 0.00	≥ 600	710 ± 0.00
Tapioca 15%	7.40 ± 0.04	30.09 ± 0.07	740 ± 0.02	≥ 600	740 ± 0.02
Tapioca 20%	7.28 ± 0.00	30.31 ± 0.03	770 ± 0.03	≥ 600	770 ± 0.03
Tapioca 25%	7.21 ± 0.01	32.98 ± 0.19	760 ± 0.02	≥ 600	760 ± 0.02
Wood bark waste biomass (dust) + 10% tapioca in the presence of various amounts of glycerol					
Glycerol 5%	7.67 ± 0.02	2.93 ± 0.12	690 ± 30	≥ 600	690 ± 30
Glycerol 10%	7.66 ± 0.02	3.04 ± 0.01	660 ± 10	≥ 600	660 ± 10
Glycerol 15%	7.67 ± 0.06	3.12 ± 0.05	680 ± 10	≥ 600	680 ± 10
Glycerol 20%	7.70 ± 0.03	3.03 ± 0.26	730 ± 20	≥ 600	730 ± 20

Note: average and standard deviation from triplicate samples; fuel pellet was prepared from 500g mixture of bark waste feedstock and additives

3.2. Chemical properties and calorific value of fuel pellet

Chemical properties of fuel pellet from Acacia bark biomass was evaluated. Evaluation was done by assesment of moisture content (MC), ash content, volatile matter and fixed carbon of pellet in the presence and absence of tapioca and biodiesel side product, glycerol into the mixtures.

From this trial production we found the distribution of moisture content (MC) of the samples were 6.39% to 8.29% and its was good to meet with the global standard quality of pellet products. Li and Liu [12] reported that a good quality pellet has MC ranging between 6% and 12%. Further, other studies also found that wood pellets having MC ranging between 9% and 14% are most durable and resistant to abrasion [13,14].Water has a crucial role in the pelletizing process and, along withlignin content, the MC of the feed is one of the most importantparameters determining pellet durability and itsalso strong influence on heating value, combustion efficiency, and bulk density[15,16].

Table 3. Chemical properties of *A. mangium* bark waste biomass fuel pellet in various amounts of tapioca and glycerol

Compositions	Chemical Properties of Acacia Bark Fuel Pellet			
	Moisture Content (%)	**Ash Content (%)**	**Volatile Matter (%)**	**Fixed Carbon (%)**
Wood bark waste biomass (dust) + various amounts of tapioca in the absence of glycerol				
Tapioca 5%	7.52 ± 1.50	5.91 ± 0.27	88.51 ± 1.70	5.59 ± 1.96
Tapioca 10%	7.07 ± 1.75	5.06 ± 0.94	89.66 ± 1.48	5.29 ± 2.43
Tapioca 15%	7.13 ± 1.78	5.10 ± 0.37	89.46 ± 2.53	5.44 ± 2.90
Tapioca 20%	6.43 ± 1.59	4.63 ± 0.20	90.37 ± 1.77	5.01 ± 1.96
Tapioca 25%	6.39 ± 2.65	4.81 ± 0.15	89.23 ± 2.41	5.95 ± 2.26
Wood bark waste biomass (dust) + 10% tapioca in the presence of various amounts of glycerol				
Glycerol 5%	8.29 ± 2.53	5.63 ± 0.17	88.43 ± 2.47	5.94 ± 2.64
Glycerol 10%	7.79 ± 2.21	5.04 ± 0.27	88.84 ± 1.98	6.12 ± 2.25
Glycerol 15%	7.56 ± 3.31	4.45 ± 0.53	89.37 ± 3.19	6.18 ± 3.72
Glycerol 20%	7.42 ± 3.37	4.30 ± 0.81	89.50 ± 3.03	6.20 ± 3.84

Note: average and standard deviation from triplicate samples; fuel pellet was prepared from 500g mixture of bark waste feedstock and additives

The ash content, volatile matter dan fixed carbon of fuel pellet produced from Acacia bark were evaluated and the results werecompared to the global pellet standard.The results showed that the ash content of Acacia bark pellets is still quite high at 4.30% - 5.91% (>1%). The higher ash content of pellet was believed due to silica, the minor element commonly found in tropical climatic plant [16]. Silica is one of the ash compounds other than potassium, calcium and magnesium. Concerning the effect of bark on the quality of biofuel,Filbakk et al [17] reported that thepresence of bark in solid biofuels is related to increased tendency towards sintering duringcombustion and is therefore related to combustion problems. Concerning ash, [18]reported that increasing ash content results to lower heating value of the biofuel, impliesthe risk of sintering and negatively affects processing equipment.

The bark pellets produced has also lower carbon content ranging from 5.01% to 6.20%. The lower fixed carbon of pellets were believed due to high ash and volatile matter of bark material used.The results also showed that the fixed carbon values was positively increased when the number of glycerol added into the mixture increased.

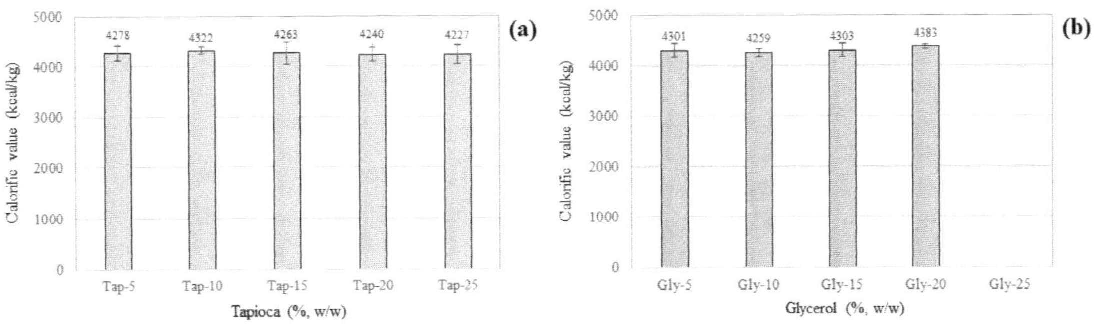

Figure 4. Calorific values of Acacia bark fuel pellet (a) in the presence various amounts of tapioca (Tap), and (b)10% of tapioca and various amounts of glycerol (Gly) as binder and additive

The calorific value of bark pellets were also evaluated to find out the highest energy potency from pelleting process. Among the data collected, we found the highest calorific value was obtained from the mixture of bark dust material and 10% tapioca in the presence of 20% of glycerol to reached 4,383 kcal/kg of heating value. On the other hand, in the absence of glycerol the heating values of bark fuel pellet were only 4,227 to 4,322kcal/kg. The heating value of pellet was increased trends linearly to correspond the increase number of glycerol added into the feedstock. In term of this, again we found that glycerol was clearly affect on the heating value of the pellet. This phenomenon was related with the properties of glycerol that known as a good source of energy and when it was used as an additive in fuel will be able to maintain the burning stability of the fuel, and increasing the heat of the fuel [10,19,20].

Table 4. General quality of fuel pellet produced from *A. mangium* waste bark biomass in the presence of tapioca and biodiesel side product, glycerol

Parameters	Status of quality	Global Standard Quality
Physical properties		
Fuel pellet density	Qualified	(SS 18 71 20)
Calorific Value	Qualified	(SS 18 71 20), (DIN 51731)
Chemical Properties		
Moisture content	Qualified	(ONORM M7135), (SS 18 71 20), (DIN 51731)
Ash content	Qualified	(ONORM M7135) (ITEBE)
Volatile matter	-	
Fixed carbon	-	
Total Sulphure	Qualified	(ONORM M7135),(SS 18 71 20), (DIN 51731)

Potential sulphur emission released on firing of fuel pellet from *A. mangium* bark waste biomass was also measured, especially those obtained in the presence of 10% tapiocaand 20% of glycerol into the mixture. The results showed that the sulphur emission level was 0.07%, which still meets the global standard of pellet products ($0.07 \leq 0.08\%$). The low sulphur levelof bark fuel pellets will positively reduce the emission rate in the air.Due to the low sulphur emission obtained, we also strong believed that bark fuel pellet will be more environmentally friendly compared to coal and other fossil fuel energy.

4. Conclusion

Finally, we concluded that *A. mangium* waste bark biomass was potentially used as alternative feedstock for the fuel pellet production. Successful manufacture of fuel pellets from Acacia bark waste biomass was achieved for all combinations studied. The optimum mixture compositions among bark waste biomass, tapioca and glycerol was found at 10% of tapioca used in the presence of 20% glycerol to reached the highest calorivic value 4,383 kcal/kg. Generally, the quality of fuel pellet produced from mix of bark waste biomass, tapioca and glycerol was satisfy to met the international standard quality of fuel pellet such as ONORM standard (Austria), SS (Sweden), DIN (Germany), EN (European) and ITEBE (Italy). Moreover, the modified common animal feed pellet press machine equipped with rotating roller-cylinders was effective used to produced fuel pellet in the small scall production.

Acknowledgments

This work was financially supported by the Directorate of Higher Education, Ministry of Research, Technology and Higher Education of Indonesia through the research scheme of Mulawarman University Excellent program (PUPT). Special thanks addressed to PT. Sarana Bina Semesta Alam (PT. SBSA), Kutai Kertanegara, Indonesia.

References

[1] MekhilefS, Saidur R, Safari A, Mustaffa WESB2011Biomass Energy in Malaysia: Current State and Prospects. *Renewable and Sustainable Energy Reviews* **15** pp 3360-70. doi: 10.1016/j.rser.2011.04.016.

[2] Dzurenda L, Geffertova J, Hecl V2010 Energy Characteristics of Wood-Chips Produced from *Salix viminalis* - Clone ULV. *Drvna Industrija* **61(1)** pp 27-31.

[3] KubíkL,Kažimírová V 2015 Mechanical properties of pellets in compression. *Research Agriculture Enginering*, **61**: S1–S8. doi: 10.17221/17/2015-RAE.

[4] Naoto M2011 Yield Prediction for *Acacia mangium* Plantations in Southeast Asia. *FORMATH*,**10** pp 295–308.

[5] Obernberger I, and Thek G 2004 Physical characteristic and chemical composition ofdensified biomass fuels with regard to their combustion behavior. *Biomass andBioenergy* **27**, pp 653–69. doi: 10.1016/j.biombioe.2003.07.006.

[6] Sultana A, Kumar A, Harfield D 2010 Development of agri-pellet production cost and optimum size. *Bioresour Technol* **101**, pp 5609–21. doi: 10.1016/j.biortech.2010.02.011.

[7] Carrillo T, Casanova P, SolísK 2016 Effect of densification conditions on physical propertiesof pellets made from sawmill residues. *American Journal of Engineering Research (AJER)*, **5(6)** pp 198-207.

[8] NasrinAB, Ma AN, ChooYM, Mohamad S, Rohaya MH, Azali A, Zainal Z 2008 Oil Palm Biomass As Potential Substitution Raw Materials For Commercial Biomass Briquettes Production. *American Journal of Applied Sciences (AJER)* **5(3)** pp 179-183.

[9] Chaiyaomporn K, and Chavalparit O 2010 Fuel pellets production from biodiesel waste. *Environment Asia*,**3(1)** pp 103-110.

[10] Shyamalee D, Amarasinghe ADUS, Senanayakaet NS2015 Evaluation of different binding materials in forming biomass briquettes with saw dust. *International Journal of Scientific and Research Publications(IJSRP)*, **5(3)** pp 1-8.

[11] Li,Y and Liu H2000 High-pressure densification of wood residues to form an upgraded fuel. *BiomassBioenergy*, **19(3)** pp 177-186.

[12] Kaliyan N, Morey VR 2009 Factors affecting strength and durability of densified biomass products. *Biomass Bioenergy*, **33(3)** pp 337-359.

[13] Kaliyan N, Morey RV 2010 Natural binders and solid bridge typebinding mechanisms in briquettes and pellets made from cornstover and switchgrass. *BioresourceTechnol*,**101** pp 1082–1090.

[14] Samuelsson R, Thyrel M, Sjöström M, Lestander TA2009 Effect of biomaterialcharacteristics on pelletizing properties and biofuel pellet quality. *Fuel ProcessTechnology (FPT)*,**90(9)** pp 1129–34.

[15] Eldoma A. and Awang K1999 Site adaptability of *Acacia mangium*, *Acacia auriculiformis*, *Acacia crassicarpa* and *Acacia aulacocarpa*. APAFRI.Publication Series No.3, Kuala Lumpur, Malaysia.

[16] Filbakk T, Jirjis R, Nurmi J, Høibø O 2011 The effect of bark content on quality parameters ofScots pine (*Pinussylvestris L.*) pellets.*Biomass and Bioenergy*, **35(8)** pp 3342-49. doi: 10.1016/j.biombioe.2010.09.011.

[17] Lehtikangas, P2001 Quality properties of pelletised sawdust, logging residues and bark. *Biomass and Bioenergy*,**20** pp 351–60. doi: 10.1016/S0961-9534(00)00092-1.

[18] Melero, J.A., vanGrieken, R., Morales, G., and Paniagua, M2007 Acidic mesoporous silica for the acetylation of glycerol: Synthesis of bioadditives to petrol fuel. *Energy Fuels*,**21(3)** pp 1782–91.

[19] Brandy, S. Tam, K., Leung, G., Salam, C 2009 Zero waste biodiesel: Using glycerin and biomass to create renewable energy. *UCR Undergraduate Research Journal* pp 5-11.

[20] Whittaker, C. and Shield, I 2017 Factors affecting wood, energy grass and straw pellet durability – A review. *Renewable and Sustainable Energy Reviews* **71** pp 1-11.

Response surface methodology to simplify calculation of wood energy potency from tropical short rotation coppice species

M T Haqiqi, Yuliansyah, W Suwinarti and R Amirta*

Faculty of Forestry, Mulawarman University. Jl. Ki Hajar Dewantara, Samarinda,East Kalimantan, Indonesia

*Corresponding Author: r_amirta@yahoo.com

Abstract. Short Rotation Coppice (SRC) system is an option to provide renewable and sustainable feedstock in generating electricity for rural area. Here in this study, we focussed on application of Response Surface Methodology (RSM) to simplify calculation protocols to point out wood chip production and energy potency from some tropical SRC species identified as *Bauhinia purpurea*, *Bridelia tomentosa*, *Calliandra calothyrsus*, *Fagraea racemosa*, *Gliricidia sepium*, *Melastoma malabathricum*, *Piper aduncum*, *Vernonia amygdalina*, *Vernonia arborea* and *Vitex pinnata*. The result showed that the highest calorific value was obtained from *V. pinnata* wood (19.97 MJ kg^{-1}) due to its high lignin content (29.84 %, w/w). Our findings also indicated that the use of RSM for estimating energy-electricity of SRC wood had significant term regarding to the quadratic model (R^2 = 0.953), whereas the solid-chip ratio prediction was accurate (R^2 = 1.000). In the near future, the simple formula will be promising to calculate energy production easily from woody biomass, especially from SRC species.

1. Introduction

Nowadays, the worldwide energy consumption is constantly growing and the utilization of primary fossil energy sources is already approaching their natural limits however mostly fossil-fuel based energy generation is linked with emissions of large quantities of pollutants, causing serious negative environmental effects [1]. Every year, the combustion of fossil fuels produces around 21.3 billion tonnes of CO_2[2]. Bioenergy, a promising alternative to subtitude fossil fuel, is the largest contributor in the world to the renewable and sustainable energy sector, and it plays a significant role in various energy industries, such as heating and cooling, electricity and power, and transportation fuel [3].

Biomass is one of the most important sources to increase the production of energy based on renewable energy sources [4]. Biomass is largely developed, by representing 10.4 % of the world's total primary energy supply or 77.4 % of global renewable energy supply [5].As a country which is located on the equator, Indonesia has big opportunity in the forest biomass resourcesutilization.Forest also contains biomass diversity richness associated in various community forestand is potential for energy feedstock.Moreover, Yuliansyah [6] explained that wood of waste biomass from the logging activities and other residues are also potential for generating electricity. According to the previous report[7], power generation using short rotation coppice (SRC) species as energy feedstock was suggested. Its sustainable systemcan support the government policies to increasebioenergy sector for prosperities in the remote area. Some SRC species such as poplar (*Populus* spp), willow (*Salix* spp), *Acacia* spp and *Eucalyptus* spp have been grown and developed in many countries in the World

[8].However, there is no list aboutSRC species implementation in forest plantation for energy production in Indonesia as far.

Many techniques and formulas to determine the energy potency from wood biomass were previously introduced [9,10]. However, the simple formula is more promising for searching and establishing the potential feedstock for short rotation coppice cultivation. The aim of this study is to promotenew alternative method for calculating energy-electricity from quadratic model ofresponse surface methodology. Further, we also analyzedtheirwood chemical composition.We suggest that using new methods will be able to give agreat option for easy calculation for wood electricity generation application in the near future.

2. Materials and Method

2.1. Short rotation coppice wood biomass

Some tropical short rotation coppice species which were potentially to be utilized for feedstock of electricity production particularly in remote area including *Bauhinia purpurea* var. corneri de Wit., *Bridelia tomentosa* Blume, *Calliandra calothyrsus* Meissner, *Fagraea racemosa* Jack ex Wall., *Gliricidia sepium* (Jacq.) Kunth ex Walp., *Melastoma malabathricum* L., *Piper aduncum* L., *Vernonia amygdalina* Delile, *Vernonia arborea* Buch.-Ham., and *Vitex pinnata*(L.) Kuntze were used in this study. The collected wood sampels were debarked, chipped, air dried and powdered.

2.2. Wood chemical properties and energy analysis

The TAPPI standard method [11] was used to determine the klason lignin content from the wood. The holocellulose and α-cellulose content were also determined according to Wise's chlorite method [12] and the TAPPI standard method [13]. In addition, the wood moisture content and their calorific value were analyzed using ASTM [14] D 7582-12 and ASTM D 5865-123, respectively. The equation proposed by Francescato [10] was used to calculate energy potency and wood chip production.

2.3. Experimental design

Response surface methodology (RSM) was used to modify and simplify the formula of energy potency, especially for the simple measurement from short rotation coppice woodspecies. The wood density (A) and moisture content (B) were choosen as potential factors with two levels (-1, +1). The experimental design with the total of 16 trials was generated by D-optimal design on Design expert® 7.0.0 software. The calculation technically used adding random calorivic value from each wood coppice species. The interaction of two factors was investigated and displayed by Analysis of Variance (ANOVA).

3. Results and Discussion

3.1. Wood Chemical properties and calorific value

The preliminary analysis on suitability of 10 selected Short Rotation Coppice (SRC) species as materials for green electricity production was done by evaluation of their wood chemical composition. The amount of holocellulose, cellulose and lignin were discussed. We found that the highest holocellulose content was obtained from *B. tomentosa* (78.99 %) followed by *M. malabathricum* (78.04 %), *C. calothyrsus* (76.94 %), *F. racemosa* (76.22 %) and *V. amygdalina* (74.91 %). On the other hands, we also found that the lowest cellulose content was obtained from *P. aduncum* (41.69 %). According to the value, the other tropical woody SRC species contained cellulose content higher than some chlones of willow SRC planted in Japan [15]. Moreover, we also found thatthe highest proportion of lignin content was obtained from *V. pinnata* (29.84 %) followed by *V. arborea* (29.64 %),*B. purpurea* (28.89 %), *G. sepium* (26.17 %) and *C. calothyrsus* (25.09 %). The examined wood chemical compositions have similar result with other tropical fast growing woody species previously studied [16]. The wood biomass composition depends on various factors such as plant species, soil conditions, climate and competition [17,18].

Tabel 1. Chemical component of wood from some tropical short rotation coppice species.

No	Plant biomass species		Holocellulose (%)	Cellulose (%)	Lignin (%)
	Latin name	Local name			
1	*B. purpurea*	Daun kupu-kupu	73.40 ± 0.35	51.51 ± 0.12	28.89 ± 0.86
2	*B. tomentosa*	Berduri	78.99 ± 0.86	53.83 ± 1.16	21.49 ± 0.52
3	*C. calothyrsus*	Kaliandra	76.94 ± 0.27	50.74 ± 0.10	25.09 ± 0.01
4	*F. racemosa*	Kopi-kopian	76.22 ± 1.03	52.25 ± 1.67	23.55 ± 0.09
5	*G. sepium*	Gamal	73.91 ± 1.30	44.07 ± 1.83	26.17 ± 0.01
6	*M. malabathricum*	Karamunting	78.04 ± 0.69	45.56 ± 0.92	20.79 ± 0.62
7	*P. aduncum*	Sirih hutan	72.82 ± 0.33	41.69 ± 0.71	23.77 ± 0.17
8	*V. amygdalina*	Sambung nyawa	74.94 ± 0.71	44.19 ± 0.23	24.17 ± 1.50
9	*V. arborea*	Hamirung	71.98 ± 1.57	49.37 ± 1.79	29.64 ± 0.09
10	*V. pinnata*	Laban	72.15 ± 0.72	44.87 ± 0.30	29.84 ± 0.11

The gross calorific value among 10 SRC species had been published at the previous paper [7]. In this paper, we analyzed the Higher Calorific Value (HCV) produced during their wood combustion process of each biomass plant species. The highest HVC was obtained from *V. pinnata*wood (19.96 MJ kg^{-1}), whereas the lowest HCV was obtained from *M. malabathricum*wood (18.12 MJ kg^{-1}). Generally, lignin content influences the obtaining calorific value from combustion of woody biomass. The highest lignin content strongly contributed to the higher calorific value of *V. pinnata*while *M. malabathricum* was the lowest from those plant species. These findings were also in line with the previous report [19,20].

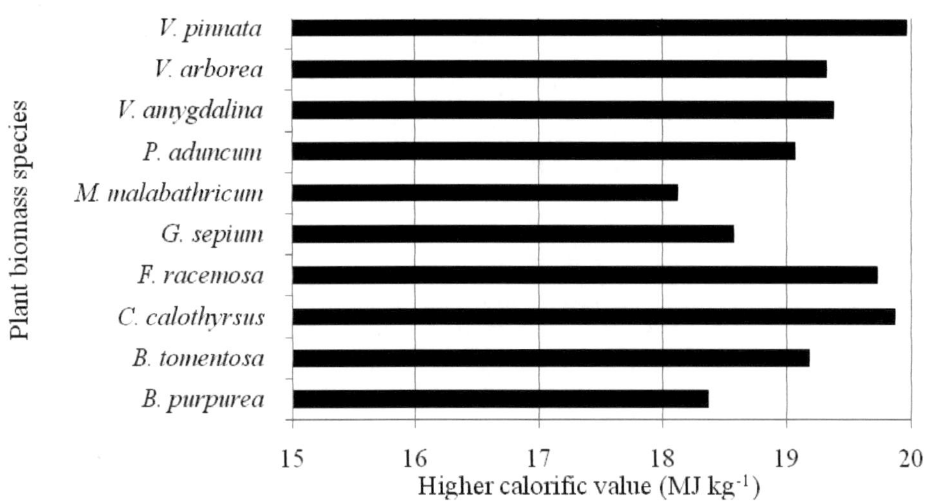

Figure 1. Higher calorivic value of wood from some tropical short rotation coppice spesies

3.2. Experimental design

The wood density and moisture content of 10 SRC plant species was analyzed on our previous report [7]. However, the results of wood density and moisture content measurement were used as variables in this study to simplify common calculation of energy potency and wood chip production using

Response Surface Methodology (RSM). Response surface plot was displayed (Figure 2) with the generated formula as follows:

$$Y_A = 3.647 - 1.032\,A - 0.386\,B + 0.614\,AB \quad (1)$$

$$Y_B = 436.719\,A + 4.36719\,AB \quad (2)$$

Where Y_A is wood energy potency (MWh/m^3), Y_B is wood chip production (kg/m^3), A is wood density (g/cm^3), and B is wood moisture content (%).

This model fitted the data with R^2value equal to 0.962and 1.000 for the response of energy potency and wood chip production, and the predicted R^2value for the same response were 0.953 and 1.000, respectively. In this case, interaction of*A* and *B* wasin significant terms. The quadratic model was evaluated using ANOVA for analyzing the effects of the factors on the response. The *p*-values which was lower than an α-level of 0.05 indicated that the quadratic model was also significant. The linear terms were significant for model and it was mean that the variables had an important influence on the energy potency and wood chip production.

The formulawas used to estimate the energy potencyand wood chip production from tropical SRC which R^2 was 0.953 and 1.000, respectively. In our approach, we eliminated the use of calorific value, an important factor on energy assesment to calculate energy electricity from woody biomass. RSM formula was also compared to common formula used for wood energy calculation [9]. Our findings proved that calculation using RSM formula had high similarity result from common formula used. It was desribed that RSM had an ability to simplify calculation from long formula to the short and easy formula.In this study, we investigated that among 10 SRC species, the highest energy potency was obtained from *F. racemosa* with the value of 3.16 MWh/m^3 while the wood chip production from its plant species was 304.91 kg/m^3.We suggest that the formula generated by RSM was useful for simple calculation of energy electricity from SRC speciesor other woody biomass species in tropical area.

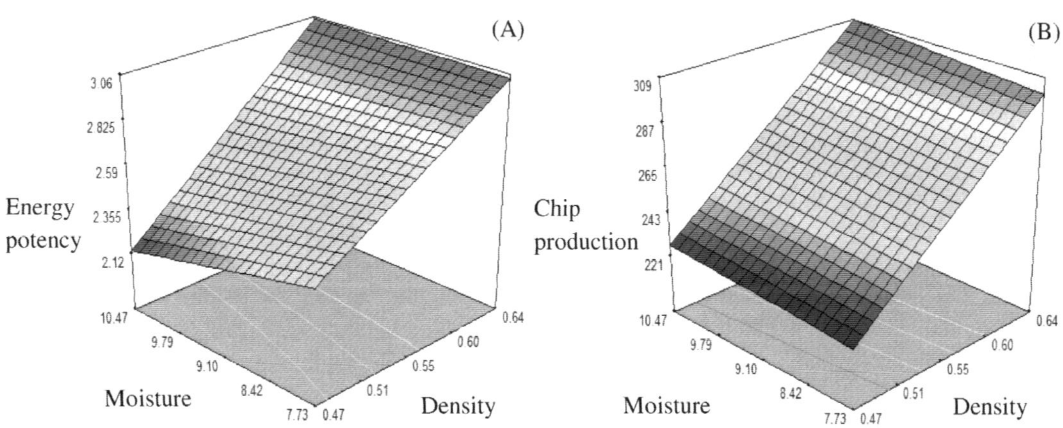

Figure 2. Response surface plot of wood energy potency (A) and wood chip production (B) from some short rotation coppice plant species

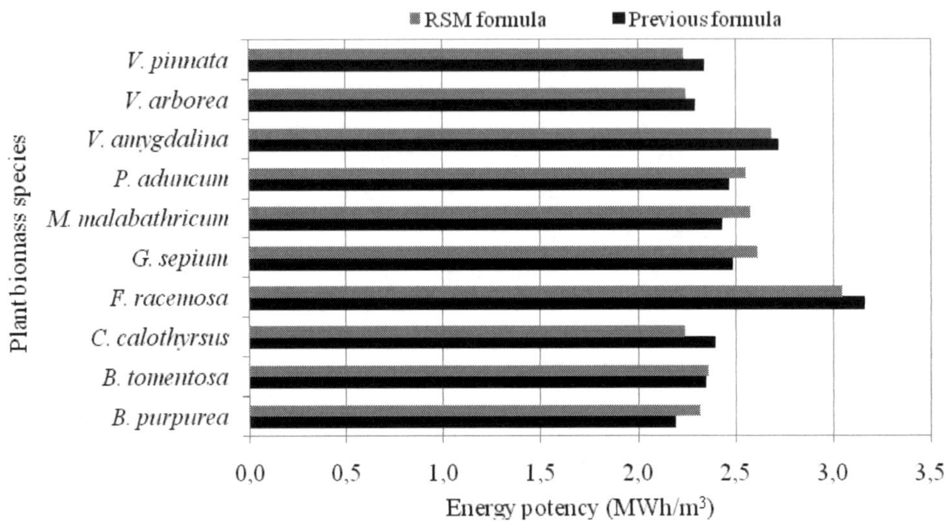

Figure 3. The high similarity results of wood energy potency between calculations using common formula and prediction using RSM formula

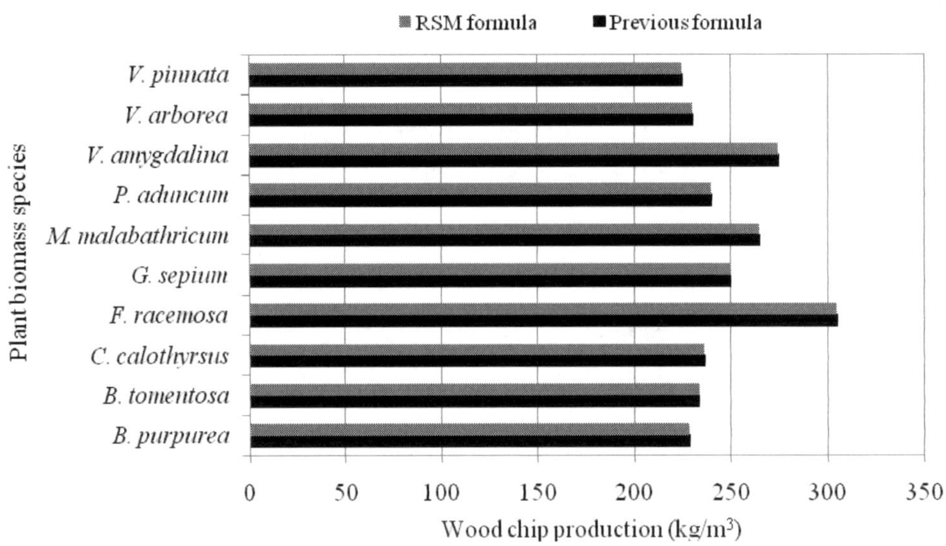

Figure 4. The high similarity results of wood chip production between calculations using common formula and prediction using RSM formula

4. Conclusion

In conclusion, we found that the use of response surface methodology quadratic model resulted in simple formula of potency energy and wood chip production calculation from short rotation coppice plant species with the significant interaction between wood density and wood moisture content. Based on wood chemical component characteristics, the high lignin content also gave the high wood calorific value production from those SRC species.

Acknowledgments

Authors were grateful to Industrial Biotechnology Laboratory, Faculty of Forestry, Mulawarman University. This work was financially supported bythe project from Islamic Development Bank (IDB) of Mulawarman University (for RA – FY 2017).

References

[1] Strzalka R, Schneider D, and Eicker U 2017 Current status of bioenergy technologies in Germany *Renewable and Sustainable Energy Reviews* **72** pp 801–820 doi:10.1016/j.rser.2017.01.091

[2] Kazulis V, Muizniece I, Zihare L, and Blumberga D 2017 Carbon storage in wood products. *Energy Procedia* **128** pp 558–63 doi:10.1016/j.egypro.2017.09.009

[3] Alsaleh M, Abdul-Rahim A S, and Mohd-Shahwahid H O 2017 An empirical and forecasting analysis of the bioenergy market in the EU28 region: Evidence from a panel data simultaneous equation model *Renewable and Sustainable Energy Reviews* **80** pp 1123–37 doi:10.1016/j.rser.2017.05.167

[4] Erol M, Haykiri-Acma H, and Kucukbayrak S 2010 Calorific value estimation of biomass from their proximate analyses data *Renewable Energy* **35** pp 170–73 doi:10.1016/j.renene.2009.05.008

[5] Carlos R M and Khang D B 2008 Characterization of biomass energy projects in Southeast Asia *Biomass and Bioenergy* **32** pp 525–32 doi:10.1016/j.biombioe.2007.11.005

[6] Yuliansyah and Amirta R 2016 Physico-chemical properties and energy potency of wood waste biomass from logging activities to generate electricity in East Kalimantan, Indonesia AIP Conf. Proc, 1755.040001-6 doi: 10.1063/1.5958476

[7] Amirta R, Yuliansyah, Angi E M, Ananto R B, Setiyono B, Haqiqi M T, Septiana H A, Lodong M, and Oktavianto R N 2016 Plant diversity and energy potency of community forest in East Kalimantan, Indonesia: Searching for fast growing wood species for energy production *Nusantara Bioscience* **8**(1) pp 22-31 doi: 10.13057/nusbiosci/n080106

[8] Marron N 2015 Agronomic and environmental effects of land application of residues in short-rotation tree plantations: A literature review *Biomass and Bioenergy* **81** pp 378-400 doi:10.1016/j.biombioe.2015.07.025

[9] Kofman P D 2010 *Units, conversion factors and formulae for wood for energy* Ireland: COFORD

[10] Francescato F, Antonini E, and Bergomi L Z 2008 *Wood fuels handbook: Production, quality requirement, trading*. Legnaro, Italy: AIEL – Italian Agroforestry Energy Association

[11] TAPPI 1988 *Acid-insoluble lignin in wood and pulp* Technical Association of the Pulp and Paper Industry, T 222 om-98. New York: Technical Assosiation of the Pulp and Paper Industry

[12] Wise L E, Murphy M, and D'Addieco A A 1946 Chlorite holocellulose, its fractionation and bearing on summative wood analysis and on studies on the hemicelluloses *Paper Trade J,* **122**(2) pp 35-43

[13] TAPPI 1988 *Alpha-, beta- and gamma cellulose in pulp and wood*. Technical Association of the Pulp and Paper Industry, T 203 om-93. New York: Technical Assosiation of the Pulp and Paper Industry

[14] ASTM International 2012 *Standart test method for gross calorific value of refuse-derived fuel by the bomb calorimeter* ASTM standart E711-87 American Society for Testing Materials

[15] Mitsui Y, Seto S, Nishio M, Minato K, Ishizawa K, and Satoh S 2010 Willow clones with high biomass yield in short rotation coppice in the southern region of Tohoku district (Japan) *Biomass and bioenergy* **34** pp 467–73 doi:10.1016/j.biombioe.2009.12.010

[16] Amirta R, Mukhdlor A, Mujiasih D, Septia E, Supriadi, and Susanto D 2016 Suitability and availability analysis of tropical forest wood species for ethanol production: a case study in East Kalimantan *Biodiversitas* **17**(2) pp 544-52 doi: 10.13057/biodiv/d170222

[17] Premjet S, Pumira B, and Premjet D 2013 Determining the potential of inedible weed biomass for bio-energy and ethanol production *Bioresources* **8** pp 701-16

[18] McKendry P 2001 Energy production from biomass (part 1): Overview of biomass *Bioresource Technology* **83**(1) pp 37-46 doi:10.1016/S0960-8524(01)00118-3

[19] Demirbas A 2003 Relationships Between Heating Value and Lignin, Fixed Carbon, and Volatile Material Contents of Shells from Biomass Products *Energy Sources* **25** pp 629–35 doi: 10.1080/00908310390212336

[20] White R H 1987 Effect of lignin content and extractives on the higher heating value of wood *Wood and Fiber Science* **19**(4) pp 446-52

Characterization and ethanol potential from giant cassava (*Manihot esculenta*) stem waste biomass

E Septia, Supriadi, W Suwinarti and R Amirta*

Faculty of Forestry, Mulawarman University, Samarinda, East Kalimantan, Indonesia

* Corresponding Author: r_amirta@yahoo.com

Abstract. *Manihot esculenta* stem waste biomass is promising material for ethanol production since it is unutilized substance from cassava production. Nowadays, cassava is the most common food in Indonesian society. The aims of this study were to identify availability and characteristic of giant cassava (*M. esculenta*) stem waste biomass for ethanol feedstock. In term of that, four plots with the size of 5m x 5m were made to calculate the total stem biomass obtained after harvesting process. In this study, various concentrations of alkaline were used to degrade lignin from the substrate. The effects of alkaline pretreatment were investigated using TAPPI method and the ethanol yield was estimated using modified NREL protocol. The results showed that the potential dry stem waste biomass from harvesting of *M. esculenta* was approximately 10.5 ton/ha. Further, alkaline pretreatment of stem waste biomass with 2% of NaOH coupled with the enzymatic saccharification process using meicelase was showed the highest production of sugar to reach of 38.49 % of total reduction sugar and estimated potentially converted to 2,62 L/ha of ethanol. We suggested *M. esculenta* stem waste biomass could be used as sustainable feedstock for ethanol production in Indonesia.

1. Introduction

As the human populations and industries increases, energy consumptions in general also increasing gradually, resulting in increased levels of pollution. Nowadays, until 2050 it is estimated that the world's population will increase by more than two billions or more people [1]. Concerning of energy consumption increased as the beginning of the development of bioethanol and biodiesel products resulting from food crops [2,3]. During this time, the production of bioethanol is more directed to the ingredients of starch and sugar, such as sugar cane, cassava root, and corn. Though these materials are basically a potential source of food, so the development of bioethanol from food in the future will be able to cause new problems due to competition for the fulfillment of food demands of the community. Therefore, it is necessary to develop bioethanol from materials that are not a food source of the community, namely lingo-cellulosic materials that allow forwards to the utilization of agricultural waste and industrial waste for bioenergy. One of kind the potential raw material source developed in Indonesia is cassava. Cassava is one of the most important root crops known and used in many countries of Africa, Latin America, and some Asian countries. Though it has its origin in South America and some Asian countries. The crop is widely grown as staple food and animal feed in these regions with a total cultivated area over 18 million ha [4,5]. In Indonesia, especially in East Kalimantan have been cultivating giant cassava (*Manihot esculenta*) that have superior quality than other cassava. Seeing the promising prospect of *M. esculenta* with the potential of large tubers, of course, many stem wastes that is not utilized. So far, few reported has been done regarding the utilization of *M. esculenta* stem biomass as the raw material for bioethanol manufacture and the characteristic properties and process technology that can be used simply to produce it. In this study, we characterized the ethanol potential of cassava stem waste into renewable energy sources and environmental benefits.

2. Methods

2.1. Giant cassava stem materials
The plant material was wood biomass from Kota Bangun, Kutai Kartanegara, East Kalimantan, Indonesia. Amount of 1% sampling intensity plot was used in this study to estimate total biomass per hectare. The 4 random sampling plots with the size of 5 m x 5 m was made. Giant cassava is commonly harvested at 12 months with the planting distance of 1 m x 1 m.

2.2. Alkaline pretreatment
Alkaline pretreatment of stem biomass was carried out for 60 min at 170ºC using a liquid-to-solid ratio 4: 1 (w/w) and NaOH concentration between 2, 4, and 6%based on dry weight of the woody biomass, respectively. The reactions were carried out using a rotary digester equipped with a controller for pressure, rotary speed, and temperature. After the reaction, the pulp fraction was separated by filtration and washed extensively with tap water until neutral pH.

2.3. Chemical component analysis
The Klason lignin content was determined by the TAPPI standard method [6]. The holocellulose and α-cellulose contents were determined according to Wise's chlorite method [7] and the TAPPI standard method [8], respectively. The reducing sugar content was determined by the Somogyi–Nelson method [9].

2.4. Enzyme activity
Filter paper unit activities were assayed in reaction mixture containing 50 mg (w/v) Whatman filter paper number 1, 50 mM tartrate buffer, pH 4.5 and the enzyme. After incubation at 50ºC for 30 min., the reducing sugars produced were determined by Somogyi-Nelson method [9]. One unit (U) of each enzyme activity is defined as the amount of enzyme, which produce 1 μmol reducing sugar as glucose in the reaction mixture per minute under above specific condition.

2.5. Saccharification of stem M. esculenta
The wet pulp fraction was hydrolyzed with a commercial cellulose preparation, meicelase from *Trichoderma viride* (Meiji Seika Co., Ltd., 224 filter paper units (FPU)/g, β-glucosidase activity 264 IU/g). The cellulase enzyme loading was an 8-FPU/g substrate. Enzymatic hydrolysis was performed at a substrate concentration of 2% in 0.05 M sodium citrate buffer (pH 4.5) containing 0.02% sodium azide at 45ºC on a rotary shaker (NTS-4000C, Rikakikai, Japan) at 140 rpm for 48 h [10]. The saccharification ratio per pulp was calculated according to the NREL LAP-009 procedure [11]. The sugar yield per wood is based on the weight percentage of the reducing sugars to the original wood. The overall yield of sugars per wood is calculated by multiplying the saccharification ratio per pulp and the pulp yield. All enzymatic hydrolysis experiments were performed in triplicate.

2.6. Estimation of ethanol production from stem M. esculenta
Potential ethanol production from Macaranga wood was estimated based on the amount of hexose sugar (HXTEL) in the lignocellulosic material obtained by enzymatic saccharification of the insoluble pulp fraction. Due to the high content of glucose in the pulp fraction, HXTEL was approximated by

the amount of reducing sugars obtained from the pulp fraction (equation 1). The ethanol yields (EtOHBIO) based on the weight of original biomass was calculated from equation (2).

$$HEXTEL = HEX \times a\ (mg/Kg) \quad (1)$$
$$EtOHBIO = HEXTEL \times Ye.\frac{h}{b}\left(\frac{mL}{Kg}\right) \quad (2)$$

Where, HEX is the hexose (D-glucose) yield upon saccharification from hexosan (w/w), of original wood basis), a is the weight of substrate (1000 mg, 1kg), Ye.h is the theoretical ethanol yield from hexose (D-glucose) (0.511), and b is the ethanol density (0.789 kg/L) (modified from Premjet) [12].

3. Results and Discussion

3.1. Measurement of cassava (M. esculenta) biomass

Overall, the number *M. esculenta* every unit produced 81 stems with total population (*M. esculenta*) per hectare obtained as many as 8,100 stems/hectare. The results showed that the percentage of *M. esculenta* in each part are root 59.39%, stem 31.84%, tuber 5.96%, and leaves 2.80%, respectively (Figure 1). Acquisition of this value is based on the post-harvest results of *M. esculenta* conducted on each plot.

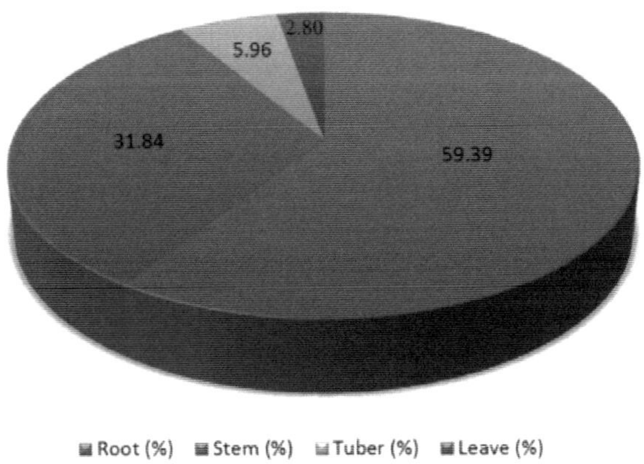

Figure 1. Percentage of *M. esculenta* in each part

Based on percentage results of *M. esculenta*, particularly the stem portion shown potential results when developed into bioethanol feedstock compared to root and other parts. According to Sumada [13], the utilization of *M. esculenta* stem wastes is not yet optimal because only 10% of the stems can be used for replanting and the remaining 90% are untapped waste. The dry biomass value obtained in *M. esculenta* stem was 10,533 kg/ha. Calculation of biomass values related to moisture contents that part of physic characteristic of *M. esculenta* stem.

3.2. Lignocellulosic components of wood biomass

The chemical properties of wood are related to the chemical content in the wood. Chemical timber or chemical components of wood are required for their existence in the chemical industry that processes wood (rayon, pulp, and paper industries) [14] and determine the timber harvest for the purpose of producing certain products. In this research, we evaluated lignin, holocellulose, and cellulose contents of wood biomass to point out their correlations to saccharification and estimated yield of ethanol. We found that the lignin content was 23.34%, holocellulose 67.26%, and cellulose 56.47% (Table 1), respectively.

Table 1. Lignocellulosic compositions (original wood components)

Lignin (%)	Holocelllulose (%)	Cellulose (%)
23.34 ± 0.39	67.26 ± 0.83	56.47 ± 0.58

3.3. Effects alkaline pretreatment on lignocellulosic components of M. esculenta stem

Effects of alkaline pretreatment on the change components were also analyzed. We found the alkaline pretreatment was effectively defibrillated and delignified the woody biomass that marked by the residual pulp fractions and decreased of Klason lignin content (Table 2,3).In general discussion, we found two different types of the residual pulp properties, hard fiber, and soft fiber. The formation of residual pulp fraction was really dependent on the concentration of alkaline and also biomass species used (Table 2, Figure 2).At the remaining 2% and 4% alkali concentrations the pulp properties are fiber with hard fiber, whereas at the remaining 6% alkali concentration the pulp properties is soft fiber. The formation of the remaining properties of the pulp depends on the concentration of alkaline and also the type of biomass used [15].

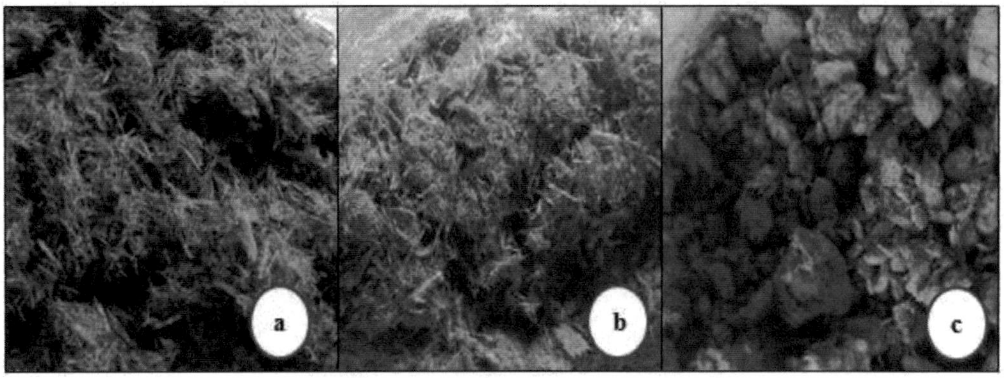

Figure 2. Pulp properties (a) 2% NaOH (b) 4% NaOH (c) 6% NaOH

Table 2. Pulp yield and pulp properties of *M. esculenta* stem

Alkaline concentration	Pulp yield (%)	Pulp properties
2%	65.81	hard fiber
4%	59.54	hard fiber
6%	40.16	soft fiber

Table 3. Residual lignin, holocellulose, and cellulose of *M. esculenta* stem

Alkaline	Lignin (%)	Holocellulose (%)	Cellulose (%)
2%	11.31 ± 0.12	43.85 ± 0.22	50.44 ± 0.01
4%	12.84 ± 0.14	41.79 ± 0.18	47.44 ± 0.21
6%	12.95 ± 0.13	36.04 ± 0.15	33.98 ± 0.37

The results showed that the pulp yield value at cooking conditions 2% was greater than the cooking conditions of concentration 4 and 6%. Differences in pulp yield produced in various types of raw materials, caused by differences in chemical components, especially holocellulose and lignin and reactions that occur during the pulping process. According to Fengel and Wegener [16] and Sjöström [17], suggesting that high levels of active alkali concentrations will cause the pulping process to proceed rapidly, but as process speed increases, lignin degradation is unavoidable and cellulosic damage will be even greater.

The highest decrease of lignin content was obtained when 6% NaOH applied at 170°C to give the value of 12.95%. Based on the values obtained in accordance with the opinion of Dewi [18], which states that the lignin content increases along with the increase of Sodium Hydroxide (NaOH) concentration. This occurs because the depolymerization process of lignin in the pulp material whose polymer turns into monomers, then the monomers react with the polymer still contained in the pulp and produce a new polymer or new lignin.

3.4. Enzymatic saccharification of M. esculenta stem

Lignocellulosic biomass consists of carbohydrates (cellulose and hemicellulose), lignin and other components (proteins, lipids, and inorganic substances) [19]. The presence of lignin greatly inhibits the process of degradation of cellulose and hemicellulose into glucose. Therefore, lignin must be eliminated either chemically or enzymatically which is a delignification process and thereafter can be fermentation process for bioethanol production. This delignification process is needed to break the long polymer chain into shorter polymer chains, increase the amorphous area (decrease the degree of crystallinity) and separate the lignin portion of the holocellulose.

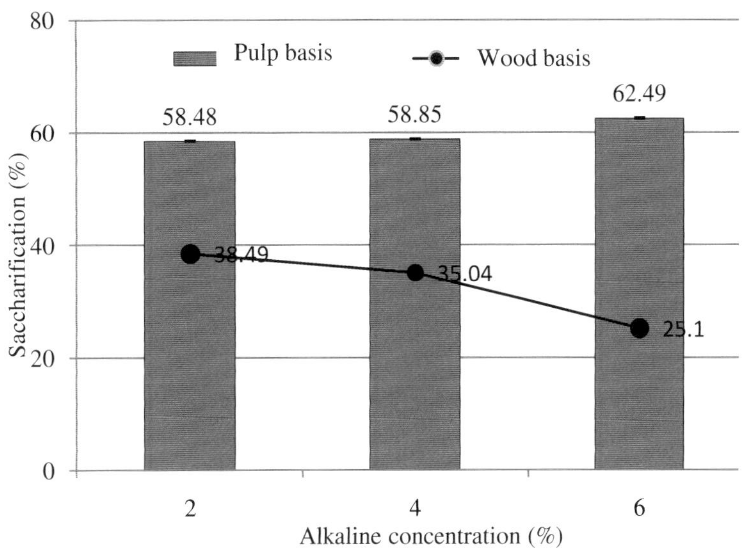

Figure 3. Saccharification yield of *M. esculenta* stem

The figure above reflects the calculation of the saccharification result of *M. esculenta* stem from the alkaline treatment concentration of 2, 4, and 6%, respectively. The higher alkaline concentration given will increase the saccharification result on the wood pulp. At the concentration of 2% alkaline addition, the saccharification result shows a number of 58.48% reducing sugars which can be produced from the pulp fraction. Meanwhile, the addition of 4% alkali concentration was able to produce reduced sugar as much as 58.85% while at 6% concentration able to reduce sugar up to 62.49% from total of pulp fraction. The effect of lignin on the saccharification process can inhibit the performance of the enzyme in breaking the glycoside bond in cellulose to produce simple sugars. According to Ko [20], lignin is a physical barrier that can limit the access of enzymes to cellulose. Lignin is able to bind to the active side of the enzyme thereby decreasing and causing unproductive enzyme performance. This causes a decrease in the activity of enzymes in degrading cellulose resulting in reduced productivity of reducing sugars.

If we look closely together, based on the overall base of the wood actually shows the opposite result. The higher the concentration given causes the decreasing of sugar yield. It also shows that the cassava wooden stem structure that is not too hard only requires preliminary treatment with mild concentration. The result of this research is obtaining optimum point of utilization of *M. esculenta* stem waste as raw material of ethanol with alkaline treatment of 2% NaOH addition with wood saccharification result 38.49%. As reported by Amirta [21], some tropical wood species such as macaranga require concentrations of up to 5% to find their optimum use point as ethanol feedstock. Moreover, this result is interesting because the used of concentrations of chemicals had suppress the production cost of the use of biomass cassava stems in the future.

3.5. Theoretical ethanol production from wood biomass

According to Premjet [12], the basic theory of ethanol calculation described in Figure 4, the highest ethanol production value was obtained from 2%, 4% and 6% alkali concentration of 249 L/ton, 227 L/ton and 163 L/ton, respectively. According to Aveline [22], the properties of biomass and plant species vary to be one of the causes of different ethanol yields at each given alkali concentration.

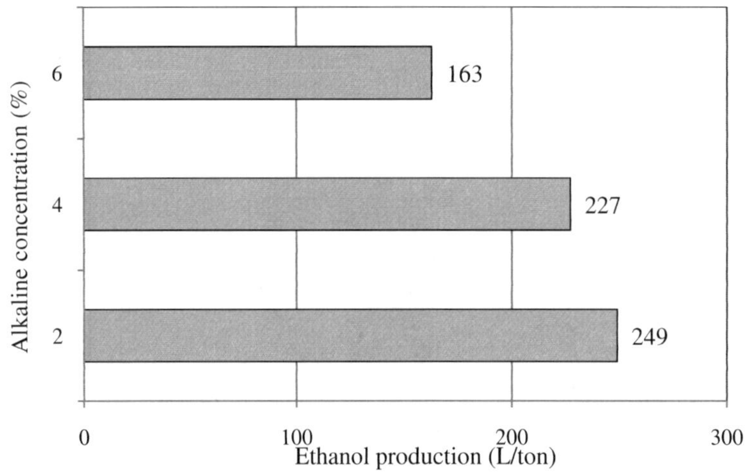

Figure 4. Potential ethanol production from wood biomass

Furthermore, the ethanol production from utilization of *M. esculenta* waste is higher than some tropical timbers from previous studies, such as *A. moluccana* (241 L/ton), *G. arborea* (239 L/ton), *A. cadamba* (233 L/ton), *A. altilis* (222 L/ton) and *P. falcataria* (220 L/ton) [15], respectively. The tropical wood species use too high concentrations of chemicals (NaOH above 3.5%), while *M. esculenta* stem waste use only 2% alkaline concentrations.

4. Conclusion

In conclusions, we found that production of ethanol from *M. esculenta* stem can be promoted as one of the more friendly process for environment, besides also explore the cassava waste stem utilization which has been widely wasted in society.

Acknowledgments

This work was financially supported by Faculty of Forestry, Mulawarman University and Laboratory of Biotechnology Industry. Authors were grateful to Mr. Yuliansyah and Ir. Edi Sukaton for handling wood component analysis and valuable discussion on the basic properties of wood biomass.

References

[1] Granham R 2011 *Beyand food versus fuel Nature* **474** pp 2-6
[2] Fargione J, Hill and Tilman D 2008 *Land clearing and the biofuel carbon debt Science* **319** pp 1235–3
[3] Cheng J J and Timilsina G R 2011 Status and barriers of advanced biofuel technologies: a review *Renewable Energy* **36** pp 3541–8
[4] Baguma Y, Sun C and Ahlandsberg 2003 *Expression patterns of the gene encoding starch branching enzyme II in the storage roots of cassava (Manihot esculenta Crantz) Plant Sci* **164** pp 833–9
[5] Baguma Y, Sun C and Borén M 2008 Sugarmediated semidian oscillation of gene expression in the cassava storage root regulates starch synthesis *Plant Signal Behav* **3(7)** pp 1–7
[6] TAPPI 1998 *Acid-insoluble lignin in wood and pulp. Technical Association of the Pulp and Paper Industry* (New York: T 222 om-98 Technical Association of the Pulp and Paper Industry)
[7] Wise L E, Murphy M and D'Addieco A 1946 Chlorite holocellulose, its fractionation and bearing on summative wood analysis and on studies on the hemicelluloses *Paper Trade J* **122(2)** pp 35-43
[8] TAPPI 1988*Alpha-beta-and gamma cellulose in pulp and wood* (New York: T 203 om-93Technical Association of the Pulp and Paper Industry)
[9] Somogyi M 1952 Notes on sugar determination *J Biol Chem* **195** pp 19-23
[10] Itoh H, Wada M, HondaY, Kuwahara M and Watanabe T 2003 Bioorganosolve pretreatments for simultaneous saccharification and fermentation of beech Wood by ethanolysis and white rot fungi *J Biotechnol* **103** pp 273-80
[11] BrownL and Torget R 1996 *Chemical Analysis and Testing Task, Laboratory Analytical Procedure, Enzymatic Saccharification Lignocellulosic BiomasNREL Laboratory Analytical Procedure #009* (Washington DC:US Department of Energy)
[12] Premjet S, Pumira B and Premjet D 2013 Determining the potential of inedible weed biomass for bio-energy and ethanol production *Bioresources* **8** pp 701-16
[13] Sumada K, Tamara P E and Alqani F 2011 Isolation study of efficient α-cellulose from waste plant stem *Manihot esculenta* crantz *Journal of Chemistry* **5** pp 434–4
[14] Kasmudjo 2010 *Forest Product Technology* (Yogyakarta: Cakrawala Media)
[15] Amirta R, Mukhdlor A, Mujiasih D, Septia E, Supriadi and Susanto D 2016 Suitability and availability analysis of tropical forest wood species for ethanol production: a case study in East Kalimantan *Biodiversitas Journal of Biological Diversity* **17(2)** pp 544-552
[16] Fengel D and Wegener G 1995 *Wood Chemistry, Ultrastructure, Reactions* (Yogyakarta: Gajah Mada University Press)
[17] Sjöström E 1995 *Wood Chemistry Fundamentals and Applications*(Yogyakarta: Gadjah Mada University Press)
[18] Dewi T K and Dandy Akbar W 2010 Effect of Concentration NaOH , Temperature, and Time on Making Pulp from Jute Stalk with Soda Process *Journal of Chemistry* **17(2)**
[19] Kim T H 2013 Pretreatment of lignocellulosic biomass In: Yang, S.T El-Enshasy H.A Thongchul, N Martin, Y *Bioprocessing Technologies in Integrated Biorefinery for Production of Biofuels Biochemicals and Biopolymers from Biomass* (New York: Wiley-USA) pp 91–109
[20] Ko JK, Ximenes E, Kim Y and LadischM R 2014 Adsorption of enzym onto lignins of liquid hot water pretreated hardwoods *Biotech Bioeng* **999(30)** pp 1-10
[21] Amirta R, Nafitri S I and Wulandari 2016 Comparative characterization of Macaranga species collected from secondary forests in East Kalimantan for biorefinery of unutilized fast growing wood *Journal of Biological Diversity* **17(1)** pp 116-123
[22] Avelin A, Skvarila J and Aulin R 2014 Forest biomass for bioenergy production-comparison of different forest species *Energy Procedia* **61** pp 1820-3

Effect of drying on the porosity of the hydroxyapatite and cellulose nata de coco compositeas bone graft candidate

S Anitasari[1,2], A Mu'ti[3,4], YO Hutahaean[5]

[1]Department of Dental Material, Dentistry Program, Faculty of Medicine, Universitas Mulawarman, East Kalimantan, Indonesia
[2]Department of Microbiology, Faculty of Medicine, UniversitasMulawarman, East Kalimantan, Indonesia
[3]Department of Biochemistry, Faculty of Medicine, UniversitasMulawarman East Kalimantan, Indonesia
[4]Department of Radiology, Faculty of Medicine, UniversitasMulawarman, East Kalimantan, Indonesia
[5]Department of Neurology, Faculty of Medicine, UniversitasMulawarman, East Kalimantan, Indonesia

E-mail:carabelli74@yahoo.com

Abstract. Bone graft is used to replace bone parts damaged by illness and accident. As a bone replacement material, the bone graft should be able to stimulate the process of the osteogenesis. The process of osteogenesis is influenced by the osteoconductive properties of a biomaterial, that porosity affects this process. The shells of blood scallop (*Anadaragranosa*) are producing hydroxyapatite (HAp),having high compressive strength, biocompatibility and osteoconductive properties, but low porosity while cellulose nata de coco (Cnc) have low compressive strength but high porosity. Therefore, the combination of two biomaterials are expected to produce composite that have high osteoconductive properties. The purpose of this research wasknowing the porosity of HAp/Cnc composite which wasbeingprecipitated for 5 hours, 15 hours, 25 hours and wasdried for 24 hours, 48 hours and 72 hours. This research usedwise drop technique to synthesis HAp powder and cellulose immersion technique for synthesis of HAp/Cnc. Results of this research, there was difference in porosity between HAp/Cnc that was precipitated for 5 hours, 15 hours and 25 hours, as well as was dried for 1 day, 2 days and 3 days. The conclusion, the synthesis of HAp/Cncwasuseful as bone graft candidate.

1. Introduction

Periodontal disease is an irreversible periodontal attachment loss, destruction of periodontal ligament and alveolar bone. In this situation bone repair and augmentation are needed [1]. Bone repair and augmentation involve the use of biomaterial and can be transplanted with biomaterial-based on scaffold for tissue regeneration. For this, biomaterial is used as a scaffold which should have specific features such as chemical, biological and mechanical properties [2].

The biological properties of scaffold must make osteoblast can be seeded, cultured and then implanted to the bone to induce the growth of new bone. The scaffold must be osteoconductive and preferably, osteoinductive which it can be induce efficiently osteogenesis process[3]. The ability of osteoblast attached to the material is influenced by the porosity, pore size and interconnectivity. One of several materials which use as bone graft is hydroxyapatite (HAp) that combine with synthetic polymer like cellulose nata de coco[1,4].

The hydroxyapatite from *Anadaragranosa* is a kind of calcium phosphate compound and it has beenused as a ceramic biomaterial because of its excellent bioactive, compressive strength, biocompatibility and osteoconductive ability, but HAp undesirable material for augmentation[2,5,6,7].

The nata de coco is bacterial cellulose (BC) that is produced from *Acetobacterxylinum* with excellent porosity but low compressive strength. Morphologically, the fibrous structure of cellulose nata de coco (Cnc)is similar to the collagenous fibrils of bone. That morphology support Cnc as a scaffolding material to supporting the growth and differentiation of human bone. The cellulose nata de coco is desirable scaffold for tissue engineering because it is biocompatibility and high porosity and accessible surface area/volume ratios [8-11]

However, study of HAp/Cnc composite material as a bone regeneration are limited. The purpose of this study was a comparison time of precipitation for 5 hours, 15 hours and 25 hours and time of drying for 24 hours (1days), 48 hours (2 days) and 72 hours (3 days) as a basis to the development a new biomaterial that can be a new bone graft candidate.

2. Materials and methods

2.1. Material sampling

The study conducted at October 2017. The place of sampling of material conducted at Samarinda, East Kalimantan. Processing of material sample fromblood scallop shell and cellulose nata de coco conducted at the Research Laboratory, Faculty of Medicine, Universitas Mulawarman. The study protocol was approved by the Research Ethics Committee on Faculty of Medicine Universitas Mulawarman.

2.2. Calcination of blood scallop shell (A. granosa)

The preparation of blood scallop shells included the process of collection, cleaning, drying, and calcination. Preparation began with the process of cleaning shells from the shells of dirt and then was dried on the air. After cleaning was continued with calcination and it was done twice time, at the temperature 800 ^{0}C for 18 hours to get calcium [12].

2.3. Hydroxyapatite synthesis

The manufacture of 1 M $Ca(OH)_2$ solution by inserting 24.81 grams of CaO into a beaker glass and added with 600 ml of distilled water while stirring at 150 rpm. Then, it was added 300 ml of phosphoric acid 1.8M and was heated above hot plate with a temperature of 40^{0}C and was stirring using a magnetizer with a speed of 300 rpm for 60 minutes. The solution was aging for 24 hours. The result was filtered by whatman membrane and was dried at 100^0 C for 3 hours. The sintering process was carried out at 900^{0}C for 5 hours with a temperature rise rate of 50 C / min [12].

2.4. Synthesis of HAp/Cnc

The synthesis of HAp/Cnc was performed by *cellulose immersion technique*. First step, purifying Cnc and making HAp suspension distilled water. Cnc was purified in distilled water for 2 days and it replaced the distilled water every 5 hours. We immersed the cellulose with 1 M NaOH solution for 2 hours. The result was a product of salt deposited in the pellicle, it was removed the salt on Cnc by distilled water until the pH is neutral and odorless.

Making HAp suspension from1 gram of HAp powder added with 30 ml distilled water. The suspension was being sonication for 30 minutes. The purified Cnc was immersed in suspension while stirring using a magnetizer with a speed of 200 rpm for 5 hours,15 hours and 25 hours. After completion of immersion, the pellicle was removed and was dried into remove its water content with a 50^0C incubator temperature [12].

2.5. Porosity test

Porosity test is given by the following equation [13]

$$Porosity(\%) = \frac{Ww-Wo}{Vf-Vo} x\ 100\% \qquad (1)$$

Five samples off all composites were equally dry weighed (Wo) and initial volume (Vo) and then immersed in distilled water, these were final volume (Vf). After 24 hours, each sample was taken out from the tube and this was weighed of wet (Ww).

2.6. Statistical analysis

Data are expressed as the mean ± standard deviation for five independent samples. Statistical comparisons between groups were performed with one-way ANOVA and Kruskal Wallis according to SPSS 23.00 software, with significant being $p<0.05$ or less.

3. Results

The Time of precipitation for 25 hours and the time of drying for 24hours (1 day) showed lowest average porosity (0,17 ±0,05%) compare time precipitation for 5 hours (0.234± 0.03%) and 15 hours (0.315 ±0.11%) as were illustrated figure 1a. Analyse of Kruskal Wallis statistical showed that the average porosity of the HAp/Cnc composite which time of precipitation for 5 hours, 15 hours and 25 hours and were dried for 24 hours were significant different ($p<0.05$)

But, there were difference of result on drying of HAp/Cnc for 48 hours. There was not difference between average porosity which time of precipitation for 5 hours, 15 hours and 25 hours as were shown in figure 1b. Based on one-way ANOVA statistical showed that were not significant different between average porosity of the HAp/Cnc composite which time of precipitation for 5 hours, 15 hours and 35 hours and were dried for 48 hours.

The best results were found for 72 hours drying at illustrated in figure1c. The average of porosity of HAp/Cnc was precipitated for 25 hours had average porosity (0.09 ± 0.02%) approaching average porosity of bone (0.03± 0.01%). Analyse by one-way ANOVA statistical showed that were significant different between time of precipitation for 5 hours, 15 hours and 25 hours ($p<0.05$) and post-hoc analyses by Tukey's HSD showed that precipitation time for 25 hours had the best average porosity.

The time of precipitation for 5 hours and the time of drying 24 hours, 48 hours and 72 hours showed that were not difference average porosity of time of drying for 24 hours (0.234± 0.03%), 48 hours (0.205 ± 0.05%) and 72 hours (0.20 ± 0.08%) as illustrated in figure 2a. One-way ANOVA statistical showed that was not significant different ($p<0.05$).

The average porosity HAp/Cnc was precipitated for 15 hours and was drying 72 hours (0.15 ± 0.02) % had lowest compare with time of drying 24 hours (0.315 ±0.11)% and 48 hours(0.194 ± 0.06) % (Figure 2b). Based on Kruskal Wallis statistical analysis showed that were significant different ($p<0.05$).

The samples were precipitated for 25 hours and were drying for 72 hours had average porosity (0.09 ± 0.02%) highest compare with a drying for 24 hour (0.17 ±0.05%) and 48 hours (0.207 ± 0.03%) as illustrated in figure 2c. Analysis data for all samples which time of precipitation were 5 hours, 15 hour and 25 hour and were compare with time of drying for 24 hours, 48 hours and 72 hours, the best average porosity was 25 hours precipitation and was drying for 72 hours (Table 1). Analyse by one-way ANOVA statistical, there were significant different ($p<0.05$) between the drying for 24 hours, 48 hours and 72 hours. Post-Hoc analyses by Tukey's HSD, the best average porosity were precipitation time 25 hours and drying for 72 hours.

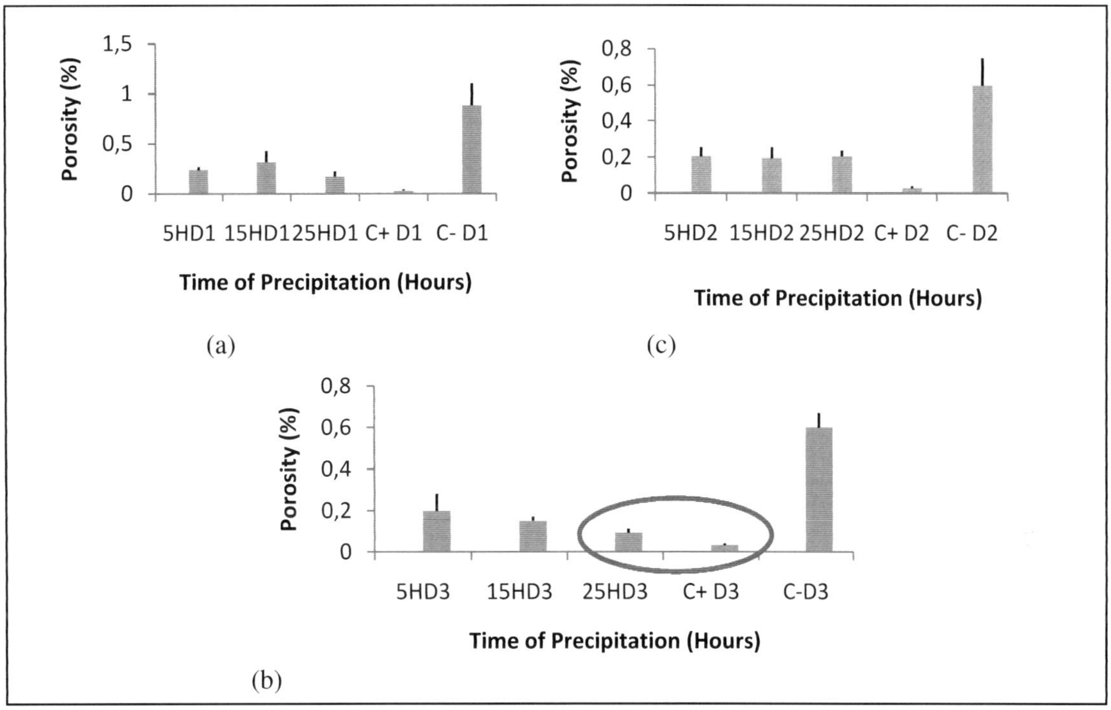

Figure 1. The average porosity of HAp/cellulose nata de coco with different time of precipitation.Data are presented as mean ±SD (n=5)

4. Discussions

Among various synthetic polymers, Bacterial Cellulose (BC) has gained attention due to its purity and abundance. Character of BC was biocompatibility, non- toxicity, non-allergic and high swelling ability make BC a potential scaffold for tissue engineering [10]. The BC scaffold have been shown to support the ingrowth of human chondrocytes and human smooth muscles cells in vitro because morphology structure of BC is similar to the collagen fibrils of human bone, but BC has too small pore size to allow the ingrowth of cells into the scaffold after in vivo implantation [5,10]. Many BC composites such as BC/polyacrylamide, BC/polyester and BC/collagen has been used as biomaterial in dentistry. One of exciting BC is cellulose nata de coco derived from coconut water fermentation and *A. xylinum*. This cellulose was used in this study besides hydroxyapatite from *A. granosa*.

Table 1. The average porosity of HAp/cellulose nata de coco with different time of precipitation and drying process. Data are presented as the mean ±SD (n=5)

Precipitation Time (Hours)	Drying for 24 hours	Drying for 48 hours	Drying for 72 hours
	Mean± SD (%)	Mean ± SD (%)	Mean ±SD (%)
5	0.234± 0.03	0.205 ± 0.05	0.20 ± 0.08
15	0.315 ±0.11	0.194 ± 0.06	0.15 ± 0.02
25	0.17 ±0.05	0.207 ± 0.03	0.09 ± 0.02
C+	0.03± 0.01	0.03 ± 0.01	0.03± 0.01
C-	0.88 ± 0.22	0.60 ± 0.15	0.60 ±0.07

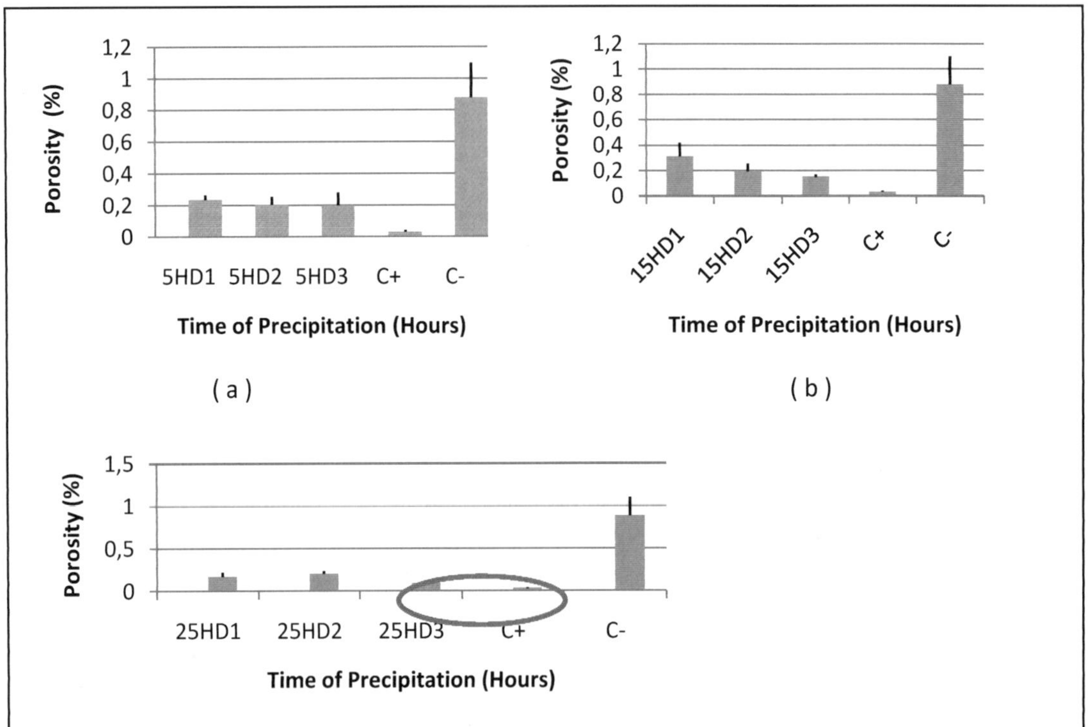

Figure 2. The average porosity of HAp/cellulose nata de coco with different time of drying. Data are presented as the mean±SD (n=5)

The Hydroxyapatite (HAp) enhances osteoblast cell growth, osteoprogenitor attachment and differentiation to promote a new bone[10]. HAp/BC has shown better biocompatibility and osteo-conductivity compared combination HAp/Polyester and HAp/Collagen [10,14]

In this study, manufacturing HAP/ BC from cellulose nata de coco (Cnc) composites with difference time of precipitation and time of drying influence the result of porosity. The comparison of manufacture based on time of precipitation shows that time of precipitation for 25 hours has lowest porosity due to it produces more HAp deposits on the surface and penetrate into Cnc fibrils.

As we know, the Cnc is hydrophobic material that has high porosity which causes ability to absorption the water was very high. While HAp is hydrophilic material with low porosity but big pore size. The combination HAp/Cnc is expected have the ability of water absorption as well as human bone [15,16]. If HAp/Cnc has ability as well as human bone, this material can make osteoblast, chondrocytes, endothelial cells, and smooth muscle cells migration and attached on the scaffold [17].

Besides, result of precipitate influences porosity composite, drying which causes shrinkage and crack in pore size, damaging interconnected porous network and porosity then makes decreasing the porosity. In this study, the composites of HAp/Cnc was dried for 72 hours had lowest porosity because it made the pore of composites was damaged and it made porosity composite decrease approaching porosity human bone. The porosity composite that approaching with porosity human bone are good candidate for new bone graft due scaffold that result from this material can attached by osteoblast which can trigger osteogenesis process and process of making new capillary [18].

5. Conclusions
The good porosity of biomaterial HAp/Cnc was precipitated of 25 hours and drying for 72 hours (3 days) because it had porosity near the bone porosity so it was feasible as bone graft candidate

Acknowledgements
Financial support provided by Faculty of Medicine Universitas Mulawarman is gratefully appreciated.

References

[1] Farah A, Giogio P, Sophia PP,Aldo B G, William V, Giannobile and Giulio R 2016 *Int. Journal of Dentistry* pp 1-15
[2] Sung J A, Young M S, Se E K, Sung I J, Jin O J, Jong S P, Hui J G, Da E S, Young C N, Seong S K, Chong Y K, Jung B H and Youn M L 2015 *Biotechnology and Bioprocess Engineering.* **20** pp 948-55
[3] Neftaha T, Ze Z, Younes M, Luciana A L, Lisineia M Z, Dennis L and Mahmoud R 2012 *AAMB Express.* **2** pp 1-61
[4] Kolos E C, Ruys A J 2014 J.Mater Sci: Mater Med. 25 1801-17
[5] Pelagie M F, Sandra P O, Mukta K, Ming G, Lucia A, Thomas J W 2016 *J. Cellulose.* **23** pp 1263-82
[6] Tetteh G, Khan A S, Delaine-Smith R M, Reilly G C, Rehman I U 2014 *J Mech Behav Biomed Mater.***39** pp 95-110
[7] Dean M L 1997 *Journal of Material Science: Materials in Medicine.* **8** pp 227-32
[8] Ozgiir K, Nicholas P B, daniel H S and Keith J Bowman 2015 *AAPS Pharm Sci Tech.***16** pp 1455-64
[9] Christina C, Ana B, Angeles B, Carlos N 2016 *Cellulose.***23** pp 57-91
[10] Deepachitra R, Thotapalli P S 2014 *J. Cellulose* **21** pp 3585-95
[11] Pericles N M, Sheila C R, Oduvaldo C M P J, Viciany E F, Sara L R L, Joao F D L N and Fernanda D C L A 2009 *Acta Veterinaria Scandinavia.* **51** pp 1-12
[12] Sugandi 2014 Sintesis Hidroksiapatit Berpori Dari Cangkang Telur Ayam Dengan Matriks Selulose *Tesis* (Bogor: Bogor Institute of Agriculture)[Indonesian]
[13] Moran, Michael J, Shapiro, Howard N 2004 *Termodinamika Teknik* Ed 4 (Jakarta: Erlangga)
[14] Kim S, Ku S H, Lim S Y, Kim J H, Park C B 2009 *Adv mater.***23** pp 2009-14
[15] Reddy N, Yang Y 2015 Bacterial cellulose Fibers (*Berlin Heidelberg)* (Berlin-Heidelberg: Springer-Verlag) pp 307-29
[16] Murr L E 2015 Handbook of Materials Structure, Properties, Processing and Performance (Switzerland: Springer International Publishing)
[17] Shin PL, Iris L C, Jeffrey M C, Je R L, Ali D, Kuan C C 2013 *Cellulose.* **20** pp 2191-219.
[18] Zhijiang C, Jaehwan K 2010 *Cellulose.* **17** pp 83-91

Prospect and policy of palm oil mill effluents for future electricity in east kalimantan (utilization of pome as renewable energy)

M I Aipassa [1*], R Kristiningrum [2], V Y Tarukan [3]**

[1,2] Faculty of Forestry Mulawarman University
[3]Head of Electric and the Energy Use Divison, East Kalimantan Mining and Energy Agency

Corresponding author: *marlon.ivanhoe@gmail.com, **citra_lestari2008@yahoo.co.id

Abstract. East Kalimantan economy for four decades was mainly based on natural resources extraction and dominated by primary sectorwith the six highest GDP in 2013. But, the contribution of oil and gas were decreasing production due to the absence of new wells.One of the mission was create natural resources and renewable energy based economic people oriented. The Goverment of EK Province chose a strategy of socio-economic transformation based on renewable natural resources. This strategy has been applied in the regional development plan by mainstreaming climate change issues. Data related to energy source and its potential, remote rural electrification, bioenergy feedstock, etc including from the Palm Oil company was collected and subsequently analized in line with the EK Governor Letter. Currently (2014) available of Biogas-Pome as bioenergy feedstock is 162 million m^3year^{-1}, where as currently utilized is only 22 millionm^{3}year^{-1}. Power demand supply status in January 2015 indicated as available capacity is 467 MW where the peak demand is 444 MW. About 22% of households without electricity are difficult to be electrified without breakthrough efforts. About 215 thousand households are un-electrified, with more power need about 150 MW in total capacity. As business opportunity, high demand for rural electrification, particularly in Kutai Kartanegera, Kutai Timur, Kutai Barat, Berau and Paser.

1. Introduction

Energy often known as the catalyst for development, is an important element to support the population growth, urbanization, industrialization and tourism. Fossilized energy has contributed the most impact of the economic development in the world through the transportation sector and energy conversion sector. Palm oil mill effluent (POME), agro-based industries and farming industries are identified as potential sources of biogas. It was studied that more than 500 kt of biomethane could be produced yearly if all the POME is treated anaerobically[1]. The utilization of biomethane has remained unexplored for its injection into natural gas grid. Middle East region has supplied about half of oil production in the world and the rest come from the central Asia region and America continent [2]. The economics of depending mostly on imported fuels have grown in recent years as oil prices have become unstable, doubling in less than two years the price of 2004 and reaching in 2006 oil prices [3]. Moreover, in the environmental side, the use of fossilized energy especially in the transportation where at least 25% has contributed to the increasing of toxic gas and affected and greenhouse emission. With the growing awareness of energy-pollution and a climate change consequence, as well as complying to the Kyoto Protocol that targets the use of less energy pollution from emitting energy source which is a clear message to the global for a change towards more sustainable energy production and conservation [4 - 6].

There are astonishing efforts to promote the use of solar power, hybrid with less consumption of fuel, hydrogen fuel cell, biofuels and others to the people. Still not many people are ready to the complete changing where the cost is still expensive in terms of cost, not truly practical and lots more researches needed to improvise those methods [7]. Nevertheless, the suitable replacement and quite similar to the fossilized fuels is the biodiesel fuels that extracted from the plants and animal fats in term of characteristics, properties and without any modification of the engine. In recent years, there are two main types of biofuels which refer to first of generation biofuels called bioethanol, which is derived from starch or sugar such as cane, corn and sugar beet, and second generation biofuels is biodiesel, derived from animal fats and vegetable oils, for example from palm oil, jathropa, soy bean, rapeseed oil, neem oil and others [7] [8].

Significant sustainability concerns are also associated with palm oil production at plantations and mills. The environmental impacts are mainly due to open lagoon treatment of palm oil mill effluent (POME) and fertilizer and pesticide use. The potential use of palm oil residues was reviewed for the first time in 2006. Quantified the potential GHG emissions and benefits from residue use and concluded that use of residues may significantly improve the environmental performance of palm oil production. Socioeconomic aspects of sustainability, such as plantation worker safety and health and land tenure rights have also been researched [9].

Indonesia has the potential of renewable energy sources such as biofuel, biomass, geothermal, water, solar wind, ocean waves to tidal energy of sea water. The president of the republic of Indonesia keeps the ambition so that renewable energy can be maximized as possible to release the dependence of fossil energy. Therefore in 20125 the use of renewable energy is expected to reach 23% of total energy. For now, fossil energy still dominates the supply of energy needs. Noted, fossil energy contributes 95% of the total available energy 47% of which are petroleum, 24% natural gas and 24% coal. While the use of renewable energy has only reached %[10].

Palm oil was as biofuel potential of renewable energy have waste when they process. There are two types of waste generated from oil palm plantations in the form of liquid waste which is the residue of the process of producing CPO or in the form of solid waste derived from empty bunches or from palm shells. Indonesia currently has 850 oil palm plantations annually capable of producing liquid waste reaching 28.7 million tons and 15.2 million tons of solid waste. If a palm oil plantation capable of processing 30 to 45 tons of fresh fruit bunches per day will be able to generate 1 megawatt of electricity from liquid waste and 3 megawatts of solid waste. The palm oil industry is ready to utilize renewable energy through biogas power plant (PLTBg). This is to welcome the government's plan which will require the waste of palm oil to be utilized for power plant raw material [11].

In this review, a brief outlook has been conducted on the potential of palm oil mill effluents for future electricity in East Kalimantan Province (utilization of pome as renewable energy). The survey cover general overview of East Kalimantan prospect of POME, business opportunity, rational for pome to electricity investment, key challenges, business risk and multi stakeholder cooperation to debottleneck investment pome to electricity.

2. General Overview East Kalimantan and Power Supply Crisis

As a developing country, Indonesia has been constantly trying on its toes to meet with the ever increasing energy demand by finding source of energy like Palm oil. East Kalimantan Province on utilization of Palm Oil Mill Waste for electricity in the context of policy and its investment opportunity in EK. As you know that East Kalimantan is the second largest island in Indonesia after Papua with total area of 12.7 million Hectares, with population approx. 3,3 million people. We have approx 12 million hectare of area, and blessed with biodiversity and resources that support our economic growth in the previous decades. Whereas total area of East Kalimantan was 12.532.799,52 ha, with total population 3.3 million and have potency of natural resources like forest, oil and gas, Mineral Mining, Ecosystem Services, etc. For four decades, East Kalimantan economy was mainly based on natural resources extraction and dominated by primary sector. With this natural resources, we develop our province and country and its been made East Kalimantan Province as province with the siz highest GDP in Indonesia in 2013. But, the contribution of oil and gas were decreasing due to the absence of new wells and degreasing due to the absence of new wells and the last decide the role of non- oil sector in East Kalimantan has increased, particulary driven by coal mining, while the contribution of renewable resources into the economy such as agriculture and service are remain small. With structure we can see that our economic growth rate was decrease sharply [10].

East Kalimantan Province have power supply crisis. Power demand-supply status in East Kalimantan per January 2015indicated available capacity is 467 MW where the peak demand is 444 MW. PLN power sales growth in Kalimantan 2009-2014 is 10.5% (PLN 2015)22% of households without electricity are difficult to be electrified without a breakthrough efforts , About 215 thousand households are un-electrified, with more power needs about 150 MW in total capacity (Source: Electrification Ratio 2015, East Kalimantan Mining and Energy Agency) [11].

Remote rural electrification is the key to reach the 80% electrification target.Most of non-electrified households are in remote rural areas. Cheap coal power plant cannot help remote rural areas because coal power plant is only feasible in a big scale operation, i.e. >10 MW; low number of inhabitants in remote rural areas, thus unable to absorb a big scale power plant and the distance between urban and remote rural areas is too far for stretching power grid resulting significant power loss.

The standard solution for electrifying rural areas is diesel generator but high operation cost (IR 2,000 - IR 4,000 per kwh) and High GHG emissions (1 kwh = 0.9 kg CO^2) and increase the pressure on fiscal (fuel subsidy) and trade balance.

3. Potency and Prospect POME

The potential of palm oil plantation waste as renewable energy includes palm oil shell of 5-6% is used as boiler plant, 60%-65% POME shifts have not been used optimally and empty bunches of 20-22% as fertilizer or mulch. POME (palm oil mill effluent) contains methane gas (50-75%) and carbon dioxide (25-25%). POME processing in open ponds causes methane gas. The effect of methane on global warming is 21 times that of CO2. Methane is a flammable gas and can be used as a fuel for renewable energy.

The benefits of POME for EBT/rural electricity for plantation companies are to become new business opportunities ie electric energy with feed-in tariffs, lower diesel / diesel consumption, support government regulation in achieving ISPO criteria and local government policies, increase social support with low-cost electricity supply by cooperation with PT PLN and improving the bargaining position of CPO domestically and the downstream industry.

The benefits of POME for EBT / rural electrification for governments and communities is a low-cost solution for improving rural electrification, promoting rural economic activity, saving fuel subsidies, reducing the value of fuel imports, helping achieving East Kalimantan energy targets by 3% by 2020, GHG emission reduction targets from industry and industrial waste sectors and mitigate business disruptions and plantation conflicts [10].

3.1 Resource Potential and Current Development of Bioenergy Development

Based on Table 1 we can explained that potential resources of biogas and biomass. Bioenergy feedstock biogos from POME, but bioenergy feedstock biomass from palm oil solid by product and vegetation on ex-mine land. Bioenergy from biogos-POME have currently available 162 million m^3 $year^{-1}$ but currently utilized just 22 million m^3 $year^{-1}$. There are mean that currently unutilized for 162 million m^3 $year^{-1}$. It means having power generation potential amount 34 MW. While palm oil solid by product have currently available 3.9 millionton $year^{-1}$ but currently utilized just 2.8 million ton $year^{-1}$. There are mean that currently unutilized for 1.1 million $tonyear^{-1}$. It means that palm oil solid having power generation potential amount 117MW. Bioenergy feedstock from vegetation on ex-mine land have currently available 234,000 ton biomass potential $year^{-1}$ but none are used for bioenergy. There are mean that currently unutilized for 31 MW.

Table 1. Bionergy development as resource potential

Re- Resources	Bioenergy feedstock	Currently available (2014)	Currently utilized (2014)	Currently Unutilized	
				Volume	Power Generation Potential (MW)
Biogas	Biogos-POME*	162 million M3/year	22 million M3/year **	140 million M3/year (2014)	34 MW (2014)***
Biomass	Palm oil solid by-products (shell, Fiber, EFB)*	3.9 million ton/year	2.8 million ton/year	1.1 million ton/year(2014)	117 MW (2014)
	Vegetation on ex-mine land (available land)*	36,000 ha → 234,000 ton biomass potential/year	None are used for bioenergy	36,000 ha (2014)	31 MW (2014)

Sources:
* GE-LAMA-I 2016
** utilized by 3 POMs
*** including 13 potential POMs

While current development, from available data could be seen at following Table 2.

Table 2. Bioenergy development as current development.

Feedstock Resources	Feeding to PLN		Self-used (biogas)
	Biomass	Biogas	
Palm Oil (60+ mils)	4 units, 0.8 MW (operating) 1 unit, 3 MW (construction)	1 unit, 4 MW (operating) 2 unit, 4 MW (construction)	2 units, 1.8 MW (operating)
Forest plantation	1 Unit, 9.5 MW (plan)		

Note : Data Biomass provided from some Palm Oil Company, as followsn : 1. PT Tanjung Buyu Perkasa; 2. PT Teladan Prima Sawit; 3. PT Telen; 4. PT Hutan Hijau Mas (Operasional), than Biogas: REA Kaltim (sold to PLN) dan PT Prima Mitrajaya Mandiri + PT Indonesia Plantation Synergy (self used).

Table 2 explains that feedstock resources from palm oil and forest plantation. Palm oil have feeding to PLN from biomass as much 4 units operating with 0.8 MW and 1 unit construction with 3 MW. But from biogas provide 1 unit operating with 4 MW and 2 unit construction with 4 MW. While self - used (biogas) as much 2 units operating with 1.8 MW.

3.2. Business Opportunity

Business case for POME to electricity could be figured out as follows : (1) potential for power generation: over 34 MW (based on table 1), (2) viable investment: IRR from 12 to 20%, (3) high demand for rural electrification, especially in Kutai Kartanegara, Kutai Timur, Kutai Barat, Berau, and Paser[10] [12][13].

3.3. Rationale for Pome to Electricity Investment

Regulatory and market pressure to electricity investment were (1) Indonesia Sustainable Palm Oil (ISPO) standard compliance requires methane capture; (2) draft of EK sustainable plantation regulation/Raperda, 2016: Palm oil mills must develop renewable energies for rural electrification; (3) market pressure for 100% RSPO-certified sustainable supply chains and (4) early mover sustainability premium.

CSR have advantages like electrification for rural areas, secure social license to operate, government relation can support to EK Government efforts towards rural electrification and GHG emission reduction.

GHG emission reduction can contribute to the regional GHG emission reduction ambitions East Kalimantan regional target of 19% reduction in GHG emission by 2020 Production of 1 tonne of CPO can generate over ½ tonne of CO2eq if methane is not captured

Financial viability must operational efficiency, continued electricity generation (currently electricity is produced only when the mill operates), reduced diesel consumption, generation of new revenue streams, potential IRR estimated between 12 to 20% depending on capacity and flexibility regarding financing needs and risk exposure (EPC versus IPP models).

3.4. Key Challenges and Business Risk

Key challenges for this program were regulatory risk (Burdensome and uncertain licensing procedures), PLN off-take (Uncertainty regarding off-take: is there enough demand in the area? Is the mill close enough from the grid or Will PLN commit to buy or not.

Oil palm companies are exposed to technical and financial risks. The main risks are due to insufficient human resource capacity for operation and long-term survival risks due to inadequate maintenance practices, as well as the lack of engine supplier support for individual site support.

Business risk this program were the development of a POME to electricity facility requires significant investment and electricity generation is not palm oil companies' core business.

3.5. Multi Stakeholder Cooperation to Debottleneck Invesment Pome to Electricity

Regulatory risk, strong political commitment and multi-stakeholder cooperation:

a) Governor letter No 671.2/4753/Distamben dated 13 May 2013 instructing Bupati to facilitate the construction of biogas power plant in their region to increase rural electrification services
b) POME Biogas Partnership signed on 22 May 2014 between Provincial Government, PLN, District Governments, GAPKI, and GIZ to cooperate in promoting investment in POME to electricity in 3 Districts
c) Governor letter No 050/2726/Bapp/2015 dated 17 April 2015 on Proposed policy to four ministries to regulate access to POME-biogas
d) Support districts to draft a regulation to optimize the palm-oil waste utilization
e) Coordination with PLN to identify priority locations for POME to electricity
f) 20 mills have been identified for further assessment
g) Provincial Government will provide assistance to conduct Pre-Feasibility Studies

Business risk, The Provincial Government, through its partnership with GIZ [13] [14] and GGGI [15] has mobilized technical and financial assistance to conduct Pre-Feasibility studies in priority locations:

a) Palm Oil Companies can be accompanied all the way through project development, from Pre-FS to financial close
b) Detail engineering design (DED) and power grid construction;
c) Proactive promotion to the potential POMs;
d) Business-to-business partnership facilitation

4. Conclusion

Sustainable palm oil development in east kalimantan has become a fact as a driver of locomotive economy in the region. The development of palm oil and its derivatives and its products are encouraged to produce environmentally friendly products. Every palm oil plantation company in east kalimatan is expected to support the greenhouse gas emission reduction program, one of which is a liquid waste processing system (POME) for electricity. In the long run to become a new business source that can supply PLN as a source of electricity through the sale and purchase cooperation of electric energy with excess power scheme

References

[1]Hoo PY, Patrizio P, Leduc S, Hashim H,Kraxner F, Tan ST, Ho WS 2017 *J.Energy Procedia***105** pp 562-9

[2] Chin MJ, Poh PE, Tey BT, Chan ES, Chin KL 2013 *J. RenewSustainEnergy***26** pp 717-26

[3] Oosterveer P 2014. *J. Jclepro*(2014)019

[4] Millo F, Debnath BK, Vlachos T, Ciaravino C, Postrioti L, Buitoni G 2015 *J. Fuel* **159** pp 614-27

[5]Sadeghinezhad E, Kazi SN, Badarudin A, Oon CS, Zubir MNM, Mehrali M2013 *J. Renew Sustain Energy* **38** 410-24

[6] Aurélio M, Lagnier B, Ferreira G, Guilherme L, Lamare A, Murta S, et al 2017 *J.Fuel***194** pp 144-56

[7] Yasin MH, Mamat R, Najafi G, Ali OM, Yusop AF, Ali MH 2017 *J. Renewable and Sustainable Energy Reviews* **79** pp 1034-49

[8] Özener O, Yüksek L, Ergenç AT, Özkan M 2014 *J. Fuel* **115** pp 875–83

[9] Hasen SB, Padfieldb R, Syayutib K, Eversc S, Zakariah Z, Masturad S 2015 *J. Jclepro* (2015)051

[10] Herdiyanto H 2016 *Development of PLTB from POME to EBT*(Samarinda:Plantation Service East Kalimantan Province Report) pp 1-23

[11] Sawit Indonesia 2015 *Palm Waste Must be Processed into Electricity*(Samarinda: Sawit Report) pp 1-60

[12] Flim2014 *Employment Contract GIZ GE-LAMA-I (consultant) East Kalimantan*(Samarinda: Flim) 8

[13] GIZ LCORE2014 *POM Mapping for POME Biogas Power Plant in East Kalimantan* (Samarinda: GIZ) pp 10-5

[14] GIZ, GELAMAI 2014 *Pre-FS at 6 locations of Palm Oil Factory*(Samarinda: GIZ Report) 20

[15]Mining and Energy 2016 *Potential Condition of Electricity and Renewable Energy Development in East Kalimantan Province*(Samarinda: Mining and Energy Report) 8

CO_2 adsorption on SBA-15: A molecular modelling

R R D J N Subagyono[a], A L Chaffee[b]

[a] Chemistry Department, Faculty of Mathematics and Natural Sciences, Mulawarman University, Samarinda, East Kalimantan, Indonesia 75242
[b] School of Chemistry, Monash University, Clayton, Victoria 3800, Australia

*corresponding author: dirgarini_julia@yahoo.co.id

Abstract. A study on molecular modelling of CO_2 adsorption on SBA-15 has been conducted. A model of SBA-15 with 5.6 nm pore diameter and 3.9 OH/nm^2 was constructed and used for the molecular studies of CO_2 adsorption. Adsorption of CO_2 molecules on SBA-15 occurred as a result of the formation of hydrogen bondings between CO_2 molecules and hydroxyl groups on the silica surface. In the presence of water molecules, the interaction between CO_2 molecules and the silica surface slightly decreased.

1. Introduction

SBA-15 (the acronym of Santa Barbara Acid 15) is one ordered mesoporous materials with a variety range of application. Due to the unique characteristics of SBA-15, this mesoporous material is applicable as catalyst [1] and adsorbent[2, 3]. SBA-15 is a mesoporous material with an ordered hexagonal structure that has a large pore size (5-30 nm), high surface areas (687-910 $m^2.g^{-1}$), large pore volumes (0.5-1.2 $cm^3.g^{-1}$), and micro pore volumes [2, 4, 5]. The micropores increase the surface areas and surface roughness that are useful for adsorption. Furthermore, these micropores presented result in a significant higher pore volumes and surface area [5]. In comparison to MCM-41 and HMS, which have a hexagonal structure, SBA-15 is hydrothermally much more stable due to the thicker pore walls of SBA-15 (2.7 nm) [6]. It appears that SBA-15 is promising for use as a CO_2 capture agent considering the pore size and the high hydrothermal performance, even in the presence of moisture.

Over the years SBA-15 has been studied in the context of CO_2 adsorption. However, there has been a limited study of the relationship between molecular structure of SBA-15 and CO_2 diffusion within the structure on an atomistic scale. Computational chemistry, based on quantum mechanics, provides features to allow investigation of molecular structures on nanoscale and reaction mechanisms within the structures [7] . Among other computational chemistry methods, molecular modeling or molecular dynamic (MD) is a powerful tool for providing a better insight of CO_2-SBA-15 interaction, especially dynamic behaviors up to an atomistic level.

The models for SBA-15 have been proposed with different methods. Most recently, Bhattacharya, *et.al* [4] constructed the realistic model for mesoporous silica SBA-15 including the micropores in the pore walls. The initial structure for the model was a group of triblock copolymer surfactants which each surfactant consists of 11 lattice units. The surfactant was constructed to mimic the experimental surfactant P123, with 5 hydrophobic units and 3 hydrophilic units on each surfactant molecule.

In this work, a model for the structure of SBA-15 is proposed. This model is used to investigate the dynamic behaviours of partially dehydroxylated SBA-15 as opposed to fully hydroxylated SBA-15. The study also aims to study the CO_2 diffusion behaviour within SBA-15 in the presence of water molecules.

2. Materials and method

Molecular dynamic studies was conducted using Material Studio® 4.4 (Accelrys; San Diego)[8] combined with the molecular simulation program DISCOVER, FORCITE and the COMPASS as forcefield. The initial model of SBA-15 was constructed and then was subjected to further preparation to get a good agreement with the experimental material. The study focused on evaluating the thermal stability of the hydroxyl groups on the surface of SBA-15 and the dynamic behaviour of SBA-15. SBA-15 was constructed with the similar procedure as reported by Chaffee [9]. The initial structure used for the molecular dynamic studies was generated from a three dimensional infinite lattice of α-quartz (a=4.913, b=4.913, c=5.4052, α=90, β =90, γ=120). The pore diameter for the model is 5.6 nm, which corresponds to the experimental materials to allow proportional comparison. After geometry optimization, the fully hydrated material was then subjected to subsequent different temperatures, namely 373 K, 473 K, 573 K. This dehydration steps were performed via an iterative sequence of MD runs followed by water removal from Si-O-Si linkages.

The partially dehydrated model was used to investigate the CO_2 diffusion behaviour, in the presence of water molecules. Dynamic simulation was conducted by placing 15 molecules of CO_2 in arbitrary positions within the structures. Calculation of the diffusion behaviour of CO_2 in the presence of H_2O was achieved with the same procedure by adding 15 CO_2 molecules and 10 H_2O molecules. The model was subjected to dynamic simulations at 348 K at constant volume and temperature (NVT). The simulations controlled by Anderson method were carried out using 110,000 steps (110 ps) and 1 fs time steps. After which, the partially hydrated model is analysed with mean square displacement to get plots of interactions between gas and the silica surface.

3. Results and discussion

3.1. SBA-15 modeling

A model of SBA-15 has been prepared with 5.6 nm pore diameter. The initial model of α-quartz was used to construct a supercell with a=7.36 nm, b=7.36 nm, c=0.54 nm, α=90, β =90, γ=120. The supercell was carved to construct a whole within the structure and was followed by hydroxylation to Si and O atoms which have less than 4 bonds and 2 bonds, respectively. The fully hydroxylated SBA-15 was subjected to geometry optimization to get a reasonable structure with the lowest energy.

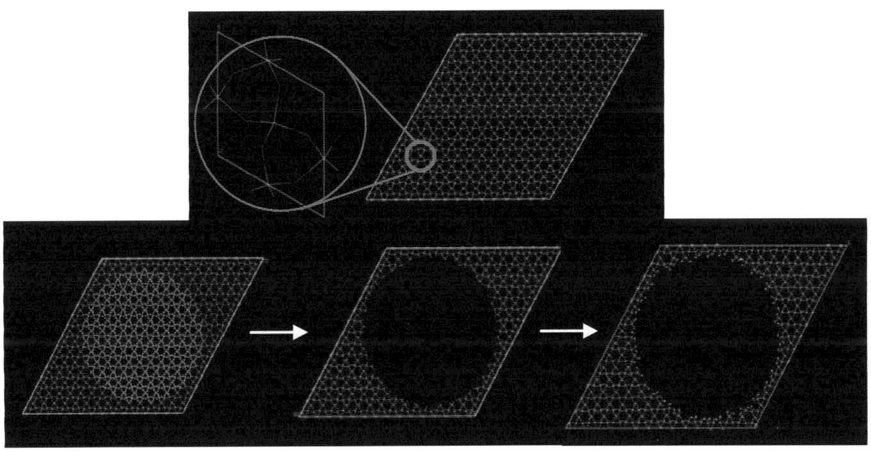

Figure 1 Construction steps of fully hydroxylated SBA-15. Three dimensional structure of α-quartz was used as initial structure. All atoms within radius 2.6 nm were selected to form a whole within the structure. The fully hydroxylated model was achieved by adding O-H or H.

Molecular dynamic simulation was conducted to determine which bonds are close enough and form hydrogen bonds. 72 sets of adjacent OH groups were edited prior to length distribution analysis. One set is the distance between O of one OH group and H of the adjacent OH group (**Figure 2**). Each set was analysed with length distribution. The result from the analysis is a plot of length distribution versus frequency, as can be seen at **Figure 3**.

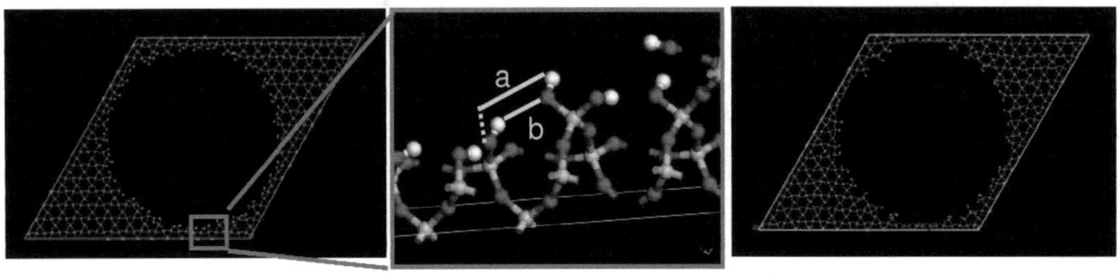

Figure 2 Selection of sets for length distribution analysis. The removal of water molecules is determined from length distribution plots resulting in partially dehydrated SBA-15.

The hydroxy pair 14 is close enough to form hydrogen bond whilst hydroxy pair 19 is not close enough to form hydrogen bond. Thus, pair 14 was chosen for dehydration by removing H_2O molecules from Si-O-Si linkages. At 373 K, there are 16 pairs of sets that are close enough to form hydrogen bonds. These all sets were dehydrated by removing water molecules. After which, the partially dehydrated model was subjected to geometry optimization with Fine quality using Smart method. Molecular dynamic simulation at higher temperature was performed followed by length distribution analysis to perform dehydration steps continuously. The iterative replication of this process, result in a reduction of silanol numbers from 8.2-3.9 OH/nm^2.

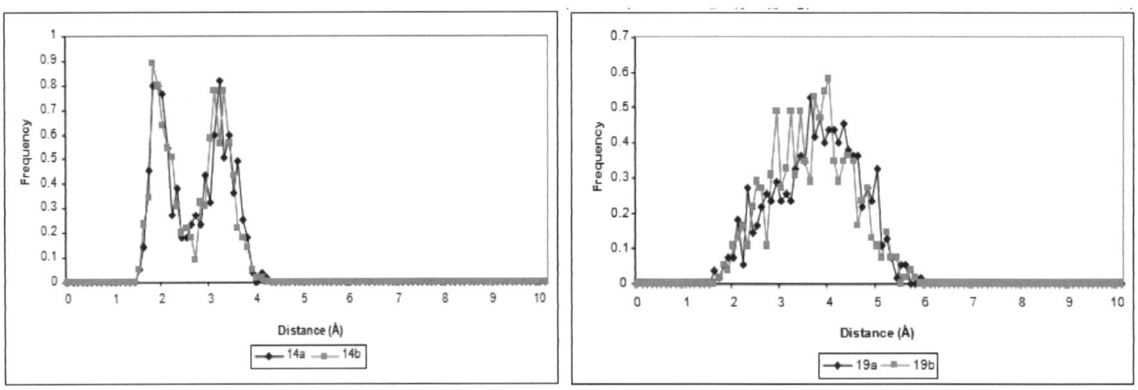

Figure 3 Length distribution plots for hydrogen bonds the hydroxy pairs 14 and 19. The maximum distance for hydrogen bond formation is 2.5 Å.

Within fully hydroxylated SBA-15, there are various types of hydroxyl groups on the silica surface, namely free (isolated) OH, vicinal (bridged) OH, geminal OH, and siloxane bridge (**Figure 4**). The dehydroxylation of silica surface occurs in several steps. Therefore, the molecular dynamic simulation was carried out at progressive temperatures. According to Voort *et.al* [6], at 373 K, below 400 K, the main dehydration occured is the removal of physisorbed water. Yet, in the fully hydroxylated model, there is no physisorbed water molecules modeled. However, at this temperature there is a significant decrease in the amount of free hydroxyl group by 50%, from 51 to 24. At higher temperature, 473 K, the amount of free OH was constant compared to that at 373 K. Yet, the number of geminal OH and vicinal OH decline slightly to 5 and 3 pairs, respectively. Approximately 27 free OH are present on the SBA-15 surface at 573 K. The models resulted from the dynamic simulation at this temperature was chosen as a final SBA-15 structure containing 3.9 OH per nm^2. At higher temperature, the distance between O of one OH and H of another OH is getting further, so that it is more unlikely for both O and H to create hydrogen bond. The increase in temperature, from 373 K to 573 K, increases the amount of free OH but decreases the amount of bridged and geminal OH. Above 673 K, it is predicted that there will be only free OH left on silica surface. The results have a good agreement compared to silica gel with pre-treatment temperature reported by Voort *et.al* [6]. Surface hydroxyl group concentration and amount of various types of OH at different temperatures are shown at **Table 1**.

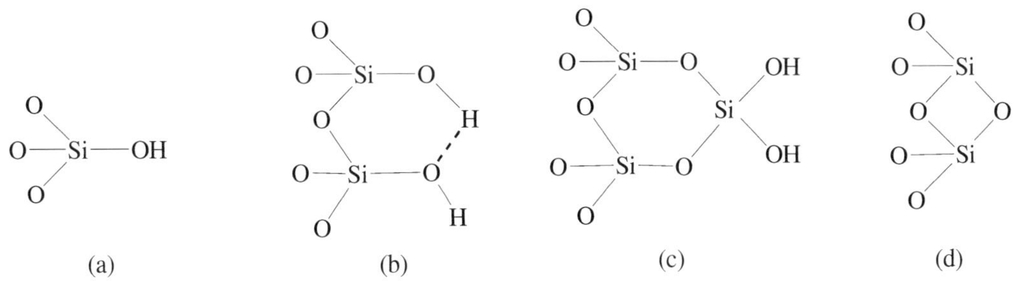

Figure 4 (a) free OH, (b) vicinal OH, (c) geminal OH, (d) siloxane bridged

Table 1 Various types of hydroxyl groups on SBA-15 models

Model	α_{OH} (nm^{-2})	Free OH	Geminal OH (pairs)	Vicinal OH(pairs)
Fully hydrated	8.2	51	11	8
Partially dehydrated				
373 K	4.9	24	8	5
473 K	4.2	24	5	3
573 K	3.9	27	4	2

3.2. CO_2 interaction within SBA-15 structure

Interaction between CO_2, in the presence and in the absence of water molecules, and the silica surface is mainly based on the formation of hydrogen bond. The formation of hydrogen bond between oxygen of CO_2 and hydrogen of hydroxyl group on silica surface bring CO_2 molecules move closer to silica surface. After geometry optimization, 10 out of 15 CO_2 molecules within fully hydrated model get closer to silica surface, whilst 5 other CO_2 molecules stay freely within the pore.

In partially dehydrated model, similar to fully hydrated model, there are 5 free CO_2 molecules within the pore after geometry optimization. However due to less hydroxyl groups this model possessed, it is assumed that there is a competition among CO_2 molecules to form hydrogen bonds, especially on the surface containing hydroxyl group. In the presence of water molecules, the formation of hydrogen bond is not only based on the interaction between O of CO_2 and H of hydroxyl group, but also is created from the interaction between O of CO_2 and H of H_2O, and between H of H_2O and O of hydroxyl groups.

Molecular dynamic simulation was performed after geometry optimization. The dynamic simulation was performed using FORCITE and the COMPASS as a forcefield at 75 ^{o}C and was followed by mean square displacement analysis. The dynamic simulation was achieved after running the model at 110 ps with 1 fs step. Plots of simulation time versus mean square displacement of CO_2 molecules are shown at Figure 5. Three CO_2 molecules interact stronger with hydroxyl groups on silica surface compared to other molecules. These three molecules, denoted as j,o and n, are molecules with low coefficient diffusions. Other molecules with higher coefficient diffusions interact much weaker rather than j,o, and n. These molecules inclined to interact with H_2O molecules and to move actively within the pore. It appears that the presence of water hinder interaction between CO_2 and hydroxyl groups on silica surface.

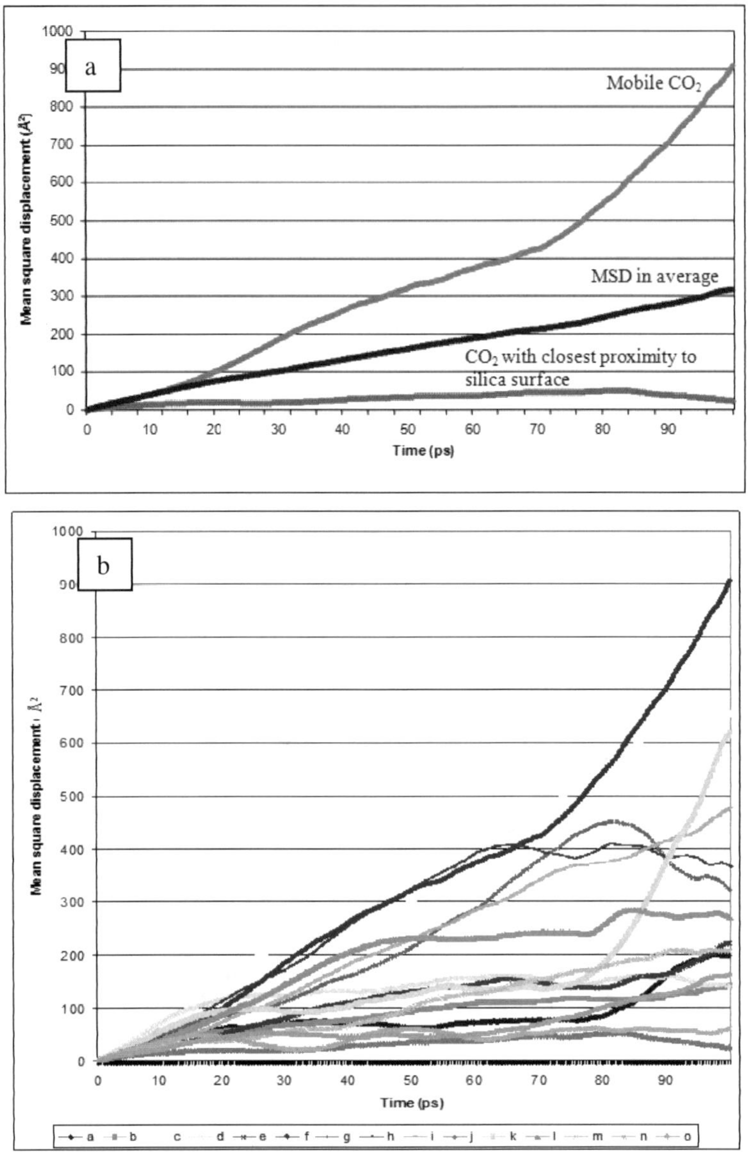

Figure 5 Mean square displacement for (a) CO_2 molecules and (b) each 15 CO_2 molecules within partially dehydrated SBA-15 in the presence of water at 75°C

Coefficient diffusion (D_α) of CO_2 molecules within three dimensional structure of partially dehydrated SBA-15 (3.9 OH/nm^2) is measured by equation below:

$$D_\alpha = \frac{1}{6N_\alpha} \lim_{t \to \alpha} \frac{d}{dt} \sum_{i=1}^{N_\alpha} \left[\left(\vec{r}_i(t) - \vec{r}_i(0)^2 \right) \right] \tag{1}$$

which $\lim_{t \to \alpha} \frac{d}{dt} \sum_{i=1}^{N_\alpha} \left[\left(\vec{r}_i(t) - \vec{r}_i(0)^2 \right) \right]$ is a slope of mean square displacement plot. In Figure 6, the slope of MSD plot is 2.9618, so that the coefficient diffusion for CO_2 molecules in average within the model is 0.4936. Calculations of coefficient diffusion of each molecule were conducted by applying equation

1. However, some mean square displacement plots are not linear, so that the slopes of the graph are determined from linier trend of the graph. For instance, to calculate coefficient diffusion of CO_2 (h), the slope is determined from the linear trend in range of 0-60 ps (figure not shown). Above 60 ps the plot is not linear so it can be neglected in gradient calculation. Coefficient diffusion of each molecule is presented in Table 2.

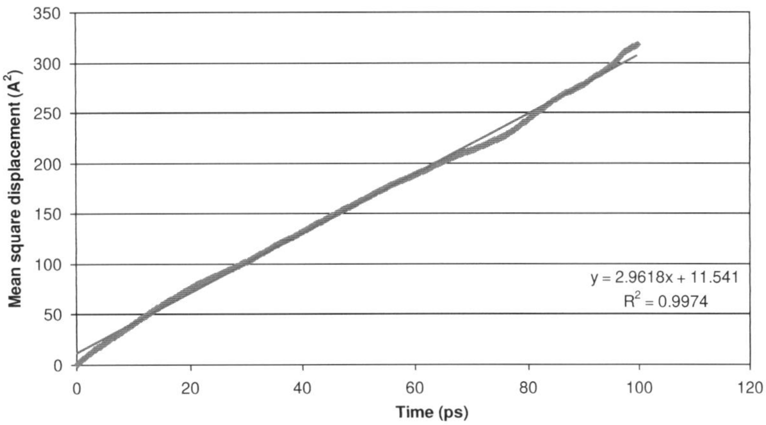

Figure 6 Mean square displacement of 15 CO_2 molecules in average

From the data presented in Table 2, it is clear that CO_2 (j), (o),and (m) are interacted stronger with silica surface compared to other CO_2 molecules. CO_2 (d) with the highest coefficient diffusion, has the greatest mobility compared to other molecules.

Table 2 Coefficient diffusion (D_α) of each molecule within partially dehydrated SBA-15 (the picture presented shows the position of each atom)

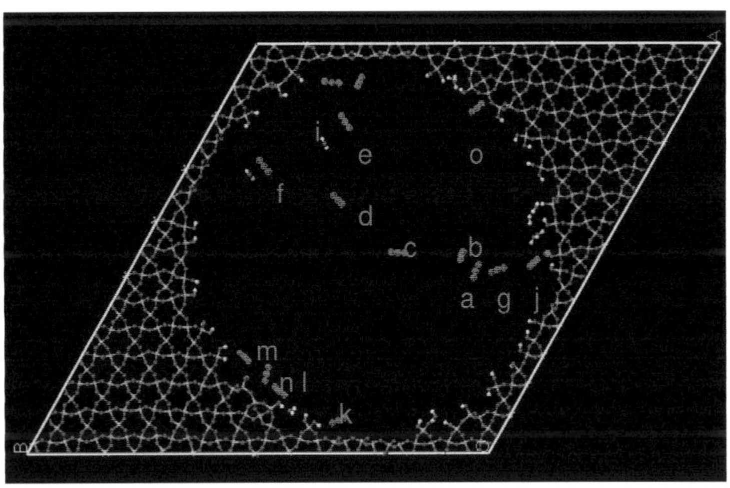

CO_2	(D_α)
a	0.0917
b	0.1232
c	0.2264
d	0.2409
e	0.0928
f	0.2179
g	0.1304
h	0.2234
i	0.1627
j	0.0442
k	0.1649
l	0.1738
m	0.0743
n	0.1024
o	0.0687

Study of CO_2 interaction in the absence and in the presence of water within fully hydrated material is worthy for future work. However, calculation time for quantum mechanical calculation increases with the size and the complexity of the structure. In this mini project, some calculations are still running. Thus, the data for these calculations cannot be reported. These calculations are geometry optimization for fully hydrated model with CO_2 and H_2O molecules within the structure and dynamics simulation for fully hydrated models.

Another future work required is a comparison study of CO_2 interaction within fully hydrated material and CO_2 interaction within partially hydrated material. It is projected that the formation of hydrogen bonds are more likely to occur within a model with more hydroxyl groups as opposed to a model with less hydroxyl groups. The presence of water molecules is also critical to be considered in correspondence to the selectivity of silica surface towards CO_2 and H_2O. It appears that the selectivity of silica surface will decrease in the presence of water.

4. Conclusions

A model of SBA-15 with 5.6 nm pore diameter has been constructed. The model has 3.9 OH/nm^2. The increase in temperature on dehydration process results in a decrease in vicinal and geminal OH, yet results in a rise in the amount of free OH and siloxane bridged. The visualisation of CO_2 interaction within the model gives a better understanding of the dynamic behaviour up to an atomistic level. CO_2 molecules have a tendency to form hydrogen bond with hydroxyl group on silica surface. However, the existence of water molecules slightly decreases the interaction between CO_2 and SBA-15.

Acknowledgments

We thank the Australian Government through Australia Award Scholarship for financial funding during the postgraduate study.

References

[1] Lee J, Park Y, Kim P, Kim H and Yi J 2004 Preparation of NaCl-incorporated plugged mesoporous silica using a cost-effective precursor and applications to the hydrodechlorination of chlorinated hydrocarbons *Journal of Materials Chemistry* **14** 1050-6

[2] Hiyoshi N, Yogo K and Yashima T 2005 Adsorption characteristics of carbon dioxide on organically functionalized SBA-15 *Microporous and Mesoporous Materials* **84** 357-65

[3] Yue M B, Sun L B, Cao Y, Wang Z J, Wang Y, Yu Q and Zhu J H 2008 Promoting the CO_2 adsorption in the amine-containing SBA-15 by hydroxyl group *Microporous and Mesoporous Materials* **114** 74-81

[4] Bhattacharya S, Coasne B, Hung F R, Gubbins K E, P.L. Llewellyn F R-R J R and Seaton N 2007 *Studies in Surface Science and Catalysis*: Elsevier) pp 527-34

[5] Zelenak V, Badanicova M, Halamova D, Cejka J, Zukal A, Murafa N and Georigk G 2008 Amine-modified ordered mesoporous silica: Effect of pore size on carbon dioxide capture *Chemical Engineering Journal* **144** 336-42

[6] Voort P V D, Ravikovitch P I, Jong K P D, Benjelloun M, Bavel E V, Janssen A H, Neimark A V, Weckhuysen B M and Vansant E F 2002 A New Templated Ordered Structure with Combined Micro- and Mesopores and Internal Silica Nanocapsules *The Journal of Physical Chemistry* **106** 5873-7

[7] Silva E F d and Svendsen H F 2007 Computational chemistry study of reactions, equilibrium and kinetics of chemical CO_2 absorption *International Journal of Greenhouse Gas Control* **1** 151-7

[8] Perego C and Bosetti A 2011 Biomass to fuels: The role of zeolite and mesoporous materials *Microporous and Mesoporous Materials* **144** 28-39

[9] Chaffee A L 2005 Molecular modeling of HMS hybrid materials for CO_2 adsorption *Fuel Processing Technology* **86** 1473-86

The Application of Electrolysis Method to Reduce Ammonia Content in Liquid Waste of Tofu

S Prabowo*[1], Nurlaili[2], Muflihah[2], R A Tindangen[2], Sukemi[2,3]

[1] Faculty of Agriculture, Mulawarman University, Samarinda, Indonesia
[2] Chemical Education Department, Faculty of Teacher Training and Education, Mulawarman University, Samarinda, Indonesia
[3] SMA N 2 Samarinda, Samarinda, Indonesia

*Corresponding Author: sprabowo@faperta.unmul.ac.id

Abstract. Ammonia (NH_3) is known as an important chemical in industrial sector. It is also known as harmful pollutant. Ammonia is a weak base, a gas in room temperature and has 330°C of BP. The aims of research were to investigate the effect of voltage (4 to 12 volt), time (1 to 30 min.), concentration of ammonia (0.01 to 0.05 M) and potassium hydroxide concentration on the ammonia content in aqueous solution by using electrolysis method with platinum as electrodes. The ammonia content was analysed by using UV-Vis spectrophotometer. The result showed that an increment in the voltage, time and potassium hydroxide concentration could increase the amount of converted ammonia. The optimum condition to reduce the ammonia content by using electrolysis method was 10 V of electrical voltage, 25 min. of electrolysis time and 0.04 M of potassium hydroxide concentration. At the optimum condition, the electrolysis method could decrease 81.13% of ammonia content in liquid waste of tofu.

1. Introduction

Environmental pollution is a crucial issue in efforts to establish a healthy environment. One of the sources of environmental pollution is waste from industrial activities. Based on observation and interview, it is known that some industries in Samarinda dispose waste to river. One of industries that continuously dispose the waste to river is tofu industry. Tofu industry produce solid and liquid wastes and they are usually disposed to river. The liquid waste of tofu contains protein.Decomposition of the protein in the river producesammonia.

Liquid waste of tofu has ammonia content around 2.21 – 16.00 mg/l. It exceeds the minimum content of the ammonia; 0.50 mg/l. Ammonia can increase the oxygen demand in gill and damaged tissue of fish and decrease ability of blood to transfer oxygen. In chronic condition an increment of ammonia content cause illness and decrease growth.

Many ways have been done to handle the tofu's liquid waste such as an aerobe process and convert the waste to biogas and fertilizer. Another way is application of electrolysis on the treatment of the waste. Electrolysis is a decomposition process of chemicals by using electrical energy. Electron flows from and to electrical source, but is not between the electrodes. Cation goes to cathode and is neutralized with the addition of electron at cathode. Anion goes to anode and release electron to be a neutral. The application of electrolysis on the waste treatment is depend on the potential electrode. By using electrolysis method, the ammonia in liquid waste can be decomposed and the ammonia content can be decreased [4, 5, 6].

Some researches on the application of electrolysis methods to reduce the ammonia content have been done, especially at developed countries. The usage of electrolysis methodis not only focus on the decrement of the ammonia content, but also focus on the usage of hydrogen as the result of ammonia decomposition as fuel. Some studies have been done on the application of electrolysis to decompose ammonia. An increment in electrical voltage that used in the electrolysis of aqueous ammonia solution increase the amount of converted ammonia solution. On the other hand, anincrement of potassium hydroxide concentration in the electrolysis of aqueous ammonia solution decreases the ammonia content in the solution [7,8, 9].

According to the reasons above, this research was created to find the optimum condition; electrical voltage, time, potassium hydroxide concentration and electrode types on the treatment of tofu's liquid waste.

2. Methods

2.1. Sample and Chemicals

Sample on this research was the liquid waste of tofu. The waste was collected from home industry that is located in Samarinda city, East Kalimantan, Indonesia. The waste was centrifuged for 10 min. Then the mixture was separated and the clear part of the waste was collected and used as the sample. All the chemicals that used in this research were analytical grade.

2.2. Electrolysis procedure

20 ml of certain concentration sample was loaded in the electrolysis container (beaker glass). 3 drops of certain concentration of KOH was added. After the electrodes were loaded in the solution, the electrolysis was started for certain time and at the certain voltage.

Determination of the optimum electrical voltage was done at the conditions; 100 ppm of initial ammonia concentration, 2 min. of electrolysis time, 0.02 M of potassium hydroxide concentration and 4, 6, 8, 10 and 12 volt of electrical voltages.

Determination of the optimum timewas done atthe conditions; 100 ppm of initial ammonia concentration, 0.02 M of potassium hydroxide concentration, 10 volt of electrical voltage, and 10, 20, 25 and 30 min. of electrolysis time.

Determination of the optimum potassium hydroxide concentrationwas done at the conditions; 100 ppm of initial ammonia concentration, 25 min. of electrolysis time, 10 volt of electrical voltage, and 0.01, 0.02, 0.03 and 0.04 potassium hydroxide concentration.

2.3. Determination of ammonia concentration

The sample was loaded into 50 ml beaker glass. 1 ml Nessler reagent was added. The mixture was homogenized and leaving for 10 min. Then the absorbance of the mixture was measured by using UV-Vis spectrophotometer at 495 nm of wavelength. The ammonia concentration was calculated by plotted the absorbance data to the regression linear of standard curve. Series standards of ammonia were made by using aqueous ammonia solution.

3. Result and discussion

3.1 Determination of optimum electrical voltage

The effect of electrical voltage on the remaining ammonia concentration in the solution by using the electrolysis process is shown in table 1.

Table 1. Remaining concentration of ammonia in the solution after treated by using electrolysis method (initial ammonia concentration = 100 ppm, electrolysis time = 2 min., concentration of KOH = 0.02 M)

Electrical voltage (Volt)	Ammonia concentration (ppm)	
	Initial	Remaining
4	100	23.81
6	100	19.96
8	100	15.15
10	100	11.31
12	100	11.98

According to table 1, it can be stated that an increment on the electrical voltage decreases the remaining concentration of ammonia; increases the converted ammonia content. This phenomenon is caused by an increment on the electrical voltage increases electric current, an increment on the electric current increases electric charge and an increment on electric charge in electrolysis process increases mass of electrolysis product.This result is in accordance with previous research. Riwayati (2010) found that an increment on electric current density decreases the remaining ammonia concentration; increases the converted ammonia content [10]. On the other hand, and increment the electric current density increases the electric current. It means an increment on the electrical voltage decreases the remaining ammonia concentration; increases the converted ammonia content. It seems that the optimum electrical voltage was 10 volt.

3.2 Determination of optimum time

The effect of time process on the remaining ammonia concentration in the solution by using the electrolysis process is shown in table 2.

Table 2. Remaining concentration of ammonia in the solution after treated by using electrolysis method (initial ammonia concentration = 100 ppm, concentration of KOH = 0.02 M, electrical voltage = 10 volt)

Time (min.)	Ammonia concentration (ppm)	
	Initial	Remaining
10	100	30.19
15	100	26.92
20	100	24.31
25	100	20.27
30	100	20.81

Table 2shows that an increment on the time process of electrolysis decreases the remaining concentration of ammonia; increases the converted ammonia content. This phenomenon is in accordance with Faraday's Law; an increment on the time process increases mass of electrolysis product[11]. When the mass of the product increases, the mass of reactant decreases; the concentration of the reactant decreases. The optimum electrolysis time process was 25 min.

3.3 Determination the optimum potassium hydroxide concentration

The effect of concentration of KOH on the remaining ammonia concentration in the solution by using the electrolysis process is shown in table 3.

Table 3. Remaining concentration of ammonia in the solution after treated by using electrolysis method (initial ammonia concentration = 100 ppm, electrolysis time = 25 min., electrical voltage = 10 volt)

Concentration of KOH (M)	Ammonia concentration (ppm)	
	Initial	Remaining
0.01	100	13.35
0.02	100	13.19
0.03	100	12.88
0.04	100	12.42

From table 3it can be concluded that an increment on the concentration of potassium hydroxide insignificantly decreases the remaining concentration of ammonia; insignificantly increases the converted ammonia content.The KOH generates alkaline environment of ammonia electrolysis. An increment on the alkalinity decreases the ammonia oxidation overpotential, shifts the ammonia oxidation potential negatively.This study is in accordance with previous study conducted by Yao and Cheng (2007), Zhou and Cheng (2008) and Riwayati (2010) [10,12,13].

The optimum conditions (10 volt of electrical voltage, 25 min. of electrolysis time, and 0.04 M of KOH concentration) were applied on the electrolysis of the real liquid waste of tofu.The method was applicable to decrease ammonia content in liquid waste of tofu and decrease 81.13% of the ammonia content.

4. Conclusion

We have investigated the optimum conditions of electrolysis process to reduce the ammonia content in the aqueous solutions. The optimum conditions were 10 V of electrical voltage, 25 min. of electrolysis time and 0.04 M of potassium hydroxide concentration. The electrolysis method was applicable to decrease ammonia content in liquid waste of tofu. 81.13% of ammonia content in the real sample could be reduced by using electrolysis method at the optimum condition.

Acknowledgment

The authors gratefully thank to Ministry of Research, Technology and Higher Education that was funded this research.

References

[1] Husni H and Esmiralda 2010, Uji Toksisitas Akut Limbah Cair Industri Tahu Terhadap Ikan Mas (*Cyprinus carpio* Lin), Studi Kasus: Limbah Cair Industri Tahu "SUPER", Padang), *Thesis* (Padang: Faculty of Enginering of Andalas University) Available at http://repo.unand.ac.id/3005/

[2] Meade JW 1985, Allowable Ammonia for Fish Culture, *The Progressive Fish-Culturist*, **47** p 135

[3] Snieszko SF 1974, Fhe Effects of Environmental Stresson Outbreaks of Infectious Diseases of Fishes, *J. Fish Biol.*, **6** p 197

[4] Yasin NHM, Mumtaz T, Hassan MA, and Rahman NAA2013, Food waste and food processing waste for biohydrogen production: A review, *J. Environ. Manage.***130** p 375

[5] Levina E 2016, Biogas from Tofu Waste for Combating Fuel Crisis and Environmental Damage in Indonesia, *APEC Youth Scientist Journal*, **8** p 16

[6] Putnarubun C, Rettob M, Gunam IBW, and Balubun SK 2015, Utilization of Tofu Waste As a Cultivation Media for *Tetraselmis sp* Microalgae in Preliminary Study of Biodiesel Production, *International Journal of Basic & Applied Sciences*, **15** p 13

[7] Bonnin EP, Biddinger EJ, and Botte GG 2008, Effect of Catalyst on Electrolysis of Ammonia Effluents, *Journal of Power Sources*,**182** p 284

[8] Muflihah, Tindangen RA, and Noor 2015. Penentuan Kadar Ammonia Pada Berbagai Variasi Tegangan Dengan Menggunakan Elektroda Grafit.*Book Research* (Samarinda: Faculty of Teacher Training and Education of Mulawarman University)

[9] Noor and Nurlaili2015,Penentuan Kadar Ammonia Pada Berbagai Variasi Waktu Dengan Menggunakan Elektroda Grafit,*Book Research* (Samarinda: Faculty of Teacher Training and Education of Mulawarman University)

[10] Riwayati I 2010. Waste To Energy: Recoverydan Elektrolisa Amoniadari Limbah Pabrik Pupuk Untuk Menghasilkan Gas Hidrogen.*Momentum*.**6** p 27

[11] Ratnawati, Sumarno, and Nugroho A 2010, Konversi Elektrokimia Amonia Menjadi Hidrogen, *Teknik*, **31**98

[12] Yao K and Cheng YF 2007, Electrodeposited Ni–Pt Binary Alloys as Electrocatalysts for Oxidation of Ammonia, *Journal of Power Sources*,**173** p 96

[13] Zhou l and Cheng YF 2008, Catalytic Electrolysis of Ammonia on Platinum in Alkaline Solution for Hydrogen Generation,*International Journal of Hydrogen Energy*, **33** p 5897

Biaccumulation and tolerance of heavy metals on the tropical earthworm, *Allobophora sp.* after exposed to contaminated soil from oil mine waste

Suhendrayatna[1]*, Darusman[2], Raihannah[1] and D Nurmala[1]

[1] Department of Chemical Engineering, Faculty of Engineering, Syiah Kuala University, Banda Aceh, Indonesia
[2]Soil Science Department, Syiah Kuala University, Banda Aceh, Indonesia

*Corresponding Author: suhendrayatna@unsyiah.ac.id

Abstract. In this study, the impact of contaminated soil from oil mine waste on survival, behavior, tolerance, and bioaccumulation of heavy metals by the tropical earthworm, *Allobophora sp.* has been quantified. Earthworm was isolated from heavy metals-contaminated soil, cultured in laboratory condition, and exposed to contaminated soil from oil mine waste for a couple of months. The behavior and response of earthworms to contaminated soil was monitored for 28 days and evaluated by the response criteria was expressed in scale index (SI) referred to Langdon method. Resistance test of the earthworm (LC_{50}) to heavy metals also conducted with variation soil concentrations of 100%, 50%, 25%, 12.5%, and 6.25%, and 0% (Control). Results showed that contaminated soil extremely affected to the earthworm live, especially length and their body weight. The Lethal Concentration 50% (LC_{50}) of earthworm against contaminated soil was 19.05% (w/w). When exposed to contaminated soil, earthworm accumulated chromium, barium, and manganese at the concentration of 88; 92.2; and 280 mg/kg-DW, respectively. Based on these results, earthworm *Allobophora sp.* has potential to reduce heavy metals from contaminated soil in the field of bioremediation process.

1. Introduction

Excessive anxiety about the presence of heavy metals in environment is caused by very high poisoning levels of heavy metals in all aspects of life [1, 2]. U.S. Environmental Agency (USEPA) has listed 13 heavy metal elements, which are the main element of dangerous pollution, such as Sb, As, Be, Cd, Cr, Cu, Hg, Ni, Pb, Se, Ag, and Zn [2]. These heavy metals in fact harmful to human health and continuity of life in the neighborhood. Although such concentration is low, the effects of heavy metal ions can take effect directly to accumulate in the food chain [3]. Mobilization of heavy metals depend on the characteristics of microorganism and heavy metals in environment [4]. As well as the sources of other environmental pollutants, heavy metals can be transferred within reach very far in the environment, potentially disrupt environmental biota and ultimately affect human health although in a long period of time and far from sources of pollution.

Natural processes such as natural cycles change resulted rocks and volcanoes contribute a very large to environment. Besides, the entering of heavy metals into the environment are derived from other sources which include; mining, oil, gold, coal, hydroelectric, pesticide, ceramics, metal smelting, fertilizer factories, and other industrial activities. In some Asian countries, heavy metal contamination has spread widely as reported by the survey team from National Agriculture and Food Research Organization and Tokyo University of Agriculture and Technology in Japan and University of Madras and Manipal Institute of Technology [5]. Our preliminary research reported that soil contaminated oil as by-product from oil explorations activities in Aceh Province contain a number of heavy metals in high concentrations such as Pb, Cr, Mn, Fe, Zn, and Ba. The problem will occur if soil piled up without being processed or without reducing the heavy metals content. These phenomena will be facing to the world in the future considering activities of oil and natural gas exploration running massively.

In line with these problems, research on the effects of heavy metals on the lives of organisms in environment will continue to be carried out. In the recent years, many researches have focused to study on the acute toxicity effects to earthworms of inorganic and organic pollutants such as the nickel, and so on [6-8]. They are also concerning heavy metals accumulation in earthworms and most are on lead, zinc, and cadmium [9-11]. Earthworms are the representative species of soil animals with the largest terrestrial faunal biomass and constitute a major component in soil ecosystem. They play important roles in element geochemistry cycle since their contribution on organic matter decomposition. In the environment, pollutants in soil such as organic pollutants and metals easily influence earthworms. They are very suitable for monitoring the contaminations effects in field with their limit mobility [12]. Earthworms have a potential as degrading organisms of heavy metals in contaminated soil by bioremediation process. Langdon et al conducted a study of earthworms *L. rubellus* and *L. terrestris* which were exposed on arsenic contaminated soil [13]. Arsenic accumulation by *L. rubellus* in uncontaminated soil smaller compared to the treatment of contaminated soils. Concentrations of arsenic found in the cells of *L. terrestris* and *L. rubellus* was 230 mg-As/kg and 92 mg-As/kg, respectively. Leveque et al reported that metal pollution has a direct impact on earthworm communities (proportion of juveniles, abundance, and diversity) [14]. The response of earthworms, *P. antipodarum* against metal arsenic trivalen (As III) faster compared to arsenic pentavalen (As V), this was caused by metal toxicity levels As III higher compared to As V [15].

Based on the explanation above, few data are available on acute toxicity effects of earthworms to heavy metals from soil contaminated. This paper describes the influence of heavy metals contaminated soil and hydrocarbons on the lives of Earthworm, *Allobophora sp.* isolated from heavy metals contaminated soil from oil mine waste.

2. Materials and Methods

2.1 Contaminated soil

Contaminated soil obtained from oil and gas exploration in East Aceh. The results of preliminary analysis of contaminated soil was used show that soil contains Ba (0.08 mg/g), Cr (0.018 mg/g), Hg (0.65 (g/g), Pb (0.057 mg/g), and Zn (0.314 mg/g), while Co and Cd found a very small concentration in the soil. Meanwhile, hydrocarbon compounds on the contaminated soil found on the average of C between C_{12} to C_{16}. Preliminary analysis results shows that deposits of hydrocarbons in the contaminated soil was hydrocarbons with contains C_{12} (5.28%), C_{13} (6.39%), C_{14} (3.60%), C_{15} (20.77%), and C_{16} (63.96%).

2.2 Growth medium of earthworms

Earthworms require a growing medium containing rich of organic matter for their growth. In this study, growth medium of earthworms was cultivated so that soil conditions can meet the requirements as their natural habitat with the following criteria.

1. Consists of fibrous organic materials that have weathered between 50-60% and has not released the unwanted gas by earthworms;
2. Able to maintain the stability of humidity with a good moisture level (35-50%) for the growth of earthworms. If the water content in the medium is too high, the earthworm body will be pale and become dead;

3. Media is always loose and not easily solid or porous;
4. Media is easy to decompose; and
5. Media temperature is in range 27-30°C with pH about 6.5 – 7.2.

2.3 Isolation of earthworms and resistance test
Preliminary research was conducted to find a species of heavy metal resistant earthworms. Earthworm was collected from contaminated site of Banda Aceh City (such as the area of Darussalam, Kajhu, and Ketapang Dua). Earthworms were kept in soil collected from the same site in brown glass tubes in field and brought to laboratory. Earthworms were picked out with forceps, as soon as possible thoroughly rinsed with distilled water, and put into the brown tubes with one filter paper in, and distilled water were added in few drops to maintain them moist. Earthworms were kept at room temperature (± 30°C) for 7 days to reach the adaptation condition and to empty their guts, then weighted, counted and identified, and their body lengths were measured.

Resistance test of the earthworm (LC_{50}) to heavy metals referred to the methods conducted by [13]. Two steps tests were conducted, namely preliminary and actual test. Preliminary test intended to explore and predict the range of concentration test that could lead to the animals' death. While the actual test based on preliminary test results that could be estimated the toxic and nontoxic concentrations, then the actual test of soil concentrations arranged. Concentrations of heavy metals contaminated soil were varied 100%, 50%, 25%, 12.5%, and 6.25%. Control (0%) were also prepared to compare the conditions of earthworm's exposure to the contaminated soil. Six tails of earthworm from stock tubes were separated and transferred into tube of experiment and exposed to contaminated soil already varied over the next 96 hours. The experiment condition was prepared on the same treatments with adaptation phase in tube of shelter. During treatment, clinical symptoms and behavior of earthworm were observed, as well as the number of earthworm live and dead were recorded to earthworms mortality calculation.

Behavior and response of earthworms to contaminated soil was monitored for 28 days and evaluated by response criteria that was expressed in scale index (SI) referred to Langdon, et al [13] as follows: (S=2) good muscle tone, the worm's response to stimulation was very fast, (SI=1) weak muscle tone, normal walking response or good muscle tone, but slow in response, and (SI=0) weak muscle tone, no response to stimulation.

2.4 Heavy metal analysis
Heavy metals concentration in soil and earthworm analysed after destructive based on Toxicity Characteristic Leaching Procedure by using Atomic Absorption Spectroscopy [16]. Heavy metals ion to be specified using gas, air pressure, Hollow Cathode, and specific current lamps.

3. Result and Discussion

3.1 Earthworm characteristics isolated from contaminated soil
Based on the preliminary results, from many places that were taken (Darussalam, Kajhu, and Ketapang Dua), only earthworms from Kajhu area could survive against heavy metal toxid compared to earthworms isolated from other areas. Earthworms isolated from Darussalam area were death after 1 day exposure. The resistance of earthworms against heavy metals toxic are strongly influenced by habitat or earthworms origin place [17,18]. The analysis results showed that earthworms isolated from Kajhu area accumulate manganese 187 mg Mn/kg-WD. This was due to the soil where earthworms live contain low concentration of manganese that not cause toxic to earthworms. The accumulation of manganese in low concentration indirectly influenced to earthworm resist against toxic heavy metals. Physical characteristics of earthworms isolated from contaminated soil have outer and inner segments, hair, having no outer frame; cuticle (outer skin) protects their body, having no motion, and no eyes. To be able to move, earthworms must use the muscles of their long body and thick that circle their body. The presence of mucus produced by epidermal glands of their body can facilitate their movement in

dense and rough places. The mucus can also lubricate their body making a hole in soil to easily get in and out of the hole. In addition to these functions, the mucus can be used to defend their self. On their body, there is an organ called seta, which is found on every segment of hair and relatively hard and short-sized. The adhesiveness of these organs is strong so earthworms can cling tightly to surface of the object. Seta can also help earthworms during marriage. Based on these investigation results, the classification of isolated earthworm was Allobophora *sp.*

3.2 Effect of contaminated soil on earthworm behaviour and response

Physical characteristics and behavior of earthworms, *Allobophora sp.* observed were length, weight, color, shape or condition of body, movement, and response to stimuli. The observations conducted before and after earthworms were exposed to contaminated soil. The conditions of earthworms exposed to contaminated soil gradually degenerate and eventually die. Pathologic indications that have been observed were slow responses to stimuli, swelling around the body, red color fades, non-directional movements, and at 100% concentrations most of earthworm conditions were worsening, while earthworm in the control treatment were in good condition.

Furthermore, examination of the effect of heavy metal contaminated soil to earthworms refers to method developed by Langdon et al [13] by setting the Index Score (IS) to the condition of earthworms. Based on the observations, earthworm conditions at various treatments illustrated in Figure 1.

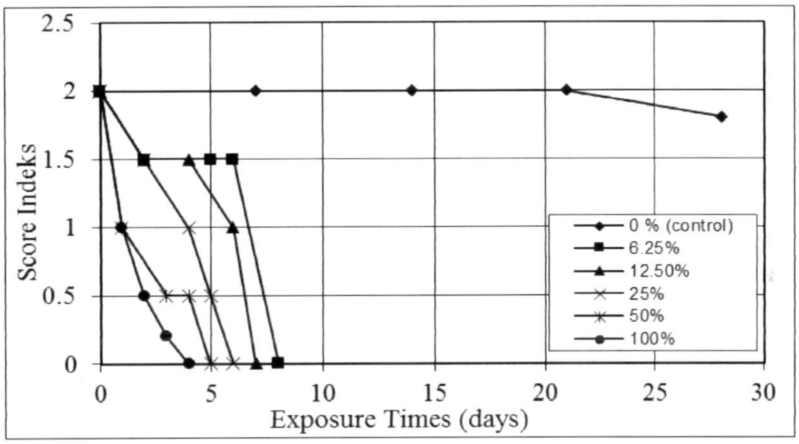

Figure 1. Earthworm conditions at various concentrations of contaminated soil

Figure 1 shows that earthworms exposed to contaminated soil were worsening in line with the increasing of contaminated soil concentration. At the concentrations of 100%, the earthworms only survived in 4 days of exposure because of acute poisoning of heavy metal toxins, whereas at concentrations of 6.25%, earthworms were still able to survive for 8 days of exposure because they have chronic or sub-chronic toxicities. Furthermore, the earthworms on control treatment were still survive in good condition for 28 days. When observed from the influence of exposure time to earthworm conditions, the conditions of these animals appeared weak, slow response, not directed movement since the first day of exposure. This indication was observed at the concentration of 100% (SI=1), whereas at lower concentrations of 6.25% the earthworms can still catch on the quick response, directed movement, although the condition was rather weak (SI=1.5). The second day of exposure for concentration of 100%, two earthworms were found dead and the others were in worsening condition. For concentration of 6.25% on the second day of exposure, no earthworms found died but the condition was weak with slow response and the directional movement (SI=1.5). Further observations conducted on earthworm conditions in line with the length of exposure time indicated that earthworm conditions became worse and eventually die, except for the control treatment was still survive in good condition on 28 days of exposure (SI=2). The results were consistent with the research conducted by Langdon et al [13] that focused on the effect of arsenic contaminated soil against two species of earthworms, *L. rubellus* and *L. terrestris.* They reported that on the first day of exposure, no earthworms died and after 5 days of exposure, an earthworm, *L. terrestris* was died. Furthermore, three earthworms, *L. terrestris* died after 10 days of exposure and six *L. terrestris* and two *L. rubellus* still survived but they were in a weak condition. While earthworms in the control treatment was still survive in good condition on 28 days of exposure (SI=2).

3.3 Toxicity value (LC_{50}) of contaminated soil to earthworms

The methods conducted by Langdon, et al [13] was applied to obtain toxicity value (LC_{50}) of contaminated soil to earthworms. Based on the calculation, LC_{50} was found 19.05%. Its mean that 50% of earthworms died when exposure to contaminated soil with the concentrations of 19.05%.

3.4 Effect of contaminated soils on weight and length of earthworms

To find out Influence of contaminated soils on weight and length of earthworms and to investigate the relationship between concentrations of contaminated soil, the research data were evaluated using statistical methods. Test results show that higher concentrations of contaminated soil cause earthworm weight increases. It was indicated by body weight of earthworms decreased larger at concentration of 100% compared to 50%. Similarly, results occurred to the length of earthworms before and after treatment. Calculation results using statistical methods concluded that contaminated soil greatly affect weight and length of the earthworms, which indicated by $F_0 > F_{Table}$. In the source of variation in length and weight, calculation result for F_0 was 2176.9 while the results of F_{Table} ($F_{0.1}$ and $F_{0.01}$) were 3.94 and 6.92, respectively.

3.5 Accumulation of heavy metals in earthworm after exposed to contaminated soil

Earthworm was exposed to contaminated soil with a concentration of 100; 50; 25; 12.5; 6.26%; and 0% (control). After three days of exposure, earthworm was taken, cleaned, and oven dried (48 h at 60 °C), ground to homogeneous powder in a quartz bowl, and heavy metal accumulated in cell was analyzed by AAS Shimadzu 630. The results are tabulated in Table 1 and illustrated in Figure 2.

Table 1. Bioccumulation of heavy metals in earthworm after exposed to contaminated soil

Contaminated soil concentration (%)	**Heavy metals accumulated in earthworm (mg/kg-DW)**							
	Cr	**Pb**	**Mn**	**Ba**	**Cd**	**Hg**	**Co**	**Zn**
Control (0)	trace	trace	187	trace	trace	trace	trace	trace
6.25	18	trace	190	18	trace	trace	trace	trace
12.5	88	trace	160	92.2	trace	trace	trace	trace
25	86	trace	280	24	trace	trace	trace	trace
50	27	trace	230	39.6	trace	trace	trace	trace
100	50	trace	220	69.9	trace	trace	trace	trace

Results showed that when earthworms were exposed to contaminated soil, concentration of heavy metals accumulated in earthworm did not correspond to the increasing rate of contaminated soil concentration. Some of accumulated metals were excreted to the soil so the toxic contents of heavy metals present in cell were reduced and toxicant may be in the form of origin, as a metabolite or as a conjugate. Heavy metals that accumulated by the earthworms were Mn, Cr, and Ba with the concentration of 88; 92.2; and 280 mg/kg DW, respectively. Other metals such as Pb, Cd, Hg, Co, and Zn found as trace in earthworm's cell. These results were in line with those reported by Lu [19]. Furthermore, earthworms exposed to control treatment were found accumulate manganese in the concentration of 187 mg-Mn/kg-DW. This was due to soil where the earthworm isolated contains manganese. Due to the presence of manganese in soil from where the earthworm isolated, accumulation of manganese higher compared to other metals. These results support research reported by Langdon et al [13,17]. Accumulation of arsenic on earthworms, *L. rubellus* exposed to the uncontaminated soil was less (<1 mg-As/kg) than that of contaminated soil (230 mg-As/kg).

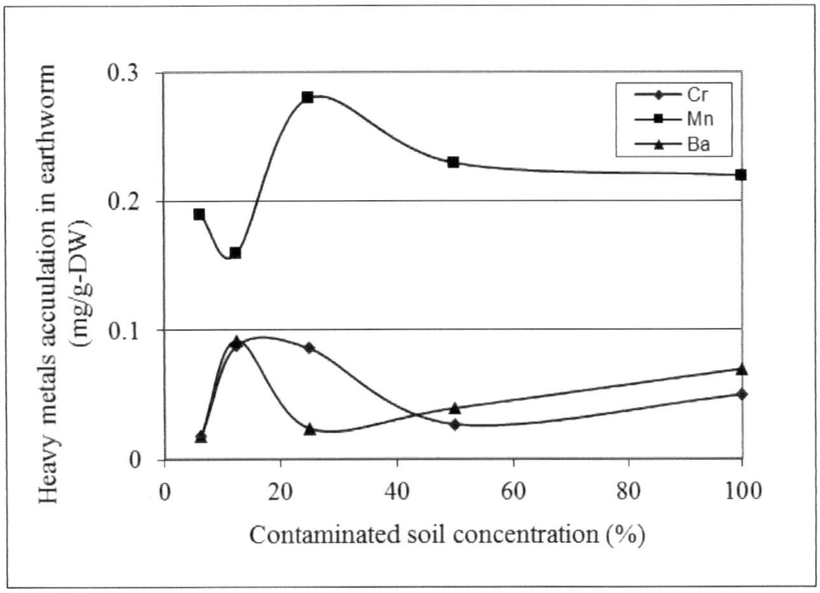

Figure 2. The relationship of heavy metal concentrations accumulated in the earthworm to the concentration of contaminated soil.

4. Conclusions

Based on the results and discussion presented, this research leads some conclusion.

1. The influence of heavy metals and hydrocarbons contaminated soil on the life of earthworms, *Allobophora sp.* increased with the increasing of contaminated soil concentrations.
2. Contaminated soil concentration causing 50% mortality (LC_{50}) on earthworm, *Allobophora sp.* was found 19.05%.
3. Contaminated soils greatly affect to the weight and length of earthworms, Allobophora *sp.* Earthworm, *Allobophora sp.* was able to accumulate heavy metal compounds from contaminated soil such as Cr, Ba, and Mn, with the concentration of 88; 92.2; and 280 mg/kg-DW, respectively.
4. These results reached to conclusion that earthworm, *Allobophora sp.* has a potential for the remediation application and can be effective to remove heavy metals from contaminated soil.

Acknowledgments

Grateful are extended to Raihanah and Nurmala Dewi for assistance in field data collection and Darusman for contribution of the results discussion.

References

[1] Kavehei A Hose GC Gore DB 2017 *Environmental Pollution* **xxx** 1-7
[2] Novotny V 1995 *Diffise Sources of Pollution by Toxic Metals and Impact on Receiving Waters,* Allan R Forstner U and Salmons W (eds.), Springer 34-64
[3] Suhendrayatna, Ohki A, and Maeda S 2001 *Applied Organometallic Chemistry* **15** 277-284
[4] Jing XB He N Zhang Y Cao YR Xu H 2012 *Can. J. Microbiol.* **58** 45-53
[5] Nilesh V 2017 *The New Indian Express* **4** 1-2
[6] Drouin M, Bradley R, and Lapointe L 2016 *Forest Ecology and Management* **64** 113–121

[7] Koen L and Colin RJ 2002 *Chemosphere* **46** 197–200
[8] Vandecasteele B, Samyn J, Quataert P, *et al* 2004 *Environ. Pollut.* **129** 363–375
[9] Kamitani T and Kaneko N 2007 *Ecotoxicol. Environ. Saf.* **66** 82–91
[10] Ma WC 2004 *Bull. Environ. Contam. Toxicol.* **72** 663–670
[11] Vliet P, Zee S M W C 2005 *Environ Pollut* **138** 505–516
[12] Zhang ZS Zheng DM Wang QC Lv XG 2009 *Bull. Environ. Contam. Toxicol.* **283** 937-42
[13] Langdon CJ Pierce TG Black S and Semple KT 1999 *Soil biol. and biochem.* **31** 1963-1967
[14] Lévêque T Capowiez Y Schreck E Mombo S Mazzia C Foucault Y Dumat C 2015 *Science of the Total Environment* **511** 738-746
[15] Golding LA Timperley MH and Evans CW 1997 *Environ. Monitoring and Assessment* **47** 239-254
[16] Eaton AD, Epps AA 1995 *Standart Methods for the Examination of Water and Wastewater*, 19th Ed. APHA AWWA and WEF Baltimore MD 341-364.
[17] Landon, CJ, Piearce GT Meharg PA and Semple KT 2001 *Eviron. Toxicol. and Chem.* **20** 2336-2341.
[18] Rodriguez-Campos J Dendooven L Alvarez-Bernal D Contreras-Ramos SM 2014 *Applied Soil Ecology* **79** 10–25.
[19] Lu FC 1994 *Basic Toxicology* University of Indonesia Press Jakarta-Indonesia 54-63.

Degradation of methylene blue using pillared TiO_2 on de-oiled spent bleaching clay

N Hindryawati*, A S Panggabean, N D Fadillah, Erwin and Daniel

Faculty of Mathematic and Natural Sciences, Mulawarman University, Samarinda, East Kalimantan, Indonesia

*Corresponding author: ienwati@yahoo.com

Abstract. Degradation of methylene blue (MB) using pillared TiO_2 onto spent bleaching clay has been conducted. Activation of deoiled spent bleaching clay (DSBC) has been done using acid, followed by pillarization with TiO_2 using *rarasaponin* from Klerak fruit as surfactant. From the X-ray diffraction results show the mineral on DSBC is rectorite with dioctahedral mica layer and dioctahedral smectite with ratio 2:1. This molecule have formula $Na.Al_4(Si, Al)_8.O_{20}.(OH)_4.H_2O$ and after calcinations the pattern TiO_2 was appearance at 2θ: 27.4460°, 36.0850°, 54.3216° and 56.6403°. In order to test the catalytic performance of Ti-DSBC for photodegradation of MB under UV light was conducted under several reaction conditions. The highest degradation of MB was 90 % within 50 minutes and Ti-DSBC can be reused until 5 cycles with percent degradation MB was 84 %.

1. Introduction

Indonesia is one of the world's largest producers and exporter of palm oil. In terms of agriculture, Indonesian Palm Oil Producers Association has been recorded palm oil is the most important industry and the total area of oil palm plantations in Indonesia is currently around 11.9 million hectares [1]. On the palm oil process, there are three steps in the refining of edible oils. There are degumming, neutralization, bleaching and deodorization. In the bleaching process palm industries are using imported natural or acid activated clays. The local consumption of bleaching clay is estimated to be 100,000 tons per year, with annual production of crude palm oil is 18 million tons/years [2]. The residual of bleaching clay will transformed to spent bleaching clay (SBC) that containing 20-40 % oil and it may cause pollutions hazard due to the substantial oil content in clay [3]. Based on their structural as smectite, SBC possess certain properties which make them as choice as low cost adsorbent or support material because there high mechanical and chemical stability, availability, affordability, ion exchange capability and environmentally friendly material [4].

Based on their performance pillared clay has been reported that dispersion TiO_2 particles into layer clays could improve catalyst performance because their stability properties. In general, TiO_2 pillared clays will substitute several ions in the interlayers of clay [5]. Photocatalysis, in which clean and safe energy of the sun can be harnessed for sustainable, non-hazardous and economically viable technologies. These processes are all characterized by the same chemical features such as production of hydroxyl radicals (OH) and superoxide anion, which are generated when a semiconductor catalyst absorbs radiation when it is in contact with water and oxygen and among of all semiconductor photocatalyst, titanium oxide has proven to be the most widely used do to its strong oxidizing power, non-toxicity and long-term photostablity [6,7]

Before SBC used as template for TiO_2, there are several treatment to activate the SBC. Solvent extraction processes for recovery of oil from spent bleaching clay have been suggested. This is because they give a higher yield and better quality [8]. There are some extractions methods such as soxhlet but in the last research report ultrasonic extraction give more advantage such as efficiency of time and solvent. This technique used ultrasound wave to impressed the molecules [9]. The deoiled spent bleaching clay (DSBC) also needs activation to open their active side and high surface area. Kheang has reported that activation use acid treatment has a good result [10].

Otherwise, DSBC had been materials there is some report the use of high molecular weight such as polyoxypropylene (POP) or ethyl Trimethyl Ammonium Bromide (CTAB) can increase surface area of montmorrilonite [5]. But, utilizing synthetic like those as modifying agent can also cause environmental hazardous because they are non biodegradable and toxic, so that an unused synthetic surfactant need to processed after modification takes place which increases operational costs [11]. Kurniawan have been reported that rarasaponin from Klerak can be performed as natural surfactant and show similar results such as surfactant synthetic [12]. Rarasaponin is sugar derivatives compound that may be found in plants, to characterize the rarasaponin could by forming foam when dissolved in water and the most widely used method to obtain rarasaponin is solid–liquid extraction. The others solvent can be used to extract rarasaponin from its solid matrices such as ethanol, methanol, and sodium methoxide [11].

In the present study, TiO_2 pillared DSBC were prepared from SBC and Titanium oxide in the presence of polymeric surfactant *rarasaponin* and the as prepared materials were characterized by X-ray diffraction (XRD), Fourier transform infrared spectra (FTIR) and scanning electro microscopy (SEM). The influence of optimum concentration and contact time of this material photocatalyst was investigated by evaluated methylene blue degradation. The fundamental information derived from this study is important in developing catalyst or adsorbent of pore structure from layered clays.

2. Experimental

2.1. Material and chemical

The spent bleaching clay was purified by dispersion in water, decantation and ultrasonic extraction of the oil and sieved with a particle size 100 mesh. Klerak was supplied form Yogyakarta. Klerak will be extracted using polar solvent to obtained *rarasaponin* as surfactant. Titanium oxide (II) pro analyst was obtained from sigma Aldrich, sulphuric acid (H_2SO_4 (98 %)), methanol 95 %, n-hexane were supplied by Sigma-Aldrich

2.2. Preparation technique

Modification DSBC was conducted by dissolving 0.5 g of rarasaponin in 25 mL of distilled water. Subsequently, 5 g of DSBC powder was added to the solution. The mixture was then stirred for 24 hours at 26°C. The solid was then separated from the mixture, repeatedly washed using distilled water and dried at 110°C for 24 hours. Furthermore, pillarization of TiO_2 was prepared with dissolve TiO_2 in 6 mL ethanol 98 % that contain 1 g of DSBC, the mixture was stirred for 24 hours and dried in oven at 120°C for 5 hours. Thereupon, the material crushed until have similar size and calcinated for 5 hours at 550°C and labeled as Ti-DSBC.

2.3. Characterization

X-ray diffraction (XRD) patterns of the resulting organoclays were obtained on Shimadzu 700 diffractometer using Cu Kα radiation at 30 kV and 30 mA. The Fourier transform infra red (FT-IR) spectra were recorded on a spectrum 501 Shimadzu. Scanning electron microscope (SEM) (JEOL JSM

501 LA) was used to observe the crystal microstructure of as-prepared pillared clays. Surface analysis was used to observe the surface area of material.

2.4. Photocatalytic experiment

Photocatalytic activities of these materials were also evaluated on their performance as catalyst in the photocatalytic oxidation of methylene blue in water. Photocatalytic reaction was carried out in a beaker glass reactor of 40 mL capacity attached to an inner radiation type 100 W ultraviolet lamp. Aqueous suspensions of methylene blue (20 mL, with three different concentrations: 5, 10 and 15 ppm) and Ti-DSBC (1 g) were placed in beaker glass. Prior to irradiation, the suspensions were shaked with different time (10, 30, 70 and 90 minutes) to establish an adsorption/desorption equilibrium between dye and the Ti-DSBC. At given intervals of illumination, a specimen (1 mL) of the suspension was collected and centrifuged. The filtrates were analyzed by UV-Vis spectroscopy at 664.97 nm.

3. Results and Discussion

3.1. Characterization of DSBC and Ti-DSBC

The powder XRD patterns of DSBC and Ti-DSBC are shown in figure 1. The diffractograms of (fig. 1A) showed that the intense diffraction peaks found at 20,85 to 50,1375° correspond to the Miller indices [100], [101] and [110] were confirmed as SiO_2, and [112] was confirmed as Al_2O_3 as the framework of DSBC. The type of clay was confirmed as rectorite. Wang (2005) had reported that the thickness of a single sheet of rectorite is the sum of the thickness of the smectite layer and the thickness of the mica layer [13]. Additionally, the original rectorite single sheet is about 40 μm lengths and 5 μm in width whereas monmorillonite (MMT) is about 1 μm length as well as in width. As a result, the separable layer thickness and layer aspect ratio (width/thickness) of rectorite are larger than those of regular MMT [14]. The XRD patters of Ti-DSBC show the appearance of peak at 2θ of 27.4460°, 36.0850°, 54.3216° and 56.6403° corresponding to Miller indices [101], [101], [211] and [220] were assigned as the diffraction peak of rutile. The diffraction peaks found at 2θ = 42,4490°, 68,3161 ([111] and [311]) indicates the presence of silica. The crystalinity of Ti-DSBC (72,5014 %) was increase compared to DSBC (55,8946 %.)

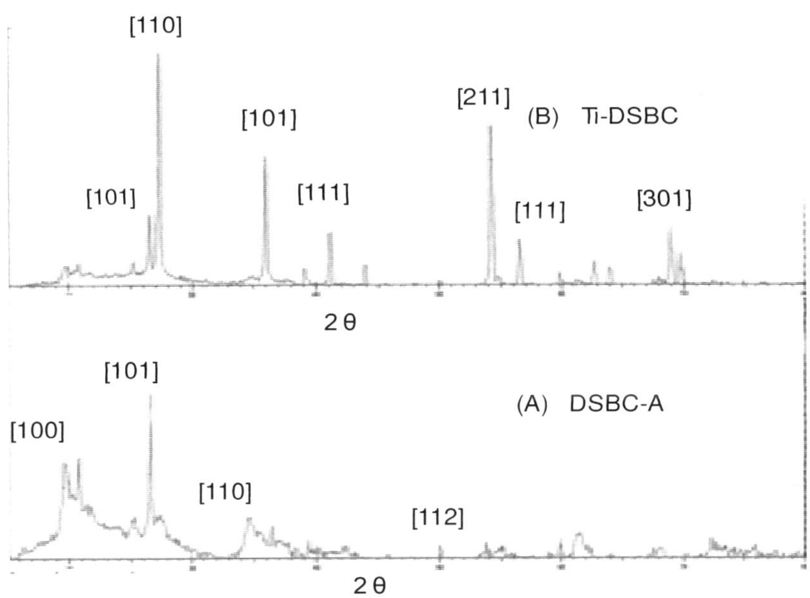

Figure 1. XRD results of Ti-DSBC and DSBC-A

The Fourier Transforms Infrared (FTIR) spectra of Ti-DSBC and DSBS-S (DSBC-Surfactant) spectrum are shown in figure 2. The FTIR spectrum of DSBC-S shown the difference of the peaks detected for DSBC after pillared with TiO_2 in wave numbers 2,900, 1,700 and 1400 cm^{-1}, which are C–H stretching, C–H bending and C=O stretching, respectively. These functional groups belong to carbon chain and carbonyl functional groups of rarasaponin. The presence of these functional groups is strong indication that intercalation of rarasaponin molecules. The weak –OH stretching vibration of water at 3570.24 cm^{-1} and 3408.22 cm^{-1} corresponding to the stretching vibration of hydroxyl group and interlayer water molecules from DSBC, is observed, the adsorption peak at 1462.02 cm^{-1} appears only after adding surfactant rarasaponin, this wave number is assigned to the asymmetrical stretching vibration of C-O carbonyl group from rarasaponin. The stretching vibration C=O indicates that carbonyl group from rarasaponin was intercalated to DSBC [12]. Furthermore, the spectra of Ti-DSBC and DSBC-S shows peak at 462.92 cm^{-1} and 470.63 cm^{-1} that belongs to Si-O stretching [4].

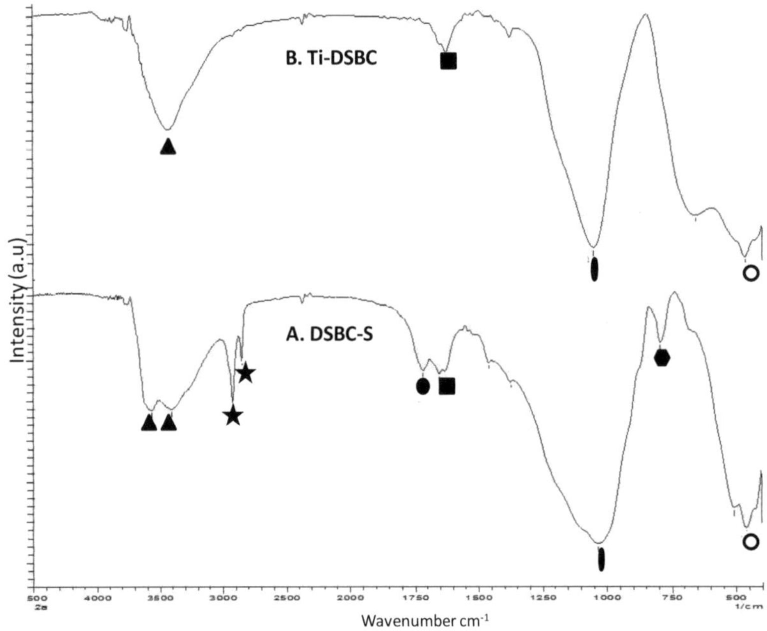

Figure 2. FTIR spectra of (A) DSBC-S and (B) Ti-DSBC; ▲:O-H; ★ :C-H stretching; ●:C=O stretching; ■:H-OH bending;⬢ : Si-O-Si;❚ : Si-O ; ○: Al-O-Si

The same chemical groups present in DSBC-S and Ti-DSBC are identified by the FTIR spectra as shown in fig. 2. The characteristic absorption peaks at 486 cm^{-1} are attributed to the vibration of the Al–O-Si bond structure, and the characteristic absorption band Si–O are shown at 858 cm^{-1}, Si–O stretching are shown at 1101 $cm^{-1}$and band at 1641 cm^{-1} shows the presence of bending vibration of water molecules bound to the material.

Figure 3 (A) and (B) show the field emission scanning electron microscopy (FESEM) images of (A) DSBC-A and (B) Ti-DSBC. Advance enlargement of the image (10000 times) discovered the existence of aggregate particles. It showed that DSBC-A particles were having porous whereas after pillarized with TiO_2 showed a different morphology occurred in Ti-DSBC, the small spherical crystals shown in the surface Ti-DSBC.

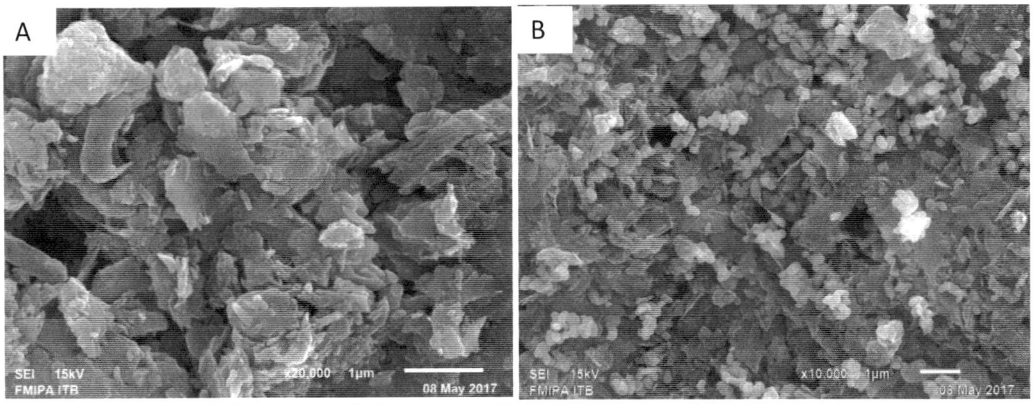

Figure 3. SEM pictures of (A) DSBC-A and (B) Ti-DSBC

3.2 Photocatalytic performance

This study investigated the effect of the contact time and the effect of methylene blue concentration in catalytic performances of Ti-DSBC under UV light is shown in figure 4. The contact time was varied in the range 10-90 min and concentration methylene blue was varied in the range 5-15 ppm.

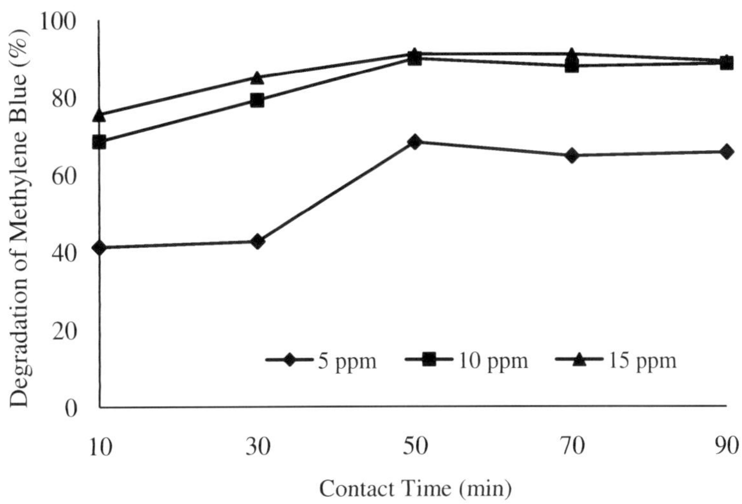

Figure 4. Photocatalytic performance of Ti-DSBC under UV light with different time

From fig.4 it can be seen that for 1 g of catalyst the photodegradation methylene blue (MB) increased within the first 10 min and reached as high as 40% , 66% and 75% for 5, 10, 15 ppm, respectively. Further, with an increase in reaction duration more than 30 min, the photodegradation of MB increased, this finding is mainly due to for longer reaction duration will increase the contact opportunity between catalyst and reactant and increase the production of electron-hole pairs on the surface of TiO_2 and the high reactive hydroxyl radicals was improved [15]. After 50 min percent degradation MB remained almost constant for 90 % at 10 and 15 ppm of MB.

The effect of catalyst dosage on degradation of MB was investigated by varying the catalyst amount (0.25 – 1.5 g) with optimal conditions 10 ppm of MB for 50 min. As shown in Fig. 5, the percent degradation MB increased as the catalyst amount increased from 0.25 to 1 g. The results obtained revealed that photocatalytic degradation performance was substantially improved with increasing the amount of Ti-DSBC and reached a plateau at a dosage of 1 g due to the enhancement of production of hydroxyl radicals.

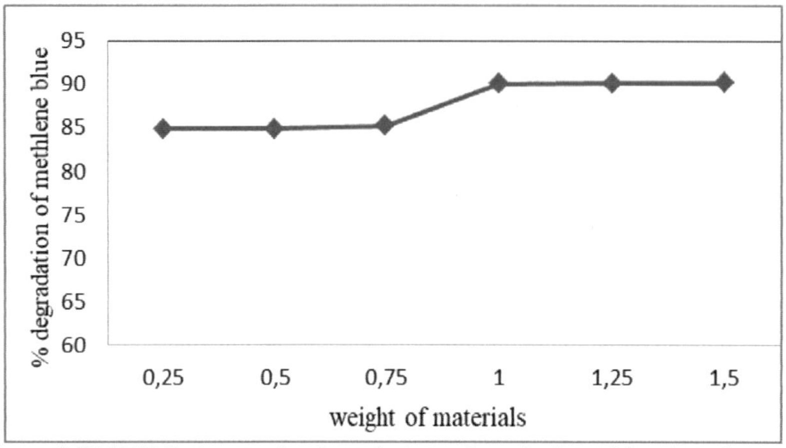

Figure 5. Effect of Ti-DSBC dosage in degradation of methylene blue by photocatalytic, initial concentration 10 ppm for 50 min reaction duration

However, the dosage of Ti-DSBC more than 1 g was not increase the photocatalytic activity due to reach the equilibrium or the agglomeration and sedimentation of TiO_2 may also occur under large amount of loadings which leads to lower surface area [15].

Reusability is one of the factors in the economical application of Ti-DSBC as the photocatalyst. The material was reused without any further activation. In order to test the reusability of the catalyst, the repeated batch experiments using the photocatalytic under UV light with optimized conditions was applied (catalyst content of 1 g, MB concentration of 10 ppm and reaction time 50 min). The results provided in Figure 6, show that the Ti-DSBC can be used up to five cycles while maintaining the activity at considerable degradation of methylene blue (84 %). This study shows that catalytic activity of catalyst has good reproducibility.

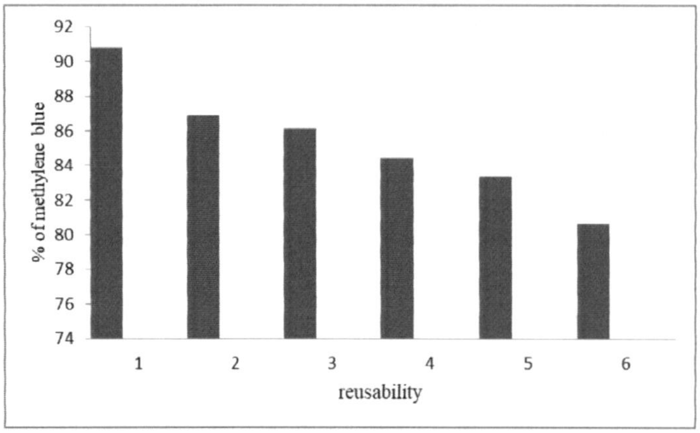

Figure 6. Reusability of Ti-DSBC

4. Conclusion

In this work, TiO_2 was pillared on DSBC in the presence of polymer *rarasaponin* as an surfactant. The XRD result proved that pillarization of TiO_2 on DSBC is a success. The morphology of Ti-DSBC shows the some spherical crystal was distributed on the pore of DSBC. The optimum reaction condition of photodegradation methylene blue (MB) were : 1 g catalyst, 10 ppm of MB and 50 minutes reaction duration with % degradation of MB 90%. The Ti-DSBC was easily to separated and could reused five cycle while maintaining 84% degradation MB.

Acknowledgement
The authors acknowledge BOPTN Grant from the Faculty of Mathematic and Natural Sciences, Mulawarman University.

References

[1] GAPKI 2017 *Minyak Kelapa Sawit* (Indonesia Invesment) [Online]. Available: Indo. [Accessed: 30-Jul-2017].
[2] Hindryawati N Erwin Daniel and Maniam G P 2017 *Preparation of Spent Bleaching Earth-Supported Calcium from Limestone as Catalyst in Transesterification of Waste Frying Oil* (J Bahan Alam Terbarukan vol 6 no 1) pp 68–75.
[3] Nursulihatimarsyila A W Cheah K Y Chuah T G Siew W L and Choong T S Y 2010 *Deoiling and regeneration efficiencies of spent bleaching clay* (Am J Appl Sci vol 7 no 3) pp 434–437.
[4] Bel H Galvez M E Da P and Ben M 2014 *TiO_2 / clay as a heterogeneous catalyst in photocatalytic / photochemical oxidation of anionic reactive blue 19* (Arab J Chem).
[5] Chen D Zhu Q Zhou F Deng X and Li F 2012 *Synthesis and photocatalytic performances of the TiO_2 pillared montmorillonite* (J Hazard Mater vol 235–236) pp 186–193.
[6] Herrmann J 1999 *Heterogeneous photocatalysis : fundamentals and applications to the removal of various types of aqueous pollutants* (Cat Today vol 53) pp 115–129.
[7] Chong M N Yang Tneu Z. Poh P E Jin B., and Aryal R 2015 *Synthesis, characterisation and application of TiO_2–zeolite nanocomposites for the advanced treatment of industrial dye wastewater* (J. Taiwan Inst Chem Eng vol 50).
[8] Daous M M and Zahrani A Al 2000 *Reycycling Of Spent Bleaching Clay And Oil Recovery* (Trans Ichem vol 78).
[9] Fuadi A 2012 *Ultrasonik sebagai alat bantu ekstraksi Oleoresin Jahe* (J Teknol vol 12 no 1) pp 14–21.
[10] Loh S K James S Ngatiman M Cheong K Y Choo Y M and Lim W S 2013 *Enhancement of palm oil refinery waste – Spent bleaching earth (SBE) into bio organic fertilizer and their effects on crop biomass growth* (Ind Crop Prod vol 49) pp. 775–781.
[11] Suwandi A C Indraswati N and Ismadji S 2012 *Adsorption of N-methylated diaminotriphenilmethane dye (malachite green) on natural rarasaponin modified kaolin* (Desalin Water Treat vol 41 no 1–3) pp 342–355.
[12] Kurniawan A Sutiono H Ju Y H Soetaredjo F E Ayucitra A Yudha A and Ismadji S 2011*Utilization of rarasaponin natural surfactant for organo-bentonite preparation: Application for methylene blue removal from aqueous effluent* (Microporous Mesoporous Mater vol 142 no 1) pp 184–193.
[13] Wang Y Zhang H Wu Y Yang J and Zhang L 2005 *Preparation, structure, and properties of a novel rectorite/styrene-butadiene copolymer nanocomposite* (J Appl Polym Sci vol 96 no 2) pp. 324–328.
[14] Thamaphat K Limsuwan P and Ngotawornchai B 2008 *Phase Characterization of TiO_2 Powder by XRD and TEM* (Nat Sci vol 42) pp 357–361.
[15] Gar M Taw A and Ookawara S 2016 *Enhancement of photocatalytic activity of TiO_2 by immobilization on activated carbon for degradation of pharmaceuticals* (J Environ Chem Eng vol 4) pp 1929–1937.

Potential of using tropical brown macroalgae *sargassum cristaefolium* meal in the diets for juvenile white shrimp (*litopenaeus vannamei*)

A Sudaryono[1*], P Sukardi[2], E Yudiarti[3], E H Hardi[4], S Hastuti[1], T Susilowati[1]

[1] Aquaculture Department, Faculty of Fisheries and Marine Science Diponegoro University, Semarang, Indonesia
[2] Faculty of Fisheries and Marine Science, University of General Soedirman, Purwokerto
[3] Marine Science Department, Faculty of Fisheries and Marine Science Diponegoro University, Semarang, Indonesia
[4] Aquaculture Department, Faculty of Fisheries and Marine Science Mulawarman University, Samarinda, Indonesia

Corresponding author: A. Sudaryono, agungsoed@yahoo.co.id

Abstract. The objectives of this study were to evaluate the effect of different dietary levels of tropical marine brown macroalgae, *Sargassum cristaefolium* meal (SCM) on growth performance and feed utilization efficiency of juvenile white shrimp, *Litopenaeus vannamei* (mean initial weight 2.65±0.1 g). The algae were collected from Bandengan Coast, Jepara, Indonesia. The study used a completely randomized design with 5 treatments of dietary different *S. cristaefolium* meal levels (0, 10, 20, 30, and 40 g algae meal/kg feed) in triplicates. The results showed that the different levels of dietary SCM did not significantly affect (P>0.05) average daily growth (ADG;.204-224 mg/day) and survival (80-97%) of the shrimp after a 42-day feeding period. However, the test diets significantly affected (P<0.05) feed intake (FI), feed conversion ratio (FCR), and protein efficiency ratio (PER). Better FCR (2.13) and PER (1.34) was found by the diet with 2% SCM compared to the control diet (no SCM) (FCR 3.42 and PER 0.81). This finding indicates that the supplement of dietary 2% SCM (20 g/ kg feed) can increase the feed utilization efficiency up to 62% or reduce the FCR up to 38% without any adverse in growth performance.

1. Introduction

The white shrimp, *Litopenaeus vannamei* is currently to become the major species being cultured worldwide [1, 2]. Bacterial and viral diseases outbreaks are a major problem faced by penaeid shrimp aquaculture industry especially penaeid shrimp, *L. vannamei*. In recent decade, the industry has got many serious economic losses due to the microbiological diseases causing high mortality in all producing countries [3, 4, 5). Reducing environment quality of commercial intensive shrimp aquaculture may have resulted in increased microbiological diseases.

Reducing the use of antibiotics due to the problems of antibiotic residual in food and resistant pathogens spreading in aquaculture environment [3, 6] has resulted in more attention for the use of immuno stimulants [7, 8]. The use of immuno stimulating substances such as carrageenan, laminaran, alginate and fucoidan derived from marine seaweeds origin extract that could enhance immune response and promote growth in some penaeid shrimp species has been studied by many workers [1-4, 9-13]. However, only a few information regarding the evaluation of marine macro algal polysaccharides on shrimp immune response, disease resistance and growth performance has been available so far. So that this work is needed to conduct.

The present study was conducted to evaluate the potential of using tropical brown marine macroalgae *Sargassum cristaefolium* meal in diets for growth performance and feed efficiency of white shrimp *Litopenaeus vannamei* juveniles.

2. Methods

This study was carried out from March to May 2017 at Coastal Eco-Development Laboratory, Faculty of Fisheries and Marine Science, Diponegoro University, Semarang, Central Java Province.

2.1. Preparation of marine brown algae meal

Marine brown algae *Sargassum cristaefolium* were collected from Bandengan Coast, Jepara, Central Java Province, Indonesia. After collection, the algae were washed thoroughly by freshwater to remove the debris and then dried under room temperature and powdered using an electric grinder. The meal was packed and stored until use.

2.2. Test shrimp and diets

White shrimp *L. vannamei* postlarvae (PL-12) were obtained from a commercial hatchery (PT. CPP Hatchery, Sluke, Rembang, Central Java Province, Indonesia) and transferred to the indoor Nutrition Laboratory, the Center for Development of Brackish Water Aquaculture, Ministry for Marine and Fisheries Affairs, Jepara. To make sure that the shrimp were in good health condition, the shrimp were tested PCR negative (nested PCR). The shrimp were reared in six tons capacity concrete tank as stock tank containing 3000L of aerated seawater at salinity 32.5-35.0 ppt, with pH 8.0-8.2 and temperature 27.5-30.0^{o}C for a period of 45 days until they reached juveniles size approximately 1.5-3.0 g before starting the feeding trial. During the nursery, the shrimp were fed a commercial *L. vannamei* feed (Starter I pellets 933S Gold Coin) containing crude protein minimum 36%, crude fibre maximum 4%, crude lipid minimum 5%, ash maximum 15% and moisture maximum 12%. A 20% daily water exchange was maintained during the nursery.

The commercial *L. vannamei* feeds as used in the nursery were used as basal diets in this study with the composition of test diets as shown in Table 1. Wheat flour was used to adjust to 100%% total proportion in the test diets.

Five experimental dry pellet diets containing similar approximately 36% crude protein (dry weight basis) were formulated to contain 0 (as a control), 10, 20, 30, and 40 g *S. cristaefolium meal* /kg diet. The procedure to prepare the test dry pellet diets followed the method conducted by [2]. The dried test diets were packed in plastic bags and stored in the refrigerator until used for feeding trial.

Table 1. Composition (g/k g) of experimental diets for *L. vannamei* juvenile

Ingredient	Composition (g/kg)				
	SD0	SD1	SD2	SD3	SD4
S. cristaefolium meal (0-4%)	0	10	20	30	40
CMC (binder) (0.5%)	5	5	5	5	5
Wheat flour	40	30	20	10	0
Basal diet* (95.5%)	955	955	955	955	955
Total (100%)	1000	1000	1000	1000	1000
Proximate composition (% dry weight basis)**					
Crude protein	35.6	36.9	36.2	36.5	35.8
Crude lipid	8.1	8.8	9.0	7.9	8.6
Crude ash	11.8	12.6	11.7	11.4	11.1
Fibre	6.5	6.2	5.8	5.7	5.9
NFE (calculated by difference)	38.0	35.5	37.1	38.5	38.6

* Commercial *L. vannamei* feed (Starter I Pellets 933S Gold Coin)
** Values are mean of triplicate samples.

2.3. Feeding trial and evaluation parameters

A total of acclimated one hundred and fifty shrimp juveniles with an initial individual mean wet weight of 2.65±0.07 g were randomly selected from the stock tank and transferred to 25-L capacity black round plastic experimental tanks filled with 20-L seawater for the feeding trial. Shrimp were randomly stocked at 10 individuals per tank in triplicates per treatment in 15 experimental tanks. The experimental tanks were substrate-free flat-bottom plastic tanks equipped with aeration and a black plastic net as a cover to minimize disturbances and avoid the shrimp from jumping out. During a period of feeding trial for 42 days, water quality in the experimental tanks were monitored with salinity ranged from 32-35 ppt, water temperature from 27.5-30°C, dissolved oxygen from 3.70-6.50 mg/L, and total ammonia <0.01 mg/L. A 12 h light: 12 h dark photoperiod was maintained throughout the trial.

The feeding protocol of this work followed the method conducted by Sudaryono et al. [2]. The test diets were given to the juveniles three times a day *ad libitum* at 8:00 (30%), 13:00 (30%) and at 17:00 (40%) on a 8% initial feeding rate of total body weight for a 42 days period. Fecal wastes and uneaten diets were collected from the tanks using siphon every before the feeding time. Uneaten diets were dried and calculated to determine feed intake and feed efficiency ratio. Every 14 days interval, the shrimp were weighed to adjust the amount of feed given and counted to monitor the survival in each experimental tank. The molting shrimp were monitored daily to determine any effect of the treatment diets on molting frequency of the shrimp. At the end of the feeding trial, growth performance measured as weight gain (WG), average daily growth (ADG), protein retention (PER), feed conversion ratio (FCR), feed intake and survival were calculated as previously described by Sudaryono et al. [2].

2.4. Biochemical and statistical analyses

Crude protein of test shrimp carcass after feeding treatments and test diets were determined using by the Kjeldahl method (Tecator Kjeltec System 1007, Denmark). Crude lipid was determined by Soxhlet extraction method, ash and moisture contents were determined according to [14].

All data obtained from this study was statistically analyzed by one-way analysis of variance (ANOVA) and Duncan's multiple comparison test using a program of SPSS version 21 for windows. Probability values of 0.05 were considered significant. Covariate ANOVA was used to make sure that there was no effects of initial shrimp weight on the calculated parameters.

3. Results And Discussion

Growth performance, feed intake , FCR, PER, survival and molting frequency of shrimp as effect of feeding treatment with diets containing different levels of brown algae *S. cristaefolium* meal for 42 days are summarized in Table 2. Initial shrimp wet weight of test shrimp were similar (P>0.05) ranging from 2.62-2.69 g/juvenile with an average of 2.65±0.07g. There were no effects (P>0.05) of the diet treatments on final shrimp weight (11.2-12.0 g), weight gain (8.6-9.4 g), ADG (204-224 mg/day), survival (80.0-96.7%), and molting frequency (36.0-44.3%) after the feeding trial for 42 days. However, feed intake (FI), FCR, and PER were significantly affected (P<0.05) by different inclusion levels of dietary *S. cristaefolium* meal and shrimp fed with 2% *S. cristaefolium* meal based diet (SD2) resulted in the best feed efficiency performance. In addition, all shrimp fed with the control diet (SD0) gave worse responses (P<0.05) in terms of FI, FCR, and PER than those with the diets containing the brown algae meal more than 2%. Similar nutrient utilization performances (P>0.05) (FCR and PER) were also shown by shrimp fed 3% and 4% *S. cristaefolium* meal containing diets.

Dietary *S. cristaefolium* supplementation levels had no significant effects on shrimp carcass composition. There were no significant differences in shrimp carcass protein, lipid, and ash contents after shrimp were fed with different test diets for 42 days (Table 3).

Disease outbreak up to now is still to be a major problem in shrimp production. The efforts to overcome and prevent shrimp disease are still difficult and the methods to support these efforts are still poorly understood. Lack of the specific immune system in crustaceans has caused vaccines for crustaceans are practically not available. Therefore, more attention is focused on the benefit of using immuno stimulants as one of problem solvers in improving immune response in crustaceans [15]. Use of various immuno stimulants such as glucan, peptidoglycans, fucoidan, alginate, lippolysaccharides, live bacteria, killed-bacteria supplemented in diets has been reported effective in strengthening the non-specific defense mechanism and protecting shrimp against disease [1, 2, 3, 4, 9, 10, 11, 12, 13, 15]. In addition, polysaccharides containing active compounds derived from marine brown algae (*Sargassum* spp.) extract and meal forms have been widely studied and documented having the ability to enhance disease resistance and promote growth in penaeid shrimp *Penaeus monodon* [3, 4] and *L. vannamei* [2, 11, 16].

Table 2. Growth performance, nutrient utililization efficiency, survival and molting number of *L. vannamei* juveniles after feeding with experimental diets supplemented with graded levels of *Sargassum cristaefolium* meal for 42 days

Paramater	Dietary *S. cristaefolium* meal supplementation levels (g/kg diet)				
	0 (SD0)	10 (SD1)	10 (SD1)	10 (SD1)	10 (SD1)
Initial weight (g/shrimp)	2.68 ± 0.02^{a}	2.62 ± 0.05^{a}	2.69 ± 0.04^{a}	2.63 ± 0.09^{a}	2.62 ± 0.12^{a}
Final weight (g/shrimp)	11.6 ± 0.4^{a}	11.2 ± 0.2^{a}	11.5 ± 0.5^{a}	12.0 ± 0.5^{a}	11.9 ± 0.5^{a}
Wight gain (g/shrimp)	8.9 ± 0.7^{a}	8.6 ± 0.5^{a}	8.8 ± 0.5^{a}	9.4 ± 0.5^{a}	9.3 ± 0.4^{a}
ADG (mg/day)	212 ± 17^{a}	204 ± 3^{a}	211 ± 13^{a}	224 ± 11^{a}	220 ± 10^{a}
Feed intake (g)	30.4 ± 1.9^{a}	28.6 ± 1.0^{ab}	18.3 ± 1.2^{d}	25.2 ± 1.6^{c}	25.8 ± 2.4^{bc}
FCR	3.4 ± 0.3^{a}	3.4 ± 0.1^{a}	2.1 ± 0.1^{c}	2.7 ± 0.1^{b}	2.8 ± 0.4^{b}
PER	0.81 ± 0.02^{c}	0.85 ± 0.04^{c}	1.34 ± 0.07^{a}	1.07 ± 0.04^{b}	1.03 ± 0.13^{b}
Survival (%)	96.7 ± 5.8^{a}	83.3 ± 5.8^{a}	93.3 ± 5.8^{a}	80.0 ± 10.0^{a}	90.0 ± 10.0^{a}
Moulting (no. of shrimp)	37.0 ± 7.0^{a}	39.0 ± 2.0^{a}	41.0 ± 4.0^{a}	36.0 ± 4.0^{a}	44.0 ± 2.0^{a}

Values are the means of triplicate group. Mean values having the same superscripts are not significantly different (P>0.05).

Table 3. Proximate composition of *L. vannamei* juveniles carcass fed with test diets supplemented with graded levels of *S. cristaefolium* meal for 42 days

Inclusion levels of brown algae meal	Crude protein (%)	Crude lipid (%)	Crude ash (%)
Carcass			
0	68.23±0.20^{a}	6.63±0.12^{a}	18.96±0.07^{a}
10	68.10±0.15^{a}	6.80±0.16^{a}	20.23±0.52^{a}
20	69.31±0.59^{a}	6.73±0.09^{a}	18.35±0.45^{a}
30	68.18±0.47^{a}	6.82±0.04^{a}	19.06±0.13^{a}
40	68.44±0.37^{a}	6.75±0.13^{a}	18.75±0.64^{a}

Values are the means of triplicate group ± SD and calculated as % of the dry matter

Previous diet studies showed that dietary brown marine algae *Sargassum wightii* supplementations were found effective in supporting growth and survival of *P. monodon* [4]. Similarly, [3] also found that growth performance (weight gain, specific growth rate, FCR and PER) and survival of *P. monodon* postlarvae enhanced when the shrimp fed the 500-2000 mg fucoidan derived from brown algae *Undaria pinnatifida*/kg diet. In the present study, in contrast with the previous studies, supplementation of brown algae *S. cristaefolium* meal in the diet did not enhance growth (final weight, ADG, weight gain, total molting) and survival of *L. vannamei*. This present result is in agreement with the earlier work of [11] who reported that survival and weight gain of *L. vannamei* juveniles were not affected by addition of dietary sodium alginate at different levels (0, 0.5, 1.0 and 2.0 g/kg diet). Similar results were also shown by [16] that no significant differences in survival and weight gain of *L. vannamei* were observed when they consumed diets containing different levels (0, 5, 10, 15%) of

Sargassum illicifolium. Addition of *S. cristaefolium* extract at a dosis of 0, 200, 600, 1000, 1400 mg/kg diet did not significantly affect on the final weight and survival of *L. vannamei* during a 42-day feeding trial period [2]. This indicates that differences in growth performance and survival are influenced by differences in shrimp species, brown algae species and the type of immunostimulants.

However, quite interestingly that addition of *S. cristaefolium* meal in the diet improved FCR and PER of *L. vannamei* in the present study. A dose of 20 g *S. cristaefolium* meal/kg diet in the present study resulted in the best FCR (2.13) and PER (1.34) compared to doses of 0-10 g and 30-40 g *S. cristaefolium* meal/kg. This indicates that 20 g algae meal/kg supplementation is adequate in promoting nutrient utilization efficiency in this study. At that concentration, shrimp showed an optimum nutrient utilization level so that it resulted in reducing FCR and increasing PER. In fact, addition of 20 g algae meal/kg diet significantly promotes better nutrient utilization efficiency. The lowest feed intake is apparent at 20 g/kg supplementation with no differences in growth as affected by increasing the levels of dietary algae meal. Lower feed intake to produce similar growth in the present study is a proof that enhancement of nutrient digestibility might have occurred, resulting in more efficient nutrient utilization including protein. Similar performance was also shown by earlier work of [16], better FCR (1.15-1.17) were obtained in *P. monodon* fed diets containing dietary increased algae levels for 45 days. Feed utilization efficiency enhancement (lower FCR) was also observed in *L. vannamei* juveniles fed *S. cristafolium* extract-supplemented diets [2]. Despite these reports, the mode by which these bioactive compounds promote feed utilization efficiency (FCR) and protein utilization rate (PER) in white shrimp *L. vannamei* is not yet fully explained.

The process of efficient nutrient utilization through digestion and assimilation caused by the activation of fixed phagocytes in the hepatopancreas to produce lytic enzymes upon stimulation might have been attributed to feed utilization enhancement effects of dietary *S. cristaefolium* meal [10]. Enhancement of health status caused by increased active immunological defense and stress tolerance from opportunistic microbial pathogens might have been attributed to the bioavailability of active compounds (immunostimulants) contained by polysaccharides derived from brown marine macroalgae *Sargassum* spp. consumed by the cultivated species [3, 11, 16]. Moreover, the present findings show that dietary *S. cristaefolium* supplementation can promote nutrient utilization efficiency of white shrimp *L. vannamei* juveniles. With limitation and restriction in using antibiotics in aquaculture, utilization of tropical brown marine macroalgae *S. cristaefolium* meal as a feed supplement can be one of problem solvers to overcome disease problem in shrimp production. However, how the active compounds of polysaccharides derived from tropical brown marine algae *S. cristaefolium* meal can enhance feed utilization efficiency in *L. vannamei* juveniles have not been understood yet and this should be addressed for further investigations.

Acknowledgments
Thanks to Diponegoro University that support funding to this work under RPI scheme project 2016/2017. Special thanks to anybody who assist this research. The authors are thankful to the Research Centre for Brackish water Aquaculture Development, Jepara, Indonesia for providing laboratory facilities.

References
[1] Chang, C., Gao, L., Cheng, W. 2013. Effect of dietary administration of hot-water extracts of *Sargassum cristaefolium* on the immunity of white shrimp *Litopenaeus vannamei* and its resistance against *Vibrio alginolyticus*. *Journal of Fisheries Society Taiwan* **40** (1) pp 27-41.
[2] Sudaryono, A., Haditomo, A.H.C., Isnansetyo, A. 2015. Evaluation of dietary supplementation of aqueous extract of brown algae *Sargassum cristaefolium* on growth performance and feed utilization of juvenile white shrimp *Litopenaeus vannamei*. *AACL Bioflux* **8**(2) pp 142-7.
[3] Traifalgar, R.F., Serrano, A.E., Corre, V., Kira, H., Tung, H.T., Michael, F.R., Kader, Md.A., Laining, A., Yokoyama, S., Ishikawa, M., Koshio, S. 2009. Evaluation of dietary fucoidan supplementation effects on growth performance and vibriosis resistance of *Penaeus monodon* post larvae. *Aquaculture Science* **57**(2) pp 167-74.
[4] Huxley, V.A.J., Lipton, A.P. 2010. Immunomodulatory effect of *Sargassum wightii* on *Penaeus monodon* (Fab.). *The Asian Journal of Animal Science* 4 (2):192-196.

[5] Kumaran, T., Citarasu, T. 2016. Isolation and characterization of vibrio species from shrimp and *Artemia* culture and evaluation of the potential virulence factor. *Intel Prop Rights* **4**(1) p 153. doi:10.4172/2375-4516.1000153
[6] Sahul Hameed, A.S., Rahaman, K.H., Alagan, A., Yoganandhan, K. 2003. Antibiotic resistance in bacteria isolated from hatchery-reared larvae and post-larvae of *Macrobrachium resenbergii*. *Aquaculture* **217** pp 39-49.
[7] Sakai, M. 1999. Current research status of shrimp immunostimulants. *Aquaculture* **172** pp 63-92.
[8] Huynh, T.G., Yeh, S.T, Lin, Y.C.Shyu, J.F., Chen, L.L., Chen, J.C 2011 White shrimp *Litopenaeus vannamei* in seawater containing *Sargassum hemiphullum* var. chinese powder and its extract showed increased immunity and resistance against *Vibrio alginolyticus* and white spot syndrome virus. *Fish and Shellfish Immunology* **31** pp 286-93.
[9] Chotigeat, W., Tongsupa, S., Supamataya, K., Phongdara, A. 2004. Effect of fucoidan on disease resistance of black tiger shrimp. *Aquaculture* **233** pp 23-30.
[10] Azad, I.S., Panigrahi, A., Gopal, C., Paulpandi, S., Mahima, C., Ravichandran, P. 2005. Routes of immunostimulantion vis-à-vis survival and growth of *Penaeus monodon* post larvae. *Aquaculture* **248** pp 227-34.
[11] Cheng, W., Liu, C.H., Kuo, C.M., Chen, J.C. 2005. Dietary administration of sodium alginate enhances the immune ability of white shrimp *Litopenaeus vannamei* and its resistance against *Vibrio alginolyticus. Fish and Shellfish Immunology* **18** pp 1-12.
[12] Rodriguez, J., Espinosa, Y., Echeverria, F., Cardenas, G., Roman, R., Ster, S 2007 Exposure to probiotics and β-1,3/1,6-glucans in larviculture modifies the immune response of *Penaeus vannamei* juveniles and both the survival to White Spot Syndrome Virus challenge and pond culture. *Aquaculture* **273** pp 405-15.
[13] Yeh, S.T, Chen, J.C 2008 Immunomodulation by carrageenans in the white shrimp *Litopennaeus vannamei* and its resistance against *Vibrio alginolyticus. Aquaculture* **276** pp 22-8.
[14] AOAC (Association of Official Analytical Chemists) 1990 *Official methods of analysis, 15th edition. Association of Official Analytical Chemists* (Washington,D.C:AOAC) pp 1094
[15] Tung H.T., Koshio, S., Teshima, S., Ishikawa, M., Yokoyama, S., Ren, T., Hirose, Y., Phuong, N.D.T 2009 Effects of heat-killed *Lactobacillus plantarum* supplemental diets on growth performance, stress resistance and immune response of juvenile kuruma shrimp *Marsupenaeus japonicus* Bate. *Aquaculture Sci* **57**(2) pp 175-84.
[16] Hafezieh M., Ajdari, D., Ajdehakosh, A.P., Hosseini, S.H 2014 Using Oman Sea *Sargassum illicifolium* meal for feeding white leg shrimp *Litopenaeus vannamei. Iranian Journal of Fisheries Sciences* **13**(1) pp 73-80.

Growth of locally isolated microalga in POME to produce lipid as alternative energy sources

Elvitriana*, E Munir, Delvian and H Wahyuningsih

Natural Resources and Environment Management Study Program, University of Sumatera Utara, Medan, Indonesia.

*Corresponding author: yatna66@yahoo.com

Abstract. Purpose of this study was to find the best growth of locally isolated microalgae that produce lipids from Palm Oil Mill Effluent (POME) as an alternative energy source. Microalgae was cultivated in POME in glass vessel at room temperature using a lighting intensity of 13,000 lux and continuously aeration for 24 and 12 hours, respectively. Biomass of microalgae were analyzed daily to get their growth by spectrophotometry at 624 nm wavelength, whereas Modified Bligh and Dyer method determined lipid content. Results show that the best growth occurred at 10% inoculum with lighting cycle and aeration of 24 hours (on/off) and resulting highest biomass content of 0.99 g dry weight/L followed by the decrease of organic substances in POME. The percentage reduction of COD, BOD, TSS, and oil at POME reached above 92%, while phosphate concentration reached 89.2%. Cultivation of microalgae in POME for 12 days showed its ability to reduce organic substances and nutrients in POME and produced biomass with lipid content of 35%. These results reached to the conclusion that locally isolated microalgae has an ability to treat POME safely for environment and POME can be used as a growing medium of microalgae that produces lipids.

1. Introduction

Environmental pollution has been a big issue in recent years caused by the continuous disposal of wastewater in large volumes, resulting a decrease in water quality such as decreased dissolved oxygen in water, increased nutrients, and odors, even causing the surrounding water to be unusable. In some developing countries, palm oil industry continues to grow rapidly followed by the increased wastewater from Palm Oil Milling Effluent (POME), which can have a negative impact on the environment. Generally, the palm oil mill entrepreneurs consider that wastewater treatment is very expensive, requires complicated technology, and has no economic value. Various renewable technologies can be applied dealing with water pollution issues that continue to degrade due to disposal of industrial wastewater, especially from palm oil industries. The increasing POME wastewater from palm oil industry is also one of the causes of greenhouse gases; therefore, POME wastewater treatment is needed for the environment and more effective and sustainable.

The use of microalgae for wastewater treatment offers several advantages over traditional waste treatment, including cost-effectiveness in removing BOD, phosphors, nitrogen, and removal of pathogenic bacteria compared to active sludge systems. Through the process of wastewater treatment using microalgae will produce biomass in large numbers [1]. Naturally, microalgae works to reduce concentration of nitrogen and phosphors in wastewater [2]. In the uncontrolled state, residual wastewater entering the waters can lead algal growth that known as algal bloom and eutrophication. A very common method for eliminating nitrogen is through denitrification process that reduces nitrates to gases nitrogen and released it into the atmosphere [3].

Liquid waste from palm oil industry is generally rich in nitrogen, phosphor, carbon, and sulfur nutrients which are nutrients for microalgae growth [4]. Processing ponds of microalgae have potential to reduce nitrogen and phosphorus nutrients through assimilation into microalgae biomass followed by harvesting. The performance of microalgae is better than that of bacteria in lowering COD concentration because microalgae can grow in water contains low carbon. This is due to its ability to utilize dissolved carbon from the air [5]. Wang et al. [6] demonstrated the feasibility of cultivation of Chlorella sp. to treat wastewater from settlement that Chlorella sp. can adapt well without experiencing lag phase. Algae growth is increasing rapidly in the middle side of processing pond due to the high content of nitrogen, phosphor, and COD. Algae biomass is also highly efficient in degrading waste into carbon (zero waste) and producing high density of liquid energy [7]. The use of microalgae for wastewater treatment offers several advantages, namely microalgae provide an effective way to consume liquid waste nutrients and provide enough oxygen for aerobic bacteria through photosynthesis.

Those studies have prompted researchers to examine locally isolated microalgae growth in POME wastewater. Based on the concept of sustainable and environmentally development, POME wastewater can be utilized as nutrients for the growth of microalgae in reducing organic substances and microalgae biomass utilized as alternative energy sources in an effort to overcome problems of impending energy deficit. In an effort to cost reduction of biodiesel production, it requires use of low-cost raw materials that produce alternative energy and at the same time reduce the potential environmental pollution due to waste disposal [8-10].

This paper describes the growth of locally isolated microalgae in reducing organic compounds in POME wastewater as an effort to obtain alternative energy source. It is expected that this research can contribute as a reference of POME treatment system in obtaining alternative energy sources, so the realization of sustainable and integrated industrial development can solve the problem of energy deficit in the future.

2. Materials and Methods

2.1 Materials

POME liquid waste samples obtained from the oil palm industry in Nagan Raya District, Aceh were precipitated and screened before being used in the treatment. The concentration of POME prepared for this research was 25% (LP_{25}). Local microalga used was a stock culture of microbiology laboratory that has been isolated from the open pound of Banda Aceh City.

2.2 POME characteristics analysis

Characteristics of POME was evaluated by measuring several key parameters such as COD, BOD, TSS, oil content, and NH_4, NO_3 (total-Nitrogen), and PO_4 using standard methods APHA [11]. At the end of culturing, the inoculum was re-analyzed to see the decrease of organic compounds in POME with the same parameters.

2.3 Cultivation of locally isolated microalgae

Ten percent of microalgae inoculum ($v_{inoculum}/v_{media}$) was cultivated in POME (LP_{25}) in 1500 ml glass vessel at room temperature using a lighting intensity of 13,000 lux and continuously aeration for 24 and 12 h, respectively. Biomass of microalgae were analyzed daily to get their growth by determining the content of algae biomass (in dry weight), whereas characteristic POME wastewater and lipid content were analyzed at the end of cultivation.

2.4 Culture analysis

Microalgae growth curve was obtained by determining the weight of dry biomass. The growth of microalgae was analyzed by spectrophotometry at 624 nm wavelength, whereas lipid content was determined by modification method of Bligh and Dyer [12]. Microalgae cells were extracted with 2:1 (v/v) chloroform and methanol solution, the solution would be separated into a chloroform and methanol layer by addition of methanol and water to give the final solvent ratio of chloroform:methanol:water (1:1:0.9). Chloroform layer was washed with 20 mL of 5% NaCl, evaporated to dryness, and lipid total was determined gravimetrically.

3. Results and Discussion

3.1 Influence of inoculum concentration on biomass growth of microalgae in POME wastewater

Locally isolated microalgae was applied in POME wastewater to reduce organic substances contained in POME in order to dispose of the wastewater safely into the environment. Figure 1 shows the characteristics of locally isolated microalgae growing in POME wastewater during 12 days cultivation. The growing microalgae in POME wastewater have the ability to adapt to highly concentrated POME and high organic content.

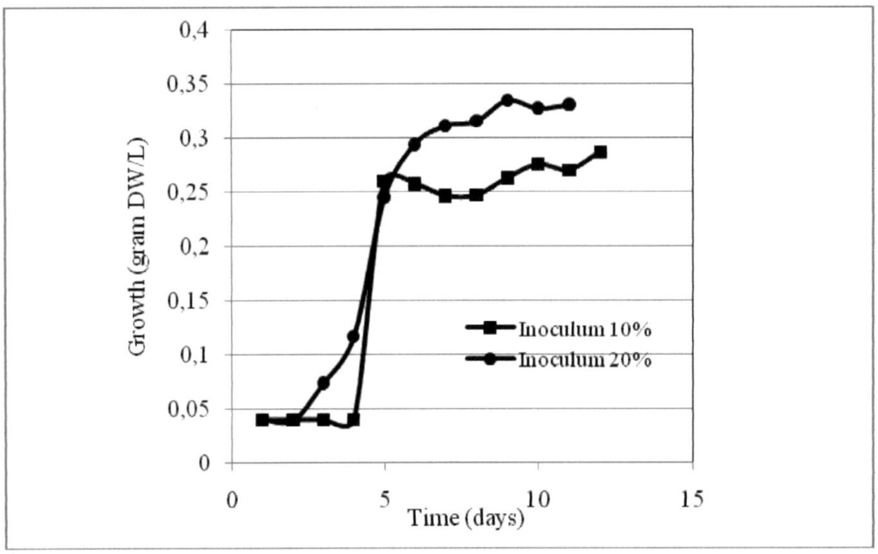

Figure 1. Effect of inoculum concentration on biomass growth of microalgae in POME

Figure 1 shows that the best microalgae growth was obtained from a 10% inoculum concentration in the 24-hour lighting and aeration cycle (on/off) which produced highest biomass of 0.34 g-dry weight/L on day 10, whereas the highest biomass content at 20% inoculum concentration reached only 0.26 g-dry weight/L. That was due to the inoculum of microalgae not acclimatized, so the growth of microalgae was very slow in adjusting to concentrated POME concentration with high organic matter content, takes long time for next growth phase, and yields less biomass. Growth of microalgae was strongly influenced by growing media used, acclimatization of inoculum needs to be done so that POME can be a good growth medium for local isolate microalgae which has potential as alternative media of mass algae growth. According to Bertoldi, et al. [13], alternative media culture has been evaluated for microalgae cultivation such as industrial and agricultural wastewater containing nutrient-rich residues as nutrients for aquaculture biomass growth. Wastewater concentration is very influential on the growth of microalgae, higher concentration of wastewater occurs growth of microalgae in the lag phase slower because it needs to take adaptation time to grow.

IOP Conf. Series: Earth and Environmental Science **144** (2018) 012050 doi:10.1088/1755-1315/144/1/012050

3.2 Effect of lighting cycle and aeration on microalgae growth in POME wastewater

Growth of microalgae was strongly influenced by carbon dioxide, lighting cycles, nutrients, and cultivation time. Figure 2 shows that microalgae biomass content was higher within 24 hours lighting and the aeration than 12 hours lighting and aeration with the same lighting intensity of 13000 lux. Furthermore, Figure 2 shows that the best microalgae growth was obtained from a 10% inoculum concentration in the 24-hour lighting and aeration cycle (on/off) which produced highest biomass of 0.99 g dry weight/L on 5 day. The next day, biomass growth gradually decreased to 12 day. The reduced of lighting cycles and nutrient content in wastewater caused the process of photosynthesis running slowly and producing less microalgae biomass.

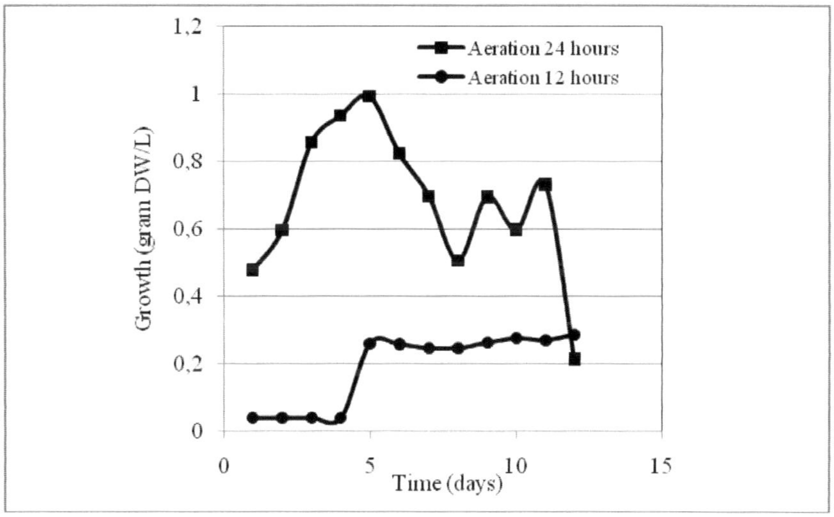

Figure 2. Effect of lighting cycle and aeration on microalgae growth in POME

3.3 Reduction of organic substance and nutrient on POME by microalgae

POME wastewater with high content of organic substances began to be degraded well by microalgae after 2 days of cultivation. In the adaptation phase, microalgae can survive to lag phase so nutrients as their food sources in the wastewater gradually decreased. Reduction of nutrients in wastewater were in line with the decreasing of organic substances in POME.

Yunardi [5] reported that the performance of microalgae is better than that of bacteria in lowering COD concentration because microalgae can grow in water contains low carbon. This is due to its ability to utilize dissolved carbon from the air. Table 1 shows the ability of microalgae to remove organic substances and nutrients in POME wastewater for 12 days. The analysis result shows that microalgae have the ability to decrease organic and nutrient substances in POME reach 92%. It was indicated from the decreasing of COD, BOD, TSS, and oil concentration, whereas phosphate concentration reach 89.2% and ammonium concentration was only 32.4%. Cultivation of microalgae in POME wastewater at LP_{25} concentration for 12 days showed its ability in reducing organic, nutrient, and yielding biomass with a total lipid content of 35%. Algae biomass was very efficient in transforming waste into carbon (zero waste) and producing high-density liquid energy [7]. Culturing process and destruction of biomass cells was very influential on the release of lipids contained in microalgae biomass. Lipid extraction with Modified Bligh Dyer [12] can extract more lipids contained in microalgae biomass because the solvent ratio can help the extraction of all lipids contained in the biomass.

Table 1. Characteristic of POME after treatment using locally isolated microalgae

No	Parameter	Before treatment (mg/L)	After lighting and aeration treatment			
			12 hours		24 hours	
			mg/L	%	mg/L	%
1	BOD	5713	1660	70.9	335.5	94.1
2	COD	38883	6532	83.2	2820	92.7
3	Suspended Solids	16800	7850	53.3	185.0	98.9
4	Phosphate (PO_4)	536.29	75.5	86.0	58.0	89.2
5	Ammonium (NH_4)	77.4	52.3	32.4	121.7	(57.2)
6	Oil contents	256.0	33.0	87.1	1.6	99.4
7	Nitrate (NO_3-N)	49.0	80.0	(63.3)	56.0	(14.3)

Condition: Inoculum 10%, light intensity 13.000 lux, POME concentration 25% (LP_{25})

4. Conclusions

The best locally isolated microalgae growth was obtained at 10% inoculum concentration with lighting cycle and aeration of 24 hours (on/off) and resulting highest biomass content of 0.99 g dry weight/L followed by the decrease of organic substance in POME wastewater. The percentage reduction of COD, BOD, TSS, and oil at POME reached above 92%, while phosphate concentration reached 89.2%. Cultivation of microalgae in POME wastewater at LP_{25} concentration for 12 days showed its ability to reduce organic substances and nutrients and produced biomass with a total lipid content of 35%.

Authors Contribution: Erman Munir, Delvian, and Hesti Wahyuningsih contributed to the discussion of the results.

References

[1] Woertz IC 2008 *Master's Theses and Project Reports* digitalcommons.calpoly.edu
[2] Johnson MB and Wen Z 2009 *Energy Fuels* **23**(10) 5179–5183
[3] Metcalf & Eddy FL Burton FL Stensel HD and Tchobanoglous G 2003 *Wastewater Engineering: Treatment and Reuse* McGraw-Hill International New York
[4] Sheehan J, Dunahay T, Benemann J, Roessler P 1998 *Al Look Back at the US Department of Energy's Aquatic Species Program - Biodiesel from Algae* National Laboratory of the US DEO
[5] Yunardi 2003 *Assessment of Algae Bioreactor Performance to the Provision of Nitrogen Compounds from Wastewater* Unsyiah Digital Library
[6] Wang L, Min M, Li Y, Chen P, Chen Y, Liu Y, Wang Y and Ruan R 2009 *Appl. Biochem. Biotechnol.* 1-13
[7] Widjaja A Chien CC, Ju YH 2009 *J. of Taiwan Institute of Chem. Eng.* **40** 13-20
[8] Daboor SM, Budge SM, Ghaly AE, Brooks SL and Dave D 2010 *Am. J. Biochem. & Biotech.* **6** (4) 239-263
[9] Lee SJ Kim SB Kim JE Kwon GS Yoon BD Oh HM 1998 *Lett. in Appl. Microbiol.* **27** 14-18
[10] Lee JY Yoo C Jun SY Ahn CY Oh HM 2010 *Bioresource Technol.* **101**(1) S75-S77
[11] APHA 1999 *Standard Method for the Examination of Water and Wastewater* Washington DC US
[12] Bligh EG and Dyer WJ 1959 *Can. J. of Biochem. and Phys.* **37**(8) 911-917
[13] Bertoldi FC, Sant Anna E, da Costa Braga MV and Oliviera JGB 2006 *Grasas y Aceites* **57**(3) 270-274

IOP Conf. Series: Earth and Environmental Science **144** (2018) 012051 doi:10.1088/1755-1315/144/1/012051

Reassessment of forest area and its scoring as a permanent production forest

T Sudarmadji* and W Hartati

Faculty of Forestry, Mulawarman University. Jl. Ki Hajar Dewantara, Samarinda, East Kalimantan, Indonesia

*Corresponding Author: triyonosudarmadji@gmail.com

Abstract. Along with the increasing demand for tropical timber, the need for wood raw materials is not enough just to rely on natural forests, and therefore in the early 1990s began to be developed timber estates concessions. The designation of forest areas can basically be altered using procedures established by the Ministry of and Forestry. PT Permata Borneo Abadi (PBA) as the holder of IUPHHK-HT covering 38.680 ha intends to propose changing of forest function from limited production forest (HPT) and convertible production forest (HPK) into permanent production forest (HP) to be developed as timber estates with fast growing species. Reassessment of forest function is intended to formulate official documents aimed to reevaluate IUPHHK-HT especially HPT and HPK. The assessment is based on established criteria and standard values (scores) covering topographic, soil and vegetation conditions as well as regional rainfall distribution. The assessment results indicate that there is an area with a slope of >40% of 840.18 ha with score >175 which must therefore as protected areas. The area of HPT that scored between 125-174 remains HPT is 12.287,77 ha. The area of HPT that scored <125 is 27.637.80 ha and therefore is possible to be converted into HP. The results can be used as an important basis for intensive forest management synergically with conservation efforts.

1. Introduction

Exploitation of natural forests outside Java and Madura is mandated by government act No. 5/67 regarding basic forestry provisions. However long time before that, the exploitation of natural forests in Kalimantan has been done traditionally by cutting down groups of floating commercial species, assembled in the river for further transported downstream to wood processing sites (wood industry). Over the time, selective logging techniques are then widely practiced with the involvement of more advanced technologies using various heavy equipments [1].

Along with the increasing demand for tropical timber in the world for various purposes including pulp and paper industries, the need for wood raw materials [1] is no longer deemed sufficient simply by relying on wood that grows naturally in natural forests. Thus, in the early 1990s a forest concession was developed that planted fast growing species to supply the timber industry and was known as timber estates or industrial plantation (HTI). Wood species such as *Acacia mangium*, *Eucalyptus sp* or *Gmelina arborea*, are widely cultivated by HTI concession holders (IUPHHK-HT).

In Indonesia, the boundaries between forest and non forest areas and between different forest areas of status and function have been set forth in the map of forest use agreement (TGHK) to determine the allocation of forest area and their functions [2, 3]. Designation of forest function principles can be changed and legally regulated in government regulations [4].

Reassessment of forest function was intended to provide complementary official document to the propose changes in the function of forest submitted to the Minister of Environment and Forestry as a determinant of national forest management policy [5]. The objective of this research was to re-assess and to find out wether there are any areas of HPT and/or HPK that might be functionally converted to HP using the criteria and scoring standards set up in the applicable government rules and regulations.

2. Method

2.1. Location and fieldworks

The study was conducted in the concession of IUPHHK-HT PT PBA, which is administratively located in East Kutai Regency, East Kalimantan Province covering 49.680 ha (figure 1). Fieldworks were conducted in 14 sample plots of field observations focused on forest stands, soil profile and soil sampling, and several weather station records in the area and its surroundings.

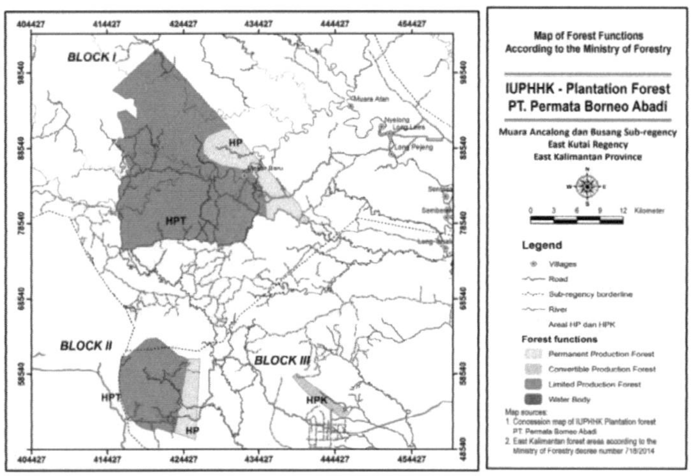

Figure 1. Map of forest functions of IUPHHK-HT PT PBA concession

2.2 Procedures

The initial environmental condition of IUPHHK-HT PT PBA was studied in the documents of forest product utilization work plan and environmental impact assessment. Basic data of lanslope, rainfall depth and rainfall intensity, soil type and other supporting data were used for assessing forest area function determination. Furthermore, the determination of proportional weight and scores were set up to obtain total scores in each observed location to find out proper recommendations for areas that are technically possible to be functionally changed from HPT and/or HPK into HP [6-8].

2.3. Analysis

Data analysis was conducted by applying the assessment criteria that HPT is a region with total score of slope, soil type, rainfall intensity of 125-174, HP designation area has the largest number of scores 124, whereas HPK is determined by criteria that the largest number of scores is 124 and/or an area if converted is still capable of maintaining environmental carrying capacity. The determination of each score follows technical guidance in the decree of Minister of Agriculture No. 837/Kpts/Um/11/80 on criteria and procedures for the establishment of protection forest [6] and No. 683/Kpts/Um/8/81 [7] on the criteria and procedure of production forest determination (table 1). All of environmental factors were analyzed to obtain units of land with unique values which are a combination of the three environmental factors [9]. The sum of these unique values generates a total score that gives direction to whether an area will be designated as HPT, HL or HP. The analysis of land cover and land system was done to justify forest function change.

Table 1. Scores of landslope, soils, and rainfall for reassessment forest functions

Class	Landslope			Soils			Rainfall		
	Description	Slope (%)	Score	Erodibiliy	Type	Score	Description	Intensity (mm/day)	Score
1.	Gentle	0 – 8	20	Low	Alluvial, Gley soil , Planosol, Grey hidromorf, Laterit	15	Low	< 13.6	10
2.	Moderate	8 – 15	40	Moderate	Latosol	30	Moderate	13.6 – 20.7	20
3.	Moderately steep	15 – 25	60	High	Cambisol, Mediteran, Brown Forest (*Inceptisol*), Non Calcic Brown (*Inceptisol*)	45	High	20.7 – 27.7	30
4.	Steep	25 – 40	80	Very High	Vertisol, Andosol (Andisol), Grumusol (*Molisol*), Laterit (*Oxisol*), Podzol (Spodosol), Podozolic (*Ultisol*)	60	Very high	27.7 – 34.8	40
5.	Extremely steep	> 40	100	Extremely High	Litosol, Organosol, Rendzina, Regosol	75	Extremely high	> 34.8	50

The classification of slopes is made to determine the condition of area associated with the potential erosion-sedimentation and conservation efforts. Soil observations were carried out in the soil profile on each land cover to determine soil characteristics for scoring soils. Subsequently, monthly ranfall and rain days were used to determine the rainfall intensity and its occurence probability events as the basis for determining proportional rainfall scores.

3. Results and Discussion

3.1. Climate condition

The average monthly air temperature ranges from 26.8 – 27.5°C, with the highest air temperature occurring in April and the lowest in July. Relative humidity is 71.3 – 82.2% with an average of 79.0%. Average duration of solar radiation intensity is 43.7% with a range of 33.7 – 50.9%. Mean annual rainfall is 1.707 mm varies relatively small or evenly throughout the year. Highest rainfall occurs in January - March and December and the lowest at September with monthly rainfall days between 11 - 17 days. Based on the Schmidt and Ferguson Climatic Classification System (1951), the study location is generally under influence of B climatic type which means having significant low rainfall periods of rainfall (table 2).

Table 2. Average values of climate elements at study sites (1986 - 2014)

Climate elements	Monthly average											
	Jan	Feb	Mar	Apr	May	Jun	Jul	Aug	Sep	Oct	Nov	Dec
Temperature (°C)	27.1	27.0	27.4	27.5	27.4	27.1	26.8	26.9	27.1	27.2	27.3	27.3
Humidity (%)	80.2	81.0	81,2	79.9	79.2	80.2	80.3	75.6	77.4	71.3	80.9	82.2
Sunshine (%)	29.7	33.7	34.3	47.8	39.2	42.2	51.9	59.8	44.3	50.9	45.8	45.0
Rainfall (mm)	185	158	200	174	165	131	117	96	90	128	170	185
Rainfall (Days)	13	11	13	12	13	12	12	10	10	11	13	13

Sources: Weather Station: Kecamatan Sangkulirang (2003 - 2014), PT Indominco Mandiri (2003 - 2012), BWS-III Sangatta (1986 - 2010), UPTD Dinas Pertanian (2002 - 2011), BWS-III Muara Ancalong (1981 - 2008).

3.2. Geology and physiography

The area consists of 7 land systems composed of sedimentary rock types that old age as formed in the Neogen era. Old sedimentary rocks would form soil with low fertility, predominantly dominated by soft and coarse textured sedimentary rock types. Physiographic conditions vary from plain to hilly and most are flat (60.1%), and are in the range of 34 - 644 m above sea level (asl).

3.3. Soils characteristic

The study area is composed of four land associations namely ultisols - inceptisols, inceptisols - ultisols - oxisols, inceptisols - entisols, and inceptisols - ultisols - molisols. The association of ultisols - inceptisols occupies about 71.61% of the area (29.194,39 ha) while inceptisols - ultisols - oxisols assosiation occupies 12.41% (5.060,50 ha), and the rest of 7.73% (3.151,71 ha) and 8.24% (3.359,15 ha) are the territories of inceptisols - entisols association and inceptisols - ultisols - molisols. Each land association of more than 70% of its territory is in a flat to steep area except the territory of inceptisols - ultisols - molisols asssociation which mostly extends from a moderate to steep slope [10].

3.4. Vegetation cover

PT PBA forest concession is currently dominated by pioneer vegetation types such as Jabon (*Antocephalus cadamba*), Mahang (*Macaranga sp.*) and group of ferns. Originally, the area was covered by a type of lowland tropical rainforest rich in species from the Dipterocarp family with very high standing stock potential. However, the high standing stock potential triggers the setting of high production targets. Therefore, intensive timber extraction from natural forests causes the productivity of forests in East Kalimantan dramatically decline. Record of two major forest fire incidents in 1982/83 and 1997/1998 helped speed up the process of losing forest resources so it is no longer economical to cultivate in the second rotation. Excessive forest exploitation transforms primary forest cover into secondary forest with 50% or more damage [11]. Deforestation greater than that number due to various factors resulted in poorer land cover classes [12, 13]. The land cover map confirms that the shrubs as the dominant land cover class in the PT PBA work area (figure 2).

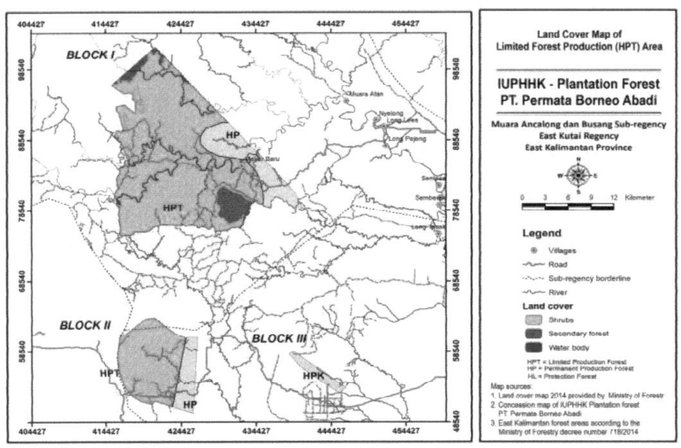

Figure 2. Vegetation map of forest concession of PT PBA

3.5. Forest function assessments

Study of changes in the function of forest area from HPT and/or HPK to HP reffers to the analysis of representative environmental factors i.e. landslope, soil type and rainfall intensity; which are considered to be represent the endogenous and exogenous energies that play a role in the formation of nature in which humans and other biological creatures [14-17]. Determination of sample plot location is based on land cover conditions, slope class, soil type and rainfall. A total of 14 sample plots were made in the field to represent some of the physical characters of the land. Thirteen sample plots measuring 50 x 50 m (0.25 hectares) and one plot measuring 30m x 60m (0,18 ha).

The class of land slopes is considered to be a major limiting factor in the process of forest area convertion [15, 17] due to its great potential to negatively impact upon the existence of living things when there was a change over it as a result of development activities [18-20]. The situation with very limited technology and access to forests, not all places can be visited to validate the accuracy of the topographical map created. This situation leads to a mismatch between the direction map function with real conditions in the fields. Even flat areas are classified as limited production forest areas (HPTs) that should be designated for special areas [21] with slopes of more than 40% or altitudes of more than 2.000 m above sea level. Similarly, vice versa in the field precisely into the area of HP.

The slope classes used in this study were obtained from the interpretation of SRTM (Shuttle Radar Topography Mission) images with spatial resolution of 30m x 30m. Currently SRTM is the only available global level elevation image even for free with high enough spatial resolution. Figure 3 is a visualization of the digital eleveation model of the SRTM image, while figure 4 is a class of PT PBA concession slope that has been derived from the SRTM digital elevation data.

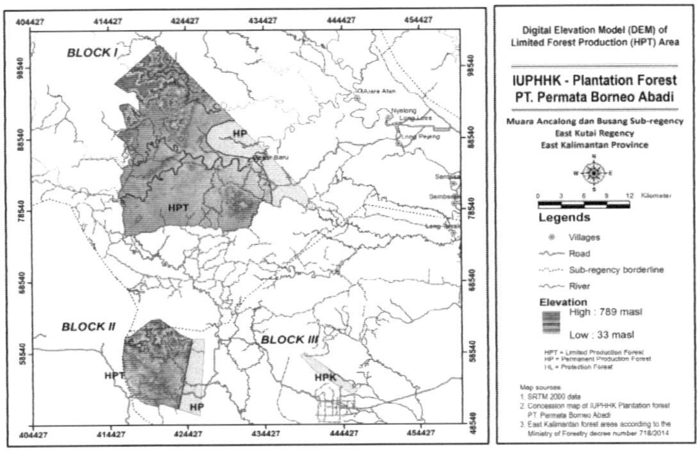

Figure 3. Map of digital elevation model of PT PBA forest concession

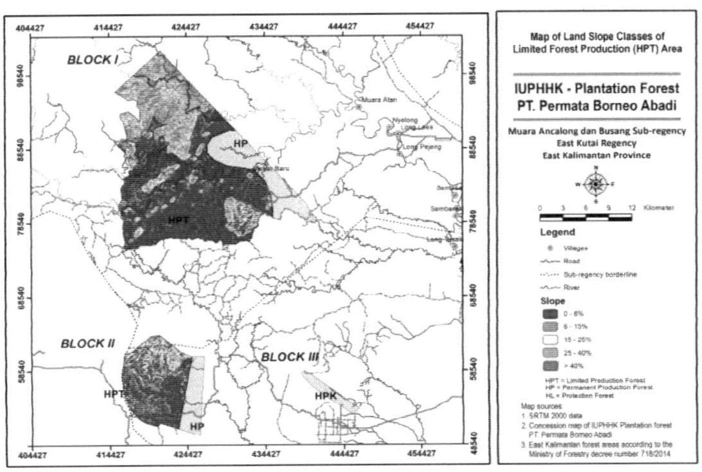

Figure 4. Map of land slope classes of PT PBA forest concession

Soil characteristic at the study site indicates the association of ultisols - inceptisols (candiudults, paleudults, dystropepts or podzolic candic, podzolic chromic, cambisol district), inceptisols - ultisols - oxisols (dystropepts, paleudults, candiudoxs or cambisol district, podzolic chromic, oxisol haplic), Association of inceptisols - entisols (tropaquepts, fluvaquents, dystropepts or district gleisols, alluvial gleics, district cambisol) and inceptisols - ultisols - molisols (dystropepts, paleudults, palehumults, argiudols - district cambisol, chromic podzolic, umbric latosol, mediterran).

Based on the soil classification results from the representative profile of ultisols - inceptisols association either on the slope class 0 - 8% or 8 - 15% there are 2 (two) soil types namely inceptisols and ultisols. The score for the cambisol soil type (inceptisols) is 45 whereas for podsolic (ultisols), but for location with a complex soil type, the class is the same as the class of soil type whose sensitivity to class erosion is determined by the soil class with highest erosion sensitivity i.e. podzolic with a score of 60. Details of the scores from each land association with various slope classes at the study sites based on the classification of the representative land profiles can be seen in table 3.

Table 3. Scoring result of soil factor based on soil profile calssification

No	Soil association	Slope class (%)	Representative soil profile and type	Soil classes	Score
01.	Ultisols - Inceptisols (Candiudults, Paleudults, Dystropepts - Podzolic candic, Podzolic chromic, Cambisol distric)	0 – 8	PBA-01 (Podzolic) PBA-03 (Podzolic) PBA-05 (Podzolic) PBA-06 (Cambisol) PBA-11 (Podzolic)	3,4	60
		8 – 15	PBA-02 (Podzolic) PBA-07 (Cambisol) PBA-12 (Cambisol)	3,4	60
02.	Inceptisols - Ultisols - Oxisols (Dystropepts, Paleudults, Candiudoxs - Cambisol distric, Podzolic chromic, Oxisol haplic)	8 – 15	PBA-04 (Latosol)	2	30
03.	Inceptisols - Entisols (Tropaquepts, Fluvaquents, Dystropepts-Gleisol distric, Alluvial gleic, Cambisol district)	0 – 8	PBA-09 (Podzolic) PBA-13 (Podzolic)	4	60
		8 – 15	PBA-10 (Podzolic) PBA-14 (Podzolic)	4	60
		15 – 25	PBA-08 (Litosol)	5	75

Rainfall intensity is estimated by using rain and rain distribution both monthly and yearly. Based on the result of map study and orientation and field observation, the data collection is done at 5 (five) observation locations. Data on the distribution of rainfall depth and rainy days are used to describe the rainfall intensity characteristics. The monthly average within 13 years (2002 - 2014) is presented in figure 5. Average annual rainfall ranges from 1,309 - 2,355 mm and monthly rain huge between 103 - 212 mm. The highest rainfall possibility occurs in December while the lowest is in August. In general, rainfall is relatively low - below the average of some other parts of East Kalimantan, but the trend is relatively similar that it has 2 (two) peaks (March and December) based on polynomial trend - order 3 with R2 = 0.9382 (r = 0.97). The number of annual rainfall days ranges from 92 - 157 and monthly between 6 - 14 days, the largest in December while the smallest is in August.

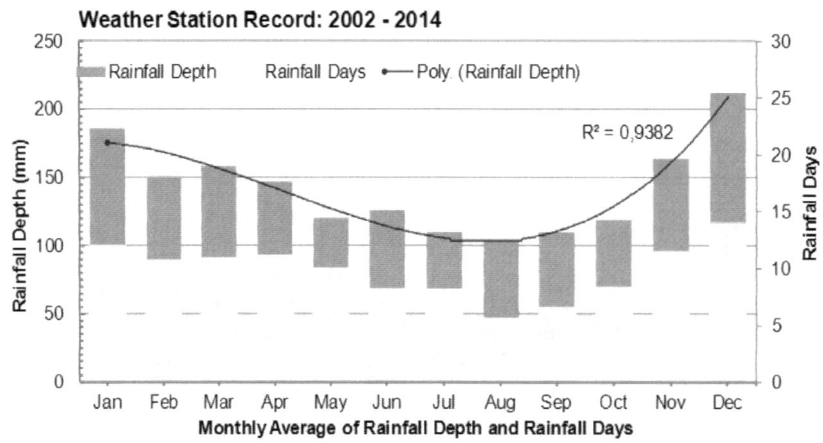

Figure 5. Distribution of rainfall depth and rainfall days at study sites

Average rainfall intensity per day (HH) in the four rainfall observation locations is presented in figure 6. The mean rainfall intensity for each location is between 5.0 - 18.5 mm/HH, 6.6 - 17.2 mm/HH, 4.5 - 15.3 mm/HH, 7.2 – 14.9 mm/HH, and 17.4 - 39.7 mm/HH. The four locations have a magnitude of 6.7 - 19.8 mm/HH. This means that the range is Very Low (<13.6 mm/HH) to Medium (20.7 - 27.7 mm/HH). Although the above monthly rainfall intensity range is between Very Low to Medium, it is important to detect in more detail whether there are other high rainfall intensities of High and Very High [22].

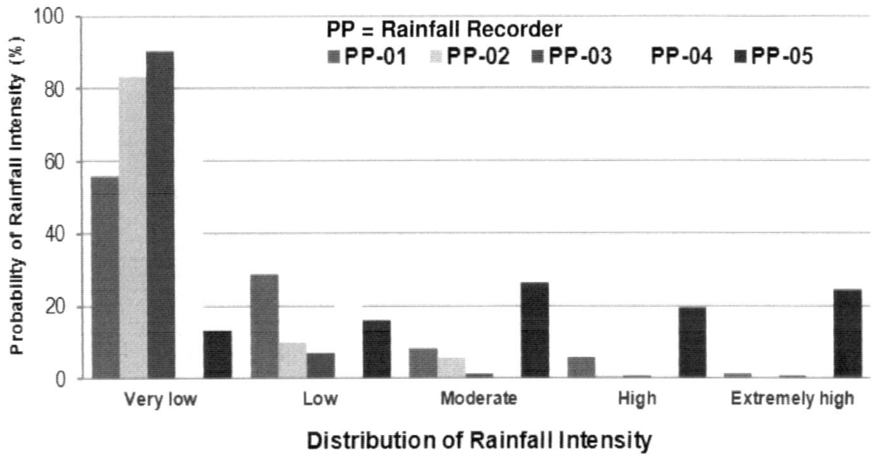

Figure 6. Distrution of monthly avarage rainfall intensity per-day occurences

Most of the possibilty of rainfall intensity are at very Low and Low prices. Decree of Minister of Agriculture No. 837/Kpts/Um/11/1980 stipulates that rainfall intensity is divided into Very Low (<13.6 mm/HH), Low (13.6 - 20.7 mm/HH), Medium (20.7 - 27.7 mm/HH, High (27.7 - 34.8 mm/HH), and Very High (> 34.8 mm/HH) , and then multiplied by the value of the scales according to the relative magnitude of the impact of the erosion event. The proportional scale value of rain intensity is 10 (ten). Given that the dynamics of rain and rainfall days of the year can not always be accurately predicted, the determination of rainfall intensity score is calculated by counting all of rainfall occurrences. Weighted score of rainfall intensity possibility for each location was multiplied by the scale of the rainfall intensity for each class (Very Low, Low, Medium, High, Very High) is 6.2, 3.3, 2.4, 2.8 , and 3.5 respectively. Thus, the total rainfall intensity score for the entire study area is 18.1.

The process of overlapping was done on each classified data except the rainfall intensity, in the form of slope and land cover maps of which each class get a weighting value. Land slope was divided into five classes of slopes (0-8%, 8-15%, 15-25%, 25-40% and >40%) superimposing to the soil type associations present in the PT PBA concession ie associations of Ultisols - Inceptisols, Inceptisols - Ultisols - Oxisols, Inceptisols - Entisols, and Inceptisols - Ultisols - Molisols. The superimposing process produced unique land units as a combination of the above three factors. These land units ware then given a new score to deteremine appropriate forest functions. In overall the superimposing process of the three data produced 261 units of land units (figure 7).

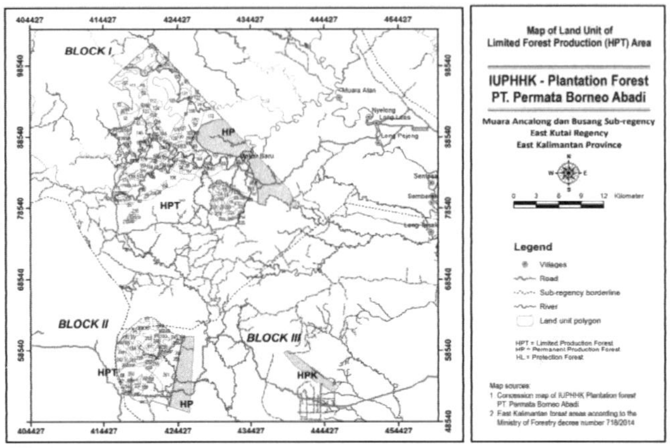

Figure 7. Land units distribution based on landslope, soil type and rainfall intensities

Each land unit has a unique identity and scores of three environmental used data. Referring to Minister of Forestry regulation No. P.34/Menhut-II/2010 on Forest Area Function Change Procedures, the function HP was determined with score below 125 while the score between 125-174 become HPT function. Score quantities greater than or equal to 175 were designated as protected forest areas. For this forest function assessment, table 4 displays the data attributes each score of land 261 land units generated from the analysis of superimpose based on standard values of forest functions regulated in Minister of Forestry regulation No. P.34/Menhut-II/2010 on the procedures for change forest function. An area of 70.53% or approximately 28,753.30 ha of total HPT area in PT PBA concession could be converted into HP. Further, n area of 840.18 ha (2.06%) should be a protected area because it has a very steep slope (>40%), and the remainder is reserved for HPT function (Figure 8).

Table 4. Forest function based on re-scoring results for effective and efficient forest management

Total score criteria	Forest function based on MoF decree P.34/Menhut-II/2010	Area (Ha)	Area (%)
<125	Permanent Production Forest (HP)	27.637,80	67,80
125-174	Limited Production Fortest (HPT)	12.287,77	30,14
≥175	Protection Forest	840,18	2,06
	Total	40.765,75	100,00

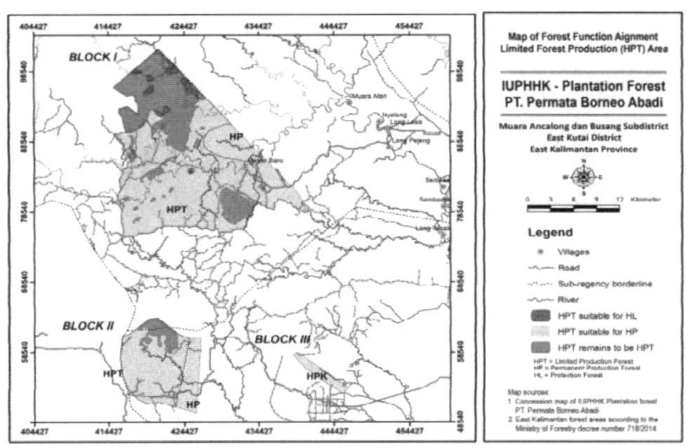

Figure 8. Map of forest function for effective and efficient forest management

Reassessment of forest function provides important direction for the function of forest areas but still need to be harmonized for the benefit of effective, efficient and compact forest management. The result of reassessment shows significant changing of forest function. From earlier designationof IUPPHHK-HT function covering 49,000 ha, the area of HPT would decrease enough from 40,765,75 ha to 12,287.77 ha. Conversely, the area of permanent production forest (HP) would increase from 7,336.25 ha to 34,974.05 ha. Consequently, HP's previously small areas would be dominant i.e. 71.37% of the total area of the previous designation forest function.

4. Conclusion

In conclusion, the area of HPT in the PT PBA concession is a region with a slope of > 40% (840.18 Ha) must therefore serve as a protected area. The area of HPT which scores between 125-174 and therefore remains HPT is 12.287,77 Ha. The HPT area that scored a new <125 score is 27,637.80 Ha and therefore allows to be converted to HP. Based on these conclusions, PT PBA may use the results of this study as the basis for full forest management in its concessions including conservation efforts while referring the laws and regulations governing the designation of protected areas and conservation.

Acknowledgments

Sincere thanks are expressed to PT Permata Borneo Abadi (PBA), East Kalimantan Indonesia for facilitating this research. The authors would also like to acknowledge all of reviewers for critical review of the manuscript.

References

[1] Sist, P., Sheil D., Kartawinata K., Priyadi, H. (2003). Reduced-impact Logging in Indonesian Borneo: some results confirming the need for new silviclutural prescriptions. Forest Ecology and Management: 179. pp. 415-427.
[2] Government of Indonesia. (2007). Government act No. 6/2007: Forest planing, forest management, and forest utilization [in Indonesian].
[3] Government of Indonesia. (2009). Government act No. 32/2009: Protection and Management of Environment [in Indonesian].
[4] Government of Indonesia. (2010). Government act No. 10/2010: Procedure of designation and changing forest function [in Indonesian]
[5] Government of Indonesia. (1999). Government act No. 41/1999: Forestry [in Indonesia].
[6] Ministry of Agriculture. (1980). Decree No. 837/Kpts/Um/11/80: Criteria and procedure of forest protection designation.
[7] Ministry of Agriculture. (1981). Decree No. 683/Kpts/Um/8/81: Criteria and procedure of forest production designation.
[8] Ministry of Forestry. (2014). Decree No. P.16/Menhut-II/2014: Guidance of forest tenurial utilization [in Indonesian].
[9] Ministry of Forestry. (2011). Decree No. P.4/Menhut-II/2011: Guidance of forest reclamation [in Indonesian].
[10] Pritchett, W. L. and Fisher, R. F. (1987). Properties and management of forest soils. Second edition. John Willey & Sons. New York. 494 p.
[11] Primack R. B., Supriatna J., Indrawan M,, Kramadibrata P. (1998). Conservation biology. Jakarta: Yayasan Obor Indonesia [in Indonesia].
[12] Morgan, R. P. C. (1996). Soil erosion and conservation, 2nd ed. Longman Group Lim. London.
[13] Sudarmadji, T. and Hartati, W. (2016). The process of rehabilitation of mined forest lands toward degraded forest ecosystem recovery in Kalimantan, Indonesia.bVolume 17, Number 1, April 2016 E-ISSN: 2085-4722. Pages: 185-191 DOI: 10.13057/biodiv/d170127.
[14] Agassi, M. (1996). Soil Erosion: Conservation and rehabilitation. Marcel Dekker, Inc. NY.
[15] Arsyad, S. (2006). Soil and water conservation. IPB Press. Bogor [in Indonesia].

[16] Bergen S. D., Bolton S. M, Fridley J. L. (2010). Design principles for ecological engineering. J Ecological Engineering 2001;18:201-210.

[17] Bradshaw, A. D. (2002). Introduction and philosophy. Perrow and A. J. Davy (eds). Handbook of Ecological Restoration. Vol. 1. Cambridge Univ. The Edinburgh Bldg, CB2 2RU, UK.

[18] Kartasapoetra, A. G. (2000). Soil and water conservation technology. Rineka Cipta. Jakarta [in Indonesian].

[19] Kusumandaria A, Nugroho P. (2014). Land capability analysis based on hydrology and soil characteristics for watershed rehabilitation. The 5th Sustainable Future for Human Security (SustaiN 2014). Procedia Environmental Sciences 28 (2015) 142-1447.

[20] Lee, R. (1978). Forest microclimatology, Columbia University Press, New York.

[21] Laurila-Panta, M., Lehikoinenb, A., Uusitaloc, L., Venesjärvib. R. (2015). How to value biodiversity in environmental management ?. Ecol. Indicators 55: 1-11

[22] Government of Indonesia. (2010). Government act No. 24/2010: Forest land use and utilization [in Indonesian].

The economic benefits of vegetation in the upstream area of Ciliwung watershed

T R Saridewi*and Nazaruddin
Bogor Agricultural Extension College Bogor West Java Indonesia

*Corresponding author: trsdewi74@yahoo.com

Abstract. Ciliwung watershed has strategic values since its entire downstream area is located in the Special Administrative Region of Jakarta (DKI Jakarta), the capital of Indonesia. This causes forest and farmland areas are converted into open areas or built-up areas. The existence of these areas provides enormous environmental and economic benefits. Economic benefit values are very important to be considered in developing a policy development plan, but they have not been calculated yet. This study aims to determine the economic benefits provided by trees and other vegetation anddevelops a development policy that takes into account simultaneously ecological and economic aspects. The study is conducted in the upstream Ciliwung watershed, by using land cover patterns in 1989, 2000, 2010 and 2014, and employs GIS and CITY green analysis. The results show that conversion of forest and farmland areas reduces the ability of Ciliwung upstream watershed to store water. Therefore, its ability to reduce the flow of surface has been decreased. This creates a decrease in the cost savings of annual stormwater, from US$ 15,175,721 in 1989 to US$ 13,317,469 in 2014. The Environmental Services Payment Policy (PES) for upstream community groups managing the watershed has been considered as a fairly effective policy.

1. Introduction

The development activities in the Ciliwung watershed are growing rapidly due to the influence of Jakarta city as the capital of Indonesia which is situated in the downstream area of the Ciliwung watershed. The upstream area which is located in Bogor Regency and Municipal is a tourism destination area that caused rapid development of recreation facilities. Meanwhile, the midstream area is located in Bogor Regency and Municipal, and Depok Municipal, and the entire downstream area is located in the Special Administrative Region of Jakarta (DKI Jakarta) which has dense settlement, public facilities, and other urban solid areas. This rapid growth and progress lead to various problems in the Ciliwung watershed such as worsen flooding and the deterioration of river quality [1].

Flooding is caused by the lack of water catchment areas because land cover in the Ciliwung watershed area is dominated by built-up areas. Flood has disrupted various public economic and social activities and caused huge losses. Great floods which occurred in 2002 have led to a loss of Rp 9.8 trillion (US$7 billion) [2]. Great floods in January 2013 claimed 20 lives, and 33,502 people were forced to flee, with an estimated loss of Rp. 20 trillion (US$20 billion) [3] with an average annual loss around US$ 321 million [2].

The government policy in flood control is dominated by a structural policies, such as the construction of dam, ponds, lake as well as green well and water body normalization. This policies turned out to be inefficient [4] and is not cost-effective for ecological improvement [5]. Spatial plan is one of the non-structural approaches to flood control [6], since inappropriate changes in land use and land management are the causes of flooding [7]. Actually, spatial plan has already been stipulated in the Spatial Plan (RTRW) of regency/municipal, but its implementation is often not in line with the rapid development activities, including in the Ciliwung watershed area. Increased demand for space is driven by the increase in population and the development of various economic sectors. Up to now, the demand for space is being fulfilled by converting agricultural and forest areas into open areas or even built-up areas. This happens because the economic appreciation of forest and agricultural land is lower than that for other uses.

The low economic appreciation occurs because we have not had a tool to interpret the economic benefits of forest and other vegetation areas. A study on economic contribution of forest and other vegetation areas should be conducted to be used as a reference by the government to prepare development policies. The forest'sfunction in slowing down rainwater runoff, increase air quality and reduce energy used to cool down air temperature can be interpreted as real saving cost to improve urban population health and welfare [8]. Based on this situation, this research intends to investigate the economic and environmental benefits provided by trees and other vegetation and to prepare policy development plan that would take into account both ecological and economical aspects.

2. Method

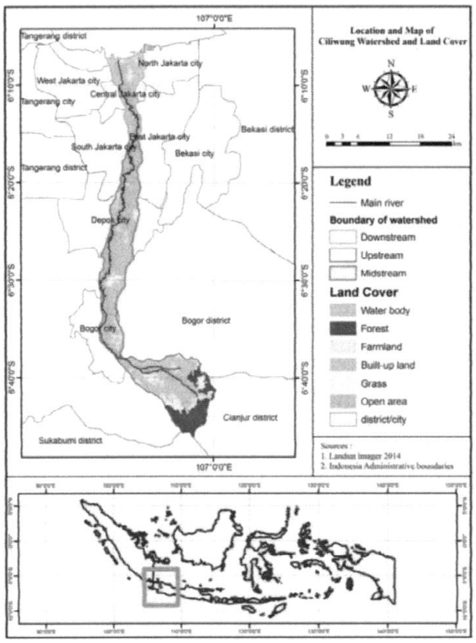

Figure 1. Ciliwung watershed location

This research is conducted in the upstream area of the Ciliwung watershed in Bogor Regency and Municipal. The Ciliwung Watershed covers an area of 386.10 km^2 spans along a 117 km main river. The watershed extends from West Java Province in the south to Jakarta Province in the north with an overall river length of 807.93 km.

The study used secondary data consisting of land cover map in 1989, 2000, 2010, and 2014 [9], administrative boundary map, watershed boundary map, annual rainfall data, and construction cost [10]. The study was conducted in two-stage analysis, i.e.the change in land cover and CITYgreen analysis. Analysis of land cover pattern change used data from landsat 7+ETM satellite image in 1989, 2000, 2010, and 2014 which identified (interpreted) land cover pattern in the upstream area of the

Ciliwung watershed using ArcGIS 10.1 [9]. Based on image classification, land cover is grouped into seven patterns which include water body, built-up areas, grass areas, open areas, dryland farming areas, forest areas, and paddy field areas in the upstream areas. CITYgreen is a comprehensive tool to calculate the environmental benefits provided by urban trees and to model the impact of various construction and planning scenarios [8]. CITYgreen 5.4 is an extension of ArcView GIS software that can calculate the environmental benefits of a particular site; one of which is to reduce the impact of surface current from rainwater (stormwater control analysis). The output of CITYgreen used in this research is the economic benefit summary.

3. Results and Discussions

3.1. Economic benefits

Ciliwung watershed area has become a densely populated area since a long time ago, because since 1989 the land use pattern is dominated by built-up areas. Ciliwung has been essential to human life since the beginning of the history of Indonesia, and since the 10th century Kalapa harbor has been established. As time passes, the number of people residing around the Ciliwung River continues to grow and develop. This is why the community needs for housing and other facilities continues to grow until now. Based on the analysis of land cover patterns from 1989 to 2014, the built-up areas in the upstream areas has increased from 6,269.74 ha in 1989 to 7,844.23 ha in 2014. Most of the land that can serve as water catchment areas have largely been reduced, the extent of forest area decreased from 4,614.16 ha in 1989 to 4,241.21 ha in 2014. The land cover patterns from 1989 to 2014 in the Ciliwung watershed area are presented in table 1.

Table 1. Land use pattern in upstream Ciliwung watershed

Land use pattern	Year 1989		Year 2000		Year 2010		Year 2014	
	Ha	%	Ha	%	Ha	%	Ha	%
Water body	35.94	0.24	78.08	0.51	96.75	0.64	64.09	0.42
Built-up areas	6269.74	41.21	6334.67	41.64	6665.10	43.81	7844.23	51.56
Grassland area	124.12	0.82	116.92	0.77	148.86	0.98	113.10	0.74
Open areas	187.97	1.24	276.49	1.82	186.35	1.22	224.70	1.48
Dryland farming areas	3257.09	21.41	3158.62	20.76	3009.34	19.78	1914.30	12.58
Forest areas	4614.06	30.33	4521.74	29.72	4390.38	28.86	4241.21	27.88
Paddy field areas	725.82	4.77	728.21	4.79	717.96	4.72	813.11	5.34
Total	15214.74	100.00	15214.74	100.00	15214.74	100.00	15214.74	100.00

Within this 25-year period, the proportion of built-up areas has increased from 41.21% of the total area in 1989 to 51.56% in 2014, or up by 10.35 %. In 1989, the proportion of dryland farming areas was 21.41% and was reduced to about 12.58% in 2014, or down by 8.83%. There was a slight increase in paddy field proportion of 0.57%, from 4.77% in 1989 to 5.34% in 2014. The proportion of grassland areas was reduced by 0.08% , while open areas and water body was increased by 1.24% and 0.18%

Based on the land use pattern, the proportion of dryland farming areas suffered the largest decline compared to other use pattern as water catchment areas. Agriculture land owners can easily convert their agricultural land because the economic benefits provided by agricultural land is lower than that of other uses. The benefits of land that is widely understood until now is only as a medium for production, while other benefits including as ecological balancer is often overlooked. Meanwhile, the presence of paddy fields in the watershed areas have three major benefits, namely (1) as a medium for paddy production, (2) as a buffer of environmental quality and ecological balancer, (3) provide nutrients to ground water, maintain the river flow, reduce runoff, and weather stabilization [11]. Thus, a policy to preserve agricultural land is needed.

CITYgreen is able to analyse the statistical results of the site which consists of land cover distribution and the extent of each of land cover, as well as the ecological benefits consisting of Current Number (CN), run off, storage volume needs, construction cost, and annual stormwater saving cost. In this research, the analysis is focused on economic benefits. There are two types of CN produced by CITYgreen, namely CN and CN Without trees. CN is actual CN value based on real land cover, while CN Without Trees is CN value based on a scenario of zero tree in the study area.

The result of the analysis shows that there was an increase in CN from 2000 (82) to 2010 (88), and a decrease in CN from 2010 (88) to 2014 (85). This means that in the upstream area of the Ciliwung watershed, there was an increase in run-off from 2000 to 2010, and a decrease in run-off from 2010 to 2014. The storage volume capacity of the upstream area of the Ciliwung watershed decreased from 1989 to 2014. This occurred because there was a change in the extent of canopy land cover (forest and agriculture land) to urban solid areas, which was demonstrated by the Curve Number (CN) value that tends to increase each year. The bigger the water storage volume capacity in the upstream area of the Ciliwung watershed, the higher the annual stormwater saving cost will be. The tabulation of the result of CITYgreen analysis is presented in table 2.

Table 2. The result of CITYgreen analysed

Information	**Year**			
	1989	**2000**	**2010**	**2014**
CN (Current)	82	82	88	85
CN (Without trees)	88	88	93	90
Run off (Current, cm)	11.48	11.48	14.65	13.03
Run off (Without trees, cm)	14.65	14.65	17.61	15,81
Storage volume needed (m3)	1.90E+15	1.90E+15	1.78E+13	1.66E+12
Construction cost (USD)	$174,064,321	$174,064,321	$163,407,322	$152,750,322
Annual stormwater saving cost (USD)	$15,175,721	$15,175,721	$14,246,595	$13,317,469

The decrease in storage volume needed from 1.90E+15 in 2000 to 1.78E+13 in 2010, and to 1.66E+12 in 2014 can be interpreted as a decrease in the capability of the upstream areas of the Ciliwung watershed as a water volume storage to reduce surface current. The decline reduced the economic value of benefit which is stated as annual stormwater saving cost, where annual stormwater saving cost is the annual cost based on financing of more than 20 years with 6% interest rate. In 2000, the annual stormwater saving cost was $15,175,721; it decreased to $14,246,595 in 2010, and further decreased in 2014 to $13,317,469.

The economic value of benefit is the dollar value of benefit calculated by multiplying storage volume needed of stormwater reduced by trees by a local cost per cubic foot for mitigation [8]. For example, the cost for building retention ponds, building additional stormwater management facilities or treating water. Additional information used to obtain economic value of benefits is the unit construction cost using general cost unit standard in 2015 and construction material cost in 2015 [12].

The decrease in economic value of benefit from 1989 to 2014 was directly proportional with the decrease inthe extent of forest and agriculture areas. Therefore, adjustment of efficient land use pattern is needed to harmonize urban economic development and urban ecological function to attain efficiency of more environmentally friendly land use [13]. Land use types influence the ecological benefit of a city, with the highest benefit value being land use as green open space, residential areas, roads, and industrial squares [14]. Hence, city forest and green open space need to be developed through fiscal planning process, systematic design, and management to give benefits to urban areas [15].

The Ministry of Forestry used their budget fund for Jakarta flood handling program in 2014 amounted to USD 114,543,000. The program's activities and their respective allocations are as follows: permanent vegetation (0.01%), agroforestry (0.2% 0), gully plug (0.02), controlling dam (0.54%), retaining dam (0.37%) and green well (98.85%). Compared to the value of annual stormwater, the allocation for maintaining vegetation carried out by the Ministry of Forestry is bigger reaching USD 24,054,030. Compared to the budget allocation for physical development, the budget allocation for maintaining vegetation is much smaller. With continuous outreach activities on the economic value of benefits of forest and other vegetation areas, the budget allocation for maintaining vegetation is expected to increase.

3.2 Development policy
Watershed is classified as common pool resources (CPRs) based on its rivalness and exclusivity nature. Substractibility or rivalness means that individual's consumption of resources will reduce the ability of others to use those resources. Exclusivity means that there are costs spent to limit the access to resources for others that can be designated as beneficiaries. Overuse and free rider are both problems and characters of CPR resources; thus, mechanism and institutionalized system are needed to prevent or avoid them. Therefore, watershed management in the development countries has to be started from institutional arrangement through a detailed mechanism to overcome imbalance of access to water and the recognition of the right to access water for poor population [16].

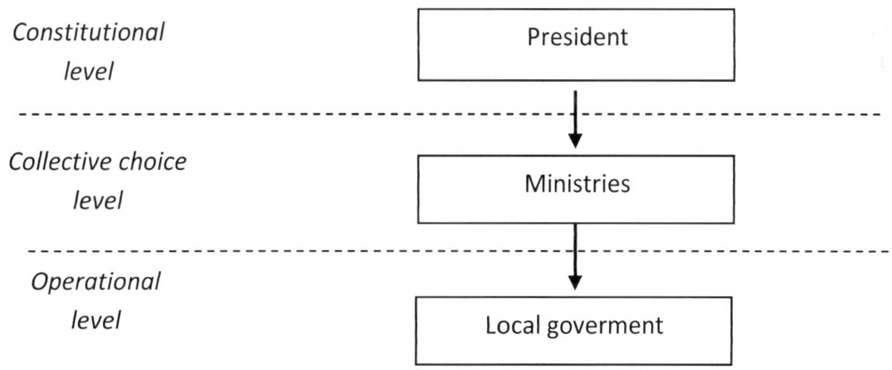

Figure 2. Government institution position in Ciliwung watershed management [17]

Payment for Environmental Service (PES) mechanism can be done to prevent farmers from doing land fragmentation and land conversion through economic incentives [18]. PES needs to be implemented as joint activities between the government and the public, starting from the transfer of knowledge on the economic benefits of forest and agriculture areas. Institutional approach for Ciliwung watershed management can be done by referring to Ostrom's *Institutional Analysis and Development* (IAD) [19]. IAD framework is a mapping of various concept levels as a means to systematically understand the policy setting process, to analyze institution that prepares collective action as a basis to determine activities and result accomplishment. PES mechanism implementation needs a collective action involving many institutions, both governmental and non-governmental. The rules at collective choice level consist of several laws and regulations that are under the authority of some ministries presented in figure 1.

The ministries responsible at collective choice level are the Ministry of Environment and Forestry, the Ministry of Public Works and Publc Housing, Ministry of Spatial Plan and Agrarian Affairs, and the Ministry of Agriculture. Regulations at operational level is Regional Regulation on Spatial Plan issued by each Regency or Municipal. Institutions responsible for this are Local Governments of Regency or Municipal consisting of Bogor Regency and Municipal, Depok Muncipal, and DKI Jakarta.

Efforts to reduce Ciliwung discharge rate to a safe condition can be conducted by implementing four scenarios, PES payment, PES payment in the form of access to market, reforestation, and land conversion prevention [20]. Institutions related to these scenarios are at collective choice level and operational level. Incentive mechanism is in the form of compensation and environmental service compensation. PES policy can gradually maintain agriculture area and reduce river current discharge if PES policy is implemented in the form of guarantee of access to markets, and run along with measures on land conversion and tree reforestation policy [20]. Reforestation policy can be implemented without compensation mechanism because the status of forests in the upstream area is a state forest. However, PES must be given to the community or people who lose their access due to the designation of the forest area [21]. For example, the community living around theforest in Ejidatarios and Pobladores Mexico receives compensation with some limitations such as, ban of forest clearance, hunting and habitat change, and the obligation to limit the number of livestock, conduct patrol, and control of erosion. The Mexican experience can serve as an example to implement compensation in forest area in the upstream area of the Ciliwung watershed. The community living in the upstream area are the most effective watershed managers. The next challenge is how PES can increase poor population's welfare in the upstream area of the watershed.

4. Conclusions

The economic value of benefit is the dollar value of benefit calculated by multiplying storage volume needed of stormwater reduced by trees by a local cost per cubic foot for mitigation. This value is the annual stormwater saving cost value that tends to decrease during this research. In 2000, the economic value of benefit of forest and agriculture areas was $15,175,721; it decreased to $14,246,595 in 2010, and further decreased in 2014 to $13,317,469. The economic benefit should be made known to the government and the public in order to understand the importance of forest and agriculture areas. The most effective way to maintain forest and agriculture areas is the implementation of PES mechanism for citizens who manage the watershed in the upstream area. PES policy needs a collective action involving many institutions, both governmental and non-governmental.

Public participation in watershed management should be on the top of the government's agenda. Communication and information dissemination related to the importance of payment for environmental services should become the government's agenda, by involving educational institution, research and public organization to accelerate the people's understanding process.

Reference

[1] Poerbandono P, Ward PJ and Julian MM 2009 Set up and calibration of a spatial tool for simulating river discharge of Western Java in recent decades : preliminary result and assessments Journal of Engineering and Technological Sciences **41**(1): 50-64
[2] Budiyono Y, Aerts J, Brinkman J, Marfai MA and Ward P 2015 Flood risk assessment for delta mega-cities: A case study of Jakarta *Natural Hazard* 75(1) pp. 389-413 American Forest 2002 CITYgreen 5.0 user manual American Forest Washington DC
[3] Kompas 2013http://forum.kompas.com/megapolitan/231945-*'inilah-kerugian-akibat-banir-jakarta-dari-tahun-ke -tahun* [Are these economic loses of Jakarta flooding ?]', html, 23-01-13 Retrieved on May 1,2015.Jakarta – Indonesia
[4] Fulazzaky MA 2014 Review challenges of Integrated Water Resources Management in Indonesia *Water* **6**:2000-20
[5] Cui B, Wang C, Tao W and You Z 2009 River channel network design for drought and flood control: A case study of Xiaoqinghe River basin, Jinan City, China *Journal of environmental management* **90**(11): 3675-3686
[6] Kodoatie R J and Sjarief R 2010 *Tata Ruang Air* [Water Land Use] ANDI Publisher Yogyakarta Indonesia
[7] Woube M 1999 Flooding and sustainable land–water management in the lower Baro–Akobo river basin, Ethiopia *Applied Geography* **19**(3): 235-51
[8] American Forest 2002 CITYgreen 5.0 user manual American Forest Washington DC

[9] Saridewi TR 2015 *Instrument pembayaran jasa lingkungan dalam mendukung penataan ruang DAS Ciliwung* [Payment for environmental services for supporting land use planning in Ciliwung Watershed] Disertation Bogor Agricultural University Indonesia

[10] Central Management of Citarum Ciliwung Watershed (Balai Pengelolaan Daerah Aliran Sungai Citarum Ciliwung)] 2011 *Penyusunan rancangan peningkatan fungsi dan daya dukung ekologi dalam pengelolaan DAS Ciliwung* [Drafting improved functionality and ecology in the management capability Ciliwung Watershed] Bogor, Indonesia

[11] Wu R-S, Sue W-R, Chien C-B, Chen C-H, Chang J-S and Lin K-M 2001 A simulation model for investigating the effects of rice paddy fields on the runoff system *Mathematical and computer modeling* **33**(6): 649-58

[12] Hasanah A 2016 *Kajian hubungan perubahan penutupan lahan dan kapasitas pengendalian aliran permukaan pada lanskap DAS Ciliwung hulu* [A relation study between landcover change with stormwater runoff control capacity in Ciliwung Upstream Watershed landscape] Thesis Bogor Agricultural University Indonesia

[13] Zhang K, Zhang J, Chen Y and Zhu Y 2006 Urban greenbelt eco-service value of Hangzhou City under effects of land use change : An evaluation with CITYgreen model *Chinese Journal of Applied Ecology* **17**(10):1918-22

[14] Peng, L, Chen, S, Liu, Y and Wang, J 2007 Application of CITYgreen model in benefit assessment of Nanjing urban green space in carbon fixation and runoff reduction *Chinese Journal of Applied Ecology* **18**(6):1293-98

[15] Wolf, KL 2003 Ergonomics of the city : Green infrastructure and social benefits In C. Kollin (ed) Engineering Green *Proc of the 11th National Urban Forest Conference* American Press Washington DC

[16] Molle F, Wester P and Hirsch P 2010 River basin closure: Processes, implications and responses *Agricultural Water Management* **97**(4): 569-77

[17] Ananda J and Proctor W 2013 Collaborative approaches to water management and planning: An institutional perspective *Ecological Economics* **86**: 97-106

[18] Muradian R, Corbera E, Pascual U, Kosoy N and May PH 2010 Reconciling theory and practice: An alternative conceptual framework for understanding payments for environmental services *Ecological economics* **69**(6) pp 1202-08

[19] Ostrom E and Cox M 2010 Moving beyond panaceas: a multi-tiered diagnostic approach for social-ecological analysis *Environmental Conservation* **37**(04) pp 451-63

[20] Saridewi TR, Hadi, S, Fauzi A and Rusastra IW 2017 Payment for Environmental Services Approach to Reduce Flood in Ciliwung Watershed *Biotropia* **24**(2) pp 127-39

[21] García-Amado LR, Pérez MR, Escutia FR, García SB and Mejía EC 2011 Efficiency of payments for environmental services: equity and additionality in a case study from a biosphere reserve in Chiapas, Mexico *Ecological Economics* **70**(12) pp 2361-68

Shoreline change after 12 years of tsunami in Banda Aceh, Indonesia: a multi-resolution, multi-temporal satellite data and GIS approach

S Sugianto[1]*, Heriansyah[2], Darusman[3], M Rusdi[1] and A Karim[1]

[1]Remote Sensing and Cartography Laboratory, Faculty of Agriculture, Syiah Kuala University
[2]Graduate Student, Integrated Coastal Management, Syiah Kuala University
[3]Soil Science Department, Syiah Kuala University, Banda Aceh

*Corresponding author: sugianto@unsyiah.ac.id

Abstract. The Indian Ocean Tsunami event on the 26 December 2004 has caused severe damage of some shorelines in Banda Aceh City, Indonesia. Tracing back the impact can be seen using remote sensing data combined with GIS. The approach is incorporated with image processing to analyze the extent of shoreline changes with multi-temporal data after 12 years of tsunami. This study demonstrates multi-resolution and multi-temporal satellite images of QuickBird and IKONOS to demarcate the shoreline of Banda Aceh shoreline from before and after tsunami. The research has demonstrated a significant change to the shoreline in the form of abrasion between 2004 and 2005 from few meters to hundred meters' change. The change between 2004 and 2011 has not returned to the previous stage of shoreline before the tsunami, considered post tsunami impact. The abrasion occurs between 18.3 to 194.93 meters. Further, the change in 2009-2011 shows slowly change of shoreline of Banda Aceh, considered without impact of tsunami e.g. abrasion caused by ocean waves that erode the coast and on specific areas accretion occurs caused by sediment carried by the river flow into the sea near the shoreline of the study area.

1. Introduction

Banda Aceh was one of the worst cities hit by tsunami on December 26, 2004. However, now Banda Aceh is totally in recovery after 12 years of rebuilt. The tsunami has affected the morphology unit, especially in the shoreline, established since Holocene period [1]. The impact of tsunami can contribute seawater submerged into the coastal plain. This can change the depositional and erosional process in the shoreline. This fact shows that the big tsunami and land subsidence effects not only occur instantaneously, but also causes morphological changes in proportion to the change, which usually occurs at a time scale of centuries to millennia.

The most powerful earthquake occuring in this area in the last 40 years was identified approximately as a 9 on the Richter scale. The earthquake caused deadly giant waves which stormed nearly a dozen countries around the world and killed ten thousand peoples [2]. The giant tsunami on 26 December 2004 has affected the shoreline along the Banda Aceh beaches, especially near the Uleulue seaport, a famous and historic seaport of Banda Aceh. The impact of the tsunami in this area has caused the change of shoreline from few meter to hundred meters [3]. The extent of the change of the shoreline along the coastline of Banda Aceh can be traced back using remote sensing data combined with Geographical Information System (GIS). Satellite data can provide precision (cm even to mm), high spatial resolution, and wide coverage measurements of surface deformation, reducing the use of in-situ traditional techniques [4, 5]. Remote sensing data combined with GIS analysis can provide the compressive result of geomorphological change. As a GIS is regarded as a toll to collect, store, manage, manipulate, analyze and visualize geographic or spatial data, it can be regarded as a special case of information systems where the database consists of observations on features distributed spatially. Using GIS involves capturing the spatial distribution of features by means of measurement or maps [6].

Digital Globe launched the QuickBird satellite sensor in October 2001, as one of the first commercial satellites that has high image spatial resolution. It has contributed to Indonesian authority to quickly access virtually and monitor the damage caused by tsunami in 2004 by providing image of the affected area. The QuickBird had observed earth for more than a decade with a four-band multispectral spatial resolution which is about 2.4 m and 0.65 m in the panchromatic band. QuickBird has been chosen for this study because of the high spatial resolution result. In line with the QuickBird image, space imaging (now owned by GeoEye) had launched the IKONOS satellite on September 1999. IKONOS, on the other hand, has a 4-meter multi-spectral spatial resolution, possibly its panchromatic band sharpened to one meter. QuickBird and IKONOS from digital Globe have been utilized to classify the five species of saltmarsh vegetation, soil and water around Venice Lagoon, Italy and give clear and accurate result [7]. Thus, with a fine spatial resolution, these datasets would be useful result to analyze the shoreline change part of Banda Aceh coastal area.

Mapping shoreline change is important aspect for Banda Aceh as its development is in progress after it was struckted by huge tsunami in 2004. The use of GIS as mapping and analysis tool for spatial data can be realised to obtain high quality of the shoreline change. The result can be used as guidelines for the local government planning in relation to shoreline policy in the future. The use of GIS to analyze for shoreline change has been implemented by Bouchahma and Wanglin in 2014 [8] at the coastal area in Djerba Island and produced a good map of coastal change in the area. The analysis of shoreline is needed not only for a good consideration of spatial and temporal analysis but also for the integration of planning in the area. Remote sensing and GIS are commonly used in coastal geomorphology since they provide a better source in studying the changes of shoreline in the long term use [9]. With the ultimate useful and high resolution, this study aims to provide multi-resolution and multi-temporal satellite image of QuickBird (by digital Globe) and IKONOS to demarcate the shoreline change in Banda Aceh shoreline before and after tsunami without considering the change of tide and sea surface during the period of acquiring and analyzing images.

2. Materials and Methods

The location of this study is along the shoreline of Banda Aceh (Figure 1). Banda Aceh is located at 95^{0}16’27,68”-95^{02}2’47,40” Longitude and 5^{0}30’47,77”- 5^{0}36’47,54” Latitude. Selected samples are located at 95^{0}18’52.62” – 95^{0} 21’ 19.40” Longitude and 5^{0}35’8.29” – 5^{0}36’39.33” Latitude. The main source of image data for this study was IKONOS with the data acquired in 2004 and 2005 and QuikBird images with the data acquired in 2009 and 2011. The data were supported by Geospatial Data Centre of Aceh Province, Indonesia. Maps related to the study were collected from National Land Resources Agency of Aceh.

Due to lack of data availability consecutively years for both IKONOS and QuickBird for the years of image acquired, we used different date of acquisition for IKONOS and QuickBird as a temporal analysis and different multi-resolution images.

Due to the lack or unavailability of image data required in consecutively years from both IKONOS and QuickBird, we used different date of acquisition for IKONOS and QuickBird as a temporal analysis and different multi-resolution images. Pre-processing images of IKONOS and QuickBird were carried out.

Selection of sample area was carried out by choosing the worst area after the tsunami which has not been built up or has been being developed. Thus, we only focus on detail calculation of shoreline change for Syiah Kuala, sub-district of Banda Aceh. The other four sub-districts, such as Meuraxa, Kuta Raja and Kuta Alam, are also included for initial analysis of a shoreline change of Banda Aceh coastal area.

2.1. Image analysis, registration and shoreline digitation

Image registration, sharpening, cutting, classification and overlaying process over the area were applied. Sharpening the image, enlarging the scale of the images in order to distinguish between the sea and shoreline, and re-digitalizing through the polyline method to determine the shoreline, were carried out and overlaid images was conducted for before tsunami image e.g. 2004 image, and after tsunami images of 2005, 2009, and 2011. Shoreline digitation was carried out through the polyline method on the scale of 1:50.000 for all images. The QuickBird and IKONOS images were good enough, allowing to determine the boundaries between the sea and land. Digitizing on the base of the 1:50.000 map scale has produced clear images with new digitized boundaries of the shoreline.

2.2. Overlaying images

Overlay images were carried out to distinguish the shoreline changes of the study area among years of images acquired. The IKONOS images acquired in 2004 and 2005 were utilized to distinguish before and just after tsunami. Meanwhile QuickBird images acquired in 2009 and 2011 were also utilized to spatially analyze the change of shoreline of Banda Aceh, not only for before and after tsunami but also along the years of change since 2004, in the time of catastrophic disaster of tsunami. Overlaid maps were done to demarcate the changes of shoreline images in 2004, 2005, 2009 and 2011. On this stage, overlay was carried out over the shoreline of digitized polylines of IKONOS and QuickBird images before tsunami in 2004 to after the tsunami; for example, 2005, 2009 and 2011.

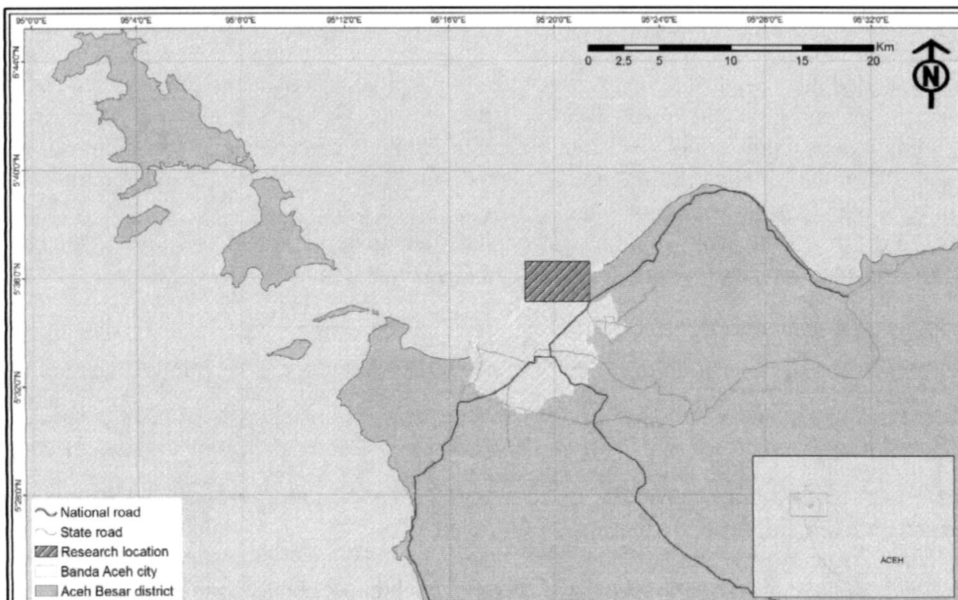

Figure 1. Map of Banda Aceh with the inset for selected sub district investigated for shoreline change analysis

3. Result and Discussion

The four sub-districts out of the nine sub-districts in Banda Aceh, i.e.: Meuraxa, Kuta Raja, Kuta Alam, and Syiah Kuala, were directly hit by tsunami in 2004 and the hitting has directly impact on shoreline change with the average change of 12.15 to 309.9 meters. The shorelines of these four sub-districts have changed after the tsunami in 2004 with various extents and sizes (figure 2), and the worst area was Syiah Kuala sub district, as sample point used for the multi-temporal and multi resolution images.

The result from IKONOS image analysis for 2004 and 2005 shows a variation of shoreline changes starting from north to south of Banda Aceh. The greatest change was at Deah Raya Village shoreline of Syiah Kuala Sub-district. QuickBird image analysis for 2009 and 2011 shows a slowly change of shoreline of Syiah Kuala sub-district. This is due to abrasion and accretion in this area. This change shows the dynamic transformation of shoreline along the coastline of Banda Aceh city after the tsunami. In the short and longtime, the changes of shoreline should be analyzed [10] and mapped as those are important for navigation safety, coastal resource management, environment maintenance, development of coastal resources, and city planning. The shoreline changes are important information for mapping and deformation of coastline.

3.1. The shoreline of Syiah Kuala before tsunami

The shoreline of Syiah Kuala sub-district before tsunami was undamaged, as shown in figure 2. It clearly indicates the shoreline of Banda Aceh before tsunami. The red line on the map indicates the shoreline of Syiah Kuala which was undamaged. While, the dotted line on the map indicates the shoreline and part of rives canal which toward to the sea.

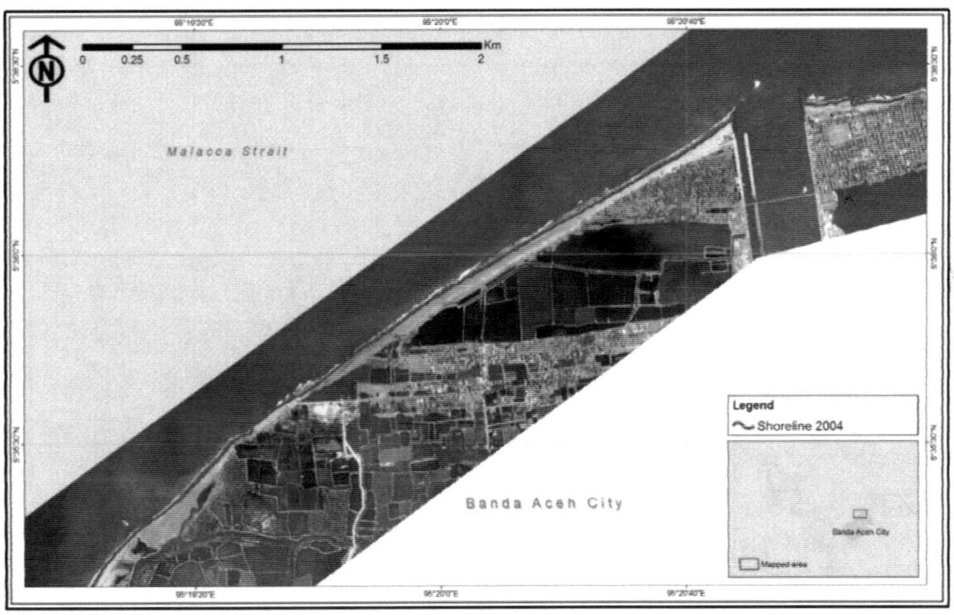

Figure 2. Shoreline map of Syiah Kuala Sub-district, Banda Aceh before tsunami in 2004

3.2. The Shoreline after tsunami

The shoreline of Syiah Kuala after tsunami in 2004 was severe damaged where it suffered high enough coastal erosion so that the level of sea water was so big enough to immerse the land area along the shoreline of Banda Aceh. For more detail, the impact of tsunami on shoreline of Syiah Kuala sub district can be traced back as described in figure 3.

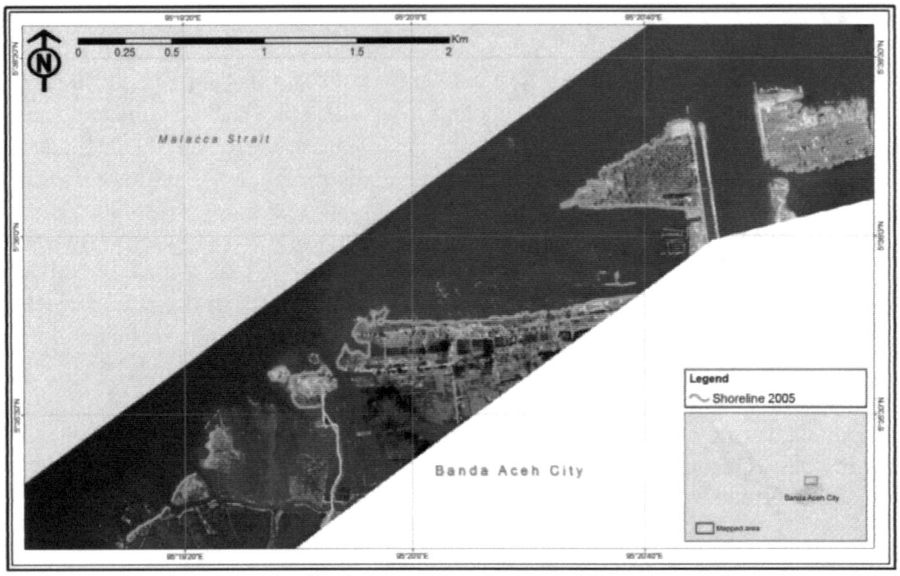

Figure 3. Syiah Kuala-sub-district map showing shoreline change for January 2005

After tsunami, the shoreline of Syiah Kuala sub district changed in certain areas, as noted with the disconnected green line in Figure 3. This change is caused by the sea flooding the land has diminished some part of the shoreline. The long term effects of tsunami on the shorelines could vary to be due to the natural erosion and the flood. However, tracing back the damage can be retrieved again as long as the image data are available.

3.3. The changes of shoreline after tsunami

The result analysis of the shoreline of Syiah Kuala area before and after tsunami based on overlaid result of image using GIS tool for 2004 and 2005 shows various changes (figure 4). Figure 4 illustrates the changes of Syiah Kuala shoreline between 2004 and 2005.

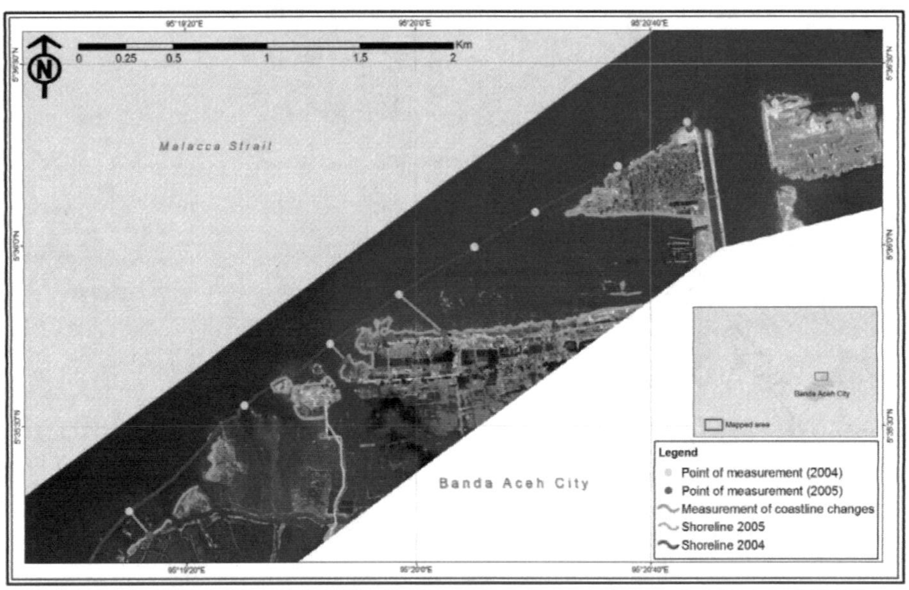

Figure 4. Map of the Syiah Kuala shoreline changes between 2004 and 2005, note: redline represent previous shoreline

The changes of shoreline at Banda Aceh between 2004 and 2005 were totally worst in which shoreline on 2005 had been disappeared caused by tsunami. On 2004 to 2005 the shoreline along Syiah Kuala had been seriously damaged (abrasion) started from the lowest level that were 12.15 meter to 309.92 meter. In certain area, the shoreline was also totally wrecked and this made flood flew into pond of residents near the area. To see the rate of shoreline changes after tsunami at Syiah Kuala subdistrict, it was conducted an analysis between 2004 and 2009 (figure 5). Figure 5 shows several changes of shoreline at Syiah Kuala with different wide of coastal owing to both abrasion and accretion. Between 2004 and 2005, we checked for nine dots as points for GPS points recorded to see the changes with the lowest (55.1 meter) and the highest (178.77 meters) changes.

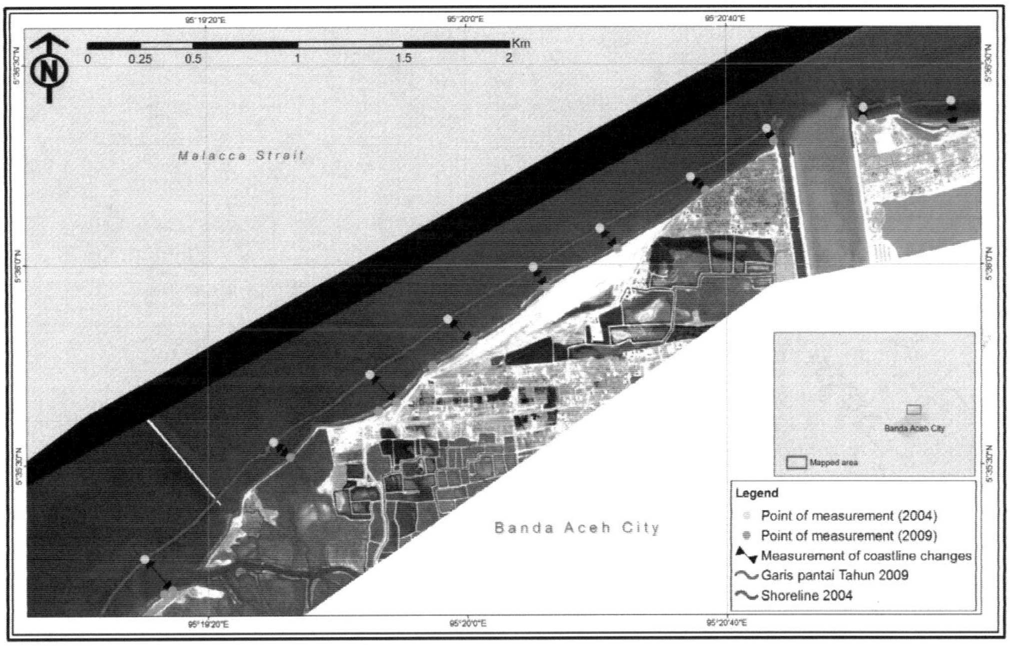

Figure 5. Map of Syiah Kuala's shoreline changes in 2004-2009

Analyzing the changes of shoreline of Syiah Kuala in the 2009 and 2011 images without considering the tsunami effects in 2004 give the assumption that the changes are not caused by tsunami, but rather the dynamic of geomorphology process in the area. The shoreline changes of Syiah Kuala shoreline between 2009 and 2011 could be due to abrasion along the shoreline area investigated. Figure 6 clearly shows the changes of shoreline. The changes occuring between 2009 and 2011 were dominated by the abrasion of sediment carried by rivers flowing into its sub-districts. The size of the abrasion was in the range of 1.08 and 32.69 meters. Meanwhile, the size of the abrasion happening only at Tibang area, a village name in Syiah Kuala sub-district, was in the range of 10.24 and 14.87 meters and this abrasion might be due to the waves and tides, as revealed by the multi-temporal image analysis.

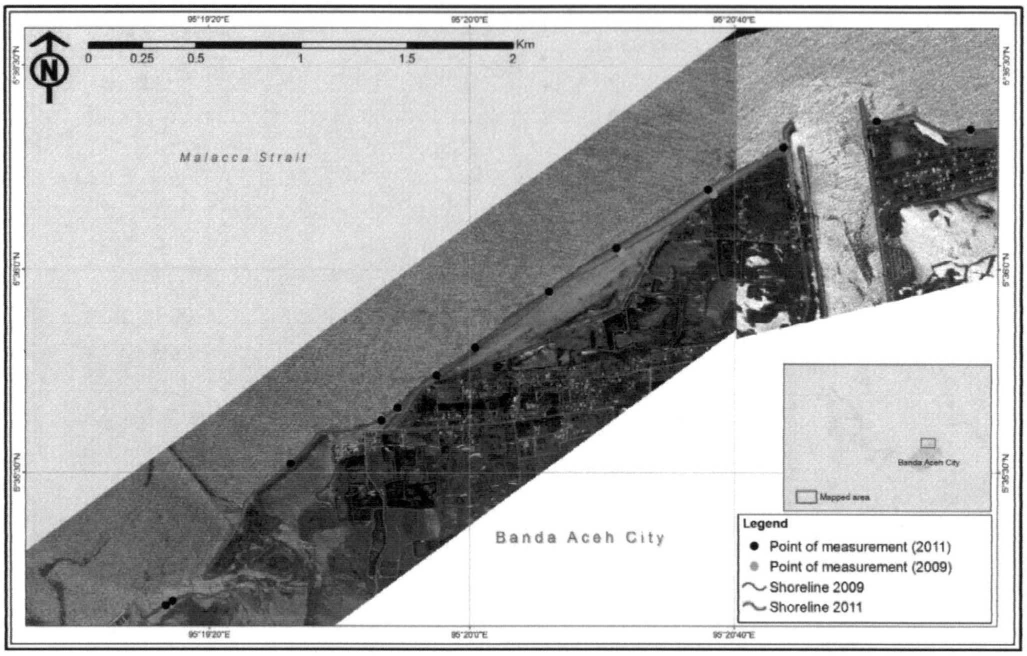

Figure 6. Map of Syiah Kuala changes on its shoreline between 2009-2011

4. Conclusions

Multi-temporal and multi resolution image analysis has been performed to demonstrate its ability to identify the extent of shoreline change in Banda Aceh, from few meters to hundreds of meter changes after 12 year of tsunami.

IKONOS images was used to show the changes of shoreline in Banda Aceh between 2004 and 2005, accounting for 12.15 to 309.92 meters from abrasion, while for accretion counting for 15.39 to 33.57 meters.

QuickBird images show an abrasion shoreline of Banda Aceh coastal area from 2004 to 2009 from 18.3 to 194.93 meters. The changes of Syiah Kuala shorelines sample point for Banda Aceh between 2009 and 2011 not only caused by tsunami, but it also could be from the abrasion and geomorphological dynamic process of this area during the period after tsunami, e.g.: from 2005 to current date.

Acknowledgment

Thanks to Heriansyah and Muhammad Rusdi for field data collection and mapping and also for GIS Analysis. Thank you due to Planning Board of Aceh for providing satellite image.

References

[1] Meilianda E 2009 Past, Present and Future Morphological Development of a Tsunami-Affected Coast. A Case Study of Banda Aceh. Disertasi. University of Twente. Enschede, The Netherlands

[2] Parape C D, Premachandra C, Masayuki T, Bari A, Ranjith D, Duminda W, Shengye J, and Masami S 2013 Building Damage and Business Continuity Management in the Event of Natural Hazards: Case Study of the 2004 Tsunami in Sri Lanka. *Sustainability*, **5** pp 456-477

[3] Heriansyah, 2016 Analisis Perubahan Garis Pantai Kota Banda Aceh Pasca Tsunami Dengan Menggunakan Sistem Informasi Geografis (Sig). Thesis. Syiah Kuala University

[4] Yang Z, Li Z, Zhu J, Yi H, Hu J, Feng G 2017 Deriving dynamic subsidence of coal mining areas using InSAR and logistic model. Remote Sens. **9** 125

[5] Zhang L, Lu Z, Ding X L, Jung H S, Feng G C, Lee C W 2012. Mapping ground surface deformation using temporarily coherent point SAR interferometry: Application to Los Angeles Basin. Remote. Sens. Environ, **117** pp 429–439
[6] Xuan Z 2014. GIS and Urban Mining, *Resources* **3** pp235-247
[7] Matthew J, McCarthy and Joanne N H 2014 Habitat Mapping and Change Assessment of Coastal Environments: An Examination of WorldView-2, QuickBird, and IKONOS Satellite Imagery and Airborne LiDAR for Mapping Barrier Island Habitats. *ISPRS Int. J. Geo-Inf,* Vol **3**pp 297-325
[8] Bouchahma M, Wangli Y 2014 Monitoring shoreline change on Djerba Island using GIS and multi-temporal satellite data *Arab J Geosci.* 7 p 3705–8
[9] Al-Hatrushi S M 2013 Monitoring of the shoreline change using remote sensing and GIS: a case study of Al Hawasnah tidal inlet, Al Batinah coast, Sultanate of Oman *Arab J Geosci.* **6** p 1479-5
[10] Taha L G E, Elbeih S F 2010 Investigation of fusion of SAR and Landsat data for shoreline super resolution mapping: the northeastern Mediterranean Sea coast in Egypt. *Appl Geomat* **2** pp 177–186

IOP Conf. Series: Earth and Environmental Science **144** (2018) 012054 doi:10.1088/1755-1315/144/1/012054

Coastal women empowerment in improving enterprise of fish product processing in Sanga-Sanga Districts

G Haqiqiansyah and E Sugiharto
Faculty of Fisheries and Marine Science, Mulawarman University, Samarinda

*Corresponding Author : gustihaqi@gmail.com

Abstract. This research was conducted to identify and scrutinize women empowerment of fish product processing group in the District of Sanga-Sanga on 2017. The method used was survey method, which is direct observation and interview to respondent. Data were collected in the form of primary and secondary data. Collected data then processed, tabulated, and displayed in the table and graph. The measurement of women empowerment degree was measured by Likert Scale on 3 level, that are score 1 = low, score 2 = less, and score 3 = high. The result of research demonstrated that the rate of empowerment women group of fish product processor was high (score 42,75). Partially, awareness level or willingness to change of processing enterprise group which indicate empowerment indicator categorized as high (91,67%). The level of capability to increase the chance of acquiring access was high (66,67%), the level of capability to overcome an obstacle tend to categorized as less (50%) and the level of capability to collaborate was high (66,67%). It means that the level of coastal women empowerment could be reliable to do a reformation.

1. Introduction

1.1. Background

In the context of coastal societies, the gap of status and role between women and men in family is commonly occurred. Women status and role are highly respected in one hands, on the other hand however women are placed as men subordinate. A lot of women status is related to household management, children raising, family finance management. Really regrettable due to their time are wasted which is actually could be utilized to increase family economy by empowering women. In the other side, fishery potential is relatively large, therefore there are a lot of chance for developing a various alternative enterprise by utilizing local resources potential. Recently, a lot of empowerment program have been implemented however it has not tackled the problem faced by societies yet. Dependencies attitudes of coastal communities are high, it mostly happened when no fishing activity (due to season affect, meanwhile non fisheries activities had not developed yet. High dependencies on ponggawa or capital owner is undeniable due to the general condition faced by fishermen community, they require aids to fulfill their needs. As this phenomenon goes by, it is difficult for societies to disentangle from poverty traps. The chain of poverty as if twisted them without stopping. Responding to this condition, it is needed to find systematized, focused and planned solution. Empowerment action of coastal women communities is obviously needs to do.

The research results on women public role in coastal societies could be found in Sinjai Regency. Processing enterprise which has been done by housewives in East Sinjai was smoked fish. The problems emerged on input subsystem of smoked fish are : the lack of sufficient amount of fish as raw material , the difficulties to acquire fire woods, the use of additives food coloring is inappropriate with food safety standard, fish container could damage easily, work force wage is unavailable [1].

The research result of demonstrate that women who live in coastal area of Sanga-Sanga District have prospective potential and chance to be empowered through various activities consisting of productive enterprise and accompaniment activities [2].

1.2. The Aim of Research
The goal to be achieved is to identify and assess the empowerment of women processing group of fishery products in Sanga-Sanga District

2. Method of Research

2.1. Location and Time of Research
This research was conducted in Sanga-Sanga District in The Regency of Kutai Kartanegara on April to August 2017.

2.2. Data Gathering Method
Survey method by direct observing is used in this research, in which direct observation is conducted in the location. Data embrace two sorts of data, i.e. primary and secondary data. In detail, required data are: a. Collecting primary and secondary data. Primary data is gained from main sources, whether through direct observation or interview to respondent refer to questionnaire which has been compiled according to research aim. Respondent who has been chosen is a group of women conducting activity of fish processing enterprise. Secondary data is required as supporting data, it is gained from village office, Marine and Fisheries Department, and reports of similar research. Collecting of secondary data is conducted by literature study and relevant reference to this research. Secondary data were included profile of area, potential and opportunity of fish enterprise developing and business capital, b. Conducting Focus Group Discussion. It is conducted to community leaders, consisted of district envoy, village leader, fishermen group and group of fish processing entrepreneur.

2.3. Data Analysis Method
Data collected will be processed in the table and graph and will be analyzed descriptively. To measure the level of empowering on women group of fish product processing using Likert Scale with three level, i.e. score 1 = low, score 2 = less, score 3 = high.

Table 1. Measurement criteria of empowering level

No	Interval Class	Criteria
1.	17,00 - 28,33	Low
2.	28,34 - 39,66	Less
3.	39,67 - 51,00	High

Source: Primary data processed, 2017

3. Results and Discussions

3.1. General Description of Research Location
Sanga-Sanga Districted is geographically located in the equator between of 117° 01'East Longitude – 117° 17' East Longitude and 0° 35' South Latitude – 0° 45' South Latitude, with the area of 233,4 km^2. The boarder area administered as follow:

North Side	: Anggana District
East Side	: Anggana District
South Side	: Muara Jawa District
West Side	: Palaran District/Kota Samarinda

Sanga-Sanga District consist of five villages: Jawa, Sanga-Sanga Dalam, Pendingin, Sarijaya, and Sanga-Sanga Muara. The river which flows in Sanga-Sanga District is Sanga-Sanga River. Jawa Village is a biggest village (60,63 km^2), and the smallest is Sarijaya Village (22,81 km^2).

The amount of dweller of Sanga-Sanga District in 2015 is greater than in 2014, it increased from 19.419 to 19.828 (increased about 0,03%). Dweller composition consist of 10.651 male (53,72%) and 9.177 female (46,28%), it distributed in 5 village. The average of dweller density in Sanga-Sanga District is around 84 inhabitant/km^2. Inhabitant distribution is uneven among village.

3.2. Fishery Potential

The amount of fishery household in Sanga-Sanga District on 2015 are 512 households consisted of 197 household of marine fisheries, 51 household of pond culture fisheries, and 264 keramba fish culture.

In Sanga-Sanga District, the vast majority of village have potential on marine fisheries unless Jawa Village. The amount of motor boat with 0-5 GT capacity used for catching fish in the sea are 51 pieces. Whereas, the amount of fishing machine boat used in the public waters are 234 pieces. Total production of marine fisheries are 148,556 ton, ponds are 0,605 ton and karamba 0,3 ton.

The most common fishing boat used by fishermen is attached motor, and motor boat in less than 5 GT capacity. It's a kind of small boat, therefore it affect on catching yield. This condition needs a consideration, thus the capacity of catching could be increased and its cruising capacity will be wider [3].

3.3. Respondent Characteristics

According to the result of interview to fish product processor (amplang and fish crackers) it is known that respondent ages are ranging from 37 to 67 years old. The vast majority of respondent ages are ranging from 37 – 46 years old (58%), 47 – 56 years old (17%), and age group of >56 years old is 25%. Refer to age composition, it is varied from youth to retired people. However, amplang or crackers processor who are retired is still enough. It demonstrates that occupation as fish product processor is a good alternative as a household livelihood. Respondent's age is provided in this picture 1.

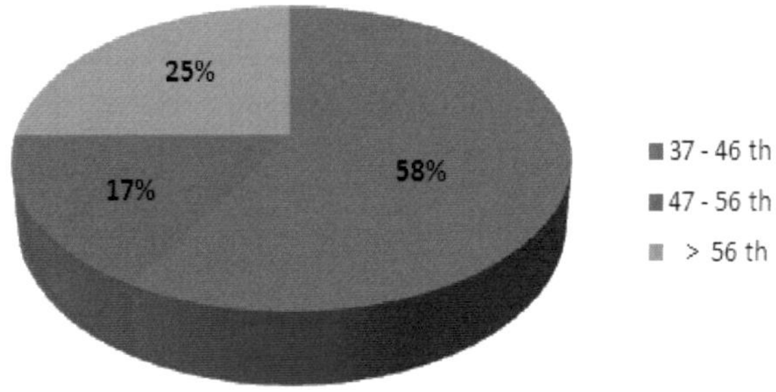

Picture 1. The percentage of respondent based on ages group

Education is an important thing to increase societies welfare. Education will increase their knowledge and will advance their thought. According to survey results, it could be known that education level is range from elementary school to senior high school. All of women processor have gained formal education. Some women processor had ever studied at non formal education such as training or courses. In detail it can be seen in picture 2.

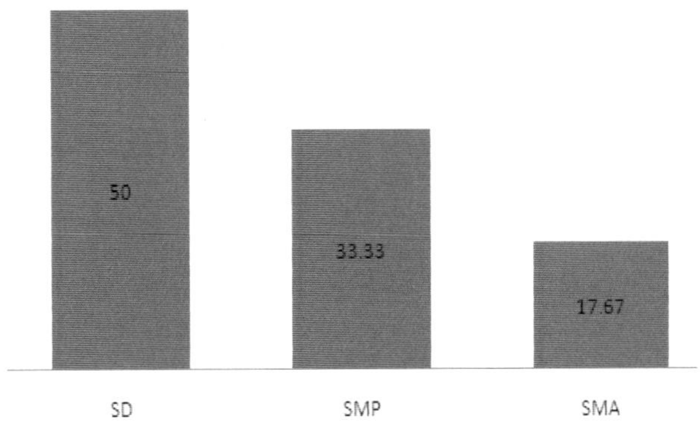

Picture 2. The percentage of respondent based on education level

3.4. Women Empowerment Indicator

Empowerment is transformation of power relation between men and women on different strata, i.e. family, society, market and state. Empowerment concept could be understood in two context. First, the power in decision process making focused on the importance of women role. Second, empowerment in term of women empowerment and its impact on men in the diversity of societies [4].

The result of study of found that there are many of poor women in Wedy District. They require special treatment for disentangling form poverty. To be seen from the profile of poor women in Wedi District, it is known that they live in unfeasible home to live, do not have education access and low income. This condition make them trap in poverty circle. Based on the result of study, it can be concluded that all of respondents (poor women) have high potential in doing entrepreneurship profession or entrepreneur candidate [5]. In line with that, the result study confirmed that the concept of policy on developing productive economy of fish product processing business sustainably using training and accompanying methods to fishermen families is able to increase fishermen societies welfare [6].

Women empowerment is a bottom-up process to transform gender relation both individually or collectively due to increasing awareness of women subordination and the creation of ability to face it. The substance of women empowerment concept is an understanding of the power itself. Women empowerment does not mean women take over the control which previously men have, however it is more related to a need to transform the value of power relation. The power could be understood as 'power within' or self confidence, 'power with' or the ability to organize collective aim collaboratively, and 'power to' emerging a change and making decision by others [7] .

As an effort to improve the strength of societies empowerment on coastal women, this research should consider its actor itself. The results of analysis and measurement of indicators used to measure women empowering level is provided in this following table.

Table 2. Analysis result of women empowering level

No	Indicators	Gained score
1	Awareness of desire to change (*Power Within*)	13.42
2	Capacity of getting access (*Power To*)	12.08
3	Capacity of overcoming threat (*Power Over*)	7.08
4	Capacity of cooperating (*Power With*)	10.17
	Jumlah	42.75

Sources : Primary data processed, 2017

From the table 2 it can be known that the level of empowerment on women group of fish product processing in Sanga-Sanga District is categorized as high (score 42,75 in 39,67 – 51,00 score range). It indicate that this group of women have high level of empowering to be able to do a change. Those condition could be utilized to conduct an event more intensively by accompanier, as originally this women group is willing to do a change.

Partially, the level of women group empowering based on empowerment indicators are follows:

3.4.1. *The awareness or the willingness to change (Power Within)* As known, community development activity is an effort to increase capacity and potential belong to a society, thus they are able to actualize their identity and prestige optimally to defend and develop themselves independently. It is intended to disentangle society from poverty trap and backwardness. Economic empowerment means the effort of income and wealth increasing focused on economic strength itself, hence the society is able to fulfill the need of living independently.

According to data analysis acquired the awareness level or willingness to change processing business group could be categorized into two part i.e. high (91,67%) and less (8,33%). It indicates that those two groups have willingness to improve their condition.

Table 3. The level of awareness or willingness to change (*Power Within*)

No	Interval Class	Percentage	Description
1.	5,00 - 8,33	0,00	Low
2.	8,34 - 11,66	8,33	Less
3.	11,67 - 15,00	91,67	High
		100,00	

Source : Primary data processed, 2017

Women ability to see and identify her potential has been classified into two categories such as the ability to identify natural resources potential surrounding their milieu and the ability to develop its potential. Those ability are is highly important due to the higher is the awareness of actors is becoming a trigger for developing its potential. It could be implemented if it is followed by many activities which is able to rise willingness up to change. Accompaniment and extension activities need to undertake systematically. In line with research result state that poverty allevation of village through women development will change women group therefore they are motivated to gain income [8]. Through empowerment, village women is able to develop its potential to create local based productive economic enterprise therefore societies reliant is set, eventually village women empowerment could reduced poverty.

3.4.2. *Capacity of getting access (Power To).* Capacity of getting access consisted of an opportunity to acquire information, market access, technology access, and access to income. Research result demonstrate that the ability to improve opportunity to get access can be categorized into three: high (66,67%), less (25%) and low (8,33%. In detailed it is displayed in table 4.

Table 4. Capacity of getting access (*Power To*)

No	Interval Class	Percentage	Description
1.	5,00 - 8,33	8,33	Low
2.	8,34 - 11,66	25,00	Less
3.	11,67 - 15,00	66,67	High
		100,00	

Source : Primary data processed, 2017

The table demonstrate that the vast majority of fish product processor state that they are able to get a chance to access information related to technology, market, information and income related to group of fish product processing or processing business. Information in terms of technology is acquired mostly from extension agent and university. Whereas, information and market opportunity related to the product is quite abandoned however, business actors still get difficulties to develop it. They said that there are many factors or constraint such as the lack of a place to promote product, the lack of capital, product continuity is uncertain due to the lack of raw material, administration requirements such as BPOM (Organization of Medicine and Food Controller), place rent, and product continuity.

Market is an attractor factor in which it will be a main constraint if it does not develop well. therefore, opening market access is a way for developing enterprise. An enterprise will not develop if there is no market. One of the ways to develop market for coastal societies product is facilitating people to big companies that export fishery product. For doing so, hence trading contract between people and company must be organized [9]. This kind of cooperation is beneficial to assure market and price for societies. People are taught about product quality control and are given capital aid for enterprise developing by companies.

3.4.3. Capacity of overcoming threat (Power over). Constraint and problem are certainty in the business teamwork context. They have to overcome and manage every single impediment faced, otherwise it will affect on teamwork and business activity. Based on research result known that the ability of fish processor to tackle an impediment could be categorized as less (50%), low (16,67%) and high (33,33%). It indicate that they did not overcome a constraint optimally. Considering to those fact, it is important for extension agent to improve their role maximally. The detail of the ability to tackle impediment displayed on table 5.

Table 5. The percentage of the ability to tackle a constraint

No	Interval class	Percentage	Description
1.	5,00 - 8,33	16,67	Low
2.	8,34 - 11,66	50,00	Less
3.	11,67 - 15,00	33,33	High
		100,00	

Sources : Primary data processed, 2017

3.4.4. Capacity of cooperation/Solidarity (Power with). Wikipedia define solidarity term as integration, level and integration type which is shown by society or group to other people or their neighbor. It refer to societies relationship or social relationship which bond to each other. This term is commonly used in sociology and other social discipline.

According to table 6 known that the vast majority of respondent state that they are able to cooperate well is 66.67%. Whereas, respondent who state that they have less ability to cooperate is 25% and 8.33% of respondent have low ability to cooperate. Strong cooperation or solidarity of each group members has a great impact on group development.

Table 6. the percentage of ability to cooperate/solidarity

No	Interval Class	Percentage	Description
1.	5,00 - 8,33	8,33	Low
2.	8,34 - 11,66	25,00	Less
3.	11,67 - 15,00	66,67	High
		100,00	

Source : Primary data processed, 2017

3.5. Women participation in inclining household economic of fishermen

Societies participation on many aspects of development is highly important. Society is not an object of development, however they can be also development subject. Therefore, societies participation will impact on development process itself. Societies will have sense of responsibility and motivation to implement development collectively. According to M. Rogers in [10] , participation terminology is involvement level of social system members in decision making process. If we scrutinize it however, participation is not only related to decision making, but also it has boarder definition, including of planning, decision making, actuating, evaluating process and benefiting development outcomes. The study result demonstrates that the income of fishermen housewife contribute to 24,04% of household total income. The income of fishermen housewife as fish processor depend on her husband's catching yields, since generally speaking that fishermen housewife only process fish with low economic value [11].

This research find that contribution of coastal women is quite good in Sanga-Sanga District. It can be seen on some activities or business mainly on fish product processing enterprise. dominant activities are amplang, crackers (kerupuk), terasi and petis processing. Those activities have grown since 15 years ago, and nowadays is still continuing.

4. Conclusion

Coastal women empowerment on processing enterprise group can be concluded from empowerment indicators as follows: the empowerment level of women group on fish product processing categorized as high (gained score is 42,75), and partially, based on empowerment indicators, it demonstrates that is the capacity level to increase a chance to gain access is about 66,67% categorized as high, 25% categorized as less, and 8,33 categorized as low, the capacity level of overcoming an impediment tend to categorized as less (50%), while 16,67% categorized as low, and 33,33% categorized as high. And the capacity level of cooperating for 66,67% is high, 25% is less and 8,33% is low.

References

[1] Amirah M, Andi A A, Yusran N I 2013 Household economic empowerment of coastal-based coastal communities in Tongke-Tongke Village Sinjai District http://pasca.unhas.ac.id/jurnal/files/61f108ed10e029141482e6ae867cf18e.pdf.

[2] Haqiqiansyah, G. Fidhiani, D. Sugiharto, E 2012 *Strengthening women's entrepreneurial capacity in Sanga-Sanga District of Kutai Kartanegara Regency* National strategic grant research report Research Institute of Mulawarman University Samarinda p 70

[3] Central Bureau of Statistics 2016 *Sanga-Sanga District in number* p 123

[4] Zakiyah 2010 Empowerment of women by Lajnah women *J. of Assessment of Religious Social Issues* Vol XVII p 44

[5] Agung U, Titin H H 2014 Model of empowering poor women through training of local potential-based entrepreneurship in District of Wedi Klaten Regency *J.of Humanities Research* 19(2) pp154-167

[6] Nugroho M 2015 Fishermen empowerment in Pasuruan District : Study of development of human resource empowerment model in coastal area. *J.of Food Technology* 6(1) pp 19-26

[7] Edi Suharto Social Assistance in the Empowerment of the Poor : Conceptions and Strategies. http://www.policy.hu/suharto/modul_a/makindo_32.htm

[8] Saptatiningsih R I, Tri S N, Sri R 2015 Village Women Empowerment To Reduce Poverty. Paper presented at National Seminar of PGRI University Yogyakarta *Proc. ISBN 978-602-73690-3-0* pp 512-524

[9] Ambo T H 2011 *Management of Coastal and Marine Ecotourism* Surabaya Brilliant International

[10] Levis L R 1996 *Rural Awareness Communications* Bandung Citra Aditya Bakti p 262

[11] Maulana F, Rikrik R 2015 Role of fisherman's wife in household improvement (case study in Penjajab Village Pemangkat District Sambas Regency) *J. of Socioekonomic Marine and Fisheries* **10** (2) pp. 241-249

Setulang forest conservation strategy in safeguarding the conservation of non-timber forest products in Malinau District

T R Hutauruk[1,2,*], A M Lahjie[2], B D A S Simarangkir[2], M I Aipassa[2] and Y Ruslim[2,]**

[1]Department of Management, Sekolah Tinggi Ilmu Manajemen Indonesia Samarinda, Indonesia.
[2]Faculty of Forestry, Universitas Mulawarman, Indonesia.

Corresponding author: *thomas_huta@yahoo.com, **yruslim@gmail.com

Abstract. Some research on human relationships with forests shows that human activities for the forests are sufficient to meet demand or to meet demand. Both will directly or indirectly change the perceptions of the people who exploit them against the forests being utilized. Setulang community one of the community groups that intensively utilize NTFP as one of the source of fulfillment and fulfill the demand of handicraft product. For needs and demand of livelihood the people of Setulang choose to make the existing forest in the Tane Olen area into. The analysis method used in this research with Stength Weakness Opportunity and Threat (SWOT). The results of this study show the extent to which the commitment of Setulang community and the conservation efforts of Setulang community both self-help and donor assistance and the local government, as well as what strategic steps can be taken by the stakeholders to conserve so that the village forest can provide benefits in the short or long term. Strategic measures need to be set up save Setulang State Forest from degradation and deforestation occurring around the village.

1. Introduction

1.1. Background

Community life in a region is closely related to the availability of nearby natural resources. Such conditions also occur in communities living in or around the forest. Communities in or around the forest will utilize forest products in the form of timber, not wood and environmental services [1,2,3, 4]. Such close dependence of the community on forests, raises awareness to keep the survival of the resources being utilized. Attitudes towards forests of local communities around the forest will be very different to those who only use forest products for commercialization but have no direct attachment to forests.

Based on the empirical data on the number of villages directly related to forest areas stated by the Ministry of Forestry in the Strategic Plan of the ministry of Forestry 2010-2014, there are 31,957 villages, consisting of 1,305 villages within the forest area, 7,943 villages on the edge of the forest area, and 22,709 villages in around the forest area (Ministry of Forestry Strategic Planning 2010-1014). Specifically, in East Kalimantan (including North Kalimantan), there are 285 villages in protected forest area, 628 villages are in Production Forest area, and 378 villages in Limited Production Forest area [5].

Setulang is one of the villages directly related to the forest in it. Setulang villagers facilitated by several donors (among others: CIFOR, GIZ-Forclime) are committed to preserving the forest in their village. The rationale for the choice of Setulang are for CIFOR's work was that the area had high environmental values, a significant population of poor, forest-dependent people, and that it was located at the forest frontier [6][7].

The commitment is manifested in the form of Village Forest. Setulang State Forest officially gained state recognition with the issuance of Forestry Ministerial Decree No. 526 / Menhut-II / 2003 on the Establishment of Limited Production Forest Area as a Setulang Forest Management Area of 4,330 hectares (hereinafter revised to Decree of Minister of Forestry Number 755 / Menhut-II / 2014 regarding Setting of Setulang Forest Working Area of 4,415 hectares in Forest Protection Area (HL) and Limited Production Forest (HPT) in KecamatanMalinau Selatan, North Kalimantan Province).

Of the 905 people in Setulang, 32.15 percent of them earn a living from the agricultural sector. Those who have livelihoods from the agricultural sector also use the Village Forest as a source of Non-Timber Forest Products (NTFP) for raw materials for handicrafts and suppliers of food and medicines. Non-wood forest products (NWFPs) or Nontimber Forest Product (NTFPs) such as nuts, mushrooms, herbs, spices, aromatic plants and gamehave been used for food, health and cultural purposes for millennia, yet there is a tendency to underestimatetheir role because they are poorly represented in international statistics, as in most cases their useand trade are confined to the informal sector. Recent studies show that NWFPs still form the basis of livesand livelihoods in many parts of the world and play a much more significant role in food and nutritionalthan previously thought [8,9,10,11]. The main reasons are the role of these products in providing livelihood and food for rural communities to reduce poverty along with supporting various environmental goals such as conservation of biodiversity. Traditional conservation schemes seek to ensure the provision of environmental services by restricting the rights of rural communities to access and use natural resources [12]. Nontimber forest products (NTFPs) are accepted as a veritable means of achieving poverty alleviation because of their role in sustaining this livelihood, for food security and environmental objectives such as biodiversity conservation. Given the importance of forests for community life, it is necessary to devise a conservation strategy that can ensure the sustainability of Village Forest by involving local communities as the main actors

1.2. Reserach Purposes
This research purposes are: 1) Know the factors that strengthen and weaken the existence of village forest in the face of threats and opportunities, and 2) Establish a conservation management strategy of Setulang Forest Village Malinau District based on current conditions

2. Methodology

2.1. Study Area
This research focuses on Conservation strategy in Setulang Forest Village, Sub-district of Malinau Selatan Hilir, Malinau District, North Kalimantan Province. It is approximately ± 32 km from the Capital city of Malinau District. The boundaries of the village, covering the northern border with the village of Sentaban, the south bordering Setarap Village, the east bordering TanjungLapang Village, and the West bordering with Paking Village. SetulangForest Village Location of Malinau Selatan Sub-district TheMalinau District is located at the coordinates of $03^0 20$ 'North Latitude – $03^0 30$' North Latitude and $116^0 24$ 'East Longitude – $116^0 29$' East Longitude (Figure 1).

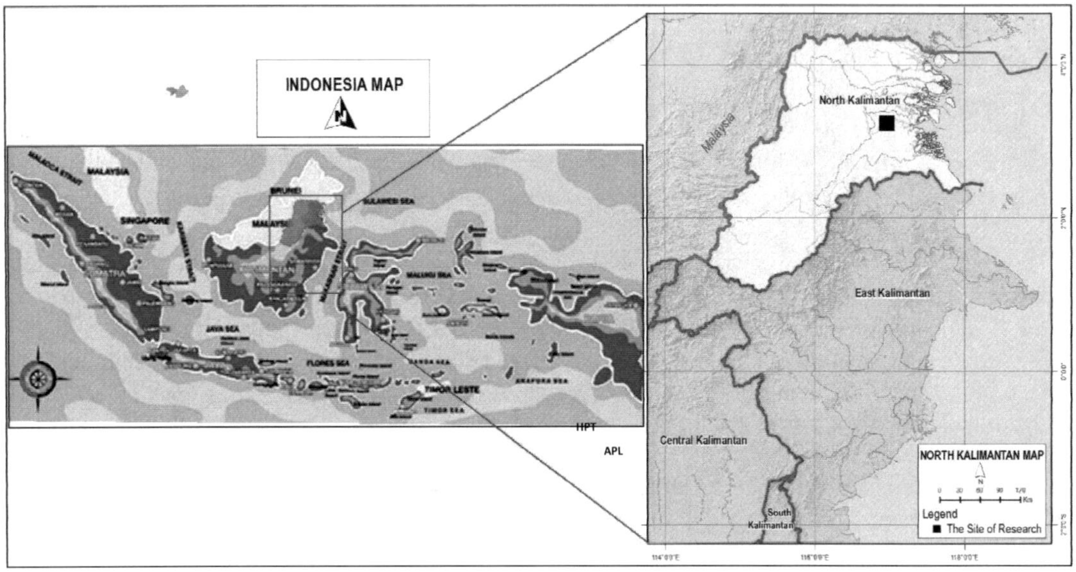

Figure 1. Setulang forest village location of Malinau Selatan Hilir sub-district (■).

2.2. Research Procedure

This research was conducted by collecting primary and secondary data. Primary data was obtained by interviewing selected informants by purposive sampling and tracing the location of Village Forest. While the secondary data collected in the form of previous research results and position maps. The collected data is tabulated and analyzed, to be further interpreted.

2.3. SWOT Analysis

The Analytic Hierarchy Process as a method of measurement with ratio scales and illustrate it with two examples. The use of SWOT as an analytical tool to strategize has been widely used by researchers. This is due to what is presented in the columns of SWOT is a form of interpretation of the results of a researcher's assessment both quantitative and qualitative.

Conducted SWOT analysis in identifying strategies for community development. Based on this exercise, the following five categories were identified and prioritized by villagers for the development of their village in future: 1. Conservation and utilization of natural resources (with reference to water and forest). 2. Development of the wasteland, agriculture and livestock sector. 3. Promotion of livelihoods resources and human resource development. 4. Promotion of health, cleanliness and education. 5. Development of village institutions [13]. SWOT analysis indicates a framework for helping the planners to identify the strategies of achieving goals. It is a technique used to analyze the strengths, weaknesses, opportunities and threats of businesses [14]. The SWOT analysis is complemented by AHP [15] to determine the priority of decision making in the formulation of a Village Forest management strategy. Assessing performance involves criteria [16] and priorities that are determined by user requirements and preferences as well as the characteristics of the individual institutes. AHP allows group decision-making. The AHP is now used as common tool in multi criteria decision making process as a part of operation research and management science because of it's both technically validity and practical usefulness [17, 18]. In the process of weighting in the AHP, determined the validity of paired data using the following formula: The ratio is obtained by dividing the Eigen Factor value of each criterion by its weight. Consistency Index (CI) is obtained through the equation:

$$CI = \frac{Average\ Ratio - n}{n-1} \quad (1)$$

CI = Contingency index
Average ratio = Average ratio are ratio of paired comparisons matrix total withcriteria total.
n = Sum of criteria

If the value of CI equal to zero, then the matched pairwise matrix is said to be consistent. The inconsistency limit set using Consistency Ratio (CR), ie comparison of consistency index (CI) with random index value (RI). The Consistency Rate can be formulated as follows:

$$CR = \frac{CI}{RI} \quad (2)$$

CI = Consistency index
RI = Random index

Random index (RI) is a random consistency index of each number of criteria. The compiled Random index as follows:

Table 1. Random index (RI)

n	**1**	**2**	**3**	**4**	**5**	**6**	**7**	**8**	**9**	**10**
RI	0.00	0.00	0.58	0.90	1.12	1.24	1.32	1.41	1.45	1.51

Note: If the value of CR <0.1, then the value of CR or inconsistency value is still acceptable. However, if the value of CR> 0.1; then the assessment matrix must be repeated.

3. Results And Discussion

3.1. Internal and External Factor Analysis

Measures related to conservation efforts by the UmaLongh Community of Setulang Village have been carried out since their origin (Long Saan). One form of conservation efforts that are done is with the Tane Olen area which until now still maintained its existence. Tane Olen's own history has been clearly. Nevertheless, conservation efforts by UmaLongh community of VillageForest, along with time shift experienced obstacles and threats. On the one hand, the establishment of Tane Olen as a Village Forest so that the Setulang people have NTFP needs providers, but on the other hand, the opening of Setulang status as Tourism Village is also increasingly open relationship with the outside world which has the potential of inclusion of threats to the preservation of Village Forest. The following will present the Setulang Forest conservation strategy prepared based on the actual conditions.

Test results on consistency of paired data obtained by CR = 0.085012 (CR <0.10); shows consistency is still acceptable.Based on EFI analysis, it is known that the total (score of x weight) obtained is 2.74. Weights on internal factors are presented in the following table 2:

Table 2. Evaluation of internal factor matrix (EFI)

No.	Internal Factor	Priority	Rating	Score
A. Strengths				
1	Forest Village in Setulang has a high diversity of biological and animal species	0.11	3	0.33
2	Boundary Clarity of Village Forest Area	0.22	4	0.88
3	Uma LonghSetulang community participation is quite high	0.15	3	0.45

4	The existence of a special institution responsible for the management of Village Forest area	0.08	3	0.24
5	Gets the attention of the world community	0.10	3	0.30
	Subtotal	0.66		2.20
B. Weaknesses				
1	Large budget uncertainties available in managing Village Forest	0.11	2	0.22
2	The dependence of donor agencies who are willing to provide guidance and technical assistance in the management of Village Forest	0.10	2	0.20
3	Most young people are less interested in taking care of the Village Forest	0.06	1	0.06
4	Most of the people of UmaLongh Setulang who are domiciled around the Village Forest do not have the ability to prevent forest destruction.	0.03	1	0.03
5	The UmaLonghSetulang community does not have the capacity to cope with the possibility of forest and land fires	0.04	1	0.04
	Subtotal	0.34		0.55
	Total	1.00		2.74

Test results on consistency of paired data obtained by CR = 0.065393 (CR <0.10); shows consistency is still acceptable. Based on EFE analysis, it is known that the total (score of x weight) obtained is 2.63. Weights on external factors are presented in the `following table 3:

Table 3. Evaluation of external factor matrix (EFE)

No.	External Factor	Priority	Rating	Score
C. Opportunities				
1	Village Forest is one of the government's priority programs in the development of social forestry	0.12	3	0.36
2	The only model of Village Forest in Prov. Kaltara who has got the Decision Letter	0.16	4	0.64
3	Positive perceptions of local communities	0.11	3	0.33
4	The attraction for visitors from different countries to see directly Forest Village	0.14	3	0.42
5	Third party support (donor agencies)..	0.08	3	0.24
	Subtotal	0.60		1.98
D. Threats				
1	The emergence of illegal plantation activities owned by individuals	0.13	2	0.26
2	The potential for horizontal and vertical conflicts.	0.13	2	0.26
3	Environmental conditions around the Tane'Olen area are more open.	0.04	1	0.04

4	Accessibility to the Village Forest location is increasingly accessible to outsiders.	0.05	1	0.05
5	Instability of socioeconomic conditions outside Setulang	0.05	1	0.05
	Subtotal	0.40		0.65
	Total	1.00		2.631

3.2. Conservation Strategy

There are several forest conservation efforts that can involve local communities, among others: All sites are forest areasconsidered to be "under threat", where communities can harvest nontimberforest products for their own consumption. As in upstream areas in otherparts of Asia, average areas of household cultivable land are <2 hectares, and mostsites are located in undulating upstream areas [19]. This means that conservation here there are efforts to prevent damage as well as recovery efforts. In support of forest restoration activities by IUCN and others, here we present a comparative analysis of ecosystem service assessment tools that can quantify ecosystem service tradeoffs. In the following sections, we present an overview of ecosystem service models relevant to restoration decision making and characterize the tools based on five key criteria. These criteria include their ability to quantify the benefits of restoration activities and scenarios in a timely and cost-effective manner across different geographic locations and scales, ecosystem service types, decision contexts, and under varying levels of uncertainty [20].

From the result of the analysis to the internal aspect, it is seen that the biggest score becomes the strength in the clarity factor of Village Forest boundary (0.87), while the highest weakness score is on the big uncertainty factor available in managing the village forest (0.22). Then on the external aspect the largest score that becomes an opportunity lies in the factor of the only model of Village Forest in Prov. North Kalimantan that has got the Decision Determination (0.66), while the biggest threat to factor the emergence of illegal plantation activities owned by individuals (0.25). Based on the largest weight gain on both internal and external aspects, the village forest management strategy depends heavily on: (1) clarity of boundaries and land / ownership status, (2) budget availability, (3) government policy support, and (4) community participation. Boundary clarity is very important because one source of conflict that occurs in the community is the ambiguity of the boundary area / land. The encounter between internal and external factors will represent the current condition of management as shown in Figure 2. Point X is obtained by (Total strength – Total Weakness) / 2 = (2.20 – 0.55)/2 = 0.82. Whereas point Y is obtained by (Total Opportunity – Total Threat) / 2 = (1.98 – 0.65)/2 = 0.67. When presented in Cartesian diagram it will appear that the coordinate position obtained lies in quadrant I.

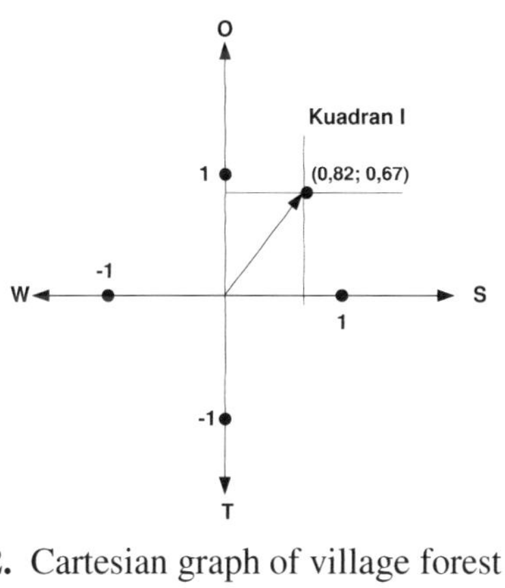

Figure 2. Cartesian graph of village forest position

In Figure 2 it is apparent that the Setulang Forest position at the time of study is in quadrant I, indicating that the Village Forest is in a very favorable situation. Village forest management has the opportunity and strength of the UmaLonghSetulang Community to take advantage of existing opportunities with the utilization of NTFPs from the village forest. The strategy to be implemented under these conditions is to support the policy of taking NTFPs to meet family needs and market demand. Here is a conservation strategy concept in sustainable Setulang Forest management considering the strengths and weaknesses (IFE) as well as the opportunities and challenges (EFE) can be arranged as follows:

Table 4. Conservation strategies in village forest management

Strategiy of S-O	**Strategy of W – O**
Involvement of local communities in intensive conservation activities in KawasanForestVillage (S1O1, S2O1, S3O1, S2O2, S2O3, S3O3,S2O4)	Sufficient budget available from APBN, APBD and / or APBDes. On the other policy it reduces the dependence of aid from donor agencies (W1O1, W2O2, W2O3, W3O3, W1O4, W1O5)
Prevent the entry of illegal activities that could disrupt the continuity of Village Forest. (S5O1, S2O2)	Strengthen the capacity of the Village Forest Management Institution and complete the management office with comprehensive information related to Village Forest. (W2O1, W4O2, W4O3, W2O4,W2O5, W3O5)
Conduct ongoing socialization related to the protection and utilization of Village Forest as a source of NTFP and sustainable environmental services. (S1O2, S4O2, S3O4, S5O4)	Provide opportunities and provide capital assistance to young children in the villages around ForestVillage in order to utilize NTFPs and Ecotourism potential. (W3O1, W3O2, W1O3, W3O4, W4O4, W3O5)
Encourage the growth of civic economic activities for local communities. (S4O3, S5O3, S1O4, S4O4)	Preparing local communities to anticipate forest degradation and forest and land fire hazards (W5O1, W5O2, W5O3, W5O4, W5O5)
Optimizing the role of trained communities in empowering local communities around VillageForest. (S4O1, S1O5, S2O5, S3O5, S4O5, S5O5)	Make the community groups trained as pioneers in every activity in contact with forest management and its follow-up. (W4O1, W1O2, W4O5)
Strategy of S –T	**Strategy of W –T**
Prevent the expansion of illegal plantation activities by giving reprimands, legal sanctions (adat), to report to the authorities. (S1T1, S4T1, S2T1, S1T2, S2T2, S1T3, S5T4, S1T5)	There is a sufficient budget for the maintenance and restructuring of the village forest area. (W1T2, W1T3, W1T4)
Enable Village Forest patrols to conduct area monitoring by involving local communities. (S3T1, S5T1, S3T2, S2T3, S3T4, S4T4, S4T5)	Local community assistance intensively as needed. (W4T1, W4T3, W2T4, W4T4, W2T5)
Establish surveillance posts along access to the Village Forest area. (S4T2, S3T3, S4T3, S2T4)	Involvement of young people in securing Village forest area. (W3T1, W3T2, W3T3, W3T4, W3T5)
Limiting the entry of outsiders to village forest areas, except those licensed by the village forest management agency (S1T4, S2T5, S3T5)	Encouraging Setulang Village as a unique wicker craft center by using NTFP. (W2T3, W1T5, W4T5)
Maximizing business opportunities generated by the community from the utilization of NTFPs. (S5T2, S5T3, S5T5)	Minimize the influx of negative impacts that can prompt societal frictions with social approaches effectively (W1T1, W2T1, W5T1, W2T2, W4T2, W5T5, W5T3, W5T4, W5T5)

4. Conclusion

From the result of the research, it can be concluded that: Setulang community has done conservation effort in a consistent way, that is only utilizing ForestVillage as a producer of NTFP and environmental services in limited amount, so as not to cause risk of damage. On the other hand, although some of the short-term threats can still be controlled, long-term if left to disrupt the sustainability of the village forest, the decline of biodiversity (flora and fauna) and the decrease of carbon stocks.

Acknowledgments

We appreciate the support received for facility and funding ofthis publication from the Faculty of Forestry Universitas Mulawarman, Department of Management Sekolah Tinggi Ilmu Manajemen Indonesia Samarinda, and GIZ-Forclime Programme. We are thanks to the Editor and reviewers and we thanks also to Umbar Sujoko for his help in creating the map of study site.

Reference

[1] Wollenberg E 1998 *Method for Assesing the Conservation and Development of Forest Product: What We Know and What We Have Yet to Learn. In Incomes From The Forest; Methods for The Development and Conservation of Forest Products for Local Communities.* Editor Eva Wollenberg and Andrew Ingles (Bogor: Center for International Forestry Research) p 227.

[2] Bishnu H, Pandit A A and Chetan K 2008 *Improving Forest Benefits for The Poor Learning From Community-Based Forest Enterprises in Nepal* (Bogor: Center for International Forestry Research).

[3] Sardjono M A 2004 *Mosaik Sosiologi Kehutanan Masyarakat Lokal, Politik dan Kelestarian Sumberdaya* (Yogyakarta: Debut Wahana Sinergi) p 300.

[4] Garnett S T, Sayer J, and Toit J 2007 *Improving the Effectiveness of Interventions to Balance Conservation and Development: a Conceptual Framework.* Ecology and Society **12**(1): 2

[5] Dewan Kehutanan Daerah Kalimantan Timur 2012 *Hutan Desa; Proses dan Pembelajaran.* (Jakarta: GIZ Forest and Climate Programme (FORCLIME) dan WWF Indonesia) p 82.

[6] Boedhihartono A K., Gunarso P, Levang P, and Sayer J 2007 *The principles of conservation and development: do they apply in Malinau?* Ecology and Society **12**(2): 2

[7] Wunder S, Campbell B, Frost P G H, Sayer J A, Iwan R, and Wollenberg L 2008 *When Donors Get Cold Feet: the Community Conservation Concession in Setulang (Kalimantan, Indonesia) that Never Happened. Ecology and Society* **13**(1): 12.

[8] Ickowitz A, Powell B, Salim M, and Sunderland 2014 *Dietary quality and tree cover in Africa.* Global Environmental Change, **24** pp 287–294

[9] Rowland, D Ickowitz, A Powell, B. Nasi, R and Sunderland T C H 2016. *Forest foods and healthydiets: quantifying the contributions.* Environmental Conservation (available at: http://www.cifor.org/library/6261/forest-foods-and-healthy-diets-quantifying-the-contributions/

[10] Basit M A 1995 *Non-Wood Forest Products from the Mangrove Forests of Bangladesh.* In P.B. Durst and A. Bishop (eds.) Beyond Timber: Social, Economic and Cultural Dimensions of Non-Wood Forest products in Asia and the Pacific p 193 - 200 in Proceedings of a Regional Expert Consultation held in Bangkok, 28 November to 2 December 1994. FAO / RAP, Bangkok.

[11] Moko H 2008 *Promoting Non Timber Forest Products as a Superior Product.* Technical Information **6** (2) pp 35 – 45.

[12] Chifamba E 2012 *Environmental Services Payments And Markets In Dry Regions Of Mahenye Campfire, Zimbabwe.* Journal of Sustainable Development in Africa **14** (6) pp 1 – 18.

[13] Singh N 2010 *SWOT Analysis – A Useful Tool for Community Vision A concept paper of central Himalayan village.* Res., **2**(9) pp 16-18.

[14] Ommani A R 2011 *Strengths, weaknesses, opportunities and threats (SWOT) analysis for farming system businesses management: Case of wheat farmers of Shadervan District, Shoushtar Township, Iran.African.* Journal of Business Management **5**(22) pp 9448-9454

[15] Saaty, T L 2008 *Decision Making With The Analytic Hierarchy Process.* International Journal of Services Sciences **1**(1) pp 83-98.

[16] Mateo J R and San C 2012 *Multi-Criteria Analysis in the Renewable Energy Industry*, 7 Green Energy and Technology DOI: 10.1007/978-1-4471-2346-0_2 (Verlag London: Springer)
[17] Dijana O 2012. *Strategy development by using SWOT* – AHP. TEM Journal **1** (4) pp 283 – 291.
[18] Kurttilaa M, Mauno P, Jyrki K, and Miika K 2000 *Utilizing the analytic hierarchy process _AHP. in SWOT analysis - a hybrid method and its application to aforest-certification case.* Forest Policy and Economics **1** pp 41- 52.
[19] Leimona B 2009 *Can rewards for environmental services benefit the poor? Lessons from Asia.* International Journal of the Commons **3**(1) pp 82–107.
[20] International Union for the Conservation of Nature (IUCN) 2016 *Forest Landscape Restoration.* Available at https://www.iucn.org/about/work/programmes/forest/fp_our_work/fp_our_work_thematic/fp_our_work_flr/, accessed 8 November 2017.

IOP Conf. Series: Earth and Environmental Science **144** (2018) 012056 doi:10.1088/1755-1315/144/1/012056

Study of optimizing water utilization in Benanga reservoir for irrigation and fresh water purposes

Tamrin[1*] and E Retati[2]

[1]Department of Civil Engineer Mulawarman University, Samarinda, East Kalimantan, Indonesia
[2]River Basin Organization III Samarinda, Kalimantan, Indonesia

*Corresponding author: *fts_tamrin@yahoo.com*

Abstract. Benanga dam was built in1978 an irrigation weir but currently it was developed into a multipurpose dam. However, based on the capacity curve measurement in 2015, the capacity curve measurement has been changed to get below. The runoff rate is calculated by using NRECA method, andwater reservoir volume is calculated by using penman modification method. The cropping pattern that has been implemented by the farmer of Lempake sincein Februaryis Paddy-Paddy-Fallow While the proposed cropping pattern in Benanga reservoir started on December, that proposed is based on the service ability for both raw water demands like irrigation and fresh water and if early planting is started besides these two months the elevation of benanga reservoir will not reach the normal elevation effective storage which is the condition pattern of reservoir operation.

1. Introduction

East Borneo Province is a province which has huge potential on water resources. However, this province has serious problems related to food production and decreased clean water supply day by day. based on temporary number area from BPS East Borneo Province in 2015, known that production of paddy decreased amout 4.08% ie previous in 2014 is 426.170 ton become 408.781 in 2015. one reason why it not able to reach the goal production is because there was planting area be reduced as an impact of long dry weather condition. Beside that there was also cropping delay because of time shift which is not effective.

For supporting food production in Samarinda, the government has built has some dam that has a function for irrigation and clean water supply such as Benanga dam. This dam is one of the dam which is located in Samarinda city stemmed Karangmumus River. This dam has been 30 years old, at the first time this dam was a dam that only has a main function for irrigation water needs. But more than the last 10 years this dam also functionalized for fresh water recervoir, although it's function has not optimal yet In order to fulfill it's function to serve irrigation dan raw water is necessary to review related to effectiveness of effective storage Benangan Dam. This review become really important so that this multipurpose dam could functioned according by early planning design. Therefore a study needed to optimize the water reservoir in the Benanga dam for two purposes ie for irrigation and clean water.

IOP Conf. Series: Earth and Environmental Science **144** (2018) 012056 doi:10.1088/1755-1315/144/1/012056

2. Theoretical Basic

2.1 Hidrology Concept

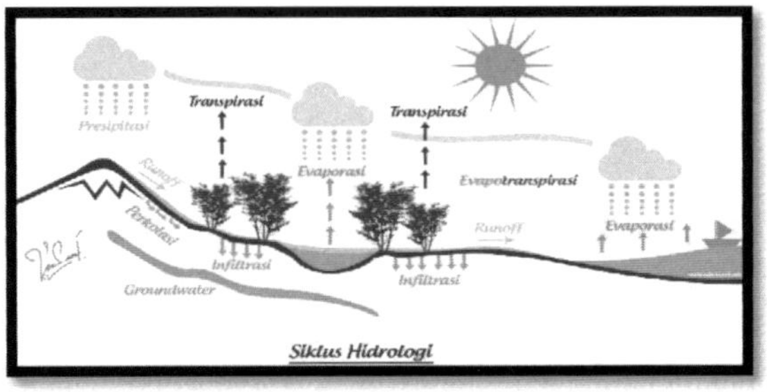

Figure 1. Hidrology Cycle
Source: *Siklus – hidrologi dan Jenis Pengairan.html*

a. Evapotranspiration

Evapotranspiration is a combination from two terms which is described physics process water transfer into the atmosphere, these are evaporation from overland flow and transpiration through the plant. The calculation evapotranspiration for irrigation is really important, because this method can give prediction total water need per day that plants need. There are a lot of method which has been developed to estimate amount of evapotranspiration such as 1) Water Balance Method, 2) Energy Balance Method, 3) Mass Transfer Method, 4) Combination of energy transfer and heat method such as Penman Method, 5) Prediction Method, 6) Method for spesific plant.

b. Defendable Rainfall

Defendable Rainfall is a rainfall that is expected to be happen in certain chance. The higher the probability of occurance that expected, the height of rainfall that occurance will be smaller and vice versa. The big probability of occurance really depends on level of need to be planned.

c. Irrigation Water Requirements

Irrigation water requirementsis the amount of water flow that is used to irrigate the land in irrigation area to know whether water that is available to irrigate existing rice field needs water balance calculation in related irrigation area. This calculation is implemented by comparing available water and existing water needs.

d. Availability Water Calculation

The availability water calculation is amout water (flow) that estimated always exist in a location (dam or other water building) with the certain amout and in certain time (period) (*Direktorat Irigasi, 1980).* Epedable flow is a minimum river flow with the certain quantities which has posibilities occurance that can be used for that needs. For irrigation needs, minimum flow that required is 80% *(Hidrologi Terapan, 2008).*

e. Water Balance Analysis

The water balance analysis can be calculated by:

$$S_t + 1 = St + lt + Rt + Et + Qt - Q_t \qquad (1)$$

where :

S_{t+1} = Water storage on period to t+1
St = Water storage on period to t
lt = inflow to reservoir on period to t
Rt = rainfall which entering into reservoir storage on t periode

Et =water loss because evaporation in reservoir on t periode
Ot = Reservoir outflow t periode
Qst = Reservoir spill out t periode

3. Analysis and Discussion

Dam Location can be seen in the following picture :

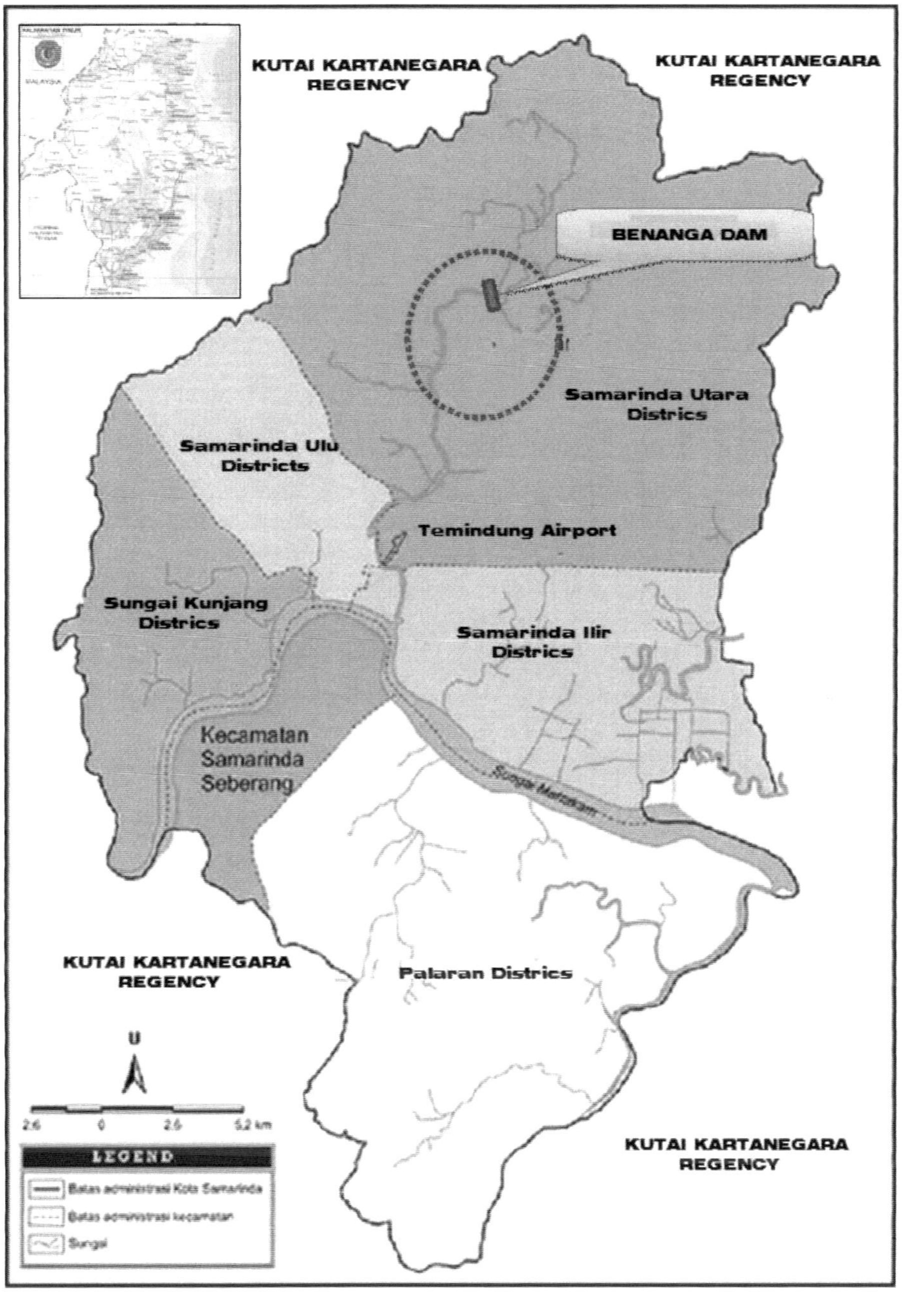

Figure 2. Location of Benanga Dam

This dam planned to fulfill irrigation water needs about 350 Ha and 210 l/seconds for clean water need. But nowdays, this dam is also utilized to accommodate water from upstream Karang mumus watershed before entering Samarinda city. So that, in data analysis the steps will btaken are :

1. Availability Water Analysis by Nreca Method
2. Irrigation Water Needs Analysis
3. Analysis of Waduk Benanga Inundation Map by reservoir capacity curve
4. Simulation of Benangan Reservoir

3.1. Evapotranspiration calculation Result

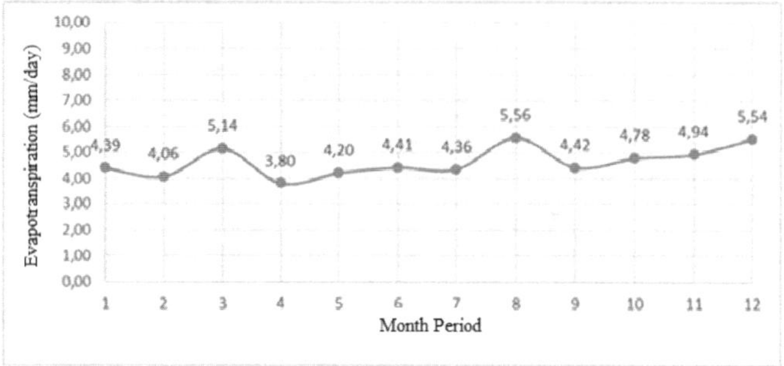

Figure 3. Evapotranspiration in Benanga Reservoir

From chart above can be seen that highess evapotranspiration in Benanga reservoir theoccurance in month 8 (August) where the value of evapotranspiration of reference plant (ETo) reaches value about 5.56 mm/days. This thing show that the biggest evaporation occurance on that month that is because of air speed about 2.06 m/dtk. *Dependable Rainfall Calculation Result 80% and 50%.*

In dependable rainfall calculation that occurance in Benangan Reservoir used the data of monthly rainfall that obtained from Badan Meteorologi Klimatalogi dan Geofisika (BMKG)that occurance between this 15 years since 2000 until 2015. This is the chart of dependable rainfall 80%:

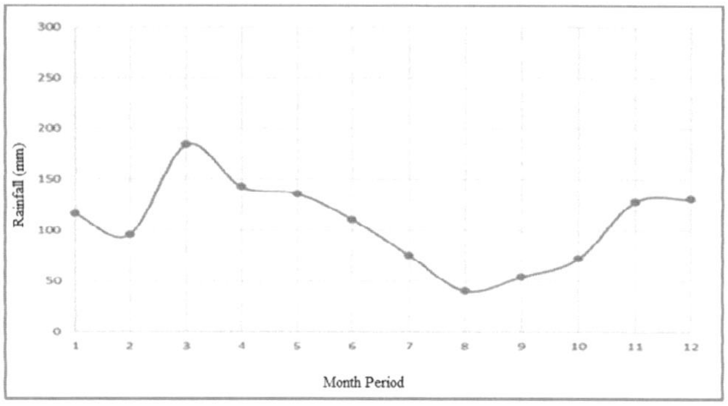

Figure 4. Dependable Rainfall Characteristics 80%

From that chart can be seen that dependable rainfall 50% characteristics which is occurance in Benanga reservoir where the highess dependable rainfall occurance on month 3 (March) which is about 183.9 mm.

IOP Conf. Series: Earth and Environmental Science **144** (2018) 012056 doi:10.1088/1755-1315/144/1/012056

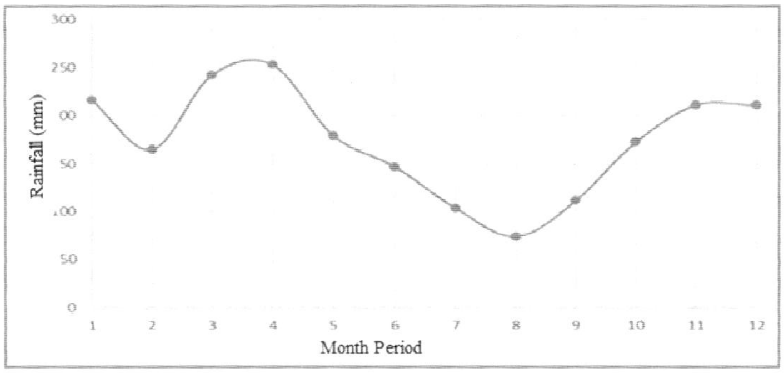

Figure 5. Dependable Rainfall Characteristic 50%

3.2. Calculation of Irrigation Water Requirements

Most of Irrigation water needs be fulfilled by overland flow. Irrgation water needs affected some factors such as climatology, soil condition, plant coefficient, cropping pattern, water supply, group system, cropping schedule, and ect. Some field condition related to water needs to agriculture varies towards time and space as mentioned in these following factors:

1. Types and varieties of cropping by farmer
2. Coeffisien varieties plants, depends on type and plants growing steps
3. When the preparation of land started (group)
4. Cropping schedule that is used by farmers, incluiding water supply related on land preparation, nurseries, and fertilization.
5. Irrigation system status and irrigation efficiency
6. Type of soil and agroclimatology factors

Des		Jan		Feb		Mar		Apr
I	II	I	II	I	II	I	II	I
LP	PD-I	PD-I	PD-I	PD-I	PD-I	PD-I	LP	LP
LP	LP	PD-I	PD-I	PD-I	PD-I	PD-I	PD-I	LP

Apr	Mei		Jun		Jul		Agust		Sep		Okt	
II	I	II	I	II	I	II	I	II	I	II	I	II
PD-II	PD-II	PD-II	PD-II	PD-II	PD-II	PLW	PLW	PLW	PLW	PLW	PLW	LP
LP	PD-II	PD-II	PD-II	PD-II	PD-II	PD-II	PLW	PLW	PLW	PLW	PLW	

Figure 6. December Cropping Pattern Planning

Legend :

LP = Land Preparation
PD-I = First Cropping Pattern
PD-II =Second Cropping Pattern
PLW = Cropping Palawija

3.3. Simulation Reservoir

To determine step that must be taken so need to besimulated with 3 alternatives such as:

1. Arrange cropping pattern for farmer only
2. Do dredging as deep 1.2 m
3. increasethe height of spilways of the dam

a. Alternative 1 (without dredging and increase spillways)
In this alternative the result of reservation simulation can be shown, as follows:

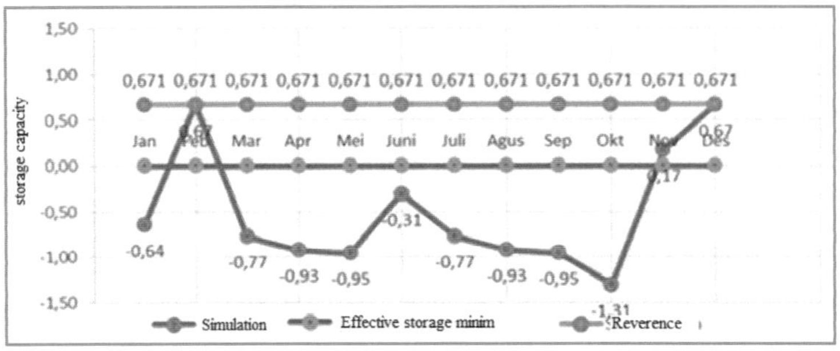

Figure 7. Result of Water Balance Simulation RTT

In chart above shows that December is the month which can be selected forthe farmers in Benanga reservoir in order to start planting period. Advantages of cropping pattern Alternative Simulation in December until January are irrigation area which can be served is larger, more raw water needs which can be fulfilled, abundant amount of water less.

b. Alternative 2 (dredged reservoir as deep 1.2 m)
In this alternative, the simulation will dredged as deep 1.2 m appropriate with the height of the main Benanga's spillway.

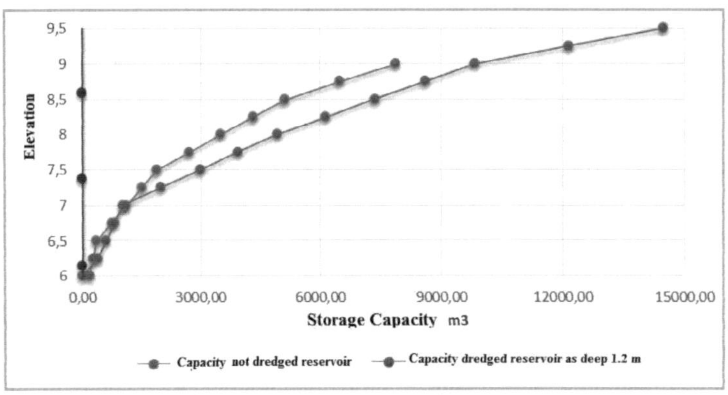

Figure 8. Simulation Result of Capacity Curve

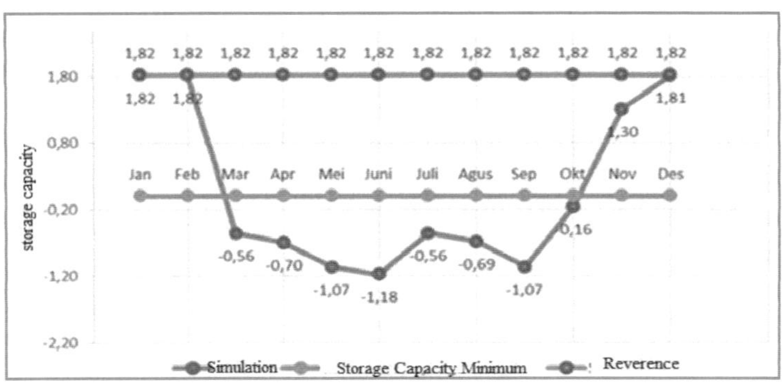

Figure 9. Dredging Simulation Result of Benanga

Figure 9 shows that if the reservoir dredged for 1.2 m. There are 3 alternative for early planting that can be used such January, February, and December and among these 3 months early planting in December month is the most profitable cropping pattern alternative where in this early planting Benanga reservoir can serve irrigation needs about 350 Ha, clean water about 210 l/second, and mantaining river about 50 l/second.

c. Alternative 3 (increase the height of Reservoir's Spilways)
Benanga's spilways nowdays is in +7.20 elevation based on water balace simulation result shows with the elevation position in +7.20 Benanga Reservoir can only serve raw water needs (irrigation water and clean water) if early cropping pattern start from February month and December with the maximum service in December month where in December month, it can serve irrigation needs about 350 Ha, Clean water 133 l/second, and mantaining river about 50 l/seconds. So elevation increasing for it's elevation which aim to waduk Benanga can fulfill raw water needs just like in it's planning 2014. Based on simulation result for increasing top elevation spilway there is in elevation +7.75 with the recaptulation of water balance values as follows :

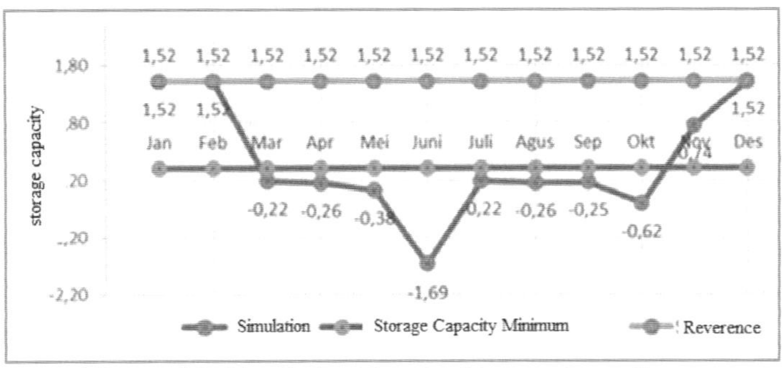

Figure 10. Simulation Result of increasing spilway

In chart above shows that if increasing the spilway has done until in 7.75 elevation then there are 3 alternative early cropping that can be used such as January, February, and December and among these three months early cropping in December month is the most profitable early cropping pattern alternative where in elevation +7.75 Benanga Reservoir able to fulfill irrigatin needs about 350 Ha, clean water about 210 l/second, for river mantaining about 50 l/second.

4. Conclusion

1. From the result known that simulation of reservoir besides February and December can not be choose because it can not reached back +7.20 elevation the storage capacity is under minimum elevation of effective storage (<6.00) so simulation of early cropping pattern should be implemented appropriate with the requirement of reservoir operation system that has done in February and December.
2. Based on simulation of water balance that appropriate with the requirement of reservoir operation system are obtained 2 alternative that fulfill the requirement there are in February and December.

References

[1] Usman 2004 *Analisa Kepekaan Beberapa Methode Penggunaan Evaporasi Transporasi Potensial Terhadap Perubahan Iklim* (Riau: Universitas Riau)
[2] Triadmojo Bambang 2008 *Hidrologi Terapan* (Yogyakarta: Beta Offset)
[3] Soedibyo 2003 *Teknik Bendungan* (Jakarta: Pradnya Paramita)

[4] Alhinduan ND dan Indrawan Ivan 2013 *Analisa Efesiensi dan Optimalisasi Pola Tata Tanam Pada Daerah Irigasi Timbang Deli Kabupaten Deli Serdang* (Medan: Universitas Sumatera Utara)

[5] Siagian SY dan Tarigan M A P 2013 *Analisa Neraca Air Daerah Irigasi Panca Arga di Kabupaten Asahan* [Medan: Universitas Sumatera Utara]

[6] Mekiuw, Yosehi 2013 *Studi Neraca Air Waduk Lapangan di Desa Semangga Jaya Kabupaten Merauke*: (Merauke: Universitas Musamus)

[7] Kementerian Pekerjaan Umum dan Perumahan Rakyat, Direktorat Jenderal Sumber Daya Air, *Balai* Wilayah Sungai Kalimantan III 2014 *Konsep Laporan Akhir Paket Pekerjaan Kajian Peningkatan Bendungan Benanga Kota Samarinda*. (Samarinda: Laporan Akhir Balai Wilayah Sungai Kalimantan III)

[8] Departemen Pekerjaan Umum, Direktorat Jenderal Sumber Daya Air. 1986. *Buku Petunjuk Perencanaan Irigasi* (Jakarta: Badan Penerbit Pekerjaan Umum)

[9] Departemen Pekerjaan Umum, Direktorat Jenderal Sumber Daya Air 1999 *Panduan Perencanaan Bendungan Urugan* (Jakarta: Direktorat Bina Teknik)

[10] Departemen Pekerjaan Umum, Direktorat Jenderal Sumber Daya Air 1986 *Standar Perencanaan* Irigasi *Kriteria Perencanaan Bagian Jaringan Irigasi (KP – 01) (*Jakarta: Direktorat Jenderal Pengairan)

[11] Departemen Pekerjaan Umum, Direktorat Jenderal Sumber Daya Air 1986 *Standar Perencanaan Irigasi Kriteria Perencanaan Bagian Bangunan (KP– 04)* (Jakarta: Direktorat Jenderal Pengairan)

Evaluation of plant species composition after thirteen years post coal mining rehabilitation in East Kutai District of East Kalimantan, Indonesia

L L Komara*, V Murtinah and Arbain

Forestry, East Kutai School of Agriculture, Jl. Sukarno Hatta No 1, East Kutai, East Kalimantan, Indonesia

*Corresponding author: liskomara@yahoo.co.id

Abstract. Coal mining leaving large overburden which needs to be rehabilitated by conducting reclamation and re-vegetation. This study objective was to evaluate plant species composition after thirteen years of re-vegetation activities in a coal mining rehabilitation site in KutaiTimur District, East Kalimantan, with coordinate 00°33'23"-00°38'17" NL and 117°23'55"-117°23'20" EL. This study compared rehabilitation site with the natural forest conditions sites. There were found 28 plant species in the rehabilitation site, consisting of 19 wood species (*Cassia siamea* and *Ficus uncinata*) and 9 non-wood species (*Nephrolepis biserata* and *Miscanthus javanica*).In comparison, 36 species were found in the natural forest condition, consisting of 25 woody species (*Nephelium eriopetalum* and *Macaranga hypoleuca*)and 11 non-wood species (*Fordia splendidisima* and *Saurauia umbellata*). Woody species diversity indices in the rehabilitation site after 13 years post mining (i.e., 2,21) was lower than in the natural forest sites (3,01); while the diversity indices for non-wood species were relatively similar (1,48 and 1,96 in the rehabilitation and natural forest sites respectively). Species richness of non-wood species was low, but it has a high coverage per species in this site.To restore rehabilitation site woody species to its assumed natural conditions, 22woody species should be planted.

1. Introduction

The composition and plant diversity depends on a number of factors, such as elevation, slope, temperature, humidity, nutrition, sunlight, topography, bedrock geology, soil characteristics, canopy structure and land use history [1, 2]. Kalimantan is the center of species diversity, it is about 138 Shorea species were pound in and 91 of them is an Kalimantan endemic [3].

The Borneo plant diversity were disturbed by coal mining activity. Coal mining activity disturbs the function of soil profile [4] and geological structure permanently [5]. Large area of mine overburden waste land is generated annually due to open cast mining and dumped on nearby area. The overburden area has a low organic contents and physics chemistry characteristics which become unfavorable for plants growth. Considering the impact of mining to the environment, the company or people conducting mining activities has to be able to guarantee the re-productivity of land affected by mining activities [6]. The overburden area need to be restored to its pre-mining condition, therefore post-mining area needs to be rehabilitated [7].

Rehabilitation methods, such as topsoil removal have been implemented to restore abiotic conditions suitable for high diversity and threatened species call as reclamation, and it may successful by using biological tools: tree seedlings, native shrubs, and grass species which call as revegetation [8]. The surrounding of post mining area are less fertile to be planted, in this type of location, plants grown have the tendency not to grow and develop well [9, 10]. It is important to note that mining area generally is an area that has a high biodiversity, conservation and economic [9]. Indicators of rehabilitation success is what on the ground, one of them is vegetation [11]. Post-mining area rehabilitation is considered success if the vegetation has 90% grown, close to the pre-mining condition [12]. Therefore it is important to know the reclamation success. Species richness is one of the criteria for evaluating the success of restoration which is very important in ecology [13]. In ecology, diversity in general aim to species diversity so the measurement of diversity is conducted by determining the number of species and species relative abundance in a community. The species diversity in area does not only rely on the species number found in the area but also on specify of species in it, whether the species are endemic for certain area of habitat. The plant criteria such as species composition and rate of growth is monitored, this modest vegetation measurement apart from being simple to be conducted is also sensitive [14, 15]. After rehabilitation are conducted, it is necessary to have evaluation on the success of the activity, thus making it important to conduct a study to find out the biodiversity of mining areas which have been revegetated. The objectives of the study were to evaluate plant species composition after thirteen years of revegetation activities in the coal mining rehabilitation site in East Kalimantan.

2. Method

2.1. Study area

The study was conducted at a post coal mining area in the lowlands (approximately 0-100 m above sea level) of East Kutai District, East Kalimantan Province (figure 1.A).

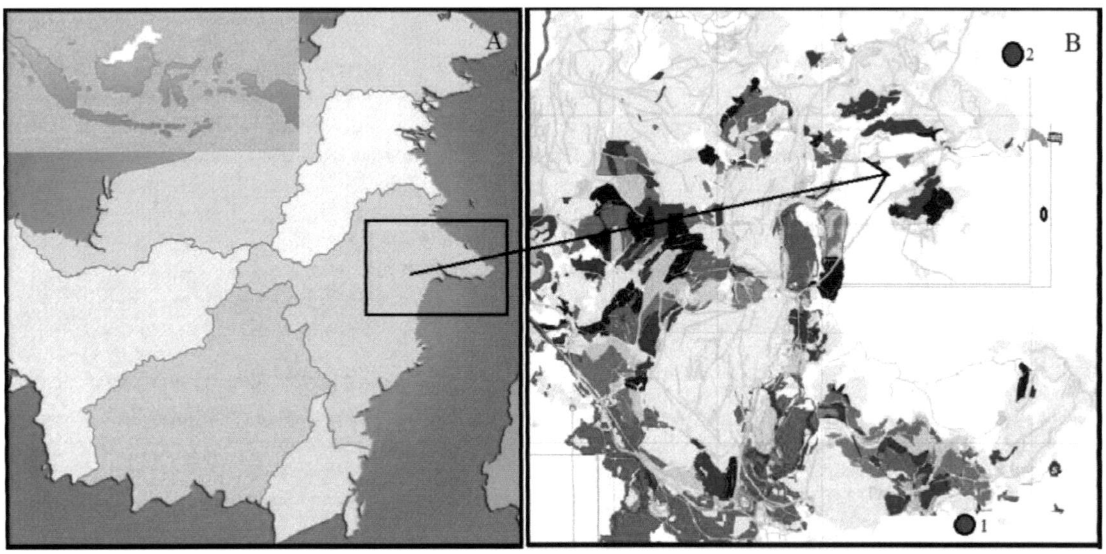

Figure 1. Location of the study area in East Kalimantan A. Blue circle 1 is the natural site and B. Blue circle 1 is the natural site and B; Blue circle 2 is rehabilitation site location.

Location of the study area in East Kalimantan at 00°33'23"-00°38'17" NL and 117°23'55"-117°23'20" EL. Borneo island. The research site was in Sangatta, East Kalimantan which has a tropical climate, with rainfall intensity characteristics varying from low to dense in a short time and can happen in a long time. average Rainfall ranges from 1543 mm per year with air temperatures minimum temperature is 26.68°C and maximum is 30.50°C, with a ±3°C daily temperature fluctuation. The average relative humidity per month between 80% - 90% with a range of 70% on afternoon and 90% in the morning. The average evaporation per month in the mine area is 4 mm/day [16].

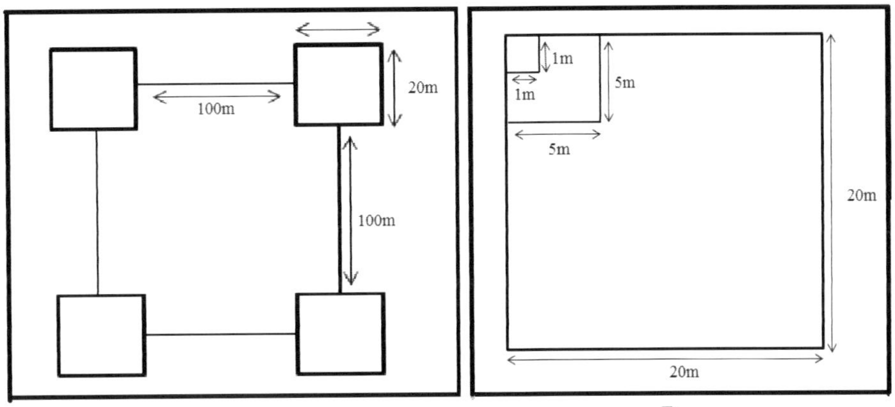

Figure 2. Plots of research in coal mining sites A .Plots in natural and rehabilitation sites, each plot measuring 20x20m^2 with 100 distance between plots; B. Plot measuring 20x20 m^2 for woody species, 5x5 m^2 for non-tree woody plants and 1x1 m^2 for herbaceous plants.

The coal mining area was previously a production forest, i.e., a forest concession area, which loaned from the Ministry of Forestry. This study was conducted in two sites: 1) a natural area and 2) a rehabilitation site which already reclamated and revegetated. Plant species composition in both the natural and rehabilitation sites was studied for a period of months (in October 2017). In both of natural and rehabilitation area, the vegetation sampling was conducted in 20 plots measuring 20x20 m^2 transects [14] with 100m^{2}distance between (fig.2). The woody species were sampled in 20x20 m^{2}plots, the non-woody plants is sampled in 5x5 m^{2}subplots and herbaceous plants were sample in 1x1 m^2 subplots. The non-tree woody plants and herbaceous plants were grouped together. The tree and herbaceous species were collected, and if there were unknown species, it identified using references in Herbarium of East Kutai School of Agriculture, East Kalimantan, Indonesia.

The data were analyzed for frequency, density and abundance [17], their relative values were calculated as follows:

$$\text{Relative density} = \frac{\text{Number of individuals of species}}{\text{Total number of inviduals}} \text{x } 100 \qquad (1)$$

$$\text{Relative dominance} = \frac{\text{Dominance of a species}}{\text{Dominance of all species}} \text{x } 100 \qquad (2)$$

$$\text{Relative frequency} = \frac{\text{Frequency of species}}{\text{Frequency of all inviduals}} \text{x } 100 \qquad (3)$$

$$\text{IVI} = \text{relative density} + \text{relative dominance} + \text{relative frequency} \qquad (4)$$

The species diversity indeces (H') were following Shannon-Wiener [7] i.e., as follows:

$$H' = -\sum_{i=1}^{S}(pi \ln pi) \qquad (5)$$

Where, H' = observed species diversity, S = the number of species; pi = the proportion of individuals or the abundance of the ith species expressed as a proportion of total cover; ln = log basen.

2.2. Data analysis
The study was conducted at a post coal mining area in the lowlands (approximately 0-100 m asl) of East Kutai District. East Kalimantan Province (figure 1).

3. Results and Discussion

3.1. Plant diversity in natural site
In the natural area we found 36 plant species, consisting of 25 woody species and non-wood plant species (tabel 1). The importance value indices of plant species in this area varied from 0.64% to 46.08%. Three species of woody plant species had importance value indices higher than 20%, specifically tree species were dominated by *Nephelium eriopetalum* (61.01%), *Macaranga hypoleuca* (22.60%) and *Croton argyratus* (21.41%) (figure 3.a). While 9 species of non-wood plants had indices higher than 15% (Figure 3), dominated by *Fordia splendidisima* (62.47%), *Saurauia umbellata* (38.21%), *Calicarpa longifolia* (37.49%), *Clidemia hirta* (27.73%) and *Homalanthus populneus* (26.95%) (figure 3.a).

3.2. Plant diversity in rehabilitation site
After 13 years of rehabilitation, we found 28 plant species which consist of 19 woody species and 8 non-wood species (tabel 1). The importance value indices of non wood plant species in this area varied from 5.29 % to 77.10%. The top 9 species with highest importance value indices were higher than 11.16% for woody species and more than 10.61% for non woody species) are presented in figure 3. Woody species were dominated by *Cassia siamea* as shown by the highest importance value index of 58.30%, followed by *Ficus uncinata* (45.49%), *Molucana falcataria* (31.52%), *Glocidion rubrum* (23.65%) and *Delonix regia* (21.26%). Non-wood species were dominated by *Nephrolepis biserata* with importance value index of 77.97%, followed by Miscanthus javanica (54.81%), Leea indica (47.97%), *Scleria purpurescens* (32.49%), *Timonius koordersii* (24.94%) and *Mesophleebion motleyanum* (21.76%).

Tabel 1. Diversity of woody species and non-wood species at natural and reclamation site (√ for present)

Woody Plant species			**Non-wood plant species**		
Plant species	**Natural**	**Rehabilitation**	**Plant species**	**Natural**	**Rehabilitation**
Actinodaphne bancana	√	×	*Calicarpa longifolia*	√	×
Alseodaphne nigrescens	√	×	*Clidemia hirta*	√	×
Arcidendron clipiera	√	×	*Curculigo latifolia*	√	×
Artocarpus campeden	×	√	*Ficus uncinata*	√	×
Barringtonia macrostachya	√	×	*Fordia splendidisima*	√	×
Beilschmiedia madang	√	×	*Homalanthus populneus*	√	×
Borasodendron boornensis	√	×	*Koilodepas brepipes*	√	×
Bridelia glauca	√	×	*Leea indica*	×	√
Cananga odorata	√	×	*Melastoma malabatrichum*	×	√
Canarium denticulatum	√	×	*M. motleyanum*	×	√
Cassia siamea	×	√	*Miscanthus javanica*	×	√
Cratoxylon sumatranum	**√**	**√**	*Nephrolepis biserata*	×	√
Croton argyratus	√	×	*Rubus molucanus*	×	√
Delonix regia	×	√	*Saurauia umbellata*	√	×

Dillenia excelsa	×	√	*Scleria purpurescens*	×	√
Dryobalanops aromatica	×	√	*Sentotecha lapaecea*	×	√
Eusideroxylon zwageri	√	×	*S. jagorianum*	√	×
Ficus uncinata	×	√	*Syzigium lineatum*	√	×
Glocidion macrophyllum	√	×	*Timonius koordersii*	×	√
Glocidion macrophyllum	×	√	*Trema canabina*	√	×
Glocidion rubrum	×	√			
Guensia petandra	√	×			
Litsea lancifolia	√	×			
Macaranga conifera	√	×			
Macaranga gigantea	√	×			
Macaranga hypoleuca	**√**	**√**			
Macaranga motleyana	√	×			
Mallotus paniculatus	×	√			
Melicope glabra	√	×			
Memecylon boornensis	√	×			
Molucana falcataria	×	√			
Nephelium eriopetalum	√	×			
Polyosma longipes	×	√			
Ptenandra echinata	√	×			
Ptenandra rostrata	×	√			
Samanea saman	×	√			
Switenia mahagoni	×	√			
Syzygium lineatum	×	√			
Vernonia arborea	√	×			
Vitex pinnata	**√**	**√**			

Table 2. Plant species diversity in natural and rehabilitation site

	Natural Site		**Rehabilitation sites**	
	Number	**H'**	**Number**	**H'**
Woody species	25	3.01	19	2.21
Non-woody species	11	1.96	9	1.48
Total Species	36		28	

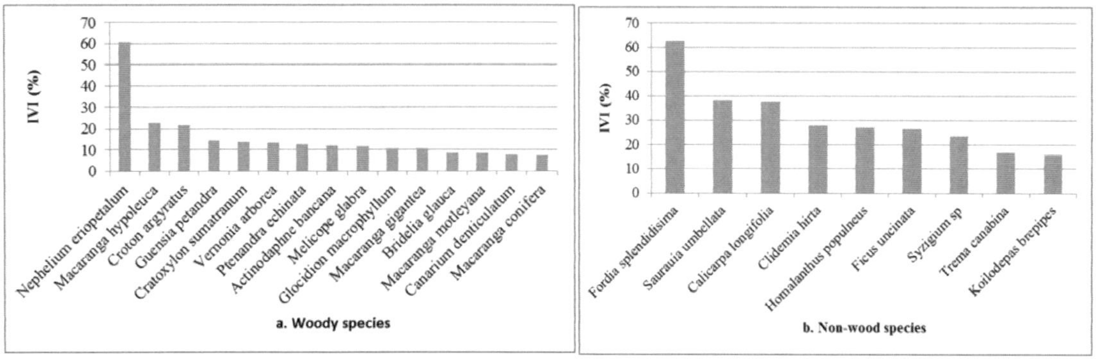

Figure 3. The importance value indices (IVI) of (a). 15 species of woody species and (b). 9 species of non-wood species in natural site

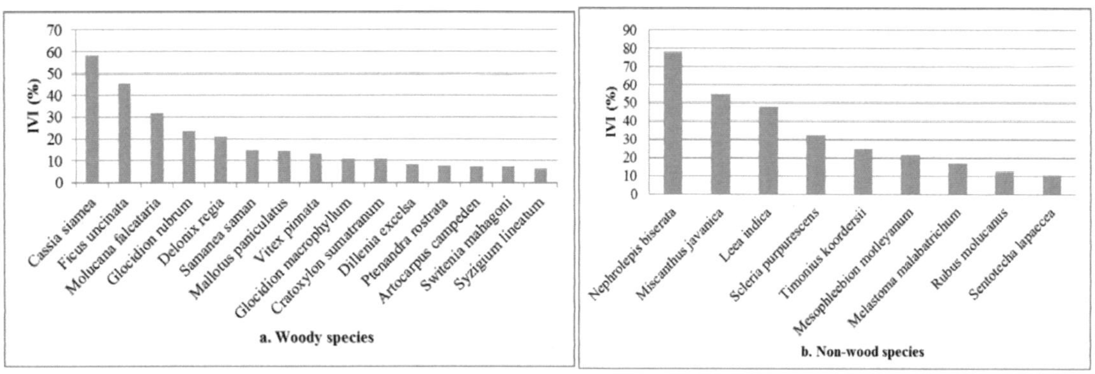

Figure 4. The importance value indices (IVI) of (a). 15 species of woody species and (b). 9 species of non-wood species in rehabilitation site

3.3. Comparison between natural and rehabilitation site

After 13 years of rehabilitation, plant diversity in the rehabilitation site was still lower than in the natural site which was used as reference. Comparison of species diversity indices (H') between the two sites indicate a significant different in woody species, however, diversity indices for non-wood plants species were similar between the two sites. The diversity index for woody species in natural area was 3.01 and in reclamation site was 2.21. Table 2 presents species richness and diversity indices in the two sites. It is interesting that between woody species of natural site and rehabilitation site is only three species the same, these are *Cratoxylon sumatranum*, *Macaranga hypoleuca* and *Vitex pinnata* (tabel 1). Between non-woody species of natural site and rehabilitation site is no same species.

3.4. Discussion

As shown by its highest importance value index, natural site were dominated rambutan hutan (*Nephelium eriopetalum*), which is a local species. It has a high value indeces perhaps because of its easy to grow. *Macaranga hypoleuca* was the second dominant species in the natural site, it is known as fast growing species. *Macaranga hypoleuca* is a sub-canopy tree growing up to 30 metres tall, it can be 30 – 40 cm in diameter. Many of *Macaranga* sp is tolerance to open areas. In the rehabilitation site, the dominant species was johar (*Cassia siamea*). However, *Cassia siamea* was quite abundant because the company planted a seedlings, which became one of the factors contributing to the importance value indices of these plant species is the high number of seedlings of local species planted in the rehabilitation site. There is only three woody species from the natural site can grows in the rehabilitation site (*Cratoxylon sumatranum* (13.72%), *Macaranga hypoleuca* (22.60) and *Vitex pinnata* (6.27)). There is three possibility first because it planted by the company, the second one is because of seeds already present in the top soil during reclamation process, and its grow when the condition is suitable and the third its because of the seed dispersal. The dominant non-wood species in natural site is *Fordia splendidisima* with the highest importance value index (62.47%) and the dominant non-wood species in the rehabilitation site was the *Nephrolepis biserata* with an importance value index of 54.81%. *Fordia splendidisima* extremely tolerant to minimum light intensity (shade plant). *Nephrolepis biserata* is tolerant to high light intensity. The non-wood species found in the rehabilitation site was not found in the natural sites (*Nephrolepis biserata* and *Miscanthus javanica*). The ability of plants to grow in the rehabilitation site can be attributed by seed dispersal. Perhaps dispersed by wind from the forest system, except *Leea Indica* which may dispersed by animal. The distance between the rehabilitation site to the nearby vegetation source and animal species and will affect the plant species dispersion e.g., the nearest forest that will affect the distribution of the species (Traveset et al. 2014).

Based on the number of plant species at rehabilitation site, to restore plant diversity in the rehabilitation site to its assumed natural conditions, it should be planted with 22 woody species. The species importance value indices can be considered a to select the woody tree species from the natural area. In the natural area, only ten local species had importance value indices higher than 10%, and 9 of them was not in the rehabilitation plant species e.g., *Nephelium eriopetalum* (61.01%), *Croton argyratus* (21.41%), *Guensia petandra* (14.59%), *Vernonia arborea* (13.35%), *Ptenandra echinata* (12.78%), *Actinodaphne bancana* (12.11%), *Melicope glabra* (11.64%), *Glocidion macrophyllum* (10.72%) and *Macaranga gigantean* (10.65%). The availability of seedlings and the ability of the plant to grow successfully can be considered in the rehabilitation program of coal mining rehabilitation condition.

4. Conclusions

After 13 years of rehabilitation, 28 plant species were found in the rehabilitation site, consisting of 19 woody species and 9 herbaceous species. In comparison, data from the natural area indicate the presence of 36 plant species, consisting of 25 local woody species and 11 non-woody species. Woody species diversity index in the rehabilitation site after 13 years post mining (i.e., 2.21) was lower than in the natural area (3.01); while the diversity indices for non-woody plants were relatively similar (1.48 and 1.96 in the rehabilitation and natural sites respectively). Although its species richness of non-wood species was low, but it has a high coverage per species in this site. To restore plant diversity in the rehabilitation site to its assumed natural conditions, it should be planted with 22 woody species.

Acknowledgements
Many thanks to PT. Kaltim Prima Coal for all the facilities during this study.

References

[1] Kennedy T S S 2008 Soil and Vegetation Change on Coal Mine 15 Years Reclamation in Aspen Parkland of Alberta Thesis. Master of Landscape Architecture degree
[2] Traveset A, Heleno R and Nogales M 2014 *Seeds: The Ecology of Regeneration in Plant Communities*. 3rd ed. (London: CAB International)
[3] Tribun kutim 2013 61 Jenis Flora Kalimantan Terancam Punah. http://kaltim.tribunnews.com
[4] Kent M and Coker P 1992 *Vegetation Description and Analysis. A practical approach* (New York: John Wiley and Sons)
[5] Soerianegara I and Indrawan 1998 *Forest Ecology of Indonesia. Forestry Faculty*. (Bogor: Institut Pertanian Bogor)
[6] Shrestha R K and Lal R 2011 Changes in physical and chemical properties of soil after surface mining and reclamation *Geoderma* **11** 268-276

[7] Spies T and Turner M 1999 The macro approach, managing forest landscapes: Dynamic forest mosaics, In: *Maintaining biodiversity in forest ecosystems*, Hunter, M.L.Jr. (Ed.)(Cambridge University Press) 665 pp. 95-160.

[8] Sheoran V and Sheoran A S 2009 Reclamation of abandoned mine Land *J Mining Metallurg* 45 **1** 13-32

[9] Dale V H and Beyeler S C 2001 Challenges in the development and use of ecological indicators *Ecol Indicat* **1** 3-10

[10] Hutchinson T F, Boerner R E J, Iverson L R, Sutherland S and Sutherland E K 1999 Landscape patterns of understory composition and richness across a cross a moisture and nitrogen mineralization gradient in Ohio USA *Quercus* forests. *Plant Ecol* **144** 177-189

[11] Cooke J A and Johnson M S 2002 Ecological restoration of land with particular reference to the mining of metals and industrial minerals: A review of theory and practice. *Environ Rev* **10** 41-71

[12] Sen S and Kumar V 2016 Evaluating soil quality an bioefficacy study of Cajanus cajan L in coal mine-degraded land *Turkish Journal of Agriculture and Forestry* **40** 499-511

[13] Regulation of State Minister for The Environment no. 4/2012 about Environment Friendly Indicators for Business and/or Activity of Open Coal Mining. [Indonesian]

[14] Hazarika P, Talukdar N C and Singh Y P 2006 Natural colonization of plant species on coal mine spoils at Tikak Colliery, Assam *International Soc Trop Ecol* 47 **1** 37-46

[15] Perrow M R and Davy A J 2002 *Handbook of Ecological Restoration, Vol. 1. Principles of Restoration*. (Cambridge, UK: Cambridge University Press)

[16] Wardana W 2008 Evaluation of soil potention and vegetation biomass in reclamation area of PT. Kaltim Prima Coal. Sengata. East Kutai. Thesis. Universitas Mulawarman, Samarinda. Indonesia.

[17] Ludwig J A, Hindley N and Barnett G 2003 Indicators for monitoring mine site rehabilitation: trends on waste-rock dumps, northern Australia *Ecol Indicat* **3**143-153.

Antagonistic evaluation of *Trichodermaviride* and *Aspergillusflavus* against wood-decay fungus *Pleurotusostreatus*

Erwin, E Iskandar, N Husien

Faculty of Forestry, Mulawarman University
Kampus Gunugn Kelua Jl. Penajam,Samarinda (75119), Kalimantan Timur, Indonesia

E-mail: erwin@fahutan.unmul.ac.id

Abstract. *Pleurotusostreatus* cultivation is often limited by the presence of antagonistic microorganisms that can also grow together on a variety of substrates that are used as growth media of *P. ostreatus*. As a preliminary study, this research aims to evaluate the antagonism of*Trichodermaviride* and *Aspergillusflavus* in interacting with the *P.ostreatus* fungus by in-vitro. An antagonism test using a dual culture method of these fungi with 1 month incubation time on Poteto Dextrose Agar (PDA) media was performed. Evaluation of the antagonistic properties of these fungi is conducted through macroscopic and microscopic observations. Macroscopically, the growth of *T. viride* and *A. flavus* mycelium showed that both colonies were more dominant than *P. ostreatus* colonies. Microscopically, hyphae of *T. viride* and *A. flavus* indicate the presence of very strong micoparasitic properties that characterized by attachment and convolution of both hyphae to *P. ostreatus* hyphae. It was concluded thatthe growth of *T. viride* and *A. flavus* fungi was more dominant and had potential as an inhibiting agent of *P. ostreatus* fungal growth, respectively, in dual culture on PDA media.

1. Introduction

Pleurotusostreatus, has long been recognized as one of the most widely used wood fungi (mushroom) for food because of its high levels of protein, minerals and vitamins [1]. Naturally, this fungus is not only grows on dead wood but also easily cultivated on various agricultural industry and residual material substrates, and other agriculture-cellulosic materials [2] because they produce oxidative enzymes [laccase and manganese peroxidase (MnP)] and hydrolytic (cellulase, xylanase and tannase) [3].The use of different substrates is able to effect on biological and economic yield of *P. ostreatus* (oyster mushroom) cultivation [4].

However, it is common to find problems in the cultivation of the *P. ostreatus* mushroom, one of which is the existence of contamination on the growth substrate media which regularly found, so that variations of heat treatment are required to eliminate the growth of other fungi such as *Trichoderma* sp., *Altemaria* sp., *Aspergillus* sp., *Fusarium*sp ., *Monilia* sp., *Mucor* sp., *Rhizopus* sp. etc.[5]. The growth of these other fungi as a competitor often leads to impaired growth of *P. ostreatus* so that it can reduce the production of its cultivation.

In the same habitat, it is well known that mushrooms have mutually supportive relationships and inhibit the growth of one another in obtaining nutritional sources for its survival. Schmidt[6] notes that the inhibiting growth relationship or better known as the antagonism (competitive reciprocal effect) is based on production of toxic metabolites,on mycoparasitism, and on nutrient competition.

In the effort of handling *P. ostreatus* growth disorders, the initial information required on how the characteristics of interdependent or antagonistic relationship between *P. ostreatus* with competitors in a substrate. Therefore, antagonistic studies of fungi were performed using two species of antagonistic fungi, *Trichodermaviride* and *Aspergillusflavus* to find out the interaction pattern of each of these fungi in inhibiting the growth of *P. ostreatus* fungus by in-vitro.

2. Materials and methods

2.1. *Fungal isolates and culture media*

The fungal isolates used were *Trichodermaviride*, *Aspergillusflavus* and *Pleurotusostreatus* stored in the collecting of the Forest Protection Laboratory of Faculty of Forestry, Mulawarman University, Samarinda.

Potato Dextrose Agar (PDA) material is made aseptically in laboratory following the procedure ofHiMedia[7]. PDA is used as a culture media for fungal growth of *P. ostreatus*, and antagonistic test as well.

2.2. *Antagonistic test*

An antagonistic test using a dual culture method of Bruce and Highley[8] on PDA media was performed on *T. viride* and *A. Flavus* against *P. ostreatus*, respectively, with an incubation period of 1 month.

Materials and equipment used in the antagonistic test were conducted aseptically in Laboratory of Wood Biology, Faculty of Forestry, Mulawarman University, Samarinda.

2.3. *Observation*s

Interaction of two fungal colonies in vitrodual culture on PDA media wasexaminedon macro-and microscopic observations.

*2.3.1. Macroscopic observation.*In macroscopic observation,NIKON SMZ645 Stereoscopic Microscope at 10-30x magnification was used to evaluate mycelium growth and the contact limit between the two fungal colonies growing on the PDA media plates.

2.3.2. Microscopic observation. By using NIKON Eclipse E400 Light Microscope at 40-400x magnification, microscopic observation was made on the growth and interaction of the two fungal hyphae growing on the PDA media plates.

3. Results and discussion

3.1. *Growth inhibition of Pleurotusostreatus by Trichodermaviride*

Macroscopically, in the dual culture method, the growth of both fungi in one Petri dish at 1-2 weeks incubation showed *Tricodermaviride* colony with mycelium and greenish spores more dominant than *Pleurotusostreatus* colony with whitish mycelium (Figure 1a).

Since the beginning contact, *T. viride* colony dominates almost thoroughly the surface area of PDA media. On the contrary, the growth of *P. ostreatus* mycelium with very limited growth area,cover the surface of PDA media, even tend to grow in the opposite direction from the contact zone. In the contact zone, there is no clear boundary separating between the two fungal colonies (Figure 1b).

The dominance of *T. viride* colonies is not only from the fungal growth fully covering the surface of the PDA media but also the fungus grows on top of the *P. ostreatus* mycelium (Figure 1). This condition shows *T. viride*plays a role micoparasite[9] by utilizing the mycelium of *P. ostreatus* as a source of nutrients other than obtained from the media. It is certain that with the dominant colonization, *T. viride* is antagonistic fungus to *P. ostreatus*.

The interaction stage of both colonies can also be shown by microscopic observations (Figure 2) in which appears to be attachment and convolution of *T. viride* hyphae on the *P. ostreatus* hyphae. The attachment is an early stage of theconvolution process, which, according toKlein Dand Eveleigh[10], is a frequent response between the *Trichoderma* spp. with other fungi in case of contact.

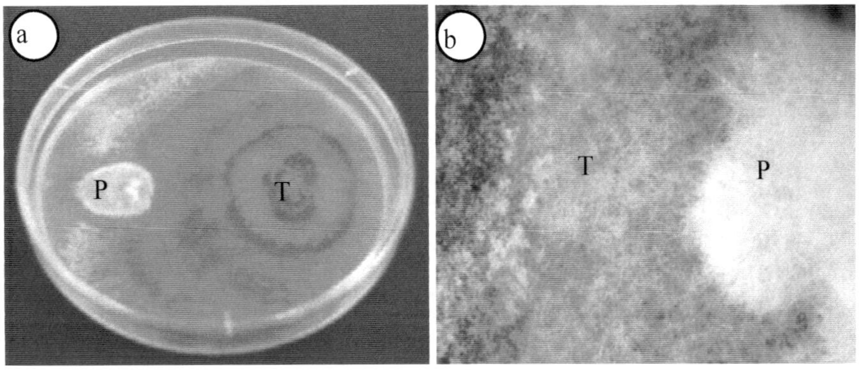

Figure 1. The dominant growth of the *Tricodermaviride* colony(T) against *Pleurotusostreatus* (P) on PDA dual culture media: (a) The overall surface of the media in Petri dishes; (b) At the contact boundary of the colonies' encounter [28 x magnifications].

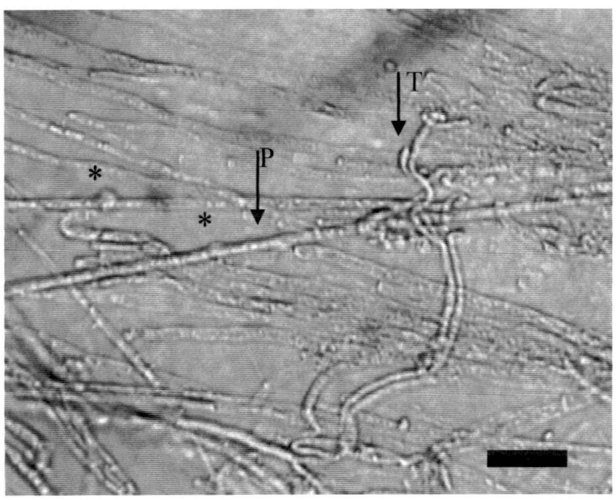

Figure 2. Attachment and convolution of*Tricodermaviride*hyphae (T) on*Pleurotusostreatus* hyphae (P) on dual culture of PDA media. Sign * denotes clamp connection of *P. ostreatus*hyphae, which is known as main characteristic of basidiomycetefungi [Bar = 20 μm].

Indication of the micoparasite process of *T. viride* as an antagonist fungus, according to Ranasingh *et al*[9], itis usually followed by releasing process of various enzymes such as chitinase, glucanase and pectinase.

Basically, the antagonistic properties are due to the ability of *T. viride* to produce extracellular enzymes capable of metabolizeon polycyclic aromatic hydrocarbons (PAHs) such as naphthalene, phenanthrene, chrysene, pyrene, and benzo[a]pyrene [11], which is indicate his dominating ability and even the destructive structure of the *P. ostreatus* mycelium. Such event was also observed in dominating of other Tricoderma species, i.e. *T. citrinoviride,T. harzianum and T. Atroviride* associated with the *Camptomyia* pestagainst the mycelial growth of ashiitake mushroom, *Lentinulaedodes* [12].

3.2. Growth Inhibition of Pleurotusostreatus *by* Aspergillusflavus

The *Aspergillusflavus* mycelial colony also showed growth dominance fulfilling the PDA media against *Pleurotusostreatus* (Figure 3a). The condition also shows that the *P. ostreatus* mycelium could not develop as it could grow in mono-culture media.

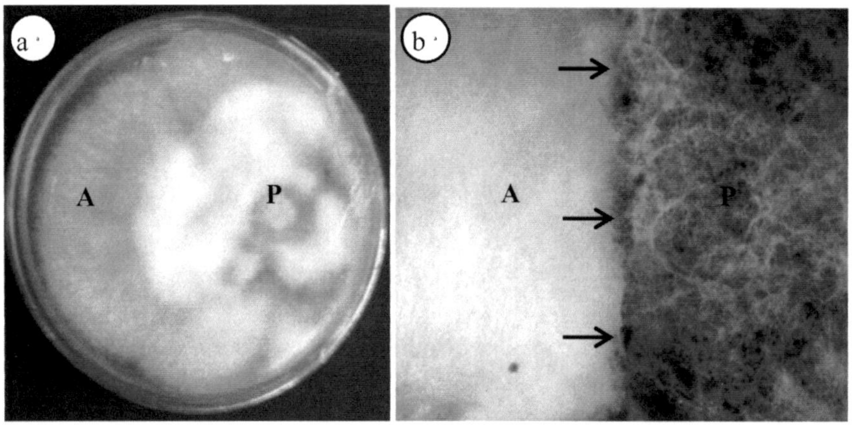

Figure 3.*Aspergillusflavus* colony (A) Dominates the surface of the media and covers the mycelium of *Pleurotusostreatus* (P) on PDA dual culture media method: (a) The overall surface of the media in Petri dishes; (b) At the contact boundary of the colonies (arrows) [28 x magnifications].

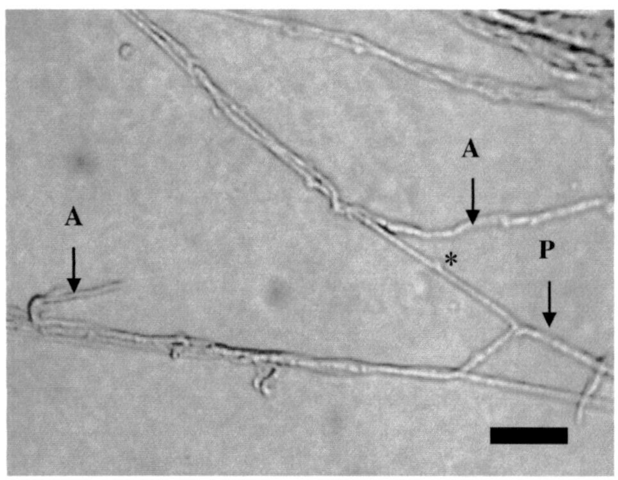

Figure 4. Attachment and convolution of *Aspergillusflavus* hyphae (A) to the *Pleurotusostreatus* hyphae (P) on dual culture of PDA media Sign * denotes clamp connection of *P. ostreatus* hyphae, which is known as main characteristic of basidiomycete fungi [Bar = 20 μm].

Covering of PDA media by *A. flavus* mycelium was also characterized by presence of a limiting zone that forms like a long line span on the second encounter of mycelium (Figure 3b). The properties of *A. flavus*micoparasite on dual-culture media were demonstrated by exploiting *P. ostreatus*mycelium as a source of nutrients beside those that obtained from PDA. This could be indicative that for the fact that *A. flavus* has in common with *T. viride*, which acts as an antagonist fungus against *P. ostreatus*.

Microscopically, the antagonistic featuresare seen that *A. flavus* hyphae also attached and convoluted *P. ostreatus* hyphae (Figure 4). The antagonistic ability of *A. flavus*hyphae is related to the extracellular xylanase enzymes that released [13], which may play a role as a metabolic toxin [5] and also has the greatest cellulolytic ability [14] against *P. ostreatus* hyphaeso that its growth in PDA media becomes inhibited. Meanwhile, aflatoxinB_1 (AFB_1)as a secondary metabolite that produced by several members of Aspergillus fungi could be degraded by ligninolytic enzymes of *P. ostreatus*strains [15,16].

4. Conclusions

Based on results of this study, we conclude that*Trichodermaviride* and *Apergillusflavus* mycelial colonies dominate the *Pleurotusostreatus* mycelia colony in utilizing nutrient in dual culture of PDA media. Both of the antagonistic fungi are very powerful microparasite against *P. ostreatus* hyphae, thereforethe preventionand inhibition oftheirgrowth in media of *P. ostreatus* cultivation must be undertaken early.

Acknowledgements

The authors are grateful to Dr. Djumali Mardji of the Forest Protection Laboratory, Faculty of Forestry, Mulawarman University for valuable suggestions in microscopic analysis of fungi.

References

[1] Wahid M, Sattar A and Khan S 1988 Composition of wild and cultivated mushrooms of Pakistan *Mushroom in Trop.***8** 47
[2] Sa'nchez C 2010 Cultivation of *Pleurotusostreatus* and other edible mushrooms *Appl. Microbiol. Biotechnol.***85** 1321
[3] Luz J M R, Nunes M D, Paes S A, Torres D P, Silva M C S and Kasuy M C M 2012 Lignocellulolytic enzyme production of *Pleurotusostreatus* growth in agroindustrial wastes *Braz.J.Microbiol.*1508
[4] Girmay Z, Gorems W, Birhanu Gand Zewdie S 2016 Growth and yield performance of *Pleurotusostreatus*(Jacq. Fr.) Kumm (oyster mushroom) on different substrates *AMB Expr.* **6** 87
[5] Yildiz S, ü. Cafer, Derya-Gezer E and Temiz A 2002 Some lignocellulosic wastes used as raw material in cultivation of the *Pleurotus ostreatus*culture mushroom *Process Biochem.***38** 301
[6] Schmidt O 2006 *Wood and Tree Fungi:Biology, Damage, Protection, and Use*(Berlin: Springer-Verlag) p 334
[7] HiMedia 2016 *Potato Dextrose Agar M096* (Mumbai: HiMedia Laboratories Pvt. Ltd) p 2
[8] Bruce A and Highley T L 1991 Control of growth of wood decay basidiomycetes by *Trichoderma* spp. and other potentially antagonistic fungi *Forest. Prod. J***41** 63
[9] Ranasingh,Subhat A and NeduchezhiyanM2006 Use of trichodermain disease management. *Orissa Rev.* 68
[10] Klein Dand EveleighD E1998Ecology of trichoderma*Trichoderma and Gliocladium* (*Basic Biology, Taxonomy and Genetics* vol 1) ed CP Kubicek and GE Harman (London: Taylor & Francis Ltd) chapter 3 pp 57-74
[11] Zafra G and Espinosa DVC 2015 Biodegradation of polycyclic aromatic hydrocarbons by Trichoderma species: a mini review *Environ. Sci. Pollut. Res.* **22** 1942
[12] Kim J Y, Kwon H W, Yun Y H and Kim S H 2016 Identification and characterization of Trichoderma species damaging shiitake mushroom bed-logs infested by Camptomyia pest*J. Microbiol. Biotechnol.***26** 909
[13] Bhushan B, Pal A, Kumar S and Jain V 2015 Biochemical characterization and kinetic

comparison of encapsulated haze removing acidophilic xylanase with partially purified free xylanase isolated from*Aspergillus flavus*MTCC 9390 *J.Food.Sci. Technol.***52** 191
[14] Handayani T and Purwantisari S 2015 Isolation and identification of mold contaminants on mushroom growing medium (bag log) and their cellulolytic performance test *J. Sains dan Matematika***23** 55
[15] Das A, Sourav B, Muthusamy P and Jayaraman A 2015Aflatoxin B1 degradation during co-cultivation of *Aspergillusflavus* and *Pleurotusostreatus* strains on rice straw*3 Biotech.***5**279
[16] Jackson L W and Pryor B M 2017 Degradation of aflatoxin B_1 from naturally contaminated maize using the edible fungus *Pleurotus ostreatus AMB Expr* **7** 110

Diversity of tree vegetation on different slopes in Sangkulirang – Mangkalihat exokarst area

R A Suwasono*, P Matius, Sutedjo

Program Study of Forestry Science, Faculty of Forestry, Mulawarman University, Kampus Gn. Kelua, Samarinda 75123, Indonesia

*Coresponding Author: raharjo.nara.luki@gmail.com

Abstract. The Karst ecosystem in East Kalimantan is predominantly located in the Sangkulirang-Mangkalihat covering an area of 1,867,676 hectares. The exokarst are all features that may be found on a surface karst landscape. The objective of this study was to determine the diversity of tree vegetation (diameters >10 cm) on different slopes. Six study locations were selected as replications where each location consisted of the different of slopes. The sample plot was set up 15 plots in each location on quadrants of 10 m x 10 m. 538 individuals had been found in the study sites consisting of 163 species, 100 genera and 43 family. The Dipterocarpaceae was dominant on slopes and the upper ridges, while *Shorea* sp. has dominated on the upper ridges. The highest diversity index (H') of 4.04were found on the slopes and valley while the Species Richness Index (R) and Evenness Index (e) were high in all three slopes. The highest Similarity Index (ISs) of41.06was in the slopes and valley, while the highest Decimilarity Index (ID) of 67.30were in the slopes and upper ridges. Meanwhile, the overall diversity of species in the Sangkulirang-Mangkalihat exokarst area is high.

1. Introducing

Karst ecosystem is one of the unique ecosystems because it has underground river flows and caves.Formed in the past of calcium carbonate ($CaCO_3$), which is secreted by coral animals brachiopods and lilies krinoidea on ancient coral reefs under the sea in prehistoric times.When the tectonic movement on the Sunda Dangkalan created the archipelago of Nusantara, these ancient coral reefs along with other sedimentary rocks were pressed and lifted up and became a large part of Borneo Island [1].

Karst in general terms of the Decree of the Minister of Energy and Mineral Resources No. 17 year 2012 on the Designation of the Karst Landscape Area is a landscape formed by the dissolution of water on limestone and / or dolomite.The notion of karst landscape area is karst which shows certain exokarts and endokarts. Exokarst is karst on the surface while endokarst is karst on the lower surface.

Lime forest is a forest grown on limestone rocks containing calcium / calcite carbonate that is easily dissolved by rainwater,which causes the formation of cracks and tunnels that resemble reliefs, thus forming a typical morphology [2]. This area is a producer for millions liters of water for the various lives of flora, fauna and humans [3].

East Kalimantan Governor Regulation No. 67 Year 2012 stated that the karst ecosystem in East Kalimantan largely contained Sangkulirang Peninsula, extending to the Tanjung Mangkalihat with an area of 1,867,676 hectares. Administratively, this area is included in two districts, Berau and East Kutai.

This area is the headwaters of five major rivers namely Sungai Bengalon, Authorship, Tabalar, Lesan and Pesab [3]. In addition, this area also stores a variety of potential, among others: water underground rivers, caves, lakes, biodiversity, and minerals [4]. Furthermore this Karst ecosystem has at least three important values in human life [3].First, the scientific value is related to earth science, lithology, geological and mineral structures, fossil sites, archeology and paleontology, as well as shelter for endemic flora and fauna. Second, the socio-cultural value, which includes the spiritual aspect especially concerning the existence of caves associated with ritual interests that are also valued aesthetics, recreation and development of science. The third value is the high economic value because it becomes the source of underground river water, swallow producers, tourism and cement materials.

One of the government's efforts in protecting and preserving Karst Sangkulirang Mangkalihat area is by the issuance of Regulation of the Minister of Energy and Mineral Resources (ESDM) which is then followed up by the Provincial Government of East Kalimantan by issuing East Kalimantan Governor Regulation No. 67 of 2012 on Protection and Management of Karst Sangkulirang-Mangkalihat in Berau and East Kutai.

Research have been done by many research institutes from within and abroad, but more to explore the potential of prehistoric culture contained therein. Research on biodiversity potency especially flora has also been done one of them by Lembaga Ilmu Pengetahuan Indonesia (LIPI) and The Nature Conservancy (TNC) in 2004 in Merabu Village area. However, because the Karst Sangkulirang-Mangkalihat is area is very wide, not all locations can be explored its potential.

This study aims to determine the diversity of tree-level vegetation in limestone forest communities, especially inSangkulirang-Mangkalihat exocarst area.

2. Method

This research was conducted in Sangkulirang – Mangkalihat Exokarst Area which is included in the administrative area of Berau and East Kutai Regency.Sampling was conducted at 6 points of observation as repetition, covering Suaran, Ulu Bias, Lobang Kelatak and Biduk-biduk – TelukSulaiman in Berau and Tondoyan and GunungGergaji in East Kutai. At each location were made 15 observation plots divided by slopes i.e. valley, slope and upper ridge.

To determine the study site and location purposive sampling method was used, and then continued with making plot of sample sized 10 m x 10 m.The sketch of the research plot can be seen in **Figure 1**.

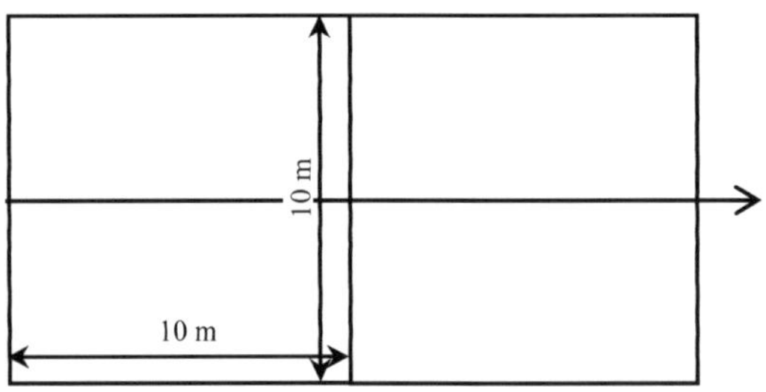

Figure 1. Transect / Path Method Design

Measurements and data collection for tree level (diameter>10 cm) were conducted for names, individual number, stem diameter and total height, chest height (1.3 m above ground) or 20 cm above the banir using phiband. Clinometer was used to measure the total height and 4 meter auxiliary rod.

Plant materials from the unknown species were taken in accordance with the procedure of sampling to make herbarium. Herbarium identification was done in Dendrology and Forest Ecology Laboratory, Faculty of Forestry, Mulawarman University.

The collected data were then analyzed quantitatively usingthe calculation of NPJ (Important Type Value), Richness Index (R) Margalef, Diversity Index (H') Shannon and Wiener, Evenness Index (e) Pielou, Similarity Index (ISs) and Decimilarity Index (ID) Sorensen.

3. Results and Discussion

After two month explorations, datahas been collected for the potential of biological richness in the form of tree level vegetation. Sampling was done on 6 points of observation i.e. Suaran, Tondoyan and Gergaji in the first month and Biatan Ulu, Lobang Kelatak and Biduk-biduk – Teluk Sulaiman in the second month (Figure 2). The results of this study described the condition of limestone forest area especially exokarst area and more specifically represented the habitat of "lowland limestone forest" Sangkulirang - Mangkalihat.

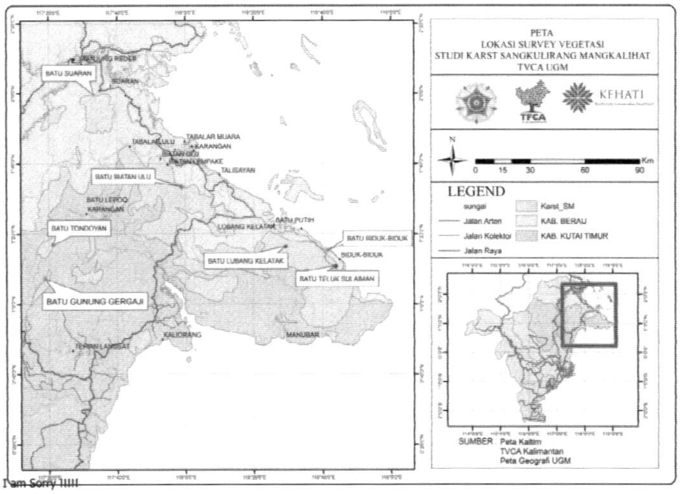

Figure 2. Map of Vegetation Survey Karst Sangkulirang MangkalihatStudy

3.1. Floristic Composition

Results of vegetation data collection in 90 plotsscattered in 6 research sites recorded as many as 538 individual tree-level vegetation belonging to 163 species, 100 genera and 43 families. In the Upper Ridge area it was recorded that the most widely vegetationfound, as many as 228 individuals, was belonging to 82 species, 59 genera and 43 families in 30 plots of samples.In the slope region, there were 155 individuals belonged to 74 species, 50 genera and 24 family (**Table 1.**).

Table 1. Individual, species, genera and family number in each slopes

	Upper ridge	Slope	Valley
Individual number	228	155	155
Species number	82	77	74
Genera number	59	53	50
Family number	31	28	24

This condition indicates that in 3 research sites have fewer species compared to similar research ever conducted by [5], which reported that in Karst Merabu area it was found 247 species belonging to 57 families. This is reasonable, however, because the research was conducted at all growth rates and more focused in an area with broader coverage of the sampling area. However, if it wasviewed as a whole at six locations with an area of 0.9 ha, it potentially have higher species richness than the extent of 4 ha in Merabu.

The main familypossessing the highest NPS in Upper Ridge is Dipterocarpaceae family with NPS of 73.51%,whiletwo main families in Slope are Meliaceae and Dipterocarpaceae with NPS of 37.93%

and 36.88% respectively. Invalley, the Dipterocarpaceaehas the highest value with NPS of 41.61% (**Table 2**).

Table 2. Top 10 highest NPS in every slopes

Upper Ridge	NPS (%)	Slope	NPS (%)	Valley	NPS (%)
Dipterocarpaceae	73.51	Meliaceae	37.93	Dipterocarpaceae	41.61
Malvaceae	29.63	Dipterocarpaceae	36.88	Meliaceae	35.05
Myrtaceae	28.00	Ebenaceae	31.98	Lauraceae	28.12
Meliaceae	26.67	Euphorbiaceae	29.58	Euphorbiaceae	27.12
Sapotaceae	17.41	Annonaceae	20.11	Annonaceae	26.57
Calophyllaceae	15.78	Lauraceae	16.00	Ebenaceae	22.56
Euphorbiaceae	12.12	Moraceae	15.01	Phyllanthaceae	16.34
Ebenaceae	8.74	Phyllanthaceae	14.30	Sapotaceae	16.34
Lamiaceae	8.54	Myristicaceae	12.87	Lythraceae	12.24
Actinidiaceae	7.05	Leguminosae	7.92	Lecythidaceae	11.98

*Shorea*sp. species predominated in Upper Ridge with NPJ of 46.49%, while in slope was dominated by *Cephalomappamalloticarpa*J.J.Sm. with NPJ of 23.48% and in Valley, the presence of tree level vegetation that has the highest NPJ is *Aglaia* sp.with NPJ of 15.83% (Tabel 3).

Table 3. Top 10 highestNPJ in every slopes

Upper Ridge	NPJ (%)	Slope	NPJ (%)	Valley	NPJ (%)
*Shorea*sp.	46.49	*Cephalomappamalloticarpa*J.J.Sm.	23.48	*Aglaia* sp.	15.83
Microcosfibrocarpa(Mast.) Burret	26.33	*Shorea*sp.	17.84	*Shorea*sp.	13.83
*Vaticarassak*Blume	16.78	*Horsfieldia*sp.	11.02	*Duabanga moluccana* Blume	12.17
*Calophyllum lanigerum*Miq.	13.08	*Diospyrosborneensis*Hiern.	10.84	*Macarangabancana*(Miq.) Müll.Arg.	11.96
Vaticaodorata (Griff.) Symington	10.43	*Shoreamujongensis*P.S.Ashton	10.69	*Dipterocarpustempehes*Sloot.	10.75
Syzygiumtawahense (Korth.) Merr. &L.M.Perry	9.57	*Diospyros*sp.	10.66	*Eusideroxylonzwageri*Teijsm. &Binn.	10.32
*Syzygium*sp.	9.30	*Aglaia* sp.	8.83	*Planchoniavalida* (Blume) Blume	8.83
Aglaia sp.	8.16	*Duabanga moluccana*Blume	7.76	*Shoreaguiso*(Blanco) Bl.	8.83
*Callicarpapentandra*Roxb.	7.93	*Polyalthiarumphii* (Bl.) Merr.	6.91	*Cephalomappamalloticarpa*J.J.Sm.	8.69
Palaquium hexandrum(Griff.) Baill.	7.12	*Prainea limpato* (Miq.) Beumee ex K.Heyne	6.56	Sp*5*.	7.73

The dominance of the presence of the Dipterocarpaceae family is accordance with what has been written by MacKinnon, that under 800 m in the not so steep slopes, the forests are higher and mainly consist of Dipterocarpaceae[1]. The survey results in the Karst Merabu area were also reported that the Dipterocarpaceae family was dominant in the area [5].

In addition to important values, also it was also performed calculations to determine the value of other indexes. The calculation of some index values were presented in **Table 4**.

Table 4. Diversity Index (H'), Richness Index (R) andEvenness Index (e) in every slopes

	Upper ridge	Slope	Valley
Richness Index (R)	14.92	15.07	14.47
Diversity Index (H')	3.80	4.04	4.04
Evenness Index (e)	0.86	0.93	0.94

R values on upper ridge, slope and valley were 14.92, 15.07 and 14.47 respectively. Based on the classification of R values made by Magurran, the three slopes has a wealth of species because it has a value of R>5.0 (Table 4).

Slope and valley were known to have a higher H'value rather than in upper ridge with H' value of 4.04, but when it was viewed from its value at all sites it has a high diversity with an H' value above 3.

The evenness index for all locations was classified as *Almost Equal*which had e value on upper ridge, slope and valley is 0.86, 0.93 and 0.94 respectively. This shows that all tree species in all locations have nearly equal numbers of individuals, or no species that have a very dominant number of individuals.

[6]mentioned that slope has low diversity due to soil leaching and faster dry land conditions, and lack of appropriate vegetation.

3.2. *Forest Structure*

In 90 plots with sized 10 m x 10 m or an area of 0.9 ha was recorded as 538 individual tree levels with diameter>10 cm.

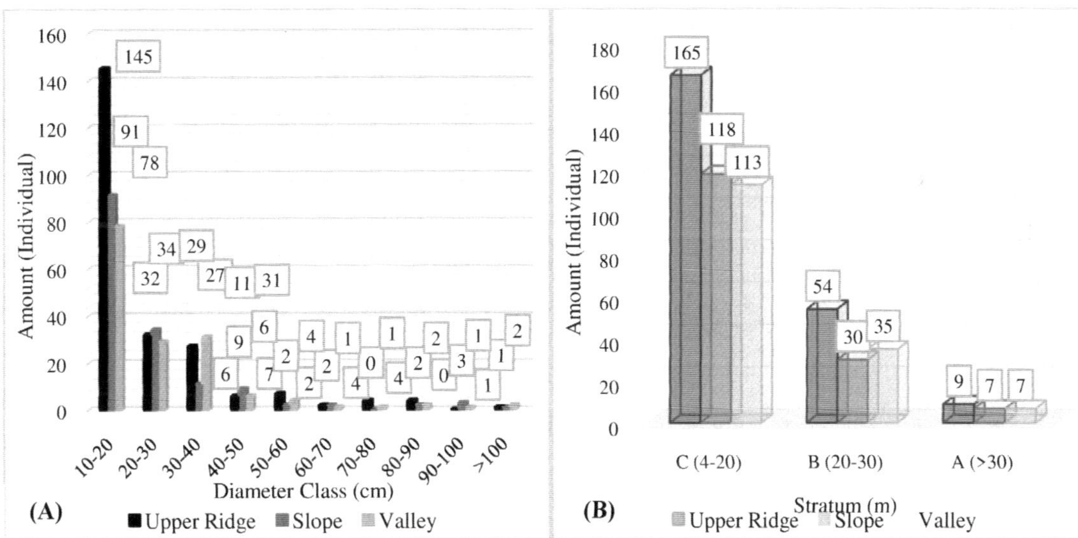

Figure 3. Distribution of class diameter **(A)** and stratum tree on each slope**(B)**

All locations, upper ridge, slope and valley,were dominated by the presence of small trees with a diameter of 10-30 cm, but there was, in few number, still found trees with a diameter of over 100 cm.Species with stem diameter more than 100 cm was *Shorea*sp. in upper ridge, *Horsfieldia*sp. in slope and in the valley it was found the species of *Duabangamoluccana* Blume and *Shorea*sp.

The distribution of individual trees composition in each grade of stem diameter in the slopes is presented in Figure 3 (A). Based on tree height and grouping based on stratum, there are 3 stratumsin all locations. All three sites were both dominated by trees with a height between 4-20 m and belonging to the stratum C(Figure 3 B).

If the points on the diagram above are connected with the line,thus is forming a curve and it will look like a curve to form of letter J but upside down. This illustrates that at the beginning of the growth, the amount of vegetation is quite a lot. But over time, these individuals experienced growth so that there is competition and natural selection resulting in a kind or different typesamong individuals. Naturally, this competition leads to a reduction in the number of individuals who survive at each grade level [7]. The distribution of such diameter class showing the potential of regeneration is quite good, because it has a diameter pattern the smaller number of individuals and more and more [8].

3.3. *Simiarity Index (ISs) and Decimilarity Index (ID)*

The Similarity Index (ISs) represents the similarity of the composition with comparable type. The highest type of similarity between slope and valley with ISs value is 41.06%, then between upper ridge and valley ISs is 38.46%.The lowest equation between upper ridge and slope with ISs value is only 32.70%. While the Decimilarity Index or Inequality Index is the opposite of the similarity index (**Table 5**).

Table 5. Similarity Index (ISs) andDecimilarity Index (ID) among location.

Level	IndeksSimilaritas (ISs)(%)		
	Upper Ridge	Slope	Valley
Upper Ridge		32.70	38.46
Slope	67.30		41.06
Valley	61.54	58.94	
DecimilarityIndex(ID)(%)			

There is no ISs value more than 50%,therefore it can be interpret that the similarity of types among sites is low and illustrates that each location has different characteristics (Table 5). [9]suggest that species composition among different locations is suspected because the influence of abiotic factors such as differential light level, nutrient availability, availability of water, wind and temperature. [10]also argued that the environmental factor that is very influential to the distribution of tree species is the height.

4. Conclusion

From the descriptionabove, it can be concluded that the Dipterocarpaceae is the main family of the natural forest of exokarst with the distribution of diameter and height of the multilevel tree and form an inverted J curve. All study sites have medium and low category of INPs, which means there is no dominant species at the study site. All locations have high Diversity Index (H ') and the Evenness Index (e) is almost evenly distributed as well as the type of similarity is low with less than 50% ISs, indicating that each location has its own uniqueness.

5. Acknowledgement

Thank you to TFCA, Kehati Foundation and UGM Karst Study Center which supports the implementation of this research. Mrs. Rita Diana and Fenny Putri M. S. who helped a lot on writing.

References

[1] MacKinnon K, Hatta G, Hakimah H, Mangalik A and Kartikasari S N 2000 *Ekologi Kalimantan* (Jakarta: Prenhallindo) p 273
[2] Achmad A2011 *Rahasia Ekosistem Hutan Bukit Kapur* (Surabaya: Brilian Internasional) p 38
[3] Suprihanto 2014 *Laporan Akhir Studi Sosial Ekonomi Masyarakat Di Sekitar Kawasan Karst Kabupaten Kutai Timur* (Sangatta: Bappeda Kabupaten Kutai timur) p 3,6
[4] Tyas N T, Vitdiawati R, Nusantari R 2016 konservasi dan pemanfaatan berkelanjutan kawasan karst gunung sewu sebagai bagian geopark untuk mempertahankan fungsi ekologi *Prosiding*

Symbion (Symposium on Biology Education) Prodi PendidikanBiologi FKIP Universitas Ahmad Dahlan **27 Agustus 2016** 316

[5] Sasmirul A, Sudiyanto, Jasari, Purnomo, Setiawan P and Haryono E 2013 *Laporan Biodiversity Kawasan Karst Merabu Kecamatan Kelay Kabupaten Berau* (Berau: The Nature Conservancy, Balai Lingkungan Hidup Kabupaten, Balai Konservasi Sumberdaya Alam Propinsi, Lekmapuri) p 41

[6] Karami R, Mehrabi H R and Ariapoor A 2015 the effect of altitude and slope in the species diversity of herbaceous plants (case study: watershed miandarqarootaggilangharb) *Journal of Applied Environmental and Biological Sciences* **5(7)** 201

[7] Abdurachman 2008 struktur tegakan pada hutan alam bekas tebangan *Info Teknis Dipterokarpa* **2(1)** 59

[8] Sunarti S 2011 penelitian vegetasi hutan di wilayah cadas gantung dan cadas pagar, cagar alam gunung tukung gede, serang *Berkala Penelitian Hayati* **Edisi Khusus: 5A** 109

[9] Sutomo, Fardila D 2013 floristic composition of groundcover after the 2010 pyroclastic fire on mount merapi *Jurnal Manajemen Hutan Tropika* **XIX(2)** 88

[10] Kurniawan A and Parikesit2008 persebaran jenis pohon di sepanjang factor lingkungan di cagar alam pananjung pangandaran, Jawa Barat *Biodiversitas* **9(4)** 278

IOP Conf. Series: Earth and Environmental Science **144** (2018) 012060 doi:10.1088/1755-1315/144/1/012060

Correlation between soil physicochemical properties and vegetation parameters in secondary tropical forest in Sabal, Sarawak, Malaysia

K Karyati[1], I B Ipor[2], I Jusoh[2] and M E Wasli[2]

[1*]Faculty of Forestry, University of Mulawarman, Kampus Gunung Kelua, Samarinda, East Kalimantan, 75119, Indonesia.
[2]Faculty of Resource Science and Technology, Universiti Malaysia Sarawak, 94300, Kota Samarahan, Sarawak, Malaysia.

[*]Corresponding Author : karyati@fahutan.unmul.ac.id

Abstract. The tree growth is influenced by soil morphological and physicochemical properties in the site. The purpose of this study was to describe correlation between soil properties under various stage secondary forests and vegetation parameters, such as floristic structure parameters and floristic diversity indices. The vegetation surveys were conducted in 5, 10, and 20 years old at secondary tropical forests in Sarawak, Malaysia. Nine sub plots sized 20 m × 20 m were established within each study site. The Pearson analysis showed that soil physicochemical properties were significantly correlated to floristic structure parameters and floristic diversity indices. The result of PCA clarified the correlation among most important soil properties, floristic structure parameters, and floristic diversity indices. The PC1 represented cation retention capacity and soil texture which were little affected by the fallow age and its also were correlated by floristic structure and diversity. The PC2 was linked to the levels of soil acidity. This property reflected the remnant effects of ash addition and fallow duration, and the significant correlation were showed among pH (H_2O), floristic structure and diversity. The PC3 represented the soil compactness. The soil hardness could be influenced by fallow period and it was also correlated by floristic structure.

1. Introduction

Swidden fallow secondary forests provide rotating habitats for succession species in a primary forest matrix thus, enhancing biodiversity. Due to the rapid re-growth in secondary forest ecosystem, the watershed and soil properties of this primary-secondary forest landscape are ecologically at par with each other [1]. The diameter increment of dominant tree species determine growth rate in the secondary forest [2]. Forest succession alters the chemical, physical, and biological properties of the soil through their occupancy of the area, and it is likely that these alterations contribute to the relative changes in the abundance of the dominant plant species that characterizes succession aspect on the land [3].

A number of soil characteristics affect plant growth and its well-being. These include soil texture (size distribution of particles), structure (arrangement of soil particles), and soil depth. Soil reaction primarily influences plant growth indirectly through its effects on the solubility of ions and the activity of microorganisms [4]. In addition, five easily observable properties of soil (texture, structure, color, depth, and stoniness) can be used to infer a great deal about how a particular soil influences plant growth [3]. It is the ability of the soil to supply nutrient elements in the amounts, forms, and proportions required for maximum plant growth. The plant growth depends on the physicochemical properties and organic matter content of the soil [5].

There are associations between species and soil characteristics under homogenous parent rock and elevation range [6] as well as correlation among topography, soil nutrient, and floristic [7]. In addition, several soil properties showed positive and negative correlation to floristic parameters ([8] [9]). On the other hand, several studies reported that there was not significant correlation between soil nutrient availability and tree growth pattern under forest succession process in the tropics [10] [11] [12]. The purpose of this study was to describe correlation between soil physicochemical properties and vegetation parameters. The information on correlation between soil properties and vegetation growth is important for addressing future management and sustainability of secondary forests.

2. Materials and methods

2.1. Study sites

The study was carried out in three stages of fallows or period of abandonment such as 5, 10, and 20 years old of secondary forests in Sabal, Sri Aman, Sarawak, East Malaysia (figure 1). The geographic locations of these sites are 01°04'43.3"N 110°59'02.0"E, 01°03'55.9"N 110°55'51.4"E, and 01°03'59.3"N 110°53'34.4"E, respectively. The study plots at Sabal were located approximately 110 km southeast of Kuching along the Kuching-Sri Aman Road and 5 to 15 km from the Sabal Agroforestry Centre. The study sites had the similar histories of abandonment after shifting cultivation based on interviews with the land owner as well as confirmation by staff of Sabal Agroforestry Centre.

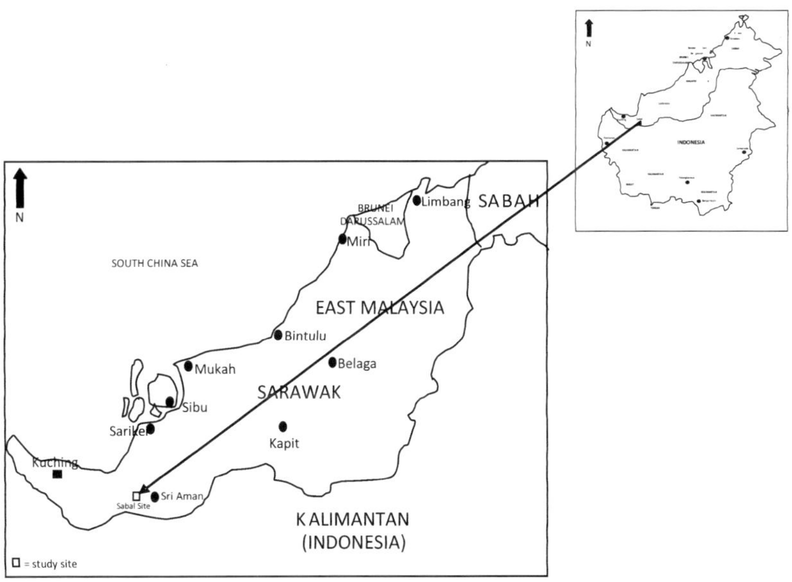

Figure 1. Map of study area in Sabal, Sarawak, Malaysia.

2.2. Data collection

The surveys of 5, 10, and 20 years old at secondary forests were conducted during 6 months from January - July 2013. Nine plots sized 20 m × 20 m were established within each study site. The diameter at breast height (DBH) and total height of all woody trees with DBH of ≥ 5 cm within the plot were enumerated and their species were identified.

2.3. Data analysis

Individual basal area (BA) and volume (V) were measured by using formulas [13]:

$$\text{Individuals BA} = \pi\,(\text{DBH}/2)^2.\ 10^{-4} \quad (1)$$

$$\text{Individuals V} = \tfrac{1}{4}\,\pi \times \text{DBH}^2.\ 10^{-4} \times \text{H} \times f \quad (2)$$

where: DBH is diameter at breast height (cm), 'H' is tree height (m), and 'f' is form factor.

Four diversity indices of Shannon-Wiener's diversity index (H'), Simpson's dominance index (D_s), Pielou's evenness index (J'), and Margalef's richness index (R) were used to measure species diversity of standing tree in each community [14]:

$$H' = -\sum_{i=1}^{s}\left(\frac{n_i}{N}\right)\ln\left(\frac{n_i}{N}\right) \quad (3)$$

$$D_s = \sum_{i=1}^{s}\left(\frac{n_i}{N}\right)^2 \quad (4)$$

$$J' = \frac{H'}{\ln(S)} \quad (5)$$

$$R = \frac{(S-1)}{\ln n} \quad (6)$$

As stated here, n_i = number of individuals of the i- th species, N = total number of all the individuals in a unit area, and S = number of species in each plot.

The correlations between soil physicochemical properties and vegetation parameters both floristic structure and diversity were determined using Pearson correlation analysis [15]. The soil physicochemical properties in the study sites had been reported by Karyati et al. [16]. The floristic structure parameters included average DBH, average height, total BA, and total volume, while floristic diversity indices were H', Ds, J', and R. Principal Component Analysis (PCA) of soil physicochemical properties were performed to prove and confirm correlation between soil properties and vegetation parameters in each study sites. All statistical tests were conducted using SPSS version 18 for Windows (SPSS Inc. 2012).

3. Results and discussion

3.1. Correlation between soil physicochemical properties and vegetation parameters

The result of the Pearson's correlation analysis between soil properties and vegetation parameters of the study sites was presented in Table 1. The soil pH (H_2O) showed positive correlation to H, BA, H', and R (P value <0.01) and to age and V (P value <0.05). A negative correlation (P value < 0.01) was also shown by the relationship between pH (H_2O) and Ds. In contrast, relationship between pH (KCl) and Mg showed no correlation to all vegetation parameters. The soil EC value showed negative correlation with H' and J' (P value < 0.05) and positive correlation to Ds (P value < 0.05). The soil T-C showed positive correlation to age stand of the study sites (P value < 0.01), DBH, H, and V (P value < 0.05) and negative correlation to N' (P value < 0.05). The negative correlation were shown by the relationship between T-N and N' (P value < 0.01) as well as H' (P value < 0.05).

The soil CEC level showed positive correlation to age, DBH, H, BA, and V (P value < 0.01) and Ds (P value < 0.05). There was negative correlation between CEC and N' (P value < 0.01) and CEC to H' and R (P value < 0.05). The exchangeable Ca showed positive correlation to BA (P value <0.01) and V (*P* value < 0.05). The positive correlation was shown by the relationship between exchangeable K to DBH, BA, V, *Ds* (*P* value < 0.05), and age (*P* value < 0.01). While the relationship between exchangeable K to *H'* and *R* showed negative correlation (*P* value < 0.05). The exchangeable Na showed positive correlation to age, DBH, H, BA, V (*P* value < 0.01) and negative correlation to *N'* (*P* value < 0.01). The CEC showed negative correlation to diameter growth rate (DGR) as well as relationship between Ca and basal area growth rate (BAGR). Meanwhile, Mg and K had negative correlation to DGR and BAGR [9].

The strong positive correlation (*P* value < 0.01) was shown by the relationship between percentage of clay to age, DBH, H, BA, and V. The positive correlation was also shown by the relationship between percentage of silt to age, DBH, H, V (*P* value < 0.01), and BA (*P* value < 0.05). The percentage of silt showed positive correlation to H (*P* value < 0.01) and negative correlation to *N'* (*P* value < 0.01). However, the percentage of sand showed a negative correlation (*P* value < 0.01) to age, DBH, BA, V, and positive correlation (*P* value < 0.05) to *N'*. The percentage of sand showed positive correlation (*P* value < 0.01) to average DBH, BA, V, and negative correlation (*P* value < 0.01) to H. In Toledo et al. [9] study, the percentage of silt had negative relationship to DGR. The soil texture had major effects on forest growth, but these effects are indirect, manifested through the effect of texture on features such as water-holding capacity, aeration, and organic matter retention [3]. The soil bulk density showed negative correlation to age (*P* value < 0.01), DBH, H, and V (*P* value < 0.05) of the study sites. In contrast, positive correlation was showed by relationship between porosity to DBH, V (*P* value < 0.05) and age and H (*P* value < 0.01). The relationship between soil hardness to age, H, BA, V, and biomass showed negative correlation (*P* value < 0.05).

Table 1. Correlation between soil properties and vegetation parameters.

Soil properties	Age (years)		N		Average DBH (cm)		Average H (m)		BA (m^2)	
	r	*P* value	*r*	*P* value	*r*	*P* value	*R*	*P* value	*r*	*P* value
pH (H_2O)	0.48*	<0.05	0.31	0.114	0.32	0.107	0.51**	<0.01	0.49**	<0.01
pH (KCl)	-0.09	0.655	-0.07	0.743	-0.04	0.861	-0.22	0.262	-0.11	0.593
EC (μS)	0.00	0.993	-0.16	0.437	0.22	0.272	0.13	0.511	0.23	0.255
T-C (%)	0.50**	<0.01	-0.41*	<0.05	0.46*	<0.05	0.39*	<0.05	0.30	0.130
T-N (%)	0.37	0.058	-0.53**	<0.01	0.32	0.107	0.14	0.501	0.06	0.766
C/N ratio	-0.21	0.302	0.42*	<0.05	-0.16	0.431	0.03	0.882	0.07	0.739
CEC	0.68**	<0.001	-0.54**	<0.01	0.73**	<0.001	0.51**	<0.01	0.53**	<0.01
Exch. Ca	0.37	0.055	0.36	0.066	0.34	0.086	0.34	0.08	0.53**	<0.01
Exch. Mg	0.32	0.099	0.22	0.267	0.23	0.255	0.25	0.205	0.35	0.071
Exch. K	0.49**	<0.01	-0.22	0.278	0.46*	<0.05	0.30	0.123	0.39*	<0.05
Exch. Na	0.92**	<0.001	-0.52**	<0.01	0.84**	<0.001	0.69**	<0.001	0.66**	<0.001
Sum of exch. bases	0.93**	<0.001	-0.33	0.088	0.85**	<0.001	0.71	<0.001	0.75**	<0.001
Clay (%)	0.81**	<0.001	-0.32	0.109	0.72**	<0.001	0.62**	0.001	0.60**	<0.01
Silt (%)	0.77**	<0.001	-0.59**	<0.01	0.70**	<0.001	0.50**	<0.01	0.44*	<0.05
Sand (%)	-0.83**	<0.001	0.46*	<0.05	-0.74**	<0.001	-0.59**	0.001	-0.55**	<0.01
BD	-0.63**	<0.001	0.13	0.518	-0.44*	<0.05	-0.511**	<0.01	-0.36	0.067
Porosity	0.65**	<0.001	-0.13	0.516	0.45*	<0.05	0.53**	<0.01	0.38	0.054
Hardness	-0.39*	<0.05	-0.11	0.595	-0.34	0.084	-0.44*	<0.05	-0.47*	<0.05

Soil properties	V (m^3)		*H'*		*Ds*		*J'*		*R*	
	r	*P* value	*r*	*P* value	*r*	*P* value	*R*	*P* value	*r*	*P* value
pH (H_2O)	0.45*	<0.05	0.56**	<0.01	-0.50**	<0.01	0.36	0.069	0.54**	<0.01
pH (KCl)	-0.07	0.715	-0.33	0.097	0.26	0.200	-0.15	0.466	-0.35	0.071
EC (μS)	0.21	0.296	-0.41*	<0.05	0.42*	<0.05	-0.47*	<0.05	-0.37	0.058
T-C (%)	0.42*	0.031	-0.10	0.635	0.10	0.635	0.07	0.743	-0.06	0.771
T-N (%)	0.23	0.248	-0.39*	<0.05	0.35	0.074	-0.16	0.439	-0.35	0.077
C/N ratio	-0.07	0.713	0.44*	<0.05	-0.38	0.053	0.23	0.247	0.41*	<0.05
CEC	0.65**	<0.001	-0.48*	<0.05	0.47*	<0.05	-0.25	0.201	-0.42*	<0.05
Exch. Ca	0.42*	<0.05	0.14	0.489	-0.03	0.877	-0.06	0.754	0.19	0.346
Exch. Mg	0.26	0.189	-0.12	0.55	0.23	0.255	-0.31	0.113	-0.05	0.808
Exch. K	0.41*	<0.05	-0.46*	<0.05	0.47*	<0.05	-0.37	0.06	-0.41*	<0.05
Exch. Na	0.80**	<0.001	-0.14	0.493	0.11	0.574	0.05	0.808	-0.11	0.603
Sum of exch. bases	0.84**	<0.001	-0.11	0.575	0.13	0.513	-0.02	0.909	-0.06	0.759
Clay (%)	0.69**	<0.001	-0.19	0.331	0.20	0.326	-0.06	0.752	-0.14	0.500
Silt (%)	0.59**	0.001	-0.34	0.088	0.28	0.155	-0.06	0.776	-0.30	0.126

Sand (%)	-0.68**	<0.001	0.27	0.174	-0.25	0.216	0.06	0.752	0.22	0.270
BD	-0.42*	<0.05	-0.23	0.246	0.21	0.303	-0.24	0.224	-0.27	0.180
Porosity	0.44*	<0.05	0.25	0.214	-0.22	0.276	0.25	0.213	0.28	0.150
Hardness	-0.39*	<0.05	-0.23	0.243	0.16	0.420	-0.05	0.813	-0.28	0.158

Note: Soil properties were at the depth of 0-10 cm (n=27); Calculation of vegetation parameters were done according to the 20 m × 20 m subplots (n=27); N=number of individuals (tree); DBH=diameter at breast height (cm); H=height (m); BA=basal area (m^2); V=volume (m^3); *H'*=Shannon-Wienner diversity index; *Ds*=Simpson dominance index; *J'*=Pielou evenness index; *R*=Margalef richness index; EC=electrical conductivity; T-C=total carbon; T-N=total nitrogen; CEC=cation exchange capacity ($cmol_c\ kg^{-1}$); Exch. Ca, Mg, and K=exchangeable Ca, Mg, and K ($cmol_c\ kg^{-1}$); Clay, Silt, and Sand=percentage of clay, silt, and sand; BD=bulk density ($g\ mL^{-1}$); Soil hardness (mm) was measured using a Yamanaka-type penetrometer; *r* are Pearson's correlation coefficients; *P* values of correlation are shown; * and ** correlations are significant at the 0.05 and 0.01 level (2-tailed).

In general, soil physicochemical properties showed positive and or negative correlation to one or more vegetation parameters, except pH (KCl) and exchangeable Mg. This result was contrast with reported by Buschbacher et al. [17]. According to Toledo et al. [9], non-significant or even weak negative correlation is found between growth rates and individual and composite soil variables. Similarly, no significant correlations between soil nutrient availability and tree growth variation were found in Borneo [10] [18], Costa Rica [11] and Brazil [12].

3.2. *Principal component analysis (PCA)*

Factor loading of the surface soil properties by Principal Component Analysis (PCA) was shown in table 2. The three components of PC1, PC2, and PC3 explained 74.66% of the total variability and each component represents a series of variables which simplifies the analysis and interpretation. The first principal component score (PC1) was defined as cation retention capacity and soil texture, such as CEC, exchangeable K, clay, silt, and porosity exhibited high positive factor loading, while sand and bulk density to a lesser degree. The second principal component (PC2) showed a high positive factor loading for T-N and a high negative factor loading for pH (H_2O), reflecting the soil acidity. Meanwhile, the third principal component (PC3) was related to soil compactness as soil hardness showed a high positive factor loading (table 2). The result indicated that the factor analysis provides statistical evidence of the ability of the three principal component scores to integrate soil physicochemical properties within the components.

Figure 2 shows the correlation between PC1 and PC2 scores from PCA in the study sites. The figure showed that the selected soil physicochemical properties grouped at each site following fallow ages. The positive factor loading of PC1 were CEC, exchangeable K, clay, silt, and porosity. The negative factor loading showed for sand and bulk density in PC1. The clear differences were observed for several soil properties which categorized into cation retention capacity, such as CEC, exchangeable K, and clay content related to PC1 based on fallow period. Hence, clay content of the soil was one of the determining factors that influenced the nutrient retention capacity such as the soil CEC level and the amount of exchangeable K. These three soil properties were affected by the succession process during fallow periods. In addition, no clear tendencies were showed for several soil physical properties, such as silt and sand contents, porosity, and bulk density related to PC1. However, the soil CEC and exchangeable K correlated significantly to floristic structure parameters and diversity indices. Meanwhile, soil texture (contents of clay, silt, and sand), porosity, and bulk density had significant correlations with several floristic structure parameters, but no correlation showed between these soil properties and floristic diversity indices of the study sites (table 1).

Figures 3 and 4 show the correlation among PC1 scores from PCA, floristic structure parameters, and floristic diversity indices of the study sites. The results indicated that soil properties may influence to both floristic structure and diversity. The analysis showed that the floristic structure parameters and floristic diversity indices were grouped according to the soil properties and fallow period of the study sites. The floristic structure parameters, such as DBH, H, BA, and V increased with increasing fallow period in three study sites. This trend was also showed by several soil properties that were included in PC1, such as soil CEC, exchangeable K, and clay content, particularly for surface soils at 0-10 cm depth. In addition, many soil physicochemical properties were significantly correlated with floristic structure parameters, except pH (KCl), electrical conductivity, and exchangeable Mg as shown in table

1. Although the correlation among floristic diversity indices and soil properties which were included in PC1 resulted the similar pattern among the study sites (figure 4), most of the soil properties such as pH (KCl), exchangeable Ca, sum of exchangeable bases, soil texture, bulk density, porosity, and hardness showed no correlation with floristic diversity indices as presented in table 1.

The result of PC1 showed that the study sites which consisted of soils with high CEC level, amount of exchangeable K, higher clay and silt contents, and higher soil porosity possessed consist high values of floristic structure parameters such as DBH, H, BA, and V. Meanwhile, the study sites with the low sand content and bulk density in soil may affect the values of floristic diversity indices, such as *H'*, *Ds*, and *R*. The lower bulk density showed the higher porosity of soil and it was an advantage for plant growth because it will provide higher water infiltration rates which reduce runoff, then improve the water availability and retention in subsoil.

Table 2. Factor loading of surface soil physicochemical properties by Principal Component Analysis.

Variables analyzed	pH (H_2O), T-C, T-N, CEC, Exch. Ca, Exch. Mg, Exch. K, Clay, Silt, Sand, BD, Porosity, and Soil hardness			
	Value	PC1	PC2	PC3
Variables with a high positive factor loading (> 0.70)	+	CEC, Exch. K, Clay, Silt, and Porosity	T-N	Hardness
Variables with a high negative factor loading (> 0.70)	-	Sand and BD	pH (H_2O)	ND
Contribution name of PC axis		43.28%	19.13%	12.26%
		Cation retention capability & soil texture	Soil acidity	Soil compactness

Note: (+)=factor loading with a positive value; (-)=factor loading with a negative value; ND=not determined; T-C=total carbon; T-N=total nitrogen; CEC=cation exchange capacity; Exch. Ca, Mg, and K=exchangeable Ca, Mg, and K; Clay, Silt, and Sand=percentage of clay, silt, and sand; BD=bulk density.

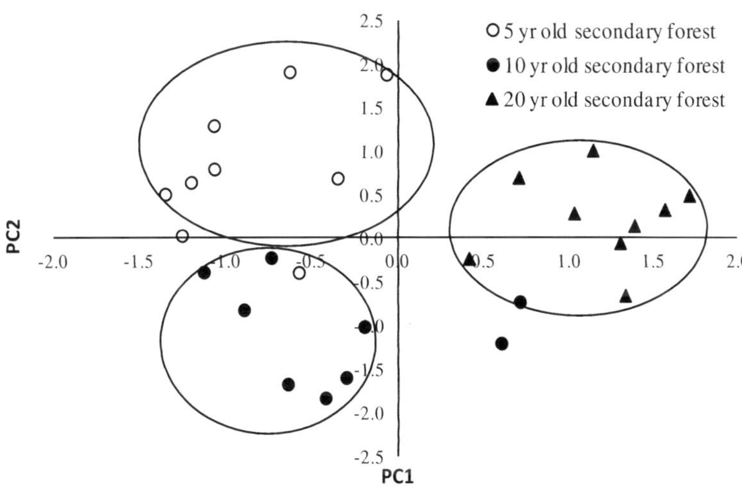

Figure 2. Correlation between PC1 and PC2 scores from Principal Component Analysis (PCA) in the study sites.

The factor loadings of the PC2 showed high positive factor for T-N and negative factor for pH (H_2O). In term of the fallow periods of the study sites, there was no clear difference for T-N and pH (H_2O) related to PC2. In addition, the analysis showed that both T-N and pH (H_2O) were correlated well to several floristic structure and diversity parameters as shown in table 1. This was mainly related to variation of soil texture and the content of exchangeable bases under similar parent materials. The high values of floristic structure and diversity probably tended to increasing T-N.

For factor loading of PC3, the correlation analysis showed that soil hardness was significant correlated to floristic structure parameters, but no correlation showed between soil hardness and floristic diversity indices as presented in table 1. The floristic structure parameters increased with decreasing soil hardness, particularly at surface soils (0-10 cm depth) following fallow period in the study sites. The soil hardness related to soil aggregates and compaction. Fisher and Binkley [3] stated that soil physical properties profoundly influence the growth and distribution of trees through their effects on soil moisture regimes, aeration, temperature profiles, soil chemistry, and even the accumulation of organic matter.

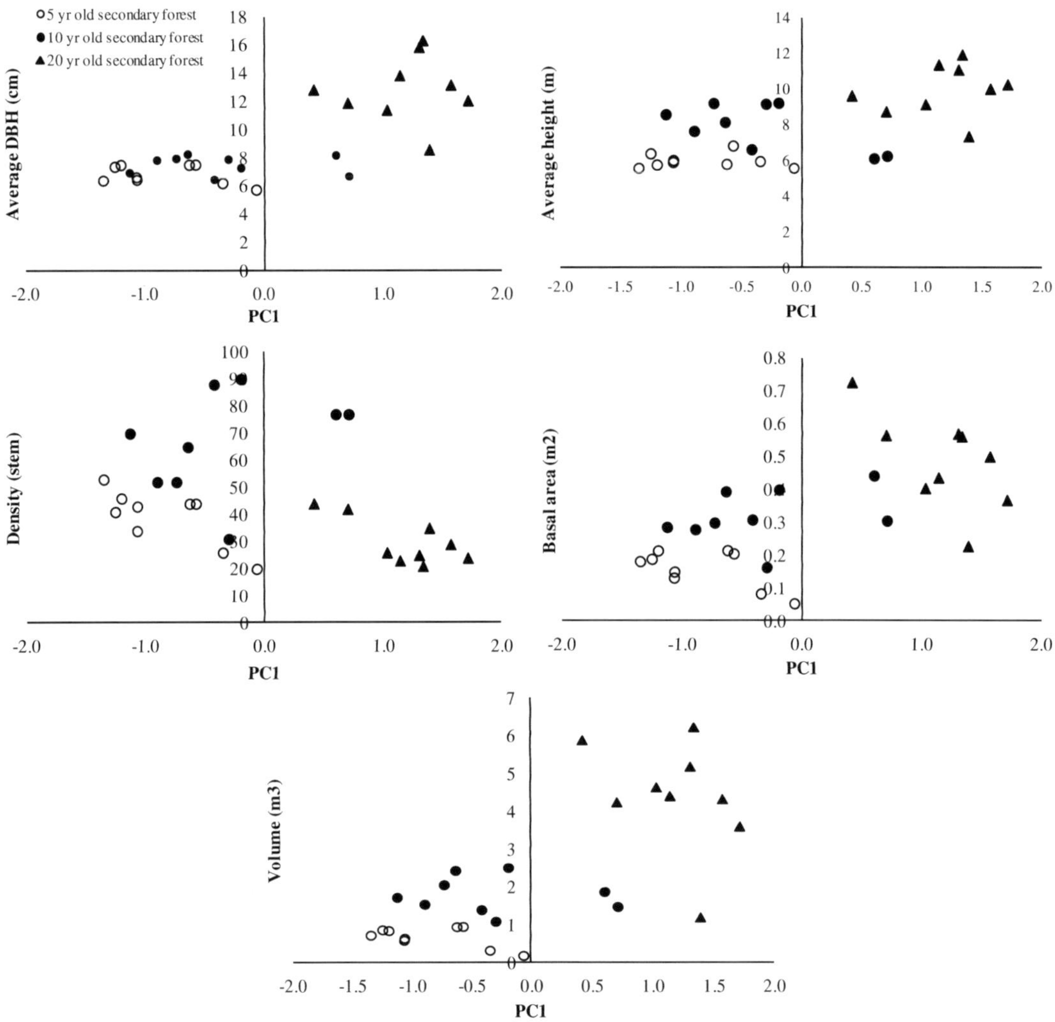

Figure 3. Correlation between PC1 scores from PCA and floristic structure parameters in the study sites.

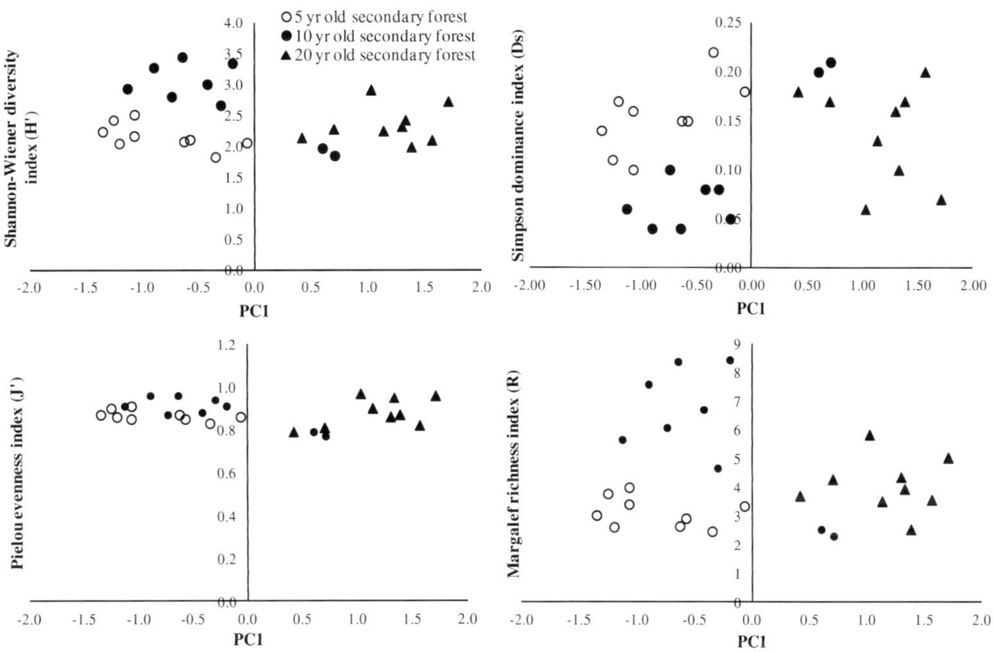

Figure 4. Correlation between PC1 scores from PCA and floristic diversity indices in the study sites.

4. Conclusion

We conclude that soil physicochemical properties were significantly correlated to floristic structure parameters and floristic diversity indices. Three principal component scores integrating and grouping soil physicochemical properties and floristic structure parameters as well as floristic diversity indices in terms fallow age within the same components. The development of floristic structure during early succession process may probably effect to improve soil physicochemical properties.

Acknowledgements

We would like to extend our sincere thanks to En. Hidir Marzuki, En. Sekudan Tedong, En. Salim Arip, and En. Muhd Najib Fardos for full support and helping in the fieldwork.

References

[1] Chokkalingam U, Smith J and de Jong W 2001 A conceptual framework for the assessment of tropical secondary forest dynamics and sustainable development potential in Asia *J. Trop. For. Sci.* **13**(4) pp 577-600

[2] Karyati, Ipor I B, Jusoh I and Wasli M E 2017 The diameter increment of selected tree species in a secondary tropical forest in Sarawak, Malaysia *Biodiversitas* **18**(1) pp 304-11

[3] Fisher R F and Binkley D 2000 *Ecology and Management of Forest Soils, 3rd Ed* (USA: John Wiley & Sons, Inc) p. 35

[4] Harris R W 1992 *Arboriculture: Integrated Management of Landscape Trees, Shrubs, and Vines* (New Jersey: Prentice Hall Career & Technology) pp. 88-117

[5] Hazra P and Som M G 2006 *Vegetable Science* (New Delhi: Kalyani Publishers) pp. 37-108

[6] Nizam M S, Norziana J, Sahibin A R and Latiff A 2006 Edaphic relationships among tree

species in the national park at Merapoh, Pahang, Malaysia *Jurnal Biosains* **17**(2) 37-53

[7] Potts M D, Ashton P S, Kaufman L S and Plotkin J B 2002 Habitat patterns in tropical rain forests: A comparison of 105 plots in northwest Borneo *Ecology* **83**(10) pp 2782-97

[8] Kumar J I N, Kumar R N, Bhoi R K and Sajish P R 2010 Tree species diversity and soil nutrient status in three sites of tropical dry deciduous forest of Western India *Trop. Ecol.* **51**(2) pp 273-79

[9] Toledo M, Poorter L, Peña-Claros M, Alarcón A, Balcázar J, Leaño C, Licona J C, Llanque O, Vroomans V, Zuidema P and Bongers F 2011 Climate is a stronger driver of tree and forest growth rates than soil and disturbance *J. Ecol.* **99** pp 254-64

[10] Ashton P S and Hall P 1992 Comparisons of structure among mixed Dipterocarp forests of North-Western Borneo *J. Ecol.* **80** pp 459-81

[11] Clark D B, Clark D A and Read J M 1998 Edaphic variation and the meso-scale distribution of tree species in neotropical rain forest *J. Ecol.* **86** pp 101-12

[12] Vieira S, de Camargo P B, Selhorst D, Da Silva R, Hutyra L, Chambers J Q, Brown I F, Higuchi N, dos Santos J, Wofsy S C, Trumbore S E and Martinelli L A 2004 Forest structure and carbon dynamics in Amazonian tropical rain forests *Oecologia* **140** pp 468-79

[13] Husch B, Miller C I and Beers T W 1982 *Forest Mensuration* (New York: John Wiley and Sons Publishing) 402 pp

[14] Odum P E 2005 *Fundamentals of Ecology 5th Ed* (Belmont CA: Thomson Brooks/Cole) 598 pp

[15] Gogtay N J, Thatte U M 2017 Principles of correlation analysis *Journal of The Association of Physicians of India* **65**(3) pp 78-81

[16] Karyati K, Ipor I B, Jusoh I and Wasli M E 2014 Soil properties under various stages of secondary forests at Sarawak, East Malaysia *Journal of Tropical Forestry and Environment* **4**(01) pp 28-39

[17] Buschbacher R, Uhl C and Serrão E A S 1988 Abandoned pastures I Eastern Amazonia. II. Nutrient stocks in the soil and vegetation *J. Ecol.* **76** pp 682-99

[18] Zakaria R, Nik Rosely N F, Mansor M and Zakaria M Y 2008 The distribution of *Macaranga* Genus, (family Euphorbiaceae) in Penang Island, Peninsular Malaysia *J. Biosci.* **19**(2) pp 91-99

The potential of woody waste biomass from the logging activity at the natural forest of Berau District, East Kalimantan

D R Sari* and Ariyanto

Forestry Faculty, Mulawarman University, Samarinda Ulu, Samarinda 75116, East Kalimantan, Indonesia.

*Corresponding Author: rakhmah_sari@yahoo.com

Abstract. The fifth principles of Sustainable Forest Management of the Forest Stewardship Council (FSC) is to encourage the efficient use of forest products and services to ensure economic viability and a wide range of environmental and social benefits, and one of the criteria is to minimize logging waste. Therefore, identification, calculations, and monitoring of logging waste should be done. The purpose of this study were to know the potential of logging waste in logging area and to know the estimated volume of waste based on the Annual Allowable Cut (AAC). This research was carried out at PT Karya Lestari, Berau District, East Kalimantan. It was found that the highest percentage of waste was in the form of the main trunk canopy, followed by the branches, stump, twigs and the rest of the bucking. It was also found that there was quite strong positive relationship between the volume of felled trees with its logging waste volume, and the estimated volume of logging waste based on AAC was 40,623 m^3 per year. This result shows that the potential logging waste is high and it is recommended for the company to do logging on trees with small volumes or lower diameter class (40-89 cm).

1. Introduction

The logging waste is one of the problems that must be addressed in the process of forest harvesting in natural forests because among the objectives of the forest harvesting activities are to maximize the value of wood and optimize the supply of industrial raw materials. There are four main activities in forest harvesting: logging, skidding, loading and hauling [1], that have to go well and sequentially so that the logging operation can be successful. If the process is not done with good and proper planning, logging as the first stage in forest harvesting can be the most potent activity to produce wood waste. The success of tree felling can be measured, among others, from the level of damage to the environment and the waste produced.

In line with that, the fifth principle of Sustainable Forest Management (SFM) by the Forest Stewardship Council (FSC) is to encourage the effective utilization of various products and services from forests to ensure economic capability and environmental and social benefits, which is one of the criteria is to minimize waste of wood harvesting (logging waste). SFM itself is a global aim for guarantee timber supply, protection of forest environmental services (such as carbon sequestration), protection of forest habitats, and protection of secure place for over than 1.6 billion human being who depend on forest [2,3,4,5]. Therefore, all harvesting activities in the forest company should get more attention, one of which is to identify, calculate, and monitor the waste of timber harvesting. However, waste in good condition can still be utilized, for example for raw materials of advanced products, raw materials for sawn, core veneer, and chips, even some of the stem waste deserve to be used as logs[6], the material of sawmill industry [7], bioenergy [8,9,10,11], and bio-based polymeric materials [12].

The purpose of this research were to know the potential of logging waste that occurs in the plot of harvest due to timber harvesting and to know the estimated volume of logging waste based on the Annual Allowable Cut (AAC). Research on wood waste in timber companies was absolutely necessary. On the basis of that information, corrective action in the forest harvesting process can be done so that the wood waste that occurs can be minimized, thus increasing the efficiency and effectiveness of timber harvesting. In addition, the existing waste potential data also becomes important information in the efforts of further waste utilization.

2. Materials and Methods

2.1. Study area

The study was conducted at Karya Lestari Company, Berau District, East Kalimantan Province, Indonesia, for ± 3 months, at the 2014 Annual Work plan block. Geographically, the work area of Karya Lestari Company is located between 116°40'12.2"-116°59'46.7"East Longitude and 01°35'15.6"-01°48'4.4"North Latitude. Based on the distribution of forest groups, Karya Lestari Company is included in the Gie River Forest Group and according to government administration, it is included in Kelay Sub-district, Berau District, East Kalimantan Province, Indonesia (Figure 1). Meanwhile, based on forestry administration, Karya Lestari Company is included in Sambaliung Forestry Unit, Berau Forestry Department, East Kalimantan Provincial Forestry Office.

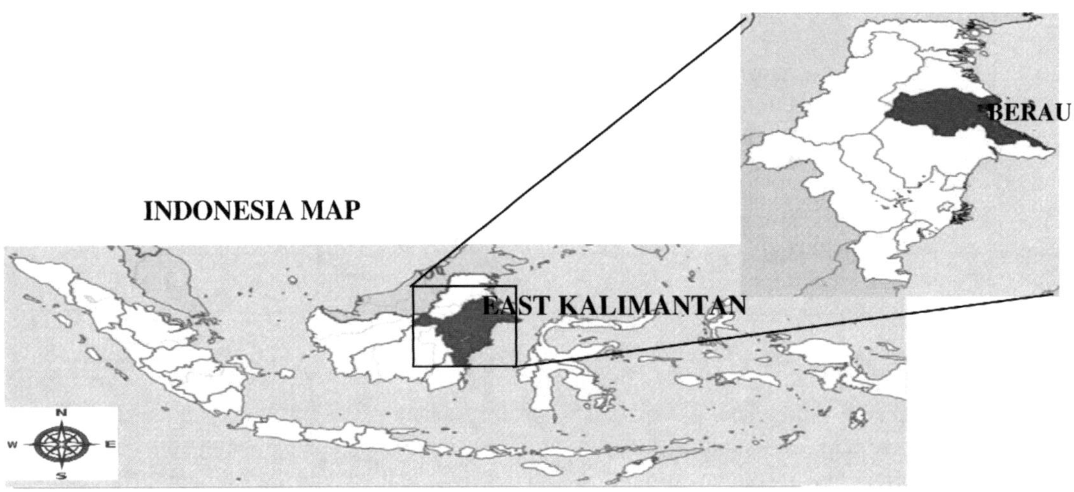

Figure 1. Study Area at Berau District, East Kalimantan Province, Indonesia

2.2. Scope of logging waste

In this research, the logging waste that we measured is the whole part of the remaining logging trees or all parts of commercial rods that left in the harvest plots consisting of stumps, the rest of the bucking, main rod of the crown, the branches and twigs. The definition of it is as follows: commercial rods are stems from the top of the stump until the first branch or stem which has been issued by forest concessions as a commercial rods; the top rod is the stem part of the first branch to the canopy which is an extension of the main (commercial) stem; branches and twigs is a canopy component of a felled tree that is above the first branch; the stump is the bottom of the tree under the notch and the notch reply; and the short cut is the stem part of the main stem containing the defect and needs to be cut [6].

In this research, we used term of the main rod of the canopy for top rod, and term of the rest of the bucking for shortcut.

2.3. Data collection and analysis
The number of sample trees measured was 60 trees, randomly selected and scattered on the logging plot. Each part of the logging waste, identified and classified as a stumps, the rest of the bucking, the main rod of the crown, branch or twig, then each piece of waste is measured in diameter and length. We calculated waste volume by using the following formula:

$$V = {}^{1}/_{4}\pi\, d^2\, L\, 10^{-3} \quad (1)$$

With V = volume of logging waste (m^3); d = diameter (cm); and L = length of logging waste (m).
To estimate the volume of logging waste based on Annual Allowable Cut (AAC)we used linear regression with following equation:

$$V_l = a+bV \quad (2)$$

With Vl = Estimated volume of logging waste based on AAC (m^3); a,b = constant and V = Volume of AAC (m^3).

3. Results and Discussion

3.1. Summary of sample trees
In this study, we observed and measured 60 sample trees with species *Shorea sp.* in the 2014 Annual Working Plan block. The smallest diameter of the sample tree measured was 45 cm and the largest diameter of 166 cm with an average diameter was 77 cm. The following table presents the sample tree distribution measured:

Table 1. Summary of sample trees

Class of Diameter (cm)	Amount of sample trees
40-49	1
50-59	12
60-69	14
70-79	12
80-89	5
90-99	8
100-109	4
110-119	1
≥120	3
Total	60

From table 1, we can see that the measured tree samples were dominated by trees with diameter of 50-79 cm that was as many as 38 trees, and the remaining 32 samples were spread over the diameter of 40-49 cm (1 tree), 80-89 cm (5 trees), 90-99 cm (8 trees), 100-119 cm (1 tree) and ≥120 cm (4 trees).

3.2. Forms of logging waste
Logging waste is log part of commercial stem, short pieces, stumps, branches and twigs [6].The forms of logging waste found in the logging area were stump(the bottom of the tree under the notch and the notch reply), the rest of the bucking(in the form of short pieces or logs, including the main stem part

that is defective or damaged and needs to be cut), the main rod of the crown(the stem part from the first branch to the canopy which is an extension of the main stem), branches and twigs(the canopy component of the felled tree that is above the first branch), as we can see at Figure 2.

Figure 2. Forms of the logging waste at the logging area: (a) stump; (b) rest of the bucking; (c) main rod of the crown and branches; and (d) twigs.

From the calculation of the volume of each waste, by using equation (1), the results obtained that the largest waste was in the form of main rod of canopy (43.43%), followed by waste in the form of branches (26.12%), stumps (13.63%), twigs (11.13%) and rest of the bucking (5.69%). Figure 3 shows the percentage of waste volume in each form.

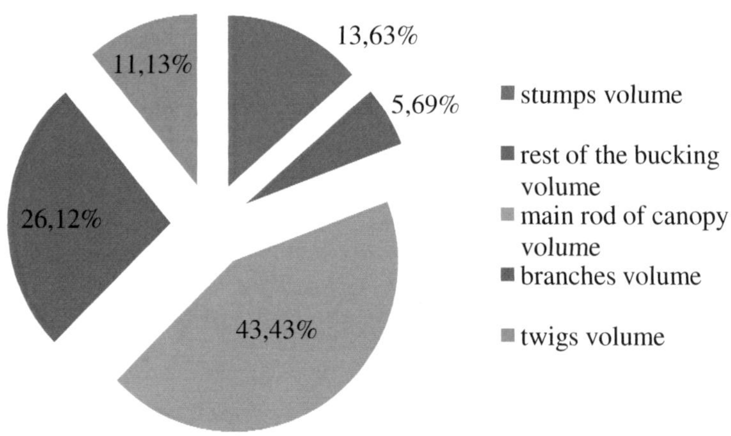

Figure 3. Percentage of waste volume in each form.

Most of the stump as the remaining logging that found on the logging block exceeded the recommended limit for natural forest, which was 50 cm above the soil surface. The stump contained in the sample tree was mostly above 1 m high and some of them were reached more than 3 m height. In addition to the physical condition of the trunk containing defects or damaged by harvesting, short pieces produced due to trimming at the base or tip also cause the residual waste bucking. This condition could actually be avoided by increasing control and monitoring of logging activities, and also by improving the logging and bucking techniques [6] such as by applying of more environmental friendly logging techniques.

3.3. Volume of logging waste

The log volume data of the felled trees is obtained from the company's data, while the volume of logging waste is calculated using formula (1). The volume calculation of felled trees and logging waste generated from 60 sampling trees obtained the following data: the total volume of felled trees was 618.7 m^3 and the total logging waste volume that occurs in logging area was 509.7 m^3, whereas the average volume of logged trees was 10.3 m^3 and the average volume of waste produced by each felled tree was 8.5 m^3. Total volume of felled trees and total volume of logging waste was shown in figure 4.

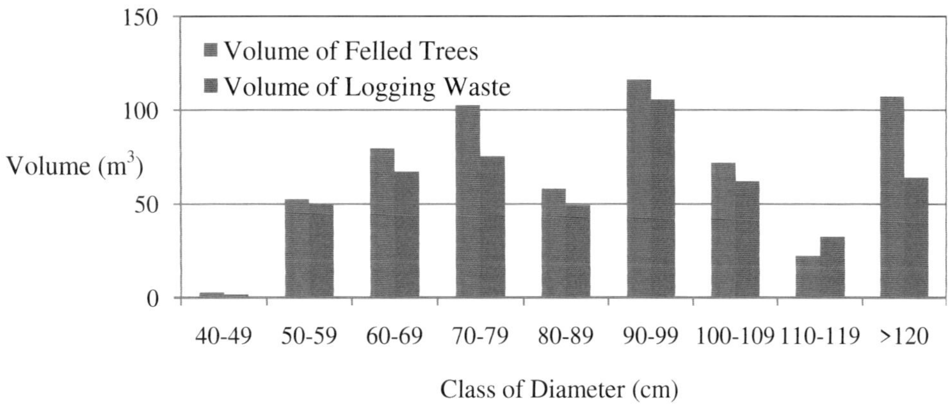

Figure 4. Total volume of felled trees and logging waste in each diameter class.

Figure 4 showed that in general, the total volume of logging waste produced is lower than the volume of the felled tree, except in the 110-119 cm diameter class, where the volume of waste generated is greater than the volume of the felled tree. It could also be seen in figure 4 that in the smaller diameter class, which were in the 60-69 cm, 70-79 cm, and 80-89 cm diameter classes, the waste produced was much smaller than the volume of the felled trees compared to the greater diameter class. The calculation of waste in the sample tree in the 70-79 cm diameter class showed the best result, where the waste produced was much smaller than the volume of the felled tree.

We also did regression analysis of the 60 data from felled trees and its logging waste volumes, which was shown in Figure 5.

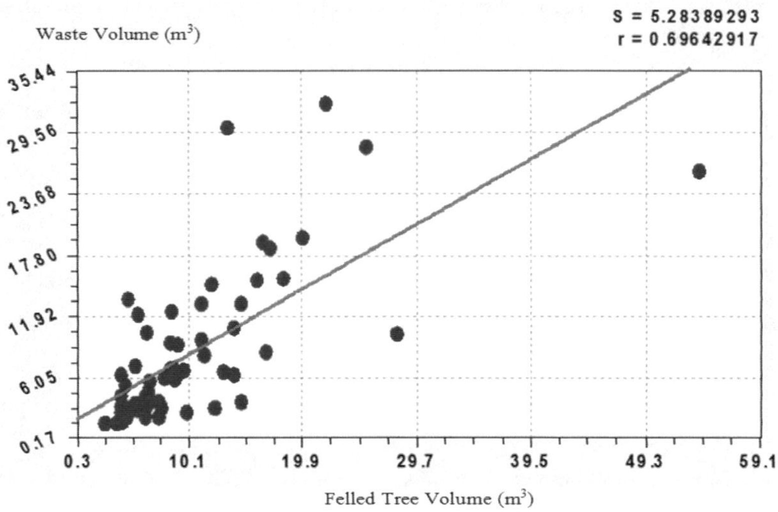

Figure 5. Graph of relationship between felled tree volume and waste volume by linear model

From Figure 5, with s value 5.283 and r value 0.696 it could be seen there was quite strong positive relationship between the volumes of felled trees with its logging waste volume. It was also could be seen the tendency that logging on smaller volumes of trees will result in a smaller volume of logging waste while logging on larger volumes will relatively result in a much larger volume of logging waste as well. From the results in figures 4 and 5, we found do more logging on smaller volume of trees or in smaller diameter class (40-89 cm) than logging on larger volume of trees or larger diameter classes can be one possible alternative to minimize the resulting of logging waste.

The result of the analysis also produces constant value a=1,939 and b=0,636, so equation (2) become Vl= 1.939+0.636V. Until year 2021, Karya Lestari Company planned to harvest the average area of the cutting block ± 1,250 ha per year and the annual harvest volume is estimated at 58,365 m^3. Thus, the volume of logging waste that will be generated was estimated at 40,623 m^3 per year which is high. The existence of waste in the forest leads to lower efficiency of timber harvesting. Several studies have shown that the efficiency of harvesting timber in natural forests ranges from 75 to 87%[13].On the other hand, potential of logging waste can also be utilized for various utilization efforts. The potential of logging waste from logged trees up to 15 cm in diameter is 57% [13]. Actually, high potentialities of plant waste for many industrial processes were overviewed [14], as the global issue of energy scarcity in the world today, whereas biomass and waste biomass must contribute to development a bioeconomy to the energy transition in combination with the other renewable energies [15]. Furthermore, utilization of woody biomass, which is produced from two specific sources: waste wood from sawmills, and logging waste left in forest land, has positive impact for regional economic development [16]. In the future, it is important to do more study on the utilization of logging waste, especially in East Kalimantan.

4. Conclusion

The potential of logging waste is high. In addition to efforts to minimize the logging waste, the potential of waste can also be utilized for various utilization efforts. There is a quite strong relationship between the volume of felled trees and the volume of its waste produced, therefore it is recommended for the company to do logging on trees with small volumes or lower diameter class (40-89 cm).

Acknowledgments
We thank to Forestry Faculty of Mulawarman University for the support, Center of Social Forestry of Mulawarman University as facilitator of this research, The Borneo Initiative as a research funder, and Karya Lestari company as a provider of research locations and necessary data. We also thank to anonymous reviewers for valuable comments.

References
[1] Conway 1976 *Logging Practices, Principles of Timber Harvesting System* (San Francisco: Miller Freeman Publication Inc) p 416
[2] Canadell J G and Raupach M R 2008 *Science* **320** pp 1456-57
[3] Chao S 2012 *Forest People: Number Across The World* (Moreton-in-Marsh: Forest People Programme) p 24
[4] FAO 2010 Global Forest Resources Assessment 2010 (Rome: FAO) p 340
[5] PutzP A et al 2012*Conservation Letters* **5** pp 296-303
[6] Budiaman 2000 *J. Teknologi Hasil Hutan* **2** pp 34-43
[7] Widarmana S 1973 *J.Hutan dan Masyarakat* **3** pp 235-346
[8] Dwivedi P, Bailis R and Khanna M 2014 *Bioenerg. Res.* **7** pp 217-31
[9] Guinée JB, Heijungs R and van der Voet E 2009 *Int J Life Cycle Assess* **14** pp 328-39
[10] Lu H Rand ElHanandeh A 2017*Clean Techn Environ Policy* **19** pp 2021-40
[11] Rios M and Kaltschmitt M 2013 *Biomass Conv. Bioref.* **3** p 239
[12] Moriana R, Vilaplana F andEk2015 *M. Cellulose* **22** pp 3409-23
[13] Dulsalam D, Tinambunan I, Sumantri and Sinaga M 2000*J.Penelitian Hasil Hutan* **4** pp 327-37
[14] Efremova SV 2012 *Russ J Gen Chem* **82** p 963
[15] Nelles M, Brosowski A, Morscheck G and Schüch A 2018 *Utilization and Management of Bioresources* Ghosh S (Singapore: Springer) pp 3-12
[16] Moon D, Isa A, Yagishita T and Minowa T 2013 *J. Wood Sci* **59** pp 321-30

The effect of clay nanoparticleon the retention and attack of drywood termite (*Cryptotermes cynocephalus* Light).

T Alex[1,2], B Winarni[3],I W Kusuma[4], E T Arung[4]and E Budiarso[4]

[1]Faculty of Forestry, Mulawarman University, Samarinda, East Kalimantan, Indonesia
[2]Department of Agricultural Technology, State Agricultural Polytechnic of Samarinda, East Kalimantan, Indonesia
[3] Department of Agricultural Management, State Agricultural Polytechnic of Samarinda, Samarinda, East Kalimantan, Indonesia
[4] Faculty of Forestry, Mulawarman University, Samarinda, East Kalimantan, Indonesia

*Corresponding Author : tamanalex2@gmail.com

Abstract. Theaplication of clay nanoparticles as wood preservative was intended to examine its resistant to infestation of drywood termite (*C. cynocephalus*Light). Loam was crushed into small pieces as clay nanoparticles, which was used as wood preservative that was soluble in water. The material was dissolved with water in a given concentration and put it into anggerung *(Trema orientalis)*, white meranti *(Shorea bracteolata*) and sengon (*Paraseriehthes falcataria*) by full cell method or impregnation. The tested samples were preserved by clay nanoparticles at the air pressure 60 psi for 2 hours and then furnace-dried, and tested by infestation of drywood termites. Results of the research showed that mortality of the drywood termites on those three preserved woods with clay nanoparticles by concentration of 2.5% and 5%, which reached 96.4% and 100% on average, respectively, in comparison without preservation by average mortality value was 27.4%. The highest retention of clay nanoparticle was 24.89 kg.m^3 was obtained by concentration of 5% for white merantiand the lowest was 9.32 kg.m^3 by concentration of 2.5% for anggerung. The application of clay nanoparticle as effective wood preservative against the infestation of drywood termite by concentration of 5% has been proven by mortality 100%.

1. Introduction

The application of clay nanoparticle for wood preservation has been known by few people, as well as academician and practitioner because clay nanoparticles are particles that derived from loam crushing process. Loam is a kind of clays that contains poor nutrient, therefore it is so-called worst clay or acrysol. The particle size of loam ranges 50-2 micron and mostly contains iron, aluminium, and silica loam [1]. According to Sutandi et al[2] explain that loam has acid property, which may inhibit the growth of crops including oil palm (*Elaeisguineensis* Jack.). Size of the loam, as mentioned above, may be become beneficial and valuable if it is processed into nanoparticle size, so that it will easily pass through the wood's cavities. Such clay nanoparticles contain more metallic elements, particularly iron, so that if such material is impregnated into the wood and fills the wood's cavities and cells, it will be poisonous for organism, which destroys the wood, particularly drywood termite (*Cryptotermes cynocephalus* Light). Anggerung (*Tremaorientalis*) belongs to Ulmaceae family.

Anggerunghas naturally low durability or less resistant to wood destroyer, and itsSpecific gravity ranges 0.33–0.4 that included as durable class V. White meranti (*S.bracteolata*) is a species of commercial timber and belongs to Dypterocarpaceae family.White meranti grows at natural forest and it mostly utilized as building materials.White meranti has average densities that range between 0.4–0.6 and belong to durable class IV [3].Sengon (*P. falcataria*) belongs to Fabaceae family. Sengon grows fast and mostly grown at timber estate and community forest in Indonesia.
Such timber can be utilized for light construction materials or non-structural. Densities of sengon range 0.28–0.37 and belong to durable class V or less resistant to drywood termite.

According to [5] drywood termite has the smallest size and belongs to Kalotermitidae family. It is a drywood termite, which lives in very low level of water. Mortality of termite is an amount of termite's death at definite time. The mortality values are determined and measured in accordance with death percentage of the drywood termites [6]. Objectives of the research were to prevent the infestation of drywood termite and to study the amount of clay nanoparticle retention on anggerung, white meranti and sengon.

2. Materials and method

Loam was taken from 80 cm depth, and then crushed with water and strained using 100 mesh strainers, and after that it was sun-dried. The dried loam was crushed into nanoparticles size using ball mill, which is so-called clay nanoparticle. Both chemical contents and pH of clay nanoparticles were analysed at Soil Laboratory of Pusat Rehabilitasi Hutan Tropika Universitas Mulawarman. The tested samples of anggerung, white meranti, and sengon were made in size of 2.5x2.5x0.5 cm from the timber, randomly and dry-aerated to achieve constant weight or balanced water content and then preserved by pressed-vacuum (impregnated) for two hour by clay nanoparticle material with concentrations 2.5% and 5%. Then, the retention values were counted. The tested samples after the preservation process and the tested samples without preservation, were furnace-dried at constant temperature of 103°C and weighed, then kept until it reached air-dried. Glass pipe by diameter of 1.8 cm and 3 cm height was attached on one side of the tested samples using paraffin (candle). Fifty healthy and active worker drywood termites taken from Wood Preservation Laboratory of State Agricultural Polytechnic of Samarinda were put into the glass pipe and covered with cotton.Then, it was put in a dark room at (25 ±2)°C temperature and (70 ±5) % humidity for 12 weeks and counted the dead termites [7]. Mortality and retention counted using equation as presented below.

2.1. Mortality equation

$$M = \frac{\Sigma dt}{\Sigma it} \text{x } 100\% \quad (1)$$

M = mortality worker drywood termite, Σdt = sum of dead worker drywood termite, Σit = sum of initial worker drywood termite.

2.2. Retention equation

$$R = \frac{W_2 - W_1}{V} \text{x } C \quad (2)$$

R = retention ($gram.cm^3$) converted to ($kg.m^3$),W_2 = weight of wood after preserved (gram), W_1= weight wood before preserved(gram), C = concentration of clay nanoparticle solutions, V = volume (cm^3)

3. Results and discussion

The observed data concerned with chemical contents and pH clay nanoparticles, mortality of termite, reduced weight of the tested samples due to infestation of such drywood termites and retention of clay nanoparticles on those three kinds of timber, such as anggerung, white meranti and sengon.

3.1. The effect of clay crushing into clay nanoparticles against chemical content and pH clay nanoparticles

Chemical content and pH loam, as well as clay nanoparticles were presented in table 1.

Table 1. Chemical content and pH of loam and clay nanoparticle.

Number	Chemical element	Material content (mg/gram)	
		Loam	Clay nanoparticle
1	Iron (Fe)	27.89	32.53
2	Lead (Pb)	0.94	0.45
3	Zinc (Zn)	0.73	0.94
4	Aluminium (Al)	0.61	0.57
5	Copper (Cu)	0.37	0.54
6	Calcium (Ca)	0.16	0.21
7	pH	3.59	4.15

The chemical contents of loam and clay nanoparticle as presented in table 1 show that of chemical content clay nanoparticle for iron (Fe) has the highest level followed by Zinc (Zn), aluminium (Al), and copper (Cu), as well as lead (Pb) and calcium (Ca). All chemical elements from loam to clay nanoparticle increase, except for (Pb) and (Al), in which (Pb) from 0.94 milligrams .gram for loam becomes 0.45 milligrams .gram for clay nanoparticle while Al from 0.61 milligrams .gram for loam becomes 0.57 milligrams .gram for clay nanoparticle. The value of pH for loam was lower than clay nanoparticle, it may be correlated with Al content. According to [8] the higher content of Al in the clay, the lower pH of the clay is and vice versa. The highest content of Fe means that the clay is very infertile. Fe, Zn, Pb, Cu, and Al are macronutrients, which are required in small amount for the plant growth, and if they are excessive, they may harm and poison the plants. According to [9] suggested that impregnation of clay nanoparticle may increase mechanical properties or strength of sengon and white meranti timbers.

3.2. The effect of clay nanoparticles on mortality of drywood termites

Mortality of drywood termites on the tested samples of three timbers, anggerung, white meranti, and sengon were presented in table 2. Drywood termite mortality on resistant test against those three species of wood, which were preserved through pressed-vacuum. Such mortality percentage shows different mortality of the drywood termites in accordance with different concentration of clay nanoparticles, in which 5% concentration of clay nanoparticles on those three kinds of timber showed that mortality of the termites reached 100%, while at concentration 2.5% on those three kinds of timber, such as anggerung, white meranti, and sengon showed mortality values for about 95.93%, 94.81% and 98.52%, respectively. For the tested samples without preservation, mortality values for those three timbers were 28.89%, 28.89% and 24.44%, respectively. It showed some differences between the treatment of preservation using clay nanoparticles and without any preservation. According to [9] stated that the treatment of preservation and concentration of the preservative have significant effect on the increase of mechanical properties of wood.

Mortality values of the drywood termites, as presented in table 2 showed that sengon wood is more resistant to drywood termite attack than white meranti and anggerung, it can be seen in wood preserved with clay nanoparticle concentration for 2.5%.In which sengon Significant difference was found for mortality of termites on preservation in accordance with analysis of variance, which is shown in table 3. Values of count from analysis of variance on mortality of drywood termites. According to[11] suggested that timbers, which are treated by wood preservatives or being preserved, are more resistant to organisms that damage the timbers in comparison with the timbers without any preservation. Further test usedFairly Significant Difference (FSD)test for concentration factor of clay nanoparticles was presented in table 4.Numbers followed by the same letter show no significant difference on Fairly Significant Difference at alpha 0.05 = 6.684

Table 2. Mortality of drywood termites on resistant test against three species of wood, which were preserved by pressed-vacuum for two hour.

Kinds of timber	Concentration (%)		
	0 (non preservation)	2.5	5
Anggerung	28.89b	95.93a	100a
White meranti	28.89b	94.81a	100a
Sengon	24.44b	98.52a	100a

Table 3. Analysis of variance on mortality of drywood termites.

Comparative material	Independent degree	Value of count
Treatment of preservation	8	122.30 **
Kinds of timber	2	0.03 ns
Concentration of clay nanoparticles	2	488.21 **
Combination between kinds and concentrations	4	0.48 ns

** = very significant difference
ns = not significant

Table 4. Fairly significant difference (FSD) test for concentration factor on mortality of drywood termites.

Concentration of claynanoparticle	Mortality
5%	100.00 a
2.5%	96.42 a
0 (without preservation)	27.41 b

Mortality of drywood termites on the tested samples from three kinds of timber along with the treatment of preservation using clay nanoparticle by concentrations 5% and 2.5% did not show any difference, but it was very different for timber without any preservation, and it meant that preservation using clay nanoparticles is effective to resist the infestation of drywood termites. The chemical content in the clay nanoparticle (Table 1) can be as toxic to drywood termite because chemicals such as iron (Fe), lead (Pb), zink (Zn), copper (Cu), aluminium (Al) and calcium (Ca) are termite chemicals. According to [14] natural resistant of the timber (without preservation) on white meranti and sengonagainst infestation of drywood termites.

3.3. The effect of clay nanoparticle on retention

The retention values of anggerung, white meranti, and sengon, which have been preserved by different concentrations of clay nanoparticles, are presented in figure 1. The retention values of anggerung, white meranti, and sengon, which have been preserved by different concentrations of clay nanoparticles, are stated in kilogram per meter cubic ($kg.m^3$). It shows that clay nanoparticles are easily passed through the timber of white meranti, by higher retention than on sengon and anggerung. According to [10] suggested that requirement of retention for wood preservation is 12 $kg.m^3$ by concentrations of 2-5%. Comparison of such retention values based on kinds of timber show that white meranti is higher than sengon by the same concentration. However, white meranti easily impregnated by clay nanoparticles. It indicates that the absorptive capacity of white meranti better than sengon and anggerung. Density measurement for white meranti is 0.47 sengon is 0,30 and anggerungis 0.37. The absorptive capacity of each timber against the preservative is different. According to [15] impregnation of preservative in timber highly depends on absorptive capacity kind of timber due to different types (kinds) of timber may have different absorptive capacity of the preservative. The higher retention values mean that the timbers are easily preserved. According to [16] concentration of solution and properties of the wood structures may determine permeability and absorption of the preservative chemical solutions in the wood. Such permeability affects retention values of the preservative. The retention values are presented in figure 1.

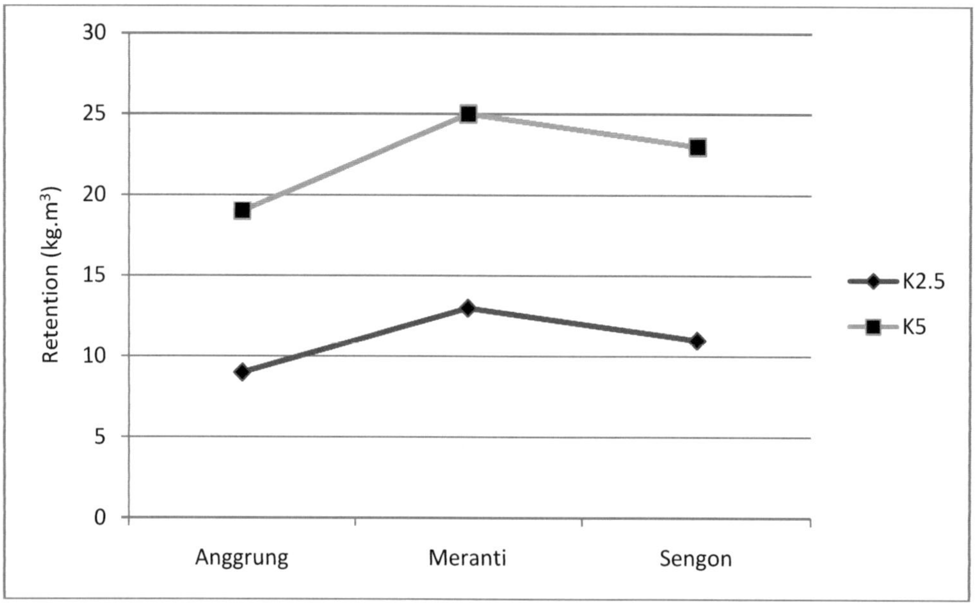

Figure 1. Graphic for retention values of anggerung, white meranti, and sengon, which are preserved by clay nanoparticles with different concentrations.

The Figure above shows that white meranti produces higher retention than anggerung and sengon, by concentrations 5% and 2.5%, and the lowest retention is found on anggerung. According to [17] stated that retention of clay nanoparticle may increase strength on sengon. Based on the description above, it can be concluded that the application of clay nanoparticle, the concentration 5% in wood preservation using pressed-vacuum (impregnation) may resist or prevent infestation of the drywood termites effectively. The lowest clay nanoparticle retention is on anggerung for about 9.32 kg.m^{3}by concentration 2.5% and the highest is on white merantifor about 24.89 kg.m^{3}by concentration 5%. High concentration may result high retention values as well. Absorption and retention are highly dependent on the void cell of the wood and solution concentration [18].

The effect of nanoparticle retention on drywood termite mortality can be attributed to its concentration, where 5% clay nanoparticle concentration provides high retention result for three species of wood and so is the mortality of drywood termite reach 100%. It can be said that the retention of clay nanoparticle has an effect on drywood termite mortality.

4. Conclusion

The application of clay nanoparticles in wood preservation in three species of wood, anggerung, white meranti and sengon can poison drywood termite (C. cynocephalus Light) mentioned in the concentration of 5% clay nanoparticles causing mortality of drywood termite to get 100%.

Acknowledgements
Special thanks to Ministry of Research, Technology and Higher Education from Directorate of Research and Public Service for funding the research through Competitive Grant in 2016 in accordance with Letter of Work Agreement/Contract Number: 25/PL.21.C/PL/2016 Date: 27th April 2016. We thanked to Suryadifor suggestions and supports.

References

[1] Anwar E K 2007 The effect of worm inoculants and the application of organic material against fertility and productivity of ultisol clay *J Trop Soil* **122** 121-30
[2] Sutandi A, Nugroho B and Sejati B 2011 Correlation between pyrite depth and some chemical properties of clay, as well as production of oil palm (*Elaeisguineensis*) *J Tanah Link.* **131** 21-4
[3] Martawijaya A and Barly 2010 *Guidance of Wood Preservation* (Bogor: IPB Press) p 107
[4] Muslich M and Ruliaty S 2016 Resistance of 45 kinds of timber in Indonesia against drywood termite *J Forest Prod Res* **342** 137-45
[5] Priadi T and Pratama A 2014 The resistance form drywood termite (*Cryptotermes cynocephalus*) after heat treatment of mindi wood (*Meliaazedarach* L) *The Utilization of Biomass from Forestand Plantation for Environment Conservation Efforts* The 6th Inter Sym of Iwors Faculty of Forestry University of Sumatra Utara Medan p 103-8 November 12-13, 2014
[6] Willeitner H and Liese W 1992 *Wood Protection in Tropical Countries a Manual on the Know How* (Deutsche GTZ GmbH Technical CooperationFederal Republic of Germany Eschbornp) p 228
[7] BSN 2014 *Resistance Test on Wood and Product of Wood against Organism that Destruct Wood SNI 7207-2014* (Jakarta: The Board of National Standardization) p 34
[8] Priyadi, Iskandar, Suwardi and Mukti R R 2015 Characteristics of heavy metals adsorption Cu, Pb and Cd using synthetieszeolite Zsm-5 *J Trop Clays* **202** 77-83
[9] Alex T, Winarni B, Kusuma I W, Arung E T and Budiarso E 2017 The clay nanoparticle impregnation for increasing the strength and quality of *sengon*(*Paraserianthesfalcataria*) and white meranti timber (*Shoreabracteolata*) timber *J Nus Biosci* **91** 107-10
[10] Persada D S, Lukmandaru G and Listyanto T 2013 Preservation of mahogany by pressure technique using deltamethrin against the infestation of drywood termites *Utilization of Renewable Natural Resources Toward Welfare and Environment Sustainability* The 5th Inter SymIwors Coordination of Private University in Kalimantan Balikpapan p 84-91 November 7-8, 2013
[11] Lebow S, Lebow P, Woodward B, Kirker G and Arango R 2015 Fifty-year durability evaluation of post treated with industrial wood preservatives *Forest Prod J.* **657** 307-13
[12] Widyorini R, Yudha A P, Lukmandaru G and Prayitno T A 2014 Mechanical and physical properties, as well as the board resistance of bamboo by citric acid adhesive against the infestation of drywood termite *The Utilization of Biomass from Forest and Plantation for Environment Conservation Efforts* Proc The 6th Inter Sym of Iwors Faculty of Forestry University of Sumatra Utara Medan p102-9 November 11, 2014
[13] Wahyudi, Ohtani Y and Ichiura H 2012 Significant feeding deterrent of berberine from talikuning (*Tinosporadissitiflora* Diels) against two subterranean termites (*Cryptotermesformosanus*Shiraki and *Reticulitermessparatus* Kolbe) *J Indon Wood Res Soc* **31** 18-22
[14] Kasmudjo 2012 *Furniture and Handicraft. Basic Theory and Application* (Yogyakarta: Cakrawala Media) p149
[15] Fatriasari W, Iswanto A H, Lestari A S, Prianto A H, Budiman I and Hadi Y S 2011 Physical mechanical, durability and electrical properties of polystyrene *Eugenia* sp. *Enhancing*

Research Based Education on Smart Processing of Inferior Timber Proc The 3rd Inter SymIwors UGM Yogyakarta 2011

[16] Siau J F 1995 *Wood: Influence of Moisture on Physical Properties*(Keene NY: Department of Wood Science and Forest Product Virginia Polytechnic Institute and State University) p 227

[17] Hidayat W, Carolina A and Febrianto F 2013 Physical, mechanical and durability properties of OSB prepared CCB treated fast growing tree species strands *J Trop wood Sci and Tec* **11** 55-61

[18] Sint K M, Adamopoulos S, Koch G, Hapla F and Militz H 2013 Impregnation of *Bombax ceiba* and *Bombax insigne* wood with a N methylal melamin compound. *Wood Sci and Tech* **471** 43-58

Conflict resolution efforts through stakeholder mapping in Labanan Research Forest, Berau, East Kalimantan, Indonesia

C B Wiati[1*], S Y Indriyanti[1,2], R Maharani[1] and Subarudi[3]

[1]Dipterocarps Forest Ecosystem Research and Development Center, Samarinda, East Kalimantan, Indonesia
[2]Master Student of Forestry Faculty, Mulawarman University, Samarinda, East Kalimantan, Indonesia
[3]Center for Research and Development on Climate Change and Policy, Bogor, West Java, Indonesia

Corresponding author: caturbudiwiati@gmail.com

Abstract. Conflict resolution in Labanan Research Forest (LRF) by the Dipterocarps Forest Ecosystem Research and Development Center (Balai Besar Penelitian dan Pengembangan Ekosistem Hutan Dipterokarpa – B2P2EHD) needs support from other parties that are also interested in such forest management. This paper aimed to presented conflict resolution in LRF through stakeholder mapping for its engagement. This research was conducted for seven months (June to December 2015) with interviews and literature study as its data collection. Collected data were analysed by a stakeholder analysis and matrix based on their interest and power levels. Two important findings were: (1) There are 19 parties having interests in the existence of LRF should be engaged; (2) Conflict resolution of LRF can be achieved: (a) ensuring key stakeholders which have high interest and high power level has same perception in existence and management of LRF, (b) establishing a partnership with primary stakeholders which have high interest and high power levels; (c) building partnerships between primary stakeholders which have high interest but low power levels, (d) building partnerships between key and secondary stakeholders which have low interest but high power levels and (e) gaining support from primary and secondary stakeholders which have low interest and low power levels. Stakeholder mapping is an important tool for tenure conflict resolution through mapping the power and interest of the conflicted parties and finding the proper parties to be approached.

1. Introduction

Conflicts of natural resource uses or tenure conflicts are often happening in Indonesia. Conflicts were becoming more frequent after Indonesia shifted to reformation and decentralization era. Wulan et al [1] reported that there were 359 evidences of conflicts in forestry sector from January 1997 to June 2003. Five main factors caused the conflict are: (i) the boundary or restriction of forest access (36%), (ii) deforestation (26%), (iii) illegal logging (23%), (iv) environmental or forest damage (12%) and (v) forest lands conversion (3%).

Conflicts also occur in Labanan Research Forest (LRF), Berau District, East Kalimantan Province, Indonesia. LRF is one of state forest area with special purpose (Kawasan Hutan Dengan Tujuan Khusus/KHDTK) belongs to the Forestry and Environment Research, Development and Innovation Agency of Ministry of Environment and Forestry (FOERDIA–MEF) and managed by the Dipterocarps

Forest Ecosystem Research and Development Center (Balai Besar Penelitian dan Pengembangan Ekosistem Hutan Dipterokarpa/B2P2EHD) as stated in Decree of the Director General (DG) of FOERDIA [2]. In the recent year, LRF has faced many conflicts in the form of illegal logging and land encroachment by local communities. In 2013, LRF security patrols in cooperation with Police Resort Berau, Section of Region I of East Kalimantan Natural Resource Conservation Agency (Balai Konservasi Sumber Daya Alam /BKSDA Kaltim) and Forestry Services of Berau District have found the encroachment in LRF area reach 2.5 Ha [3]. In 2014 also found a wood pile of mixed wood in size: 10 x 12 m (46 pcs), 5 x 20 m (43 pcs), 6 x 16 m (23 pcs),10 x 10 m (9 pcs), 5 x 10 m (8 pcs) and 5 x 7 m (13 pcs) [4]. This finding indicates that there are conflicts between LRF management with local communities.

Although the B2P2EHD is responsible for LRF management, the resolution of conflicts cannot be achieved without supporting from other stakeholders that are also interested in LRF existence. The existence of LRF is very important asset because of the presence of STREK (Silvicultural Techniques for the Regeneration of Logged Over Area in East Kalimantan) plots as a data series collected over 25 years (1990–2015). The LRF is also useful for other parties in its surrounding area. The sustainable management of LRF can be achieved by building a common understanding of all parties about the importance of LRF. This paper aimed to find conflict resolution efforts through stakeholder mapping in LRF.

2. Materials and Methods

2.1. Time and Location Research
This research was conducted for seven months (from June to December 2015) in Berau District, East Kalimantan Province, Indonesia, especially in around LRF area.

2.2 Method of Data Collection and Analysis
The parties/stakeholders were generally defined by Mitchell et al [5] as an individual or group that is able to influence the achievement of a goal and have a relationship correlated, while mapping the parties (stakeholder mapping) was defined by BSR [6] as a collaborative process that includes research, debate and discussion on determine the list of the main stakeholders of the spectrum of the parties as a whole. Mapping of the parties in this study was used as a tool to identify and assess the influence and role of a party or a group of the existence of LRF.

This study used a qualitative approached which data collected was conducted through interviews with key respondents and combined with literature study. Research stages were carried out by: (a) Parties identified and listed; (b) Parties classified based on the importance and authority; (c) Stakeholder analyzed, and (d) Priorities stages engagement of parties established in conflicts resolution of LRF. This study analysis was used a matrix of influence and interests of parties or stakeholders power – interest matrix/grid [7]. While determination of parties position in matrix was based on the authority and relevant parties views of LRF.

3. Results and Discussions

3.1 Overview of Labanan Research Forest and Surrounding Villages
LRF originated from government cooperation project between Indonesian Government and France Government in 1989–1990. STREK Plot has been developed in this project in order to found balances between benefits of tree production (wood) and its environment through tree increment measurement, annually. After this project finished, STREK Plot continued with cooperation project between Indonesian Government and European Union Government through Berau Forest Management Project (BFMP) in 1996–2001. BFMP Project was intended as a pilot project for sustainable forest management at operational level in Labanan Unit of INHUTANI I Concession (136,000 ha) to encourage Sustainable Forest Management (SFM) Certification, implementation of Reduce Impact

IOP Conf. Series: Earth and Environmental Science **144** (2018) 012063 doi:10.1088/1755-1315/144/1/012063

Logging (RIL) and Information System of Tree Topography (Sistem Informasi Topografi Pohon/ SIPTOP). Then, STREK Plot activities followed up with concession area of Labanan Unit of INHUTANI I established with covering ± 142,691 ha as research and operational activities of BFMP by Decree of Ministry of Forestry and Plantation [8]. In order to ensure legal certainty of STREK Plot research and its sustainability researches, then this area is appointed become LRF by Decree of Ministry of Forestry [9] with an area of 7,900 ha. Furthermore, according to Ministry of Forestry Decree Number: 64/Menhut-II/2012 (February 3, 2012) [10], the area was designated as KHDTK Labanan/LRF with an area of 7,959.10 ha based on definitive boundaries [11].

The administration of LRF in Berau District including 3 sub-districts namely: Teluk Bayur, Sambaliung and Kelay. There are 5 villages of 3 districts nears to LRF. From the villages bordering with LRF, the village which already have clear boundaries only Labanan Makmur and Labanan Makarti villages. This is understandable due to these two villages originated from transmigration villages. Maps of LRF can be seen in this following figure 1.

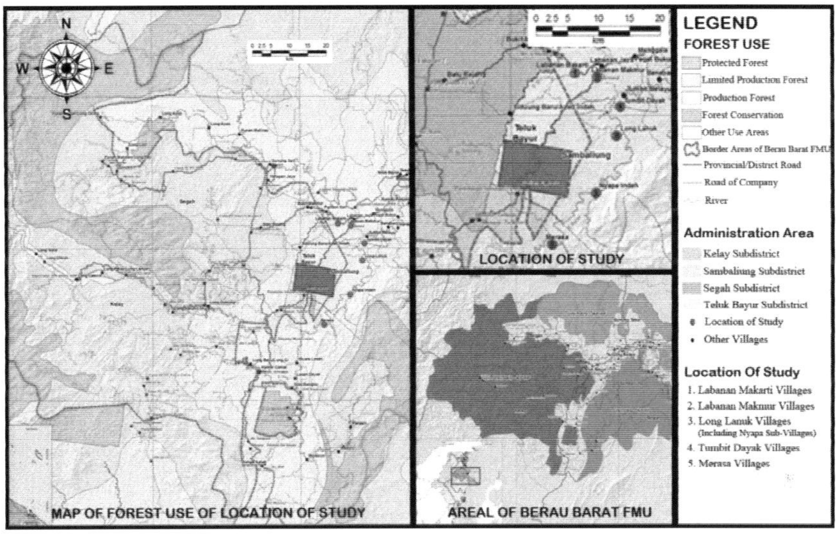

Figure 1. Map of Location Study

The profile of the villages shown in following table.

Table 1. Profiles of Villages in around Labanan Research Forest [12-19]

Sub District	Village	Position from LRF	Established year	Large (Km^2)
Teluk Bayur	Labanan Makarti[a]	North	1984	14.31
	Labanan Makmur	North	1982	9.26
Sambaliung	Long Lanuk[a] (including Nyapa Indah Subvillage)	East	Naturally	427.11
	Tumbit Dayak	Northeast	Naturally	81.19
Kelay	Merasa[a]	Southeast	1968	345.99

[a]The village directly bordering with LRF

3.2 Interest Conflicting Occured in Labanan Research Forest

Generally, LRF disallowance by village communities around before this area was designated as LRF, precisely since 2005, followed with BFMP activities finished in this area. This is evidenced by destruction of STREK camp (Km 37) and claiming of land developed by Merasa village community.

Unfortunately, B2P2EHD was not investigated the perpetrators yet due to authority limited, but it has been widely reported.

LRF disturbances by personal attacked from Merasa Village begins on July 2013. Some seasonal crops such as bananas and others were planted in that area. Land clearings massively happened on May 2015 by Merasa villagers followed by Nyapa Indah sub-village resident. Land clearings in LRF effected lost in some assets including research plots (STREK Plot and others), facilities and infrastructure that had been built. The research plots in LRF that destroyed by land clearing can be seen in following table.

Table 2. Research Plots Damaged in LRF due to Land Clearings by Local Community [20]

No	Plot Impacted	The Extent of Impact (Ha)	Total Area (Ha)	Percentage Impacted (%)
A	Merasa Village Community Group	49.19	349	
1	Bina Pilih	7.37	24	30.71%
2	Watershed Rehabilitation (Rehabilitasi DAS)	25.79	285	9.05%
3	Intensive Silviculture (Silvikuktur Intensif/SILINT)	15, 07	36	41.86%
4	STREK RKL I Plot 2 Square 3 & 4	0.96	4	24.00%
B	Nyapa Indah Sub-village Community Group	15.42	104	
1	In-Situ Keruing	2.03	12	16.92%
2	Seed Orchard of Keruing	7.71	25	30.84%
3	Seed Orchard of Meranti	4.15	50	8.30%
4	Seed Orchard of Tengkawang	1.12	5	22.40%
5	SILINT	0.41	12	3.42%
	Total damaged area	64.61	453	14.26%

Wiati and Indriyanti [21] reported some causes of the conflict in LRF as follows LRF less well known by local people around, there are some company's activities around LRF, land claims and land clearing by local people around LRF and weakness of law enforcement. The dominant conflict causes is motivation of land claims by local people and land clearing by Merasa villagers that motivated other villages to do the same also. The cause of the conflict in LRF briefly shown in the following figure.

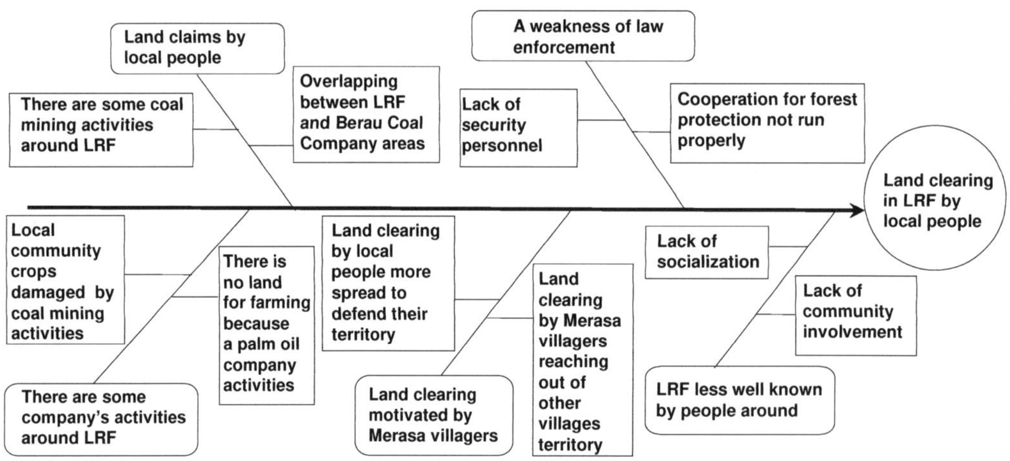

Figure 2. Diagram *Fishbone* causes of conflict in LRF

3.3 The Stakeholders in Labanan Research Forest

To better understand the conflict in LRF and find ways of conflict resolution, it is necessary to analyze all of parties to map relationship between parties and conflict between parties. However, because this research has not yet conduct Focus Group Discussion (FGD) between parties who concerned on existence of LRF, then this article just will disclose information about the parties who potentially can work together to support B2P2EHD for the resolution of conflicts in LRF. For ease of analysis, then stakeholders associated with presence and management LRF divided into 3, namely: (i) Primary Stakeholders, who having interests directly with a policy, program or project; (ii) Secondary Stakeholders (supporting), who have no direct interest, but participate and have influence on public attitudes and decisions; and (iii) Key Stakeholders, which lacks legal authority in decision-making as shown in Table 3.

Table 3. Clasification of Stakeholders Involved in the Tenurial Conflict in Labanan Research Forest

Primary Stakeholders	Secondary Stakeholders	Key Stakeholder
1. Dipterocarps Forest Ecosystem Research and Development Center (B2P2EHD) 2. Berau Barat Forest Management Unit (Kesatuan Pengelolaan Hutan Produksi/KPHP Berau Barat) 3. Hutan Sanggam Labanan Lestari (HLL) Company 4. Berau Coal (BC) Company 5. Local People around LRF 6. Center for Orangutan Protection (COP) 7. Mahakam Berau Watershed Management Agency (Balai Pengelolaan Daerah Aliran Sungai/BP DAS Mahakam Berau) – Nusantara Berau Coal (NBC) Company	1. Berau District Regional People Representative Assembly (Dewan Perwakilan Rakyat Daerah/DPRD Berau) 2. Forest and Climate Change (ForClime) Programme 3. Bestari Foundation 4. East Kalimantan Natural Resource Conservation Agency (BKSDA Kaltim) 5. Sub-District (Teluk Bayur, Sambaliung and Kelay) 6. Rimba Anugerah Kaltim (RAK) Company 7. Kaltim Jaya Bara (KJB) Company 8. Armed Battalion of 18/105 Tarik Buritkang	1. Ministry of Environment and Forestry (MEF) 2. Forestry and Environment Research, Development and Innovation Agency of Ministry of Environment and Forestry (FOERDIA–MEF) 3. Ministry of Energy and Mineral Resources (MEMR) 4. Berau District Government

Table 3 shown that key stakeholders which have legal authority for existences and management of LRF were FOERDIA–MEF, MEF and Berau District Government. Meanwhile, stakeholders related to LRF management (primary stakeholders) were included 7 parties e.g. B2P2EHD, KPHP Berau Barat, HLL Company, BC Company, local people around LRF, COP, and BP DAS Mahakam Berau – NBC Company.

3.4 The Parties Participation in Conflict Resolution in Labanan Research Forest

In order to determine the parties which may be involved in the conflict resolution of LRF, each stakeholder was divided into 4 groups based on its interest and power levels as shown in Figure 3. Group A showed stakeholder groups that have a high interest towards to LRF but have a low power on policy-making, including members of institution or person responsible for the implementation of activities but not as policy makers. Group B was a group of stakeholders that have a low interest and a low power on existences and management of LRF. Group C shown stakeholder groups that have a high degree of interest and power related to the existences and management of LRF. While group D showed a group of stakeholders that have a low interest but have a high power on the existences and management of LRF.

Interest	Power: Low	Power: High
High	**Group A** • Local People around LRF • COP	**Group C** • MEF • FOERDIA–MEF • MEMR • B2P2EHD • KPHP Berau Barat • HLL Company • BC Company
Low	**Group B** • BP DAS Mahakam Berau – NBC Company • BKSDA Kaltim • Sub-district (Teluk Bayur, Sambaliung and Kelay) • RAK Company • KJB Company • Armed Battalion of 18/105 Tarik Buritkang	**Group D** • Berau District Government • DPRD Berau • ForClime Programme • Bestari Foundation

Figure 3. Matrix grouping of stakeholders based on its interests and impacts

Figure 3 indicated that although the stakeholders can be divided into 3 groups: primary, secondary and key stakeholders, they can also be divided by their degree of interest and power levels. This division is usefull because some parties have a high interest and a low power should be involved in conflict resolution. Grouping through this matrix would much help for conflict resolution because the stakeholders would increase their understanding about the position of other parties based on their difference levels of interest and power. Level differences in interest and power should be addressed through building of common perception about management and existences of LRF in order to achieve its conflict.

Related to parties involvement, Figure 3 showed there are several steps for conflict resolution that can be performed by B2P2EHD i.e. (1). First step, to ensure that key stakeholders included in Group C (MEF, FOERDIA–MEF, MEMR) had been common perception in existences and management of LRF. Its common perception was related to policy authority including area identity changes and area permit that can be greatly impacted on management of LRF, in the future; (2). Second step, to develop cooperation between primary stakeholders were included in Group C (KPHP Berau Barat, HLL Company, and BC Company) had payed attention to the problem happened. In this steps, B2P2EHD was received fund and human resources from Group C to carrying conflict resolution. On the other hand, this cooperation can be informed to the public that LRF management activities being carried out by the B2P2EHD and other party; (3). Third step, to develop cooperation with primary stakeholders from Group A (local people around LRF and COP). In this step, B2P2EHD needs to ensure that local people around LRF and COP interest in accordance with B2P2EHD, especially LRF utilization. Important perception, although B2P2EHD has legal authority to manage LRF, its management will be considered to other parties interest, and (4). Fourth step, to develop cooperation with key and secondary stakeholders that were in group D (Berau District Government and DPRD Berau). This cooperation was very important due to LRF is forest area included in the administrative area of Berau District. This condition causes local communities around LRF more obey with Berau District Government and DPRD Berau than B2P2EHD. Besides, B2P2EHD also need to develop cooperation with For Clime Programme and Bestari Foundation that also included in Group D. Both institutions are carrying out their activities by developing a strong interaction with local people around LRF, so that they can give a mindset influences for local people. This cooperation will helped B2P2EHD can be accepted by community around LRF. Last, when the support of secondary stakeholders in Group D was considered insufficient, then next step required B2P2EHD will supported from primary and secondary stakeholders in Group C (BP DAS Mahakam Berau–NBC Company, Sub–district (Teluk Bayur, Sambaliung and Kelay), RAK Company, KJB Company, BKSDA Kaltim and Armed Battalion of 18/105 Tarik Buritkang). It was due to their work location relatively close to LRF, so it can covered B2P2EHD weaknesses, especially related to lack of funding and human resources in LRF management, particularly in security activities.

4. Conclusion

Nineteen parties have interested in existences of LRF. They can be put in the stakeholders group (as primary, secondary, and key stakeholders) and the stakeholders matrix based on their own position determined by their interest and impact to the use of LRF. This position has to be considered and addressed by the management of LRF for its conflict resolution agenda. Another agenda is to socialize the important of LRF for both the sustainable forest management program and the prosperity of local people living in surrounding LRF area. Five steps can be carried out by the B2P2EHD for conflict resolution namely: (a) ensuring key stakeholders which have high interest and high power level has same perception in existence and management of LRF; (b) establishing a partnership with primary stakeholders which have high interest and high power level; (c) building partnership between primary stakeholders which have high interest but low power level; (d) building partnership between key and secondary stakeholders which have low interest but high power level and (e) gaining support from primary and secondary stakeholders which have low interest and low power level. However, the stakeholder mapping is important tools for solving the tenure conflict and finding the parties to be approached.

Acknowledgements

The authors gratefully acknowledge to Dipterocarps Forest Ecosystem Research and Development Center (B2P2EHD); Forestry and Environment Research, Development and Innovation Agency of Ministry of Environment and Forestry (FOERDIA–MEF); Forestry Faculty of Mulawarman University and also all interviewed related parties.

References

[1] Wulan Y C, Yasmi Y, Purba C and Wollenberg E 2004 *Analisa Konflik Sektor Kehutanan di Indonesia 1997 – 2003* (Bogor: Center for International Forestry Research (CIFOR))
[2] Surat Keputusan. 2007 Badan Penelitian dan Pengembangan Kehutanan SK No. 90/Kpts/VIII/2007 tentang Penunjukan Penanggung Jawab Pengelolaan Kawasan Hutan Dengan Tujuan Khusus (KHDTK) Lingkup Badan Penelitian dan Pengembangan Kehutanan
[3] B2PD 2013 Laporan KHDTK. Balai Besar Penelitian Dipterokarpa p 7
[4] B2PD 2014 Laporan KHDTK. Balai Besar Penelitian Dipterokarpa p 9
[5] Mitchell R K, Agle B R and Wood D J 1997 Toward a theory of stakeholder identification and salience: defining the principle of who and what really counts *Academy of Management Review* **22** p 853–86
[6] BSR 2011 *Stakeholder Mapping. Business for Social Responsibility.* Diakses dari http://gsvc.org/wp-content/uploads/2014/11/Stakeholders-Identification-and-Mapping.pdf tanggal 30 Juni 2015
[7] Slaba M 2014 Stakeholder Power – Interest Matrix and Stakeholder Responsibility Matrix in Corporate Social Responsibility *Proc. The 8th International Days of Statistics and Economics* (Prague) p 1366–74
[8] Surat Keputusan 1999 Menteri Kehutanan dan Perkebunan SK No. 866/Kpts–II/1999 tentang Penetapan Areal Konsesi PT. Inhutani Unit I Labanan Seluas ± 142.691 Hektar Sebagai Areal Penelitian dan Kegiatan Operasional BFMP

[9] Surat Keputusan 2007 Menteri Kehutanan SK No. 121/Menhut-II/2007 tentang Penunjukan Kawasan Hutan Produksi Tetap seluas ± 7.900 (Tujuh Ribu Sembilan Ratus) Hektar di Kabupaten Berau Provinsi Kalimantan Timur sebagai Kawasan Hutan Dengan Tujuan Khusus untuk Hutan Penelitian Labanan

[10] Surat Keputusan 2012 Menteri Kehutanan SK No. 64/Menhut-II/2012 tentang Penetapan Kawasan Hutan Dengan Tujuan Khusus untuk Hutan Penelitian Labanan yang terletak di Kabupaten Berau, Provinsi Kalimantan Timur seluas 7.959,10 (Tujuh Ribu Sembilan Ratus Lima Puluh Sembilan dan Sepuluh Perseratus) Hektar

[11] BPKH Wilayah IV. 2009. Berita Acara Pembuatan Tata Batas Definitif Kawasan Hutan Dengan Tujuan Khusus Untuk Hutan Penelitian Labanan, Kabupaten Berau, Provinsi Kalimantan Timur. Balai Pemantapan Kawasan Hutan Wilayah IV

[12] BPS Kabupaten Berau 2015[a] *Kecamatan Sambaliung Dalam Angka* (Badan Pusat Statitistik Kabupaten Berau) p 31

[13] BPS Kabupaten Berau 2015[b] *Kecamatan Teluk Bayur Dalam Angka* (Badan Pusat Statitistik Kabupaten Berau) p 32

[14] BPS Kabupaten Berau 2015[c] *Kecamatan Kelay Dalam Angka* (Badan Pusat Statitistik Kabupaten Berau) p 35

[15] BFBP – Yayasan Bestari 2003[a] *Profil Desa Labanan Makarti; Laporan Survey Sosial Ekonomi Masyarakat Sekitar Hutan* (Berau Forest Bridging Project – Yayasan Bestari) p 1–3

[16] BFBP – Yayasan Bestari 2003[b] *Profil Desa Labanan Makmur; Laporan Survey Sosial Ekonomi Masyarakat Sekitar Hutan* (Berau Forest Bridging Project – Yayasan Bestari) p 1–3

[17] BFBP – Yayasan Bestari 2003[c] *Profil Kampung Long Lanuk; Laporan Survey Sosial Ekonomi Masyarakat Sekitar Hutan* (Berau Forest Bridging Project – Yayasan Bestari) p 1–5

[18] BFBP – Yayasan Bestari. 2003[d]. *Profil Kampung Merasa; Laporan Survey Sosial Ekonomi Masyarakat Sekitar Hutan* (Berau Forest Bridging Project – Yayasan Bestari) p 1–4

[19] BFBP – Yayasan Bestari 2003[e] *Profil Kampung Tumbit Dayak; Laporan Survey Sosial Ekonomi Masyarakat Sekitar Hutan* (Berau Forest Bridging Project – Yayasan Bestari) pp 1–3

[20] B2PD 2015 *Studi Pengembangan Model Kemitraan Kehutanan di KHDTK Labanan Kabupaten Berau* (DIREC Penelitian Dipterokarpa: Laporan Hasil Penelitian) p 25

[21] Wiati C B and Indriyanti S Y 2015 Upaya Penanganan Konflik Di KHDTK Labanan Kabupaten Berau Melalui Pembangunan Kemitraan Kehutanan *Makalah dalam Seminar Solusi Penanganan Konflik Masyarakat Hutan Melalui Upaya Pengelolaan Kawasan Hutan Secara Partisipatif* pp 9–12

Examining spectral properties of Landsat 8 OLI for predicting above-ground carbon of Labanan Forest, Berau

A Suhardiman*, B A Tampubolon, M Sumaryono

Faculty of Forestry, University of Mulawarman

*Corresponding Author: suhardiman@fahutan.unmul.ac.id

Abstract. Many studies revealed significant correlation between satellite image properties and forest data attributes such as stand volume, biomass or carbon stock. However, further study is still relevant due to advancement of remote sensing technology as well as improvement on methods of data analysis. In this study, the properties of three vegetation indices derived from Landsat 8 OLI were tested upon above-ground carbon stock data from 50 circular sample plots (30-meter radius) from ground survey in PT. Inhutani I forest concession in Labanan, Berau, East Kalimantan. Correlation analysis using Pearson method exhibited a promising results when the coefficient of correlation (r-value) was higher than 0.5. Further regression analysis was carried out to develop mathematical model describing the correlation between sample plots data and vegetation index image using various mathematical models. Power and exponential model were demonstrated a good result for all vegetation indices. In order to choose the most adequate mathematical model for predicting Above-ground Carbon (AGC), the Bayesian Information Criterion (BIC) was applied. The lowest BIC value (i.e. -376.41) shown by Transformed Vegetation Index (TVI) indicates this formula, $AGC = 9.608*TVI^{21.54}$, is the best predictor of AGC of study area.

1. Introduction

The advancement of satellite-based remote sensing technology brings the opportunity to study forests much better. Various types of satellite imagery are available in many different spatial resolution. Landsat 8 OLI (Operational Land Imager) is the latest generation of NASA (National Atmospheric and Space Administration) earth resource satellite series which was launched in 2013. This new satellite persist its 30-meter spatial resolution to match with the previous generation. The good thing about Landsat is that all image products are completely free and available worldwide through USGS website. For country like Indonesia, Landsat images have been used as main source of land cover mapping since 1990. On the other hand, for remote sensing communities, Landsat spectral properties is always interesting to be explored and studied in various fields of research and applications.

Although there are encouraging results in utilizing remote sensing data to study forest attributes such as basal area, stem volume or aboveground biomass in larger areas [1], correlation between the vegetation cover on the satellite images with forest attributes (e.g. above-ground biomass) can be very poor [2] especially in the complex tropical forest where vegetation are composed from hundreds of species and having different ages. Higher variability in tropical forest structure is major reasons for uncertainties in estimating forest attributes [3], [2]. The variation depends on factors such as soil type, topographic position [4], soil nutrients [5] and levels of human disturbance [6].

In order to increase accuracy, some studies suggested utilization of high spatial resolution image or Lidar data [7]. Nonetheless, in case of Indonesia, only a few of government bodies or private companies including forest concession holders willingly spent their money to purchase high spatial resolution image. Depends on the project scale, some high spatial resolution images were bought limited to particular location. The fact is that most of forest concessions are still using moderate spatial resolution images such as Landsat [8] to be used for forest areas mapping and monitoring activities.

Hence, the accurate estimation of structural characteristics of tropical forest vegetation remains as a major obstacle [9]. The promising results are mostly obtained from studies on conifer forest, which consist of pure stands dominated by a single tree species [10], [11], [12]. For optical remote sensing, various image transformation methods such as vegetation indices and texture analysis have been used to estimate forest attributes and so far the accuracy depends on the site, forest type as well as spatial resolution of the image. In this study, the spectral properties of relatively new Landsat 8 OLI were examined in order to predict above-ground carbon (AGC) at large scale based on pixel values and sample plots data.

2. Method

2.1. Study Area

The study area covered approximately 1,380 square kilometers (km^2)of PT. Inhutani I forest concession at Labanan, Berau District, East Kalimantan. Various land cover types were presented in this area but two most dominant are primary lowland Dipterocarp forest (51%) and secondary lowland Dipterocarp forest (23%) which is source for most valuable and commercial timber from this forest. Figure 1 below shows the situation map of study area and the distribution of sample plots.

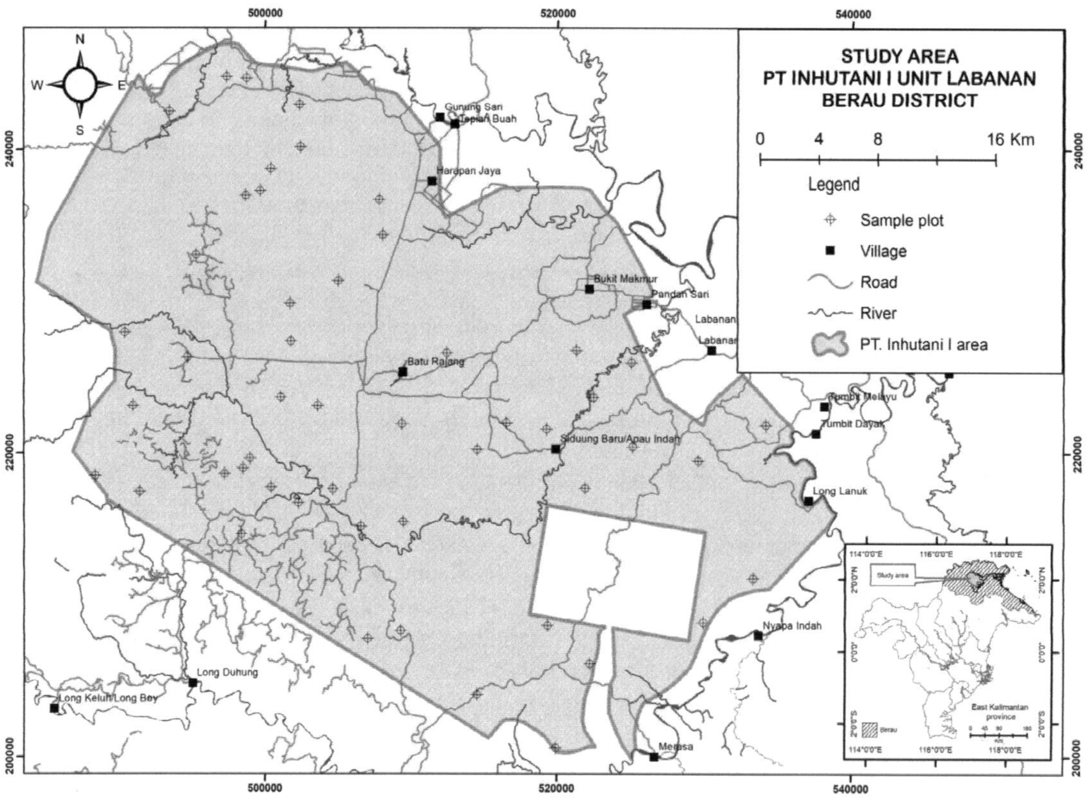

Figure 1. Study area and sample plot distribution

2.2. Materials

There are two set of data needed for this study and therefore will be explained in this section. First data related to the Landsat 8 OLI images which have been transformed into three spectral vegetation indices. Landsat imagery is continuous type of data contains grid pixel values which indicate the magnitude of spectral reflectance over objects on earth. Other dataset was AGC (ton C/ha) from sample plots which was collected in September 2016.

2.2.1. Satellite data. Landsat 8 OLI path/row 117/059 images were downloaded from http://earthexplorer.usgs.gov and two acquisition periods were selected on May and September 2016. In tropics, the persistent of cloud is major issue, especially when dealing with optical remote sensing products such as Landsat. Therefore finding images on Landsat catalog sometimes very challenging. One image scene may not be sufficient due to cloud which was cover up the interest point or location of the study.

Landsat 8 OLI brought more sensors than its predecessor. The sensing product of sensor is called band. The pixel value of band represents bit-package combination of surface, atmosphere, and sensor condition that can affect the overall usefulness of a given pixel [13]. Furthermore [14]mentioned that vegetation indices may be sensitive to the atmospheric condition. Therefore in this study, radiometric correction was applied to pixel band so that atmospheric factor can be excluded using Top of Atmosphere (ToA) formula as provide in [15].

Among nine Landsat 8 OLI spectral bands, near-infrared (NIR) and red (R) band are essential to study biomass content of vegetation. NIR and R band were used in this study to transform pixel values to vegetation index values. There are numerous vegetation index which has been developed and some of them were exhibited a positive correlation withstand or other forest properties.In this study, three vegetation indices that is Normalized Difference Vegetation Index (NDVI), Single Ratio Vegetation Index (SRVI) and Transformed Vegetation Index (TVI) were examined. Landsat pixel value was transformed to vegetation index value using these following formula:

$$NDVI = \frac{NIR - R}{NIR + R} \quad \ldots\ldots (1)$$

$$SRVI = \frac{NIR}{R} \quad \ldots\ldots (2)$$

$$TVI = \sqrt{NDVI + 0.5} \quad \ldots\ldots (3)$$

Where NIR is near infrared band and R is red band of Landsat 8 OLI. The transformation resulted a continuous layer of NDVI, SRVI, and TVI with similar format to Landsat 8 OLI unless the pixel values were altered. The new pixel value were critical in this study since it role as predictor of value from ground sample plot data.

2.2.2. Sample plot data. As many as 50 sample plots were established during field surveying September 2016. Simple random sampling were chose to pinpoint the sample plots location so that bias would be minimized. Four team of surveyor were set up to do forest inventory in the plots. GPS receiver helped teams to navigate plots. Once the survey team arrived at targeted location, a 30-meter radius of circular sample plots were then established. Sample plots were also nested to capture different strata of vegetation. Smaller radius of plots captured smaller trees. A 30-meter radius sample plot is set to collect data for big trees (dbh $\geq$ 50 cm) while the smallest subplot (radius equals to 1.5 m) were collected seedling data (juvenile trees where dbh < 2 cm and height < 1.5 meter). Figure 2 shows layout of the sample plot established on the ground.

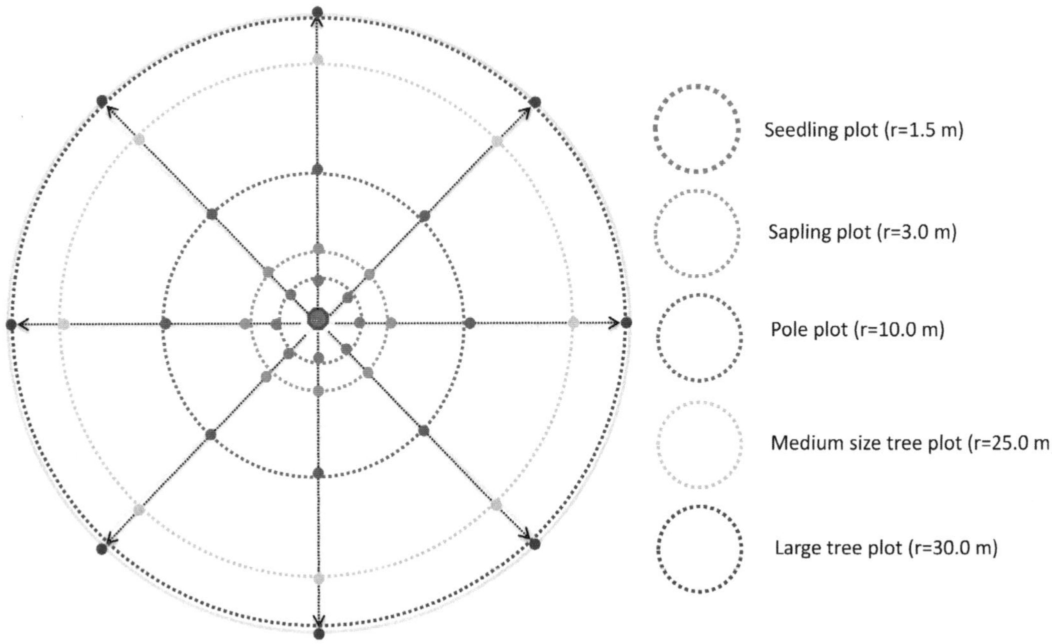

Figure 2. Layout of sample plots on the ground

Diameter at breast height (dbh) of the trees, pole, and sapling were measured in each sample plots and subplots while species/genus/family name and land cover types were carefully identified. The center of the sample plot were marked using GPS receiver in order to update the plot position on maps.Eventually, allometric equation of [16] were used to estimate above-ground biomass (AGB) content for each trees.

$$Ln(AGB) = -2.289 + \left(2.649 \cdot Ln(Dbh)\right) - \left(0.021 \cdot Ln(Dbh)^2\right). \dots\dots\dots\dots\dots\dots\dots\dots\dots\dots\dots\dots\dots (4)$$

Where AGB above-ground biomass (kg) and Dbh is diameter at breast height (cm). Conversion of AGB to AGC simply by multiplying AGB with a constant value of 0.47 [17]. The estimation of AGC (ton C ha^{-1}) for each sample plot was calculated by summing up all carbon trees and multiplied by area conversion factor.

2.2.3. Statistical data analysis. The position of 50 sample plots were overlaid with three-layer data of vegetation indices in geographic information system (GIS) environment. Position of sample plots were drew as circle polygon with radius 30 meter rather than points. Pixel values that correspond to the polygon of sample plot position were averaged. At the end, 50 carbon stocks from sample plots have their corresponding pixel value.

Furthermore, statistical analysis was carried out to denote potential relationship between carbon stock and pixel value. Different mathematical function were tested i.e. linear and non-linear (logarithmic, quadratic, power and exponential) to develop befittingrelationship models. The coefficient of Pearson correlation (r) were used to measure how well two sets of data are related to each other.Theoretically r-value closed to 1 exhibits strong positive correlation.

In this study, models with r-value greater than 0.5 were selected to be further analyzed using Bayesian Information Criterion (BIC). BIC is a tool to estimate the quality of mathematical or statistical model of a given dataset. The model with lowest BIC is preferred [18].For multi datasets, BIC is calculated using following formula.

$$BIC = \sum_{i=1}^{N} \left(n_i \cdot ln \frac{RSS}{n_i} \right) \cdot +k \cdot N \cdot \ln(n_t) \quad (5)$$

Where BIC is Bayesian Information Criterion, RSS is residual sum of square of regression, k is number of parameters used in the model, N is number of datasets, n_t= total number of data point used in the set (i.e. in all N datasets), i is index of dataset 1, 2,..., N, and n_i is the number of points in dataset 1, 2,..., N.

3. Results and discussion

3.1. Above-carbon stock of sample plots.

The characteristics of Labanan forest of PT.Inhutani I could be recognized from sample plots data. Tropical forest of Kalimantan has complex stand structure and diverse species composition. That is the reason why variation among forest properties (e.g. stand volume, AGB or AGC) is relatively high in tropics especially where forest has been disturbed by logging or any other human interference. Area of PT Inhutani I is an example of lowland dipterocarp forest that has been logged for years so that different forest cover type exist as an impact of logging activities including encroachment and et cetera.

Eighty-four percent of the total sample plots were randomly located at lowland dipterocarp forest as major land cover type over the study area. The average AGC approximately 135.18 ton C/ha with standard deviation is 57.16 ton C/ha. Hence the variation in this area is about 42.28%. The minimum AGC is 11.12 ton C/ha while the maximum one is 287.32 ton C/ha. This average carbon stock value is slightly lower than the average value of Indonesia secondary lowland forest carbon stock which is 169 ton C/ha.

3.2. Correlation of AGC and vegetation index.

Table 1 below summarizes the relationship between AGC and pixel value of three different vegetation indices. The degree of relationship denoted by r-value exhibit a promising result. All three indices showed r-value greater than 0.5. According to [19], R-values between ±0.3 and ±0.7 indicate a moderate positive or negative linear relationship. The relatively low r-value as result of this study was admittedly predicted.[2] mentioned that correlation between the vegetation cover on the satellite images with forest attributes (e.g. aboveground biomass) can be very poor especially in the complex tropical forest where vegetation are composed from hundreds of species and having different ages.Furthermore, [20] stressed and argued that low coefficient of correlation is often showed by most of optical remote sensing data analysis when it used to estimate forest attributes.

Table 1. Pearson correlation coefficient of each vegetation index

Vegetation index	Coefficient of Pearson correlation (r)	Significant (2-tailed)
NDVI (*Normalized Difference Vegetation Index*)	0.523	0.000
SRVI (*Single Ratio Vegetation Index*)	0.528	0.000
TVI (*Transformed Vegetation Index*)	0.517	0.000

Table 2 showed mathematical models of vegetation index as predictor. Different behavior were demonstrated by each model. Linear model as well as logarithmic and quadratic exhibited lower coefficient of determination (R^2) compare to power and exponential model. This finding suggested that correlation between AGC and vegetation index is adequate to be modeled by power and exponential function for all vegetation indices.

Table 2. Regression model of vegetation index as predictor using five mathematical functions

Indices	Regression / Mathematical model	R^2	R^2 Adjusted	Model
NDVI	-330.1 + 608.6*NDVI	0.272	0.257	Linear
SRVI	-50.84 + 23.97*SRVI	0.278	0.263	Linear
TVI	-1319 + 1294*TVI	0.267	0.252	Linear
NDVI	240.4 + 388.3*ln NDVI	0.255	0.239	Logarithmic
SRVI	-191.8 + 160.8*ln SRVI	0.287	0.272	Logarithmic
TVI	-26.34 + 1382*ln TVI	0.262	0.247	Logarithmic
NDVI	$300.7 - 1300*NDVI + 1411*NDVI^2$	0.295	0.265	Quadratic
SRVI	$-161 + 55.7*SRVI - 2.202*SRVI^2$	0.291	0.261	Quadratic
TVI	$7275 - 1.465e{+}04*TVI + 7379*TVI^2$	0.296	0.266	Quadratic
NDVI	$624.8*NDVI^{6.117}$	0.569	0.560	Power
SRVI	$1.026*SRVI^{2.338}$	0.546	0.536	Power
TVI	$9.608*TVI^{21.54}$	0.573	0.564	Power
NDVI	0.0952*Ln(9.3284*NDVI)	0.575	0.567	Exponential
SRVI	$9.507*1.385^{SRVI}$	0.462	0.451	Exponential
TVI	$(2.035*10^{-8})*(4.875*Ln(20.005*TVI))$	0.574	0.565	Exponential

Figure 3 below showed the scatter plot diagram as well as the trendline of dataset using power and exponential function which explore the relationship between vegetation index and AGC from sample plots. All vegetation indices showed a promising result where R^2 value slightly higher than 0,5 except SRVI model developed using polynomial function (Figure 3d). Either NDVI and TVI, both indices demonstrated narrow value range which indicate variation among pixels are a bit lower. In contrast, the variability of AGC from sample plots showed wider range that indicate higher variation of AGC value between sample plots.

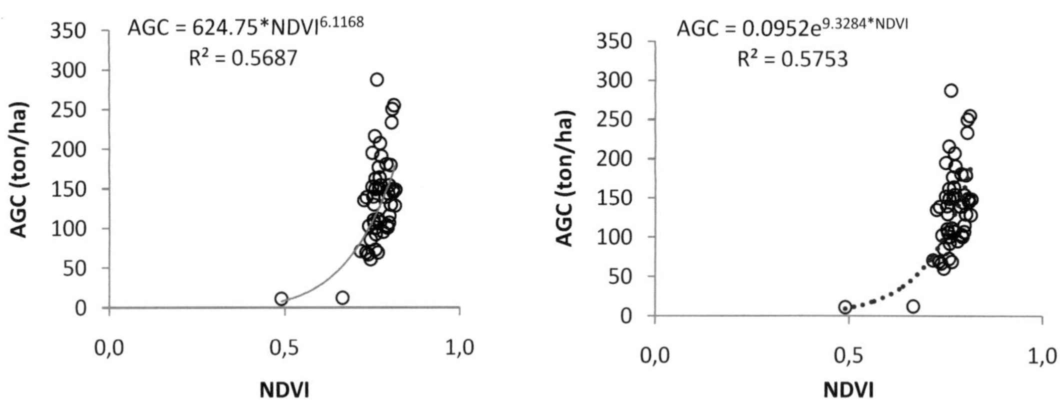

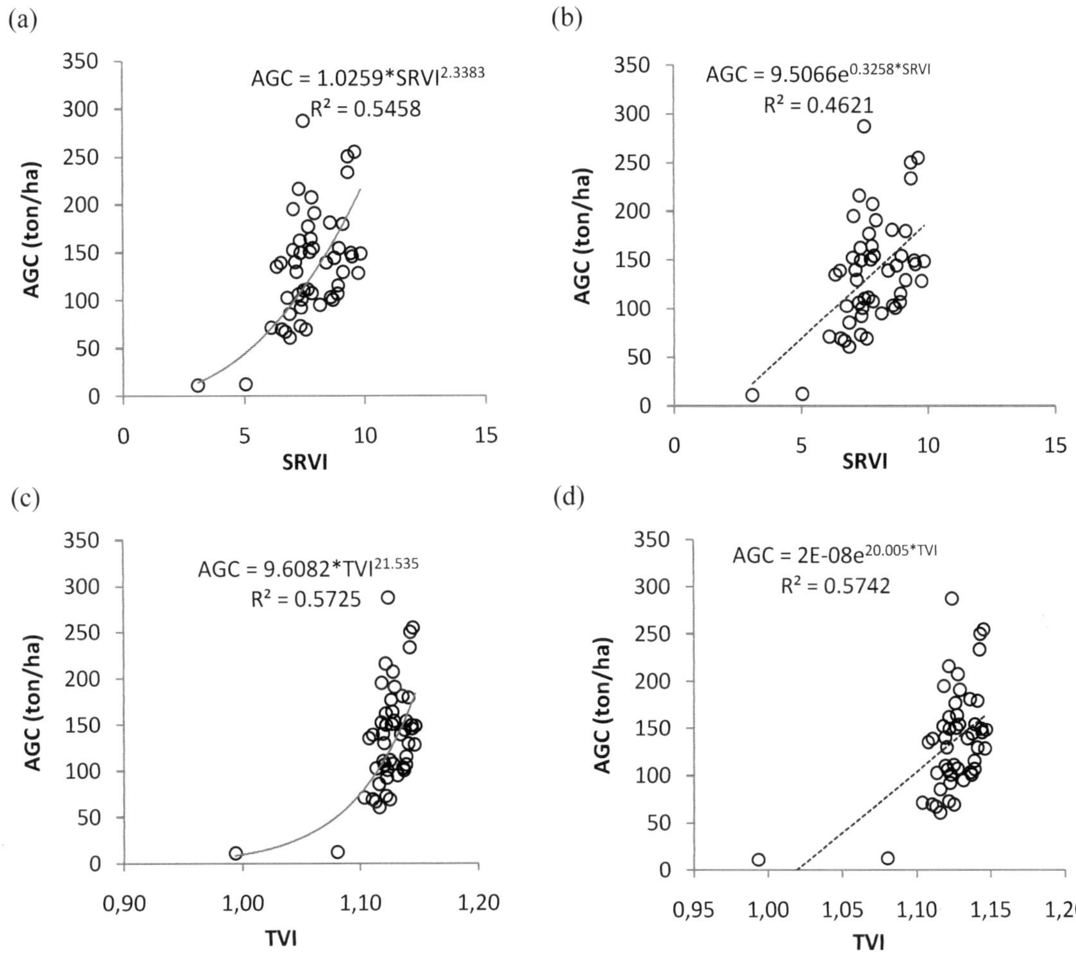

(e) (f)

Figure 3. Graphical relationship between AGC and vegetation index using power function (a, c and e) and exponential function (b, d, and f).

Table 3 below showed the calculated BIC values only for power and exponential models. Lowest BIC value is -376.41 and it belongs to the mathematical model where TVI is the predictor. Then the selected mathematical model for predicting AGC (ton/ha) in PT. Inhutani I Unit Labanan in Berau district is $AGC = 9.608*TVI^{21.54}$

Table 3. The calculated Bayesian Information Criterion (BIC) for selecting best mathematical model

	Mathematical model	N	n_i	n_t	k	RSS	BIC
Power	$624.8*NDVI^{6.117}$	3	50	150	3	0.051	-299.55
Power	$1.026*SRVI^{2.338}$	3	50	150	3	35.196	27.54
Power	$9.608*TVI^{21.54}$	3	50	150	3	0.011	-376.41
Exponential	0.0952*Ln(9.3284*NDVI)	3	50	150	3	0.050	-300.25
Exponential	$9.507*1.385^{SRVI}$	3	50	150	3	41.709	36.03
Exponential	$(2.035*10^{-8})*(4.875*Ln(20.005*TVI))$	3	50	150	4	0.011	-361.49

4. Conclusion

Landsat products are the most widely used satellite imagery in Indonesia due to its extensive coverage and available at no-cost. Therefore exploration of the capability of Landsat product especially Landsat

8 OLI to predict land-based properties is always interesting subject. This study revealed once again that spectral properties of Landsat image as reflectance of energy from the sun exhibited a moderate correlation with forest properties which is above-ground carbon stock.The complexity of forest structure is one factor preventing coefficient of correlation to meet high value i.e. 0.7 or 0.8. Landsat pixel value captures only the emergent trees canopy while co-dominant and suppressed trees are unable to be recognize. Another factor is related to the pixel size of Landsat that seems too large (30 × 30 m) to capture underlying variability of certain objects on the ground such as AGC. Therefore further study may be needed to discriminate AGC based on the forest stratum before correlation with pixel values is undertaken.

Acknowledgments
This study supported from GIZ Forclime Forests and Climate Change Programme and Forestry office of Berau District.

References

[1] Odzemir I, Karnieli A. 2011. Predicting Forest Structural Parameters Using the Image Texture Derived from WorldView-2 Multispectral Imagery in a Dryland Forest, Israel. International Journal of Applied Earth Observation and Geoinformation 13: 701-710.
[2] Saatchi S, Houghton R, Dos Santos Alvala R, Soares J, Yu Y. 2007. Distribution of Aboveground Live Biomass in the Amazon Basin. Global Change Biology 13: 816-837.
[3] Intergovernmental Panel on Climate Change. 2001. Climate Change 2001: The Scientific Basis. Contribution of Working Group I to the Third Assessment Report of the Intergovernmental Panel on Climate Change. Editor Houghton J, Ding Y, Griggs D, Noguer M, van Linden P, Dai X. Cambridge: Cambridge University Press
[4] Clark D, Clark D. 2000. Landscape-scale Variation in Forest Structure and Biomass in a Tropical Rain Forest. Forest Ecology and Management 137: 185-198.
[5] Laurence W, Fearnside P, Laurance S, Delamonica P, Lovejoy T, Rankin de Merona J. 1999. Relationship Between Soils and Amazon Forest Biomass: a Landscape-scale Study. Forest Ecology and Management 118: 127-138.
[6] Urquiza-Haas T, Dolman P, Peres C. 2007. Regional Scale Variation in Forest Structure and Biomass in the Yucatan Peninsula, Mexico: Effects of Forest Disturbance. Forest Ecology and Management 247: 80-90.
[7] Powel S, Cohen W, Healey S, Kennedy R, Moisen G, Pierce K. 2010. Quantification of Life Aboveground Forest Biomass Dynamics with Landsat Time-series and Field Inventory Data: A Comparison of Empirical Modeling Approaches. Remote Sensing of Environment 114: 1053-1068.
[8] Lu D. 2005. Aboveground Biomass Estimation Using Landsat TM Data in the Brazilian Amazon. International Journal of Remote Sensing 26 (12): 2509-2525.
[9] Drake JB, Dubayah RO, Clark DB, Knox RG, Blair JB, Hofton MA. 2002. Estimation of Tropical Forest Structure Characteristics Using Large-footprint Lidar. Remote Sensing of Environment 79: 305-319.
[10] Kayitakire F, Hamel C, Defourny P. 2006. Retrieving Forest Structure Variables Based on Image Texture Analysis and IKONOS-2 Imagery. Remote Sensing of Environment 102 (3-4): 390-401.
[11] Sprintsin M, Karnieli A, Berliner P, Rotenberg E, Yakir D, Cohen S. 2007. The Effect of Spatial Resolution on the Accuracy of Leaf Area Index Estimation for a Forest Planted in the Desert Transition Zone. Remote Sensing of Environment 109: 416-428.
[12] Wolter P, Townsend P, Sturtevant B. 2009. Estimation of Forest Structural Parameters Using 5 and 10 meter SPOT-5 Satellite Data. Remote Sensing of Environment 113: 2019-2036.
[13] Anonymous. 2017. Landsat 8 Pre-Collection Quality Assessment Band. Available online at https://landsat.usgs.gov/qualityband. Accessed on November 29, 2017

[14] Freitas S, Mello M, Cruz C. 2005. Relationship Between Forest Structure and Vegetation Indices in Atlantic Rainforest. Forest Ecology and Management 218: 353-362.
[15] Anonymous. 2017. Using the USGS Landsat 8 Product. Available online at http://landsat.usgs.gov/using-usgs-landsat-8-product. Accessed on November 29, 2017.
[16] Brown S. 1997. Estimating Biomass and Biomass Change of Tropical Forests: A primer. Rome: FAO Forestry Paper.
[17] Standard Nasional Indonesia. 2011. Pengukuran dan Perhitungan Cadangan Karbon – Pengukuran Lapangan untuk Penaksiran Cadangan Karbon Hutan (Ground based forest carbon accounting). Badan Standardisasi Nasional. Jakarta.
[18] Navakatikyan MA. 2007. A Model for Residence Time in Concurrent Variable Internal Performance. Journal of the Experimental Analysis of Behavior 87: 121-141.
[19] Ratner B. 2003. Statistical Modeling and Analysis for Database Marketing. Effective Techniques for Mining Big Data. Florida: Chapman & Hall/CRC.
[20] Gibbs H, Brown S, O'Niles J, Foley J. 2007. Monitoring and Estimating Tropical Forest Carbon Stock: Making REDD a Reality. Environ. Res. Lett. 2: 1-13.

IOP Conf. Series: Earth and Environmental Science **144** (2018) 012065 doi:10.1088/1755-1315/144/1/012065

The diversity of local sorghum (*Sorghum bicolor* L. Moench) in Nusa Tenggara Timur province

L Mukkun[1*], H J D Lalel[1], N Richana[2], M B Pabendon[3] and S R Kleden[1]
[1]Agriculture Faculty of Nusa Cendana University, Jl. Adi Sucipto Penfui, Kupang
[2]Badan Litbang Pertanian, Balai Besar Pascapanen, Jl Tentara Pelajar, Cimanggu, Bogor
[3]Badan Litbang Pertanian, Balai Penelitian Tanaman Serealia, Jl. Dr. Ratulangi No. 274 Maros

*Corresponding author: lincemukkun@yahoo.co.id

Abstract. Sorghum (*Sorghum bicolor* L. Moench) is an important food crop in the dry land including Nusa Tenggara Timur (NTT) Province. This plant has a high adaptability to drought, can produce on marginal land, and is relatively resistant to pests and diseases. The study aims to collect and identify the species of local sorghum being cultivated by farmers, and the purposes of cultivation. In addition, this study will preserve germ plasm of local sorghum by providing bank seeds for the next growing season. A collection of local sorghum samples was conducted in 7 districts using survey and observation method. A total of 53 species of sorghum were collected, with various characteristics and different local names. Based on the skin color of the seeds, the accessions were grouped into white groups (26.42%), light yellow (15.09%), black (20.75%), brown (24.52%), and red (13.20 %). Sorghum is used for complementary food for rice, consumption in times of food insecurity, fodder, and as a fence for corn and rice. It is necessary to characterize the type of local sorghum that has the potential to be developed for food, industrial raw materials, and for functional food.

1. Introduction

East Nusa Tenggara (NTT) province is characterized by low rainfall and prolonged drought, 8 - 10 months, and a short rainy season (3 - 4 months) to make sorghum one of the most suitable food crops to be developed. A 92.32 percent or about 1.5 million hectares of total agricultural land in NTT is dry land [1]. Sorghum (*Sorghum bicolor* L. Moench) is a subsistence crop for many farmers in the semi-arid tropics in Africa. The cropping systems based on sorghum used mainly local varieties which are part of their strategy to reduce the risks in the constraining areas. Local varieties have been defined as variable plant populations adapted to local agro-climatic conditions which are named, selected and maintained by the traditional farmers to meet their social, economic, cultural and ecological needs [2].

Sorghum is cultivated for human consumption.The production is mainly ensured by local varieties which are diversified and belong mostly to the botanical race guinea [3]. The guinea varieties are rustic, well adapted to low agronomic conditions and climatic uncertainties [4]. Moreover, their grain quality is well suited to the various local processing.

Sorghum is one type of food crops that has been well adapted and cultivated by the people of NTT, therefore, sorghum has a different local name each region. In East Flores, sorghum is known as *watablolo, solo corn,* or *wataholot*, in Sumba, it is called as *wataruhemutuji* and *watanruhemukadipta*, in Timor known as *benwuka* or *bennina*, Rote people call it *jagung rote*, Ende people call it *watar*, in Lembata known as *watarwoloq*. However, the existence of local sorghum began to be replaced by other food crops that have a better economic value such as corn and rice. Therefore, the collection of sorghum germ plasm is an essential activity to seek and excite their genetic diversity to prevent genetic erosion of local sorghum species.

Sorghum is basically a self-pollinated plant with an open panicle form so that the possibility of cross-species cross-breeding high and producing offspring in the form of accessions or new varieties. Germ plasm diversity can be derived from wild-type relatives (wild-type), landraces, local varieties and introduced varieties and somaclonal variations. To maintain the genetic diversity of local sorghum in NTT, conservation efforts should be made, in the form of collections of species that are still cultivated by farmers, to further develop both in-situ and ex-situ. This study aims to explore local sorghum germ plasm in NTT that is still cultivated by farmers, and reproduce it ex-situ. This preliminary study aims to re-promote sorghum as a commodity that can support food security of rural communities and increase the comparative value of sorghum.

2. Materials and Methods

2.1. Collection Area
Panicles and seeds of local sorghum samples have been collected from farmers in areas that still grow sorghum both in the yard and on the land. The location of sampling include 7 (seven) districts and 15 villages, which is determined intentionally based on the criteria that the location is still cultivated sorghum. The seven sampling sites Wet Flores Timur District (1 villages), Lembata District (3 villages), Kupang District (1 village), Sabu Raijua District (1 village), Belu District (1 village), and Sumba Timur District (8 villages).

2.2. Germ plasm Collection
The collection was preceded by a participatory diagnostic in each village. A collection of sorghum samples was performed on a population of sorghum farmers in both small and larger planting areas. Observations were made on the characteristics of panicles and seeds visually i.e. panicle, panicle density, husk color, husk nature, presence or absence of feather on seed, seed cluster, and seed thickness. In addition, it was also measured the water content using portable water meter (hygrometer). The weights of 1000 seeds were weighed using analytic scales. In addition, interviews were conducted on farmers using questionnaires to obtain data on harvested area, planting location, plant characteristics, harvest time, and sorghum utilization.

2.3. Data Analysis
Qualitative data obtained were analyzed descriptively and presented in tables and figures, while quantitative data were analyzed using mean and standard deviation.

3. Results and Discussion

3.1. Local Sorghum Type at Farmer Level
The type of sorghum identified in the study sites were 53 samples and 31 species based on the color of panicles and seeds (Table 1). It proves that the diversity of local sorghum species is quite high, although the farmers are less interested in cultivating sorghum. The results of interviews with farmers revealed that the lack of interest of farmers to grow sorghum due to post-harvest technology and complicated processing compared with other foodstuffs.

The highest diversity of local sorghum was found in East Sumba Regency which was 8 species, followed by Sabu Raijua Regency with 5 types. Variations of panicle/seed color were also varied, namely white, light yellow, brown, red, black, and mix between colors. This indicates that crossbreeding between species of sorghum produces new genotypes increasing the diversity of

IOP Conf. Series: Earth and Environmental Science **144** (2018) 012065 doi:10.1088/1755-1315/144/1/012065

sorghum germ plasm. Sorghum is basically a self-pollinated plant with high cross-pollination levels due to its open-air panicles [5]. In order to maintain the genetic diversity of the collection, conservation measures are necessary, both *in-situ* and *ex-situ*. Such genetic diversity can be utilized as genetic material and gene donors for the improvement of various plant characters in breeding programs. Breeding programs are impossible without high genetic diversity, so the availability of germ plasm collection is a critical stage of breeding programs [6].In terms of its utilization, sorghum as a staple food can also be used as a functional food ingredient, especially red and black sorghum type. It has been reported that phenol compounds contained in sorghum such as phenol, flavonoid, and tannin are antioxidant compounds that play an important role in overcoming oxidation-related diseases such as cancer, diabetics, cardiovascular disease, and obesity.The sorghum phytochemicals show high antioxidant activity against different free radicals in vitro relative to fruits and vegetables and may offer similar benefits attributed to fruits and vegetables. However, overall epidemiological evidence suggests sorghum has anti-carcinogenic properties when consumed regularly in diet [7]. The red and black sorghum contains tannins and anthocyanins 10 times higher than the white sorghum [8-9].

Table 1. Collected local sorghum during the study

Sample Origin	Number	Color of panicle/seed	No. of sample	Type	Local's name
Lembata	5	Reddish brown	3	Local	Watarwoloq
		Light yellow	3	Kawali	
		White	1	Numbu	
		Black-spotted white	1	Local	
		White, the bottom is red	1	Local	
Flores Timur	4	Light yellow	2	Kawali	-
		White	1	Numbu	-
		Black	1	Local	Watablolong hitam
		Reddish brown	1	Local	Watablolong merah
Sumba Timur	8	White, ½ the base of the seeds is brown	3	Local	Wataruhemu
		Maroon	1	Local	
		Red	2	Local	
		Brownish yellow	2	Local	
		Dark black	3	Local	
		Black	1	Local	
		White	2	Numbu	
		Reddish brown	1	Local	
Rote Ndao	3	Black	2	Local	Beladai hitam
		White with black spot	4	Local	Beladai putih
		Brownish red	1	Local	Beladai merah
Kab. Kupang	3	Reddish brown	4	Local	benwuka
		Dark black	1	Local	bennina
		Light yellow	3	Local	
Belu	3	White	1	Local	-
		Reddish brown	2	Local	
		Black	1	Local	
SabuRaijua	5	Reddish black	1	Local	Jagung rote
		Red	1	Local	
		Brown	1	Local	
		Black	1	Local	
		White	1	Local	
Total	31		53		

Local sorghum has a specific name at each location, indicating that sorghum has been known by the community for a long time and has adapted well to local conditions.

IOP Conf. Series: Earth and Environmental Science **144** (2018) 012065 doi:10.1088/1755-1315/144/1/012065

3.2. Characteristics of sorghum panicle

Characteristics of panicle such as shape, density, color, and nature of husk are specific features to identify the type of sorghum (Table 2). Sorghum with a compact panicle is preferred by farmers because it has more seeds than those with loose panicles. There are five forms of panicles: (1) inverted pyramid, (2) width at the top of the panicle, (3) symmetrical, (4) wide at the bottom of panicle, and (5) pyramid [10].Agronomic characterization can be used as a simple method to separate the accession of the cluster based on similarity or dissimilarity, then used to look at heterogenetic patterns and the genetic distance of accession [11].

Information on morphological properties of sorghum germ plasm is essential for plant breeders in the process of assembling superior varieties and improving the nature of the existing strains and varieties.

Table 2. Characteristics of Local Sorghum NTT

Sample code	Panicle shape	Panicle compactness	Grain Covered by glume	Pericarp color
Flotim 01	symmetric	compact	25% grain covered	light yellow
Flotim 02	symmetric	compact	25% grain covered	white
Flotim 03	symmetric	loose	75% grain covered	black
Flotim 04	pyramid	Semi-compact	25% grain covered	reddish brown
Kupang 01	symmetric	compact	25% grain covered	reddish brown
Kupang 02	symmetric	compact	25% grain covered	dark black
Kupang 03	pyramid	compact	25% grain covered	light yellow
Lembata 01	pyramid	loose	25% grain covered	reddish brown
Lembata 02	symmetric	compact	25% grain covered	light yellow
Lembata 03	symmetric	compact	25% grain covered	white
Lembata 04	pyramid	loose	25% grain covered	reddish brown
Lembata 05	pyramid	loose	fully covered	black
Sabu 01	*	*	75% grain covered	reddish black
Sabu 02	*	*	75% grain covered	red
Sabu 03	*	*	75% grain covered	brown
Sabu 04	*	*	75% grain covered	black
Sabu 05	*	*	25% grain covered	white
Rote 01	pyramid	loose	50% grain covered	black
Rote 02	pyramid	compact	25% grain covered	variegated
Rote 03	pyramid	loose	fully covered	brownish red
Sumba 01	pyramid	loose	75% grain covered	variegated
Sumba 02	pyramid	loose	75% grain covered	red
Sumba 03	symmetric	loose	fully covered	red
Sumba 04	symmetric	loose	75% grain covered	brownish yellow
Sumba 05	pyramid	loose	fully covered	dark black
Sumba 06	symmetric	compact	very short	white
Sumba 07	pyramid	loose	fully covered	red
Sumba 08	pyramid	loose	75% grain covered	black
Belu 01	symmetric	loose	25% grain covered	white
Belu 02	pyramid	loose	25% grain covered	reddish brown
Belu 03	pyramid	loose	75% grain covered	black

*no panicle, only seed

3.3. Seed Properties

The weight of 1000 local sorghum seeds ranges from 29 to 51 g and belongs to the category of high and very high potency of seeds. Balitsereal (1996) has classified the potency of sorghum accession by weight of 1000 seeds i.e. low (<15 g), medium (15 - 25 g), high (26 - 35 g), very high (> 36 g). The results showed that almost all species were potential to be developed into high-yielding varieties, suitable for food consumption and for food diversification. The samples namely Flotim 01, Lembata01, 02, 04, Sabu 02, 05, Rote 01, Sumba 03, 05, and 06 are the potential types used as genetic sources for developing high-yielding varieties.

Seed moisture content is one of the important factors in maintaining seed quality. The water content of sorghum at the farm level ranges from 10 to 17 percent, so it needs proper drying to achieve ideal water content for storage of 8 to 10 percent. The main requirement for increasing the productivity of sorghum is the use of seeds quality. The three important aspects related to the quality of seeds are (1) correct seed production techniques, (2) techniques to maintain the quality of distributed seeds, and (3) seed quality detection techniques [12]. The use of superior quality seed varieties contributes significantly to phenotypic appearance and yield components of plants [13].

Availability of sorghum seed both quantity and quality at farm level have not been fulfilled. The results of studies in East Nusa Tenggara show that farmers have not been able to produce superior seeds, unavailable of storage facilities, and markets uncertain of seeds affect the success of farmers or breeders in providing sorghum seed [14].

Table 3. Water content and 1000 weight seed

Sample code	Water content (%)	1000 weight seed (g)
Flotim 01	11.07	46.00
Flotim 02	10.77	29.00
Flotim 03	10.00	34.00
Flotim 04	12.40	33.00
Kupang 01	10.50	32.00
Kupang 02	11.53	31.00
Kupang 03	12.10	36.00
Lembata 01	12.37	40.00
Lembata 02	11.97	51.00
Lembata 03	16.73	31.00
Lembata 04	12.53	40.00
Lembata 05	12.30	28.00
Sabu 01	13.93	36.00
Sabu 02	14.33	40.00
Sabu 03	12.83	38.00
Sabu 04	13.40	36.00
Sabu 05	14.13	37.00
Rote 01	13.23	40.00
Rote 02	16.67	35.00
Rote 03	15.37	38.00
Sumba 01	11.43	36.00
Sumba 02	12.50	34.00
Sumba 03	12.20	40.00
Sumba 04	14.80	38.00
Sumba 05	15.57	41.00
Sumba 06	17.23	42.00
Sumba 07	15.43	32.00
Sumba 08	14.73	43.00
Belu 01	11.20	43.00
Belu 02	11.57	37.00
Belu 03	11.73	39.00
Range	10.00 – 17.23	29 - 51

4. Conclusions

This study on local sorghum variety in the seven regency of Nusa Tenggara Timur Province has shown that the diversity cultivated sorghum in the villages is dominated by inherited varieties from parents.The local sorghum in NTT has a high diversity, with 13 collected accessions. These types have distinctive features such as color, shape, and different seed size. The colors of the seeds are white, yellow, brownish red, and black indicated that sorghums are rich in phytochemical compounds that are beneficial to human health. In addition, each type has a panicle that is also different depending on the type of sorghum. Based on the results of the interview is known that the current consumption of sorghum by the population began to decrease even some who no longer consume them. Cultivation is not intensive anymore but is cultivated for the purpose of fodder or as a fence/trap plant.The type of sorghum identified in the study sites were 53 samples and 31 species based on the color of panicles and seeds. The highest diversity of local sorghum is found in East Sumba Regency which is 8 species, followed by Sabu Raijua Regency with 5 types. Variations of panicle/seed color are also varied, namely white, light yellow, brown, red, black, and mix between colors. This indicates that crossbreeding between species of sorghum produces new genotypes thus increasing the diversity of sorghum germ plasm. Characteristics of panicle such as shape, density, color, and nature of husk are specific features to identify the type of sorghum. The weight of 1000 local sorghum seeds ranges from 29 to 51 g and belongs to the category of high and very high potency of seeds. The results showed that almost all species were potential to be developed into high-yielding varieties, suitable for food consumption and for food diversification. The samples namely Flotim 01, Lembata 01, 02, 04, Sabu 02, 05, Rote 01, Sumba 03, 05, and 06 are the potential types used as genetic sources for developing high-yielding varieties. The water content of sorghum at the farm level ranges from 10 to 17 percent, so it needs proper drying to achieve ideal water content for storage of 8 to 10 percent.

Acknowledgments
We acknowledge The Agricultural Research and Development Agency, Jakarta, through KP4S Project for the financial support to complete this research. We also grateful to Rector of Nusa CendanaUniversity for giving us the chance and encouraging us to get this grand. We also express our appreciation to Mrs. Maria Lorethaand, Director of YaspenselLarantuka, for the assistance during the collection and identification of local sorghum.

References

[1] Biro Pusat Statistik 2015 Nusa Tenggara Timur dalam Angka [BPS Propinsi NTT]
[2] Teshome A, Fahrig L, Torrance JK, Lambert JD, Arnason JT, Baum BR 1999 Maintenance of sorghum (Sorghum Bicolor, Poaceae) landrace diversity by farmers' selection in Ethiopia. *Econ. Bot.* **53**(1) pp 79-88
[3] Barro-Kondombo C, Sagnard F, Chantereau J, Deu M, vom Brocke K, Durand P, Gozé E, Zongo JD 2010 Genetic structure among sorghum landraces as revealed by morphological variation and microsatellite markers in three agroclimatic regions of Burkina Faso. *Theor. Appl. Genet.* **120** pp 1511-23
[4] Kouressy M, Dingkuhn M, Vaksmann M, Heinemann AB 2008 Adaptation to diverse semi-arid environments of sorghum genotypes having different plant type and sensitivity to Photoperiod. *Agric. For.Meteorol.* **148** pp 357-71
[5] Kondombo, C. P, Barro, A, Kaboré, B. and Marc Bazié, J. 2016 On-farm diversity of sorghum [Sorghum bicolor (L.) Moench] and risks of varietal erosion in four regions of Burkina Faso
[6] Acquaah, G. 2007. Principles of plant genetics and breeding. *Blackwell Publ.* p 584
[7] Awika, J.M. and Rooney, L.W 2004 Sorghum phytochemicals and their potential impact on human health. *Phytochemistry* **65** pp 1199–221
[8] Dykes, L., and. Rooney, L.W 2006 Sorghum and millet phenols and antioxidants. *Journal of Cereal Science* **44** pp 236–51
[9] Rooney, L.W. and D.S. Murty 1982 Evaluation of sorghum food quality.Sorghum in the eighties: *Proc Symposium on Sorghum*. Patancheru 2-7 Nov 81. ICRISAT. India. pp. 571-88
[10] IBPGR and ICRISAT 1993 Descriptors for sorghum [Sorghum bicolor (L.) Moench]. Rome, Italy: International Bureau of Plant Genetic Resources; and Patancheru 502 324, Andhra Pradesh, India: International Crops Research Institute for Semi-Arid Tropics. pp 76
[11] Ali, M.L., F. Rajewski, P.S. Baenziger, K.S. Gill, K.M. Eksridge, and I. Dweikat 2008 Assessment of genetic diversity and relationship among a collection of US sweet sorghum germplasm by SSR markers. *Mol. Breeding* **21** pp 497-509
[12] Saenong, S., M. Azrai, Ramlah A, and Rahmawati 2007 Pengelolaan benih jagung. Buku Jagung. Teknik Produksi dan Pengembangan. Pusat Penelitian dan Pengembangan Tanaman Pangan. Bogor
[13] Subagio, H. and Aqil, M 2014 Perakitan dan pengembangan varietas unggul sorgum

untuk pangan, pakan, dan bioenergi. *Iptek Tanaman Pangan* **9**(1)
[14] Subagio, H and Suryawati 2013 Wilayah dan ragam penggunaan sorgum di Indonesia In: Sorgum, Inovasi Teknologi dan Pengembangan (eds.). Sumarno, Darmadjati, D.S., Syam, M. dan Hermanto. Badan Penelitian dan Pengembangan Pertanian, Kementerian Pertanian. *IAARD Press*

Sulawesi cacao (*Theobroma cacao*, l.) performances under two different agricultural system in east coast of Central Sulawesi

Muslimin[1] *, W Wijayanti[1] , A Anshary[2], Z Basri[2], A F Cruz[3], I N Suwastika[4] and T Shiina[3]

[1]Faculty of Forestry, Tadulako University, Palu 94118, Indonesia
[2]Faculty of Agricultural, Tadulako University, Palu 94118, Indonesia
[3]Graduate School of Life and Environmental Sciences, Kyoto Prefectural University, Kyoto 606-8522, Japan
[4]Biology Department Faculty Of Sciences, Tadulako University, Palu 94118, Indonesia

* Corresponding author: madjidmuslimin@gmail.com

Abstract. Cacao (*Theobroma cacao* L.) is one major agricultural commodity from central Sulawesi, which is significantly affected by a range of pest and diseases including Cacao Pod Borer (CPB), Pod Rod Diseases (PRD), and Vascular Streak Dieback (VSD). Intensive and extensive approaches, including introduction of various new superior cultivars/clones, and development of eco-friendly agricultural system were implemented in order to overtake that problem. Here, we observed the performance of three different clones (namely: Sulawesi 1, MCC01, and Tadulako-1) in two different locations (Parigi and Poso-which located along east coast of central Sulawesi province), and under two different Agricultural management system (echo-friendly-intensive and Non-Intensive). All clones tested showed their well-adapted and suitable to local conditions. The performance of each clone can be improved by intensive management system. Based on all observed parameters (number of wet bean per pot, bean count, fat and shell percentage, and percentage of infection) it convincingly showed that intensive management system was working well in improving the quality and quantity of cacao beans production, and it fit to commercial requirements. The highest rate of infection was by *Phytophthora* and no significant differences on the overall performances. Clones of Sulawesi-1 and MCC01 most likely was better than Tadulako-1.

1. Introduction

Cacao (*Theobroma cacao* L.) is not endogenous Indonesian agronomic plant. This agronomic plant was introduced from South America, which has three different ancestor called as Criolla, Forestero, and Trinitario [1].The present time we can found various cultivars of Cocoa with very different kind of phenotypic characters. It is believed that cultivars are the current plant after natural adaptation into local condition. In Central Sulawesi, there were more than 20 clones (such as Sulawesi 1, Sulawesi 2, ICCRI, MCC etc.) that have been identified and cultivated by local farmer [2].

Statistically, Central and South Sulawesi contribute more than 60% (more than 250 thousand ton) of annual cacao bean for national product [3]. Unfortunately, cocoa farming is affected by a range of pest and decreases with the estimated average loss 30-40% of total production [4]. The serious cases including; Pod Rot known as Black Pod (caused by the fungus of *Phytophthora* spp.)causes global yield loss of 20-30% and tree deaths of 10% annually), Vascular-streak dieback (VSD)(caused by the fungus of *Oncobasidium theobromae*, causes heavy loss of trees in mature plantations), and Cocoa Pod Borer (CPB)(caused by the insect of *Conopomorpha cramerella*, inflicts loss of US$ 40 million per year).These are serious threat causing vast losses in the cacao farming in Sulawesi.

Cacao farms is widely distributed over Central Sulawesi area, including in highland such as Kulawi and Palolo, or in beach areas such as Parigi and Poso (which located in Eastern coast of Sulawesi island), and also in lowland and west coast of Island, such as Dolo, Donggala, and Toli-toli. The different areas show their local conditions including microclimate. Due to these local conditions and lack of information related the resistance again pathogen of each cultivar in different local condition, here in this study we determine the status of productivity and the resistance to main pest and diseases, also the quality of yield from three different cultivars in two east coast locations (Parigi and Poso), which supply significantly in production of cacao bean. Moreover, we also describe the effect of cacao farming management (we called echo friendly intensive farming) compared to the conventional one.

2. Material and method

This observation was done continuously from February 2013 to April 2014. Sampling was taken up based on the randomized block design in Split-Plot pattern, from two locations (Poso and Parigi). The main plot was three different of clones (cultivars) i.e. Sulawesi 1, MCC01, and Tadulako-1, while as sub-plot were two different management systems i.e. echo friendly intensive and conventional (Nonintensive) Agricultural system. The subject plants as samples were clones propagated by grafting in 2008.The designed locations are listed in Table 1.

Table 1. The designed locations

No.	Management system	Location	coordinates	Altitude (m above sea level)
1.	Echo-friendly Intensive	Parigi	S 00° 56.919' E 120° 13.002'	182
2.	Echo-friendly Intensive	Poso	S 01° 18.943' E 120° 35.171'	82
3.	Non-Intensive (conventional)	Parigi	S 00° 56.453' E 120°12.981'	47
4.	Non-Intensive (conventional)	Poso	S 01° 18.903' E 120° 13.002'	48

Eco-friendly Intensive Management system that we introduced to the farmer, can be described as intensive in choosing clone for grafting, regular pruning, and using organic material for fertilizer, insecticide and pesticide applications, but during rate of pathogen attack (e.g. during wet season), farmer can use chemical commercial materials in minimum amount. While conventional system refer to non-intensive management, using commercial fertilizer, and applying insecticide, and pesticide in high dose, without combination to the organic compound.

Observed parameters were quantity and quality of yield and rate of pest and diseases infection from sample of each plot. Sampling was done during period of April and November.

3. Result and Discussion

In this study, we observed 45 plants as sample for each plot, so in total, we observed 270 plants from each location. Firstly, we observed the infection rate on each clone in those locations by measure the infection rate of CPB (Cacao Pod Borer), *Phytophthora*, and VSD (Vascular-streak dieback). The outcome can be seen in Figure 1 and 2.The management system showed significant effect in reducing infection rate of these three pathogens. CPB and VSD can be controlled until zero by this intensive system, while infection due to *Phytophthora* can be eliminated. It could be the fact that fungi are difficult to be controlled during wet season. During rainy season, the spore of *Phytophthora* easily dispersed growth and infected the cacao pot [5-6]. In both locations (Poso and Parigi), *Phytophthora* show still dominant diseases compared to CPB and VSD.

The low rate of infection here also indicates that three observed clones have high resistance again CPB, *Pythoptora*, and VSD infections. Since these three clones are resulted from breeding and selection program and it could be superior clones and suitable for the local condition of Central Sulawesi, especially in east-coast area.

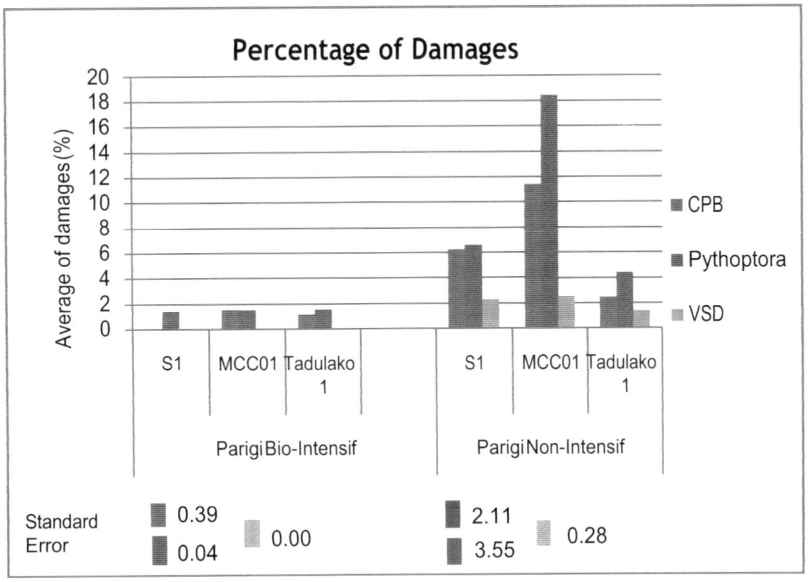

Figure 1. Infection rate of CPB (Cacao Pod Borer), *Phytophthora*, and VSD (Vascular-streak dieback) on three different cultivars of cacao) from two different management system in Parigi.

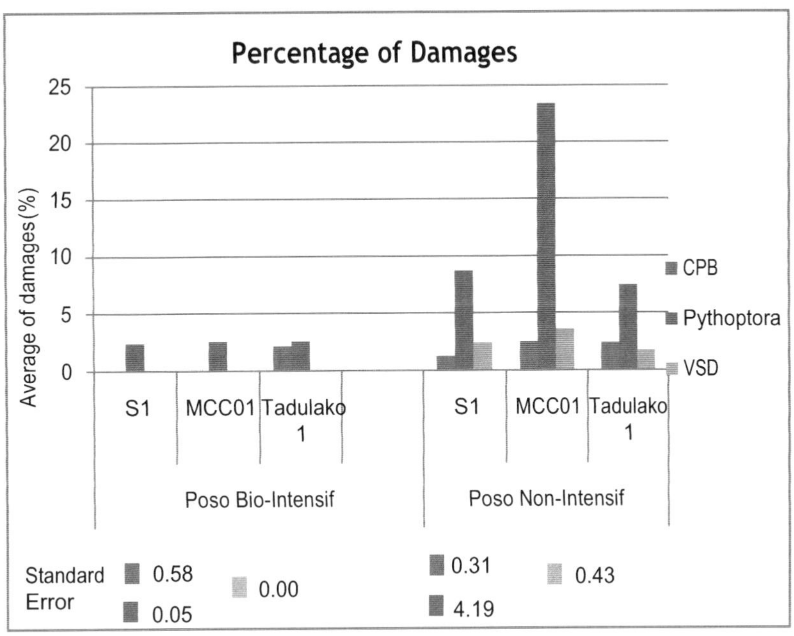

Figure 2. Infection rate of CPB (Cacao Pod Borer), *Phytophthora*, and VSD (Vascular-streak dieback) on three different cultivars of cacao) from two different management system in Poso.

The quality and quantity of yield from this area were observed based on the seed number, fresh weight of bean (wet bean), shell content, and fat content of bean from each clones and from each location. Bean count referred to number of bean in 100 g. This number was not only depend on genetic properties but also can be proved by good management system. The data of bean account is presented in Figure 3.

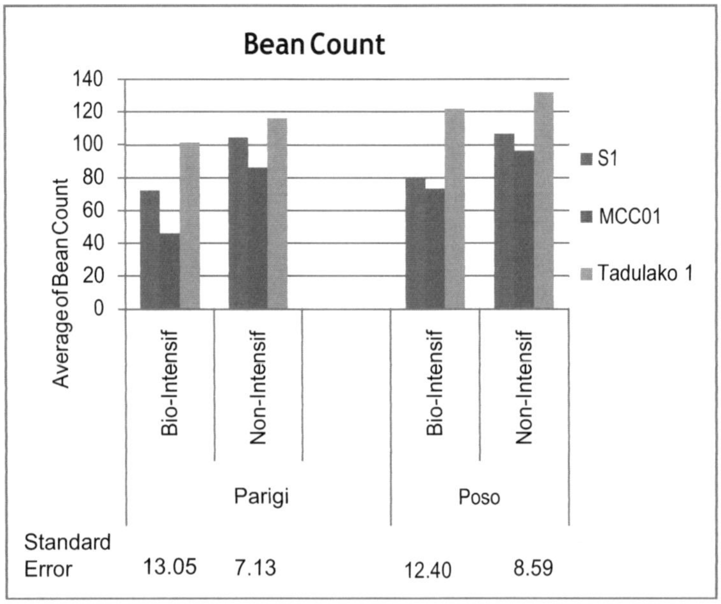

Figure 3. Number of beans on 100 g. Data was collected from each plot from Parigi and Poso farms cultivated by applying two different systems (echo-friendly intensive and non-intensive-conventional). Intensive system in the farm on those three clones, it could increase the size of produced bean.

Produced bean can be improved by applying the intensive management of cacao farming. The each bean become bigger and heavier when plant healthy. In intensive system, plant could be supplied enough nutrition, enough light for carrying photosynthesis, and less pest and pathogen infections as similar case reported in Ivory Coast [7]. As a result, the plants can grow better than it under conventional agronomic system. Figure 3 also indicates that plant growth condition in Parigi is better than in Poso. It shows that the nutrition status in Parigi is more suitable for cacao plant than that of in Poso for all clones. However, clones of tadulako-1 produces smaller/lighter of bean compared to bean of MCC01 and Sulawesi-1. Clone of MCC01 has higher sensitivity in case of bean size, compared to other clones. The bean weight of MCC01 is easily improved if we apply the echo-friendly intensive system.

Furthermore, we also observed the quality of bean by observe the shell content and fat content. Shell content on dried bean is parameter that cacao bean produced in the farming, are less of unexpected contaminants. While fat content is important parameter in case of chemical properties. Data from measurement of shell content and fat content are presented in Figure 4 and 5, respectively. Our data showed that there were no significant difference on shell content and fat content of bean produced from two different management system. The differences only can be detected in between clones. It seems that genetic factor is more dominant than environmental factors. The data also suggests that genetic potency of Tadulako-1 is lower than Sulawesi-1 and MCC01 properties.

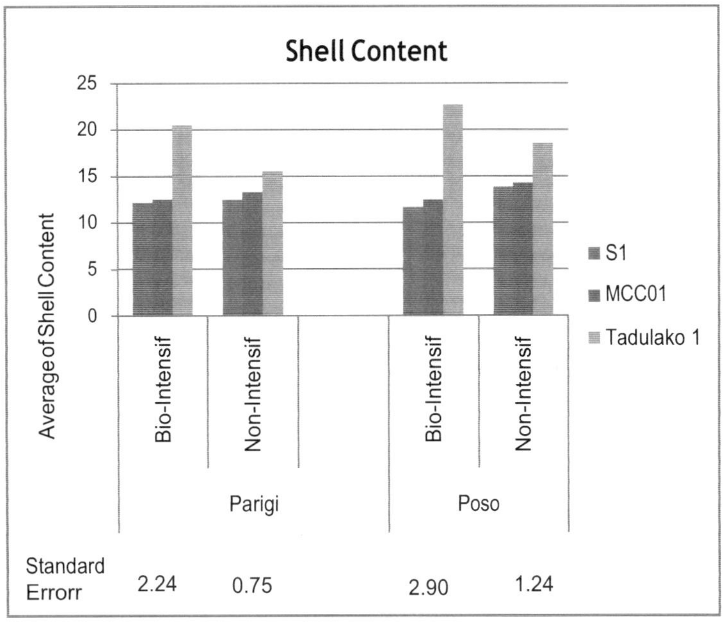

Figure 4. Percentage of shell content on dried beans produced by three different clones (Sulawesi-1, MCC01, and Tadulako-1) from farming area of Parigi and Poso implementing echo-friendly intensive and non-intensive agricultural system.

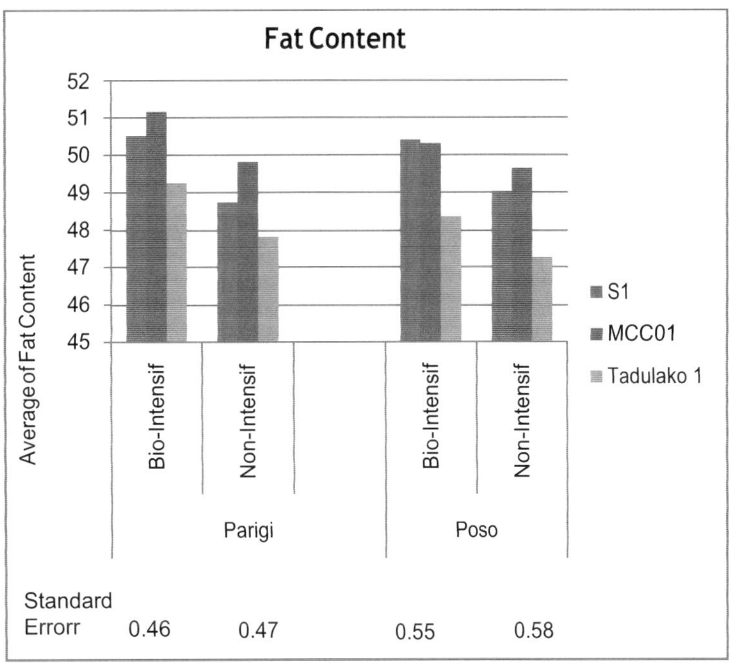

Figure 5. Fat content on dried beans of three different clones (Sulawesi-1, MCC01, and Tadulako-1) from farming area of Parigi and Poso implementing echo-friendly intensive and non-intensive agricultural system.

Altogether, it shows that clones of Sulawesi-1 and MCC01 are better than Tadulako-1 clone. Although all clones are suitable for Parigi and Poso condition, the genetic potency of Sulawesi-1 and

MCC01 can be improved by improving the ecologic condition as also reported before[8]. The tested clones give response to the ecological condition that build-up by applying good echo-friendly-intensive management system. We recommend to farmer in Parigi and Poso to use clones of Sulawesi-1 and MCC01 for cultivating (by grafting) and practicing in eco-friendly system for their cacao farming.

Acknowledgment

This research was supported by DGHE (Directorate General Higher Education), under program Dissertation grant program. It was also supported by research grant from INS and TS, under program of International Cooperation and International Publications - from the Ministry of Research, Technology and Higher Education of the Republic of Indonesia, and it also supported by JSPS (Japan Society for Promotion Science)-DGHE (Directorate General Higher Education) International grant between Japan and Indonesia.

References

[1] Motamayor J C, Lachenaud P, da Silva E M J W, Loor R, Kuhn D N, Brown J S and Schnell R J 2008 Geographic and genetic population differentiation of the Amazonian chocolate tree (Theobroma cacao L)*PloS One* **3(**10).

[2] Suwastika I N, Muslimin, Rifka, Aisyah N, Rahmansyah, Mutmainah, Ishizaky, Basri and Shiina T 2015 Genotyping Based on SSR Marker on Local Cacao (Theobroma Cacao L.) from Central Sulawesi *Procedia Environmental Sciences* **28** pp 88–91.

[3] Indonesia Directorate General of Estate Crops 2016Plantation Statistics Indonesia, December 2015.

[4] Hebbar P K 2007 Cacao Diseases : Important Threats to Chocolate Production Worldwide Cacao Diseases : A Global Perspective from an Industry Point of View*Phytopathology* **97(**12) pp 1658–63

[5] Ristaino J B, Madritch M, Trout C L, Parra G, Pathology P, Carolina N, and Carolina N 1998 PCR Amplification of Ribosomal DNA for Species Identification in the Plant Pathogen Genus Phytophthora*Applied and environmental microbiology* **64(**3) pp 948–954.

[6] Hanada R E, Pomella A W V, Soberanis W, Loguercio L L, and Pereira J O 2009 Biocontrol potential of Trichoderma martiale against the black-pod disease (Phytophthora palmivora) of cacao *Biological Control* **50**(2), pp 143–9

[7] Donald P F 2004 Biodiversity Impacts of Some Agricultural Commodity Production Systems. *Conservation Biology* **18(**1) pp 17–37

[8] Saltini R, Akkerman R, and Frosch S 2013 Optimizing chocolate production through traceability: A review of the influence of farming practices on cocoa bean quality *Food Control* **29**(1) pp 167–187

IOP Conf. Series: Earth and Environmental Science **144** (2018) 012067 doi:10.1088/1755-1315/144/1/012067

Effect of aluminum stress in early-stage growth of soybean

D Sagala[1*], E Suzanna[1], Prihanani[1], M Ghulamahdi[2], I Lubis[2] and Trikoesoemaningtyas[2]

[1]Agrotechnology Program Study, Agricultural Faculty, University of Prof. Dr. Hazairin, SH, Bengkulu, 38117, Indonesia

[2]Department of Agronomy and Horticulture, Faculty of Agriculture, Bogor Agricultural University, Jl. Meranti, Kampus IPB Darmaga, 16680, Bogor, Indonesia

Email: danner_10@yahoo.com

Abstract. The sensitivity of soybean to aluminum stress is well known. One of the abiotic stresses in tidal swamps is the aluminum toxicity. Therefore, it is necessary to find the appropriate management and cropping pattern to obtain a high yield of soybean in the tidal land. We supposed that it would be related to the development stage of soybean. This study was aimed to find the most sensitive of soybean vegetative growth stage to aluminum. Three cultivars of soybean (tanggamus, Karasumame, and M652) and four aluminum treatments (control, 10 days after planting/DAP, 20 DAP, 30 DAP) were arranged in a completely randomized design with three replications. Measurements have been made on root length, root/shoot ratio, biomass dry weight, leaves Aluminum content, leaves the area, and sensitivity index. Overall, observation results indicate that the earlier soybean experience aluminum stress, the more its vegetative growth is disturbed.

1. Introduction

Fertile soil has been converting to either non-agriculture or non-crop agriculture use for years in Indonesia. It is about 110000 ha $year^{-1}$[1]. Extensification has to be done to meet the soybean national need in addition to intensification program. Developing the tidal swamps for soybean cultivation should be one of the extensification programs. Indonesia has a large of tidal swamps. According to reference [2], there is about 9.53 million ha of tidal swamps suitable for agriculture activity. However, there are some obstacles faced in the development of the tidal swamps.

Aluminum (Al) is one of the chemical element that can be found in tidal swamps soil and inhibit the soybean root growth[3–6].Reference [7] concluded in his review that Al toxicity initially disrupted root growth. Al is absorbed by root and enters to plant tissue through apoplast and symplast system[8].Soybean Al-sensitive cultivar will experience the enlargement of plasma membrane and plasmolysis in the root[9].

There are many reports about the danger of Al to soybean root growth. However, it is lack of information about which stage of the soybean development is the most sensitive to Al stress. The damage to the soybean root caused by Al should affect the entire soybean growth. Therefore, this recent work will show the soybean vegetative growth at Al stress and the most affected soybean growth stage. This information will be important to the development of tidal swamp for soybean cultivation in Indonesia. It is expected to be useful, especially, for saturated soil culture studies of soybean done by references[10–12].

2. Materials and Method

The experiment tested 4 levels of Al application time treatment for 3 soybean cultivars. The Al application time treatments were control (No Al treatment), 10, 20, and 30 days after planting (DAP).

The three soybean cultivars were Tanggamus (Indonesian cultivar), Karasumame (Nepali cultivar), and M652 (Indian cultivar). They were arranged in a completely randomized design. The experiment was replicated three times.

The experiment was conducted in a greenhouse. A nutrient solution was used as growing medium of the soybean. The use of nutrient solution as a growing medium was intended to control the Al concentration and to ensure the homogeneity of all environmental factors other than the treatment. The end of the experiment was at age of 40 DAP i.e. 10 days after 30 DAP treatment.

2.1. Nutrient Solution Composition

A nutrient solution composition of reference[13] was referred in this experiment. Some chemical compounds consisted of1.5 nM $Ca(NO_3)_2.4H_2O$; 1.0 mM NH_4NO_3; 1.0 mM KCl; 0.4 mM $MgSO_4.7H_2O$; 1 mM KH_2PO_4,;0.50 ppm $MnSO_4.4H_2O$; 0.02 ppm $CuSO_4.5H_2O$; 0.05 ppm $ZnSO_4.7H_2O$; 0.5 ppm H_3BO_3;0.01 ppm $(NH_4)_2MO_7O_{24}.4H_2O$ and 0.068 mM $FeSO_4.7H_2O$ were dissolved in water so that the volume of the nutrient solution was 2 l each pot. The concentration and form of Al were referred from reference[14] study. They found that soybean started to be stressed at 0.7 nM $AlCl_36H_2O$ (169 ppm).

Soybean was sowed in sand medium until the age of 5 days. The experimental pots were filled with the 2 l nutrient solution. The 5-days seedling was transferred to the pot. Each pot was connected to installed aerator for oxidation need (Figure 1).

Figure 1. Arranged pots of the experiment

2.2. Data Collecting and Analysis

Data were collected on root length, root/shoot ratio, biomass dry weight, leaves Aluminum content, leaves the area, and sensitivity index. They were measured at the end of the experiment except for leaves area. Leaf area was observed 1 week after each Al treatment. Sensitivity index formula was referred from reference[15] as follows:

$$SI = \frac{1-(\frac{y}{p})}{1-(\frac{x}{xp})} \quad (1)$$

SI = Sensitivity Index
y = mean value of cultivar treated by Al
yp = mean value of cultivar not treated by Al (control)
x = mean of all cultivars treated by Al
xp = mean of all cultivar not treated by Al (control)
Criteria:SI< 0.5 = tolerant; 0.5 <IS ≤ 1 = moderate; SI ≥ 1 = sensitive

The variants of the data were analyzed. The Duncan Multiple Range Test probability 5% was used to see the mean comparison. This statistical analysis was done by using SPSS 22.

3. Result

Root length was affected by the presence of Al. Figure 2 (a) shows that the data pattern of root length follows the pattern of the Al treatment. Control is insignificantly different from 30 DAP but significantly different from 20 DAP, and so on up to 10 DAP. The root of soybean control is longer than those treated by Al. The shortest root is found in 10 DAP treatment and significantly different from control.

Figure 2 (b) shows the root/shoot ratio of all soybean cultivar according to the time of Al application. These root/shoot ratio values show that the root dry is lighter than shoot dry weight for all of the values are less than 1. The time of Al application insignificantly affects the root/shoot ratio although the root/shoot ratio pattern tends to be the same as the pattern of root length.This implies that the root growth affects the shoot growth. The development of root in soybean control tends to be faster than the other.

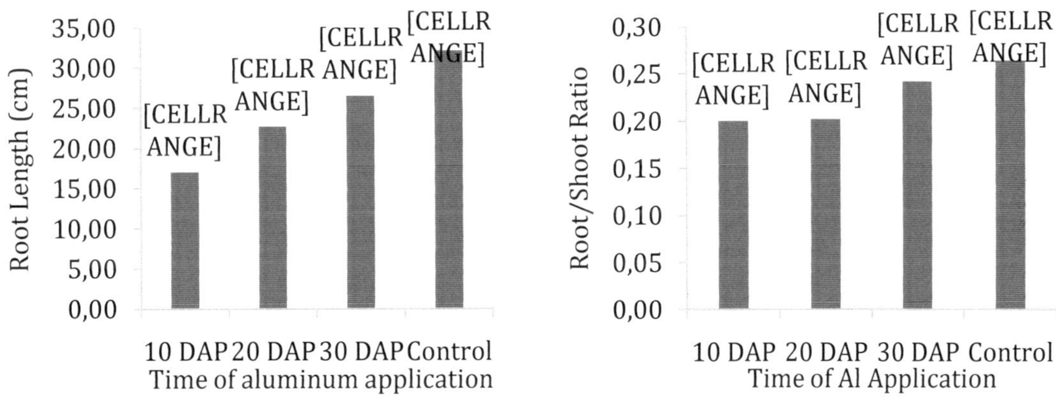

(a) (b)

Figure 2. Root length (a) and root/shoot ratio(b) of all soybean cultivar at Al treatment. DAP is the day after planting. Values followed by a different letter are significant at $P<0.05$.

Interaction of cultivar and time of Al application affected total dry weight. Figure 3 (a) shows the total dry weight of different Al application time in each cultivar. The pattern of dry weight value also tends to be the same to root length and root/shoot ratio pattern. The dry weight of Tanggamus treated Al at 10 and 20 DAP is significantly lighter than those treated Al at 30 DAP and control. The dry weight of Karasumame treated by Al earlier is lighter than those treated later. The dry weight of M652 treated Al at 10 DAP is significantly different from the other. The dry weight of M652 treated Al at 20 DAP is insignificantly different from 30 DAP but they are significantly different from 10 DAP and control.

Figure 3 (b) shows that cultivar M652 is significantly the lightest in 10 and 30 DAP treatment. Cultivar Tanggamus and Karasumame are statistically similar at all of the Al treatment.M652 are the lightest than the other although insignificantly different from Tanggamus and Karasumame at 20 DAP and control.

Interaction of cultivar and time of Al application affected the Al content. Al content was derived from soybean leaves. Figure 4 (a) shows the Al content of different Al application time in each cultivar. The Al content of Tanggamus at 10 and 20 DAP are significantly different from 30 DAP and control. The Al content of Karasumame at 10 DAP, 20 DAP, and control are statistically different from 30 DAP. The Al content of M652 at 30 DAP is the highest but insignificantly different from 10 and 20 DAP.

Figure 4 (b) shows that the highest Al content is found in M652 leaves at 30 DAP. The Al content of Tanggamus, Karasumame, and M652 are statistically similar at 10 DAP, 20 DAP, and control. The high Al content of M652 at 30 DAP influenced the dry weight of M652 at 30 DAP (Figure 3). The dry weight of M652 at 30 DAP is low while its Al content is high.

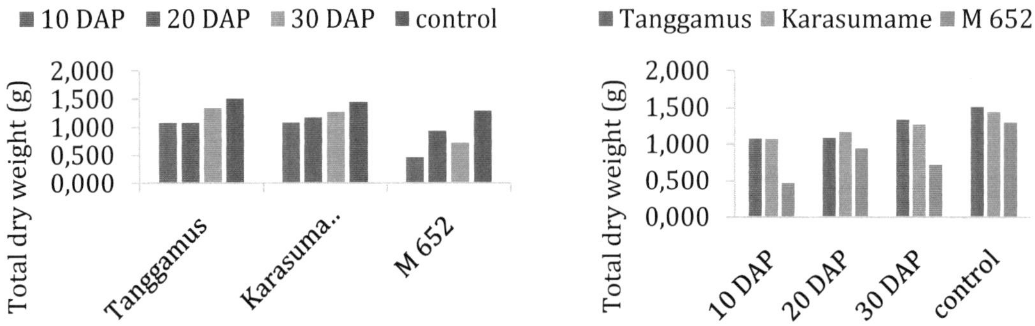

(a) (b)

Figure 3. Interaction of cultivar and Al treatment on the total dry weight of soybean biomass. (a) Mean comparison of Al treatments in each cultivar. (b) Mean comparison of cultivars in each Al treatment. DAP is the day after planting. Within each cultivar (a) or each Al treatment (b), values followed by a different letter are significant at P<0.05.

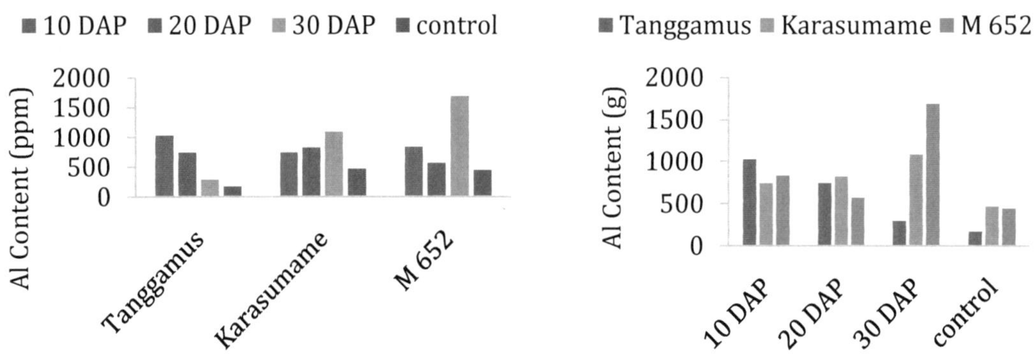

(a) (b)

Figure 4. Interaction of cultivar and Al treatment on Al content of soybean. (a) Mean comparison of Al treatments in each cultivar. (b) Mean comparison of cultivars in each Al treatment. DAP is the day after planting. Within each cultivar (a) or each Al treatment (b), values followed by a different letter are significant at P<0.05.

Figure 5 shows the difference of leaves area of control and Al treatment at 10 DAP (a), 20 DAP (b), and 30 DAP (c). All cultivar leaves are larger in control than those at Al treatment. The differences leave area of Tanggamus between control and Al treatment 10 DAP, 20 DAP and 30 DAP respectively are 46%, 15%, and 6%. The differences of leaves area of Karasumame between control and Al treatment 10 DAP, 20 DAP and 30 DAP respectively are 59%, 15%, and 14%. The differences of leaves area of M652 between control and Al treatment 10 DAP, 20 DAP and 30 DAP respectively are 73%, 65%, and 24%. The higher the difference means the more disrupted the soybean. The difference of leaves area is higher in 10 DAP treatment and lowest in 30 DAP treatment.

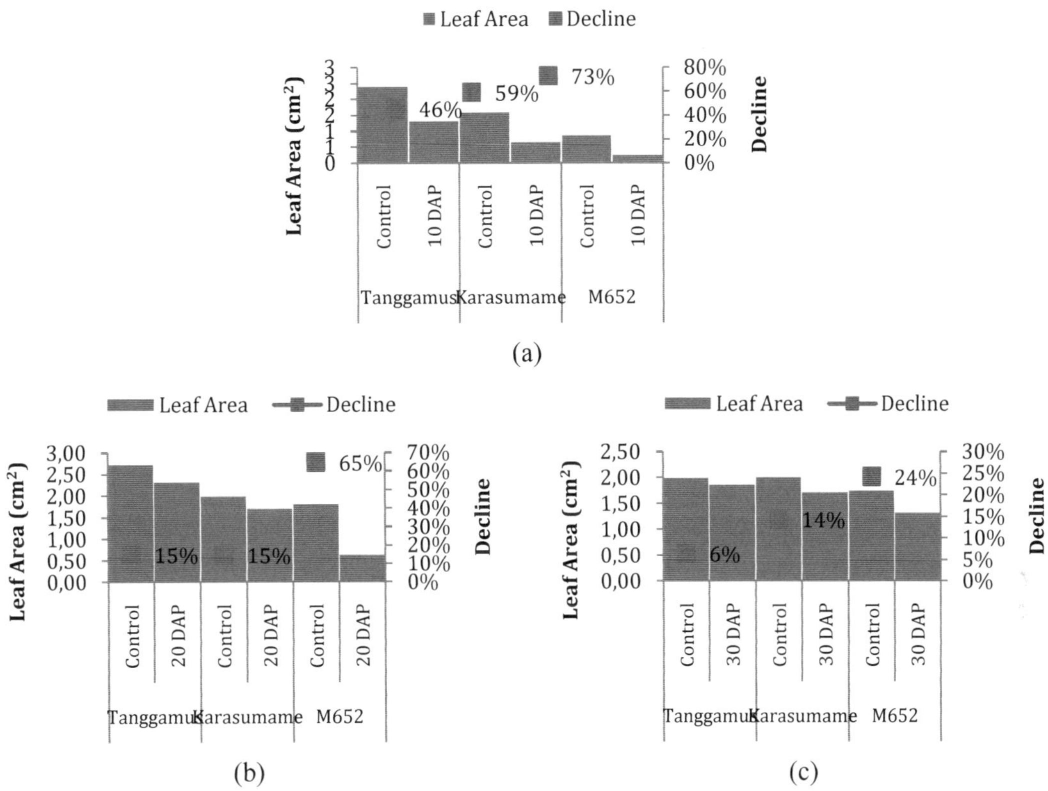

Figure 5. The difference of soybean leaf area between control and Al treatment in each cultivar. (a) leaf area of soybean for 10 DAP treatment. (b) leaf area of soybean for 20 DAP treatment. (c) leaf area of soybean for 30 DAP treatment. The observation was done at 5 days after each Al treatment, for example, the leaf area of (a) was measured at 15 DAP (5 days after 10 DAP treatment). DAP is the day after planting.

Figure 6 shows that sensitivity index of the three cultivars studied at 10 DAP is more than 1. It means that soybean at the age 10 DAP is sensitive to Al toxicity. The sensitivity index of Tanggamus tends to decrease with the time of Al application. Sensitivity index of Karasumame is more than 1 at 10 and 20 DAP but decline sharply at 30 DAP. The sensitivity index of M652 is more than 1 at 10 DAP and decline at 20 and 30 DAP.

4. Discussion

Soybean growth was disrupted due to Al toxicity. Al affected soybean growth since Al was still inside the roots. Root growth was disrupted by cellular damage occurring within the plant tissue. Physiological processes within the roots were disrupted due to cellular damage. Al also affected the physiology process directly due to plasmolysis that occurs in the presence of Al [9].

Sensitive Soybean cultivars experience growth disturbance to all plant organs. We found a relationship between inhibitions of root and shoot growth in this study. The root growth pattern over the time of the al application tends to be same as the root/shoot ratio pattern. This suggests that impaired root growth also results in a disruption to shoot growth [14].

The three cultivars studied in this experiment absorbed Al to leaf. This indicates that there is no Al-detention mechanism at the root for not getting into the leaf[16]. The higher the concentration of Al in the leaves the greater the growth disturbance experienced by the Tanggamus. This does not occur in the two other cultivars. It was not found a regular pattern in both varieties.

It was found that the Al treatment at age 10 DAP made the worst growth of soybean compared to other treatments all parameters observed. Therefore, this study has found that early toxicity of Al is more dangerous than in older age.

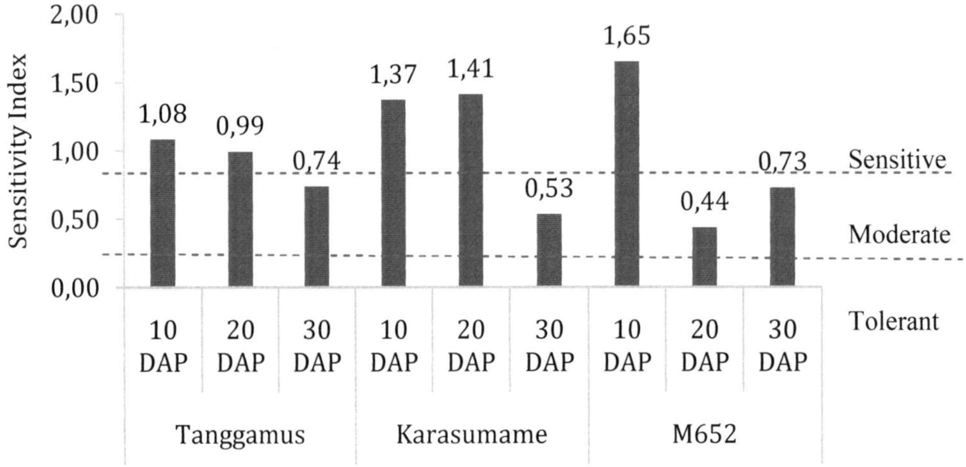

Figure 6. Sensitivity index of soybean on all Al treatment. SI < 0.5 = tolerant; 0.5 <IS ≤ 1 = moderate; SI ≥ 1 = sensitive. SI was counted using the formula of [15].

5. Conclusion

Al affected all of the soybean vegetative organs observed. Cultivar M652 was the most vulnerable cultivar to Al toxicity. All cultivar were sensitive at 10 DAP. So, we conclude that the earlier all soybean cultivar experienced Al, the more sensitive all of the soybean cultivar to Al

Acknowledgement

We are grateful for the financial support for this research from the Directorate of Research and Community Service, Directorate General for Research and Development, Ministry of Research, Technology and Higher Education for the provision of the grant of the Applied Product Research Program.

References

[1] Ministry of Agriculture 2015*Strategic Plan of the Ministry of Agriculture 2015-2019*(Jakarta: Ministry of Agriculture, Republic of Indonesia) p 364

[2] Haryono 2013*Lahan Rawa Lumbung Pangan Masa Depan Indonesia*(Jakarta: Badan Penelitian dan Pengembangan Pertanian) p 142

[3] Wijanarko A and Taufiq A 2016 Effect of lime application on soil properties and soybean yield on tidal land *Agrivita J. Agric. Sci.***38** pp 14–23

[4] Muhrizal S, Shamshuddin J, Fauziah I and Husni MAH 2006 Changes in iron-poor acid sulfate soil upon submergence*Geoderma***131** pp 110–22

[5] Anda M, Siswanto AB and Subandiono RE 2009 Properties of organic and acid sulfate soils and water of a "reclaimed" tidal backswamp in Central Kalimantan, Indonesia*Geoderma* **149** pp 54–65

[6] Minh L 1998 Soil and water table management effects on aluminum dynamics in an acid sulfate soil in Vietnam*Agric. Ecosyst. Environ.***68** pp 255–62

[7] Zheng SJ 2010 Crop production on acidic soils: overcoming aluminum toxicity and phosphorus deficiency*Ann. Bot.***106** pp 183–4
[8] Horst WJ, Wang Y, and Eticha D 2010 The role of the root apoplast in aluminum-induced inhibition of root elongation and in aluminum resistance of plants: a review *Ann. Bot.* **106** pp 85–97
[9] Yu HN, Liu P, Wang ZY, Chen WR and Xu GD 2011 The effect of aluminum treatments on the root growth and cell ultrastructure of two soybean genotypes *Crop Prot.* **30** pp 323–8
[10] Ghulamahdi M, Melati M and Sagala D 2009 Production of Soybean Varieties under Saturated Soil Culture on Tidal Swamps *J. Agron. Indonesia.***37** pp 226–32
[11] Ghulamahdi M, Melati M, Sagala Dand Sahuri 2011 The effect of water depth and bed width on the production of soybean (Glycine max L. Merr) under saturated soil culture on tidal swamps *J. ISSAAS* **17** pp 258–9.
[12] Sagala D, Ghulamahdi M, Melati M 2011 Pola Serapan Hara dan Pertumbuhan Beberapa Varietas Kedelai dengan Budidaya Jenuh Air di Lahan Rawa Pasang Surut *J.Agroqua* **9** pp 1–10.
[13] Sopandie D 1990 *Studies of plant responses to salt stress* (Okayama: Okayama University)
[14] Pujiwati H, Ghulamahdi M, Yahya S, Aziz SA and Haridjaja DO 2016Tanggap kedelai hitam terhadap cekaman aluminum pada kultur hara*J. Penelit. Pertan. Tanam Pangan* **35** pp 149–53
[15] Fischer R and Maurer R 1978 Drought resistance in spring wheat cultivars I. Grain yield responses *Aust. J. Agric. Res.***29** pp 897-912
[16] Sopandie D 2014*Fisiologi Adaptasi Tanaman Terhadap Cekaman Abiotik pada Agroekosistem Tropika* (Bogor: IPB Press) p 228